$A \cap B$	$A \cap B$ = ... set intersection (see Appendix A)
$A \subseteq B$	A is a subset of B (see Appendix A)
$\emptyset$	$\emptyset = \{\ \} =$ empty set (see Appendix A)
$\sim A$	$\sim A = \{x \in U \mid x \notin A\} =$ set complement (see Appendix A)
$A \sim B$	$A \sim B = \{x \in A \mid x \notin B\} =$ set difference (see Appendix A)
$|A|$	The number of elements (or the cordinality) of a set A (see Appendix A)
$\delta(x, y)$	Kronecker delta: $\delta(x, y) = 1$ if $x = y$, else $\delta(x, y) = 0$
$\mathscr{I}(\mathrm{STR1,STR2})$	Index of coincidence of two strings
$P(E)$	Probability of an event E (see Appendix B)
$P(E \mid F)$	Conditional probability of an event E given that event F has occurred (see Appendix B)
$E[X]$	Expectation (or expected value) of a random variable X (see Appendix B)
$\mathscr{P}$	Plaintext space
$\mathscr{C}$	Ciphertext space $\mathscr{K}$
$\kappa \in \mathscr{K}$	κ is a key belonging to the keyspace $\mathscr{K}$
E or e_κ	Encryption mapping
D or d_κ	Decryption mapping (inverse function of E or e_κ)
$(a_1, a_2, \cdots, a_k)$	A cycle (= cyclic permutation)
$A \oplus B$	The XOR of two bit strings
$\pi(x)$	The number of prime numbers p satisfying $p < x$
$\phi(n)$	Euler's phi function
$\mathrm{ord}_n(a)$	The order of a relative to n (or order of a mod n)
$\mathbb{Z}_n^\times$	$\mathbb{Z}_n^\times = \{$all invertible elements of $\mathbb{Z}_n\}$
E_g	$E_g : \mathbb{Z}_p^\times \to \mathbb{Z}_p^\times$, defined by $E_g(\ell) \equiv g^\ell \pmod{p}$ (Modular exponentiation function—inverse of discrete logarithm, see Chapter 9)

INTRODUCTION TO CRYPTOGRAPHY

WITH MATHEMATICAL FOUNDATIONS
AND COMPUTER IMPLEMENTATIONS

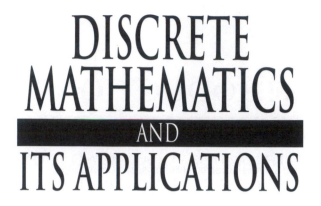

DISCRETE MATHEMATICS AND ITS APPLICATIONS

Series Editor

Kenneth H. Rosen, Ph.D.

Juergen Bierbrauer, Introduction to Coding Theory

Francine Blanchet-Sadri, Algorithmic Combinatorics on Partial Words

Richard A. Brualdi and Dragoš Cvetković, A Combinatorial Approach to Matrix Theory and Its Applications

Kun-Mao Chao and Bang Ye Wu, Spanning Trees and Optimization Problems

Charalambos A. Charalambides, Enumerative Combinatorics

Gary Chartrand and Ping Zhang, Chromatic Graph Theory

Henri Cohen, Gerhard Frey, et al., Handbook of Elliptic and Hyperelliptic Curve Cryptography

Charles J. Colbourn and Jeffrey H. Dinitz, Handbook of Combinatorial Designs, Second Edition

Martin Erickson, Pearls of Discrete Mathematics

Martin Erickson and Anthony Vazzana, Introduction to Number Theory

Steven Furino, Ying Miao, and Jianxing Yin, Frames and Resolvable Designs: Uses, Constructions, and Existence

Randy Goldberg and Lance Riek, A Practical Handbook of Speech Coders

Jacob E. Goodman and Joseph O'Rourke, Handbook of Discrete and Computational Geometry, Second Edition

Jonathan L. Gross, Combinatorial Methods with Computer Applications

Jonathan L. Gross and Jay Yellen, Graph Theory and Its Applications, Second Edition

Jonathan L. Gross and Jay Yellen, Handbook of Graph Theory

Darrel R. Hankerson, Greg A. Harris, and Peter D. Johnson, Introduction to Information Theory and Data Compression, Second Edition

Darel W. Hardy, Fred Richman, and Carol L. Walker, Applied Algebra: Codes, Ciphers, and Discrete Algorithms, Second Edition

Daryl D. Harms, Miroslav Kraetzl, Charles J. Colbourn, and John S. Devitt, Network Reliability: Experiments with a Symbolic Algebra Environment

Silvia Heubach and Toufik Mansour, Combinatorics of Compositions and Words

Leslie Hogben, Handbook of Linear Algebra

Derek F. Holt with Bettina Eick and Eamonn A. O'Brien, Handbook of Computational Group Theory

Titles (continued)

David M. Jackson and Terry I. Visentin, An Atlas of Smaller Maps in Orientable and Nonorientable Surfaces

Richard E. Klima, Neil P. Sigmon, and Ernest L. Stitzinger, Applications of Abstract Algebra with Maple™ and MATLAB®, Second Edition

Patrick Knupp and Kambiz Salari, Verification of Computer Codes in Computational Science and Engineering

William Kocay and Donald L. Kreher, Graphs, Algorithms, and Optimization

Donald L. Kreher and Douglas R. Stinson, Combinatorial Algorithms: Generation Enumeration and Search

C. C. Lindner and C. A. Rodger, Design Theory, Second Edition

Hang T. Lau, A Java Library of Graph Algorithms and Optimization

Elliott Mendelson, Introduction to Mathematical Logic, Fifth Edition

Alfred J. Menezes, Paul C. van Oorschot, and Scott A. Vanstone, Handbook of Applied Cryptography

Richard A. Mollin, Advanced Number Theory *with Applications*

Richard A. Mollin, Algebraic Number Theory

Richard A. Mollin, Codes: The Guide to Secrecy from Ancient to Modern Times

Richard A. Mollin, Fundamental Number Theory with Applications, Second Edition

Richard A. Mollin, An Introduction to Cryptography, Second Edition

Richard A. Mollin, Quadratics

Richard A. Mollin, RSA and Public-Key Cryptography

Carlos J. Moreno and Samuel S. Wagstaff, Jr., Sums of Squares of Integers

Dingyi Pei, Authentication Codes and Combinatorial Designs

Kenneth H. Rosen, Handbook of Discrete and Combinatorial Mathematics

Douglas R. Shier and K.T. Wallenius, Applied Mathematical Modeling: A Multidisciplinary Approach

Alexander Stanoyevitch, Introduction to Cryptography with Mathematical Foundations and Computer Implementations

Jörn Steuding, Diophantine Analysis

Douglas R. Stinson, Cryptography: Theory and Practice, Third Edition

Roberto Togneri and Christopher J. deSilva, Fundamentals of Information Theory and Coding Design

W. D. Wallis, Introduction to Combinatorial Designs, Second Edition

Lawrence C. Washington, Elliptic Curves: Number Theory and Cryptography, Second Edition

DISCRETE MATHEMATICS AND ITS APPLICATIONS

Series Editor KENNETH H. ROSEN

INTRODUCTION TO CRYPTOGRAPHY

WITH MATHEMATICAL FOUNDATIONS AND COMPUTER IMPLEMENTATIONS

Alexander Stanoyevitch

California State University
Carson, California, U.S.A.

CRC Press
Taylor & Francis Group
Boca Raton London New York

CRC Press is an imprint of the
Taylor & Francis Group, an **informa** business

A CHAPMAN & HALL BOOK

Chapman & Hall/CRC
Taylor & Francis Group
6000 Broken Sound Parkway NW, Suite 300
Boca Raton, FL 33487-2742

© 2011 by Taylor and Francis Group, LLC
Chapman & Hall/CRC is an imprint of Taylor & Francis Group, an Informa business

No claim to original U.S. Government works

Printed in the United States of America on acid-free paper
10 9 8 7 6 5 4 3 2 1

International Standard Book Number: 978-1-4398-1763-6 (Hardback)

Library of Congress Cataloging-in-Publication Data

Stanoyevitch, Alexander.
   Introduction to cryptography with mathematical foundations and computer implementations / Alexander Stanoyevitch.
      p. cm. -- (Discrete mathematics and its applications)
   Includes bibliographical references and index.
   ISBN 978-1-4398-1763-6 (hardcover : alk. paper)
   1. Coding theory. 2. Cryptography--Data processing. 3. Cryptography--Mathematics. 4. Data encryption (Computer science) I. Title. II. Series.

QA268.S693 2010
005.8'2--dc22   2010010970

**Visit the Taylor & Francis Web site at
http://www.taylorandfrancis.com**

**and the CRC Press Web site at
http://www.crcpress.com**

Contents

Preface. xiii
About the Author. xv
Dependency Chart. xvii
Acknowledgments. xix

1   An Overview of the Subject 1

Basic Concepts   1
Functions   4
One-to-One and Onto Functions, Bijections   5
Inverse Functions   7
Substitution Ciphers   8
Attacks on Cryptosystems   12
The Vigenère Cipher   15
The Playfair Cipher   18
The One-Time Pad, Perfect Secrecy   25
Chapter 1 Exercises   28
Chapter 1 Computer Implementations and Exercises   35
   Vector/String Conversions   35
   Integer/Text Conversions   36
   Programming Basic Ciphers with Integer Arithmetic   38
   Computer-Generated Random Numbers   39

2   Divisibility and Modular Arithmetic 43

Divisibility   43
Primes   44
Greatest Common Divisors and Relatively Prime Integers   46
The Division Algorithm   47
The Euclidean Algorithm   48
Modular Arithmetic and Congruences   52
Modular Integer Systems   58
Modular Inverses   60
Extended Euclidean Algorithm   61
Solving Linear Congruences   64
   Summary of Procedure for Solving the Single
   Linear Congruence (Equation 2.2)   66
The Chinese Remainder Theorem   67
Chapter 2 Exercises   71
Chapter 2 Computer Implementations and Exercises   85

3   The Evolution of Codemaking until the Computer Era 91

Ancient Codes   91
Formal Definition of a Cryptosystem   94
Affine Ciphers   96
Steganography   100
Nulls   102
Homophones   105
Composition of Functions   109
Tabular Form Notation for Permutations   110
The Enigma Machines   111
Cycles (Cyclic Permutations)   114
Dissection of the Enigma Machine into Permutations   119
Special Properties of All Enigma Machines   126
Chapter 3 Exercises   127
Chapter 3 Computer Implementations and Exercises   136
    Computer Representations of Permutations   140

4   Matrices and the Hill Cryptosystem 145

The Anatomy of a Matrix   145
Matrix Addition, Subtraction, and Scalar Multiplication   146
Matrix Multiplication   147
Preview of the Fact That Matrix Multiplication Is Associative   149
Matrix Arithmetic   149
Definition of an Invertible (Square) Matrix   151
The Determinant of a Square Matrix   153
Inverses of 2×2 Matrices   155
The Transpose of a Matrix   156
Modular Integer Matrices   156
The Classical Adjoint (for Matrix Inversions)   159
The Hill Cryptosystem   162
Chapter 4 Exercises   166
Chapter 4 Computer Implementations and Exercises   174

5   The Evolution of Codebreaking until the Computer Era 181

Frequency Analysis Attacks   181
The Demise of the Vigenère Cipher   187
    The Babbage/Kasiski Attack   188
    The Friedman Attack   192
The Index of Coincidence   193
Expected Values of the Index of Coincidence   193
How Enigmas Were Attacked   201
    German Usage Protocols for Enigmas   202
    The Polish Codebreakers   203
    Rejewski's Attack   203
Invariance of Cycle Decomposition Form   205
Alan Turing and Bletchley Park   206
Chapter 5 Exercises   208
Chapter 5 Computer Implementations and Exercises   214

Programs to Aid in Frequency Analysis   214
Programs to Aid in the Babbage/Kasiski Attack   215
Programs Related to the Friedman Attack   218

6   Representation and Arithmetic of Integers in Different Bases 221

Representation of Integers in Different Bases   221
Hex(adecimal) and Binary Expansions   224
Addition Algorithm with Base b Expansions   229
Subtraction Algorithm with Base b Expansions   231
Multiplication Algorithm in Base b Expansions   234
Arithmetic with Large Integers   237
Fast Modular Exponentiation   239
Chapter 6 Exercises   241
Chapter 6 Computer Implementations and Exercises   248

7   Block Cryptosystems and the Data Encryption Standard (DES) . . . 251

The Evolution of Computers into Cryptosystems   251
DES Is Adopted to Fulfill an Important Need   252
The XOR Operation   254
Feistel Cryptosystems   255
A Scaled-Down Version of DES   258
DES   265
The Fall of DES   272
Triple DES   273
Modes of Operation for Block Cryptosystems   274
    Electronic Codebook (ECB) Mode   274
    Cipherblock Chaining (CBC) Mode   275
    Cipher Feedback (CFB) Mode   276
    Output Feedback (OFB) Mode   278
Chapter 7 Exercises   279
Chapter 7 Computer Implementations and Exercises   286

8   Some Number Theory and Algorithms 293

The Prime Number Theorem   293
Fermat's Little Theorem   295
The Euler Phi Function   298
Euler's Theorem   300
Modular Orders of Invertible Modular Integers   301
Primitive Roots   302
    Existence of Primitive Roots   304
    Determination of Primitive Roots   304
Order of Powers Formula   305
Prime Number Generation   308
Fermat's Primality Test   309
Carmichael Numbers   311
The Miller–Rabin Test   312
The Miller–Rabin Test with a Factoring Enhancement   315

The Pollard $p - 1$ Factoring Algorithm   316
Chapter 8 Exercises   319
Chapter 8 Computer Implementations and Exercises   325

9   Public Key Cryptography **331**
An Informal Analogy for a Public Key Cryptosystem   331
The Quest for Secure Electronic Key Exchange   332
One-Way Functions   333
Review of the Discrete Logarithm Problem   334
The Diffie–Hellman Key Exchange   336
The Quest for a Complete Public Key Cryptosystem   337
The RSA Cryptosystem   338
Digital Signatures and Authentication   343
The ElGamal Cryptosystem   345
Digital Signatures with ElGamal   347
Knapsack Problems   349
The Merkle–Hellman Knapsack Cryptosystem   352
Government Controls on Cryptography   356
A Security Guarantee for RSA   357
Chapter 9 Exercises   360
Chapter 9 Computer Implementations and Exercises   369

10   Finite Fields in General, and *GF*(2⁸) in Particular. **377**
Binary Operations   377
Rings   378
Fields   381
$\mathbb{Z}_p[X]$ = the Polynomials with Coefficients in $\mathbb{Z}_p$ 385
Addition and Multiplication of Polynomials in $\mathbb{Z}_p[X]$ 386
Vector Representation of Polynomials   387
$\mathbb{Z}_p[X]$ Is a Ring   388
Divisibility in $\mathbb{Z}_p[X]$ 389
The Division Algorithm for $\mathbb{Z}_p[X]$ 391
Congruences in $\mathbb{Z}_p[X]$ Modulo a Fixed Polynomial   395
Building Finite Fields from $\mathbb{Z}_p[X]$ 396
The Fields GF(2⁴) and GF(2⁸)   399
The Euclidean Algorithm for Polynomials   404
Chapter 10 Exercises   406
Chapter 10 Computer Implementations and Exercises   411

11   The Advanced Encryption Standard (AES) Protocol **417**
An Open Call for a Replacement to DES   417
Nibbles   419
A Scaled-Down Version of AES   421
Decryption in the Scaled-Down Version of AES   429
AES   432
Byte Representation and Arithmetic   432
The AES Encryption Algorithm   435

The AES Decryption Algorithm   439
Security of the AES   440
Chapter 11 Exercises   441
Chapter 11 Computer Implementations and Exercises   445

12   Elliptic Curve Cryptography. **451**
Elliptic Curves over the Real Numbers   452
The Addition Operation for Elliptic Curves   454
Groups   458
Elliptic Curves over $\mathbb{Z}_p$ 460
The Variety of Sizes of Modular Elliptic Curves   462
The Addition Operation for Elliptic Curves over $\mathbb{Z}_p$ 463
The Discrete Logarithm Problem on Modular Elliptic Curves   466
An Elliptic Curve Version of the Diffie–Hellman Key Exchange   467
Fast Integer Multiplication of Points on Modular Elliptic Curves   470
Representing Plaintexts on Modular Elliptic Curves   471
An Elliptic Curve Version of the ElGamal Cryptosystem   473
A Factoring Algorithm Based on Elliptic Curves   475
Chapter 12 Exercises   477
Chapter 12 Computer Implementations and Exercises   483

Appendices. **489**

Appendix A: Sets and Basic Counting Principles **491**
Concepts and Notations for Sets   491
Two Basic Counting Principles   495

Appendix B: Randomness and Probability **501**
Probability Terminology and Axioms   501
Conditional Probability   507
Conditioning and Bayes' Formula   509
Random Variables   511

Appendix C: Solutions to All Exercises for the Reader **515**
Chapter 1: An Overview of the Subject   515
Chapter 2: Divisibility and Modular Arithmetic   517
Chapter 3: The Evolution of Codemaking until the Computer Era   522
Chapter 4: Matrices and the Hill Cryptosystem   526
Chapter 5: The Evolution of Codebreaking until the Computer Era   530
Chapter 6: Representation and Arithmetic of Integers in Different Bases   536
Chapter 7: Block Cryptosystems and the Data Encryption Standard (DES)   540
Chapter 8: Some Number Theory and Algorithms   545
Chapter 9: Public Key Cryptography   550
Chapter 10: Finite Fields in General, and $GF(2^8)$ in Particular   554
Chapter 11: The Advanced Encryption Standard (AES) Protocol   560
Chapter 12: Elliptic Curve Cryptography   563

Appendix D: Answers and Brief Solutions to Selected Odd-Numbered Exercises **569**

Chapter 1   569
Chapter 2   572
Chapter 3   581
Chapter 4   587
Chapter 5   592
Chapter 6   595
Chapter 7   599
Chapter 8   601
Chapter 9   604
Chapter 10   608
Chapter 11   609
Chapter 12   611

Appendix E: Suggestions for Further Reading **615**

Synopsis   615
History of Cryptography   615
Mathematical Foundations   615
Computer Implementations   616
Elliptic Curves   616
Additional Topics in Cryptography   616

References **619**

Index of Corollaries, Lemmas, Propositions, and Theories **623**

Index of Algorithms **625**

Subject Index **627**

Preface

This book is an entirely self-contained sophomore-level text on the exciting subject of cryptography, which is the science of achieving secure communication over insecure channels. It was carefully designed to accommodate a wide variety of both mathematics and computer science courses, ranging from a ground-level course with minimal prerequisites to a follow-up course to a standard number theory course. The author began teaching cryptography as a special topic in more general courses that included both mathematics and computer science majors (for example, in discrete structures, in abstract algebra, and even in finite mathematics) starting about 12 years ago. Rather than aiming at an encyclopedic treatment of almost all topics in cryptography, this book provides a focused tour of the central and evergreen topics and concepts of cryptography, following the exciting history of its development from ancient times to the present day. Cryptography is a very important subject in both mathematics and computer science, and it serves as an ideal melting pot, where the subjects, students, and practitioners can draw on each other and become stronger in the process. The fact that cryptography is heavily relied on by business, government, and industry, coupled with the increasing new technologies for transferring data, guarantees that it will play a permanent role in research and development, as its architects and hackers continue to battle for the upper hand.

Each chapter is written in an engaging and easy-going, yet rigorous style. Important concepts are introduced with clear definitions and theorems, the proofs are written in a style that students find appealing. Numerous examples are provided to illustrate key points, and figures and tables are used to help illustrate the more difficult or subtle concepts. The first chapter gives an overview of the subject, provides a road map for the rest of the book, and lays out some terminology of the subject. Some of the concepts and definitions can only be informally defined in such a preliminary chapter, but the subsequent chapters follow a more rigorous path, and additional details for some topics introduced in the overview chapter will be revisited later on in the text with more formal treatments. Chapters 2 through 12 basically develop cryptography in chronological order, with mathematical concepts being developed as they are needed. The text proper of each chapter is punctuated with "Exercises for the Reader" that provide the readers with regular opportunities to test their understanding of their reading (and help them to become more active readers). The exercises for the reader range in difficulty from routine to more involved, but appendices at the end of the book include detailed solutions to all of them. In addition to these exercises for the reader, each chapter has an extensive and well-crafted set of exercises that range in difficulty from routine to nontrivial. An appendix contains answers or brief solutions to most of the odd-numbered exercises. A separate instructor's manual has been prepared for the corresponding solutions of the even-numbered exercises. Every chapter concludes with a Computer Implementation and Exercises section, which guides interested readers through the process of writing their own programs for the cryptographic concepts of each chapter.

There is more than enough material included in this text for a one-semester course, and a variety of different courses could be created from it. Since the computer implementation material is separated from the main text, the book can be used without computers. A useful set of platform-independent applet pages has been created to perform many of the core algorithms of this book; these can be freely downloaded along with other relevant materials from the book's Web page. Rather than print the URL of the Web page, which may change as servers get updated or as (the author's physical or electronic) addresses may change, the easiest way to access this page is through the author's homepage, which can be obtained by a simple Web search of the author's last name. The Web page should also be navigable via the publisher's Web site. There is a sufficient amount of number theory in the book to satisfy the number theory credential requirement for students studying to be secondary math teachers in the State of California, providing an exciting option to replace the standard number theory course with a more applied version. After all, cryptography is largely responsible for reenergizing the traditionally very pure subject of number theory into a practical one. The book also contains a decent amount of abstract algebra, presented in a concrete and applied fashion, which should help students to better understand the motivations and purpose of this subject. A chapter dependence chart is provided following this Preface to help instructors design their own courses.

About the Author

Alexander Stanoyevitch completed his doctorate in mathematical analysis at the University of Michigan–Ann Arbor, has held academic positions at the University of Hawaii and the University of Guam, and is presently a professor at California State University–Dominguez Hills. He has published several articles in leading mathematical journals and has been an invited speaker at numerous lectures and conferences in the United States, Europe, and Asia. His research interests include areas of both pure and applied mathematics, and he has taught many upper-level classes to mathematics students as well as computer science students.

Dependency Chart

The following chart should be helpful to readers or instructors aiming to plan courses with this book. Major dependencies are indicated with solid arrows, minor ones with dashed arrows. The two minor dependencies stemming from Chapter 3 rely on only the general function/cryptosystem concepts, not on any of the specific cryptosystems that are introduced in that chapter.

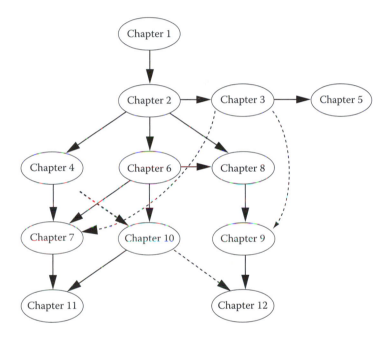

Figure 1 The material surrounding Algorithm 6.2 through Algorithm 6.4, although helpful in gaining understanding of the concepts, is not used in subsequent chapters.

Acknowledgments

Creating such a substantial book could not have been done so effectively without the help and encouragement of many individuals. I am generally grateful to my departments for encouraging and allowing me to work cryptography into my classes, first as topics, and eventually as complete courses. I am grateful to my students, who signed up for these classes although they were not always required to, for their useful feedback on early drafts. I am particularly grateful to my former student Eric Flior; he has carefully read over the entire book and provided some very useful suggestions. Moreover, he has designed the very nice platform-independent applet pages that correspond to several of the core chapters of this book and are available on the author's Web page. I am enormously grateful to my colleague George Jennings, who has also carefully read over the whole book and has provided many useful suggestions and scholarly observations. In addition, George has done a superb job with the task of preparing an instructor's solutions manual that contains completely worked-out solutions for all of the even-numbered exercises (instructors should contact CRC Press to obtain a copy of this manual). I am also grateful to my neighboring colleague Will Murray at California State University–Long Beach, who has read over portions of this book and has provided useful feedback. I thank Yumi Nishimura for creating some of the technical drawings. I thank my mother, Christa Stanoyevitch, for her continued encouragement and very careful proofreading of the entire book. I am grateful to my editor, Sunil Nair, who has provided much insightful guidance, advice, and support throughout the project. Tara Nieuwesteeg, the project editor of this book, has been extremely conscientious and a pleasure to work with throughout the production process. I also wish to express my gratitude to CRC series editor Ken Rosen, who has read over much of the book and has provided me with some very illuminating comments and suggestions. Initial feedback from the anonymous reviewers of the book has also been very helpful. Librarian Rene Stein of the National Cryptographic Museum (located at the NSA headquarters) has provided me with a great deal of historical information and assistance with photographs. I am delighted to express my sincere appreciation to the following illustrious cryptographers who have either provided me with personal photographs or helped me to secure permission for other's photographs of them: Leonard Adleman, Clifford Cocks, Whit Diffie, Joan Daemen, and Vincent Rijmen. I extend special thanks also to the widow of James Ellis for providing me with several photographs.

An Overview of the Subject

In this chapter we introduce some key definitions and general concepts of cryptography that will be used throughout the text. A few simple examples of cryptosystems are provided to help better illustrate the topics. The chapter also provides a road map of the various parts of the subject and where they will be developed in the book. Cryptography has a fascinating history, elements of which punctuate this chapter as well as latter portions of the text. We point out that the general tone of this chapter is informal, and some of the preliminary definitions given here are developed more rigorously in later chapters, after the needed mathematical concepts have been considered.

Basic Concepts

Cryptography is the science of protecting data and communications. One of its main components involves communicating messages or information between designated parties by changing the appearance of the messages (or data) in ways that aim to make it extremely difficult or impossible for other parties to eavesdrop on or interfere with the transmission. Other important aspects include *authentication*, which allows receiving parties the means to ascertain that the communication really does come from the designated sender, and *integrity*, which, among other things, ensures that the message received has not been altered. The subject of cryptography is as old as written languages. There have always been situations where it is important to convey a confidential message. A spy's life could depend on certain messages not being compromised; launch codes for nuclear and other weapons of mass destruction, if cracked, could cause the demise of a whole city, a country, or even the world. Keeping data and messages confidential has become an essential and almost daily issue for almost all of us in our high-tech society. When anyone sends out a personal e-mail, he or she certainly would like to know all who might be able to read it. A supervisor or even a curious coworker may have easy access. Cryptography is vital to electronic commerce, for otherwise it would not be possible to make credit card purchases over the Internet or to wire money from a bank to another location. As our point of departure into this exciting subject, we consider Figure 1.1, which shows the basic idea behind most cryptosystems.

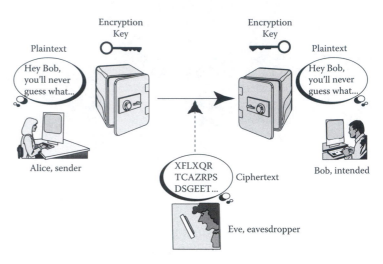

Figure 1.1 A basic reference illustration for a cryptosystem. Alice, the sender, wishes to send Bob, the intended recipient, a confidential message. On Alice's end, the message gets encrypted before it is sent to Bob, who, as the designated recipient, will be able to decrypt the message. Eve, the eavesdropper (a hacker), tries to intercept this message but will not have the key to decode it.

The characters **Alice** and **Bob** (the communicating parties) and **Eve** (the evil eavesdropper) of Figure 1.1 have become standard in cryptographic literature and are used throughout this book.* Another character—Mallory (the malicious manipulator)—will be later introduced as one who tries to fabricate messages or otherwise corrupt communications. As with any sort of security system, greater levels of security that are desired or required need more sophisticated systems to prevent breaches by intruders who must employ more sophisticated means to breach the system. The development in this book is given in essentially chronological order, with the math tools being introduced as they are needed. One general trend is that as cryptosystems have developed over several centuries, the mathematical foundations on which they rely have become increasingly sophisticated.

Any cryptographic system must anticipate **attacks** by **hackers** who might try to break the code of the transfer and thus compromise the data integrity. The advent of high-speed computers has had a tremendous impact on the standards for what are considered to be effective **ciphers** or **cryptosystems**, which are algorithms for rendering messages unintelligible except to the designated recipients. A cryptosystem has two parts: **encryption**, which is done at the sender's end of the message and means to put the actual **plaintext** (original message) into **ciphertext** (secret code), and **decryption**, which is done at the recipient's end and means to translate the ciphertext back into the original plaintext message. Encryption and decryption are done using a **key**, or perhaps two keys (one to "lock" the message in the encryption stage and the other to "unlock" it during decryption), along with algorithms that

---

* The names in our picture have become folk tradition in cryptography circles. Later, we will examine a related problem where a different sort of hacker tries to send phony messages to Bob, while attempting to make him think they came from Alice. Such an individual is called a "Mallory."

can perform the encryption and decryption. It can usually be assumed that hackers will be able to determine the cryptosystem that is used, but without complete knowledge about the corresponding keys, the system should remain secure. Thus, certain details of the keys must be kept secret.[*]

Depending on the usage, the consequences of an unauthorized break-in, and the skill level of anticipated hackers, a cryptosystem may be simple or may be very sophisticated. High-speed computers have made it possible to implement extremely sophisticated cryptosystems, but at the same time, hackers can use powerful computing tools to help them in being able to crack cryptosystems. Cryptography is a huge industry with many public and private companies working hard to keep the technology state-of-the-art and to keep one step ahead of hackers. The latest technologies in the field depend heavily on many mathematical tools ranging from abstract algebra and number theory to probability. The U.S. federal government is, of course, a big user and consumer of cryptography. Usage comes not only from the defense and intelligence industries (Pentagon, CIA, FBI, and so forth) but also from financial and technology industries. The branch in the U.S. government that is solely dedicated to cryptography is the *National Security Agency* (NSA). The NSA constantly and actively recruits people with mathematics and computer science degrees (from bachelor's degrees to PhDs).

Cryptosystems can be implemented on any alphabet. An **alphabet** is any finite set of symbols. Any ordered sequence of letters from a certain alphabet is called a **string** (from the alphabet).[†] For example, QXUZTKM is a string of length 7 (since it has seven alphabet characters) in the alphabet of the 26 uppercase English letters {A, B, C, ..., Y, Z}. **Binary strings** (also called **bit strings**) are strings from the **binary alphabet** {0, 1}; for example, 011100 is a bit string of length 6 (since it consists of six characters). The individual digits in a binary string are called **bits**. The plaintext and ciphertext may be represented in different alphabets. Thus, although it is most convenient to input a plaintext message in a familiar alphabet (such as English letters and digits), the ciphertext produced by the computer would probably be formed in an alphabet that is efficient for computer architecture and manipulations (such as the binary alphabet).

All cryptosystems require algorithms and/or functions to accomplish the encryption and decryption processes. **Algorithms** are simply lists of instructions (or programs or procedures) designed to accomplish certain tasks. The concept of a function is also very general, involving rules or formulas that show how to get an associated output for each permissible input value. Functions can be described in many ways, using graphs, tables, formulas, or algorithms. For example, a table giving the daily high temperatures (rounded to the nearest degree Fahrenheit) at the Los Angeles International Airport for every day over the past five years is a function. The inputs are the days over the past five years, and the outputs are the corresponding high temperatures. To find the output of this function for a

---

[*] Traditionally, cryptography referred to the design of cryptosystems, cryptanalysis to methods of attacking them, and cryptology to both of these tasks. Increasingly, cryptography is replacing cryptology as the main descriptor of the field, and we will adhere to this convention.

[†] Like sets, strings can be empty; but since empty strings will be unusual in our work, our default assumption will be that strings are nonempty unless explicitly stated otherwise.

given day (over the past five years), we simply look up the temperature on that day in the table. Since functions are used throughout the subject of cryptography, we will now provide a formal definition.

Functions

Definition 1.1

A **function** (or **mapping**) **from a set** A (= the set of inputs) **to a set** B (= a set containing all possible outputs) is a rule, formula, or algorithm that assigns to each element $a \in A$ (an input) a unique element $f(a) \in B$ (the corresponding output). The element $f(a)$ is also called the **image of** a **under** f, and if $f(a) = b$, we say that f **maps** a to b. The notation $f : A \to B$ is used to indicate that f is a function from the set A to the set B. The set A is called the **domain** of the function, and the set B is called the **codomain**. The set of all outputs of f is called the **range of** f and is denoted as $f(A)$. Note that the range is a subset of the codomain, that is, $f(A) \subseteq B$.

It is helpful to visualize a generic function with a diagram such as the one shown in Figure 1.2.[*]

In contrast with calculus courses, almost all of the functions that are dealt with in cryptography have domains and codomains that are either finite sets or discrete infinite sets.[†] For functions involving small domains and codomains, diagrams can easily be drawn describing their actions; the following example demonstrates this idea.

Example 1.1

Which of the three diagrams in Figure 1.3 represent(s) functions from the domain {a, b, c} to the set {1, 2, 3}?

Solution: The rule F is not a function since the input b is assigned to have two outputs. The other two rules specify functions, since each element of the domain {a, b, c} is assigned exactly one output in the codomain.

---

[*] In lower-level mathematics classes, students are sometimes taught that a function is just a formula such as $f(x) = x^2$. What is usually intended is that the domain is taken to be the largest possible subset A of real numbers for which the formula makes sense (in this case A = {real numbers}), and so $f: A \to$ {real numbers}.

[†] Unlike continuous infinite sets such as the set of real numbers that contains whole intervals of numbers, discrete infinite sets can be formed by taking a union of an infinite sequence of finite sets. For example, binary strings of any fixed length form a finite set. But the set of all binary strings of finite length, the union of all of the sets of binary strings of length 0, 1, 2, 3, and so forth, is an example of a discrete infinite set. Here is another distinction. It is always possible to represent any element of a discrete set with a finite string (in some alphabet), whereas for continuous infinite sets, this typically cannot be done. For example, there are many real numbers whose decimal expansions are nonending and nonrepeating, and would require an infinite string of decimal digits to write down.

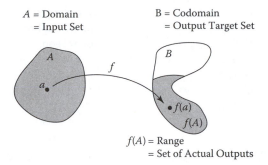

Figure 1.2 Schematic diagram of a function $f: A \to B$ with the output $f(a)$ of an element in the domain A. The range $f(A)$ (shaded on the right) is a subset of the codomain B.

We point out two important observations about the functions G and H of Example 1.1. First, note that $G(a) = G(b) = 1$, that is, two inputs are assigned a single output; a function is allowed to do this (but not the other way around). The function H does not do this: different outputs are assigned different inputs. Second, note that not every element of the codomain of G is an actual output: 3 does not occur as an output, but the function H does actually realize each element of its codomain as an output. These two properties of H are very important, and they are the given official designations in the following definition.

One-to-One and Onto Functions, Bijections

Definition 1.2

Suppose that $f : A \to B$ is a function.

  (a) We say f is **one-to-one** if different inputs are always assigned different outputs; in other words, if two elements $x, y \in A$ have the same outputs (under f): $f(x) = f(y)$, then they must be the same: $x = y$.

  (b) We say f is **onto** if every element of the codomain B occurs as an actual output; in other words, if b is an element of the set B, then there exists an element a of the domain such that the output of a is b: $f(a) = b$. In other words, the range equals the codomain, i.e., $f(A) = B$.

  (c) We say f is **bijective**, or a **bijection**, if it is both one-to-one and onto.

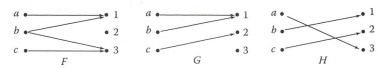

Figure 1.3 Three diagrams assigning elements of the set $\{a, b, c\}$ to elements of the set $\{1, 2, 3\}$.

Note that the temperature function mentioned earlier (with domain being the days over the past five years, codomain being the set of integers $\{\cdots, -3, -2, -1, 0, 1, 2, 3, \cdots\}$, and rule specified by a table giving the high temperature at the LAX airport each day) is not a one-to-one function since there are different days (over the past five years) that had the same high temperature (that is, two inputs share the same output). This function is also not onto (Why?).

Example 1.2

Determine whether the following functions are one-to-one and/or onto:

(a) $F : \{a, b, c, \cdots, x, y, z\} \rightarrow \{0, 1, 2, \cdots, 23, 24, 25\}$ defined by the rule

$F(i\text{th letter of the alphabet}) = i - 1.$

This rule is an abbreviation for writing out each of the 26 input/output relations:

$F(a) = 0,\ F(b) = 1,\ F(c) = 2,\ F(d) = 3$, etc.

(b) The function G : {length 4 binary strings} → {length 3 binary strings} defined by the rule

$G(b_1 b_2 b_3 b_4) = b_1 b_2 b_4$

(i.e., the third binary digit is deleted from the input to produce the output; for example, $G(0010) = 000$).

(c) The function H : {length 3 binary strings} → {length 4 binary strings} defined by the rule

$H(b_1 b_2 b_3) = b_1 b_2 b_3 b^*$

where the final bit of the output, b^*, is taken to be 0 if the first three bits add up to an even number, and 1 if they add up to an odd number. For example, if $b_1 b_2 b_3 = 101$, then $b_1 + b_2 + b_3 = 1 + 0 + 1 = 2$, which is even, so $b^* = 0$, and thus $H(101) = 1010$. Similarly, $H(100) = 1001$.

Solution: Part (a): This function F is both one-to-one and onto; it merely codes each letter (input) into its uniquely defined place in the alphabet, less one (output). Thus, no two letters are assigned the same output (one-to-one), and the codomain consists exactly of all of the outputs (onto).

Part (b): This function G is onto but not one-to-one. To see why it is onto, consider any length-3 binary string $c_1 c_2 c_3$ where $c_i = 0$, or 1 (that is, any element of the codomain), and notice that it will be the output of G applied to either of the length-4 binary strings $c_1 c_2 0 c_3$ or $c_1 c_2 1 c_3$. The fact that these two different inputs have the same output shows also that G is not one-to-one.

Part (c): This function H is one-to-one but not onto. To see that it is one-to-one is easy: if $H(b_1b_2b_3) = H(c_1c_2c_3)$, this means that $b_1b_2b_3b^* = c_1c_2c_3c^*$. But for two strings to be the same, each of the corresponding components must be equal. Just looking at the first three gives $b_1 = c_1$, $b_2 = c_2$, $b_3 = c_3$, which is tantamount to $b_1b_2b_3 = c_1c_2c_3$. To see that H is not onto, notice that since the fourth bit, b^*, of the output $H(b_1b_2b_3) = b_1b_2b_3b^*$ is completely determined by the input bits, only one of the two strings $b_1b_2b_30$, $b_1b_2b_31$ will be an output. For example, since $H(101) = 1010$, the string 1011 will not be an output.

Given any alphabet A, the set of all finite strings in A includes all strings with characters in A of length 1, 2, 3, and so on, and also includes a single **empty string** that contains no characters and so has length 0. We denote this empty string as $\varnothing$. Given two strings σ_1 and σ_2 in A, having respective length ℓ_1 and ℓ_2, their **concatenation** $\sigma_1 \cdot \sigma_2$ is the string of length $\ell_1 + \ell_2$ obtained by pasting the string σ_2 at the right end of string σ_2.

Exercise for the Reader 1.1

(a) Is function C: {finite length binary strings} $\rightarrow$ {finite length binary strings} defined by $C(\sigma) = 1010 \cdot \sigma$ one-to-one? Is it onto?

(b) Determine whether the following function is one-to-one:

D : {length 3 binary strings} $\rightarrow$ {length 3 binary strings} defined by $D(b_1b_2b_3) = d_1d_2d_3$, where $d_1 = b_1$ and

$$d_2 = \begin{cases} 1, & \text{if } b_1 + b_2 \text{ is odd} \\ 0, & \text{if } b_1 + b_2 \text{ is even} \end{cases} \quad \text{and} \quad d_3 = \begin{cases} 1, & \text{if } b_1 + b_2 + b_3 \text{ is odd} \\ 0, & \text{if } b_1 + b_2 + b_3 \text{ is even} \end{cases}$$

Inverse Functions

The one-to-one property of a function is very important when we use functions in cryptosystems because their processes can be reversed. Since there is only one input for each realized output, the association can be reversed; think of a function diagram as in Figure 1.2: when a function is one-to-one, the arrows can be reversed. If a function $f : A \rightarrow B$ is also onto (so a bijection), then every element of the codomain is an output that corresponds to a unique input, and so we can define a function from B to A by associating each element $b \in B$ the corresponding input under f whose output is b. We call this function the **inverse function of** f, and it is denoted as $f^{-1} : B \rightarrow A$. Thus, $f^{-1}(b) = a$ if, and only if, $b = f(a)$. The inverse function simply "undoes" what the function does; see Figure 1.4.

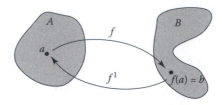

Figure 1.4 Illustration of the inverse function $f^{-1}: B \rightarrow A$ of a bijection $f: A \rightarrow B$.

Example 1.3

(a) Letting F be the bijection of Example 1.2(a), determine the string $F^{-1}(6)F^{-1}(4)F^{-1}(13)F^{-1}(4)F^{-1}(21)F^{-1}(0)$.

(b) Consider the function $F: \{1, 2, 3, 4\} \rightarrow \{1, 2, 3, 4\}$ defined by the rule that for each $a \in \{1, 2, 3, 4\}$, $F(a)$ is the remainder when a^3 is divided by 5.[*] Determine whether the inverse function F^{-1} exists, and if it does, explain how it works.

Solution: Part (a): It is helpful to draw a table for the values of F; see Table 1.3 later in this chapter (but change letters to lowercase). From such a table, we can easily identify the given string to be "geneva."

Part (b): The domain is small enough so that we can compute all of the values of F rather quickly: Since $1^3 = 1 = 5 \cdot 0 + 1$, we get $F(1) = 1$. From $2^3 = 8 = 5 \cdot 1 + 3$, we get $F(2) = 3$. Similarly, the equations $3^3 = 5 \cdot 5 + 2$ and $4^3 = 12 \cdot 5 + 4$ lead us to $F(3) = 2$ and $F(4) = 4$. We now can see that F is a bijection, and the inverse function $F^{-1}: \{1, 2, 3, 4\} \rightarrow \{1, 2, 3, 4\}$ can be described by reversing the inputs and outputs. But this clearly results in the same function, that is, $F^{-1} = F$.

Substitution Ciphers

We are nicely prepared to define our first cipher, known as a substitution cipher. Many of us have some experience with such ciphers going back to our days in elementary school when we wanted to pass notes to some of our classmates in such a way that if the note was intercepted by the teacher (or another unintended student), he or she would not be able to read it.

Definition 1.3

A **substitution cipher** is simply a function F from a plaintext alphabet P to a ciphertext alphabet C, that is both *one-to-one* and *onto*. Thus, for

---

[*] The remainder when we divide a positive integer b by 5 is the unique integer r, with $0 \leq r < 5$, such that $b = 5q + r$, for some integer q. This is just the usual remainder in long division that one learned in grade school; we give a much more thorough account of this topic in the next chapter.

every plaintext letter $p \in P$, the function associates a unique ciphertext letter $c = F(p) \in C$, such that:

(i) *One-to-one condition.* Different plaintext letters will always be associated with different ciphertext letters; i.e., if $p_1 \neq p_2$ (two letters in P), then $F(p_1) \neq F(p_2)$ (two letters in C).

(ii) *Onto condition.* Every ciphertext letter is associated with a plaintext letter; i.e., if $c \in C$, then there is an associated plaintext letter $p \in P$, with $c = F(p)$.

The function F (that specifies the correspondence between plaintext and ciphertext letters) is called the *key* of the substitution cipher. More generally, a **key** in a certain cryptosystem is some parameter that is sufficient to completely describe the encryption and/or decryption mapping of any particular instance of the cryptosystem. In some situations, as with a general substitution cipher, the key and the encryption and/or the decryption mapping are synonymous, because it is difficult to describe a general substitution cipher with anything less than a specification of the encryption mapping. Once the key is known, it is straightforward to encode plaintext messages, which are strings in the plaintext alphabet P, into ciphertext, and to decode ciphertext messages back into plaintext. Thus, the key should be made available only to the sender of the messages (who needs it in order to encrypt the plaintext message to the ciphertext message) and the intended recipient (who needs it to decrypt the ciphertext message back to its original plaintext form).

We give a simple example of a substitution cipher in which both plaintext and ciphertext alphabets consist of the set of 26 English letters. For added clarity, we will let the plaintext alphabet be the set of lowercase letters: $P = $ {a, b, c, ..., x, y, z}, and the ciphertext alphabet be the set of uppercase letters: $C = $ {A, B, C, ..., X, Y, Z}.* In cases where the plaintext and ciphertext alphabets are (essentially) the same, a substitution cipher corresponds to a *rearrangement* (or *permutation*) of the letters of the alphabet. A special case is where each letter is shifted a certain number of letters down the alphabet (where the ciphertext letters A, B, C, ... cycle back after Z). Such substitution ciphers are called **shift ciphers**. The following example describes a shift cipher that was used by the Roman emperor Julius Caesar (100 b.c.–44 b.c.), and has come to be known as the *Caesar cipher.*

Example 1.4: The Caesar Cipher

Consider the substitution cipher determined by the permutation of the 26 (uppercase) letters of the alphabet obtained by shifting each plaintext letter three letters down in the alphabet

---

* All of the ideas that we present would work equally well for larger alphabets, and in practice all contemporary encryption devices are able to deal with plaintext involving upper- and lowercase letters, numbers, punctuation marks, and other symbols. Most modern computer-based cryptosystems (that are discussed after Chapter 7 of this book) process plaintext and ciphertext as binary strings (sequences of zeros and ones), integers, or even objects in more abstract number systems.

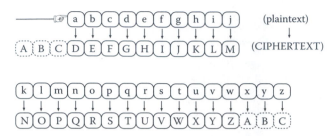

Figure 1.5 Schematic diagram of the shift permutation associated with the Caesar cipher. Each (lowercase) plaintext letter is simply shifted three letters forward in the alphabet to obtain the corresponding ciphertext.

(and recycling back to the beginning of the alphabet when we pass Z). Thus, the ciphertext letters of a, b, and c are D, E, and F, respectively. This entire shift permutation is shown in Figure 1.5.

If Caesar used this cipher to encode his famous quote:

i came, i saw, i conquered

the corresponding ciphertext would be (omitting spaces and commas):[*]

LFDPHLVDZLFRQTXHUHG

To decrypt this ciphertext, an intended recipient would simply need to shift each ciphertext letter backward three letters (see Figure 1.5). The intended recipient would be privy to the cipher, and so would be easily able to perform the decryption.

Notice the key of the Caesar cipher—namely, the shift permutation shown in Figure 1.5—can be used to both encrypt plaintext messages into ciphertext and also to decrypt ciphertext messages back into plaintext. More generally, for any substitution cipher determined by a one-to-one and onto function $F : P \to C$, the decryption procedure simply uses the inverse substitution function $F^{-1} : C \to P$, a table that can be obtained by simply reversing the arrows of the table for F (or in Figure 1.5 by reversing the arrows). Notice that the resulting inverse function of the forward shift by three letters is a backward shift by three letters. If we know we are dealing with a shift cipher, the key can be abbreviated simply by giving the number $\kappa > 0$ of letters that we shift the plaintext letters down the alphabet to obtain the corresponding ciphertext letters. Thus, the key for the Caesar (shift) cipher is $\kappa = 3$.[†]

---

[*] Preserving spaces, either directly or by means of an additional ciphertext character, would be easily detected and would render any substitution cipher much less secure since it would convey complete information on word lengths of the plaintext.

[†] Note that when $\kappa = 13$, the shift cipher is its own inverse (i.e., it is self-decrypting). This is the famous "rot13" cipher that was used in the early days of the Internet. It was discovered in 1999 that this low-security system was actually used by a major international e-mail provider to store user passwords.

Figure 1.6 A Confederate cipher disk.

Shift ciphers and generalizations of it are most naturally described in terms of modular integers and their arithmetic, and these concepts are developed in the next chapter. More sophisticated ciphers and the ability to program them for computers rely on arithmetic in other number systems such as matrices (Chapter 4), various bases (Chapter 6), finite fields (Chapter 10), and most recently, an interesting arithmetic involving certain points that lie on special curves known as elliptic curves (Chapter 12).

Before the age of computers, mechanical devices and machines were created for the sole purpose of encrypting and decrypting messages with respect to particular cryptosystems. Thus, rather than simply exchanging keys for the cryptosystem, designated parties would all have the same cryptographic devices (which, of course, were kept very secure). A very simple shift cipher device was created by the Confederates for encryption/decryption during the U.S. Civil War, showing that devices such as Caesar's cipher remained in serious use for nearly two millennia. A photograph of a Confederate cipher disk is shown in Figure 1.6.* Only five such original devices are known to exist today; one is on display at the NSA museum in Fort Meade, Maryland. Such mechanical cryptosystems reached their pinnacle with the notorious German Enigma machines, which were extremely sophisticated mechanical and electric devices. We return to this interesting era of history in Chapter 3.

Exercise for the Reader 1.2

(a)   Find the ciphertext for the plaintext message: "Meet the iceman at noon," using the shift cipher with a shift of 12 letters down the alphabet.

(b)   The following ciphertext was encrypted using the shift cipher of part (a):

VQZWUZEUEMFGDZOAMF

Find the original plaintext message.

---

* We kindly acknowledge the Confederate Secret Service Camp 1710 for permission to include this photograph (http://home.earthlink.net/~cssscv/).

Attacks on Cryptosystems

Let us now briefly digress to the other side of the game. How would eavesdropper Eve be able to crack a substitution cipher? Generally, it is safe to assume that the intruder has some information about the type of cryptosystem used, for example, a substitution cipher. Depending on what else Eve knows, there are several different approaches. Some common approaches to eavesdropping, or **passive attacks**, in a cryptosystem are described in general in the following definition. We remind the reader that there are other ways that a cryptosystem can be compromised, for example, by attempting to modify messages or sending an encrypted message pretending to be from someone else. Such an intrusion would be called an **active attack**, since it attempts to change or corrupt the data, and would be done by a Mallory (rather than an Eve). We address active attacks later in this chapter.

Definition 1.4   Types of Passive Attacks on a Cryptosystem

We differentiate the various attacks that Eve can make depending on what information she has about the cryptosystem.

(a) If Eve has only a string (or strings) of ciphertext, her attack would be termed **ciphertext only**.

(b) If Eve has both a string (or strings) of ciphertext and the corresponding plaintext, it is called a **known plaintext** attack.

(c) In a **chosen plaintext** attack, Eve would have temporary access to the encryption system, be able to use it to encrypt some plaintext strings of her choice, and see the corresponding ciphertext strings.

(d) In a **chosen ciphertext** attack, Eve has temporary access to the decryption machine and could use it to decrypt some ciphertext strings of her choice (perhaps ones that she has previously intercepted).

Example 1.5: Passive Attacks on a Substitution Cipher

We discuss how each type of passive attack could be implemented on a substitution cipher.

Since substitution ciphers are **monoalphabetic** ciphers, meaning that each plaintext character is always encrypted to the same ciphertext character, a chosen plaintext or a chosen ciphertext attack could easily reveal the whole system. For example, in a chosen plaintext/ciphertext attack, if we simply encrypted the string "abcd ... xyz," we would have the entire key. If the system was a forward shift cipher (and we had this information), we would only need to encrypt/decrypt a single letter—say, "a"—to determine the key. We will soon introduce ciphers that are **polyalphabetic**, meaning that plaintext characters may encrypt to different ciphertext characters at different instances. Such a naïve approach as above will not suffice for a chosen plaintext/ciphertext attack on polyalphabetic ciphers.

For a general substitution cipher, a known plaintext attack would tell Eve exactly how the letters appearing in the known plaintext are encrypted. If we were dealing with a shift cipher, then, as above, the information about a single plaintext character would determine the entire key.

Finally, we move on to discuss ciphertext-only attacks, which are typically the most difficult. If it is known, however, that we are dealing with a shift cipher, then since there are only 26 keys (25 actually), the **brute-force approach** of simply trying each of them to decode a given ciphertext (until it produces something that makes sense) could easily be implemented (on a computer), and this would completely determine the key. For a general substitution cipher, however, there are too many possibilities to check for a brute-force approach to be feasible, even using supercomputers. Indeed, to see how many different one-to-one and onto substitution functions $F:P \rightarrow C$ there are from the 26-letter English plaintext alphabet P to another set C (the ciphertext alphabet) of the same size, we note that there are 26 choices for $F(a)$, and after this is specified—say, $F(a) = Q$—there will then be 25 choices for $F(b)$ (that is, all ciphertext letters except Q, since it was already used); once one is specified, there will be 24 choices for $F(c)$, and so on, until we get to $F(z)$, when there will be only one remaining choice. It follows from the multiplication principle,[*] that the total number of substitution functions $F:P \rightarrow C$ is $26 \cdot 25 \cdot 24 \cdots 3 \cdot 2 \cdot 1$. This product is abbreviated as 26! and is read as "26 *factorial*."[†] Since $26! = 4.0329\ldots \times 10^{26}$, even if we had a computer that could check 1 trillion permutations per second, since there are "only" 3.1536×10^7, seconds in a year, it would require over 10 billion years—over twice the age of the Earth, to have this (fast) computer check through all permutations.

A much more effective tool in a ciphertext-only attack, or to use after one has already made use of a known plaintext attack but still has not completely determined the cipher, is **statistical frequency counts**. The idea of statistical frequency counting methods relies on the fact that some letters tend to occur more frequently in written English than others. Many tables have been

---

[*] The multiplication principle is a very useful principle for counting. In its general form, it states that if we have a process involving a finite sequence of choices: choice #1 has k_1 possible options, choice #2 has k_2 possible options, choice #3 has k_3 possible options, and so on, then the total number of outcomes of this sequence of choices is the product of the numbers of options: $k_1 \cdot k_2 \cdot k_3 \cdots$.

[†] In general, if n is any positive integer, $n!$ (n factorial) is defined to be the product of all positive integers that are less than or equal to n; i.e., $n! = n \cdot (n-1) \cdot (n-2) \cdots 3 \cdot 2 \cdot 1$.

TABLE 1.1 Frequencies of the Letters of the English Alphabet

Letter	Probability	Letter	Probability
a	.082	n	.067
b	.015	o	.075
c	.028	p	.019
d	.043	q	.001
e	.127	r	.060
f	.022	s	.063
g	.020	t	.091
h	.061	u	.028
i	.070	v	.010
j	.002	w	.023
k	.008	x	.001
l	.040	y	.020
m	.024	z	.001

published on this; for example, Table 1.1 shows the frequencies that were computed by Beker and Piper [BePi-82].[*]

Thus, *e* is by far the most frequently appearing letter (nearly 13% of all characters encountered in written English tend to be *e*'s). Although it is possible to create exceptional passages which violate these frequencies,[†] they tend to be very useful in ciphertext-only attacks. Thus, in a long ciphertext (from a substitution cipher), if a certain character appeared most often—or better yet, close to the 12.7% frequency of *e*—we would predict that this letter is the encryption of *e*. We could continue "guessing letters" in this fashion. Setting it up like a game of hangman, we could sometimes guess new letters simply by completing words. Apart from single letters, we can also use the fact that certain two-letter and three-letter combinations occur more frequently than others. For example, the most common two-letter combinations are (more common items listed first) *th*, *he*, *in*, *er*, *an*, *re*, *ed*, *on*, *es*, *st*, *en*, *at*, and *to*, and the most common three-letter combinations are *the*, *ing*, *and*, *her*, *ere*, *ent*, *tha*, *nth*, and *was*. Larger portions of ciphertext tend to make such statistical methods more effective.

The first polyalphabetic ciphers were created in the 14th and 15th centuries. Since cryptosystems were in constant use for military and diplomatic issues, new developments were sometimes kept as carefully guarded secrets by the ruling governments. Scientists who worked in the

---

[*] Of course, there will be variations in frequencies depending on the text corpus being examined. For example, the distributions in e-mails, brief text messages, and computer codes would each have distinguishing characteristics. But for most written English that is not completely informal, the distribution given in Table 1.1 works remarkably well.

[†] In 1939 an entire novel, *Gadsby*, was written by Ernest Vincent Wright and did not contain the letter e; it had over 50,000 words. Unfortunately, Wright died (at age 66) on the day his book was published, so he never saw it in print.

Figure 1.7 Blaise de Vigenère (1523–1596), French diplomat and cryptographer.

field were, of course, made to understand that even if they were to make a groundbreaking discovery, they could not expect to enjoy any fame, let alone any public recognition for it.

The Vigenère Cipher

A prototypical story exhibiting such characteristics concerns the so-called **Vigenère cipher**. Blaise de Vigenère[*] (Figure 1.7) described this cipher in his authoritative book on cryptography, *Traicté des Chiffres ou Secrètes Manières d'Escrire*[†] (first published in 1586). In it he explained that in the development of his cipher, many of the ingredients came from prominent cryptographers of the recent past; the table was invented by German Johannes Trithemius (1462–1516) and the keyword idea was introduced in a 1553 pamphlet by Italian Giovanni Battista Bellaso. Vigenère's additional contribution to the method had to do with the way in which the key was implemented. Nonetheless, it was through Vigenère's influential book that the method became widely known and hence attributed to him. The Vigenère cipher was easy to implement and many practitioners became confident in its security; it was used extensively up through the mid-19th century. In fact, it earned the name *le chiffre indéchiffrable* ("the unbreakable cipher"). It took a full three centuries for the Vigenère cipher to finally meet its demise. We now explain how the cipher works. In Chapter 5 we show its vulnerability to an ingenious ciphertext-only attack.

---

[*] Vigenère was born in the town of Saint-Pourçain, the son of a French nobleman. He received his primary education in Paris, after which at age 17 he began his diplomatic career as an assistant to the secretary of state of Francis I. His interest in cryptography began during some long-term diplomatic visits to Italy, beginning at age 26, where he met several prominent Italian cryptographers and began reading books on the subject. After retiring as a diplomat at age 47, he spent much of his retirement working on cryptography and he wrote over 20 books on the subject.

[†] The title of Vigenère's book is in old French (before the Academie Française codified spelling), akin to Shakespearian English. Translation: *Treatise on Numerals and Secret Ways of Writing*.

Definition 1.5   The Vigenère Cipher

The Vigenère cipher is determined by a key that can be any string of letters of the English alphabet, along with the **Vigenère tableau**, which is shown in Table 1.2. To encode a plaintext message, we work our way from left to right. For each plaintext character, we use the corresponding character of the key, and locate the key character's row (the key row) of the Vigenère tableau. The corresponding ciphertext character will be directly below the plaintext character in this key row. If and when the key characters are used up, we recycle back to the start of the key and continue until the plaintext is encoded.

To decode a ciphertext message, we also work from left to right using one key character for each ciphertext character. This time, the key tells us the key row in which we locate the ciphertext character, and the corresponding plaintext character will be the letter on the top of the corresponding column.

Example 1.6

(a) Use the Vigenère cipher to encode the message "Vive la France," using the keyword "money."

(b) Given that the Vigenère cipher of part (a) was used to produce the ciphertext:

NFVREAIGXFQUHMJXCGMLQ

find the original plaintext message.

Solution: Part (a): To encode the first plaintext letter v, the key row would be the m-row (first letter of the key), and the corresponding ciphertext character would be directly below the v-column, that is, H. (This process is shaded in Table 1.2.) Similarly, to encode the second letter i of plaintext, we look in the o-row under i to get W. We continue in this fashion. Note that when we get to the sixth plaintext character a (and again at the 11th plaintext character c), we would recycle back to the beginning of the keyword (so use the m-row). The complete encryption is thus:

```
plaintext:  v i v e l a f r a n c e
keyword:    m o n e y m o n e y m o
ciphertext: H W I I J M T E E L O S
```

Notice that the repeated instances of plaintext letters v, e, and a encrypt to different letters; this is in sharp contrast to substitution ciphers!

Part (b): To decode the first ciphertext letter N, we search for the location of N in the m-row of Table 1.2 (m is the first letter of the key "money"). Since N appears in the b-column of Table 1.2, the first plaintext letter is b. In the same fashion since the second ciphertext letter F appears in the r-column of the o-row of Table 1.2,

TABLE 1.2 Vigenère Tableau

Plaintext Letters

	a	b	c	d	e	f	g	h	i	j	k	l	m	n	o	p	q	r	s	t	u	v	w	x	y	z
a	A	B	C	D	E	F	G	H	I	J	K	L	M	N	O	P	Q	R	S	T	U	V	W	X	Y	Z
b	B	C	D	E	F	G	H	I	J	K	L	M	N	O	P	Q	R	S	T	U	V	W	X	Y	Z	A
c	C	D	E	F	G	H	I	J	K	L	M	N	O	P	Q	R	S	T	U	V	W	X	Y	Z	A	B
d	D	E	F	G	H	I	J	K	L	M	N	O	P	Q	R	S	T	U	V	W	X	Y	Z	A	B	C
e	E	F	G	H	I	J	K	L	M	N	O	P	Q	R	S	T	U	V	W	X	Y	Z	A	B	C	D
f	F	G	H	I	J	K	L	M	N	O	P	Q	R	S	T	U	V	W	X	Y	Z	A	B	C	D	E
g	G	H	I	J	K	L	M	N	O	P	Q	R	S	T	U	V	W	X	Y	Z	A	B	C	D	E	F
h	H	I	J	K	L	M	N	O	P	Q	R	S	T	U	V	W	X	Y	Z	A	B	C	D	E	F	G
i	I	J	K	L	M	N	O	P	Q	R	S	T	U	V	W	X	Y	Z	A	B	C	D	E	F	G	H
j	J	K	L	M	N	O	P	Q	R	S	T	U	V	W	X	Y	Z	A	B	C	D	E	F	G	H	I
k	K	L	M	N	O	P	Q	R	S	T	U	V	W	X	Y	Z	A	B	C	D	E	F	G	H	I	J
l	L	M	N	O	P	Q	R	S	T	U	V	W	X	Y	Z	A	B	C	D	E	F	G	H	I	J	K
m	M	N	O	P	Q	R	S	T	U	V	W	X	Y	Z	A	B	C	D	E	F	G	H	I	J	K	L
n	N	O	P	Q	R	S	T	U	V	W	X	Y	Z	A	B	C	D	E	F	G	H	I	J	K	L	M
o	O	P	Q	R	S	T	U	V	W	X	Y	Z	A	B	C	D	E	F	G	H	I	J	K	L	M	N
p	P	Q	R	S	T	U	V	W	X	Y	Z	A	B	C	D	E	F	G	H	I	J	K	L	M	N	O
q	Q	R	S	T	U	V	W	X	Y	Z	A	B	C	D	E	F	G	H	I	J	K	L	M	N	O	P
r	R	S	T	U	V	W	X	Y	Z	A	B	C	D	E	F	G	H	I	J	K	L	M	N	O	P	Q
s	S	T	U	V	W	X	Y	Z	A	B	C	D	E	F	G	H	I	J	K	L	M	N	O	P	Q	R
t	T	U	V	W	X	Y	Z	A	B	C	D	E	F	G	H	I	J	K	L	M	N	O	P	Q	R	S
u	U	V	W	X	Y	Z	A	B	C	D	E	F	G	H	I	J	K	L	M	N	O	P	Q	R	S	T
v	V	W	X	Y	Z	A	B	C	D	E	F	G	H	I	J	K	L	M	N	O	P	Q	R	S	T	U
w	W	X	Y	Z	A	B	C	D	E	F	G	H	I	J	K	L	M	N	O	P	Q	R	S	T	U	V
x	X	Y	Z	A	B	C	D	E	F	G	H	I	J	K	L	M	N	O	P	Q	R	S	T	U	V	W
y	Y	Z	A	B	C	D	E	F	G	H	I	J	K	L	M	N	O	P	Q	R	S	T	U	V	W	X
z	Z	A	B	C	D	E	F	G	H	I	J	K	L	M	N	O	P	Q	R	S	T	U	V	W	X	Y

Note: The 26 columns (labeled a through z) correspond to plaintext characters, the 26 rows (labeled a through z) correspond to key characters, the uppercase letters inside are ciphertext letters. For example, the shaded v-column corresponds to the plaintext letter v, the shaded m-row corresponds to the key character m, and where they intersect gives the corresponding ciphertext character H.

the second plaintext letter is r. Continuing this process, recycling the keyword "money," we arrive at the following decryption:

```
ciphertext:   N F V R E A I G X F Q U H M J X C G M L Q
keyword:      m o n e y m o n e y m o n e y m o n e y m
plaintext:    b r i n g o u t t h e g u i l l o t i n e
```

Exercise 1.3

(a) Use the Vigenère cipher to encode the message "Code blue alert," using the keyword "dijon."

(b) Given that the Vigenère cipher of part (a) was used to produce the ciphertext:

EZNOXRCCOGPQMBVJPC

find the original plaintext message.

Exercise 1.4

(a) Explain how the decryption of a Vigenère cipher can be realized as an encryption of a Vigenère cipher with another keyword.

(b) Find the keyword for the Vigenère cipher corresponding to the decryption process of the Vigenère cipher of Example 1.6.

The Playfair Cipher

Our next example is the first historically documented example of what is known as a **block cipher**. In a block cipher, letters are grouped into same-sized blocks, and these plaintext blocks are processed together to form the corresponding blocks of ciphertext in a way that changing a single letter in a plaintext block can potentially change other letters in the corresponding ciphertext block. It was created in the mid-19th century by the British scientist Sir Charles Wheatstone. It is known as the **Playfair cipher**, after Baron Lyon Playfair, who helped to promote its use by the British government in its South African (Boer) wars. It continued to be used by the British military through World War I.*

---

* Sir Charles Wheatstone (1802–1875) was a British scientist and prolific inventor most famous for developing the Wheatstone bridge, a device for measuring resistances in electric circuits. He also invented a telegraph before Samuel Morse—an achievement for which he was knighted; a musical instrument (the concertina); and a three-dimensional image display device (the stereoscope). Cryptography was one of his hobbies that he shared with his friend Baron Lyon Playfair (1818–1898), who lived across London's Hammersmith Bridge. They took Sunday walks together where they worked on cracking codes. Their dispositions were quite different. Wheatstone was so extremely shy that, although appointed as a professor, he rarely gave public lectures. In contrast, Playfair, also a scientist, was a public figure who served in an assortment of official roles including as Speaker in the House of Commons and as president of the British Association of Advancement of Science. He had direct access to many policymakers and was able to convince them to adopt the Playfair cipher.

Definition 1.6   The Playfair Cipher

We start with a key, which can be any word. To illustrate, we use "basketball" for the key. Repeated letters of the key are removed; in our example, we get "basketl." The letters of the reduced word are then deployed into a 5×5 array (starting from the upper left and proceeding in reading order), and the remaining spaces of the array are filled with the remaining letters of the alphabet, except that i and j are treated as a single letter. In our example the array would be:

$$
\begin{array}{ccccc}
b & a & s & k & e \\
t & l & c & d & f \\
g & h & ij & m & n \\
o & p & q & r & u \\
v & w & x & y & z
\end{array}
$$

Encryption scheme: Given a plaintext message, for example,

the iceman will arrive at midnight

we group the letters into adjacent pairs, but if any pair has the same two letters, we insert an x between them and regroup.

th ei ce ma nw il la rx ri ve at mi dn ig ht

In case there is an odd number of letters, we would append an additional x at the end to complete the last pair.

Each pair of letters is encrypted using the above 5×5 array depending on which of the following three cases is applicable:*

Case 1. The two letters are not in the same row or column of the array. In this case, we replace each letter with the letter in its row that is in the column of the other letter. In our example, the first pair *th* falls into this case, so *t* gets replaced by *l*, and *h* gets replaced by *g*, so the pair gets encrypted as *lg*.

Case 2. The two letters are in the same row. In this case, we replace each letter with the letter to its immediate right, cycling back to the beginning of the row if the letter is all the way on the right. In our example, the pair *ig* falls in this case, so it gets encrypted as *mh*.

Case 3. The two letters are in the same column. In this case, we replace each letter with the letter immediately below it, cycling back to the top of the column if the letter is all the way at the bottom. In our example, the pair *la* falls in this case, so it gets encrypted as *hl*.

---

* As in the Vigenère cipher, rows are horizontal segments of the table, and columns are vertical segments.

Continuing with the remaining pairs, we obtain the sequence:

lg sn fs hk hz hc hl qy qm z bbl nm fm mh gl

and thus the ciphertext is

LGSNFSHKHZHCHLQYQMZBBLNMFMMHGL

The decryption process is accomplished by reversing the above process. First go through the pairs of letters according to their cases (in cases 2 and 3, the replacements are done with the letters immediately to the left or above), then remove any redundant x's to recover the original message.

Exercise for the Reader 1.5

(a)   The Playfair cipher is used with keyword "barcelona" to encrypt the message "Meet agent Yullov at the Auberge Restaurant." Find the ciphertext.

(b)   The Playfair cipher of part (a) was used to produce the following ciphertext:

MAXHNVGLBERCCXSIHBXSGBBCACMRDERQRZ

Decode this message.

Although it is more secure than substitution ciphers, the Playfair cipher is susceptible to ciphertext-only attacks by doing statistical frequency counts of pairs of letters, since any pair of letters will always get encrypted in the same fashion. But since there are $26^2 = 676$ such ordered pairs of letters and the distinctions are less pronounced than those for single-letter statistics, a ciphertext-only attack would typically require significantly larger portions of ciphertext. Also, short keywords make the Playfair cipher much easier to crack (since the portion of the array after the keyword is much more predictable). For more details on the cryptanalysis of the Playfair cipher, the interested reader may consult [Gai-89]. More sophisticated block ciphers are often naturally developed in terms of matrices, and Chapter 4 presents all of the properties about matrices that we will need.

The 20th century saw a proliferation of ever more sophisticated block cryptosystems that required special mechanical and/or electric devices to use. These systems continued to evolve into the computer age. In Chapter 7 we develop the **Data Encryption Standard** (DES), which was a system adopted in 1973 by the U.S. government to address the growing cryptographic needs of business and industry. The encryption process of DES involved 16 complicated rounds of processing blocks consisting of binary strings (zeros and ones) of size 64. The details are quite complicated, involving various substitutions, permutations, and some other functions that we will explain later. This is a "computer-only" system that is unfeasible for hand calculations. The DES system has a high degree of

entropy, meaning that minor changes in the plaintext can produce radically different ciphertexts. The system was in widespread worldwide use for nearly 30 years. With increasing computer speeds and new cryptanalysis methods being developed, it started to become apparent that a more secure system was required, and this led to the **Advanced Encryption Standard** (AES) system in 2002. Whereas the DES, although quite complicated, relied on rather basic mathematical functions and operations, the AES is based on arithmetic in an abstract number system called a finite field. We discuss finite fields in Chapter 10 and develop the AES in Chapter 11.

All of the cryptosystems described above and all of the other ones that we did not mention dating before the 1970s shared the common disadvantage that they are so-called **symmetric key** (or **private key**) **cryptosystems**. This simply means that the decryption key and process are essentially the same as the encryption key process (perhaps with certain elements of the process being reversed). The ramification is that the key must be provided to both the sender and recipient so that secure communication can take place, and of course, the keys must be kept out of reach from any anticipated hackers.

Most experts had believed that there was no way around this symmetric key concept; in other words, if one knows how a message gets encrypted, then one should be able to figure out how to reverse the process and thus be able to decrypt any message sent under the same cryptosystem. One of the main drawbacks of all symmetric key cryptosystems is the fact that in order for such a system to be employed, the keys must be distributed to all participating parties before any secure communication can take place. This task by itself is often difficult or impractical. Such drawbacks can now be circumvented thanks to a remarkable revolution in cryptography known as **public key cryptography** or **asymmetric key cryptography** that occurred in the 1970s. The discovery was first published in a groundbreaking 1976 paper by American cryptographers Whit Diffie and Martin Hellman [DiHe-76].* Although Diffie and Hellman did not provide a complete practical implementation of a public key cryptosystem, they provided an important key exchange protocol (*the Diffie–Hellman key exchange*) by which two remote parties could establish a secure key using public (insecure) channels. Inspired by the Diffie–Hellman paper and the need for a practical cryptosystem implementation

---

* Merkle and Hellman later collaborated to develop one of the first public key cryptosystems; it is discussed, with others, in Chapter 10. Bailey Whitfield (Whit) Diffie went straight from earning his B.S. degree (1965) in mathematics at MIT to a job at the MITRE Corporation, where he became very interested in cryptography. This interest motivated him to accept a position four years later at Stanford's artificial intelligence laboratory. Martin E. Hellman earned his B.S., M.S., and Ph.D. degrees in electrical engineering from New York University. After completing postdoctoral positions at IBM and MIT, he moved on to take an academic position at Stanford in 1971, where he met Diffie. Both received numerous accolades for their pioneering work, including an honorary doctorate for Diffie from the Swiss Federal Institute of Technology. Hellman remained at Stanford until his retirement, where he had an illustrious career with continuous strong research activity and as an award-winning teacher. Diffie worked for most of the rest of his career in industry and currently serves as a vice president and chief security officer at Sun Microsystems.

Figure 1.8 American cryptographers Martin Hellman (1945–) (middle), and Whit Diffie (1944–) (right), pictured with Ralph Merkle (1952–). With permission of Chuck Painter/Stanford News Service.

of their concept, MIT scientists Ronald Rivest, Adi Shamir, and Leonard Adleman invented their *RSA cryptosystem*[*] in 1978. This has turned out to be one of the most important and widely used public key cryptosystems, and for their ingenious achievement, the three were awarded the Turing Award in 2002. The Turing Award is often referred to as the Nobel Prize in computer science.

The concept of public key cryptography had actually been discovered by British cryptographer James Ellis (Figure 1.9)[†] in the late 1960s, and the

---

[*] It was in their RSA paper [RiShAd-78] that the characters "Alice" and "Bob" were introduced as permanent fixtures in the cryptography saga.

[†] James Ellis was born in Britain and studied physics at Imperial College in London. After college, his first job was with the Post Office Research Station (which had an active cryptography team), and he was subsequently recruited in 1952 by the GCHQ (which had previously been *Bletchley Park*). His discovery of public key cryptography was made in the late 1960s, apparently motivated by his reading of a World War II-era paper on the concepts of adding/subtracting random noise to encrypt voice communications. Ellis did not have a sufficient mathematical background to adapt his concept into a practical algorithm. Clifford Cocks had a very strong mathematical background to nicely complement Ellis's strengths. He won the silver medal at the International Mathematical Olympiad as a high school student and went on to study mathematics at Cambridge, and then to do graduate work in number theory at Oxford. As a graduate student, he was recruited by GCHQ in 1973, and after learning of Ellis's public key discovery, he invented, in his first year at GCHQ, the public key cryptosystem that was later known as RSA. It was not until 1997 that the GCHQ allowed information about these discoveries to be made public. This dissemination was made through a public lecture by Cocks in that same year. The timing was unfortunate, since Ellis had passed away one month before this talk.

Figure 1.9 James H. Ellis (1924–1997), British cryptographer.

RSA implementation of it by Clifford Cocks (Figure 1.10) in 1973, while they were employed at the *Government Communications Headquarters* (GCHQ), the British analogue of the United States' NSA. The latter scientists did not receive any recognition for their discoveries until 1997, when the British government decided to declassify the information. Such stories are typical of many of the unsung heroes of cutting-edge cryptography, who often are required (by their governments) to keep a tight lid on their discoveries as matters of national security.

We will enter into the technical details of public key cryptography in Chapter 9, but it will be helpful to first give a superficial overview: Each communicating party (or individual) has two keys, a **public key** and a **private key**. Unlike with symmetric key cryptography, it is not feasible to obtain the private key from knowledge of the public key. The directory of public keys is made available to the general public (including Eve and Mallory), while all parties keep their private keys only to themselves. When Alice sends a message to Bob, she encrypts the message using Bob's public key. Only Bob, who has the corresponding private key, will be able to decrypt Alice's message. Apart from removing the prerequisite key distribution issue,

Figure 1.10 Clifford C. Cocks (1950–), British cryptographer.

public key cryptography also greatly reduces the number of keys needed. For example, if we had a network of one million parties, with a symmetric key cryptosystem, each pair would need a separate key exchanged before communications could take place. This would amount to half a trillion keys, all of which need to be securely transmitted—this is a logistical nightmare. A public key system, on the other hand, would require only two million keys, none of which would need to be securely transmitted.

Basically, public key cryptography translates the difficulty of cracking into the system (by determining a private key from public key registries) into the difficulty of solving certain notoriously difficult mathematical problems, whose "inverse" problems are much easier to solve. For example, the RSA system, to be discussed in Chapter 9, is based on the difficulty of factoring large positive integers. The inverse problem is simply multiplying large positive integers, which has always been easy. We will learn much more about prime numbers and the associated number theory in Chapter 8, which also addresses the important practical problem of generating large prime numbers (since they are needed for many public key cryptosystems). Encryption is based on the easier inverse problem, whereas unauthorized decryption would be based on the computationally infeasible problem. Using such problems that have been well known and actively researched for a long time adds to the confidence of the security of such a system. Any of these public key systems are subject to faltering upon any new discovery of efficient algorithms for the intractable problems on which they are based. Although it has not been proved, for example, that an efficient algorithm for prime factorization cannot exist, it is the general consensus that this is the case. Other public key cryptosystems are based on a very special class of intractible problems known as NP complete problems.* We will introduce knapsack cryptosystems, which are based on the NP complete knapsack problem. It is interesting to point out that because of the increased importance that such problems now have due to the widespread use of cryptosystems that are based on them, the NSA strictly regulates certain areas of research relating to such problems. American scientists who make any novel discoveries in areas relating to public key cryptography need to clear them with the NSA before announcing them to the public (or publishing).

Several other public key cryptosystems are developed in Chapter 9. In addition to the confidentiality that is provided by symmetric key cryptosystems, public key cryptosystems all provide the following additional features:

---

* There are a very large number of computational problems where there is an "efficient" way to check whether a proposed answer is correct, in that it can be done in an amount of time that is bounded by a power of the input size (this is called "*in polynomial time*") but where no known algorithm has been designed to find the solution that will also work in polynomial time. A prototypical example is the prime factorization problem. It has been established that there is a plethora of such problems that are seemingly unrelated but if a polynomial time algorithm is discovered for one of them, then polynomial time algorithms can be produced for all of them! This latter class of problems is known as the *NP complete problems*, whereas problems that can be solved in polynomial time are called *P problems*. Most scientists believe that $NP \neq P$, but the conjecture remains one of the most famous unsolved problems in mathematics and computer science. For more details on the $P = NP$ problem, the interested reader is referred to the classic but authoritative reference by Garey and Johnson [GaJo-79]. Resolving this problem is one of the seven millennium problems for which the *Clay Foundation* (http://www.claymath.org/millennium/) is offering $1 million prizes.

- **Authentication**: The intended recipient of a message will be able to verify that it came from the indicated sender.
- **Nonrepudiation**: The sender of a message will not be able to deny that he or she was the sender.

These can be achieved by so-called **digital signature schemes**. Digital signatures, unlike ordinary signatures, are unique for each sender and cannot be forged.

With all of the added advantages and high security of public key cryptosystems, a natural question thus arises: Why even bother anymore with symmetric key cryptosystems? The answer is that symmetric key cryptosystems are significantly faster and more efficient than public key cryptosystems. Thus both types of cryptosystems can continue to live a productive coexistence: public key cryptosystems can be used to securely exchange private keys, after which the faster private key cryptosystems can be used.

In the mid-1980s, a new sort of public key cryptosystem was developed using a geometrically motivated (but analytically complicated) arithmetic of points with integer coordinates on certain planar curves known as *elliptic curves*. In spite of their name, these curves are not ellipses but a more diverse family of unbounded curves. The key sizes required for a given elliptic curve cryptosystem are significantly smaller than what would be required for other typically known public key cryptosystems with the same degree of security, and this fact has made elliptic curve cryptography one of the most promising and extensively studied branches of cryptography. Elliptic curve cryptography will be studied in Chapter 12.

The One-Time Pad, Perfect Secrecy

Circumstances and needs, as well as advances in technology, fuel the constant efforts to design (and attempts to crack) evermore sophisticated cryptosystems. The eminent scientist Claude Shannon* (Figure 1.11) wrote a number of seminal papers on cryptography in which he gave two important properties that cryptosystems should possess to avoid being compromised: *diffusion* and *confusion*. *Diffusion* means that changing just a single character in the plaintext should diffuse (spread out) to affect changes in several ciphertext letters (the more the better). *Confusion* means

---

* Claude Shannon grew up in Michigan. He earned a bachelor's degree with a double major in mathematics and electrical engineering from the University of Michigan–Ann Arbor. His landmark discovery of an effective symbolism for electric circuits actually came from his master's thesis at MIT: *A Symbolic Analysis of Relay and Switching Circuits*. This thesis has had a tremendous impact on industry by changing circuit design from an art to a science. Shannon went on to earn a doctorate at MIT and continued to make valuable contributions to the electronics and communications fields during his career working at Bell Labs, where his laboratory office ceiling was adorned with a rainbow of gowns from honorary doctorates that he had received. He developed a secure cryptosystem that was used by Roosevelt and Churchill for transoceanic communications during World War II. His work in this area motivated the development of the field of coding theory, for which he is considered the founder. Coding theory studies what are called error-correcting codes, which are used in everything from CDs to routine data transmissions. We have Shannon to thank, for example, when a scratched music CD will still play perfectly well.

Figure 1.11 Claude E. Shannon (1916–2001), American applied mathematician.

that there should be no simple relationship between a cryptosystem's key and instances of its ciphertext. For example, any substitution cipher does not exhibit diffusion since changes in a single plaintext letter will affect only the corresponding ciphertext letter. Block ciphers are conceived to have good diffusion.

Another important contribution of Shannon was the concept of *perfect secrecy* that he introduced in 1949. This concept rigorously defines what it means for a cryptosystem to be "unbreakable," in the sense that seeing the ciphertext of any plaintext message (in a ciphertext-only attack) gives the hacker absolutely no information about the plaintext. There is actually a rather simple cryptosystem that exhibits perfect secrecy: the Vigenère cipher with a randomly generated key that is the same length as the plaintext; it is called a **one-time pad**. This cryptosystem is sometimes also called the **Vernam cipher**, after its inventor, Gilbert S. Vernam, a cryptographer with AT&T. It is not very practical to use because of the large keys, and the fact that once a key is used it must be thrown out. Although it had been conjectured for several decades that the one-time pad was perfectly secure, Shannon was the first to provide a rigorous proof. One-time pads have since been used for some of the most sensitive communication purposes; for example, Figure 1.12 shows a one-time pad system at the U.S. end of the Moscow–Washington hotline, in use during the Cold War era.

The next example shows how the one-time pad works.

Example 1.7: The One-Time Pad

The concept of a one-time pad involves *randomness*. By its very nature, any random process is unpredictable and this will be the key element that results in the system's being perfectly secure. There are 26 different shift operators, corresponding to the keys $\kappa = 0, 1, 2, 3, \cdots, 25$. The key for a one-time pad needs to

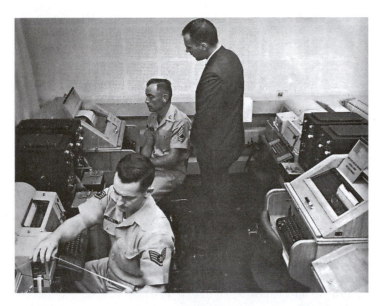

Figure 1.12 Photograph of the one-time pad machines (black) in use by the U.S. Signal Corps to support the Washington–Moscow hotline. The white machines were used to print and read plaintext messages. Photograph courtesy of the United States National Archives.

consist of a sequence of shift keys that are randomly selected from the list of 26 possible keys. Each key corresponds to how many letters down the alphabet the plaintext letter *a* (and hence all plaintext letters) gets shifted, see Table 1.3.

Suppose that we need to send a message that contains N characters. The one-time pad would require a key of length at least N. To produce the key, imagine that we label 26 identical balls with the possible key numbers 0–25 and place them in an urn; see Figure 1.13.

We shuffle the balls, randomly draw one ball, record its number, then replace it in the urn and reshuffle. We repeat this process N times to produce the one-time pad key. Although it seems contradictory, computer algorithms (which are programmed to follow a fixed set of instructions) have been designed to produce so-called *pseudorandom numbers*, which, for all practical purposes, can be assumed random.* The computer implementations given at the end of this chapter provide some schemes for producing such random numbers. For example, suppose that we needed to create a one-time pad cipher with keylength

---

* Of course, any computer algorithm runs on a specified set of instructions, so technically such a program cannot produce truly random numbers. Nonetheless, effective algorithms can be created that produce streams of numbers that satisfy all of the important statistical tests for randomness. Moreover, the programs can call on the computer clock to produce the "seed" of the generator so the algorithm will produce different streams at each call. For more details on such pseudorandom number generator algorithms, we refer the reader to Chapter 2 of [LePa-06] or Chapter 3 of [Knu-98].

TABLE 1.3 Key Values and Letters

A	B	C	D	E	F	G	H	I	J	K	L	M	N	O	P	Q	R	S	T	U	V	W	X	Y	Z
0	1	2	3	4	5	6	7	8	9	10	11	12	13	14	15	16	17	18	19	20	21	22	23	24	25

Note: The correspondence of key values κ (bottom row) and the ciphertext letter (top row) to which the plaintext letter "a" gets shifted to with a shift cipher. The value $\kappa = 0$ is not allowed as a key since it corresponds to the identity shift (that is, ciphertext letters would be identical to plaintext letters).

Figure 1.13 An urn containing 25 balls of identical size, weight, and texture can be used for the purpose of random number generation.

$N = 15$. Resorting to a random number generator, we obtained the following sequence that we will use as the key for the one-time pad:

$$\kappa = [21\ 23\ 4\ 23\ 16\ 3\ 7\ 14\ 24\ 25\ 4\ 25\ 9\ 13\ 21]$$

By consulting Table 1.3, we see that the resulting one-time pad will simply be the Vigenère cipher with keyword: vxexqdhoyzezjnv. Notice that we used lowercase letters although Table 1.3 had uppercase letters (Why?).

Chapter 1 Exercises

1. For the three diagrams shown below, indicate which specify functions. For each function, identify its domain, codomain, and range, and determine whether it is (a) one-to-one, (b) onto, or (c) bijective.

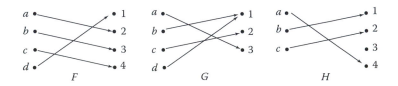

2. For these three diagrams, indicate which specify functions. For each function, identify its domain, codomain, and range, and determine whether it is (a) one-to-one, (b) onto, or (c) bijective.

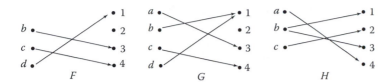

3. Consider the function f: {a, b, c, y} → {length 2 binary strings} defined by f(a) = 00, f(b) = 01, f(c) = 10, f(y) = 11.
    (a) Is f one-to-one?
    (b) Is f onto?
    (c) Determine the binary string f(a) f(b) f(b) f(y).
    (d) Suppose that the binary string 1000010111 was produced by concatenating the outputs of f under a corresponding string of input values. Determine the input string.

4. Consider the function G : {a, b, e, f, l, t, y} → {length 3 binary strings} defined by G(a) = 000, G(b) = 001, G(e) = 010, G(f) = 011, G(l) = 100, G(t) = 101, G(y) = 110.
    (a) Is G one-to-one?
    (b) Is G onto?
    (c) Determine the binary string G(b) G(e) G(l) G(t).
    (d) Suppose that the binary string 100000101010 was produced by concatenating the outputs of G under a corresponding string of input values. Determine the input string.

5. (a) Suppose that $f : A \to B$ is a function, where A and B are finite sets, and that A has more elements than B. Does f necessarily have to be onto? Can f ever be one-to-one? Explain.
    (b) Suppose that $f : A \to B$ is a function, where A and B are finite sets, and that B has more elements than A. Does f necessarily have to be one-to-one? Can f ever be onto? Explain.

6. (a) Suppose that $f : A \to B$ is a one-to-one function, where A and B are finite sets, each containing the same number of elements. Explain why f is necessarily bijection.
    (b) Suppose that $f : A \to B$ is an onto function, where A and B are finite sets, each containing the same number of elements. Explain why f is necessarily bijection.

7. Provide an example of a function from the positive integers {1, 2, 3, ⋯} to {1, 2, 3, ⋯} that is:
    (a) Neither one-to-one nor onto.
    (b) One-to-one, but not onto.
    (c) Onto, but not one-to-one.
    (d) A bijection f such that $f(a) \neq a$, for each positive integer a.

8. Provide an example of a function from the set {finite length binary strings} to the set {finite length binary strings} that is:
    (a) Neither one-to-one nor onto.
    (b) One-to-one, but not onto.
    (c) Onto, but not one-to-one.
    (d) A bijection f such that $f(\sigma) \neq \sigma$, for finite length binary string σ.

9. Consider the suffix function G: {finite length binary strings} $\rightarrow$ {finite length binary strings} defined by $G(\sigma) = \sigma \cdot 1$; i.e., $G(\sigma)$ is the concatenation of σ with the length 1 string "1." For example, $G(1010) = 10101$. (In other words, G tacks a suffix "1" onto every string.)
    (a) Is G one-to-one?
    (b) Is G onto?
    (c) In case G is a bijection, determine the inverse function.

10. Consider the reversal function H: {finite length binary strings} $\rightarrow$ {finite length binary strings} defined by $H(b_1 b_2 \cdots b_{n-1} b_n) = b_n b_{n-1} \cdots b_2 b_1$; i.e., the output of any binary string (under H) is the string of the same length, but with the bits given in the opposite order. For example, $H(1010) = 0101$.
    (a) Is H one-to-one?
    (b) Is H onto?
    (c) In case H is a bijection, determine the inverse function.

11. Consider the function f: {length 8 binary strings} $\rightarrow$ {length 8 binary strings} defined by $f(b_1 b_2 b_3 b_4 b_5 b_6 b_7 b_8) = b_2 b_4 b_6 b_8 b_1 b_3 b_5 b^*$, where $b^* = 1$ if $b_6 + b_7 + b_8$ is an even number; otherwise, $b^* = 0$. For example, $f(11110000) = 11001101$.
    (a) Is f one-to-one?
    (b) Is f onto?
    (c) In case f is a bijection, determine the inverse function.

12. Consider the function g: {length 4 binary strings} $\rightarrow$ {length 4 binary strings} defined by $g(b_1 b_2 b_3 b_4) = c_1 c_2 b_1 b_4$, where $c_1 = 1$ if $b_1 + b_2$ is an even number; otherwise, $c_1 = 0$, and while $c_2 = 1$ if $b_2 + b_4$ is an even number; otherwise, $c_2 = 0$. For example, $g(1111) = 0011$.
    (a) Is g one-to-one?
    (b) Is g onto?
    (c) In case g is a bijection, determine the inverse function.

13. (a) Use the Caesar cipher to encrypt the following strings of plaintext:
        (i) the shipment will arrive at noon
        (ii) lay low until friday
        (iii) always use the back door
        (iv) the phone is bugged
    (b) Decrypt each of the following ciphertexts that came from the Caesar cipher:
        (i) EULQJWKHLWHPWRMHQNLQV
        (ii) VHQGDJHQWSRONDVLJQDO
        (iii) LQWHUFHSWWWKHLUFDVHZRUNHU
        (iv) FKHFNLQWRWKHKRWHO

14. (a) Use the Caesar cipher to encrypt the following strings of plaintext:
        (i) two minutes until alarm sounds
        (ii) spread out your team
        (iii) reconnaissance is on schedule
        (iv) this hotel is safe
    (b) Decrypt each of the following ciphertexts that came from the Caesar cipher:

> (i) OHDYHPRQHBLQVZLVVDFFRXQW
> (ii) VWDOOWKHPIRUWKUHHKRXUV
> (iii) GRQWOHDYHZLWKRXWDJHQWGXFKRYVNL
> (iv) ERRNDIOLJKWWRSUDJXHXQGHUDOLDV

15. (a) Use the shift cipher with key $\kappa = 22$ (i.e., encryption is accomplished by shifting 22 letters down the alphabet) to encrypt each of the strings of plaintext of Exercise 13, part (a).

    (b) Decrypt each of the following ciphertexts that came from the shift cipher with key $\kappa = 18$:
       (i) OSALAFYXGJAFKLJMULAGFK
       (ii) KMTBWULZSKTGSJVWVHDSFW
       (iii) ESCWAFALASDUGFLSULSKSTMKAFWKKESF
       (iv) GHWJSLAGFZSKTWWFUGEHJGEARWV

16. (a) Use the shift cipher with key $\kappa = 6$ (i.e., encryption is accomplished by shifting six letters down the alphabet) to encrypt each of the strings of plaintext of Exercise 14, part (a).

    (b) Decrypt each of the following ciphertexts that came from the shift cipher with key $\kappa = 1$.
       (i) SFUVSOUPGJFMEPGGJDFOPX
       (ii) BTTFNCMFZPVSTUSJLFUJNFCZNJEOJHIU
       (iii) TFOEGPSBEEJUJPOBMBHFOUT
       (iv) JOGPSNBMMMPDBMDBTFXPSLFSTPGUIFQMBO

17. (a) Use the Vigenère cipher with key rocket to encrypt each of the strings of plaintext of Exercise 13, part (a).

    (b) Decrypt each of the following ciphertexts that came from the Vigenère cipher with key bluefog:
       (i) ILLVJZRXTFPGSCBTNMSULPCSSZ
       (ii) USYQJHZJYAANHNXLNWTBOTLMIYIV
       (iii) DZGIFZUOPVYYPXJYACTIXQTYGJ
       (iv) SPHXFFUPXCRYVKIZNIQAGSTARTBOOEBIK
           WLUSUVWCTETMIRSTU

18. (a) Use the Vigenère cipher with key mole to encrypt each of the strings of plaintext of Exercise 14, part (a).

    (b) Decrypt each of the following ciphertexts that came from the Vigenère cipher with key timbucktu:
       (i) VWPFABXYG
       (ii) TTXTSCMYFAMSYEIUGLDFUNRNHZGO
       (iii) UZUOAIHOKVUHBDOCLQAOAYZAEME
       (iv) RWGSUVBULXMTMZHLMQEXUSMCGORPLIHO

19. (a) Use the Playfair cipher with key diskjockey to encrypt each of the strings of plaintext of Exercise 13, part (a).

    (b) Decrypt each of the following ciphertexts that came from the Playfair cipher of part (a):
       (i) RBIABDIGTPSZ
       (ii) QMBGDTYASKCZXKPKCIDUICTPYBQM
       (iii) REBSLUMNGYXYNBLFCR
       (iv) QTBPCPSCDZLXYBQTDMYIKDTKUFGEQD
           SIYEITBQGYGDGAKW

20. (a) Use the Playfair cipher with key `crimson` to encrypt each of the strings of plaintext of Exercise 14, part (a).

(b) Decrypt each of the following ciphertexts that came from the Playfair cipher of part (a):

    (i) KFMCVFNIRAQGCFASOIEFQY

    (ii) EFFLDINGKOMCQBORGV

    (iii) YTFCGCIDIOCHINRAYTFCKCPMAVBC

    (iv) OHXNCFNERDRQFCCDBPKFIOYTKOIN
    PCAVNELBQW

21. Explain how a known plaintext attack on the Vigenère cipher would work. How much plaintext would be required for the attack to work?

22. (a) Explain how a chosen plaintext attack on the Vigenère cipher would work. How much plaintext would be required for the attack to work?

(b) Explain how a chosen ciphertext attack on the Vigenère cipher would work. How much ciphertext would be required for the attack to work?

ADFGVX Cipher

A cipher that is similar to the Playfair cipher, known as the **ADFGVX cipher**, was used by the Germans during the First World War. The ciphertexts involve only these six letters, which were chosen because of their easy distinctions in Morse code (which through telegraphs and radio was the primary means of military communications). We explain how this cipher works through a specific example. First, the method begins by randomly arranging the 26 letters of the alphabet along with the 10 digits into a 6×6 array with the rows and columns labeled with the letters ADFGVX. Table 1.4 shows such a table.

TABLE 1.4 ADFGVX Table

	A	D	F	G	V	X
A	8	p	3	d	1	n
D	1	t	4	o	a	h
F	7	k	b	c	5	z
G	j	u	6	w	g	m
V	x	s	v	i	r	2
X	9	e	y	0	f	q

Encryption: Suppose that we are given a plaintext, such as "Ambush at the Rhein."

Step 1. Replace each plaintext letter with the pair of letters in the ADFGVX table (Table 1.4) that label the plaintext letter's row and column. So a is replaced by *DV*, t by *DD*, and so on.

plaintext:	a m b u s h a t t h e r h e i n
Step 1:	DV GX FF GD VD DX DV DD DD DX XD VV DX XD VG AX

At this point, we have a substitution cipher, which at the time of the First World War would have certainly been long-outdated technology and easily hacked. The second and final step makes the plaintext much more difficult to hack.

Step 2. This part, which depends on a keyword, will permute the output string of Step 1. In this example, we use the keyword MAGIC. We create a new table with columns labeled by the keyword, and fill in the cells below it in reading order, row, by row. After this is done, we rearrange the columns of this table, so the keyword letters are in alphabetical order. The ciphertext is obtained by taking the letters of each column, from top to bottom, and taking the alphabetized columns in order.

M	A	G	I	C	A	C	G	I	M
D	V	G	X	F	V	F	G	X	D
F	G	D	V	D	G	D	D	V	F
D	X	D	V	D	X	D	D	V	D
D	D	D	D	X	D	X	D	D	D
X	D	V	V	D	D	D	V	V	X
X	X	D	V	G	X	G	D	V	X
`A	X				X				A

Reading down the columns of the second (column permuted table) gives us the ciphertext:

Ciphertext: VGXDDXXFDDXDGGDDDVDXVVDVVDFDDXXA

Decryption is performed by reversing the encryption process. Note that in addition to the keyword, the ADFGVX table (Table 1.4) is also part of the key, since it depends on how the letters and digits were randomly deployed in the 36 cells.

Historical Aside: By the time of the First World War, the French had assembled a very strong cryptography team, after having suffered an embarrassing defeat where they had lost the provinces of Alsace and Lorraine in the Franco-Prussian War of 1870. This defeat would most probably have been avoided if the French had better intelligence. Soon after the Germans began confidently using the ADFGVX cipher in 1918, as they were making plans to take over Paris, the French put their most prized cryptographer, Lieutenant Georges Painvin (Figure 1.14), to work on decrypting this new cipher. Painvin

Figure 1.14 Georges Painvin (1886–1980), French cryptographer.

worked day and night to crack it and was able to succeed with three months of hard work. His efforts were so consuming, though, that they affected his health; he lost 30 pounds in the process. Readers interested in learning more details about Painvin's ingenious attack may refer to [Kah-96].

23. (a) Use the ADFGVX cipher with key PARIS to encrypt each of the strings of plaintext of Exercise 13, part (a).

   (b) Decrypt each of the following ciphertexts that came from the ADFGVX cipher of part (a):

   (i) VVVDXDVDDXVDDD

   (ii) XXDDGADAXVVXGGXVXXGVXGXGVGGD
        DXDAGDGDDADAXAGAVAFVXVGVDXGDXA

   (iii) DVDGVGDGDFDDVDFVVXGVGDVDGDX
        VGDXVDDGDVD

   (iv) XFDDDDAXDDGXXDVVVFDDADXXDGD
        VADVDVXAVAAXXDGFDXDAGAFDGD
        DDDVGDFDG

24. (a) Use the ADFGVX cipher with key CRIMSON to encrypt each of the strings of plaintext of Exercise 14, part (a).

   (b) Decrypt each of the following ciphertexts that came from the ADFGVX cipher of part (a):

   (i) DVDDAAXDVGFGDDDXGFXFVVADVXVGFA

   (ii) VVADXDDGXGDDDDVGDADDXXVDGX
        VGVDXVXXGXXVVXVDVGGGXDVDDA

   (iii) DXGDXVDDVVXVGXXGVVGXFDFXGDGD
        FVDDFDDDAFAGXXGFVD

   (iv) DFAVVXDADVDDDDVVDGXXXDDVD
        DXXDDXVDVADDDDDDGDGXDGXDAXVD
        DDVDAXADDVDXDAD

25. Do identical adjacent pairs of plaintext typically encode to the same four-letter ciphertext strings under the ADFGVX cipher? Explain your answer.

26. Are there any problems with procedure and/or loss of security with the ADFGVX cipher if one were to use a keyword with duplicated letters (such as LONDON)? Explain your answer.

27. (a) Do we gain any new ciphers by allowing the shift ciphers to shift to the left (rather than just to the right)? Explain your answer.

   (b) Do we gain any new ciphers by allowing the shift ciphers to shift more than 25 letters to the right? Explain.

28. Suppose that we construct a cryptosystem consisting of a Vigenère cipher, followed by another Vigenère cipher, where the keywords of each have the same length. Explain how much additional security, if any, such a system would provide over a single Vigenère cipher.

29. Suppose that $n > m$ are positive integers. Discuss the differences in security of the following two cryptosystems:

   (i) Use a Vigenère cipher with a keyword of length nm.

   (ii) Use a Vigenère cipher with a keyword of length n followed by another Vigenère cipher of keyword length m.

30. (a) List all binary strings of length 0, 1, 2, and 3.
    (b) Use the multiplication principle to compute the number of binary strings of length n, where n is any positive integer.
    (c) Letting B_n denote the binary strings of length n, explain why every string in B_{n+1} can be uniquely expressed as either $0 \cdot \sigma$ or $1 \cdot \sigma$ for some length-n binary string σ.
    (d) Use the result of part (c) to give another proof of the result of part (b) using mathematical induction.

31. Discuss the secrecy of a substitution cipher that is used to send a plaintext message that consists of just a single letter.

Chapter 1 Computer Implementations and Exercises

Note: Some of the exercises below ask the reader to write programs that may not be feasible on some computing platforms or that require knowledge of certain sorts of data structures that will not be essential in later developments in this book. For example, most of the cryptosystems that we will develop after this chapter are designed to work directly on either strings or ordered lists (vectors) of numbers. The numbers will most often be integers or binary numbers (zeros and ones). Later, we will essentially assume that plaintexts will be presented in this form. In cases where the programming for particular exercises is not feasible or not important for a particular platform or use, such an exercise may be suitably improvised or even skipped without any loss of continuity.

Vector/String Conversions

Oftentimes in computer implementations of cryptosystems, it is more convenient to work with **vectors** rather than strings. A vector is simply an ordered list. This will be the case, for example, in our development of DES in Chapter 8. On the other hand, it is often more aesthetic to display binary strings rather than binary vectors. For example, the binary vector corresponding to the binary string 101100011101 might display (depending on your particular computing platform) as

        [1 0 1 1 0 0 0 1 1 1 0 1]

or as

        [1, 0, 1, 1, 0, 0, 0, 1, 1, 1, 0, 1]

Vectors are more versatile data structures than strings, since elements could be digits or any numbers. For example, the vector [32, 5] could not be so unambiguously represented as a string (325 would not do). The first two exercises below ask you to create conversion programs to pass between strings and vectors. If you need to work with strings of digits (such as binary strings), you need to know the syntax by which to enter

TABLE 1.3 Key Values and Letters

A	B	C	D	E	F	G	H	I	J	K	L	M	N	O	P	Q	R	S	T	U	V	W	X	Y	Z
0	1	2	3	4	5	6	7	8	9	10	11	12	13	14	15	16	17	18	19	20	21	22	23	24	25

Note: The correspondence of key values *κ* (bottom row) and the ciphertext letter (top row) to which the plaintext letter a gets shifted to with a shift cipher. The value *κ* = 0 is not allowed as a key since it corresponds to the identity shift (that is, ciphertext letters would be identical to plaintext letters).

them into your computing platform. For example, the number 101 is a different data structure than the binary string 101.

1. *Program for Converting Strings of Digits to Vectors of Digits.* Write a program `Vec = String2Vec(Str)` that inputs a string `Str` of digits (binary or decimal) and outputs the corresponding vector `Vec`. Thus, for example, the command `String2Vec (24821)` should produce the output `[2 4 8 2 1]`. Run your program with the following inputs, and record the outputs:
   (a) 110101111
   (b) 22953688
   (c) 9876543210

2. *Program for Converting Vectors of Digits to Strings of Digits.* Write a program `Str = Vec2String(Vec)` that inputs a vector `Vec` of digits (binary or decimal) and outputs the corresponding string `Str`. Thus, for example, the command `Vec2String([2 4 8 2 1])` should produce the output `24821`. Run your program with the following inputs, and record the outputs:
   (a) `[1 0 1]`
   (b) `[1 0 1 1 0 0 0 1 1 1 0 1]`
   (c) `[9 8 7 6 5 4 3 2 1 0]`

Integer/Text Conversions

The next four exercises ask you to develop programs that will make conversions between the integer/text correspondence of Table 1.3.

Since vectors tend to be easier to work with than strings, it is probably best (depending on your particular platform) to have programs work internally with vectors but accept inputs and/or display outputs as strings. In order to achieve conversions relating to Table 1.3, it is most obvious to first think of employing a simple lookup type code:

(using a FOR loop to go through each inputted symbol `Let`, and then)

```
IF Let == A
        SET Code = 0
ELSE IF Let ==B
        SET Code = 1
ELSE IF Let == C
        SET Code = 2
(...etc...)
```

Although this would certainly work, it would be more efficient to make use of any built-in text conversion programs that your platform may have available. Most platforms have a conversion program that converts any of the 256 standard ASCII symbols (including upper- and lowercase letters, punctuation marks, and so forth) into its unique representative as an integer from 0 to 255. The 26 uppercase/lowercase letters should be mapped to contiguous blocks of integers, so you would simply need to find out where A (or a) gets mapped in order to create a very simple program. For example, if A gets mapped to 65 (so Z would get mapped to 90), you could simply take the output of this built-in mapping function and subtract 65 to arrive at the letter-to-integer conversion of Table 1.3.

In case your platform's program can convert a whole string at once (into a vector of integers), your program could be accomplished in a single line of code.

3. *Program for Converting Uppercase Text to Integers.* Write a program Vec = UCText2Int(STR) that inputs a string STR of uppercase English letters, and outputs the corresponding vector Vec of integers as per Table 1.3. Thus, for example, the command UCText2Int(CATBIRD) should produce the output [2 0 19 1 8 17 3]. Run your program with the following inputs, and record the outputs:
    (a) JUSTDOIT
    (b) ROADTRIPTHISWEEKEND
    (c) HIGHSTAKESGAME

4. *Program for Converting Lowercase Text to Integers.* Write a program Vec = LCText2Int(str) that inputs a string str of lowercase English letters, and outputs the corresponding vector Vec of integers as per Table 1.3. Thus, for example, the command Text2Int(catbird) should produce the output [2 0 19 1 8 17 3]. Run your program with the following inputs, and record the outputs:
    (a) longlivetheking
    (b) letsgotoamovie
    (c) dinnerpartytonite

5. *Program for Converting Integers to Uppercase Text.* Write a program STR = Int2UCText(Vec) that inputs a vector Vec of integers in the range 0 to 26, and outputs the corresponding string STR of uppercase English letters, as per Table 1.3. This is simply the inverse function of the function of Computer Exercise 3. First run this program on the outputs for UCText2Int when applied to the inputs of parts (a), (b), and (c) of Computer Exercise 3 to check that your new function is really the inverse of UCText2Int. Next, run your program with the following inputs, and record the outputs:
    (a) [2 7 0 12 15 0 6 13 4]
    (b) [5 8 11 4 19 12 8 6 13 14 13]
    (c) [2 7 14 2 14 11 0 19 4 12 14 20 18 18 4]

6. *Program for Converting Integers to Lowercase Text.* Write a program str = Int2LCText(Vec) that inputs a vector Vec of integers in the range 0 to 26, and outputs the corresponding string str of lowercase English letters, as per Table 1.3 (but with lowercase letters). This is simply the inverse function of the program LCText2Int of Computer Exercise 4. What happens if you apply this program to the output of the program UCText2Int of Computer Exercise 5 to a string of uppercase letters? Check your conclusion by the evaluation of Int2LC Text(UCText2Int(CATBIRD)). Run your program with the following inputs, and record the outputs:
    (a) [15 8 2 10 20 15 19 7 4 15 8 4 2 4 18]
    (b) [0 1 14 17 19 19 7 4 12 8 18 18 8 14 13 13 14 22]
    (c) [15 17 14 2 4 4 3 22 8 19 7 2 14 13 19 8 13 6 4 13 2 24 15 11 0 13 19 22 14]

Programming Basic Ciphers with Integer Arithmetic

The programs of the preceding computer exercises should facilitate writing encryption/decryption programs for most of the basic ciphers that were introduced in this section. The basic idea to consider is that it is much simpler to work with integers rather than the letters they correspond to in Table 1.3. This simplicity will be further enhanced as we introduce new forms of arithmetic. For example, modular arithmetic of the next chapter is particularly suitable for implementing shift and related ciphers. For now, if we wanted to implement a shift cipher, say the Caesar cipher, using the integer representation of Table 1.3, we would simply add 3 (the key) to a given plaintext representative, as long as the result is less than 26. For example, the plaintext letter f is represented by 5 (in Table 1.3), adding 3 gives 8, the corresponding representative for the ciphertext letter I (in Table 1.3). In case adding 3 gives an integer greater than 25, we would subtract 26 from the result, as this would have the same effect as cycling back to the beginning of the alphabet. For example, the plaintext letter y corresponds to 24, adding 3 gives 27, and since this is greater than 25, we subtract 26 to get 1, which is the representative of the corresponding ciphertext letter B.

7. *Program for Shift Cipher.* Write a program `StrOut = Shift Crypt(str,kappa)` that inputs a string `str` of plaintext in lowercase English letters, and an integer `kappa` mod 26. The output `StrOut` should be the corresponding ciphertext (in uppercase letters) after the shift operator with key `kappa` is applied to the plaintext. Then use your program to redo the computations of Chapter Exercises 13 and 15.

   Note: In the decrypting parts, you will need to change your ciphertexts to lowercase (and choose the correct shift parameter).

   Suggestion: The programs of some of the preceding computer exercises should be useful here.

8. *Ciphertext-Only Attack on the Shift Cipher.* It is known that the following ciphertexts were encrypted using (perhaps different) shift ciphers. Decrypt these messages and determine the corresponding keys that were used.
   (a) HXDALJAANBNAEJCRXWRBDWMNACQNWJVN
       SXWNB
   (b) BCJHJCCQNARCIKDCYJATHXDALJAJCCQN
       FJUMXAO
   (c) DWQYIDMCIFBSKGDODSFOHHVSTFCBHRSGYI
       BRSFHVSBOASXCBSG
   (d) XLIHIXEMPWSJCSYVQIXXMRKAMPPFISRXLI
       WXSGOTEKIAVMXXIRMRGSHI

9. *Program for Vigenère Cipher.* Write a program `StrOut = VigenereCrypt(str,keystr)` that inputs a string `str` of plaintext in lowercase English letters and another such string `keystr` representing a key. The output `StrOut` should be the corresponding ciphertext (in uppercase letters) after the Vigenère

cipher with key `keystr` is applied to the plaintext. Then use your program to redo the computations of Chapter Exercise 17, part (a).

Note: In the decrypting parts, you will need to change your ciphertexts to lowercase (and choose the correct key).

Suggestion: The programs of some of the preceding computer exercises should be useful here. Your program should proceed character by character, using a FOR loop.

10. *Program for Decryption of Vigenère Cipher.* Write a program `strOut = VigenereDeCrypt(STR,keystr)` that inputs a string `STR` of ciphertext in uppercase English letters and another such string `keystr` representing a key. The output `strOut` should be the corresponding plaintext (in lowercase letters) before the Vigenère cipher with key `keystr` is applied to produce the ciphertext. Then use your program to redo the computations of Chapter Exercise 17, part (b).

   Suggestion: Modify your program `VigenereCrypt` by changing each individual shift to its inverse shift.

11. *Program for Playfair Cipher.* Write a program `StrOut = PlayfairCrypt(str,keystr)` that inputs a string `str` of plaintext in lowercase English letters and another such string `keystr` representing a key. The output `StrOut` should be the corresponding ciphertext (in uppercase letters) after the Playfair cipher when key `keystr` is applied to the plaintext. Then use your program to redo the computations of Chapter Exercise 19, part (a).

12. *Program for Decryption of the Playfair Cipher.* Write a program `strOut = PlayfairDeCrypt(STR,keystr)` that inputs a string `STR` of ciphertext in uppercase English letters and another lowercase string `keystr` representing a key. The output `strOut` should be the corresponding plaintext (in lowercase letters) before the Playfair cipher with key `keystr` is applied to it. Then use your program to redo the computations of Chapter Exercise 19, part (b).

Computer-Generated Random Numbers

Most computing platforms feature built-in "random number generators" that are of production quality. Recall that the text cites references that provide detailed developments of such programs, and the interested reader may wish to pursue these, but our approach will be to make the following convention.

Convention: We assume that a random number generator is available on our computing platform. We denote it by `rand`, and assume that it functions as follows: Each time `rand` is called, the output will be a pseudorandom real number (with decimals) from the interval (0,1); that is, $0 < \text{rand} < 1$.

In the language of statistics, we say that rand is uniformly distributed in the interval (0,1). This means that each time `rand` is called to generate a random number, the probability that `rand` will lie in any subinterval of (0,1) will equal the length of that subinterval. For example, the probability that `rand` (on any given call) be less than 1/2—that is, $0 < \text{rand} < 1/2$—is 1/2, and the probability that rand will be greater than 7/8—that is, 7/8 < rand < 1—is 1/8 [the length of the interval (7/8,1)]. This rand function may also have ways to reset its "seed" from its default value so that it will start off differently whenever the program is restarted; linking the seed to the computer's clock is usually a good way to accomplish this. Additional features of the rand function may include options that will allow it to produce ordered lists (vectors) of such random numbers,* and such a feature is particularly convenient for generating one-time pads. Although rand produces real numbers (with decimals) in the special range (0,1), we often need to generate random integers in a specified range. This can be done using the Algorithm 1.1, which is based on the following simple fact.

Fact: Since rand is uniformly distributed in (0,1), if N is any positive integer, then N rand will be uniformly distributed in the interval (0,N).

In order to convert real numbers to integers, we use the **floor** function (built in to most computer platforms). This is a function mapping the real numbers to the integers, which operates as follows: For any real number x, floor(x) will be the greatest integer that is less than or equal to x. For example, floor(2.1) = 2 = floor(2) = floor(2.999), floor(π) = 3, and floor(–2.6) = –3.

Algorithm 1.1: Generating Random Integers Using rand

Given two integers $\ell < k$, the number $J = \ell + \text{floor}(\lceil k - \ell + 1 \rceil \times \text{rand})$ will be a random integer in the range $\ell \leq J \leq k$.

To help better understand this algorithm (in a very relevant situation), suppose we take $\ell = 0, k = 25$. Then since $k - \ell + 1 = 26$, the fact mentioned above tells us that $[k - \ell + 1] \times \text{rand} = 26 \times \text{rand}$ is a real number that is uniformly distributed in the interval (0, 26). When we take the floor: floor($[k - \ell + 1] \times \text{rand}$), the possible integers that can arise are the integers from 0 to 25 (inclusive) and since each of these integers will occur if the previous number lies in an interval of length 1 (in the total interval of length 26), it follows that each of these 26 integers has a 1/26 chance of occurring.

---

13. *Program for Creation of Keys for One-Time Pads.*
    (a) Write a program key = OneTimePadKeyMaker (keylength) that inputs a positive integer keylength, and outputs a vector having keylength randomly chosen integers from the range {0, 1, ..., 25}. strOut should be the corresponding plaintext (in lowercase letters) before the Vigenère cipher with key keystr is applied to it. Use your program to produce a length-12 key.
    (b) Write a program having syntax LetterStr = OneTime PadKeyMaker(keylength) that functions like the one in part (a) except that the output will be a string (rather than a vector) of lowercase English letters that are determined by Table 1.3 (from the random integers that are generated). Use your program to produce a random key of length 12.

---

* If this feature is not available, ordered lists can easily be produced by using a FOR loop.

14. *Program for Random Integer Generator.*

    (a) Write a program Vec = RandIntGen(ell, k, length) that inputs three integers, the first two need only satisfy ell < k, and the third, length, is any positive integer. The output, Vec, is a vector with length elements consisting of randomly generated integers from the range $ell \leq J \leq k$. The program should be based on Algorithm 1.1.

    (b) Use your program to produce a length-20 vector of random integers from the range $26 \leq J \leq 30$. Print out this vector.

    (c) Use your program to produce a length-1000 vector of binary digits (0s and 1s). Do not print this vector, but (get your computer to) count how many of the entries are 0s and write this down. Repeat this and record the new count of the zeros.

2

Divisibility and Modular Arithmetic

In this chapter we start with some basic concepts of divisibility and primes and then move on to introduce the systems of modular integers and their associated arithmetic. Some cryptographic applications of modular arithmetic will be shown beginning with the next chapter. The modular integers are prototypical abstract number systems on which cryptosystems can be built, and they are used as the underpinnings of several more advanced systems that will be introduced in later chapters. The chapter contains a good number of important theorems and propositions; readers with less theoretical backgrounds may wish to skip some of the longer, more technical proofs. Additional topics in number theory are covered in Chapter 8.

As children, our first experience with numbers involved the set of **positive integers**

$$\mathbb{Z}_+ = \{1, 2, 3, 4, 5, \cdots\}$$

If we expand this set to include zero and negative integers, we arrive at the set of **integers**

$$\mathbb{Z} = \{\cdots -3, -2, -1, 0, 1, 2, 3, \cdots\}$$

which is simply the set of all real numbers (the numbers on the number line) that have nothing after their decimal points. There is a rich theory and structure concerning the set of positive integers (and less specifically the set of integers) called **number theory**, but many mysteries, unsolved problems, and problems that can be solved only very inefficiently remain and make the integers a virtual goldmine for building cryptosystems.

Divisibility

One of the most fundamental concepts of the integers is that of divisibility, which is first learned in grade school. Here is the formal definition:

Definition 2.1

Suppose that a and b are integers with $a \neq 0$. We say that a **divides** b (written $a \mid b$) if there is an integer c such that $b = ac$. This can also be expressed by saying a is a **factor** of b, or b is a **multiple** of a. If a does not divide b, we write $a \nmid b$.

Here are some simple examples: $3 \mid 6$, since $6 = 3 \cdot 2$. Also, $-5 \mid 15$, since $15 = (-5) \cdot (-3)$. But $8 \nmid 20$, because 20/8 is not an integer. Notice also that for any nonzero integer a, we have $a \mid a$ (since $a = a \cdot 1$), and $a \mid 0$ (since $0 = a \cdot 0$). The following theorem contains some basic yet very useful properties of divisibility.

Theorem 2.1

Let a, b, and c be integers.

   (a) *Divisibility is transitive.* If $a \mid b$ and $b \mid c$, then $a \mid c$.

   (b) If $a \mid b$ and $a \mid c$, then $a \mid (bx + cy)$ for any integers x and y.

Proof: Part (a): Since $a \mid b$ we can write $b = ae$ for some integer e. Similarly, since $b \mid c$, we can write $c = bf$ for some integer f. Substituting the former into the latter gives $c = (ae)f = a(ef)$. Since ef is an integer, we conclude that $a \mid c$.

   Part (b): The hypotheses allow us to write $b = ae$ and $c = af$ for some integers e and f. Substituting these gives us $bx + cy = aex + afy = a(ex + fy)$. Since $ex + fy$ is an integer, we conclude that $a \mid (bx + cy)$. □

Primes

Definition 2.2

An integer $p > 1$ is called **prime** if the only positive factors of p are 1 and itself. An integer $a > 1$ that is not prime is called **composite**.

The first few primes are 2, 3, 5, 7, 11, 13, 17, 19, 23, 29, 31, …. Prime numbers are the building blocks of the integers because any integer greater than 1 can always be uniquely factored into primes. This is the so-called *fundamental theorem of arithmetic*. We state this important theorem here but postpone its proof until we have developed some additional needed concepts.

Theorem 2.2: Fundamental Theorem of Arithmetic

Every positive integer $a > 1$ can be uniquely expressed as the product of primes. In other words, there exist unique prime numbers $p_1 < p_2 < \cdots < p_n$ and corresponding positive exponents $\alpha_1, \alpha_2, \cdots, \alpha_n \in \mathbb{Z}_+$ such that $a = p_1^{\alpha_1} p_2^{\alpha_2} \cdots p_n^{\alpha_n}.$[*]

In general, it becomes difficult to verify whether or not a positive integer is prime for larger integers. If a positive integer a has a nontrivial

---

[*] This important theorem is the reason why the number 1 is not considered to be prime. If 1 were prime, we would no longer have unique factorization; for example, $18 = 2 \cdot 3^2 = 1^3 \cdot 2 \cdot 3^2 = 1^{12} \cdot 2 \cdot 3^2$, and so forth.

factorization $a = bc$, with $b, c > 1$, then one of b or c must be $\leq \sqrt{a}$ (otherwise we would have the contradiction $a = bc > \sqrt{a}\sqrt{a} = a$). This means that to check if a given positive integer a is prime, we need only look for (prime) factors that are at most equal to $\sqrt{a}$. But testing primality and, more generally, determining the prime factorization of large positive integers can take an inordinate amount of time, even with the best computers and algorithms.[*]

Example 2.1

Find the prime factorizations of each of the following integers:

(a) 847
(b) 4808
(c) 6177

Solution: Part (a) Using the basic principle mentioned above, we begin checking, in order, for prime factors of 847 (knowing that we can stop after we check primes up to $\sqrt{847} = 29.1033$). Certainly $2 \nmid 847$ (since the latter is odd), and also since $847/3 = 282\ 1/3 \notin \mathbb{Z}$, we know that $3 \nmid 847$. Since 847 does not end in 0 or 5, $5 \nmid 847$. But $847/7 = 121$, and we are now reduced to looking for prime factors of 121, so we can stop when we get to $\sqrt{121} = 11$. The following diagram is often used when such factorizations are done by hand. The resulting prime factorization is thus $847 = 7 \cdot 11^2$.

$$
\begin{array}{r}
11 \\
\hline
11\,|\,121 \\
\hline
7\,|\,847
\end{array}
$$

Parts (b) and (c): Going through the same procedure, the corresponding prime factorizations are $4808 = 2^3 \cdot 601$ and $6177 = 3 \cdot 29 \cdot 71$.

One natural question arises: How many primes are there (infinitely many, or does the list eventually end)? This question was resolved a very long time ago by Euclid, the Greek mathematician who lived 325 b.c.–265 b.c. and is most famous for his timeless geometry book *The Elements*; he proved that there are infinitely many primes. This never-ending supply of primes that are as large as we could possibly want has, as we will see later, important

---

[*] To illustrate this, we point out that RSA Security (a high-tech cryptographic security company) offered a number of challenges on their company Web site that were open to the public. One of these offered a $100,000 prize to the first person to factor a certain 304-digit number (larger prizes were available for factoring larger integers). This particular challenge remained open for several years. Such challenges actually benefit the company by helping to test the security of some of its secret codes (that rest on the infeasibility of being able to factor such large or even larger integers) against potential hackers. We discuss such topics in greater detail in Chapter 9.

ramifications in cryptography. Euclid's elegant proof uses the fundamental theorem of arithmetic.[*]

Theorem 2.3: Euclid

There are infinitely many primes.

Proof: Suppose the assertion were false. Then the list of all primes would be finite: $p_1 < p_2 < \cdots < p_M$. Consider the integer $N = p_1 \cdot p_2 \cdots p_M + 1$. By the fundamental theorem of arithmetic, N can be factored (uniquely) into primes. Let p_i be (any) one of the prime factors of N. Then, since $p_i \mid p_1 \cdot p_2 \cdots p_M$, and $p_i \mid N$, it follows from Theorem 2.1(b) that $p_i \mid (N - p_1 \cdot p_2 \cdots p_M)$, that is, $p_i \mid 1$. But this is a contradiction since no prime can divide 1. □

Greatest Common Divisors and Relatively Prime Integers

Definition 2.3

Suppose that a and b are integers not both equal to zero. The **greatest common divisor** of a and b, denoted $\gcd(a,b)$, is the largest integer d that divides both a and b. We say that a and b are **relatively prime** if $\gcd(a,b) = 1$.

For a simple example, since the common factors of 12 and 20 are 1, 2, and 4, we have $\gcd(12,20) = 4$. Similarly, since the only common (positive) factor of 8 and 15 is 1, $\gcd(8,15) = 1$, and 8 and 15 are relatively prime. For integers of moderate size that can be readily factored into primes, the greatest common divisor can be easily read from the prime factorizations—simply take all common prime factors and use the corresponding minimum exponents of each prime. It is routine to verify that this product of common prime powers is the desired gcd (see Exercise for the Reader 2.2). This method is illustrated in the following example.

Example 2.2

Find $\gcd(50,165)$, and $\gcd(1960,10800)$.

Solution: The prime factorizations of the first pair of numbers are $50 = 2 \cdot 5^2$ and $165 = 3 \cdot 5 \cdot 11$; therefore, $\gcd(50,165) = 5$. Similarly, after computing the prime factorizations of $1960 = 2^3 \cdot 5 \cdot 7^2$ and $10800 = 2^4 \cdot 3^3 \cdot 5^2$, we conclude that $\gcd(1960, 10800) = 2^3 \cdot 5 = 40$.

---

[*] By contrast, the problem of whether there are infinitely many prime pairs has not yet been resolved. A prime pair consists of two primes whose difference is two; for example, 3 and 5, 5 and 7, 11 and 13, and 17 and 19 are the first few prime pairs.

Exercise 2.1

(a)   Find the prime factorizations of 16000 and 42757.
(b)   Compute gcd(100, 76), gcd(16000, 960).

Exercise 2.2

For a pair of nonzero integers a and b, the **least common multiple** of a and b, denoted lcm(a,b), is the smallest integer m that is divisible by both a and b.

(a)   Find lcm(12, 28), and lcm(100, 76).
(b)   Show that if $p_1 < p_2 < \cdots < p_n$ are the distinct primes appearing in the prime factorizations of either a or b, if $a = p_1^{\alpha_1} p_2^{\alpha_2} \cdots p_n^{\alpha_n}$ and if $b = p_1^{\beta_1} p_2^{\beta_2} \cdots p_n^{\beta_n}$, then lcm($a,b$) $= p_1^{\mu_1} p_2^{\mu_2} \cdots p_n^{\mu_n}$, where $\mu_i = \max(\alpha_i, \beta_i)$, and gcd($a, b$) $= p_1^{\sigma_1} p_2^{\sigma_2} \cdots p_n^{\sigma_n}$, where $\sigma_i = \min(\alpha_i, \beta_i)$.
(c)   Show that lcm(a,b) $\cdot$ gcd(a,b) $= ab$.

For pairs of large integers, the above procedure for finding greatest common divisors is very slow (because there is no known fast algorithm for prime factorization); a much more efficient method circumvents the need to factor by using the so-called *Euclidean algorithm*. This simple yet very useful procedure in number theory is also due to Euclid. We first formalize the procedure of dividing one integer by another nonzero integer.

The Division Algorithm

Proposition 2.4: The Division Algorithm

If a is an integer and d is any positive integer, then there exist unique integers q and r satisfying $0 \le r < d$, such that $a = dq + r$. Here, a is called the **dividend**, d is called the **divisor**, q is called the **quotient**, and r is called the **remainder**.

Finding q and r is really just the "long division" problem $a \div d$ that one learns about in grade school, but Exercise for the Reader 2.3 shows how to quickly compute q and r, if one is using a calculator (or computer). The uniqueness proof of Proposition 2.4 is routine and is left as an exercise. Although the proposition is not really an algorithm, the terminology is nonetheless standard in number theory, so we will adhere to it. In the language of modular arithmetic that we introduce later in this chapter, the result of the division algorithm can be expressed as $a \equiv r \pmod{d}$. Given the integers a and d, the dividend d is most easily expressed in terms of the following "floor" function. The floor function inputs any real number and outputs the first integer below the number. The formal

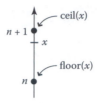

Figure 2.1 Illustration of the floor and ceiling functions.

definition of the floor function and the corresponding ceiling function are as follows.

Definition 2.4   The Floor and Ceiling Functions

The **floor function** is a function from the set of real numbers to the set of integers, defined by

$\text{floor}(x) = \lfloor x \rfloor$ = the greatest integer k that is less than or equal to x

The **ceiling function** has the same domain and codomain and is defined by

$\text{ceil}(x) = \lceil x \rceil$ = the least integer k that is greater than or equal to x

---

It is helpful to visualize the actions of these two functions using a vertical number line; see Figure 2.1. Notice that there are two different notations in use for the floor and ceiling functions; the abbreviated word *notation* is more natural and common in many computing platforms, while the symbolized notation is more compact and often used in mathematical developments. Here are some simple examples of some floors and ceilings: floor(4.75) = $\lfloor 4.75 \rfloor = 4$, ceil(4.75) = $\lceil 4.75 \rceil = 5$, $\lfloor -4.75 \rfloor = -5$, ceil(−4.75) = −4. Notice that when n is an integer, floor(n) = n = ceil(n), and conversely, if either of these two equations holds, n must be an integer.

Exercise for the Reader 2.3

(a)   Show that if the division algorithm is applied to an integer division $a \div d$, where (as usual) $d > 0$, then the quotient and remainder are given as follows: $q = \lfloor a/d \rfloor$ and $r = a - qd$.

(b)   Use part (a) to find the quotient and remainder when the division algorithm is applied to the following integer divisions:
(a) $123 \div 5$
(b) $-874 \div 15$.

The Euclidean Algorithm

The Euclidean algorithm consists of repeatedly applying the division algorithm. It is based on the following simple property:

Proposition 2.5

If a, d, q, and r are as in the division algorithm, then $\gcd(a, d) = \gcd(d, r)$.

Proof: From the equation $a = dq + r$, and Theorem 2.1(b), we see that if $e \mid r$ and $e \mid d$, then $e \mid a$. If we rewrite the equation as $r = a - dq$, then by the same token we get that if $e \mid a$ and $e \mid d$, then $e \mid r$. We have proved that the set of all common divisors of r and d equals the set of all common divisors of a and d, from which the result of the theorem directly follows. □

For the pair of integers $(100,76)$ let us observe what happens when we repeatedly apply the division algorithm by dividing all new remainders into the previous divisors:

$$100 = 1 \cdot 76 + 24$$
$$76 = 3 \cdot 24 + 4$$
$$24 = 6 \cdot 4 + 0$$

From Proposition 2.5, we see that $\gcd(100,76) = \gcd(76,24) = \gcd(24, 4) = \gcd(4,0) = 4$. In general, this procedure will always stop since the sequence of remainders is strictly decreasing. (By the division algorithm, the new remainder must be less than the previous one because the previous remainder has become the divisor.) The last nonzero remainder will be the gcd of the two starting integers. This procedure is the Euclidean algorithm. We now make a formal statement of it.

Algorithm 2.1: The Euclidean Algorithm

Input: A pair of integers a and b, not both equal to zero.
Output: The greatest common divisor, $\gcd(a,b)$.

We may assume that $a \geq b$ (if not, switch a and b). Apply the division algorithm to write $a = q_1 b + r_1$. If $r_1 = 0$, then $\gcd(a,b) = b$; otherwise, continue by dividing successive divisors by successive remainders until a zero remainder is reached:

$$b = q_2 r_1 + r_2, \quad 0 \leq r_2 < r_1$$
$$r_1 = q_2 r_2 + r_3, \quad 0 \leq r_3 < r_2$$
$$\cdots$$
$$r_{n-2} = q_{n-1} r_{n-1} + r_n, \ 0 \leq r_n < r_{n-1}$$
$$r_{n-1} = q_n r_n + 0$$

The last nonzero remainder r_n is $\gcd(a,b)$.

Since the sequence of successive remainders is strictly decreasing, $b > r_1 > r_2 > \cdots > r_n > 0$, the algorithm must eventually terminate (at most

b steps). Since Proposition 2.5 implies that $\gcd(a,b) = \gcd(b,r_1) = \gcd(r_1,r_2) = \gcd(r_2,r_3) = \cdots = \gcd(r_{n-1},r_n) = \gcd(r_n,0) = r_n$, it follows that $\gcd(a,b) = r_n$, the last nonzero remainder that is encountered in this process.

We summarize how the division algorithm data is used in going from one round to the next:

$$\text{remainder} \rightarrow \text{divisor} \rightarrow \text{dividend} \rightarrow \text{not used}$$

In practice, the Euclidean algorithm is a very efficient method for computing greatest common divisors, and it does not require any factorizations. As we will soon discover, it can also be used to solve other interesting problems, and it is readily translated into computer programs.

Exercise for the Reader 2.4

Use the Euclidean algorithm to compute $\gcd(65, 91)$ and $\gcd(1665, 910)$.

One very useful consequence of the Euclidean algorithm can be previewed by looking at the preceding example where we used it to find $\gcd(100, 76) = 4$. If we start with the second-to-last equation where this $\gcd(4)$ appeared as the remainder and work our way up, we will be able to express 4 in the form of $100x + 76y$ for some integers x and y, that is, as an *integer combination* of 100 and 76 (the integers for which we wanted to find the gcd). Here are the steps: we start with $76 = 3 \cdot 24 + 4$ and isolate the $\gcd(= 4)$ to write it as an integer combination of the two previous remainders: $4 = 3 \cdot 24 - 1 \cdot 76$. We then use the next equation up, $100 = 1 \cdot 76 + 24$, solve it for 24, and substitute the result into what we had just previously obtained: $4 = 3 \cdot 24 - 1 \cdot 76 = 3 \cdot (100 - 1 \cdot 76) - 1 \cdot 76 = 3 \cdot 100 - 4 \cdot 76$. The following theorem contains the general result.

Theorem 2.6

Suppose that a and b are integers not both equal to zero, and let $d = \gcd(a,b)$. Then there exist integers x and y such that $d = ax + by$. In the special case in which a and b are relatively prime, we can write $1 = ax + by$.

Proof: The proof is a constructive one in that it provides an algorithm for finding such an x and y. If we set $r_0 = b$ and $r_{-1} = a$, then the Euclidean algorithm consists of $n + 1$ applications of the division algorithm, and these can all be expressed as $r_{i-1} = q_i r_i + r_{i+1}$ $(i = 0, 2, \cdots n)$. Each of these is then rewritten to be solved for the last remainder: $r_{i+1} = q_i r_i - r_{i-1}$ $(i = 0, 2, \cdots n)$. Since $d = r_n$, we will start with the second-to-last equation $(i = n - 1)$ and rewrite it as $d = x_n r_{n-2} + y_n r_{n-1}$. Thus d is expressed as an integer combination of r_{n-2} and r_{n-1}. If we substitute the next equation up $(i = n - 2)$ $r_{n-1} = r_{n-3} - q_{n-2}r_{n-2}$ into our expression for d, we arrive at $d = x_n r_{n-2} + y_n r_{n-1} = x_n r_{n-2} + y_n (r_{n-3} - q_{n-2}r_{n-2}) = y_n r_{n-3} + (x_n - q_{n-2})r_{n-2} =: x_{n-1}r_{n-3} + y_{n-1}r_{n-2}$. We continue this process of successively moving up the list of division algorithm equations and substituting them into our existing integer combination of d. At the kth step $(i = n - k)$, we will have obtained an expression for d as an integer combination $x_{n-k+1}r_{n-k-1} + y_{n-k+1}r_{n-k}$. At the final step $(k = n; i = 0)$, we will have $d = x_1r_{-1} + y_1r_0 = x_1a + y_1b$, as desired. □

Although the proof is a bit technical, the idea is simple enough, and the whole scheme is nicely amenable to translate into a computer program. Later in the section we provide a very efficient implementation of this algorithm (that can be directly translated into a computer program).

Exercise for the Reader 2.5

(a) Use the procedure described in the proof of Theorem 2.6 to express gcd(65, 91) as an integer combination of 91 and 65. Similarly, express gcd(1665, 910) as an integer combination of 1165 and 910.

(b) Explain why the integers x and y in Theorem 2.6 are not unique.

Aside from its practical applications, Theorem 2.6 turns out to be very useful for obtaining new theoretical results. We demonstrate this by using it to prove the following result, which, in turn, will allow us to prove the fundamental theorem of arithmetic.

Proposition 2.7: Euclid's Lemma

(a) Suppose that p is a prime and that a and b are integers. If $p \mid ab$, then either $p \mid a$ or $p \mid b$.

(b) Suppose that p is a prime and that $a_1, a_2, \cdots, a_n$ are integers. If $p \mid a_1 a_2 \cdots a_n$, then p must divide at least one of the factors $a_1, a_2, \cdots, a_n$.

Proof: (a) Assuming that $p \mid ab$, if also $p \mid a$, we are done; so assume that $p \nmid a$. We need to show that $p \mid b$. Since p is a prime and $p \nmid a$, it follows that gcd$(a,p) = 1$. Theorem 2.6 thus allows us to write $1 = ax + py$ for some integers x and y. We multiply this equation by b to obtain $b = (ab)x + pyb$. But since $p \mid ab$, and certainly $p \mid p$, it follows from Theorem 2.1(b) that $p \mid b$, as desired.

(b) We can achieve the proof of part (b) by using the just proved special case of part (a) (when $n = 2$) to repeatedly chip away at it. Assuming that $p \mid a_1(a_2 \cdots a_n)$, part (a) tells us that either $p \mid a_1$, in which case we are done, or we get that $p \mid a_2 \cdots a_n$, which involves one less factor. Applying part (a) again to this smaller case $p \mid a_2(a_3 \cdots a_n)$, we find that either $p \mid a_2$, in which case we are done, or we get that $p \mid a_3 \cdots a_n$. If we continue this process, we will either be done or we arrive at a division involving the final two factors $p \mid a_{n-1}a_n$, to which one final application of part (a) will complete the proof. □

We are now nicely poised to prove the fundamental theorem of arithmetic.

Proof of Theorem 2.2. Part I: *Existence.* Suppose that there were positive integers greater than 1 that were not expressible as a product of primes. Let n be the smallest such integer. Since n cannot be prime (because a single prime is a product of primes), it must be composite, so we can write $n = ab$, where a and b are smaller integers with $1 < a, b < n$. But since n was chosen to be the smallest integer that cannot be written as a product of primes, both a and b must be expressible as a product of primes. Since $n = ab$, we can multiply prime factorizations of a and b to obtain a prime factorization of n. With this contradiction, the existence proof is complete.

Part II: *Uniqueness.* Suppose that a positive integer n had two different prime factorizations:

$$n = p_1^{\alpha_1} p_2^{\alpha_2} \cdots p_k^{\alpha_k} = q_1^{\beta_1} q_2^{\beta_2} \cdots q_\ell^{\beta_\ell}$$

where $p_1 < p_2 < \cdots < p_k$ and $q_1 < q_2 < \cdots < q_\ell$ are primes, and $\alpha_1, \alpha_2, \cdots, \alpha_k$ and $\beta_1, \beta_2, \cdots, \beta_\ell$ are positive exponents. If there are any primes among the p's and q's that are common, they can be divided through (cancelled) on both sides of the equation so that the lists $p_1 < p_2 < \cdots < p_k$ and $q_1 < q_2 < \cdots < q_\ell$ can be assumed to have no primes in common, and we assume that this is indeed the case. Now, since $p_1 \mid p_1^{\alpha_1} p_2^{\alpha_2} \cdots p_k^{\alpha_k} = q_1^{\beta_1} q_2^{\beta_2} \cdots q_\ell^{\beta_\ell}$, it follows from Euclid's lemma [Proposition 2.7(b)] that $p_1 \mid q_j$ for some index j. But since p_1 and q_j are both primes, it follows that $p_1 = q_j$, which contradicts the assumption that the p's and q's have no primes in common. This completes the uniqueness proof. □

Modular Arithmetic and Congruences

Among his numerous contributions to mathematics and science, the illustrious mathematician Carl Friedrich Gauss[*] (Figure 2.2) developed the extremely useful number–theoretic concepts of congruences and modular arithmetic. These concepts led to an infinite supply of abstract number systems that have turned out to play a pivotal role in an assortment

---

[*] Carl F. Gauss is widely considered to be the greatest mathematician who ever lived. His potential was discovered early, and his mathematical aptitude was astounding. While he was in second grade, his teacher, needing to keep him occupied for a while, asked him to perform the addition of the first 100 integers: $S = 1 + 2 + \cdots + 100$. Two minutes later, Gauss gave the teacher the answer. He did it by rewriting the sum in the reverse order $S = 100 + 99 + \cdots + 1$, adding vertically to the original to get $2S = 101 + 101 + \cdots + 101 = 100 \cdot 101$, so $S = 50 \cdot 101 = 5050$. This idea yields a general proof of an important mathematical series identity. Apart from his numerous groundbreaking contributions to mathematics, Gauss did significant work in physics and astronomy, as well as in other sciences. His brilliant ideas came to him so rapidly that he had a file cabinet full of them waiting to be written up for formal publication. He would often receive visits from other prominent international mathematicians who would proudly share with Gauss recent discoveries, and very often Gauss would simply reach into his file cabinet to pull out his ideas on the topic that frequently eclipsed those of the visitor. For many years until the inception of the Euro, Germany had honored Gauss by placing his image on the very common 10 Deutsche mark banknote (the value was approximately US$5). Figure 2.2 is an image of this banknote, with a drawing of Gauss's important normal (bell-shaped) curve, a cornerstone of statistics.

Figure 2.2 Carl Friedrich Gauss (1777–1855), German mathematician.

of cryptosystems. The entire framework is based on the following very simple definition.

Definition 2.5

Let m be a positive integer. We say that two integers a and b are **congruent mod(ulo)** m, and denote this as $a \equiv b \pmod{m}$, if $m \mid (a - b)$. The number m is called the **modulus** of the congruency. If $m \nmid (a - b)$, we say that a and b are **incongruent mod** m, and write this as $a \not\equiv b \pmod{m}$.

Example 2.3: Two Familiar Moduli

(a) Notice that $15 \equiv 3 \pmod{12}$, since $15 - 3 = 12$; similarly, since $27 - 3 = 24 = 2 \cdot 12$, we have $27 \equiv 3 \pmod{12}$. The reader can similarly check that $3 \equiv -9 \equiv -21 \equiv -33 \cdots$ (mod 12). Congruences mod 12 can be visualized by means of a traditional (as opposed to a digital) clock; see Figure 2.3. Two times are congruent in the clock if one can be made into the other by turning the (hour) hand of the clock a complete number of revolutions either clockwise (corresponding to adding 12) or counterclockwise (corresponding to subtracting 12).

(b) Anyone who has studied angles or trigonometry will already be familiar with 360 as a modulus, since 360° corresponds to a complete revolution angle (so adding any multiple of it results in the same angle as wherever we started). Thus the angular equalities: $-90° = 270° = 630° = \cdots$ correspond to the congruences $-90 \equiv 270 \equiv 630 \cdots \pmod{360}$. To see that $-90 \equiv 630 \pmod{360}$, for example, we note that $-90 - 630 = -720 = -2 \cdot 360$.

If we rewrite the condition $m \mid (a - b)$ for $a \equiv b \pmod{m}$ as $a - b = km$ (for some integer k), we then obtain $a = b + km$. We summarize this alternate formulation:

$$a \equiv b \pmod{m} \quad \Leftrightarrow \quad a = b + km, \text{ for some } k \in \mathbb{Z} \qquad (2.1)$$

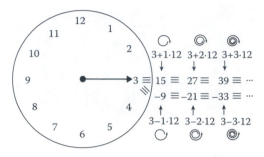

Figure 2.3 Congruence modulo 12 is like clockwork; two integers are congruent mod 12 if one can be obtained from the other by adding or subtracting a multiple of 12, corresponding to making an integral number of revolutions around the clock.

This formula is illustrated in Figure 2.3, showing how to get all of the integers that are congruent to 3 (mod 12). Congruences satisfy three basic properties that will be tacitly used throughout the remainder of this book.

Proposition 2.8: Basic Properties of Congruences

The three properties of this proposition—reflexivity, symmetry, and transitivity—can be applied to any relation between pairs of objects of a set, and if they are satisfied together the relation is called an *equivalence relation*. Thus congruence mod m is an equivalence relation.

If m is a positive integer, then congruence mod m satisfies the following properties:

(a) *Reflexivity.* $a \equiv a \pmod{m}$ for any integer a.
(b) *Symmetry.* If $a \equiv b \pmod{m}$, then $b \equiv a \pmod{m}$ for any integers a, b.
(c) *Transitivity.* If $a \equiv b \pmod{m}$ and $b \equiv a \pmod{m}$, then $a \equiv c \pmod{m}$ for any integers a, b, c.

Proof: Part (a): Since $m \mid 0 = (a - a)$, we obtain that $a \equiv a \pmod{m}$.

Part (b): Using Equation 2.1, from $a \equiv b \pmod{m}$, we can write that $a - b = km$ for some integer k. Negating both sides of this equation produces $b - a = -(a - b) = (-k)m$, which by Equation 2.1 is equivalent to $b \equiv a \pmod{m}$.

Part (c): Applying Equation 2.1 to the assumptions $a \equiv b \pmod{m}$ and $b \equiv a \pmod{m}$, we may write $a - b = km$ and $b - c = \ell m$ for some integers k, ℓ. Adding these two equations leads us to $a - c = (a - b) + (b - c) = mk + m\ell = m(k + \ell)$, which by Equation 2.1 is equivalent to $a \equiv c \pmod{m}$. □

One important consequence of Proposition 2.8 is that for any positive integer m, the integers can be broken down into m *congruence classes* of integers that are mutually congruent mod m. For example, for congruences mod 12 (clockwork), the 12 congruence classes are as follows:

$$\cdots -36 \equiv -24 \equiv -12 \equiv 0 \equiv 12 \equiv 24 \equiv 36 \cdots (\bmod 12)$$
$$\cdots -35 \equiv -23 \equiv -11 \equiv 1 \equiv 13 \equiv 25 \equiv 37 \cdots (\bmod 12)$$
$$\cdots -34 \equiv -22 \equiv -10 \equiv 2 \equiv 14 \equiv 26 \equiv 38 \cdots (\bmod 12)$$
$$\cdots$$
$$\cdots -25 \equiv -13 \equiv -1 \equiv 11 \equiv 23 \equiv 35 \equiv 47 \cdots (\bmod 12)$$

The reader should observe that (i) the sets are disjoint, (ii) all integers are accounted for (read from top to bottom, proceed to next column left or right), and (iii) the gap between successive integers in any row is always 12. In terms of the clock (Figure 2.3), this means that the rows are obtained by starting at any given hour on the clock and successively adding/subtracting 12 (corresponding to complete revolutions on the clock).

Exercise for the Reader 2.6

Show that $a \equiv b(\bmod 2)$ if, and only if, a and b have the **same parity**, that is, a and b are both even or both odd. Describe the congruence classes mod 2.

The following simple proposition gives yet another useful way to view the relation of congruence mod m.

Proposition 2.9: Congruences and Remainders

If m is a positive integer and a, b are integers, then $a \equiv b(\bmod m)$ if, and only if, a and b both have the same remainder $r \in \{0,1,2,\cdots,n-1\}$ when they are divided by m using the division algorithm (Proposition 2.4).

Proof: We use the division algorithm (Proposition 2.4) to write $a = mq + r$ and $b = mq' + r'$, where q, r, q', r' are uniquely determined integers with $0 \leq r, r' < m$.

Now, if the remainders are the same, that is, $r = r'$, then we have:

$$a - b = mq + r - mq' - r' = m(q - q') + \underbrace{(r - r')}_{=0} = m(q - q')$$

so $m \mid (a - b)$, which means that $a \equiv b(\bmod m)$.

Conversely, if we start with $a \equiv b(\bmod m)$, this means that $m \mid (a - b)$, and substituting the above expressions for a and b, we obtain $m \mid (mq + r - mq' - r')$ or $m \mid (m[q - q'] + r - r')$. Since m certainly divides $m[q - q']$, it follows [from Theorem 2.1(b)] that $m \mid (r - r')$. But since the remainders r, r' lie in the range $\{0,1,2,\cdots,m-1\}$, their difference $r - r'$ must lie in the range $\{-(m-1), \cdots, -2, -1, 0, 1, 2, \cdots, m-1\}$, and the only possible way to have $m \mid (r - r')$, would be if $r - r' = 0$, that is, the remainders are the same. $\square$

For the clockwork example, the 12 possible remainders, 0, 1, 2, ..., 11, correspond to the 12 congruence classes. Except for the fact that 0 represents 12, these numbers represent the hours on the clock. In fact, any member of a congruence class can be used to represent the whole class. What makes this concept so powerful is that the *answers we get with arithmetic operations on any integers will land in the same congruence class mod m, regardless which integers in a congruence class we use.* Before formally enunciating this important fact, we provide a motivating example.

Example 2.4

Compare the answers (mod 12) of the following arithmetic operations with the corresponding answers using instead the representatives of the integers taken from the set of possible remainders $\{0, 1, 2, \cdots, 11\}$.

(a) $56 + 81$
(b) $-23 \cdot 187 \cdot 38^4$

Solution: Part (a): $56 + 81 = 137 = 11 \cdot 12 + 5$, so $(56 + 81) \equiv 5 \pmod{12}$. On the other hand, since $56 = 4 \cdot 12 + 8$ and $81 = 6 \cdot 12 + 9$, we have (by Equation 2.1) $56 \equiv 8$ and $81 \equiv 9 \pmod{12}$. Performing the addition of the remainders gives $8 + 9 = 17 = 1 \cdot 12 + 5$, so $(8 + 9) \equiv 5 \pmod{12}$—the same answer (mod 12).

Part (b): Computing this large quantity and then applying the division algorithm leads us to

$-23 \cdot 187 \cdot 38^4 = -23 \cdot 187 \cdot 2085136 = -8968169936 = -747347495 \cdot 12 + 4$, so that $(-23 \cdot 187 \cdot 38^4) \equiv 4 \pmod{12}$.

On the other hand, since $-23 \equiv 1$, $187 \equiv 7$ and $38 \equiv 2 \pmod{12}$ when we compute the same operations with these much less unwieldy remainders, we obtain $1 \cdot 7 \cdot 2^4 = 7 \cdot 16$; if we replace, in turn, 16 with 4 (its remainder mod 12), we get $7 \cdot 4 = 28 = 4 \pmod{12}$—once again, the same answer that we obtained with the larger numbers (mod 12).

The results of the above example are not coincidental; the next result confirms this good news, establishing that modular arithmetic is easier than ordinary arithmetic.

Proposition 2.10: Validity of Congruent Substitutions in Modular Arithmetic

Suppose that m is a positive integer and that a, b, a', b' are integers with $a \equiv a' \pmod{m}$ and $b \equiv b' \pmod{m}$. The following congruences are then valid:

(a) $a + a' \equiv b + b' \pmod{m}$
(b) $-a \equiv -a' \pmod{m}$
(c) $a \cdot b \equiv a' \cdot b' \pmod{m}$
(d) $a - a' \equiv b - b' \pmod{m}$
(e) $a^k \equiv (a')^k \pmod{m}$, for any positive integer k

We will prove this result momentarily, but let us first relish some of its ramifications. In the motivating example, we previewed some of these consequences when we replaced each integer with its remainder (mod 12). Remainders are often convenient replacements, but the proposition tells us that we are free to use any replacements that we find convenient. As another example, consider the problem of computing (mod 12) the power 47^{129}. If we computed this integer directly, it would have nearly 500 digits! But since $47 \equiv 11 \pmod{12}$ (its remainder), part (e) of the proposition tells us we could instead compute 11^{129} and will get the same answer (mod 12). This number still has about 300 digits, but if we notice that $11 \equiv -1 \pmod{12}$, the proposition would tell us that we could simply compute $(-1)^{129}$, which we immediately see (by hand) is -1 (as is any odd power of -1). So, we may conclude that $47^{129} \equiv -1 \equiv 11 \pmod{12}$. In Chapter 6, we will develop an efficient scheme of computing any powers in modular arithmetic. This algorithm, known as *fast exponentiation*, will be a vital component of certain public key cryptosystems that will be considered later.

Exercise for the Reader 2.7

Working in mod 10 arithmetic, compute each of the following quantities, using representatives in $\{0, 1, 2, \cdots, 9\}$ for your final answers.

(a) $88 + 1234 + 82645$
(b) $(11!)^2$

Explain why the answers to these (and any arithmetic computations mod 10) will simply be the one's (final) digit of the corresponding answer in (ordinary) integer arithmetic.

Proof of Proposition 2.10. We begin by recasting the assumptions $a \equiv a' \pmod{m}$ and $b \equiv b' \pmod{m}$ as $a - a' = km$, and $b - b' = \ell m$, for some integers k, ℓ.

Part (a): Since $a + a' - (b + b') = (a - b) + (a' - b') = km + \ell m = (k + \ell)m$, it follows that $a + a' \equiv b + b' \pmod{m}$.

Part (b): Since $-a - (-a') = -a + a' = -(a - a') = (-k)m$, it follows that $-a \equiv -a' \pmod{m}$.

Part (c): Since $a \cdot b - a' \cdot b' = a \cdot b - a \cdot b' + a \cdot b' - a' \cdot b' = a(b - b') + (a - a')b = a(\ell m) + (km)b = [a\ell + kb]m$, it follows that $a \cdot b \equiv a' \cdot b' \pmod{m}$.

Part (d): This part follows from parts (a) and (b), since $a - b = a + (-b)$.

Part (e): This part follows from part (c), since exponentiation is a sequence of multiplications. □

Note: Students familiar with computer platforms might be familiar with the **mod function**, which depends on a parameter m (the modulus). This is a function from the set of integers $\mathbb{Z}$ to the set of possible remainders mod m: $\{0, 1, 2, \cdots, m-1\}$, for any inputted integer a, this mod function outputs its remainder when a is divided by m. It is usually denoted as mod(a,m). Thus, the output of mod(33, 12) would be 9, since 9 is the remainder when 33 is divided by 12.

Exercise for the Reader 2.8

For a given modulus m, a positive integer, is the mod function that is described in the above note a one-to-one function? Is it an onto function? Explain your answers.

Having established some basic properties of congruences and modular arithmetic, we are now poised to introduce the abstract number systems that are known as modular integers. In contrast with the system $\mathbb{Z}$ of integers, which is an infinite set, all modular integer systems are finite sets.

Modular Integer Systems

Definition 2.6

If m is a positive integer, the set of **integers modulo m**, denoted by $\mathbb{Z}_m$, is the set of possible remainders when dividing by m:

$$\mathbb{Z}_m = \{0, 1, 2, \cdots, m-1\}$$

We define the arithmetic operations of addition, subtraction, multiplication, and exponentiation on $\mathbb{Z}_m$ by performing the corresponding arithmetic operations on the integers, converting, whenever convenient, in the final answer to an element of $\mathbb{Z}_m$.[*]

Proposition 2.10 assures us that the results of such operations will always be consistent. Our next example will look at addition and multiplication tables for $\mathbb{Z}_m$ for two small values of m. Although such tables are usually not constructed in practice, examination of the tables in these small cases will help to enlighten some general properties of modular integer systems.

Example 2.5

Create addition and multiplication tables for $\mathbb{Z}_5$ and $\mathbb{Z}_6$. Do you notice any similarities or differences in the corresponding tables?

Solution: In Table 2.1 and Table 2.2, we construct addition and multiplication tables for $\mathbb{Z}_5$, and Table 2.3 and Table 2.4 give the corresponding tables for $\mathbb{Z}_6$.

Exercise for the Reader 2.9

Perform the following operations in $\mathbb{Z}_{12}$: $11+8$, $5 \cdot 8$, 11^2. Is there an element $b \in \mathbb{Z}_{12}$ such that $5b = 1$ in $\mathbb{Z}_{12}$?

---

[*] With its addition and multiplication operations, the system $\mathbb{Z}_m$ of integers modulo m inherits almost all of the nice properties of arithmetic that the system $\mathbb{Z}$ of integers possess, such as commutativity of addition and multiplication: $a+b = b+c$ and $ab = ba$, associativity of addition and multiplication: $(a+b)+c = a+(b+c)$ and $(ab)c = a(bc)$, and the distributive law: $a(b+c) = ab+ac$. The modular integers are examples of what are called *commutative rings* in abstract algebra.

TABLE 2.1 Addition Table for $\mathbb{Z}_5$

+	0	1	2	3	4
0	0	1	2	3	4
1	1	2	3	4	0
2	2	3	4	0	1
3	3	4	0	1	2
4	4	0	1	2	3

TABLE 2.2 Multiplication Table for $\mathbb{Z}_5$

×	0	1	2	3	4
0	0	0	0	0	0
1	0	1	2	3	4
2	0	2	4	1	3
3	0	3	1	4	2
4	0	4	3	2	1

The addition tables for $\mathbb{Z}_5$ and $\mathbb{Z}_6$ are quite similar in structure. The row for 0 is simply a copy of the row of the second numbers (upper column) corresponding to the fact that 0 is the *additive identity*: $0 + a = a$. The remaining rows are simply cyclic shifts of the first row; each time we shift to the left by 1 (with wraparound) from the previous row, corresponding to the next higher number being added. This simple structure is common to addition tables for any $\mathbb{Z}_m$. There is a stark difference, though, in the multiplication tables. Notice that each nonzero row (or column) of the multiplication table for $\mathbb{Z}_5$ contains each of the elements of $\mathbb{Z}_5$ (as is the case for the addition tables), but this is not the case for the multiplication table for $\mathbb{Z}_6$. For example, in $\mathbb{Z}_6$, we can get 0

TABLE 2.3 Addition Table for $\mathbb{Z}_6$

+	0	1	2	3	4	5
0	0	1	2	3	4	5
1	1	2	3	4	5	0
2	2	3	4	5	0	1
3	3	4	5	0	1	2
4	4	5	0	1	2	3
5	5	0	1	2	3	4

TABLE 2.4 Multiplication Table for $\mathbb{Z}_6$

×	0	1	2	3	4	5
0	0	0	0	0	0	0
1	0	1	2	3	4	5
2	0	2	4	0	2	4
3	0	3	0	3	0	3
4	0	4	2	0	4	2
5	0	5	4	3	2	1

by multiplying the two nonzero numbers 2 and 3. This sort of problem generally occurs in any $\mathbb{Z}_m$ when m is composite but never occurs in $\mathbb{Z}_p$ for a prime modulus p (Why?). Furthermore, when p is prime, any nonzero row in the multiplication table of $\mathbb{Z}_p$ will always contain all of the elements of $\mathbb{Z}_p$. In order to elaborate on these concepts, we first need a definition.

Modular Inverses

Definition 2.7

For any $a \in \mathbb{Z}_m$, we say that a is **invertible** (or has an inverse) if there exists another element $a^{-1} \in \mathbb{Z}_m$ such that $a \cdot a^{-1} = a^{-1} \cdot a = 1$. The element a^{-1}, if it exists, is called the (**multiplicative**) **inverse** of a.

If we have the multiplication table available, it can be determined whether an element $a \in \mathbb{Z}_m$ has an inverse simply by checking if the row of this element in the table contains a 1. For example, from Table 2.2 (multiplication table for $\mathbb{Z}_5$), we see that in $\mathbb{Z}_5$ we have $1^{-1} = 1$, $2^{-1} = 3$, $3^{-1} = 2$, and $4^{-1} = 4$. (In any $\mathbb{Z}_m$, the element 1 will always be its own inverse, and the element 0 will never have an inverse—why?) From Table 2.4, we see that in $\mathbb{Z}_6$, $1^{-1} = 1$, $5^{-1} = 5$, and no other elements have inverses. We point out that since multiplication in $\mathbb{Z}_m$ is commutative (that is, $ab = ba$ for all $a, b \in \mathbb{Z}_m$), only one of the two conditions for inverses ($a \cdot a^{-1} = 1$, or $a^{-1} \cdot a = 1$) needs to be checked. Also, there can be only one inverse of any element $a \in \mathbb{Z}_m$. [*Proof*: Suppose that both b and $a^{-1} \in \mathbb{Z}_m$ were inverses of $a \in \mathbb{Z}_m$. Then $b = b \cdot 1 = b \cdot (a \cdot a^{-1}) = (b \cdot a) \cdot a^{-1} = 1 \cdot a^{-1} = a^{-1}$. □] Inverses are important for an assortment of reasons, some of which we will discuss shortly. The following result provides a simple criterion to determine whether an element $a \in \mathbb{Z}_m$ has an inverse.

Proposition 2.11: Inverses in $\mathbb{Z}_m$

An element $a \in \mathbb{Z}_m$ is invertible precisely when $\gcd(a, m) = 1$, that is, when it is relatively prime to m. Moreover, the inverse a^{-1} can be obtained from the integer equation $1 = ax + my$ (recall Theorem 2.6), as $a^{-1} \equiv x \pmod{m}$. In particular, all nonzero elements of $\mathbb{Z}_m$ are invertible precisely when $m = p$ is prime.

Proof: If $\gcd(a, m) = 1$, then by Theorem 2.6, there exist integers x and y such that $1 = ax + my$. If we rewrite this as $1 - ax = my$, we see that $m \mid (1 - ax)$, and so $ax \equiv 1 \pmod{m}$. This implies (resetting x to be its mod m representative between 1 and $m - 1$) that $x = a^{-1}$ in $\mathbb{Z}_m$. Conversely, if a has an inverse a^{-1} in $\mathbb{Z}_m$, then $a \cdot a^{-1} \equiv 1 \pmod{m}$, or $m \mid (1 - a \cdot a^{-1})$. It follows (from the definition of divisibility) that $1 - a \cdot a^{-1} = mk$, or $mk + a \cdot a^{-1} = 1$, for some $k \in \mathbb{Z}$. From this latter equation, it readily follows that $\gcd(a, m) = 1$, because any (prime) factor of both a and m would also [by Theorem 2.1(b)] necessarily have to be a (prime) factor of 1 (which has no prime factors). The last statement easily follows from the criterion since any prime number p is relative prime to all positive integers that are less than p, that is, to all nonzero elements in $\mathbb{Z}_p$. $\square$

One important task relating to modular arithmetic concerns *solving congruences*. For example, a simple congruence such as $x + 5 \equiv 2 \pmod{12}$ can easily be solved by subtracting 5 from both sides (just like in basic algebra): $x \equiv 2 - 5 \equiv -3 \equiv 9 \pmod{12}$. From Proposition 2.10 we can always add or multiply both sides of a congruence by any number (or an equivalent representative of that number, mod m). Dividing both sides of a congruence is more difficult; in general, to divide both sides of a congruence by a number a, a needs to be invertible in $\mathbb{Z}_m$; that is (from Proposition 2.11), we must have $\gcd(a,m) = 1$. Furthermore, once it is ascertained that we can divide both sides by a, we will actually be multiplying both sides by a^{-1} rather than dividing (with real numbers) by a. This is where modular arithmetic and ordinary arithmetic are very different (unlike with multiplication and addition). This process is illustrated in the following example.

Example 2.6

Solve each of the following congruences for x:

(a) $3x + 2 \equiv 1 \pmod 5$.

(b) $5x - 4 \equiv 4 \pmod 6$.

Solution: Part (a): From Table 2.2 (or by simple trial and error multiplying 3 by each of the four nonzero elements of $\mathbb{Z}_5$ until we get 1), we see that $3^{-1} = 2$ in $\mathbb{Z}_5$. Thus, when we need to divide the congruence by 3, we will be multiplying both sides by 2: $3x + 2 \equiv 1 \Rightarrow 3x \equiv 1 - 2 \equiv -1 \equiv 4 \Rightarrow x(= 2 \cdot 3x) = 2 \cdot 4 \equiv 8 \equiv 3 \pmod 5$. This answer is easily checked in the original congruence: $3 \cdot 3 + 2 \equiv 11 \equiv 1 \pmod 5$.

Part (b): In $\mathbb{Z}_6$, 5 is invertible since it is relatively prime to 6, so we will be able to perform the necessary division by 5 to solve this congruence. From Table 2.4 (or by simple trial and error), we find that $5^{-1} = 5$, so when we need to divide this congruence by 5, we will multiply both sides by (its inverse) 5: $5x - 4 \equiv 4 \Rightarrow 5x \equiv 4 + 4 \equiv 8 \equiv 2 \Rightarrow x = 5 \cdot 2 \equiv 10 \equiv 4 \pmod 6$. Once again, this answer is easily checked.

Extended Euclidean Algorithm

We will discuss more general congruences shortly, but first we provide an algorithm for computing the inverse of an invertible integer a mod m. From Proposition 2.11, we know that a must satisfy $\gcd(a,m) = 1$, which means that there exist integers x and y, such that $1 = ax + my$. As shown in the proof of Proposition 2.11, a^{-1} can be taken to be the integer x (mod m) in this equation. We explained earlier how these integers x and y can be computed by working backwards through the intermediate steps of the Euclidean algorithm. The algorithm below is an extended version of the Euclidean algorithm that works in a more organized fashion to output the numbers x and y, along with the greatest common divisor d. The algorithm operates on ordered lists (vectors) with three components. The components of such an ordered list V will be denoted (in order) as $V(1)$, $V(2)$, $V(3)$. Thus, for example, if $V = [2, 4, 6]$, then $V(1) = 2$, $V(2) = 4$, $V(3) = 6$. The algorithm will also be multiplying ordered lists by numbers, and this is done by multiplying each of the components by the number, for example: $5 [2, 4, 6] = [10, 20, 30]$.

Algorithm 2.2: The Extended Euclidean Algorithm

Input: A pair of positive integers a and b, with $a \geq b$.
Output: Three integers, $d = \gcd(a, b)$, x, and y, that satisfy the equation $d = ax + by$.

Step 1.   Set $U = [a, 1, 0]$, $V = [b, 0, 1]$ (Initialize recordkeeping vectors)
Step 2.   WHILE $V(1) > 0$

(Tasks below will be repeated while first component of V is positive)

$\quad\quad\quad W = U - \text{floor}\,(U(1)/V(1))V$
$\quad\quad\quad$ Update: $U = V$
$\quad\quad\quad$ Update: $V = W$
END (WHILE)[*]

Step 3.   Output: $d = U(1)$, $x = U(2)$, $y = U(3)$

The ordinary Euclidean algorithm (Algorithm 2.1) had the same inputs but outputted only d. It is not so obvious that this algorithm actually does what is claimed; we will explain it and prove that it does indeed work after the following illustrative example.

Example 2.7

(a) Use Algorithm 2.2 to compute $d = \gcd(148, 75)$ and integers x and y such that $d = 148x + 75y$.
(b) If it exists, compute the $75^{-1} \bmod 148$.

Solution: Part (a):

Step 1. We initialize $U = [148, 1, 0]$, $V = [75, 0, 1]$
Step 2. Since $V(1) = 75 > 0$

---

[*] This notation means that the operations after the WHILE instruction and before its END are to be executed repeatedly until the condition indicated after the WHILE—in this case, $V(1) > 0$—fails to be valid.

we set: $W = U - \text{floor}(U(1) / V(1))V$

$$= [148, 1, 0] - \lfloor 148 / 75 \rfloor [75, 0, 1]$$

$$= [148, 1, 0] - 1 \cdot [75, 0, 1] = [73, 1, -1].$$

We update $U = V = [75, 0, 1]$, and $V = W = [73, 1, -1]$.

Since $V(1) = 73 > 0$, we repeat this with the updates:

$W = U - \text{floor}(U(1) / V(1))V = [75, 0, 1] - \lfloor 75 / 73 \rfloor [73, 1, -1]$

$$= [75, 0, 1] - 1 \cdot [73, 1, -1] = [2, -1, 2].$$

$U = V = [73, 1, -1]$, and $V = W = [2, -1, 2]$.

Since $V(1) = 2 > 0$, we again repeat these updates:

$W = U - \text{floor}(U(1) / V(1))V = [73, 1, -1] - \lfloor 73 / 2 \rfloor [2, -1, 2]$

$$= [73, 1, -1] - 36 \cdot [2, -1, 2] = [1, 37, -73].$$

$U = V = [2, -1, 1]$, and $V = W = [1, 37, -73]$.

Since $V(1) = 1 > 0$, we need one final updating:

$W = U - \text{floor}(U(1) / V(1))V = [2, -1, 1] - \lfloor 2 / 1 \rfloor [1, 37, -73]$

$$= [2, -1, 1] - 2 \cdot [1, 37, -73] = [0, -75, 147].$$

$U = V = [1, 37, -73]$, and $V = W = [0, -75, 147]$.

Step 3. Output: $d = U(1) = 1$, $x = U(2) = 37$, $y = U(3) = -73$

The resulting relationship is easily checked: $1 = 37 \cdot 148 - 73 \cdot 75$.

Part (b): The result of part (a) tells us that $\gcd(148, 75) = 1$, so from Proposition 2.11 and the equation $1 = 37 \cdot 148 - 73 \cdot 75$, and since $-73 \equiv 75 \pmod{148}$, we get that $75^{-1} = 75$ in $\mathbb{Z}_{148}$.

Exercise for the Reader 2.10

(a) Use Algorithm 2.2 to compute $d = \gcd(1155, 862)$, and integers x and y such that $d = 1155x + 862y$.

(b) If it exists, compute 862^{-1} in $\mathbb{Z}_{1155}$.

Algorithm 2.2 is really just the Euclidean algorithm in disguise, with some additional recordkeeping (hence the three-element vectors). The proof below will use the notation of the Euclidean algorithm (Algorithm 1.1), so it might be helpful for the reader to review this algorithm before reading this proof.

Proof That the Outputs d, x, and y of Algorithm 1.2 Satisfy $d = \gcd(a, b)$ and $d = ax + by$. We first point out that throughout the algorithm, any of the length-3 vectors $Z = U$, V, or W always corresponds to a valid equation:

$$Z(1) = a \cdot Z(2) + b \cdot Z(3), \text{ where } Z = [Z(1), Z(2), Z(3)]$$

To see this, note first that it is clearly true for the initial vectors $U = [a, 1, 0]$ and $V = [b, 0, 1]$. (For example, for $Z = U$, the equation becomes $a = a \cdot 1 + b \cdot 0$.) All other vectors created or updated in the algorithm are either taken to be a previously constructed vector or (in the case of a W vector) taken as a vector of the form $U + \alpha V$, where α is an integer. It suffices to show if the vectors U and V both correspond to a valid equation with the above scheme, then so will the vector $U + \alpha V$. Indeed, from the corresponding equations for U and V: $U(1) = a \cdot U(2) + b \cdot U(3)$, $V(1) = a \cdot V(2) + b \cdot V(3)$, if we add α times the second to the first, we get $U(1) + \alpha V(1) = a \cdot [U(2) + \alpha V(2)] + b \cdot [U(3) + \alpha V(3)]$, which is the (valid) equation corresponding to the vector $U + \alpha V$. With this being done, it now suffices to show that the algorithm eventually terminates, and when it does, we have (the final value of) $U(1) = \gcd(a,b)$. As in the proof of Theorem 2.6, if we set $r_0 = b$ and $r_{-1} = a$, the Euclidean algorithm can be expressed as successive applications of the division algorithm, where each one defines the next element of the remainder sequence: $r_{i-1} = q_i r_i + r_{i+1}$ $(i = 0, 1, 2, \cdots n)$. Recall that the sequence of remainders is strictly decreasing and the final nonzero remainder (r_n) is $\gcd(a,b)$. If we look at the first component of the recursive formula of Algorithm 2.2, that is, $W(1) = U(1) - \text{floor}(U(1) / V(1)) \cdot V(1)$, we see that $W(1)$ is simply the remainder when the division algorithm is applied to the integer division of $U(1)$ by $V(1)$. Since $U(1)$ starts off at a, $V(1)$ starts off at b, and at each iteration, $U(1)$ is updated to $V(1)$ and $V(1)$ to (the new remainder) $W(1)$, we see that at the ith iteration of Algorithm 2.2, $W(1)$ is exactly the value of the new remainder in the ith iteration of the Euclidean algorithm. It follows that the values of $U(1)$ are strictly decreasing integers (so the algorithm terminates) whose last nonzero value is $\gcd(a,b)$, as claimed. □

Solving Linear Congruences

We have completely described an efficient method for solving any linear congruence:

$$ax + b \equiv c \pmod{m} \qquad (2.2)$$

whenever $\gcd(a,m) = 1$, in which case there is always a unique solution. Since the first step of subtracting b from both sides (in modular arithmetic) is always easy, the heart of solving such a congruence is the (modular) division step, so we really can focus attention on the simpler equation (obtained by setting $b = 0$ in Equation 2.2):

$$ax \equiv c \pmod{m} \qquad (2.3)$$

We complete our analysis of Equation 2.3 by moving to the remaining situation where $\gcd(a,m) = d > 1$. In order for a solution to exist, we must have $d \mid c$. [*Proof*: For any solution x, we would have $m \mid ax - c$, so since $d \mid m$, we get also that $d \mid ax - c$, and since $d \mid a$, it follows that $d \mid ax - (ax - c) = c$. □] In case $d \mid c$, it turns out that the congruence (Equation 2.3) will always have d distinct solutions (mod m).

> **Algorithm 2.3 (*Procedure for Solving* $ax \equiv c$ (mod m)**
> **in the Case $d = gcd(a,m) > 1$ and $d \mid c$)**
>
> Recall that if $d \nmid c$, there are no solutions.
>
> *Step 1.* Solve the modified congruence $(a/d)y \equiv (c/d)$ (mod m/d) as explained earlier in this section. This is possible, and there will be a unique solution y_0, since $gcd(a/d, m/d) = 1$.
>
> *Step 2.* The d solutions of the original congruence are y_0, $y_0 + m/d$, $y_0 + 2m/d, \cdots, y_0 + (d-1)m/d$ (mod m).

Before we explain why this algorithm works, we give an example to illustrate its use.

Example 2.8

Find all solutions of the following congruences:

(a) $2x \equiv 7$(mod 10)
(b) $6x \equiv 12$(mod 21)

Solution: Part (a): Since $d = gcd(2, 10) = 2$ does not divide 7, there is no solution.

Part (b): Since $d = gcd(6, 21) = 3$ does divide 12, there will be 3 (= d) distinct solutions (mod 21). We use Algorithm 2.3 to find them. The modified congruence from Step 1 is $(6/3)y \equiv (12/3)$ (mod 21/3) or $2y \equiv 4$(mod 7), which has the (unique) solution $y_0 = 2$(mod 7).[*] Step 2 now gives us the set of two solutions of the original congruence: $\{2, 2 + 21/3, 2 + 2 \cdot 21/3\} = \{2, 9, 16\}$. These are easily checked to satisfy the original congruence.

Exercise for the Reader 2.11

Find all solutions of the following congruences:

(a) $123x \equiv 12$(mod 456)
(b) $15x + 4 \equiv 20$(mod 25)

We now explain why Algorithm 2.3 does its job.

Proof That Algorithm 2.3 Correctly Finds All Solutions of the Indicated Congruence. Since the d solutions indicated by the algorithm are distinct integers mod m, there are two things we need to do: (i) we must show that the d solutions indicated by the algorithm actually solve the original congruence, and (ii) there are no other solutions (mod m).

---

[*] Up to now, our method for solving the congruence $2y \equiv 4$ (mod 7) would be to first find that the inverse of 2 (mod 7) is 4 [since $2 \cdot 4 = 8 \equiv 1$(mod 7)], and then multiply both sides of the congruence to obtain $y \equiv 4 \cdot 4 \equiv 16 \equiv 2$(mod 7). Whenever a congruence can be solved by ordinary integer arithmetic, the resulting solution will also be a valid one for the congruence. This is because if two real numbers are equal, then they will be congruent modulo any m. However, this method should not be applied directly to any congruence $ax \equiv b$ (mod m), where $d > 1$, and $d = gcd(a,b) \mid c$, because it will give only one of the d solutions: For example, $5x \equiv 15$ (mod 25) has four solutions, but $x \equiv 3$(mod 25) has only one!

Part (i): The fact that $(a/d)y_0 \equiv (c/d) \pmod{m/d}$ means that $(m/d)\,|\,[(a/d) y_0 - (c/d)]$. This divisibility relation implies that $m\,|\,[ay_0 - c]$, which is equivalent to the congruence $ay_0 \equiv c \pmod{m}$. Now, since $d\,|\,a$ for any integer i, we have $a(y_0 + im/d) \equiv ay_0 + (a/d)im \equiv c + 0 \equiv c \pmod{m}$, so $y_0 + im/d$ solves the indicated congruence. In particular, so do the d indicated solutions.

Part (ii): We first observe that there can be no other solutions of the original congruence of the form $y_0 + im/d$ $(i \in \mathbb{Z})$ $\pmod{m}$ other than the d solutions indicated by the algorithm. This is because for any integer i, if r is the remainder when i is divided by d, $y_0 + im/d \equiv y_0 + rm/d \pmod{m}$. It remains to show that there can be no solutions other than these of the original congruence. Indeed, suppose that there was a solution z_0, $az_0 \equiv c \pmod{m}$, which is not of this form. Therefore, there is a unique integer i_0, such that $y_0 + i_0\, m/d < z_0 < y_0 + (i_0 + 1)m/d$. If we rewrite this double inequality as $i_0\, m/d < z_0 - y_0 < (i_0 + 1)m/d$, it is clear that $z_0 \not\equiv y_0 \pmod{m/d}$. However, the argument in part (a) shows that since $az_0 \equiv c \pmod{m}$, we have $(a/d)z_0 \equiv c/d \pmod{m/d}$, and this contradicts the fact that y_0 was the unique solution of this latter congruence. $\square$

We have thus completely described how to solve a single linear congruence of form Equation 2.3 (or Equation 2.2). We summarize the procedure:

Summary of Procedure for Solving the Single Linear Congruence (Equation 2.2)

$$ax + b \equiv c \pmod{m}$$

Step 1. Subtract b from both sides to obtain the equation $ax \equiv c - b$ $\pmod{m}$.

Step 2. First compute $d = \gcd(a, m)$.

Case 1. $d = 1$. *Unique Solution.* Use the extended Euclidean Algorithm 2.2 to compute integers e and f such that $1 = ae + mf$, to obtain $a^{-1} \equiv e \pmod{m}$. The unique solution of Equation 2.2 is given by $x \equiv a^{-1} \cdot (c - b) \pmod{m}$.

Case 2. $d > 1$ and $d \nmid c$. *No Solution.* There are no solutions of the congruence Equation 2.2 $\pmod{m}$.

Case 3. $d > 1$ and $d\,|\,c$. *Multiple Solutions.* Use the extended Euclidean Algorithm 2.2 to compute integers e' and f' such that $1 = (a/d)e' + (m/d)f'$, to obtain $(a/d)^{-1} \equiv e' \pmod{m/d}$. Use this to find the unique solution of the modified congruence $(a/d)y \equiv ([c - b]/d)$: $y_0 \equiv (a/d)^{-1} \cdot ([c - b]/d) \pmod{m/d}$. The d solutions of the original congruence are $y_0, y_0 + m/d, y_0 + 2m/d, \cdots, y_0 + (d - 1)m/d \pmod{m}$.

Exercise for the Reader 2.12

Find all solutions of the following congruences:

(a) $6x + 2 \equiv 5 \pmod{9}$
(b) $6x + 2 \equiv 3 \pmod{9}$
(c) $5x + 2 \equiv 3 \pmod{9}$

The Chinese Remainder Theorem

In many applications, including some in cryptography that we will see later on, it is necessary to simultaneously solve a system of linear congruences of different moduli[*]:

$$\begin{cases} a_1 x \equiv c_1 (\text{mod } m_1) \\ a_2 x \equiv c_2 (\text{mod } m_2) \\ \quad \vdots \\ a_k x \equiv c_k (\text{mod } m_k) \end{cases} \qquad (2.4)$$

More precisely, we would like to know when we find an integer x that solves each of the congruences in Equation 2.4. Furthermore, in cases where such a simultaneous integer solution exists, we would like to classify all of the solutions.

Puzzles that can be modeled by simultaneous congruences such as Equation 2.4 have appeared in various ancient mathematical documents, including those from the Greeks (dating back to the first century a.d.), the Chinese (dating to back to the third century a.d.), and the Hindus (dating back to the seventh century a.d.). The following is an example of such an ancient Hindu puzzle.

Example 2.9: A Hindu Puzzle from the Seventh Century a.d.

Determine a system of simultaneous congruences that models the following puzzle:

While a woman is on her way to the market, a horse steps on her basket and crushes all her eggs. The rider agrees to pay for the damage and asks how many eggs she had. She does not recall the exact number, but she knows that when she had taken them out two at a time, there was one egg left. The same thing happened when she removed them three, four, five, and six at a time, but when she took them out seven at a time, they all came out. What was the smallest number of eggs she could have had?

Solution: Letting x denote the (unknown) number of eggs that were in the woman's basket, the problem tells us that x must solve each of the following congruences:

$$\begin{cases} x \equiv 1 (\text{mod } 2) \\ x \equiv 1 (\text{mod } 3) \\ x \equiv 1 (\text{mod } 4) \\ x \equiv 1 (\text{mod } 5) \\ x \equiv 1 (\text{mod } 6) \\ x \equiv 0 (\text{mod } 7) \end{cases} \qquad (2.5)$$

---

[*] Since the reduction from a more general linear congruence $ax + b \equiv c \ (\text{mod } m)$ is simple, it suffices to assume that our linear congruences are in the form $ax \equiv c \ (\text{mod } m)$.

The problem seeks the smallest positive solution of Equation 2.5. We point out one very simple observation about congruences that will sometimes help to simplify such systems.

> **Proposition 2.12**
>
> Suppose that m_1, m_2 are positive integers with $m_1 \mid m_2$. Any solution of a linear congruence $ax \equiv c \pmod{m_2}$ will also be a solution of the same congruence $\pmod{m_1}$.

Proof: By definition, x solves the first congruence means that $m_2 \mid (ax - c)$. Since we are assuming that $m_1 \mid m_2$, it follows by transitivity of divisibility [Theorem 2.1(a)] that $m_1 \mid (ax - c)$, which means that $ax \equiv c \pmod{m_1}$. □

If we apply this proposition to the Equation 2.5, since both 2 and 3 divide 6, the congruences $x \equiv 1 \pmod 2$, $x \equiv 1 \pmod 3$ are redundant consequences of the congruence $x \equiv 1 \pmod 6$. Thus, they can be safely removed from the system to produce the following simpler, but equivalent system[*]:

$$\begin{cases} x \equiv 1 \pmod 4 \\ x \equiv 1 \pmod 5 \\ x \equiv 1 \pmod 6 \\ x \equiv 0 \pmod 7 \end{cases} \tag{2.6}$$

We will return to Equation 2.6 and the Hindu puzzle momentarily, but we first consider the problem of solving Equation 2.4. First of all, it is clear that in order for a simultaneous solution to exist, each individual congruence must have a solution, and from our development for single linear congruences, this means that we must have $d_i \mid c_i$ $(1 \le i \le k)$, where $d_i = \gcd(a_i, m_i)$. With these conditions being satisfied, in light of Algorithm 2.3, Equation 2.4 can be reduced to the simpler system:

$$\begin{cases} x \equiv b_1 \pmod{n_1} \\ x \equiv b_2 \pmod{n_2} \\ \vdots \\ x \equiv b_k \pmod{n_k} \end{cases} \tag{2.7}$$

where $n_i = m_i / d_i$ and $b_i = (a_i / d_i)^{-1} (c_i / d_i) \pmod{n_i}$. The following theorem shows that Equation 2.7 always has a solution in cases where the moduli are *pairwise relatively prime*—that is, $\gcd(n_i, n_j) = 1$ whenever $i \ne j$ $(1 \le i, j \le k)$—and it includes a uniqueness statement. This theorem has been found to date back to a Chinese mathematics book that was published in 1247 a.d. by the Chinese mathematician Qin Jiushao

---

[*] An equivalent system of equations is one that has the same solution set as the original system.

(1202–1261),[*] and has come to be known as the *Chinese remainder theorem*. We will give a constructive (algorithmic) proof of the existence of the solution, and thus provide a practical method of solving the Equation 2.7 and hence also Equation 2.4.

Theorem 2.13: The Chinese Remainder Theorem

Suppose that $n_1, n_2, \cdots, n_k > 1$ are pairwise relatively prime integers. Then for any integers $b_1, b_2, \cdots, b_k$, the system of congruences:

$$\begin{cases} x \equiv b_1 (\text{mod } n_1) \\ x \equiv b_2 (\text{mod } n_2) \\ \vdots \\ x \equiv b_k (\text{mod } n_k) \end{cases} \qquad (2.7)$$

has a simultaneous integer solution x that is unique modulo $N = n_1 n_2 \cdots n_k$.

As the following proof contains an algorithm, we will defer giving an example until after proving the theorem.

Proof: Part (a): *Existence*: (*Constructive Proof*) For each index i ($1 \leq i \leq k$), since $\gcd(N / n_i, n_i) = 1$, there is a (unique) solution e_i of the congruence $e_i (N / n_i) \equiv 1 \pmod{n_i}$. We claim that

$$x = \sum_{i=1}^{k} b_i e_i (N / n_i) \qquad (2.8)$$

is a simultaneous solution of Equation 2.7. Indeed, for each index j, we have (mod n_j) $N / n_i \equiv 0$, unless $j = i$, and this forces all terms other than the jth term in the sum of Equation 2.8 to be 0 (mod n_j). Thus $x \equiv b_j e_j (N / n_j) \equiv b_j \cdot 1 \equiv b_j \pmod{n_j}$, as desired.

Part (b): *Uniqueness*: Suppose that x' is another simultaneous solution of Equation 2.7. It follows that for each index i, $x \equiv b_i \equiv x' \pmod{n_i}$, so that $n_i \mid (x - x')$. Since the n_i's are pairwise relatively prime, it follows (from the fundamental theorem of arithmetic) that their product N also divides $x - x'$; that is, $x \equiv x' \pmod{N}$. □

---

[*] Qin Jiushao (also transliterated as Ch'in Chiu-Shao) was a tour-de-force among ancient mathematicians; his principal scientific contributions were published in his 1247 book *Shushu Jiuzhang* (*Mathematical Treatise in Nine Sections*). The first chapter of his book contained the development and proof of (what is now called) the Chinese remainder theorem. The book also included analyses of higher-order equations that modeled certain interesting applied problems such as the following (from Chapter 2 of his book): Determine the height of rainfall on level ground, given that it reached a height h in a cylindrical vessel with circular top and bottom having respective radii $a > b$. Aside from his mathematics, Qin had quite an interesting military and government career. In his youth, he fought the armies of Genghis Khan. He was notorious for his corruption and manipulations of his government posts, from which he amassed a tremendous amount of wealth. His book was actually written during a hiatus from work when he returned to his hometown to mourn the death of his mother.

Example 2.10

Solve the following system of congruences:

$$\begin{cases} x \equiv 2(\text{mod } 3) \\ x \equiv 3(\text{mod } 5) \\ x \equiv 6(\text{mod } 14) \end{cases}$$

Solution: Since the moduli are pairwise relatively prime, Equation 2.8 (in the proof of the Chinese remainder theorem) provides us with a scheme for obtaining a simultaneous solution. We first set $N = 3 \cdot 5 \cdot 14 = 210$. With $b_1, b_2, b_3 = 2, 3, 6$ and $n_1, n_2, n_3 = 3, 5, 14$, in order to use Equation 2.8, we must first determine e_1, e_2, e_3, by their defining equations: $e_i(N / n_i) \equiv 1 \ (\text{mod } n_i)$.

For e_1: $e_1 \cdot 70 \equiv 1(\text{mod } 3) \Leftrightarrow e_1 \cdot 1 \equiv 1(\text{mod } 3) \Leftrightarrow e_1 \equiv 1(\text{mod } 3)$.

For e_2: $e_2 \cdot 42 \equiv 1(\text{mod } 5) \Leftrightarrow e_2 \cdot 2 \equiv 1(\text{mod } 5) \Leftrightarrow e_2 \equiv 3(\text{mod } 5)$ [since $2^{-1} = 3 \ (\text{mod } 5)$].

For e_3: $e_3 \cdot 15 \equiv 1(\text{mod } 14) \Leftrightarrow e_3 \cdot 1 \equiv 1(\text{mod } 14) \Leftrightarrow e_3 \equiv 1(\text{mod } 14)$.
Now we have all that we need to apply Equation 2.8 to get a desired solution:

$$x = \sum_{i=1}^{3} b_i e_i (N/n_i) = 2 \cdot 1 \cdot (70) + 3 \cdot 3 \cdot (42) + 6 \cdot 1 \cdot (15)$$

$$= 608 \equiv 188(\text{mod } 210)$$

Thus 188 is the smallest positive integer solution of the original system.

Exercise for the Reader 2.13

Determine the general solution of the following system of congruences:

$$\begin{cases} x \equiv 0(\text{mod } 2) \\ x \equiv 2(\text{mod } 5) \\ 3x \equiv 4(\text{mod } 7) \end{cases}$$

Although we cannot apply the Chinese remainder theorem to the Equation 2.6 of the Hindu problem, part (b) of the following proposition, which generalizes Proposition 2.12, will allow us to convert Equation 2.6 into a form to which the theorem is applicable.

Proposition 2.14

(a) Any set of divisibility relations of the form $m_1 \mid b$, $m_2 \mid b, \cdots$, $m_k \mid b$ is equivalent to the single divisibility relation $\text{lcm}(m_1, m_2, \cdots, m_k) \mid b$.

(b) Any system of congruences of the form

$$\begin{cases} ax \equiv c \,(\mathrm{mod}\ m_1) \\ ax \equiv c \,(\mathrm{mod}\ m_2) \\ \quad\vdots \\ ax \equiv c \,(\mathrm{mod}\ m_k) \end{cases}$$

(that is, the same congruence under different moduli) is equivalent to the single congruence:

$$ax \equiv c \,(\mathrm{mod}\ \mathrm{lcm}(m_1, m_2, \cdots, m_k))$$

Note: Unlike in the Chinese remainder theorem, the moduli need not be pairwise relatively prime in this result.

The proof is similar to that of Proposition 2.12 and is left to the following exercise for the reader.

Exercises for the Reader

Exercise 2.14

Prove Proposition 2.14.

Exercise 2.15

What is the answer to the ancient Hindu problem of Example 2.9?

Chapter 2 Exercises

1. Determine whether each of the statements below is true or false.
    (a) $9 \mid 128$
    (b) $3 \mid 111$
    (c) $13 \mid 5271$
    (d) $-1 \mid a$ for any $a \in \mathbb{Z}$
    (e) $a \mid ab$ for any $a, b \in \mathbb{Z}$
    (f) $0 \mid 12$

2. Determine whether each of the statements below is true or false.
    (a) $7 \mid -49$
    (b) $2 \nmid 111$
    (c) $17 \mid 5271$
    (d) $1 \mid a$ for any $a \in \mathbb{Z}$
    (e) $a \mid a$ for any $a \in \mathbb{Z}$
    (f) $12 \mid 0$

3. If a and b are integers such that $a \mid b$, and $b \mid a$, show that $a = \pm b$.

4. If a, b, and c are positive integers such that $a \mid b$, show that $ac \mid bc$.

5. Which of the following integers is prime?
   (a) 67
   (b) 91
   (c) 893
   (d) 8671
   (e) 6581
   (f) 148,877

6. Which of the following integers is prime?
   (a) 83
   (b) 97
   (c) 893
   (d) 1229
   (e) 46,189
   (f) 12,499

7. Find the prime factorization of each of the following positive integers, as guaranteed by the fundamental theorem of arithmetic:
   (a) 24
   (b) 88
   (c) 675
   (d) 6400
   (e) 74,529
   (f) 183,495,637

8. Find the prime factorization of each of the following positive integers, as guaranteed by the fundamental theorem of arithmetic:
   (a) 52
   (b) 96
   (c) 512
   (d) 4725
   (e) 130,321
   (f) 7,817,095

9. Find the quotient and remainder when the division algorithm is applied to each of the following integer divisions:
   (a) $67 \div 2$
   (b) $108 \div 5$
   (c) $-77 \div 2$
   (d) $882 \div 13$
   (e) $1228 \div 25$
   (f) $-1582 \div 36$

10. Find the quotient and remainder when the division algorithm is applied to each of the following integer divisions.
   (a) $67 \div 3$
   (b) $180 \div 5$
   (c) $-90 \div 13$
   (d) $-564 \div 14$
   (e) $1268 \div 42$
   (f) $-8888 \div 25$

11. Using prime factorizations, compute the indicated greatest common divisors or least common multiples.
   (a) gcd(12, 36)
   (b) lcm(20, 25)
   (c) gcd(100, 56)
   (d) gcd(560, 1400)
   (e) lcm(120, 50)
   (f) gcd(121275, 5788125)

12. Using prime factorizations, compute the indicated greatest common divisors or least common multiples.
   (a) gcd(15, 40)
   (b) lcm(15, 40)
   (c) gcd(136, 86)
   (d) gcd(1925, 1568)
   (e) lcm(150, 350)
   (f) gcd(256500, 109395)

13. Use the Euclidean algorithm to compute each of the quantities in Chapter Exercise 11. For those that are lcm's, use the formula of Exercise for the Reader 2.2(c).

14. Use the Euclidean algorithm to compute each of the quantities in Chapter Exercise 12. For those that are lcm's, use the formula of Exercise for the Reader 2.2(c).

15. For each pair of integers a, b that are given, use the Euclidean algorithm (as explained in the proof of Theorem 2.6) to determine integers x and y, such that $\gcd(a, b) = ax + by$.
   (a) 12, 36
   (b) 100, 56
   (c) 560, 1400
   (d) 121275, 5788125

16. For each pair of integers a, b that are given, use the Euclidean algorithm (as explained in the proof of Theorem 2.6) to determine integers x and y, such that $\gcd(a, b) = ax + by$.
   (a) 15, 40
   (b) 136, 86
   (c) 1925, 1568
   (d) 256500, 109395

17. Determine whether each of the statements below is true or false.
   (a) $43 \equiv 1 \pmod 2$
   (b) $-43 \equiv 3 \pmod 4$
   (c) $488 \equiv 10 \pmod{12}$
   (d) $2205 \equiv 45 \pmod{360}$
   (e) $-443 \equiv 18 \pmod{22}$
   (f) $7^7 \equiv 5^8 \pmod{371}$

18. Determine whether each of the statements below is true or false.
   (a) $43 \equiv 2 \pmod 3$
   (b) $-43 \equiv 3 \pmod 6$
   (c) $2207 \equiv 11 \pmod{12}$
   (d) $11340 \equiv 90 \pmod{360}$
   (e) $-2444 \equiv -446 \pmod{666}$
   (f) $3^6 \equiv 5^2 \pmod{44}$

19. Perform the following operations in $\mathbb{Z}_{24}$:
    (a) $18 + 20$
    (b) $5 - 21$
    (c) $8 \cdot 8$
    (d) $2^8 - 3^8$
    (e) 21^{223}

    Suggestion: For part (e), compute some smaller powers first (mod 24) to build up to the final answer using laws of exponents.

20. Perform the following operations in $\mathbb{Z}_{53}$:
    (a) $39 + 47$
    (b) $25 - 36$
    (c) $18 \cdot 35$
    (d) $12^5 - 19^4$
    (e) 33^{100}

    Suggestion: For part (e), compute some smaller powers first (mod 53) to build up to the final answer using laws of exponents.

21. Create addition and multiplication tables for (a) $\mathbb{Z}_2$ and (b) $\mathbb{Z}_4$. (See Example 2.8.) By looking at the multiplication tables, determine all invertible elements.

22. Create addition and multiplication tables for (a) $\mathbb{Z}_3$ and (b) $\mathbb{Z}_9$. (See Example 2.8.) By looking at the multiplication tables, determine all invertible elements.

23. Find all invertible elements of $\mathbb{Z}_8$ and for each one find the inverse.

24. Find all invertible elements of $\mathbb{Z}_{10}$ and for each one find the inverse.

25. Find all invertible elements of $\mathbb{Z}_7$ and for each one find the inverse.

26. Find all invertible elements of $\mathbb{Z}_{11}$ and for each one find the inverse.

27. Solve each of the following congruences working in (mod 8):
    (a) $3x \equiv 5$
    (b) $7x + 2 \equiv 3$
    (c) $5x - 2 \equiv 2$

28. Solve each of the following congruences working in (mod 10):
    (a) $3x \equiv 5$
    (b) $7x + 2 \equiv 3$
    (c) $9x - 8 \equiv 7$

29. For each integer below, use the extended Euclidean algorithm (Algorithm 2.2) together with Proposition 2.11 to find the inverse in $\mathbb{Z}_{388}$, if the inverse exists.
    (a) 3
    (b) 55
    (c) 149
    (d) 97

30. For each integer below, use the extended Euclidean algorithm (Algorithm 2.2) together with Proposition 2.11 to find the inverse in $\mathbb{Z}_{299}$, if the inverse exists.
    (a) 2
    (b) 52
    (c) 80
    (d) 199

31. For each integer below, use the extended Euclidean algorithm (Algorithm 2.2) together with Proposition 2.11 to find the inverse in $\mathbb{Z}_{1353}$, if the inverse exists.
    (a) 2
    (b) 44
    (c) 886
    (d) 350

32. For each integer below, use the extended Euclidean algorithm (Algorithm 2.2) together with Proposition 2.11 to find the inverse in $\mathbb{Z}_{2555}$, if the inverse exists.
    (a) 2
    (b) 74
    (c) 98
    (d) 1972

33. Find all solutions for each of the following congruences:
    (a) $3x \equiv 59 (\mathrm{mod}\ 388)$
    (b) $149x \equiv 225 (\mathrm{mod}\ 388)$
    (c) $2x \equiv 1225 (\mathrm{mod}\ 1353)$
    (d) $886x \equiv 35 (\mathrm{mod}\ 1353)$

34. Find all solutions for each of the following congruences:
    (a) $2x \equiv 59 (\mathrm{mod}\ 299)$
    (b) $199x \equiv 99 (\mathrm{mod}\ 299)$
    (c) $2x \equiv 847 (\mathrm{mod}\ 2555)$
    (d) $1972x \equiv 363 (\mathrm{mod}\ 2555)$

35. Find all solutions for each of the following congruences:
    (a) $3x \equiv 6 (\mathrm{mod}\ 18)$
    (b) $15x \equiv 21 (\mathrm{mod}\ 51)$
    (c) $8x \equiv 12 (\mathrm{mod}\ 28)$
    (d) $8x \equiv 6 (\mathrm{mod}\ 28)$

36. Find all solutions for each of the following congruences:
    (a) $2x \equiv 6 (\mathrm{mod}\ 16)$
    (b) $6x \equiv 16 (\mathrm{mod}\ 27)$
    (c) $14x \equiv 21 (\mathrm{mod}\ 88)$
    (d) $25x \equiv 55 (\mathrm{mod}\ 95)$

37. Find all solutions for each of the following congruences:
    (a) $6x \equiv 28 (\mathrm{mod}\ 776)$
    (b) $15x \equiv 21 (\mathrm{mod}\ 1940)$
    (c) $596x \equiv 900 (\mathrm{mod}\ 1552)$
    (d) $3544x \equiv 900 (\mathrm{mod}\ 5412)$

38. Find all solutions for each of the following congruences:
    (a) $8x \equiv 16(\text{mod } 1196)$
    (b) $400x \equiv 125(\text{mod } 1495)$
    (c) $1393x \equiv 175(\text{mod } 2093)$
    (d) $17748x \equiv 6642(\text{mod } 22995)$

39. Find all solutions for each of the following systems of congruences:

    (a) $\begin{cases} x \equiv 3(\text{mod } 5) \\ x \equiv 4(\text{mod } 7) \end{cases}$

    (b) $\begin{cases} x \equiv 2(\text{mod } 3) \\ x \equiv 1(\text{mod } 5) \\ x \equiv 3(\text{mod } 11) \end{cases}$

    (c) $\begin{cases} x \equiv 2(\text{mod } 6) \\ x \equiv 1(\text{mod } 5) \\ x \equiv 3(\text{mod } 7) \\ x \equiv 1(\text{mod } 13) \end{cases}$

40. Find all solutions for each of the following systems of congruences:

    (a) $\begin{cases} x \equiv 3(\text{mod } 4) \\ x \equiv 4(\text{mod } 5) \end{cases}$

    (b) $\begin{cases} x \equiv 0(\text{mod } 2) \\ x \equiv 1(\text{mod } 5) \\ x \equiv 6(\text{mod } 9) \end{cases}$

    (c) $\begin{cases} x \equiv 1(\text{mod } 4) \\ x \equiv 2(\text{mod } 5) \\ x \equiv 3(\text{mod } 9) \\ x \equiv 1(\text{mod } 11) \end{cases}$

41. A group of 15 pirates has just looted a stash of identical and very valuable gold coins. They plan to equally divide them the next morning. During the night, one pirate, who does not trust the others, gets up to divide the coins into 15 equal parts, finds there are 8 remaining, so takes these 8 with him. But he was unknowingly followed by another pirate who then kills him. This second pirate then divides the remaining coins by 14, finds there are 11 left, and stashes these 11 (plus the 8 he got from the first pirate) under the hull of the boat, but in the process falls overboard and drowns. The next morning the remaining 13 pirates divide the coins by 13 and find there are 5 left. What is the smallest number of coins that originally could have been present?

42. Suppose that a certain computer server has less than 1GB (1 billion bytes) of memory; any time it runs jobs, it allocates an equal number of bytes of memory to each job and leaves any remaining bytes unused. Suppose that when it runs 95 jobs, it has 86 unused bytes; when it runs 98 jobs, it has 13 unused bytes; when it runs 99 jobs, it has 46 unused bytes; and when it runs 101 jobs, all bytes get allocated. Determine the exact size of the computer's memory.

Note: Chapter Exercises 43 through 45 give some applications of congruences to coding theory. **Coding theory** is an area of applied mathematics involved with the efficient transportation of information that is prone to transmission errors (for example, either through incorrectly entering the data on the sender's end or through a noisy channel). The basic idea is to transmit some additional (redundant) information that can be used to detect and sometimes even correct errors. The exercises below deal with error-detecting codes in ISBN (International Standard Book Number) numbers and in credit card numbers. Error-correcting codes have built-in mechanisms that can actually correct errors. An example is with audio CDs, which can play fine with scratches on the playing surface, or even if one were to drill a 2.5 mm hole through the playing surface. Coding theory also makes use of other areas of mathematics such as linear algebra and abstract algebra. For a good general introduction, see, for example, [LiXi-04]. A more comprehensive treatment in error-correcting codes can be found in [Moo-05].

43. *Application of Congruences: ISBN Error-Detecting Codes.* Since the 1970s, to facilitate inventory control and the ordering/selling of books, almost all books published have an attached unique ISBN number. For over 30 years the same system had been in widespread use, but as of 2007, the 10-digit ISBNs were replaced by 13-digit ISBNs. This exercise will discuss 13-digit ISBNs, and the next one will look at the previously used 10-digit system. To distinguish, we refer to each system as ISBN-13 or ISBN-10. An ISBN-13 consists of five blocks of digits. For example, the ISBN-13 of the author's book, *Introduction to Numerical Ordinary and Partial Differential Equations Using MATLAB*, is 978-0-471-69738-1. The first block always consists of three digits. Most books presently in print use 978 in this field (for the U.S. ISBN Agency). The second group is a single digit encoding the country or language of the publisher (0 indicates English), the third group of digits may range from two to seven digits and indicates the publisher (471 indicates John Wiley & Sons), the fourth block of digits has length 8 less the number in the preceding field, and indicates the publisher's assigned number for the particular book. Thus, larger publishers will be assigned smaller publisher codes to allow for larger capacities in the book field (up to six digits, or 1 million books). The fifth and final group is a single **check digit** from 0 to 9. If the digits (in order) of an ISBN-13 number are $x_1 x_2 \cdots x_{13}$, then the check digit x_{13} is determined by the equation

$$x_{13} \equiv 10 - (x_1 + 3x_2 + x_3 + 3x_4 + \cdots + x_{11} + 3x_{12}) \pmod{10} \quad (2.9)$$

For example, the right-hand side of Equation 2.9 for the above-mentioned ISBN-13 number works out to be $10 - (9 + 3\cdot7 + 8 + 3\cdot0 + 4 + 3\cdot7 + 1 + 3\cdot6 + 9 + 3\cdot7 + 3 + 3\cdot8) = -129 \equiv 1(\mathrm{mod}\,10)$. This indeed coincides with the last check digit $x_{13} = 1$. This check system (Equation 2.9) was designed to detect any single error of the following most common ones occurring in typing an ISBN number: mistyping one of the digits or switching two adjacent digits. This is important, since with a single error, the resulting incorrect ISBN could correspond to a totally different book and would otherwise go unnoticed.

(a) Each of the following is the first 12 digits of an ISBN-13 number. Find the 13th check digit for each: 978055215169, 978082482223, 978006123400.

(b) Show that if a valid ISBN-13 number has exactly one mistyped digit, then Equation 2.9 will fail to hold; i.e., the error will be detected.

(c) Show that if a valid ISBN-13 number has two different adjacent digits that were typed in the wrong order, then Equation 2.9 may fail to detect this error.

(d) Give an example of an incorrectly typed ISBN-13 number, along with two corresponding valid ISBN-13 numbers that differ from the former in exactly one digit. This shows that although the ISBN-13 system can detect common errors, it cannot correct them.

Suggestions: For part (b): If $x_1 x_2 \cdots x_{13}$ was incorrectly typed as $y_1 y_2 \cdots y_{13}$, where each $y_i = x_i$, with a single exception $y_j \neq x_j$, assume that both ISBNs checked with Equation 2.9. Then, by subtracting the corresponding two equations, we would be left with either $y_j \equiv x_j$ or $3y_j \equiv 3x_j \,(\mathrm{mod}\,10)$. But since 3 is invertible $(\mathrm{mod}\,10)$, we could multiply both sides of the latter equation by the inverse to obtain $y_j \equiv x_j (\mathrm{mod}\,10)$, which forces $y_j = x_j$—a contradiction! A similar argument works for part (c).

44. *Application of Congruences: ISBN Error-Detecting Codes.* The reader should first read Chapter Exercise 43 for some general background. From 1970 up to 2007, the ISBN-10 system was used for published books. Books published prior to 2007 needed to have their ISBN-10 numbers converted to ISBN-13. For example, the ISBN-10 number of the book *Introduction to Numerical Ordinary and Partial Differential Equations Using MATLAB* is 0-471-69738-9. The first three blocks correspond to the second through fourth blocks of the ISBN-13 numbers. The fourth and final block is a single **check digit** that is either a digit from 0 to 9, or the letter X (corresponding to 10). If the digits (in order) of an ISBN-10 number are $x_1 x_2 \cdots x_{10}$, then the check digit x_{10} is determined by the equation

$$x_{10} \equiv x_1 + 2x_2 + 3x_3 + 4x_4 + \cdots + 9x_9 \ (\mathrm{mod}\,11) \qquad (2.10)$$

For example, the right-hand side of Equation 2.10 for the above-mentioned ISBN-10 number works out to be $0 + 2\cdot4 + 3\cdot7 + 4\cdot1 + 5\cdot6 + 6\cdot9 + 7\cdot7 + 8\cdot3 + 9\cdot8 = 262 \equiv 9(\mathrm{mod}\,11)$. This indeed checks with the last check digit $x_{10} = 9$. This check system Equation 2.10 was designed to detect any single error of the following two: mistyping one of the digits or switching *any* two unequal digits. Thus, the ISBN-10 system can detect more general errors than the ISBN-13 system.

(a) Each of the following is the first 9 digits of an ISBN-10 number. Find the 10th check digit for each: 951020387, 082482223, 013014400.

(b) Show that if a valid ISBN-10 number has exactly one mistyped digit, then Equation 2.10 will fail to hold; i.e., the error will be detected.

(c) Show that if a valid ISBN-10 number has two different digits that were switched, then Equation 2.10 will fail to hold; i.e., the error will be detected.

(d) Give an example of an incorrectly typed ISBN-10 number along with two corresponding valid ISBN-10 numbers that differ from the former in exactly one digit. This shows that although the ISBN-10 system can detect common errors, it cannot correct them.

Suggestions: See the suggestions for Chapter Exercise 43 for ideas for parts (b) and (c).

Note: Comparing the error-detecting capabilities of ISBN-10 versus ISBN-13, although the latter system has a larger capacity, the former is better at detecting permutation errors [see part (c) of this and the preceding exercise].

45. *Application of Congruences: Credit Card Error-Detecting Codes.* Different credit cards use similar coding systems that include identifying information as well as a check digit. This is how Web sites can often immediately inform you if the number you keyed in is not a valid credit card number. In this exercise, we will explore the system that VISA cards use. VISA card numbers either contain 13 or 16 digits, and the first digit is always 4, indicating it is a VISA card (MasterCards always begin with 5). The second through the sixth digits identify the bank that issued the VISA card, and the seventh through the second-to-last digit give the account number. The final digit is the check digit. If the digits (in order) of a 16-digit VISA card number are $x_1 x_2 \cdots x_{16}$, then the check digit x_{16} is determined by the equation

$$x_{16} = -[2x_1 + x_2 + 2x_3 + x_4 + 2x_5 \cdots + x_{14} + 2x_{15}]$$
$$- r \ (\text{mod } 10)$$
(2.11)

where r is the number of terms in the bracketed expression that are greater than or equal to 10. For example, in the VISA card number 4784 5580 0246 1888, the bracket expression in Equation 2.11 is $[2 \cdot 4 + 7 + 2 \cdot 8 + 4 + 2 \cdot 5 + 5 + 2 \cdot 8 + 0 + 2 \cdot 0 + 2 + 2 \cdot 4 + 6 + 2 \cdot 1 + 8 + 2 \cdot 8] = [8 + 7 + 16 + 4 + 10 + 5 + 16 + 0 + 0 + 2 + 8 + 6 + 2 + 8 + 16]$ and has $r = 4$ two-digit terms and equals 108; thus, the right-hand side of Equation 2.11 equals $-108 - 4 \equiv 8 \ (\text{mod } 10)$, which coincides (as it should) with the (last) check digit x_{16}.

(a) Which of the following are valid 16-digit VISA card numbers? For those that are not, explain why: 4238 1678 1139 5207, 5602 8333 5495 1777, 4671 8899 3663 1942.

(b) Show that if a valid VISA number has exactly one mistyped digit, then Equation 2.11 will fail to hold; i.e., the
error will be detected.

(c) Show that if a valid VISA number has two different adjacent
digits that were typed in the wrong order, then Equation 2.11
will fail to hold; i.e., the error will be detected. Can an error
still be detected if the digits are not adjacent?

(d) Give an example of an incorrectly typed VISA card
number along with two corresponding valid VISA card
numbers that differ from the former in exactly one digit.
This shows that although the VISA card system can detect
common errors, it cannot correct them.

(e) Suppose that a 16-digit VISA card number was correctly
sent in but one of the digits printed out illegibly. Is it
always possible to recover the missing digit? Explain your
answer.

Suggestions: See the suggestions for Chapter Exercise 43 for
ideas for parts (b) and (c).

46. *Application of Congruences: Round Robin Tournaments.* In
sports competitions involving matches among sets of teams
(or individuals), a *round robin tournament* is a tournament
in which each team plays every other team exactly once in a
series of *rounds* (each team can play in at most one match per
round). Congruences can be used to set up a round robin tournament among N teams that will go through exactly N rounds.
Here is the algorithm:

(a) Label the teams as $1, 2, \cdots, N$.

(b) In the rth round ($1 \leq r \leq N$), match team i with team j
($1 \leq i, j \leq N, i \neq j$) if, and only if, $i + j \equiv r \pmod{N}$.

   (i) Use this algorithm to set up a round robin tournament for a competition involving $N = 4$ teams.

   (ii) Repeat the instructions of part (a) when $N = 6$.

   (iii) Prove that this algorithm will always produce a
round robin tournament.

   (iv) Note (quite obviously) that with $N = 2$ teams, only
one round is needed in a round robin tournament,
but show that with $N = 3$ teams, any round robin
tournament would need (at least) three rounds.

Suggestions for part (iii): To show that two (different) teams i
and j play each other exactly once (in the N rounds), apply the
division algorithm to the division of $i + j$ by N. The remainder
will be the unique round in which i and j are matched (except
if $r = 0$, this should mean the Nth round). Use this division to
show also that i cannot play any other team in this same round
(and similarly for j).

47. For each of the following divisibility statements, either prove
it or give a counterexample. Assume throughout that all variables represent integers.

(a) If $a \mid b$ and $a \mid (b + 1)$, then $a = \pm 1$.

(b) If n is even, then $4 \mid n^2$.

(c) If a and b are both even or both odd, then $a^2 - b^2$ is even.

48. For each of the following divisibility statements, either prove it or give a counterexample. Assume throughout that all variables represent integers.
  (a) If $a \mid b$ and $c \mid d$, then $ac \mid bd$.
  (b) If $a \mid b$ and $a \mid c$, then $a \mid \gcd(b,c)$.
  (c) If n is an integer, then $3 \mid n^3 - n$.

49. For each of the following statements, either prove it or give a counterexample. Assume throughout that all variables represent integers, unless otherwise specified.
  (a) $4 \mid [a(a+1)(a+2)]$
  (b) $4 \mid (a^4 - a^2)$
  (c) If $n > 1$ is an odd integer, then $4^n - 3$ is prime.

50. For each of the following statements, either prove it or give a counterexample. Assume throughout that all variables represent integers, unless otherwise specified.
  (a) If ab is a multiple of 4, then either a is a multiple of 4 or b is a multiple of 4.
  (b) If a is odd, then $4 \mid (a^2 - 1)$.
  (c) If a is odd, then $8 \mid (a^2 - 1)$.

51. (a) Prove the following identity: $\gcd(ab, ac) = a \gcd(b,c)$, whenever a, b, and c are positive integers.
  (b) Prove that if b and c are positive integers with $d = \gcd(b,c)$, then $\frac{b/d}{c/d}$ will be the lowest-terms representation of the fraction b/c.
  (c) Explain how the Euclidean algorithm can be used to create an algorithm for obtaining the lowest-terms representation of any fraction $\frac{b}{c}$ of positive integers. Apply your algorithm to obtain the lowest-terms representation of the fraction $\frac{1474}{39,463}$.

52. Prove the following variation of Euclid's lemma:
   If $a \mid bc$, and $\gcd(a,b) = 1$, then $a \mid c$.

   Suggestion: Use Theorem 2.6.

53. Prove that if $n \in \mathbb{Z}_+$ and $2^n - 1$ is prime, then n must be prime.

   Suggestion: Use the following (easily verified) algebraic factorization identity: $x^{ab} - 1 = (x^a - 1)(x^{a(b-1)} + x^{a(b-2)} + \cdots + x^a + 1)$.

Historical Note: The converse of Chapter Exercise 53 is false; for example, $2^{11} - 1 = 2047 = 23 \cdot 89$. Prime numbers of the form $2^n - 1$ are called *Mersenne primes*, after Marin Mersenne (1588–1648, French priest and scholar). Mersenne boldly stated (without proof) in the preface of his book *Cogitata Physica-Mathematica* (1644) that $2^n - 1$ is prime for whenever $n = 2$, 3, 5, 7, 13, 17, 19, 31, 67, 127, and 257, and composite for all other values of n. The last of these numbers has 77 digits, so the claim would have been very difficult to verify at the time, when all calculations had to be done by hand. Mersenne's conjecture initiated much activity in the area, and it was not until over a century later, in 1783, that someone discovered that Mersenne had missed one: $2^{61} - 1$ was proved to be a Mersenne prime. Taking advantage of their unique form, specialized and very efficient algorithms have been developed to check whether $2^n - 1$ is prime, and because of these algorithms, the largest known primes are Mersenne primes. Since the mid-1990s there has been an open project—*The Great Internet Mersenne Prime Search* (GIMPS—http://www.mersenne.org/)—that has scientists across the world attempting to break new records. GIMPS offers $3,000 for each new Mersenne prime that is discovered, and their website provides free specialized software programs that can help. In 2009, the first (Mersenne) prime that broke the 10-million digit benchmark was discovered on a UCLA computer setup (using GIMPS software) by Edson Smith. This prime, $2^{37,156,667} - 1$, has 12,978,189 digits, and garnered a long-standing $100,000 prize, offered by an anonymous donor for being the first to find a prime number with at least 10 million digits. The discovery of the 37th known Mersenne prime was made by Roland Clarkson, who worked independently, and at the time was a 19-year-old student at California State University–Dominguez Hills. There are still many theoretical questions that remain unanswered regarding Mersenne primes. For example, it is not yet known whether there are infinitely many Mersenne primes.

54. *Application of Congruences: Divisibility Criteria.*
  (a) Prove that for any positive integer n, $3 \mid n$ if, and only if, 3 divides the sum of the digits of n. Equivalently, if we write $n = \sum_{k=0}^{D} d_k \cdot 10^k$, where $0 \le d_k \le 9$ (the digits), $3 \mid n$
  $\Leftrightarrow n = \sum_{k=0}^{D} d_k$.
  (b) With the notation of part (a), prove that $4 \mid n \Leftrightarrow 4 \mid (d_0 + 10d_1)$.

  Suggestion: For part (a), show that $n \equiv \sum_{k=0}^{D} d_k \pmod 3$. For part (b), use the fact that $4 \mid 10^2$.

55. *Application of Congruences: Divisibility Criteria.*
  (a) Prove that for any positive integer $n = \sum_{k=0}^{D} d_k \cdot 10^k$, $11 \mid n$ if, and only if, $11 \mid (d_0 + d_2 + d_4 + \cdots - d_1 - d_3 - d_5 - \cdots)$. For example, $11 \mid 930391$ since $11 \mid (9 + 0 + 9 - 3 - 3 - 1)$.
  (b) With the notation of part (a), prove that $7 \mid n \Leftrightarrow 7 \mid (d_0 + 10d_1 + 100d_2 - d_3 - 10d_4 - 100d_5 + d_6 + 10d_7 + 100d_8 - \cdots)$. For example, $7 \mid 4001006002$ since $7 \mid (4 - 1 + 6 - 2) = (4 + 10 \cdot 0 + 100 \cdot 0 - 1 - 10 \cdot 0 - 100 \cdot 0 + 6 + 10 \cdot 0 + 100 \cdot 0 - 2)$.

56. Prove that for any odd modulus $m > 2$, we have $\sum_{k=1}^{m-1} k \equiv 0 \pmod m$.

57. Prove that for any even modulus $m > 1$, we have $\sum_{k=1}^{m-1} k \equiv m/2 \pmod m$.

58. In the notation of our (constructive) proof of the Chinese remainder theorem (Theorem 2.13), prove that $\sum_{i=1}^{k-1} e_i (N/n_i) \equiv 1 \pmod N$.

59. Suppose that a is an integer and n and m are relatively prime integers, both greater than 1.

   (a) If $x \equiv a \pmod{m}$ and $x \equiv a \pmod{n}$, show that $x \equiv a \pmod{mn}$.

   (b) Does the result of part (a) remain valid without the assumption that m and n are relatively prime? Either prove that it does or provide a counterexample.

Note: An integer a is said to have a **square root modulo m** ($m > 1$ an integer) if the equation $x^2 \equiv a \pmod{m}$ has at least one solution. Any solution is called a **square root of a mod m**. Exercises 60–63 will explore some situations in which it can be determined if a certain number a has a square root modulo a certain m, and also the problem of determining how many square roots a has, once it is known to have at least one.

60. *Square Roots Modulo a Prime.* Let p be an odd prime (positive) integer.

   (a) Prove that the integers that have square roots mod p are precisely those in the set $\{0^2, 1^2, 2^2, \cdots, [(p-1)/2]^2\} \pmod{p}$.

   (b) Show that the elements listed in the set of part (a) are all different mod p.

   (c) Show that if $p > 2$, then each of the $(p-1)/2$ *nonzero* elements listed in the set of part (a) has exactly two (distinct) square roots mod p.

   (d) Find all the numbers in $\mathbb{Z}_{11}$ that have square roots, and for each one, find all of its square roots (mod 11).

   (e) Repeat the instructions of part (d) for $\mathbb{Z}_{13}$.

   (f) What happens to the results of parts (a), (b), (c) in case $p = 2$?

   Suggestions: For part (a), if $z > (p-1)/2$, show that $w = p - z$ is $\leq (p-1)/2$ and $z^2 \equiv w^2 \pmod{p}$. For part (b), suppose that $0 \leq w < z \leq (p-1)/2$ but that $w^2 \equiv z^2 \pmod{p}$. This means that $p \mid z^2 - w^2 = (z+w)(z-w)$. Use Euclid's lemma (Proposition 2.7) to obtain a contradiction. For part (c), if z is a square root, then so is $-z (\equiv p - z) \pmod{p}$.

61. *Square Roots Modulo a Product of Distinct Primes.* Let $p < q$ both be odd primes.

   (a) Show that the equation $x^2 \equiv a \pmod{pq}$ is equivalent to the system $\begin{cases} x^2 \equiv a \pmod{p} \\ x^2 \equiv a \pmod{q} \end{cases}$. Indicate a scheme for finding all square roots of an integer a mod pq.

   (b) Using Chapter Exercise 60 and the result of part (a), obtain a result for the existence of square roots modulo a product of distinct odd primes.

   (c) Find (i) all (if any) square roots of 9 mod 35, and then (ii) compute $\sqrt{51} \pmod{493}$.

   (d) How would the results found in parts (a) and (b) change in case $p = 2$?

   (e) Find the following square roots (if they exist): (i) $\sqrt{11} \pmod{26}$, and (ii) $\sqrt{68} \pmod{86}$.

   Suggestions: Make use of the Chinese remainder theorem.

62.   *Square Roots Modulo a Prime-Efficient Algorithms.* Chapter Exercise 60 completely explained the existence and number of square roots that an integer can have modulo a prime; however, the resulting method for extracting square roots was not very much faster than a brute-force search. The following proposition gives a fast method for finding all square roots of any integer modulo a prime p in the case that $p \equiv 3 \pmod 4$.

Proposition 2.15

Assume that p is a prime number that is congruent to $3 \pmod 4$, and let $x \not\equiv 0$ be any integer $\pmod p$. Then either x or $-x$, but not both, will have two square roots $\pmod p$, and these square roots are given by $\pm w$, where $w = x^{(p+1)/4} \pmod p$.

For a proof as well as another theorem covering the remaining case where $p \equiv 1 \pmod 4$, we refer to [Coh-93]. Here we look only at the practical applications of Proposition 2.14.[*] Use Proposition 2.14 to determine all square roots of the following:

(a) $\sqrt{7} \pmod{59}$

(b) $\sqrt{142} \pmod{607}$

(c) $\sqrt{10} \pmod{2143}$

63.   *More Square Roots Modulo a Product of Distinct Primes.* Let $p \neq q$ both be prime (positive) integers. The technique for extracting square roots mod pq given in Chapter Exercise 61 can be speeded up if one of the two primes is congruent to $3 \pmod 4$, since Proposition 2.14 of Chapter Exercise 62 can then be applied. Use these ideas to find all square roots of the following:

(a) $\sqrt{5} \pmod{413}$

(b) $\sqrt{32} \pmod{22459}$

(c) $\sqrt{34} \pmod{23573}$

64.   *Wilson's Theorem.* Wilson's theorem states that if p is a prime, then $(p-1)! \equiv -1 \pmod p$.

(a)   Show that the converse of Wilson's theorem is true, i.e., show that if $n > 1$ is an integer that satisfies $(n-1)! \equiv -1 \pmod n$, then n must be prime.

(b)   Use Wilson's theorem to prove that $p \mid 2(p-3)! + 1$ for any odd prime number p.

65.   Prove that if a is any integer and n is any nonnegative integer, then a and a^{4n+1} have the same last digit.

66.   Suppose that $a > 1$ is an integer and that $k > \ell$ are positive integers.

(a)   Let r be the remainder of the division $k \div \ell$, i.e., $r = k - \text{floor}(k / \ell)$. Prove that the remainder of the division $(a^k - 1) \div (a^\ell - 1)$ is $a^r - 1$.

(b)   Prove that $\gcd(a^k - 1, a^\ell - 1) = a^{\gcd(k, \ell)} - 1$.

---

[*] More efficient implementations of Proposition 2.14 will be possible after we introduce an algorithm for fast modular exponentiation in Chapter 6. For now, when taking large powers in modular arithmetic, the reader should begin with small powers and work his or her way up, as suggested in Exercises 19 and 20.

Suggestion for part (b): Apply the Euclidean algorithm in conjunction with the result of part (a).

Chapter 2 Computer Implementations and Exercises

Note: If your computing platform is a floating point arithmetic system, it may allow you only up to 15 or so significant digits of accuracy. Symbolic systems allow for much greater precision, being able to handle 100s or 1000s of significant digits. Some platforms allow the user to choose if he or she wishes to work in floating point or symbolic arithmetic but will work in floating point arithmetic by default since operations are faster and usually sufficiently accurate for general purposes. If you have access only to a floating point system, you should keep these limitations in mind when you do computer calculations with large integers. The specific computations asked for below should be amenable to all floating point systems, which tend to be accurate to about 15 significant digits.

1. *Program for Check if a Positive Integer Is Prime.*
   (a) Write a program y = PrimeCheck(n) that inputs an integer $n > 1$, and outputs an integer y that is 1 if n is a prime number and 0 if it is not. The method used should be a brute-force check to see whether n has any positive integer factor k, checking all values of k (if necessary) up to floor($\sqrt{n}$).
   (b) Run your program with each of the following input values: $n = 30, 31, 487, 8893, 987654323, 131317171919$.
   (c) Assuming that your computing platform can perform 1 billion divisions per second [and assuming that the rest of the program in part (a) takes negligible time], what is the largest number of digits an inputted integer n could have so that the program could be guaranteed to execute in less than one minute?
   (d) Under the assumption of part (c), how long could it take for the program in part (a) to check whether a 100-digit integer is prime?
   Note: Of course, the program could execute very fast if a small prime factor is found quickly. A prime input would always take the most time since the full range of k values would need to be checked. There are some efficiency enhancements that we could incorporate into the above program: For example, after it is checked that 2 is not a factor, we need only check odd integers after 2. Such a fix could cut the runtimes essentially in half, but there are more efficient (and sophisticated) prime checking algorithms than such brute-force methods. For more on this interesting area, we refer the reader to [BaSh-96].

2. *Program for Prime Factorization of Positive Integers.*
   (a) Write a program FactorList = PrimeFactors(n) that inputs an integer $n > 1$, and outputs a vector FactorList that lists all of the prime factors, from smallest to largest, and with repetitions for multiple factors. For example, since the prime factorization of 24 is $2^3 \cdot 3$, the output of

PrimeFactors(24) should be the vector [2 2 2 3]. Starting with $k = 2$ and running k through successive integers, the method should check to see whether n has k as a factor. If it does, then the k should be appended to the FactorList output vector. We then replace n with n/k, and continue to check whether k divides into (the new) n until it no longer does, and then we move on to update k to $k + 1$. With this scheme, only prime factors will be found. (Why?) Also, we may stop as soon as k reaches the current value of floor($\sqrt{n}$).

(b)   Run your program with each of the following input values: n = 30, 31, 487, 8893, 987654323, 131317171919.

Note: Although there are faster algorithms, the factorization problem of this computer exercise is more intractable than the primality checking algorithm of the preceding exercise. Indeed, a polynomial time algorithm has been found for the primality checking problem, but it is strongly believed that there can exist no such algorithm for the prime factorization problem. We will return to this and related topics in greater detail in Chapter 8.

3.   *Program for Finding the Next Prime.*

(a)   Write a program p = NextPrime(n) that inputs an integer n > 1 and outputs the smallest prime number p ≥ n.

(b)   Run your program with each of the following input values: n = 8, 30, 32, 487, 8899, 987654321, 131317171919.

Suggestion: Call on the program of Computer Exercise 1.

4.   *Program for the Division Algorithm.*

(a)   Write a program (q, r) = DivAlg(a, d) that inputs an integer a = the dividend, and a positive integer d = the divisor. The output will be the (unique) integers q = the quotient and r = the remainder, satisfying a = dq + r, with $0 \le r < a$ (as guaranteed by Proposition 2.4).

(b)   Run your program with each of the following pairs of input values: (a, d) = (5, 2), (501, 13), (1848, 18), (123456, 321).

Suggestion: Use the idea of Exercise for the Reader 2.3.

5.   (a)   Either find a program on your computing platform that performs as follows, or write one of your own: Write a program with syntax b = mod(a,m) that will take as inputs an integer a and an integer m > 0 (the modulus). The output should be a nonnegative integer b, with $0 \le b < m$, that is congruent to a (mod m). In other words, b should be the remainder when a is divided by m using the division algorithm.

(b)   Use this function to redo the hand calculations that were asked in Chapter Exercises 19 and 20.

(c)   Use this function to check the identities of Chapter Exercises 56 and 57 with the modulus values m = 10, 11, 50, 51, 562, and 563.

Suggestion for part (b): In case your system works in floating point arithmetic (or if you are not sure) in parts (e) of Chapter Exercises 19 and 20, directly evaluating the very large ordinary integers as inputs in the mod function would result in inaccurate results. Instead, for example, to evaluate 21^{223} (mod 24), start by iteratively computing $b1 = \mathrm{mod}(21\char94 2,m)$, $b2 = \mathrm{mod}(b1\char94 2,m) \equiv 21^4 = 21^{2^2}$, $b3 = \mathrm{mod}(b2\char94 2,m) \equiv 21^8 = 21^{2^3}$, $b4 = \mathrm{mod}(b3\char94 2,m) \equiv 21^{16} = 21^{2^4}$, etc., until you get to the largest such exponent less than (or equal to) the desired exponent. Then multiply out some of the appropriate intermediate results (using the `mod` function) to obtain the desired power. This idea can be streamlined into a very fast and effective modular exponentiation algorithm, and this will be done in Chapter 6.

6. *Program for the Euclidean Algorithm.*
   (a) Write a program d = `EuclidAlg(a,b)` that inputs two positive integers a and b, and outputs d = gcd(a,b), computed using the Euclidean algorithm (Algorithm 2.1).
   (b) Check your program with the results of Exercise for the Reader 2.4, and then run it on each of the following pairs of input values: (a, b) = (525, 223), (12364, 9867), (1234567890, 0987654321), (13131717191919, 191917171313).

Note: **mod *Function on Computing Platforms*.** Most computing platforms have some sort of remainder or modular integer converter; this will be particularly useful for modular arithmetic computations. The next exercise will ask you either to find such a function on your platform or (if you cannot find one or if there is none) to write a (simple) program for one.

7. *Brute-Force Program for Finding Modular Inverses.*
   (a) Write a program with syntax $ainv$ = `modinv_bf(a,m)` that will take as inputs a modular integer a and an integer m > 1 (the corresponding modulus). If a has an inverse (mod m), the output `ainv` will be the corresponding (unique) inverse; if a has no inverse, then there will be no output for `ainv` but only a message to the effect that "there is no inverse mod m." The programming should be done by brute force, i.e., checking through all of the integers b mod m, and multiplying these by a to see if you get 1 mod m. If this ever happens, b will be the inverse, so you can output `ainv` as b; if not, then a has no inverse. You may save time in this search by using the necessary condition gcd(a,m) = 1 and gcd(b,m) = 1 for elements to be invertible (or be inverses).
   (b) Use this program to redo Chapter Exercises 29 through 31.
   (c) Use this program to determine the following inverses, if they exist: 1335^{-1}(mod 39467) and 87451^{-1}(mod 139467).

8. *Program for the Extended Euclidean Algorithm.*
   (a) Write a program with syntax $outVec$ = `ExtEucAlg(a, m)` that takes two inputs, a and b, which are positive integers such that $a \geq b$. The output will consist of a length-3 vector, `outVec`, that has three integer components, d, x, and y, where d = gcd(a,b), and x and y satisfy $d = ax + by$ (as in Algorithm 2.2). The program should follow Algorithm 2.2.
   (b) Use your program to check the results of Example 2.7(a) and of Exercise for the Reader 2.10.

(c) For each pair of integers a and b that we list here, use your program to compute gcd(a,b), and two integers x and y such that $d = ax + by$.

(i) $a = 8359$, $b = 4962$

(ii) $a = 95{,}243$, $b = 24{,}138$

(d) Use your program to solve the equation $88243x + 16947y = 1$ for integers x and y (or to determine that such a solution does not exist).

9. *Program for Finding Inverses with the Extended Euclidean Algorithm.*

(a) Write a program with syntax ainv = modinv(a, b) that has the same syntax, inputs, and outputs as the program of Computer Exercise 7, but now the programming should be using the extended Euclidean algorithm (Algorithm 2.2). Alternatively, your program can directly call on the one of Computer Exercise 8, if you have done that one.

(b) Use this program to redo Chapter Exercises 29 through 31.

(c) Use this program to determine the following inverses, if they exist: $1335^{-1}(\bmod\ 39467)$ and $87451^{-1}(\bmod\ 139467)$.

(d) Compare the performance times of modinv and mod-inv _ bf (of Computer Exercise 7) for the following pairs of inputs: $a = 967$ and

(i) $m = 10{,}001$

(ii) $m = 100{,}001$

(iii) $m = 1{,}000{,}001$

(iv) $m = 10{,}000{,}001$

10. *Program for Solving General Congruences of the Form $ax \equiv c$ (mod m).*

(a) Write a program with syntax SolVec = LinCong Solver(a,c,m) that will take as inputs the parameters a, c, and m that determine a single linear congruence $ax \equiv c$ (mod m) and will output a vector SolVec (possibly empty, if there are no solutions) containing all of the solutions of this congruence (mod m). In case gcd(a,m) > 1, the program should follow Algorithm 2.3. The program can be made simpler if it calls on the program modinv of Computer Exercise 9.

(b) Use your program to check the results of Example 2.7 and of Exercise for the Reader 2.11.

(c) Use this program to redo Chapter Exercises 33 and 35.

(d) Use this program to redo Chapter Exercises 37 and 38.

11. *Program to Check ISBN-13 Numbers.*

(a) Write a program with syntax check = ISBN13(vec), where the input is a *vector* vec of the 13 digits representing an ISBN-13 number (read Chapter Exercise 43 for the necessary background). For example, the ISBN-13 number 978-0-471-69738-1 would be inputted as the vector [9 7 8 0 4 7 1 6 9 7 3 8 1] (dashes are left out). The output check will be a string of text: either "Valid ISBN-13 number" if the vector entered does indeed correspond to a valid ISBN number (i.e., satisfies Equation 2.9) or "Invalid ISBN-13 number."

   (b) Use this program in conjunction with a for loop to redo part (a) of chapter Exercise 43.

12. *Program to Check VISA Credit Card Numbers.*

   (a) Write a program with syntax `check = VISA16 (vec)`, where the input is a *vector* `vec` of the 16 digits representing a 16-digit VISA card number (read Chapter Exercise 45 for the necessary background). For example, the VISA card number 4784 5580 0246 1888 would be inputted as the vector [4 7 8 4 5 5 8 0 0 2 4 6 1 8 8 8]. The output check will be a string of text: either "Valid VISA card number" if the vector entered does indeed correspond to a valid ISBN number (i.e., satisfies Equation 2.11) or "Invalid VISA card number."

   (b) Use this program in conjunction with a for loop to redo part (a) of Chapter Exercise 44.

13. *Program to Check the Chinese Remainder Theorem.*

   (a) Write a program with syntax `SolVec = ChineseRem (bVec, modVec)`, where the inputs are two *vectors*: `bVec` whose components are the values $[b_1, b_2, \cdots, b_k]$ of Equation 2.5, and `modVec` whose components are the values of the moduli $[n_1, n_2, \cdots, n_k]$ of Equation 2.5. The output, `SolVec`, should be the corresponding values of the solution of Equation 2.5 $[x_1, x_2, \cdots, x_k]$, as guaranteed to exist by the Chinese remainder theorem (Theorem 2.13), under the assumption that the moduli are pairwise relatively prime. Your algorithm should operate according to the algorithm specified in the proof of Theorem 2.13.

   (b) Check your program on the results of Example 2.9 and Exercise for the Reader 2.13.

   (c) Apply your program to solve the congruences of Chapter Exercises 39 and 40.

3

The Evolution of Codemaking until the Computer Era

We briefly outline the historical highlights in the evolution of codemaking from antiquity through the completion of the World War II. Whenever it is convenient, we take the liberty of explaining codes using the concepts of functions and modular arithmetic that were described in the previous two chapters. Exclusively within this chapter, we treat codemaking in a more general sense instead of deliberately changing the plaintexts, as is the proper goal in cryptography. Although some elements of codebreaking achievements are mentioned here, the corresponding evolution of code-breaking will be treated more extensively in Chapter 5.

Ancient Codes

In cryptography, the traditional distinction between a *code* and a *cipher* was that the latter changes plaintexts letter by letter, whereas a code changes whole words at a time. In recent times, codes have been expanded to encompass any form of communication that may not be understood by unintended recipients, and we shall adopt this meaning when we use the words *codes* and *codemakers*. Any form of written or spoken language can be considered as a code for expressing messages, thoughts, concepts, history, and so forth. When two people are able to speak or read the same language, thoughts and ideas can be effectively communicated by writing or speaking. But if the languages of two parties have no common roots, the writing or speaking of the language of one will appear as a secret code to the other. There are about 7,000 living languages and dialects in the world,* but this number continues to decrease as our world becomes more interconnected. Many languages of ancient societies exist only in the form of written documents and tablets that continue to get unearthed by new archaeological finds. These ancient writings often hold interesting and important keys to history and evolution, and as a consequence, the task of their decryption gives rise to some of the most important unsolved

---

* Interested readers may consult the authoritative reference [Gor-05], which contains extensive statistical details, maps, and origins of all living languages.

mysteries of our past. Being able to uncover the meaning of such historic scribes has seen some celebrated successes, by amateurs as well as professional linguists, classicists, and archaeologists. But many important unsolved codes remain and constitute the oldest unsolved codebreaking problems in the world.

One particularly notable achievement in breaking a code of a widely used ancient script was the deciphering of the Egyptian *hieroglyphics*. A hieroglyph is a character made by a graphical figure such as an animal or tool. Hieroglyphics began to be used by Egyptians about 3,000 years ago to record their history. The ability to read hieroglyphics vanished with the disappearance of ancient Egyptian society, and for many years, the hieroglyphics found in caves and on paintings and papyrus (an ancient form of paper) remained a bewildering mystery. A breakthrough occurred during the conquest of Egypt by the French forces under Napoleon Bonaparte in 1799, where the *Rosetta Stone* was discovered; see Figure 3.1.[*] The Rosetta stone contained a translation of a hieroglyphic script into ancient Greek and into demonic, a more common form of ancient Egyptian writing. Thus, the stone provided something analogous to a ciphertext and plaintext correspondence.

Through some extremely clever linguistic and archaeological forensics, the mystery of hieroglyphics was solved in the ensuing three decades. This valuable contribution, which allowed scientists to read all other hieroglyphic documents, was due to two brilliant scientists, Thomas Young (1773–1829) from England and Jean-François Champollion (1790–1832) from France. Young had laid the groundwork that he published in an 1814 paper, where he focused on some circled sequences of hieroglyphics (*cartouches*) that he hypothesized represented names of eminent pharaohs. Young subscribed to the popular opinion that the remaining hieroglyphs were *semagrams* (each representing a whole idea) and he was unable to make further progress. At the age of 10, Champollion was introduced to the problem through Jean-Baptiste Fourier, a famous mathematician who was on Napoleon's scientific staff. Fourier showed Champollion several articles with hieroglyphics and told him that no one had been able to understand this code. Champollion was so intrigued that he made a commitment that he would discover this long-standing mystery. Champollion indeed followed through with his promise; at the age of 17 he published a groundbreaking paper on the subject, in which he completely described the meanings of the hieroglyphics. Although some hieroglyphs indeed represented semagrams, most turned out to represent alphabetic characters. His discovery shattered the general consensus as postulated by Young, and it quickly earned this 17-year-old prodigy a professorship. Champollion was so overwhelmed with this news that he fainted upon hearing it.

There have been other celebrated successes of deciphering ancient scripts, but many others remain to be solved. One of these is the famous *Phaistos*

---

[*] The Rosetta Stone was so named because it was discovered near the town of Rosetta. The stone was later acquired by the English after Napoleon's defeat and now sits in the British Museum. The author gratefully acknowledges the British Museum for providing him with the source photograph for this book.

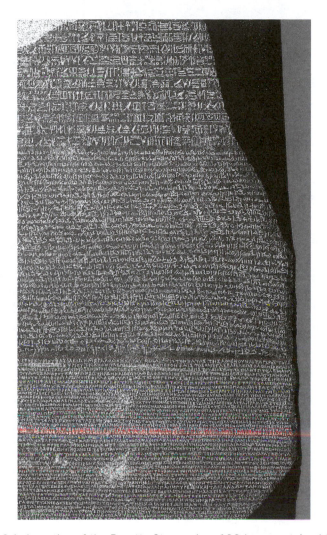

Figure 3.1 A portion of the Rosetta Stone, circa 199 b.c., contained translations of hieroglyphics into Greek and a third written language. Although badly damaged, this stone led to the decryption of hieroglyphics. On the top hieroglyphics, some cartouches (circled sequences) are visible.

disk; see Figure 3.2. The message on the disk perhaps holds some mysteries of the Minoan culture, a particularly vibrant one in Greek history.

Although such ancient alphabets were not originally intended to convey secrecy, the techniques needed to discover their original meaning require some of the common ingredients that codebreakers used to break later cryptosystems. Indeed, since many of the earlier cryptosystems were substitution ciphers, codebreakers needed to be well versed in linguistics. One language-based cryptosystem that was used as late as World War II by the United States proved to be extremely effective. The system employed members of the Native American Navajo Indian tribes who were tasked to give messages in radio communications in their native language. The Navajo language was chosen since the tribe was sufficiently large (making it easy to find recruits) and was one of the few American Indian tribes that

Figure 3.2 A drawing of the Phaistos disk, a clay disk dating back to 1700 b.c., was discovered in 1908 on the island of Crete. It contains 241 symbols, 45 of which are different. The ancient text has thus far eluded decryption.

had not been infiltrated with foreign students. Native American languages are very hard to learn and dissimilar to any other language since they were created in isolation. A Hollywood movie, *Windtalkers* (2002, starring Nicolas Cage), was made to retell this interesting story.

Formal Definition of a Cryptosystem

We next digress in order to give a general and formal definition of a cryptosystem, one that will serve us to describe almost all cryptosystems that have been developed over time as well as systems that will be devised in the future.

Definition 3.1

A **cryptosystem** consists of the following:

A **plaintext alphabet** P, a **ciphertext alphabet** C, a **plaintext space** $\mathcal{P}$, consisting of sets of strings in the plaintext alphabet P, a **ciphertext space** $\mathcal{C}$ (that is, a set of strings in the ciphertext alphabet C), a **keyspace** $\mathcal{K}$, a collection of invertible **encryption functions** $e_\kappa : \mathcal{P} \to \mathcal{C}$ (one for each key $\kappa \in \mathcal{K}$) along with the corresponding **decryption functions** $d_\kappa : \mathcal{P} \to \mathcal{C}$. The decryption functions are the inverses of the corresponding encryption functions. In cases where the key easily determines both the encryption and decryption functions, the cryptosystem is called a **symmetric key cryptosystem**.* In an **asymmetric key cryptosystem**, also known as a **public key cryptosystem**, the keys (and sometimes even the keyspaces) are different for encryption and decryption, and it is usually not feasible, or practically impossible, to compute the decryption key (or function) knowing only the encryption key.

---

* A cipher is sometimes used as a synonym for a cryptosystem, but at other times it is used to specify a particular instance of a cryptosystem. Thus, the Caesar cipher is a special instance of the shift cryptosystem/cipher, but it would usually not be referred to as the Caesar cryptosystem.

Thus any of the ancient scribes, or any language for that matter, constitutes a cryptosystem with a single key, where the encryption functions are the functions that translate our known language into the ancient or foreign language (with the alphabets generally allowed to be letters, symbols, syllables, or whole words).[*] For the remainder of this chapter we will identify the 26 letters of the English alphabet with the integers mod 26 according to Table 3.1

To help the reader appreciate the versatility of Definition 3.1, our next example gives the details of how a couple of simple ciphers from Chapter 1 fall under the umbrella of this definition, using the association of Table 3.1.

Example 3.1

(a) The shift ciphers are cryptosystems with the following parameters: Both P and C are $\mathbb{Z}_{26}$, both the plaintext space $\mathscr{P}$ and the ciphertext space $\mathscr{C}$ are the set of all finite sequence of integers mod 26, and the key space $\mathscr{K}$ is $\mathbb{Z}_{26}$. For a key $\kappa \in \mathbb{Z}_{26}$ the corresponding encryption function is

$$e_\kappa : \mathbb{Z}_{26} \to \mathbb{Z}_{26} :: e(x) \equiv x + \kappa \ (\mathrm{mod}\ 26)$$

(applied to each modular integer in the sequence of plaintext) and the corresponding decryption function is the inverse of this function:

$$d_\kappa : \mathbb{Z}_{26} \to \mathbb{Z}_{26} :: d(x) \equiv x - \kappa \ (\mathrm{mod}\ 26)$$

The Caesar cipher arises when we use the key $\kappa = 3$. This is a symmetric key cryptosystem. Although the key $\kappa = 0$ corresponds to the identity mapping on strings (not a very secret cipher), we include it for completeness.

(b) The class of all substitution ciphers on the English alphabet [for which the shift ciphers of part (a) are a special case] has the following cryptosystem components: Both P and C are $\mathbb{Z}_{26}$, both the plaintext space $\mathscr{P}$ and the ciphertext space $\mathscr{C}$ are the set of all finite sequences of integers mod 26, and the key space $\mathscr{K}$ is the set of all permutations on $\mathbb{Z}_{26}$. A permutation $\sigma \in \mathscr{K}$ can be identified by a sequence (the rearrangement) of the 26 integers mod 26 in some order: $\sigma = a_1, a_2, a_3, \cdots, a_{26}$ where $\sigma(i) = a_i$ $(1 \le i \le 26)$. For a key $\sigma \in \mathscr{K}$, the corresponding encryption function is

$$e_\sigma : \mathbb{Z}_{26} \to \mathbb{Z}_{26} :: e(x) \equiv \sigma(x)\ (\mathrm{mod}\ 26)$$

---

[*] For example, in Chinese, one of the most ancient surviving languages, single characters correspond to whole English words. As a general rule in linguistics, alphabetic written languages usually have between 20 and 40 characters (for example, Russian has 36, and there are 25 ancient Egyptian hieroglyphs), syllabic languages tend to have between 50 and 100 characters (such as the Thai alphabet with 65 characters), and semagram alphabets where characters represent concepts or ideas typically have thousands of alphabet characters (like the Chinese alphabet, which has over 5,000 characters).

TABLE 3.1 Correspondence of the English Alphabet Using the Integers mod 26 ($\mathbb{Z}_{26}$)

A	B	C	D	E	F	G	H	I	J	K	L	M	N	O	P	Q	R	S	T	U	V	W	X	Y	Z
0	1	2	3	4	5	6	7	8	9	10	11	12	13	14	15	16	17	18	19	20	21	22	23	24	25

Note: This correspondence will be a default mechanism for translating ciphertexts as well as plaintexts into vectors of modular integers.

(applied to each modular integer in the sequence of plaintext), and the corresponding decryption function is the inverse of this function

$$d_\sigma : \mathbb{Z}_{26} \to \mathbb{Z}_{26} :: d(x) \equiv \sigma^{-1}(x) \,(\mathrm{mod}\, 26)$$

This is a symmetric key cryptosystem since the inverse permutation is readily deduced from the key permutation. The key corresponding to the identity permutation again is included for completeness.

Although we rarely formally identify each of the components of cryptosystems in terms of Definition 3.1, we should nonetheless keep these notions in the back of our mind as we contemplate, develop, or analyze any cryptosystems.

Affine Ciphers

A slightly more general class of ciphers than the shift ciphers that is also easily described in terms of modular arithmetic is the class of *affine ciphers*, which is defined as follows.

Definition 3.2

Given any pair α, β of integers mod 26, where α is relatively prime to 26, the corresponding **affine cipher** is defined using the following encryption function: $\phi_{\alpha,\beta} : \mathbb{Z}_{26} \to \mathbb{Z}_{26} :: \phi_{\alpha,\beta}(x) \equiv \alpha x + \beta \,(\mathrm{mod}\, 26)$ (applied to each modular integer in the sequence of plaintext). The condition $\gcd(\alpha, 26) = 1$ is required for this function to be invertible (see Chapter 2), in which case the inverse function would be specified by $\phi_{\alpha,\beta}^{-1}(y) \equiv \alpha^{-1}(y - \beta)\,(\mathrm{mod}\, 26).$[*] The keyspace for this affine cryptosystem can simply be taken to be the set of pairs (α, β), where $\alpha, \beta \in \mathbb{Z}_{26}$, and α is relatively prime to 26.

Note: More generally, if $m > 1$ is an integer and $\alpha, \beta \in \mathbb{Z}_m$, $\alpha \neq 0$, the function $\phi_{\alpha,\beta} : \mathbb{Z}_m \to \mathbb{Z}_m :: \phi_{\alpha,\beta}(x) \equiv \alpha x + \beta \,(\mathrm{mod}\, m)$ is called an **affine function** (or **affine mapping**). It will be invertible if, and only if, $\gcd(\alpha, m) = 1$.

Exercise for the Reader 3.1

(a) Verify that the function $\phi_{\alpha,\beta}^{-1} \equiv \alpha^{-1}(y - \beta)\,(\mathrm{mod}\, m)$ is indeed the inverse mapping of the affine function $\phi_{\alpha,\beta} : \mathbb{Z}_m \to \mathbb{Z}_m :: \phi_{\alpha,\beta}(x) \equiv \alpha x + \beta \,(\mathrm{mod}\, m)$, provided that α is invertible (mod m).

(b) Compute the inverse of the affine mapping with parameters $\alpha = 21, \beta = 4$, when $m = 26$.

---

[*] In case α is not relatively prime to 26, the function $\phi_{\alpha,\kappa}$ could still encrypt a plaintext message; however, the problem is that it could not be decrypted since $\phi_{\alpha,\kappa}$ is no longer an invertible function.

Example 3.2

(a) Using the affine cipher with key $(\alpha, \beta) = (3, 11)$, encrypt the plaintext message "code blue alert."

(b) Use the above affine cipher to decrypt the following ciphertext message:

QGKXXGBNQJSXNNEBQQXU

Solution: Part (a): The plaintext "codebluealert" corresponds (under Table 3.1) to the vector [2 14 3 4 1 11 20 4 0 11 4 17 19]. The encryption mapping $\phi(x) = \phi_{3,11}(x) = 3x + 11 (\mathrm{mod}\ 26)$ gets applied to each integer in this sequence to obtain the corresponding vector of integers that represent the ciphertext. For example, since $\phi(2) = 3 \cdot 2 + 11 \equiv 17 (\mathrm{mod}\ 26)$ and $\phi(14) = 3 \cdot 14 + 11 = 53 \equiv 1 (\mathrm{mod}\ 26)$, the first two elements in the ciphertext vector are 17 and 1. Continuing to evaluate the remaining elements of the plaintext integer vector produces [17 1 20 23 14 18 19 23 11 18 23 10 16], which, according to Table 3.1, corresponds to the ciphertext RBUXOSTXLSXKQ.

Part (b): Since $9 \cdot 3 = 27 \equiv 1 (\mathrm{mod}\ 26)$, we have $3^{-1} = 9$, and the decryption mapping is (see Definition 3.2) $\phi^{-1}(y) \equiv 3^{-1}(y - 11) \equiv 9y - 99 \equiv 9y + 5 (\mathrm{mod}\ 26)$. The given ciphertext corresponds (under Table 3.1) to the following vector:

[16 6 10 23 23 6 1 13 16 9 18 23 13 13 4 1 16 16 23 20]

The decryption mapping $\phi^{-1}(y) \equiv 9y + 5 (\mathrm{mod}\ 26)$ gets applied to each integer in this sequence to obtain the corresponding vector of integers that represent the plaintext. For example, since $\phi^{-1}(16) = 9 \cdot 16 + 5 = 149 \equiv 19 (\mathrm{mod}\ 26)$ and $\phi^{-1}(6) = 9 \cdot 6 + 5 = 59 \equiv 7 (\mathrm{mod}\ 26)$, the first two elements in the ciphertext vector are 19 and 7. Continuing to evaluate the remaining elements of the plaintext integer vector produces

[19 7 17 4 4 7 14 18 19 8 11 4 18 18 15 14 19 19 4 3]

which, using Table 3.1, corresponds to the plaintext "three hostiles spotted."

Exercise for the Reader 3.2

(a) Using the affine cipher with key $(\alpha, \beta) = (5, 8)$, encrypt the plaintext message "code blue alert."

(b) Use the above affine cipher to decrypt the following ciphertext message:

CJISEIZCZRCFPCUWXCVZ

We point out that when the multiplier α in the affine cipher is taken to be 1, it reduces to the shift cipher. Our next example will show how each of the four basic passive attacks on an affine cipher can be accomplished.

Example 3.3: Passive Attacks on an Affine Cipher

Explain how each of the four types of passive attacks (see Definition 1.4) could be implemented against an affine cipher.

Solution: We assume that we know we are dealing with an affine cipher but that the key $\kappa = (\alpha, \beta)$ is unknown.

(a) *A ciphertext-only attack:* The possible values for α are the mod 26 integers that are relatively prime to 26, namely {1, 3, 5, 7, 9, 11, 15, 17, 19, 21, 23, 25}. Thus, by the multiplication principle, the number of keys $\kappa = (\alpha, \beta)$ for the class of affine ciphers is $12 \cdot 26 = 312$. This is certainly stronger than the 26-key shift cipher, but any ciphertext-only attack could be easily accomplished by using a computer to run through all of the 312 possibilities. With any ciphertext of size at least 10 or so characters, it is highly unlikely that more than one of the 312 possibilities could be a possible plaintext.

(b) *A known plaintext attack:* Any known plaintext/ciphertext character correspondence gives rise to a linear equation in the parameters α, β (mod 26). With only two known plaintext/ciphertext character correspondences, most of the time we can completely solve the two linear equations (mod 26) for α, β. We provide three specific examples that typify the three situations that could arise, given a pair of plaintext/ciphertext character correspondences.

(i) Suppose that we knew the following two plaintext/ciphertext correspondences: e $\mapsto$ X and t $\mapsto$ D. In terms of modular integers and the encryption mapping, these correspondences can be expressed as $\phi_{\alpha,\beta}(4) = 23$, $\phi_{\alpha,\beta}(19) = 3$, or

$$\begin{cases} 4\alpha + \beta \equiv 24 \ (\text{mod } 26) \\ 19\alpha + \beta \equiv 3 \ (\text{mod } 26) \end{cases} \qquad (3.1)$$

Subtracting the first equation of system 3.1 from the second gives us $15\alpha \equiv -21 \equiv 5(\text{mod } 26)$, and since $\gcd(15, 26) = 1$, this latter equation can be solved by multiplying both sides by $15^{-1} = 7(\text{mod } 26)$.[*] We thus obtain $\alpha \equiv 7 \cdot 15\alpha \equiv 7 \cdot 5 \equiv 9(\text{mod } 26)$, and substituting this into the first equation of system 3.1 gives $4 \cdot 9 + \beta \equiv 24 \Rightarrow \beta \equiv 24 - 10 \equiv 14(\text{mod } 26)$. So the key is thus completely determined.

---

[*] Computing modular inverses and solving linear equations in modular arithmetic was discussed in full detail in the preceding chapter.

(ii) Suppose instead that we knew the following two plaintext/ciphertext correspondences: $e \mapsto Y$ and $s \mapsto C$. In terms of the encryption mapping this reads as $\phi_{\alpha,\beta}(4) = 24$, $\phi_{\alpha,\beta}(18) = 2$, which produces the system

$$\begin{cases} 4\alpha + \beta \equiv 24(\text{mod } 26) \\ 18\alpha + \beta \equiv 2(\text{mod } 26) \end{cases} \tag{3.2}$$

Eliminating β from this system by subtracting the first equation from the second produces $14\alpha \equiv -22 \equiv 4 \ (\text{mod } 26)$, but since $\gcd(14, 26) = 2$, there is not a unique solution. By Algorithm 2.3, there are two solutions: α_0, $\alpha_0 + 13$, where α_0 is the unique solution of the related congruence $(14/2)\alpha_0 \equiv 4/2(\text{mod } 26/2)$, and $7\alpha_0 \equiv 2(\text{mod } 13)$. Since $7^{-1} = 2(\text{mod } 13)$, we have $\alpha_0 \equiv 2 \cdot 2 \equiv 4(\text{mod } 13)$, so we arrive at the two possible values for $\alpha = \alpha_0$, $\alpha_0 + 13 = 4,17$. Since 4 is not relatively prime to 26, $\alpha = 17$ is the only feasible option. If we substitute $\alpha = 17$ into either of the equations of system 3.2 and solve for β, we obtain $\beta = 8$. Once again, the key has been completely determined.

(iii) Suppose instead that we knew the following two plaintext/ciphertext correspondences: $e \mapsto W$, and $r \mapsto J$. In terms of the encryption mapping this reads as $\phi_{\alpha,\beta}(4) = 22$, $\phi_{\alpha,\beta}(17) = 9$, which produces the system

$$\begin{cases} 4\alpha + \beta \equiv 22(\text{mod } 26) \\ 17\alpha + \beta \equiv 9(\text{mod } 26) \end{cases} \tag{3.3}$$

Eliminating β from this system by subtracting the first equation from the second produces $13\alpha \equiv -13 \equiv 13 \ (\text{mod } 26)$, but since $\gcd(13, 26) = 13$, there is not a unique solution. By Algorithm 2.3, there are 13 solutions: α_0, $\alpha_0 + 2, \alpha_0 + 4, \cdots, \alpha_0 + 24$ where α_0 is the unique solution of the related congruence $(13/13)\alpha_0 \equiv 13/13(\text{mod } 26/13)$, or $\alpha_0 \equiv 1(\text{mod } 2)$. This gives the set of all odd integers mod 26: 1, 3, 5, $\cdots$, 25 as the solution set of $13\alpha \equiv 13 \ (\text{mod } 26)$. Unlike in the first two situations, we are unable to ascertain the key with the given information.

(c) *A chosen plaintext attack:* If we are able to encrypt just two letters of our choice, we could easily determine the key of an affine cipher by encrypting the letters a and b. These would encrypt to $\phi_{\alpha,\beta}(0) \equiv \alpha \cdot 0 + \beta \equiv \beta$ and $\phi_{\alpha,\beta}(1) \equiv \alpha \cdot 1 + \beta \equiv \alpha + \beta \ (\text{mod } 26)$, from which the key could be readily obtained.

(d) *A chosen ciphertext attack:* Since the decryption function

$$\phi_{\alpha,\beta}^{-1}(y) \equiv \alpha^{-1}(y - \beta) \equiv \alpha^{-1}y - \alpha^{-1} \cdot \beta \equiv \gamma y + \delta(\text{mod } 26)$$

is also affine, we could determine its parameters [integers γ, δ (mod 26)] by decrypting A and B. The parameters could be read off the resulting system just as in part (c). This completely determines the decryption function, which could be easily inverted to produce the encryption function and produce the key.

Steganography

We now return to the historical evolution of codemaking. Although most ancient inscriptions on sacred stones or tablets were not intentionally designed for secrecy, sometimes the scribes wished to convey a mystery or holiness so as to motivate citizens to attempt to understand and interpret these messages. The earliest documented forms of secret writing for military and state purposes date back to the fifth century b.c. These initial schemes for secret writing did not fall under the umbrella of cryptography but rather another technique called **steganography**, in which the form of the plaintext is left intact, but efforts are made to conceal its existence from unintended recipients. The ancient Chinese employed steganography by writing plaintext messages on very thin silk sheets that were then rolled up, covered in wax, and ingested by the messenger. The Greeks had also used steganography in a variety of creative ways. In one instance of historical significance during the fifth century b.c., the Persian King Xerxes had been building up his forces and planning a surprise attack on the Greeks in 480 b.c. These facts were observed by Demaratus, a Greek exiled in Persia, who felt a strong allegiance to his homeland and wanted to inform the Greeks of this impending attack. In those days, messages were sent on inscribed wax tablets, but in order to avoid detection, Demaratus scraped the wax off such a tablet, carved the message in the wood frame underneath, and then applied a fresh coat of wax, thus rendering what looked like a blank tablet. This tablet was successfully sent to the Greeks, and the message helped the Spartans defend their country from what could have been a decisive victory by the Persians.

During World War II, the Germans began using a form of steganography known as **microdots**, whereby an entire message is printed completely on top of the period at the end of a certain sentence, in an otherwise innocuous document. See Figure 3.3 for an example of a ciphertext hidden in a microdot.

Although steganography and cryptography are separate subjects and both serve to provide secrecy, the former aims to conceal the fact that a confidential message is being transmitted, whereas the latter aims to disguise the contents of the message. Both methods can be combined to achieve both goals and an additional layer of secrecy. Some of the modern steganographic techniques include disguising messages in digital pictures by encoding them as pixel values that would be insignificant to the naked eye. There have been reports in major U.S. newspapers of terrorist groups, such as al Qaeda, employing such steganographic tools to exchange messages.

Many of the early cryptosystems were simple substitution ciphers. The first leader who has been recorded to have used cryptography for both

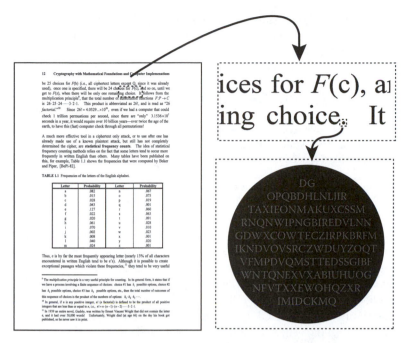

Figure 3.3 A combination of steganography (the microdot) to conceal the existence of a message with cryptography to disguise the contents of a message, even if the message is found.

military and domestic purposes was the illustrious Roman leader Julius Caesar (100 b.c.–44 b.c.). The so-called **Caesar cipher**, which we introduced in Chapter 1, is simply the shift cipher with $\kappa = 3$. Even before the Caesar cipher, the so-called **scytale**, an invention of the ancient Greeks that dates back to the seventh century b.c., was a permutation cipher. The scytale worked by wrapping a narrow and long cloth diagonally around a cylindrical rod of a particular radius (which serves as the key to the scytale cipher), and writing the plaintext message, line by line, over the cloth on the rod; see Figure 3.4. The cloth was then unwound, rendering the letters of the plaintext appearing in vertical order (the ciphertext). The intended recipient would possess another rod of the same radius and could therefore recover the plaintext. The scytale cipher is not a substitution cipher but an example of what is called a **transposition cipher** because it switches the locations of the plaintext letters (rather than substituting

Figure 3.4 Illustration of a scytale, an early transposition cipher.

the letters themselves). Substitution and transposition ciphers can be combined for greatly synergized security (an example is the ADFGVX cipher described in the exercises of Chapter 1).

Exercise for the Reader 3.3

Find the ciphertext of the message "rescue the prince from the palace" if the scytale of Figure 3.4 is used, and the cloth is wrapped four times around the rod.

The earliest writings about cryptography date back to the fourth century a.d., an example of which is the famous Indian *Kāma-sūtra* book (which has been translated into English and is still in print today) that advises women to be skilled in 64 arts. Apart from traditional skills such as cooking, also included are practical skills such as being able to determine a man's character from his features. Another is secret writing, so that women may maintain confidentiality of their secret love affairs. The first book written exclusively about cryptography was published by the Arab civilization in 725 a.d. During this era the Arab culture and civilization were thriving. There were many scholars and artists, and the governments used cryptography for maintaining the security of sensitive records and communications. The Arabs are credited with being the first to successfully develop a ciphertext-only attack on the widely used substitution ciphers. The key idea was frequency analysis, which was briefly described in Chapter 1 and will be further developed in Chapter 5. This discovery was extremely significant as it suddenly rendered the security of reams of substitution ciphertext around the world vulnerable to attack.

Whereas the Arab civilization and cryptographic techniques were flourishing, the European society was just beginning its renaissance period. When the Arabs had discovered how to attack substitution ciphers, the Europeans were just beginning to use these ciphers, and their use was mostly confined to religious parties. It was not until the 15th century that European countries began systematic use of cryptography for diplomatic purposes. This was motivated by independent developments in different countries and the associated need for secure communications and reconnaissance. Throughout the Middle Ages, the powerful Vatican had employed cryptography and developed skills to crack the cryptosystems of other countries. In 1555, the Vatican created the very important post of *Cipher Secretary to the Pontiff*, the first of whom, Benicio de Assisi, was instrumental in deciphering the codes of Spain's King Phillip, helping the Vatican to reach an important peace agreement with Spain.

Nulls

It is not known whether the Europeans learned from the Arabs how to employ frequency analysis to attack substitution ciphers or if they discovered this method on their own, but they took means to safeguard substitution ciphers from such attacks. One method was the use of so-called **nulls**, which are additional ciphertext symbols that were known to be ignored by the intended recipient and are intended to throw off frequency analysis by unintended recipients. The following is a specific example that is based on an affine mapping with a prime modulus.

Example 3.4

Consider the affine mapping $\phi : \mathbb{Z}_{31} \to \mathbb{Z}_{31}$ defined by $\phi(x) = 10x + 5 \pmod{31}$. Since the modulus 31 is prime, any affine map with a nonzero coefficient on x will be invertible. We use this mapping to encrypt plaintext letters a through z by applying this mapping to their integer equivalents of Table 3.1. The corresponding ciphertext letters will be the 26 images $\phi(0), \phi(1), \cdots, \phi(25) \pmod{31}$. The remaining 5 $(= 31 - 26)$ integers mod 31 will be nulls. A ciphertext is obtained from a plaintext by applying the mapping ϕ to (the integer corresponding to) each letter of plaintext, and punctuating the ciphertext images with the affine map's images of nulls.

  (a) Find all nulls.
  (b) Explain why the encryption mapping as described above, does not technically satisfy the conditions required by a cryptosystem (Definition 3.1). Nonetheless, show why ciphertexts can always be decrypted if they have been obtained from the above scheme.
  (c) Give an encryption of the plaintext "yes" using this cipher.
  (d) Decrypt the ciphertext: [27, 16, 20, 14, 22, 14, 27, 5, 30, 14, 0, 20, 23, 30, 21, 11, 14, 20, 30].
  (e) Provide a scheme in which the nulls could be used to make frequency analysis more difficult.

  Solution: Part (a): The nulls are $\phi(26), \phi(27), \phi(28), \phi(29), \phi(30) = 17, 27, 6, 16, 26$.
  Part (b): The set of all affine mappings $\phi : \mathbb{Z}_{31} \to \mathbb{Z}_{31}$ of the form $\phi(x) = \alpha x + \beta \pmod{31}$, where $\alpha \neq 0, \beta$ are integers (mod 31) gives rise to a cryptosystem with keyspace $\mathcal{K}$ consisting of all such pairs (α, β), but the corresponding encryption mapping should technically associate a unique ciphertext string for each plaintext string.[*] Although the nulls will be easily detected by any intended recipient and thus will cause him or her no confusion, the given description of the encryption process is not an actual function (as required by Definition 3.1). This technicality could be corrected in a number of ways by simply inserting the nulls in some deterministic fashion. As a simple example, we could insert a null after every sixth true ciphertext character. Which null? We need to be specific (to correctly define a function), but at the same time, we should aim to throw off potential attackers. One example would be to let it depend on

---

[*] This is a basic requirement in the definition of a function: Any input (in the domain) must be associated with a unique output (in the codomain).

the plaintext integer value x (between 0 and 25) of the character that immediately precedes it. We could use the $\mod(x,5) + 1$ (a number between 1 and 5) element of the ordered list of nulls 15, 12, 9, 6, 3. For example, if the sixth plaintext character is t (with integer equivalent 19), then, since $\mod(19,5) + 1 = 5$, the seventh ciphertext character would be the null 3 (which is the fifth element in the ordered list of nulls). Is it more effective to deploy the nulls in a random (nondeterministic) fashion? We will give such a strategy in the solution of part (e).

Part (c): The plaintext "yes" corresponds (via Table 3.1) to the integers 24, 4, 18, and $\phi(24), \phi(4), \phi(18) \equiv 10 \cdot 24 + 5,\ 10 \cdot 4 + 5,\ 10 \cdot 18 + 5 \equiv 28, 14, 30 \pmod{31}$. A feasible corresponding ciphertext could be formed by inserting any number of the five nulls, anywhere in this ordered list. For example: [28, 16, 4, 18].

Part (d): The nulls can be deleted from the ciphertext to reduce it to

[20, 14, 22, 14, 5, 30, 14, 0, 20, 23, 30, 21, 11, 14, 20, 30]

Since $10^{-1} \equiv 28 \pmod{31}$, we obtain $\phi^{-1}(y) \equiv 10^{-1}(y - 5) \equiv 28y - 28 \cdot 5 \equiv 28y + 15 \pmod{31}$. Applying this to each element of the above reduced ciphertext produces

[17, 4, 11, 4, 0, 18, 4, 15, 17, 8, 18, 14, 13, 4, 17, 18]

which (by Table 3.1) represents the plaintext message "release prisoners."

Part (e): This part should become more plausible after we have provided details of frequency analysis-based attacks in Chapter 5. Deterministic schemes, like the one given in the solution of part (b), have a predictability that may be exploited by potential attackers. One basic countermeasure is to mimic the distribution of prevalent letters with nulls. For example, the most commonly appearing English letters are e (12.7%), t (9.1%), a (8.2%), o (7.5%), and i (7%); see Table 1.1. Thus, one of these letters appears a total of 44.5% (on average) in plaintext messages. Accounting for adding nulls with these prevalence percentages, we should divide them by 1.445 to obtain corresponding prevalence percentage for the nulls: 15 (8.79%), 12 (6.30%), 9 (5.67%), 6 (5.20%), 3 (4.84%). This adds up to a combined total of 30.8%. One randomized encryption scheme would be to repeat the following procedure until the entire plaintext is encrypted: Spin the spinner of Figure 3.5, if it lands in the "ordinary ciphertext portion" (a sector making up 100% minus 30.8%, or 69.2%, of the circular region), we use the affine map $\phi(x) = 10x + 5 \pmod{31}$ to encrypt the next letter of plaintext; otherwise, we insert the appropriate null (each of whose circular sector areas corresponds to the prescribed percentages). As explained earlier, such nondeterministic schemes do not produce actual encryption functions (different series of spins usually produce different ciphertexts), but this causes no problems with decryption, since once the nulls are deleted, the resulting ciphertexts will always be identical.

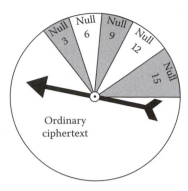

Figure 3.5 A spinner to illustrate the randomized encryption scheme of the solution to Example 3.4(d).

Such randomized spinner models are easily implemented into computer programs, as we will describe in the computer implementations supplement to this chapter.

We extend Definition 3.1 of cryptosystems to include randomized schemes, as was constructed in the previous example, whenever any actual (possibly nonunique) ciphertexts that are created can always be unambiguously decrypted into the unique corresponding plaintexts with the decryption algorithm.

Homophones

Recall that randomization was the key to the one-time pad's complete secrecy (see Chapter 1), and it is usually an important ingredient in many aspects of cryptography (both in codemaking and codebreaking), as we will see frequently throughout the rest of the book. Increasing the number of nulls makes frequency analysis more difficult, but still possible, for a ciphertext-only attack. As a general rule, the more nulls being used (in randomized schemes), the larger the ciphertext samples that would be required to distinguish the nulls from true ciphertext characters, and the actual determination of the key. Another effective way to safeguard a substitution cipher is to use so-called **homophones**, which are additional ciphertext characters used to represent more common plaintext letters. The idea is to even out the frequencies of the various ciphertext letters. The following example provides a substitution cipher with homophones that uses an affine mapping.

Example 3.5

Using the frequencies of the 26 English letters of Table 1.1, we see that they range from a low of 0.1% (for the letters x and z) to a high of 12.7% (for the letter e). Consider the affine mapping $\phi : \mathbb{Z}_{1001} \rightarrow \mathbb{Z}_{1001}$ defined by $\phi(x) = 359x + 207 \,(\text{mod } 1001)$. (Any nontrivial invertible affine mapping would do; our choice was somewhat arbitrary.) We let both the plaintext alphabet and the ciphertext alphabet be $\mathbb{Z}_{1001} = \{0, 1, 2, \cdots, 1000\}$.* Using Table 1.1 of letter frequencies, the plaintext representations of the 26 letters are obtained as follows:

(i) Since the letter *a* has a frequency of 8.2% (= 82/1000), we let the first 82 elements of $\mathbb{Z}_{1000}$ each represent *a*; that is, $a \sim \{0, 1, 2, \cdots, 81\}$. Each of these homophones corresponds to the same plaintext letter *a*.†

(ii) Since the letter *b* has a frequency of 1.5% (15/1000), the next 15 elements of $\mathbb{Z}_{1000}$ each represent *b*; that is, $b \sim \{82, 83, 84, \cdots, 96\}$.

This process of assigning homophones continues until we finish with *z*, and all the elements of $\mathbb{Z}_{1001}$ have been assigned. The affine mapping ϕ is applied to each of these to obtain the corresponding ciphertexts character. For example, $\phi(93) \equiv 359 \cdot 93 + 207 \equiv 561$ and $\phi(95) \equiv 359 \cdot 95 + 207 \equiv 278 \pmod{1001}$ are ciphertext symbols both representing the plaintext letter *b*. Tables can be created that list all ciphertext symbols for each of the plaintext letters. Randomized encryption schemes can be used to encrypt plaintexts in a way that would make each ciphertext letter occur, on average, with a frequency of 0.1%; the description of such a scheme will be called for in the following exercise.

Exercise for the Reader 3.4

(a) Develop a randomized encryption scheme to create a cryptosystem based on the homophones and affine mapping of Example 3.5.

(b) Consider the following deterministic (nonrandomized) cryptosystem: The plaintext alphabets and ciphertext alphabets are $\mathbb{Z}_{1001}$, as in Example 3.5. To encrypt an English plaintext (in the 26 lowercase letters), we use the ciphertexts of the homophones in order. For example, to encrypt the plaintext *abba*, we use the first two homophone representations of the plaintext letters *a* (0 and 1) and *b* (82 and 83), taken in order, and use the affine mapping to obtain the associated ciphertexts. Thus the ciphertext would be: $\phi(0)\phi(82)\phi(83)\phi(1) \equiv [207, 645, 4, 566] \pmod{1000}$. Is this scheme as secure as the randomized scheme that you developed in part (a)? Explain.

At first glance, a well-designed randomized homophonic cryptosystem might seem to be resistant to ciphertext-only attacks using frequency

---

* The reason we choose 1001 rather than 1000 homophones is that the sum of the frequencies in Table 1.1 is 1.001 (rather than 1.000).

† The terminology stems from the meaning of two Greek words: *homo* means "same," and *phone* means "sound." Thus any two homophones for *a*, say 22 and 64, each have the same "sound," namely *a*.

analysis. Indeed, the system indicated in the solution of Exercise for the Reader 3.4 was constructed so that each of the 1001 ciphertext symbols appears (on average) with nearly identical frequencies. Whereas single-letter frequency counts are useless for such a cryptosystem, one can look for *digraphs* (pairs of letters) and/or *trigraphs* (triples of letters). Frequency counts for such multiple letter combinations have been established for written English, and these could be applied to execute a ciphertext-only attack on a homophonic cryptosystem. But the size of a ciphertext sample needed as well as the work and ingenuity required can increase dramatically. For example, the most common digrams are (in order) *th*, *he*, and *an*. Also some digrams could never appear in written English; for example, in any digram starting with the letter *q*, it must be followed by the letter *u*. The first homophonic cipher used in Western civilization dates back to 1401 in Italy. When combined with nulls, homophonic ciphers have served as very practical and decently effective ciphers. Even after the more secure Vigenère cipher was introduced in 1553, since the encryption and decryption processes were rather slow, the homophonic substitution ciphers (with nulls) continued to be used in most practical cryptosystems. The latter types of ciphers had a great variety; for example, one could use only 100 homophones, and a look-up table could be used for quick decryption by field officers. Indeed, it took a century for the Vigenère cipher to start to catch on, and it was not until the 19th century that it rose to become the method of choice in most cryptographic circles.

We mention a related cryptosystem that also enjoyed longevity, the ultimate sign of a good design. This system was invented by one of the Founding Fathers of the United States, Thomas Jefferson (Figure 3.6). In his many important roles in early American government, he was deeply concerned with the need for private communication. Jefferson's wheel cipher, which he designed in the 1790s, is shown in Figure 3.7. The way the wheel cipher works is as follows: It consists of 36 numbered wheels, each containing a different permutation of the alphabet on its circumference. The key consisted of a specified permutation of the 36 wheels

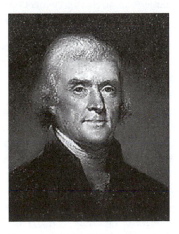

Figure 3.6 Thomas Jefferson (1743–1826), American statesman and third president of the United States.

Figure 3.7 Thomas Jefferson's wheel cipher.

by which the wheels should be loaded onto the spindle. (Thus, even if the enemy were to obtain the wheel, there are $36! > 10^{41}$ possible keys.) The way a plaintext message gets encrypted is to take the first (up to) 36 characters, and line them up together on the wheels, the corresponding ciphertext would then be the resulting string of 36 characters on any of the other 25 remaining rows of the wheels. The process continues for the next block of 36 plaintext characters, until the entire message is encrypted. Decryption is equally simple. The first 36 characters of the ciphertext are lined up on the 36 wheels, and one of the 25 remaining rows of the wheels will contain the corresponding plaintext message. Jefferson's wheel was a natural precursor to many of the mechanical and computer program-based cryptosystems that were developed throughout the 20th century, and it was used for official purposes as late as World War II by the U.S. Navy. Since his discovery remained immune from attacks for over 150 years, Jefferson has been called the father of American cryptography.

Exercise for the Reader 3.5

> Describe the design of a device that is similar to Jefferson's wheel cipher and that could serve as an encryption/decryption device for the Vigenère cipher. Provide specific design and implementation details.

Toward the beginning of the 20th century, a new scientific break-through would serve as a tremendous impetus for cryptographers to advance the subject to be able both to harvest and to protect secrets in the ensuing environment. Wireless communication was now possible by means of radio transmissions. This meant that messages could be sent instantaneously over long distances, but at the same time, many unintended recipients would be able to receive these transmissions. Prior to radio transmissions, wire telegraphs were the dominant technology for fast communications (both cryptographic and nonsecret) since the mid-19th century. But this technology required wire networks between communicating parties, which was a tremendous hindrance, for example, in military field communications. Vigenère's *chiffre indéchiffrable* had been compromised by Babbage and Kasiski. New ciphers were constructed using similar substitutions and rearrangements to what had been used in the past, but needs for efficiency and practicality for the needed hand computations put limits on the complexity of the systems. Furthermore, the amount of intercepted ciphertexts was increasing at an alarming rate. This provided hackers with plentiful supplies of ciphertext data on which to analyze cryptosystems.

Composition of Functions

As we have already mentioned, sophisticated cryptosystems can be created by combining several simpler encryption schemes into a single grand scheme. The underlying mathematical concept is a process whereby two (or more) functions are successively applied (in a certain order) to an input to result in a new function called the *composition* of the two functions (in the specified order). In order for this to work, the output of the first function that is applied must always lie in the domain of the second function. The formal definition is as follows:

The composition is illustrated in Figure 3.8.

Definition 3.3

If a function $f : A \rightarrow B_1$ has its range being a subset of the domain of another function $g : B \rightarrow C$ (that is, $f(A) \subseteq B$), then these functions can be combined in a natural way to form a new function $g \circ f : A \rightarrow C$ called the **composition** of f and g, defined by $(g \circ f)(a) = g(f(a))$.*

Example 3.6

(a) Consider the affine functions $\phi : \mathbb{Z}_{26} \rightarrow \mathbb{Z}_{26} :: \phi(x) \equiv 3x + 5$ and $\psi : \mathbb{Z}_{26} \rightarrow \mathbb{Z}_{26} :: \psi(x) \equiv 11x + 2$. Obtain formulas for the two composition functions $\psi \circ \phi$ and $\phi \circ \psi$. Is either of these composition functions affine? Are they the same function?

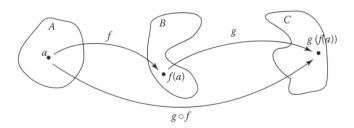

Figure 3.8 The composition $g \circ f : A \rightarrow C$ of two functions $f : A \rightarrow B_1$ and $g : B \rightarrow C$, where $f(A) \subseteq B$.

---

* We emphasize that in any composition $g \circ f$, the right function (here, f) is applied *first* to the input (domain element).

(b) For any set A, we define the *identity function* for A to be the function $i_A : A \to A$ given by the formula $i_A(a) = a$. Show that if $f : A \to B$ is a bijection, then $f^{-1} \circ f = i_A$ and $f \circ f^{-1} = i_B$.

Note: When mention of the set itself is not important, the identity function is sometimes denoted simply as *id*.

Solution: Part (a): Using the definition of composition, we obtain

$$(\psi \circ \phi)(x) \equiv \psi(\phi(x)) \equiv \psi(3x+5) \equiv 11(3x+5) + 2 \equiv 33x + 57$$

$$\equiv 7x + 5 \,(\text{mod } 26)$$

$$(\phi \circ \psi)(x) \equiv \phi(\psi(x)) \equiv \phi(11x+2) \equiv 3(11x+2) + 5 \equiv 33x + 11$$

$$\equiv 7x + 11 \,(\text{mod } 26)$$

Thus, we see that both compositions of these affine maps are affine but are different functions (that differ by a shift mapping). The fact that $\psi \circ \phi \neq \phi \circ \psi$ is rather typical. *Compositions of functions are not commutative.*

Part (b): For each $a \in A$, the image $f(a)$ will be some element b of B. By definition of the inverse function, we have $f^{-1}(b) = a$. Hence, $(f^{-1} \circ f)(a) = f^{-1}(f(a)) = f^{-1}(b) = a = i_A(a)$. This proves that $f^{-1} \circ f = i_A$. Since $(f^{-1})^{-1} = f$, the second identity follows from the first.

Tabular Form Notation for Permutations

Notation: If A is any finite set, say $A = \{1, 2, \cdots, n\}$, then a permutation of the set A—that is, a bijection $\sigma : A \to A$—can always be represented in **tabular form**:

$$\sigma : \begin{pmatrix} 1 & 2 & 3 & \cdots & n-1 & n \\ \sigma(1) & \sigma(2) & \sigma(3) & \cdots & \sigma(n-1) & \sigma(n) \end{pmatrix}$$

where the image of each element in A in the first row is listed directly below it. For a specific example, the table

$$\sigma : \begin{pmatrix} 1 & 2 & 3 & 4 & 5 & 6 \\ 4 & 6 & 2 & 1 & 5 & 3 \end{pmatrix}$$

represents the permutation $\sigma : \{1, 2, \cdots, 6\} \to \{1, 2, \cdots, 6\}$ specified by $\sigma(1) = 4$, $\sigma(2) = 6$, $\sigma(3) = 2$, $\sigma(4) = 1$, $\sigma(5) = 5$, and $\sigma(6) = 3$. This function is clearly both one-to-one and onto; hence, it is a permutation. The corresponding table for the inverse permutation can be obtained by "turning the above table upside-down"—that is, by switching the rows and then reordering the columns (as the reader should verify):

$$\sigma^{-1} : \begin{pmatrix} 1 & 2 & 3 & 4 & 5 & 6 \\ 4 & 3 & 6 & 1 & 5 & 2 \end{pmatrix}$$

Exercise for the Reader 3.6

Let $\sigma, \tau : \{1, 2, \cdots, 6\} \rightarrow \{1, 2, \cdots, 6\}$ be the permutations specified by

$$\sigma : \begin{pmatrix} 1 & 2 & 3 & 4 & 5 & 6 \\ 4 & 6 & 2 & 1 & 5 & 3 \end{pmatrix} \text{ and}$$

$$\tau : \begin{pmatrix} 1 & 2 & 3 & 4 & 5 & 6 \\ 3 & 4 & 5 & 6 & 1 & 2 \end{pmatrix}$$

Find the tabular representations for the composition $\sigma \circ \tau$ and its inverse $(\sigma \circ \tau)^{-1}$. Compare the latter with the tabular form for the composition $\tau^{-1} \circ \sigma^{-1}$.

The observations made in the previous Exercise for the Reader are generalized in the following.

Exercise for the Reader 3.7

The inverse of a composition of invertible functions is the reverse-order composition of the inverses.

Show if $f : A \rightarrow B$ and $g : B \rightarrow C$ are both bijective functions, then so is their composition $g \circ f : A \rightarrow C$, and furthermore we have $(g \circ f)^{-1} = f^{-1} \circ g^{-1}$.

The Enigma Machines

Toward the end of the World War I in 1918, a pair of German engineers, Arthur Scherbius and Richard Ritter, invented one of the first practical encryption/decryption machines that was to foreshadow many of the computer-based cryptosystems that would follow in the latter portion of the 20th century. Their device, known as *Enigma* (see Figure 3.9), was adopted by the German government and military and, through its

Figure 3.9 An Enigma machine.

modifications, became one of the most notorious and powerful crypto-graphic tools in history. There are many different models of the Enigma machine, and they were used by a number of countries. It is estimated that roughly 100,000 Enigmas had been built. In particular, both Germany and Japan used them during World War II (Japan took a German machine and modified it; the end product was called *Purple*). The Allied Forces were able to crack both machines, and this had a decisive impact on the timing of their defeat of Hitler. By some accounts, this intelligence breakthrough curtailed the war by about two years. Although the Enigma's encryption process is quite complicated, it is much easier to understand if we break things down into more basic components. We first explain the elements of the original Enigma machine. Then, in Example 3.8, we provide a detailed example of a scaled-down Enigma machine simplified to work with a six-character alphabet rather than a 26-character alphabet. This will allow all of the details to be understood with the convenience of reduced-size plaintext/ciphertext alphabets. We subsequently explain the differences that arise when we expand to a full alphabet, along with some later modi-fications. In Chapter 5, we will return to the even more fascinating history and details of how the Enigma machines were successfully attacked.

We see from Figure 3.9 that the Enigma machine looked like a fat type-writer. The keys in the middle are arranged and labeled just as the 26 letter keys on an ordinary (typewriter) keyboard. On top, there are bulbs that are arranged and labeled just as the keys below. When a key is pressed, an elec-tronic/mechanical encryption process converts the corresponding letter into a ciphertext letter, whose bulb will be lit up on the lampboard. The internal elements that worked together to produce the ciphertext consisted of

(a) **Rotors:** Rotors (originally three, later expanded to more), all lined up on a spindle, each corresponding to certain permutation of 26 inputs, but each could be rotated to any of 26 settings. Each rotor has 26 electrical contacts on each face that are equally spaced around a circle, and those on one face are wired to those of the other with respect to a specific permutation. As with Jefferson's wheel cipher, the rotors can be put on the spindle in any order.

(b) **Reflector:** This plate has 26 electrical contacts on its face adja-cent to the last rotor. These contacts are connected in pairs by a full 13 pairs of connections. After having passed through the rotors, a signal passes through this reflector, and then goes back through the rotors in reverse order.

(c) **Plugboard:** This is an exposed portion of the machine that has six pairs of plugs that can be used to interchange six pairs of letters. Plaintext inputs first pass through the plugboard before entering the rotors, and the pre-ciphertext that exits the rotors again passes through the plugboard before reaching the lampboard as ciphertext.

Figure 3.10 shows a schematic diagram of these elements, with the pro-cessing of a typical plaintext element.[*]

---

[*] Note that the keyboard/lampboard orientation in this diagram corresponds to that of the original German Enigma machines; there are some differences with the standard American English keyboards.

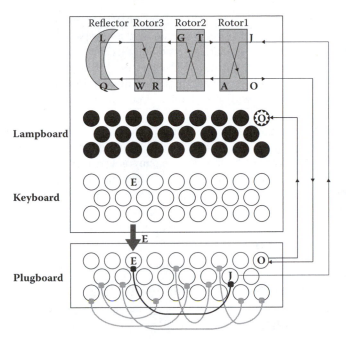

Figure 3.10 Schematic diagram for an Enigma machine.

The edge of each rotor wheel is marked with the numbers 0 through 25 (corresponding to the letters of the alphabet), equally spaced around the circumference. The rotors lie inside the machine, but lifting the rotor lid allows the user to set each of the rotors to any initial setting specified by the three numbers (from 0 to 25) that show on top. When the rotor lid is closed, the three numbers on the top are visible through the **rotor window**. The appearance is like a mod 26 odometer, but the functionality of the rotors is also very similar to that of an automobile odometer, except the Enigma rotors work mod 26. When a plaintext character is entered on the keyboard, the first rotor always rotates forward one *notch* = 1/26 of a complete rotation. So, for example, if the first rotor's initial setting was 3, when a key is entered, the first rotor would advance to 4. Every time the first rotor advances from 25 back to 0, the second rotor will advance one notch. Similarly, every time the second rotor advances from 25 back to 0, the third rotor advances one notch. (Once again, this is exactly the same mechanical functionality as an automobile odometer, except in the latter which works mod 20, a "notch" represents 1/10th of a complete rotation, rather than 1/26th for an Enigma rotor.) Now, each time a plaintext letter is entered on the keyboard, the rotors advance as just described, and the electric path determined by the rotor permutations (and their current position), the plugboard, and the reflector determines the ciphertext character. This ensures (with fixed plugboard settings) that the resulting single-character encryption functions will be different functions for the first $26^3 = 17,576$ plaintext characters.

One remarkable property of the Enigma machines is that one may decrypt any ciphertext message produced by an Enigma machine simply by entering the ciphertext into any other (same model) Enigma machine using the same initial settings (that were used to key in the plaintext). These initial settings served as the keyspace for the Enigma cipher. We give a mathematical proof of this and another interesting property of Enigma machines at the end of this section.[*] Shortly we give an example that provides a detailed description of how Enigma machines work, but with all elements reduced to smaller scales, for ease of illustration.

When the German government decided to adopt Enigma machines for military communications, it realized that some Enigma machines would be captured and probably be sold to enemy governments. But even with the possession of an Enigma machine, decryption of ciphertext was far from trivial. In order to use the machine to decrypt a ciphertext produced by it, the initial settings at the start of the encryption must be used. Let us now count the number of possible initial settings of the basic Enigma machine with three rotors.

The number of ways to arrange the three rotors on the spindle: $3! = 6$.

The number of rotor settings: $26^3 = 17,576$.

The number of ways to install six plugboard cables (each connecting a pair of letters): 100,391,791,500.[†]

Multiplying these possibilities gives the total number of initial arrangements of the basic Enigma machine to be over 10,000,000,000,000,000 (that is, over 10 quadrillion).

It is convenient to temporarily digress to introduce notations and concepts for special sorts of permutations known as cycles. Any permutation can be uniquely decomposed into its associated cycles, and these cycles are very helpful in understanding the details of a permutation, much as the prime factorizations help to understand many properties of integers. Cycles played an important role, as we will see in Chapter 5, in the breaking of the Enigma machines.

Cycles (Cyclic Permutations)

One very interesting and useful concept relating to permutations is that of a *cycle*. If we start with any element of the domain and continue to take images of it under a permutation, we will eventually "cycle back" to the element. For example, with the permutation σ of $\{1, 2, 3, 4, 5, 6\}$ specified by

$$\sigma : \begin{pmatrix} 1 & 2 & 3 & 4 & 5 & 6 \\ 4 & 6 & 2 & 1 & 5 & 3 \end{pmatrix}$$

if we repeatedly apply σ to each element of the domain until we arrive back at the original element, we obtain the following sequences (or chains):

---

[*] This also follows from the fact that an electric circuit is formed in the machine that connects the plaintext key on the keyboard to the ciphertext bulb on the lampboard.
[†] See Exercise 37 for the underlying counting scheme.

$$1 \overset{\sigma}{\mapsto} 4 \overset{\sigma}{\mapsto} 1 \qquad \Leftrightarrow \quad (1,4)$$

$$2 \overset{\sigma}{\mapsto} 6 \overset{\sigma}{\mapsto} 3 \overset{\sigma}{\mapsto} 2 \quad \Leftrightarrow \quad (2,6,3)$$

$$3 \overset{\sigma}{\mapsto} 2 \overset{\sigma}{\mapsto} 6 \overset{\sigma}{\mapsto} 3 \quad \Leftrightarrow \quad (3,2,6)$$

$$4 \overset{\sigma}{\mapsto} 1 \overset{\sigma}{\mapsto} 4 \qquad \Leftrightarrow \quad (4,1)$$

$$5 \overset{\sigma}{\mapsto} 5 \qquad\qquad \Leftrightarrow \quad (5)$$

$$6 \overset{\sigma}{\mapsto} 3 \overset{\sigma}{\mapsto} 2 \overset{\sigma}{\mapsto} 6 \quad \Leftrightarrow \quad (6,3,2)$$

Each of these ordered sequences by itself represents a special type of permutation on $\{1, 2, 3, 4, 5, 6\}$, known as a *cycle*, which leaves fixed all elements of $\{1, 2, 3, 4, 5, 6\}$ that do not appear in the sequence. So, for example, the first sequence, abbreviated by $(1, 4)$, represents the permutation τ on $\{1, 2, 3, 4, 5, 6\}$ that permutes 1 and 4: $\tau(1) = 4$, $\tau(4) = 1$, and *fixes* all other elements of the domain: $\tau(k) = k$, if $k \neq 1, 4$. Notice that this permutation is the same as the fourth, abbreviated $(4, 1)$, which is simply a cyclic reordering of $(1, 4)$. Similarly, the three ordered sequences $(2, 6, 3)$, $(6, 3, 2)$, and $(3, 2, 6)$ are all cyclic reorderings of one another (each one is obtained from the previous one by moving the number at the front of the ordered list to the end) and all represent the same permutation μ on $\{1, 2, 3, 4, 5, 6\}$ defined by $\mu(2) = 6$, $\mu(6) = 3$, $\mu(3) = 2$, and $\mu(k) = k$, if $k = 1, 4, 5$. The actions of this cycle are illustrated (in two different ways) in Figure 3.11.

The following cycle 5 really just corresponds to the identity map, which fixes all elements of $\{1, 2, 3, 4, 5, 6\}$, and is called a *trivial cycle*. Thus, the original permutation σ can be completely described in terms of the two cycles $\tau = (1, 4)$ and $\mu = (2, 6, 3)$. In fact, the reader should verify that $\sigma = \tau \circ \mu = \mu \circ \tau$. Ordinarily, we know that composition of functions is not commutative, but cycles involving disjoint elements will always commute since they act independently of one another. Because of this fact, it is common to abbreviate disjoint cycle compositions such as $\tau \circ \mu = (1, 4) \circ (2, 6, 3)$ simply as a product: $(1, 4)(2, 6, 3)$ (omitting the composition symbol). Thus we can rewrite the original permutation σ as the product of its disjoint cycles: $\sigma = (1, 4)(2, 6, 3)$. Such a product representation of a permutation is called the *disjoint cycle decomposition* of

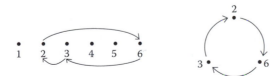

Figure 3.11 Two illustrations of the actions of the cycle (2,6,3), shown as a permutation of {1, 2, 3, 4, 5, 6}. The elements 1, 4, 5, that are not listed in the cycle are left fixed. This cycle can also be written as either of its cyclic reorderings: (6, 3, 2) or (3, 2, 6).

the cycle.[*] One final observation about cycles: If we take the cycle $(2, 6, 3)$ and consider the related cycle obtained by listing the same elements in the reverse order: $(3, 6, 2)$, this corresponds to reversing all arrows in Figure 3.11, and thus corresponds to the inverse permutation, that is, $(2, 6, 3) = (3, 6, 2)^{-1}$. All of these concepts generally hold true, but before we give the general result, we formally define cycles.

Definition 3.4

If $n > 1$ is an integer and $a_1, a_2, \cdots, a_k$ are distinct elements of the set $\{1, 2, \cdots, n\}$, then the **cycle** $(a_1, a_2, \cdots, a_k)$ is the permutation μ on $\{1, 2, \cdots, n\}$, that is defined by

$$\mu(a_1) = a_2, \mu(a_2) = a_3, \cdots \mu(a_{k-1}) = a_k, \text{ and } \mu(a_k) = a_1$$

Also, $\mu(j) = j$, if $j \notin \{a_1, a_2, \cdots, a_k\}$ (that is, μ fixes all other elements of $\{1, 2, \cdots, n\}$).

Notice that any of the cyclic permutations of the above cycle correspond to the same permutation and thus all represent the same cycle:

$$(a_1, a_2, \cdots, a_k) = (a_2, a_3, \cdots a_k, a_1) = (a_3, a_4, \cdots, a_k, a_1, a_2)$$

$$= \cdots = (a_k, a_1, \cdots, a_{k-2}, a_{k-1})$$

We say the cycle $(a_1, a_2, \cdots, a_k)$ has **length k** and call it a **k-cycle**. Two cycles, $\mu = (a_1, a_2, \cdots, a_k)$ and $\tau = (b_1, b_2, \cdots, b_\ell)$, are said to be **disjoint** if their lists of elements have nothing in common, that is, $\{a_1, a_2, \cdots, a_k\} \cap \{b_1, b_2, \cdots, b_\ell\} = \varnothing$. A cycle $(a_1, a_2, \cdots, a_k)$ is **associated with a permutation** σ on $\{1, 2, \cdots, n\}$, if $\sigma(a_i) = a_{i+1}$ for each i, $1 \le i < k$ and $\sigma(a_k) = a_1$.

The following proposition generalizes some of the interesting facts that we observed for a specific permutation and its associated cycles.

Proposition 3.1

Let σ be a permutation and $\mu = (a_1, a_2, \cdots, a_k)$ and $\tau = (b_1, b_2, \cdots, b_\ell)$ be two cycles on $\{1, 2, \cdots, n\}$.

(a) *Disjoint cycles commute.* If the cycles μ and τ are disjoint, then $\tau \circ \mu = \mu \circ \tau$. We define the product of two disjoint cycles μ and τ to be this (commutative) composition, and denote it as $\tau\mu$ or $\mu\tau$. Products of any finite number of pairwise disjoint cycles are similarly (and unambiguously) defined.

(b) *Inverse of a cycle.* The inverse (permutation) of a cycle $\mu = (a_1, a_2, \cdots, a_k)$ is the same length cycle with the elements being listed in reverse order: $\mu^{-1} = (a_k, a_{k-1}, \cdots, a_2, a_1)$.

(c) Any two cycles μ and τ associated with a permutation σ are either identical or disjoint.

---

[*] Since the trivial cycle (5) is just the identity mapping, there is no need to include it in the cycle decomposition; that is, although it would be correct to write $\sigma = (1, 4)(2, 6, 3)(5)$, the trivial cycle (5) is redundant.

(d) *Decomposition of a permutation into disjoint cycles.* If $\mu_1, \mu_2, \cdots, \mu_K$ are all of the nontrivial cycles associated with σ, then $\sigma = \mu_1 \mu_2 \cdots \mu_K$.

Notation: When dealing with permutations, it is sometimes convenient to have a notation for repeated compositions of a permutation with itself. It is common to use the *j*th power notation for the **j-fold composition** of a permutation with itself. Thus, $\sigma^2(x) \equiv (\sigma \circ \sigma)(x) = \sigma(\sigma(x))$, $\sigma^3(x) \equiv (\sigma \circ \sigma \circ \sigma)(x) = \sigma(\sigma(\sigma(x)))$, and in general $\sigma^j(x) \equiv (\underbrace{\sigma \circ \sigma \circ \cdots \circ \sigma}_{j \text{ times}})(x) = \underbrace{\sigma(\sigma(\cdots \sigma}_{j \text{ times}}(x)\cdots))$.

Proof: Part (a): Let $A = \{a_1, a_2, \cdots, a_k\}$, $B = \{b_1, b_2, \cdots, b_\ell\}$, and C be all of the remaining elements of $\{1, 2, \cdots, n\}$. These three sets are disjoint and their union is $\{1, 2, \cdots, n\}$. Let $i \in \{1, 2, \cdots, n\}$; we separate into three cases.

Case 1. $i \in A$: In this case, τ fixes i—that is, $\tau(i) = i$—so we have $(\mu \circ \tau)(i) = \mu(i)$. But since also $\mu(i) \in A$, τ also fixes $\mu(i)$, and we have $(\tau \circ \mu)(i) = \tau(\mu(i)) = \mu(i)$.

Case 2. $i \in B$: In this case, μ fixes both i and $\tau(i)$, and it follows (just as in Case 1) that $(\mu \circ \tau)(i) = \tau(i) = (\tau \circ \mu)(i)$.

Case 3. $i \in C$: In this case, both μ and τ fix i, so $(\mu \circ \tau)(i) = i = (\tau \circ \mu)(i)$.

This completes the proof of part (a) for two disjoint cycles. The proof for larger numbers of cycles is accomplished in a similar fashion, with an additional case for each additional cycle.

Part (b): We temporarily denote the purported inverse as $\omega = (a_k, a_{k-1}, \cdots, a_2, a_1)$. Since both $\mu = (a_1, a_2, \cdots, a_k)$ and ω fix all elements of $\{1, 2, \cdots, n\}$ except those in $\{a_1, a_2, \cdots, a_k\}$, we need only check that $(\omega \circ \mu)(a_i) = a_i = (\mu \circ \omega)(a_i)$ whenever $1 \leq i \leq k$. This result follows from the facts that μ takes any a_i to the next element on μ's list (a_{i+1} or a_1, if $i = k$), and $\omega = (a_k, a_{k-1}, \cdots, a_2, a_1)$ takes any a_i to the previous element on μ's list (a_{i-1} or a_k, if $i = 1$).

Part (c): Let $\mu = (a_1, a_2, \cdots, a_k)$ and $\tau = (b_1, b_2, \cdots, b_\ell)$ be cycles associated with σ. We may assume that $k \leq \ell$. If the cycles μ and τ were not disjoint then, by taking cyclic permutations of μ and τ, if necessary, we may assume that $a_1 = b_1$. Since both of these cycles are associated with σ, we have $a_2 = \sigma(a_1) = \sigma(b_1) = b_2$. By the same token, we may next infer that $a_3 = \sigma(a_2) = \sigma(b_2) = b_3$. Continuing in this fashion, we successively obtain that $a_4 = b_4$, $a_5 = b_5$, and so on, until we arrive at $a_k = b_k$. From here, we obtain $a_1 = \sigma(a_k) = \sigma(b_k)$. Since $a_1 = b_1$, it follows that $\ell = k$, and thus $\mu = \tau$.

Part (d): This will be a constructive proof that was foreshadowed by the motivating example that was given.

Initialize the set $S = \{1, 2, \cdots, n\}$.

Iterative Step: Begin by repeatedly applying the permutation σ to the elements of the set S (starting from the smallest) until $\sigma(i) \neq i$. All elements of S smaller than i are thus left fixed by σ, and so give rise to trivial cycles associated with σ. Next we form the sequence obtained by iteratively applying σ to i:

$$a_{1,1} = i, \ a_{1,2} = \sigma(i), \ a_{1,3} = \sigma(a_{1,2}) = \sigma(\sigma(i)), \ a_{1,4} = \sigma(a_{1,3})$$

$$= \sigma(\sigma(\sigma(i))), \ \cdots, a_{1,k_1+1} = \sigma(a_{1,k_1})$$

until we arrive at a duplication. Thus, in the notation above, we have $a_{1,1}, a_{1,2}, \cdots, a_{1,k_1}$ are all different, and a_{1,k_1+1} equals one of these numbers.

Claim: $a_{1,k_1+1} = i$

Proof of Claim: If instead we had $a_{1,k_1+1} = a_{1,j}$ for some integer j, $1 < j \le k_1$, then we would have $\sigma^{k_1}(i) = \sigma^{j-1}(i)$. If we repeatedly apply the inverse permutation σ^{-1} to this equation $(j - 1)$ times, we arrive at $\sigma^{k_1-j+1}(i) = i$, which contradicts the fact that $\sigma^{k_1+1}(i)$ was the first duplicated entry in the list.

In light of the claim, the cycle $\mu_1 = (a_{1,1}, a_{1,2}, \cdots, a_{1,k_1})$ is associated with σ. Now we delete from the set S each of the elements $a_{1,1}, a_{1,2}, \cdots, a_{1,k_1}$ of this cycle, as well as any elements smaller than $a_{1,1}$ that were found to correspond to trivial cycles. If S is the empty set, then $\sigma = \mu_1$, and we are done. Otherwise, we go back and apply the above iterative step again, but this time we change the labeling of the cycle elements as $\mu_2 = (a_{2,1}, a_{2,2}, \cdots, a_{2,k_2})$. We again update S by deleting each of the new cycle elements $a_{2,1}, a_{2,2}, \cdots, a_{2,k_2}$, as well any elements smaller than $a_{2,1}$ that were found to correspond to trivial cycles. This process continues until S is empty. The cycles $\mu_1, \mu_2, \cdots, \mu_K$ that were constructed are all of the nontrivial cycles associated with σ, and the construction shows that any number of $\{1, 2, \cdots, n\}$ that is left unfixed by σ will belong to exactly one of these associated cycles. By part (a), the action of σ will be the same as the action of the product $\mu_1\mu_2\cdots\mu_K$. □

Example 3.7

(a) Express the permutation σ on $\{1, 2, \cdots, 10\}$ specified by

$$\begin{pmatrix} 1 & 2 & 3 & 4 & 5 & 6 & 7 & 8 & 9 & 10 \\ 3 & 2 & 1 & 5 & 9 & 10 & 7 & 4 & 8 & 6 \end{pmatrix}$$

as a product of its disjoint cycles.

(b) Express the permutation σ^{-1} as a product of its disjoint cycles.

Solution: Part (a): Applying the constructive procedure in the proof of part (d) of Proposition 3.1, we detect the following nontrivial cycles (in order, starting with smallest elements): (1, 3), (4, 5, 9, 8), (6, 10). Forming their product yields: $\sigma = (1, 3)\,(4, 5, 9, 8)\,(6, 10)$.

Part (b): Since, by part (b) of Proposition 3.1, the inverse of any cycle is simply the reverse order cycle, it follows that the inverses of the cycles associated with σ will be the disjoint cycles associated with σ^{-1}. Thus, $\sigma^{-1} = (1, 3)^{-1}(4, 5, 9, 8)^{-1}(6, 10)^{-1} = (3, 1)(8, 9, 5, 4)(10, 6)$.

Exercises for the Reader

Exercise 3.8

Find the disjoint cycle decomposition of the permutation σ on $\{1, 2, \cdots, 10\}$ specified by

$$\begin{pmatrix} 1 & 2 & 3 & 4 & 5 & 6 & 7 & 8 & 9 & 10 \\ 1 & 4 & 7 & 2 & 5 & 10 & 3 & 9 & 8 & 6 \end{pmatrix}$$

Exercise 3.9

Let σ be the one unit shift permutation on $\mathbb{Z}_6$, that is, $\sigma(i) \equiv i + 1 \pmod 6$. Show that σ^k (the k-fold composition) corresponds to the k-unit shift permutation on $\mathbb{Z}_6$, and that the following disjoint cycle decompositions hold: $\sigma = (0, 1, 2, 3, 4, 5)$, $\sigma^2 = (0, 2, 4)(1, 3, 5)$, $\sigma^3 = (0, 3)(1, 4)(2, 5)$, $\sigma^4 = (0, 4, 2)(1, 5, 3)$, and $\sigma^5 = (0, 5, 4, 3, 2, 1) = \sigma^{-1}$. Show also that $(\sigma^4)^{-1} = \sigma^2$ and $(\sigma^3)^{-1} = \sigma^3$.

While the results of Exercise for the Reader 3.9 might seem appealing, we stress that shift permutations are most easily conceived and manipulated in terms of modular arithmetic. The following proposition documents some useful properties of general one unit shifts and their powers. The proof of this proposition is routine and is left as an exercise (Exercise 20).

> **Proposition 3.2: General One-Unit Shift Permutations and Their Powers**
>
> Given a positive integer $m > 1$, let $\sigma : \mathbb{Z}_m \to \mathbb{Z}_m$ be the unit (forward) shift operator represented by the m-cycle $(0, 1, 2, \cdots, m-1)$. Then for any positive integer k, the k-fold composition $\sigma^k : \mathbb{Z}_m \to \mathbb{Z}_m$ is the k-unit shift and can be computed by the formula
>
> $$\sigma^k(x) \equiv x + k \pmod m \qquad (3.4)$$
>
> The inverse of the k-fold shift is the k-unit backward shift, which is the same as the $(n - k)$-unit forward shift, and can be computed by the following formula:
>
> $$(\sigma^k)^{-1}(x) \equiv \sigma^{m-k}(x) \equiv x - k \pmod m \qquad (3.5)$$

One simple but important consequence of this proposition is that if $k \equiv \ell \pmod m$, then $\sigma^k = \sigma^\ell$; this simply corresponds to the fact that shifting a multiple of m units in $\mathbb{Z}_m$ is just the identity mapping. For example, since $\mod(122, 6) = 2$, we may write $\sigma^{122}(x) \equiv \sigma^2(x) \equiv x + 2 \pmod 6$. Also, by Equation 3.5, we see that $(\sigma^{122})^{-1}(x) \equiv (\sigma^2)^{-1}(x) \equiv x - 2 \equiv x + 4 \pmod 6$.

Dissection of the Enigma Machine into Permutations

We will show how to decompose the actions of an Enigma machine into a composition of permutations. Such a decomposition naturally leads to a means for easily creating programs for Enigma machines, and such tasks are explored in the computer exercises of this chapter. The following example contains the complete details for a scaled-down version of the original Enigma machine where the alphabet size is reduced from 26 letters to 6. This size reduction is only for ease of illustration; the ideas are the same and can easily be used for full-sized Enigma machines.

Example 3.8: Detailed Description of a Scaled-Down Enigma Machine

Figure 3.12 shows a schematic of a scaled-down Enigma machine with

- A plaintext/ciphertext alphabet of six characters, i.e., $P = \{a, b, c, d, e, f\}$, $C = \{A, B, C, D, E, F\}$.
- Two plugboard cables, connecting two pairs of letters.

The three rotors will correspond to the following permutations of six objects* $\{0, 1, 2, 3, 4, 5\}$:

Rotor 1: (0, 2)(3, 4, 5) Rotor 2: (0, 1)(2, 3)(4, 5) Rotor 3: (0, 2, 4)(1, 3, 5)

The initial settings of the rotors are determined by which numbers are lined up with the letter a: Rotor 1: 5, Rotor 2: 3, Rotor 3: 1. Through the rotor window, these settings would show as: |1|3|5|. Now, as soon as a plaintext letter is entered on the keyboard, as explained earlier, the first rotor (on the right) will advance one notch (here, one notch = 1/6 of a full rotation) from 5 to 0. Since the first rotor has advanced to 0, the second rotor advances one notch from 3 to 4. Thus, before the electronic portion of the plaintext to ciphertext letter takes place, the rotors have advanced from their initial setting of |1|3|5| to |1|4|0| (the reader should think of a mod 6 automobile odometer). This new setting is shown in Figure 3.12, which illustrates the encryption process of a letter from the plugboard/rotor interface, through the rotors and reflector, and back to this interface. The encryption of e from and back to the plugboard is indicated with these new rotor settings: Rotor 1: 0, Rotor 2: 4, Rotor 3: 1.

We will also use plugboard cable connections (a, f), (b, d), and the reflector will be configured with the permutation (a, f)(b, d)(c, e).

(a) Use compositions of functions to express the action of the scaled-down Enigma machine as described above on a single plaintext character with the indicated rotor and plugboard settings.

---

* *Note:* The original Enigma machines also had the rotors labeled with numbers rather than letters. We adhere to the number notation since this will help to make the underlying concepts more transparent. By contrast, the plugboard/keyboard interface and the reflector do not rotate, and their corresponding contacts will be labeled with the letters a, b, c, d, e, f.

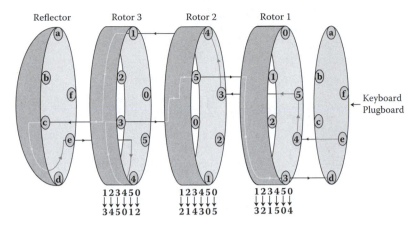

Figure 3.12 The path of the character "e," coming from the keyboard/plugboard going through the rotors and reflector and back to the plugboard rotor interface, is shown, emerging as "d."

    (b) Given that the rotation direction of the rotors is counter-clockwise when viewed from the right-side perspective of Figure 3.12 (so when Rotor 1 turns advances one notch = 1/6 of one revolution, the number 5 contact will be adjacent to the letter *a* on the plugboard/keyboard interface), generalize the decomposition of part (a) to describe the actions of this scaled-down Enigma machine on any plaintext message using the above initial settings.

    (c) With the above initial settings, how would the above scaled-down Enigma machine encode the German word *ebbe*?

    (d) Check that the ciphertext can be decrypted by feeding it into this scaled-down Enigma machine if the same initial settings are used.

    Solution: Part (a): In order to facilitate the two-letter/digit interfaces indicated on the left (reflector $\leftrightarrow$ rotor 3) and on the right (Rotor 1 $\leftrightarrow$ plugboard interface), we will need to pay a small price of introducing the following bijection:

$$F : \{a,b,c,d,e,f\} \to \mathbb{Z}_5 = \{0,1,2,3,4,5\}$$

defined by $a \mapsto 0, b \mapsto 1, c \mapsto 2, d \mapsto 3, e \mapsto 4, f \mapsto 5$

(This is an abbreviated bijection for that which was indicated in Table 3.1.) A key concept will be to understand how to represent the effect of a single rotor that may have been rotated to any position.

MATHEMATICAL DESCRIPTION OF THE PERMUTATION ASSOCIATED WITH A SINGLE (ROTATED) ROTOR

Since the rotors act independently, we will focus attention on a single rotor that is in a certain setting/position *j*. We let P denote the permutation associated with the rotor. For

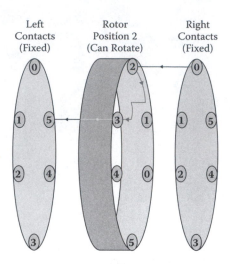

Figure 3.13 The action of a single rotated rotor.

example, it may be Rotor 2 that is in position $j = 2$, as is shown in Figure 3.13 with the fixed contact plates (that do not rotate) on the left and right. We need to describe the resulting permutation on the set {0, 1, 2, 3, 4, 5} that results from the action of the rotor as a mapping from the numbers on the right contact plate to those on the left contact plate. As can be seen from Figure 3.13 (with $j = 2$), any contact number i on the right contact plate enters the rotor as the number $i + j$ (mod 6). Let σ denote the one unit forward shift operator, so (see Proposition 3.2) σ^k is the k-unit forward shift operator, and this initial entrance permutation is simply σ^j. Figure 3.13 shows 0 on the right contact plate entering the rotor at (with $j = 2$) $\sigma^2(0) \equiv 0 + 2 \equiv 2 (\mathrm{mod}\, 6)$. Next, the rotor applies its own permutation ρ to produce its resulting output $\rho \circ \sigma^2(i)$. Figure 3.13 shows $2 = \sigma^2(0)$ being permuted to $\rho(2) = 1$ by the rotor. Finally, as the output of the rotor enters the left contact plate, since the rotated outputs (on the left side of the rotor) are still shifted to be j units more (mod 6) than those on the contact plates, we need to apply the inverse of the j-unit shift σ^{-j} so that the correct output number on the left plate is produced. In Figure 3.13, this corresponds (with $j = 2$) to converting the output 1 on the left side of the rotor, to the number 5 on the left contact plate: $\sigma^{-2}(1) \equiv 1 - 2 \equiv 5 (\mathrm{mod}\, 6)$.

In summary, we have shown that if any of the rotors is in position j and has associated permutation ρ, then the resulting permutation in going from the right contact plate numbers through the rotor to the left contact plate numbers is described by

$$\sigma^{-j} \circ \rho \circ \sigma^j = (\sigma^j)^{-1} \circ \rho \circ \sigma^j$$

Such a triple composition occurs quite often in mathematics, and is given a special name.

Definition 3.5

If ρ, τ are permutations on the same set, then the permutation $\tau^{-1} \circ \rho \circ \tau$ is called the **conjugate** (or the **conjugation**) of ρ by τ.

Note that since the inverse of a composition is the composition of the inverses in the reverse order, it follows that $(\tau^{-1} \circ \rho \circ \tau)^{-1} = \tau^{-1} \circ \rho^{-1} \circ \tau$. In other words: *The inverse of the conjugate is the conjugate of the inverse*. Also, from what was shown above, the permutation resulting from any rotor that has been rotated to position j is simply the conjugate of its permutation ρ by the j-unit shift σ^j. Since these conjugates are a core component in Enigma computations, we introduce a special notation for them: *Notation*: The conjugate of a permutation ρ (on the set $\{0, 1, 2, 3, 4, 5\}$) by a j-unit shift σ^j will be denoted as $C_j[\rho]$. Thus,

$$C_j[\rho] \triangleq \sigma^{-j} \circ \rho \circ \sigma^j$$

and since (as pointed out above) the inverse of the conjugate is the conjugate of the inverse, we may write

$$C_j[\rho]^{-1} = C_j[\rho^{-1}]$$

Let us now proceed through each step of the initial plaintext letter encryption, introducing appropriate permutations as they arise. For brevity, disjoint cycle decompositions will be used.

  (i) The plugboard: The given plugboard settings are represented by the following permutation on $\{a, b, c, d, e, f\}$: $\beta = (a,f)(b,d)$.

  (ii) For plugboard/right rotor interface we need the bijection F specified above.

  (iii) Rotor 1 is in position 0 and has its permutation $\rho_1 = (0,2)$ $(3,4,5)$, so the resulting permutation is $C_0[\rho_1]$.

  (iv) Rotor 2 is in position 4 and has its permutation $\rho_2 = (0, 1)$ $(2, 3)(4, 5)$, so the resulting permutation is $C_4[\rho_2]$.

  (v) Rotor 3 is in position 1 and has its permutation $\rho_3 = (0, 2, 4)(1, 3, 5)$, so the resulting permutation is $C_1[\rho_3]$.

  (vi) For the rotor 3 to reflector interface, we need the inverse F^{-1} of the bijection F specified above.

  (vii) Reflector action: $\tau = (a, f)(b, d)(c, e)$.

The transition back to the lampboard is affected by using the inverse functions of (i) through (vi), taken in the reverse order. Thus the complete encryption of (any) first plaintext letter corresponds to the following composition:[*]

$$\beta \circ F^{-1} \circ C_0[\rho_1^{-1}] \circ C_4[\rho_2^{-1}] \circ C_1[\rho_3^{-1}] \circ F \circ \tau \circ F^{-1} \circ C_1[\rho_3] \circ C_4[\rho_2]$$
$$\circ C_0[\rho_1] \circ F \circ \beta$$

---

[*] Since σ^0 is the identity map *id*, it is redundant in this composition and can be omitted, but we leave it in to better depict the function needed to process a general plaintext character.

Note that $\beta^{-1} = \beta$ (since β is a product of disjoint two-cycles and any two-cycle is its own inverse—Why?)

Part (b): For subsequent encryptions past the first plaintext letter, the only change needed in part (a) is the advancement of the three rotors (which together can be thought of as a mod 6 automobile odometer). We let k_1 be an integer in $\{0, 1, 2, 3, 4, 5\}$ corresponding to the initial position for the first rotor, and k_2 correspond to the initial setting of the second rotor. At the processing of the ℓth plaintext character, the rotors will each need to be rotated the following number of "notches" past their initial settings:

Rotor 1. $\ell_1 = \ell$ notches
Rotor 2. $\ell_2 = \text{floor}([k_1 + \ell] / 6)$ notches
Rotor 3. $\ell_3 = \text{floor}([k_2 + \ell_2] / 6)$ notches

In light of Proposition 3.2, the corresponding necessary changes needed in part (a) are as follows:

(iii)   Rotor 1 permutation: $C_{5+\ell_1}[\rho_1]$.
(iv)   Rotor 2 permutation: $C_{4+\ell_2}[\rho_2]$.
(v)   Rotor 3 permutation: $C_{1+\ell_3}[\rho_3]$.

When proceeding back from the reflector, the corresponding conjugations of the inverse permutations should be used.

Part (c): The first plaintext character encryption (up to the final plugboard application) is illustrated in Figure 3.12. We perform the step-by-step composition using the formula of part (a): Since the first plaintext character e is left fixed by β (the plugboard), we may compute:

$$\beta \circ F^{-1} \circ C_0[\rho_1^{-1}] \circ C_4[\rho_2^{-1}] \circ C_1[\rho_3^{-1}] \circ F \circ \tau \circ F^{-1} \circ C_1[\rho_3] \circ C_4[\rho_2]$$
$$\circ C_0[\rho_1] \circ F(e)$$

$$= \beta \circ F^{-1} \circ C_0[\rho_1^{-1}] \circ C_4[\rho_2^{-1}] \circ C_1[\rho_3^{-1}] \circ F \circ \tau \circ F^{-1} \circ C_1[\rho_3] \circ C_4[\rho_2]$$
$$\circ C_0[\rho_1](4)$$

$$= \beta \circ F^{-1} \circ C_0[\rho_1^{-1}] \circ C_4[\rho_2^{-1}] \circ C_1[\rho_3^{-1}] \circ F \circ \tau \circ F^{-1} \circ C_1[\rho_3] \circ C_4[\rho_2](5)$$

$$= \beta \circ F^{-1} \circ C_0[\rho_1^{-1}] \circ C_4[\rho_2^{-1}] \circ C_1[\rho_3^{-1}] \circ F \circ \tau \circ F^{-1} \circ C_1[\rho_3] \circ (4)$$

$$= \beta \circ F^{-1} \circ C_0[\rho_1^{-1}] \circ C_4[\rho_2^{-1}] \circ C_1[\rho_3^{-1}] \circ F \circ \tau \circ F^{-1}(0)$$

$$= \beta \circ F^{-1} \circ C_0[\rho_1^{-1}] \circ C_4[\rho_2^{-1}] \circ C_1[\rho_3^{-1}] \circ F \circ \tau(a)$$

$$= \beta \circ F^{-1} \circ C_0[\rho_1^{-1}] \circ C_4[\rho_2^{-1}] \circ C_1[\rho_3^{-1}] \circ F(f)$$

$$= \beta \circ F^{-1} \circ C_0[\rho_1^{-1}] \circ C_4[\rho_2^{-1}] \circ C_1[\rho_3^{-1}](5)$$

$$= \beta \circ F^{-1} \circ C_0[\rho_1^{-1}] \circ C_4[\rho_2^{-1}](3)$$

$$= \beta \circ F^{-1} \circ C_0[\rho_1^{-1}](2)$$

$$= \beta \circ F^{-1}(0)$$

$$= \beta(a)$$

$$= f$$

This calculation demonstrates that a machine was really the only practical way to implement such a complicated cryptosystem. The remaining three plaintext characters can be encrypted in a similar fashion, the only difference being that the rotor 1 initial setting should be changed from 0 to 1 for the second character, to 2 for the third character, and 3 for the fourth character. The corresponding composition evaluations to what was done for the first plaintext character proceed as follows for the next three plaintext characters:

Transformation of second plaintext character: b

$$\overset{\beta}{b} \mapsto \overset{F}{d} \mapsto \overset{C_1[\rho_1]}{3} \mapsto \overset{C_4[\rho_2]}{4} \mapsto \overset{C_1[\rho_3]}{5} \mapsto \overset{F^{-1}}{1} \mapsto \overset{\tau}{b} \mapsto \overset{F}{d} \mapsto \overset{C_1[\rho_3^{-1}]}{3} \mapsto \overset{C_4[\rho_2^{-1}]}{1} \mapsto \overset{C_1[\rho_1^{-1}]}{0} \mapsto \overset{F^{-1}}{0} \mapsto \overset{\beta}{a} \mapsto f$$

Transformation of third plaintext character: b

$$\overset{\beta}{b} \mapsto \overset{F}{d} \mapsto \overset{C_2[\rho_1]}{3} \mapsto \overset{C_4[\rho_2]}{1} \mapsto \overset{C_1[\rho_3]}{0} \mapsto \overset{F^{-1}}{2} \mapsto \overset{\tau}{c} \mapsto \overset{F}{e} \mapsto \overset{C_1[\rho_3^{-1}]}{4} \mapsto \overset{C_4[\rho_2^{-1}]}{2} \mapsto \overset{C_2[\rho_1^{-1}]}{3} \mapsto \overset{F^{-1}}{2} \mapsto \overset{\beta}{c} \mapsto c$$

Transformation of fourth plaintext character: e

$$\overset{\beta}{e} \mapsto \overset{F}{e} \mapsto \overset{C_3[\rho_1]}{4} \mapsto \overset{C_4[\rho_2]}{4} \mapsto \overset{C_1[\rho_3]}{5} \mapsto \overset{F^{-1}}{1} \mapsto \overset{\tau}{b} \mapsto \overset{F}{d} \mapsto \overset{C_1[\rho_3^{-1}]}{3} \mapsto \overset{C_4[\rho_2^{-1}]}{1} \mapsto \overset{C_3[\rho_1^{-1}]}{0} \mapsto \overset{F^{-1}}{2} \mapsto \overset{\beta}{c} \mapsto c$$

Thus the ciphertext is FFCC (putting it in uppercase as usual).

Part (d): We now place ourselves in the position of an intended recipient of the ciphertext FFCC. We would know to reset the rotor settings to their initial settings, and then feed the ciphertext characters into the machine. The procedure and permutation mappings are exactly the same as with the encryption. Here are the details for each letter:

Transformation of first ciphertext character: F

$$\overset{\beta}{f} \mapsto \overset{F}{a} \mapsto \overset{C_0[\rho_1]}{0} \mapsto \overset{C_4[\rho_2]}{2} \mapsto \overset{C_1[\rho_3]}{3} \mapsto \overset{F^{-1}}{5} \mapsto \overset{\tau}{f} \mapsto \overset{F}{a} \mapsto \overset{C_1[\rho_3^{-1}]}{0} \mapsto \overset{C_4[\rho_2^{-1}]}{4} \mapsto \overset{C_0[\rho_1^{-1}]}{5} \mapsto \overset{F^{-1}}{4} \mapsto \overset{\beta}{e} \mapsto e$$

Transformation of second ciphertext character: F

$$\overset{\beta}{f} \mapsto \overset{F}{a} \mapsto \overset{C_1[\rho_1]}{0} \mapsto \overset{C_4[\rho_2]}{0} \mapsto \overset{C_1[\rho_3]}{1} \mapsto \overset{F^{-1}}{3} \mapsto \overset{\tau}{d} \mapsto \overset{F}{b} \mapsto \overset{C_1[\rho_3^{-1}]}{1} \mapsto \overset{C_4[\rho_2^{-1}]}{5} \mapsto \overset{C_1[\rho_1^{-1}]}{4} \mapsto \overset{F^{-1}}{3} \mapsto \overset{\beta}{d} \mapsto b$$

Transformation of third ciphertext character: F

$$\overset{\beta}{c} \mapsto \overset{F}{c} \mapsto \overset{C_2[\rho_1]}{2} \mapsto \overset{C_4[\rho_2]}{3} \mapsto \overset{C_1[\rho_3]}{2} \mapsto \overset{F^{-1}}{4} \mapsto \overset{\tau}{e} \mapsto \overset{F}{c} \mapsto \overset{C_1[\rho_3^{-1}]}{2} \mapsto \overset{C_4[\rho_2^{-1}]}{0} \mapsto \overset{C_2[\rho_1^{-1}]}{1} \mapsto \overset{F^{-1}}{3} \mapsto \overset{\beta}{d} \mapsto b$$

Transformation of fourth ciphertext character: C

$$\overset{\beta}{c} \mapsto \overset{F}{c} \mapsto \overset{C_3[\rho_1]}{2} \mapsto \overset{C_4[\rho_2]}{0} \mapsto \overset{C_1[\rho_3]}{1} \mapsto \overset{F^{-1}}{3} \mapsto \overset{\tau}{d} \mapsto \overset{F}{b} \mapsto \overset{C_1[\rho_3^{-1}]}{1} \mapsto \overset{C_4[\rho_2^{-1}]}{5} \mapsto \overset{C_3[\rho_1^{-1}]}{4} \mapsto \overset{F^{-1}}{4} \mapsto \overset{\beta}{e} \mapsto e$$

We have thus recovered the original plaintext message *ebbe*. Note that the transformations of letters/numbers in the four decryptions are the same as those in the corresponding encryptions, but in the reverse order.

Exercise for the Reader 3.10

Find the ciphertext if the German plaintext *beef* is encrypted with the scaled-down Enigma machine of Example 3.8, but with initial rotor settings: Rotor 1: 0, Rotor 2: 2, Rotor 3: 5.

Special Properties of All Enigma Machines

We close this chapter with the fulfillment of our promise to show that decryptions from any Enigma machine can be accomplished by entering the ciphertext into the machine (using the same initial settings). This was witnessed in Example 3.8, as was the second interesting property included in the following.

Proposition 3.3: General Properties of Enigma Machines

(a) *Enigmas are self-decrypting.* Suppose that γ is a ciphertext character (i.e., a letter in the range A through Z) that lights up on an Enigma machine (with any number of rotors and any number of plugboard cables) when a certain plaintext character π is entered on the keyboard. If the same rotor (and plugboard) settings are used, and if γ is entered on the keyboard, then π will light up on the lampboard.

(b) *Enigmas never preserve any plaintext letters.* If a plaintext letter π is entered into an Enigma keyboard, and γ is the corresponding ciphertext character that lights up on the lampboard, then $\pi \neq \gamma$.

At first glance property (b) may seem like an added security feature, but it is actually helpful to codebreakers since it allows them to easily eliminate one possibility (out of 26) for each intercepted ciphertext letter. We will prove part (a) and leave the similar proof of part (b) as an exercise (Exercise 33).

Proof of Proposition 3.3(a): We will use the function notation established in Example 3.8. If the machine has N rotors (any positive integer), the ith rotor will have an associated permutation ρ_i and a particular setting j_i, where j_i is a modular integer in the relevant base (26 for an original Enigma machine, 6 for the scaled-down version of Example 3.8). The action of these N rotors (in moving toward the reflector plate) can be combined into the following function:

$$\rho = C_{j_N}[\rho_N] \circ C_{j_{N-1}}[\rho_{N-1}] \circ \cdots \circ C_{j_2}[\rho_2] \circ C_{j_1}[\rho_1]$$

The entire Enigma encryption function (with these specific rotor settings) can now be expressed as

$$E = \beta \circ F^{-1} \circ \rho^{-1} \circ F \circ \tau \circ F^{-1} \circ \rho \circ F \circ \beta$$

We need to show that this mapping is its own inverse, that is, that $E \circ E = id$ (the identity mapping). From Proposition 3.1, we know that any two-cycle has this property. Since β, τ are products of disjoint two-cycles, they too must have this property: $\beta \circ \beta = \tau \circ \tau = id$. Using these facts, together with the fact that any invertible function ϕ satisfies $\phi \circ \phi^{-1} = id = \phi^{-1} \circ \phi$, we obtain

$$E \circ E = \beta \circ F^{-1} \circ \rho^{-1} \circ F \circ \tau \circ F^{-1} \circ \rho \circ F \circ \beta \circ \beta \circ F^{-1} \circ \rho^{-1} \circ F \circ \tau \circ F^{-1}$$

$$\circ \rho \circ F \circ \beta$$

$$= \beta \circ F^{-1} \circ \rho^{-1} \circ F \circ \tau \circ F^{-1} \circ \rho \circ F \circ id \circ F^{-1} \circ \rho^{-1} \circ F \circ \tau \circ F^{-1}$$

$$\circ \rho \circ F \circ \beta$$

$$= \beta \circ F^{-1} \circ \rho^{-1} \circ F \circ \tau \circ \tau \circ F^{-1} \circ \rho \circ F \circ \beta$$

$$= \beta \circ \beta$$

$$= id \quad \square$$

Chapter 3 Exercises

1. Use the affine cipher $\phi_{\alpha,\beta} : \mathbb{Z}_{26} \to \mathbb{Z}_{26} :: \phi_{\alpha,\beta}(x) \equiv \alpha x + \beta$ (mod 26) with key $(\alpha, \beta) = (3, 3)$ (and the letter/integer correspondence of Table 3.1) to encrypt the following messages:
    (a) The laundromat is a front
    (b) Khinchine is on Noriega's payroll
    (c) Offer them all immunity
    (d) The money is under the flooring

2. Use the affine cipher $\phi_{\alpha,\beta} : \mathbb{Z}_{26} \to \mathbb{Z}_{26} :: \phi_{\alpha,\beta}(x) \equiv \alpha x +$ (mod 26) with key $(\alpha, \beta) = (17, 8)$ (and the letter/integer correspondence of Table 3.1) to encrypt the following messages:
    (a) Remain in the safehouse
    (b) The meeting place has been infiltrated
    (c) The container is full of military weapons
    (d) You can trust the vice president

3. The affine cipher with key $(\alpha, \beta) = (11, 22)$ was used to encrypt each of the following strings of the ciphertexts. Decrypt each of them.
    (a) AUIYIMXOPGXHOZUBOVUNXWBBGTOM
    (b) EOWBWBODMVGBXWJDSWF
    (c) NUUCZUBWXWNNSVGJOMOEUYWJ
    (d) XUKWGJWSSOMMIMOXVOEUBDSNOTONWJD

4. The affine cipher with key $(\alpha, \beta) = (7, 13)$ was used to encrypt each of the following strings of the ciphertexts. Decrypt each of them.
    (a) DPQQHQKPWXABQRHAHAPKHXCMNQP
    (b) MHBNQPNDPAQQLHGPCHGPCH
    (c) BKPBFQKNQQKPIHBXTPAQJNCPBHTOMPQP
         NAINXQKPAQRB
    (d) HWWPCIHXUMPONZTPAQWHCTHCPRAWHCT
         NQRHA

    Note: The scytale system can be illustrated by paper and pencil as follows: Given a plaintext message—say "Today you need to meet with me"—we choose a number of rows and columns for a table in which to insert the plaintext characters in reading order.

t	o	d	a	y	y
o	u	n	e	e	d
t	o	m	e	e	t
w	i	t	h	m	e

    Additional redundant letters (or nulls) may be inserted to completely fill out the grid. The ciphertext message is then obtained by forming the string of characters by proceeding down the columns, then moving from left to right: TOTWOUOIDNMTAEEHYEEMYDTE.

5. (a) Using the scytale scheme of the preceding note, give the ciphertext of the plaintext message "The assassination will proceed," using a three-row, eight-column grid.
    (b) Decrypt the following message that was encrypted with a scytale scheme: MTPEORSDRONEEOED.

6. (a) Using the scytale scheme of the preceding note, give the ciphertext that corresponds to the plaintext message "The emperor will dine with Agent Smith" using a four-row, eight-column grid.
    (b) Decrypt the following message that was encrypted with a scytale system:
         AOERRRCJRAOEASVNNEEKGCRIEUFNFROS.

7. In each part, a ciphertext is given, along with a portion of the plaintext. An affine cipher of the form $\phi_{\alpha,\beta} : \mathbb{Z}_{26} \rightarrow \mathbb{Z}_{26} :: \phi_{\alpha,\beta}(x) \equiv \alpha x + \beta \pmod{26}$ was used for each encryption. Indicate whether the key can be uniquely determined mathematically from the given information. Explain why this is or is not possible; in cases where it is possible, perform the decryption.
    (a) Ciphertext: FNZMTPFSEXXWKFII,
         Plaintext: ***r*c** ***t****
    (b) Ciphertext: QMBUOBPXMDDYL,
         Plaintext: **ns****** ***
    (c) Ciphertext: ZYJWEXLWMYJIFMJMNDW,
         Plaintext: **an ******a********
    (d) Ciphertext: XCPPWPIKWJHTWOREPSNIJIK,
         Plaintext: *er***********b********

8. In each part, a ciphertext is given, along with a portion of the plaintext. An affine cipher of the form $\phi_{\alpha,\beta} : \mathbb{Z}_{26} \to \mathbb{Z}_{26} ::$ $\phi_{\alpha,\beta}(x) \equiv \alpha x + \beta \,(\text{mod } 26)$ was used for each encryption. Indicate whether the key can be uniquely determined mathematically from the given information. Explain why this is or is not possible; in cases where it is possible, perform the decryption.

   (a) Ciphertext: XCLQYLKQCLSYJK, Plaintext: *ar****** **e**

   (b) Ciphertext: XMIXGVWQEGNYQERIWUE, Plaintext: *****a **s***e*****

   (c) Ciphertext: YCCBAROEBYAJRAOHB, Plaintext: **e** ***t*******h*

   (d) Ciphertext: NKLXUNPDOFSUNPNXSWLSO, Plaintext: av ******t************

9. Consider the affine mapping $\phi : \mathbb{Z}_{29} \to \mathbb{Z}_{29}$ defined by $\phi(x) = 6x + 12\,(\text{mod } 29)$. We use this mapping to encrypt plaintext letters a through z by applying this mapping to their integer equivalents of Table 3.1. The corresponding ciphertext letters will be the 26 images $\phi(0)$, $\phi(1)$, $\cdots$, $\phi(25)$ (mod 29). The remaining 3 (= 29 − 26) integers mod 29 will be nulls. A ciphertext is obtained by a plaintext by applying the mapping ϕ to (the integer corresponding to) each letter of plaintext and punctuating the ciphertext images with nulls.

   (a) Find all nulls.

   (b) Give two different encryptions of the plaintext *stop* using this cipher.

   (c) Decrypt the ciphertext [10, 12, 14, 7, 6, 10, 25, 7, 10, 27, 12, 6, 23, 2, 3, 10, 9, 19, 27, 7, 3, 9, 6, 18, 20, 7].

10. Consider the affine mapping $\phi : \mathbb{Z}_{34} \to \mathbb{Z}_{34}$ defined by $\phi(x) = 15x + 7\,(\text{mod } 34)$. We use this mapping to encrypt plaintext letters a through z by applying this mapping to their integer equivalents of Table 3.1. The corresponding ciphertext letters will be the 26 images $\phi(0)$, $\phi(1)$, $\cdots$, $\phi(25)$ (mod 34). The remaining 8 (= 34 − 26) integers mod 34 will be nulls. A ciphertext is obtained by a plaintext by applying the mapping ϕ to (the integer corresponding to) each letter of plaintext and punctuating the ciphertext images with nulls.

   (a) Find all nulls.

   (b) Give two different encryptions of the plaintext *stop* using this cipher.

   (c) Decrypt the ciphertext [30, 2, 7, 27, 2, 13, 31, 26, 20, 11, 25, 2, 2, 14, 24, 25, 18, 7, 27].

11. Let $\sigma, \tau, \omega : \{1, 2, \cdots, 7\} \to \{1, 2, \cdots, 7\}$ be the permutations specified in tabular forms by

$$\sigma : \begin{pmatrix} 1 & 2 & 3 & 4 & 5 & 6 & 7 \\ 4 & 6 & 2 & 3 & 5 & 7 & 1 \end{pmatrix}, \quad \tau : \begin{pmatrix} 1 & 2 & 3 & 4 & 5 & 6 & 7 \\ 3 & 4 & 5 & 6 & 7 & 1 & 2 \end{pmatrix}, \text{ and}$$

$$\omega : \begin{pmatrix} 1 & 2 & 3 & 4 & 5 & 6 & 7 \\ 5 & 1 & 3 & 6 & 7 & 4 & 2 \end{pmatrix}.$$

Compute the tabular form representations for the following related permutations:

(a) $\sigma \circ \tau$

(b) $(\sigma \circ \tau) \circ \omega$

(c) $\tau \circ \omega$

(d) $\sigma \circ (\tau \circ \omega)$

After you complete this exercise, read Exercise 28.

12. Repeat the directions of Exercise 11 for each of the following permutations:

(a) σ^{-1}

(b) τ^{-1}

(c) $(\sigma \circ \tau)^{-1}$

(d) $\tau^{-1} \circ \sigma^{-1}$

Compare with the result of Exercise for the Reader 3.7.

13. Express each of the following permutations on $\{1, 2, \cdots, 10\}$ as a product of disjoint cycles:

(a) $\begin{pmatrix} 1 & 2 & 3 & 4 & 5 & 6 & 7 & 8 & 9 & 10 \\ 1 & 3 & 4 & 5 & 2 & 9 & 8 & 7 & 6 & 10 \end{pmatrix}$

(b) $\begin{pmatrix} 1 & 2 & 3 & 4 & 5 & 6 & 7 & 8 & 9 & 10 \\ 10 & 9 & 8 & 7 & 6 & 5 & 4 & 3 & 2 & 1 \end{pmatrix}$

(c) $(1, 2, 5) \circ (7, 4, 2)$

(d) $(1, 2) \circ (2, 3) \circ (3, 4)$

(e) $(1, 2, 5) \circ (7, 4, 2)^{-1}$

(f) $(1, 2) \circ (2, 3, 5, 9) \circ (1, 2)^{-1}$

14. Express each of the following permutations on $\{1, 2, \cdots, 10\}$ as a product of disjoint cycles:

(a) $\begin{pmatrix} 1 & 2 & 3 & 4 & 5 & 6 & 7 & 8 & 9 & 10 \\ 8 & 5 & 9 & 7 & 2 & 4 & 1 & 3 & 6 & 10 \end{pmatrix}$

(b) $\begin{pmatrix} 1 & 2 & 3 & 4 & 5 & 6 & 7 & 8 & 9 & 10 \\ 1 & 3 & 5 & 7 & 9 & 2 & 4 & 6 & 8 & 10 \end{pmatrix}$

(c) $(10, 8, 5) \circ (6, 4, 5, 8)$

(d) $(2, 1) \circ (3, 2) \circ (4, 3) \circ (5, 4)$

(e) $(10, 9, 6)^{-1} \circ (5, 4, 6)$

(f) $(1, 2) \circ (2, 4, 1, 10) \circ (1, 2)^{-1}$

15. Explain whether the Enigma machine is a transposition cipher, a substitution cipher, or both.

16. (a) Explain why a brute-force approach to a ciphertext-only attack on the Jefferson wheel cipher would be more time consuming than a similar attack on the original (three-rotor,

six-plugboard cable) German Enigma machine, assuming the attacker possessed both devices.

(b) Given a *crib*, a matching pair of sequences of consecutive plaintext/ciphertext characters that were obtained from a Jefferson wheel cipher encryption, explain how this could be used to reduce the number of possible wheel permutations. Explain further that if the crib has a sufficient number of characters, then the key could be completely determined.

17. (a) Find the ciphertext if the plaintext *aaaa* is encrypted with the scaled-down Enigma machine of Example 3.8, with initial rotor settings: Rotor 1: 5, Rotor 2: 4, Rotor 3: 1.

(b) Find the ciphertext if the plaintext *aabbcc* is encrypted with the scaled-down Enigma machine of Example 3.8, with initial rotor settings: Rotor 1: 4, Rotor 2: 5, Rotor 3: 4.

(c) Find the ciphertext if the plaintext *eeddccbbaa* is encrypted with the scaled-down Enigma machine of Example 3.8, with initial rotor settings: Rotor 1: 5, Rotor 2: 4, Rotor 3: 5.

(d) Check that the ciphertexts that resulted in parts (a) through (c) decrypt to the original plaintexts if they are fed back into the scaled-down Enigma machine with the same initial rotor settings.

18. (a) Find the ciphertext if the plaintext *abcd* is encrypted with the scaled-down Enigma machine of Example 3.8, but with initial rotor settings: Rotor 1: 4, Rotor 2: 0, Rotor 3: 2.

(b) Find the ciphertext if the plaintext *abcdef* is encrypted with the scaled-down Enigma machine of Example 3.8, with initial rotor settings: Rotor 1: 3, Rotor 2: 3, Rotor 3: 3.

(c) Find the ciphertext if the plaintext *aabbccddee* is encrypted with the scaled-down Enigma machine of Example 3.8, with initial rotor settings: Rotor 1: 5, Rotor 2: 4, Rotor 3: 5.

(d) Check that the ciphertexts that resulted in parts (a) through (c) decrypt to the original plaintexts if they are fed back into the scaled-down Enigma machine with the same initial rotor settings.

19. Suppose that $\phi_{\alpha,\beta} : \mathbb{Z}_m \to \mathbb{Z}_m :: \phi_{\alpha,\beta}(x) \equiv \alpha x + \beta \pmod{m}$ is an invertible affine mapping (so that $\gcd(\alpha, m) = 1$). Show that $\phi_{\alpha,\beta}^{-1} : \mathbb{Z}_m \to \mathbb{Z}_m$ is the affine mapping $\phi_{\alpha^{-1}, -\alpha^{-1}\beta}$.

20. Suppose that $\phi_{\alpha,\beta} : \mathbb{Z}_m \to \mathbb{Z}_m :: \phi_{\alpha,\beta}(x) \equiv \alpha x + \beta \pmod{m}$ and $\phi_{\gamma,\delta} : \mathbb{Z}_m \to \mathbb{Z}_m :: \phi_{\gamma,\delta}(x) \equiv \gamma x + \delta \pmod{m}$ are affine mappings. Show that the composition $\phi_{\alpha,\beta} \circ \phi_{\gamma,\delta}$ is the affine mapping $\phi_{\alpha\gamma,\alpha\delta+\beta}$.

21. Suppose that we construct a cryptosystem consisting of an affine cipher, followed by a shift, and then followed by another affine cipher. Thus, the keyspace would have five parameters: $\alpha, \beta, \kappa, \alpha', \beta'$. Explain how much additional security such a system would have over a single affine cipher.

22. *A Variable Ciphertext Space Affine-Based Cipher.* Let $p > 26$ be any prime number. Consider the following affine mapping based cryptosystem: The plaintext space is the set of finite strings in the 26-letter lowercase English alphabet. The ciphertext space is the set of finite length vectors of integers mod p. The keyspace $\mathcal{K}$ consists of all triples (α, β, p), where $p > 26$ is a prime integer, and α, β are integers mod p, where $\alpha \not\equiv 0$. Let f denote the standard bijection from the plaintext alphabet (a through z) to $\mathbb{Z}_{26}$ as specified by Table 3.1, and let $\phi_{\alpha, \beta} : \mathbb{Z}_p \to \mathbb{Z}_p$ be the affine mapping defined by $\phi_{\alpha, \beta}(x) \equiv \alpha x + \beta \pmod{p}$. Note that since p is prime, $\alpha \not\equiv 0$ implies that α is relatively prime to p, so $\phi_{\alpha, \beta}$ is invertible. The encryption mapping is the composition $\phi_{\alpha, \beta} \circ f$, and decryption is accomplished using the composition $f^{-1} \circ \phi_{\alpha, \beta}^{-1}$.

   (a) Explain why, although the above system does define encryption/decryption processes (that are inverses of each other), it does not quite fall under the umbrella of Definition 3.1 of a cryptosystem.

   (b) Explain why the composition $f^{-1} \circ \phi_{\alpha, \beta}^{-1}$ is not properly defined as a mapping from $\mathbb{Z}_p$ to the plaintext alphabet (a through z), but when restricted to actual ciphertext elements of $\mathbb{Z}_p$, the mapping is indeed well defined.

   (c) Use this cryptosystem with parameters $p = 97$ and key $(\alpha, \beta) = (44, 22)$ to encrypt the plaintext message "Beware of ambush."

   (d) Decrypt the following ciphertext that was created with the cipher in part (c): [22, 73, 56, 83, 57, 82, 39, 4, 13, 83, 82, 11, 56, 48, 74, 83, 4, 73].

   (e) How many keys does this cryptosystem have corresponding to a given prime p? (The answer will depend on p.)

23. *A Variable Ciphertext Space Affine-Based Cipher.*
   (a) Use the cryptosystem of the preceding exercise with parameters $p = 53$ and key $(\alpha, \beta) = (44, 22)$ to encrypt the plaintext message "Beware of ambush."

   (b) Decrypt the following ciphertext that was created with the cipher in part (a): [10, 12, 39, 20, 39, 39, 10, 3, 11, 21, 3, 19, 19, 39, 10, 1, 46].

24. *A Variable Ciphertext Space Affine-Based Cipher.* Compare the security of the cryptosystem of Exercise 22 against frequency analyses (in a ciphertext-only attack) with that of the standard affine cryptosystem (Definition 3.2).

25. *A Variable Ciphertext Space Affine-Based Cipher.* Explain whether it would be feasible to perform a known plaintext attack on the cryptosystem of Exercise 22. In particular, how many plaintext characters would one need to encrypt to determine the key? How would things change if the prime p were known?

26. *A Digraph Block Cryptosystem Involving Affine Mappings.* The parts of this exercise will develop a block cryptosystem that processes two-character plaintext blocks using a single bijective affine function.

(a) Let $\mathcal{D}$ denote the set of all $26^2 = 676$ digraphs (pairs of letters). Let f denote the standard bijection from the plaintext alphabet (a through z) to $\mathbb{Z}_{26}$ as specified by Table 3.1. Define a mapping $F : \mathcal{D} \to \mathbb{Z}_{676}$ by $F(\alpha\beta) = 26 f(\alpha) + f(\beta)$, where α, β denote any plaintext alphabet letters. For example, $F(\mathrm{be}) = 26 f(\mathrm{b}) + f(\mathrm{e}) = 26 \cdot 1 + 4 = 30$. Compute $F(\mathrm{th}), F(\mathrm{st})$, and $F(\mathrm{ts})$.

(b) Show that the mapping F of part (a) is a bijection, and find a formula (or rule) for the inverse function $F^{-1} : \mathbb{Z}_{676} \to \mathcal{D}$. Compute $F^{-1}(0), F^{-1}(145)$, and $F^{-1}(495)$.

Using this function F, it is now easy to develop a cryptosystem as follows: The plaintext space consists of all finite strings of lowercase letters, the ciphertext space is the set of all even-length strings of uppercase letters. The keyspace $\mathcal{K}$ consists of all pairs (α, β) of integers mod 676, where $\gcd(\alpha, 676) = 1$. For a given key, we let $\phi_{\alpha,\beta} : \mathbb{Z}_{676} \to \mathbb{Z}_{676} :: \phi_{\alpha,\beta}(x) \equiv \alpha x + \beta \pmod{676}$ denote the corresponding invertible affine mapping. A plaintext is encrypted by applying the composition $F^{-1} \circ \phi_{\alpha,\beta} \circ F$ to successive pairs of plaintext characters to produce corresponding pairs of ciphertext. In case the plaintext string has an odd length, we insert an additional letter x at the end (which should cause little confusion).

(c) Explain the decryption process.

(d) Use this system with key $(\alpha, \beta) = (99, 362)$ to encrypt the plaintext message "reconnaissance."

(e) Go through the decryption process to recover the plaintext if the cryptosystem of part (d) has produced the ciphertext VOZYMEHQQMTLXPZSBI.

(f) How could a chosen ciphertext attack on this system be accomplished?

27. *A Digraph Block Cryptosystem Involving Affine Mappings.*
(a) Use this system with key $(\alpha, \beta) = (465, 182)$ to encrypt the plaintext message "reconnaissance."
(b) Go through the decryption process to recover the plaintext if the cryptosystem of part (a) has produced the ciphertext GPQUZAVOXFDHVCVKNORY.
(c) Is this cryptosystem a substitution cipher? Explain.
(d) How could a chosen ciphertext attack be organized?
(e) Suppose two pairs of corresponding plaintext/ciphertext digraphs are known. Discuss the circumstances under which the key could be mathematically determined from this information.

28. *Composition of Functions Is Associative.* Suppose that $f : A \to B$, $g : B \to C$ and $h : C \to D$ are functions. Prove that $h \circ (g \circ f) = (h \circ g) \circ f$.

29. *Orders of Permutations.* Suppose that n is a positive integer and that σ is a permutation on $\{1, 2, \cdots, n\}$. The *order* of the permutation σ is the smallest positive integer k, such that the k-fold composition σ^k is the identity permutation. It may not be immediately clear that such a positive integer exists, but this exercise will show more.

    (a) Show σ in ℓ-cycle $\sigma = (a_1, a_2, \cdots, a_\ell)$ has order ℓ.

    (b) Show that if the orders of the disjoint cycles in the cycle decomposition of σ are $\ell_1, \ell_2, \cdots, \ell_t$, then the order of σ is $\mathrm{lcm}(\ell_1, \ell_2, \cdots, \ell_t)$.

    (c) Show that the order of σ must divide $n!$.

30. *Original Enigma Machine.* One of the original 3-rotor German Enigma machines was characterized by the following permutations (presented in terms of disjoint cycle decompositions):

    Rotor1Perm = (aeltphqxru) (bknw) (cmoy) (dfg) (iv) (jz) (s)
    Rotor2Perm = (fixvyomw) (cdklhup) (esz) (bj) (gr) (nt) (a) (q)
    Rotor3Perm = (abdhpejt) (cflvmzoyqirwukxsg) (n)
    ReflectorPerm = (ay) (br) (cu) (dh) (eq) (fs) (gl) (ip) (jx) (kn) (mo) (tz) (vw)

    (a) Find the ciphertext if the German plaintext *alle* is encrypted with this Enigma machine, with the following initial settings: The rotors are inserted in standard (unpermuted) order; their settings are Rotor 1: 25, Rotor 2: 18, Rotor 3: 3; and the plugboard cables are (a,f), (c,z), (g,u), (h,i), and (s,t).

    (b) Check that the ciphertext that resulted in part (a) decrypts to the original plaintext if it is fed back into this Enigma machine with the same initial rotor/plugboard settings.

31. *Original Enigma Machine.*
    (a) Find the ciphertext if the German plaintext *auch* is encrypted with the Enigma machine described in the previous exercise, with the following initial settings: The rotors are inserted in standard (unpermuted) order; their settings are Rotor 1: 24, Rotor 2: 0, Rotor 3: 13; and the plugboard cables are (a,f) (c,z), (g,u), (h,i), and (s,t).

    (b) Check that the ciphertext that resulted in part (a) decrypts to the original plaintext if it is fed back into this Enigma machine with the same initial rotor/plugboard settings.

32. Prove Proposition 3.2.

33. Prove Proposition 3.3(b).

34. Suppose that a general Enigma machine (as described in this chapter) was allowed to have two plugboards P1 and P2 rather than one, where the signals exiting the keyboard pass through P1 and P2 in this order, and signals exiting from the rotors pass through them in opposite order (P2, then P1) before entering the lampboard. Which, if any, of the two properties of Proposition 3.3 will continue to hold for such a modified machine? Explain your answer.

35. Suppose that a general Enigma machine as described in this chapter was allowed one of the following modifications on the reflector permutation:

(a) The permutation still consists of disjoint two-cycles, but is allowed to have fewer than 13 of them in the disjoint cycle decomposition.

(b) The permutation is allowed to be a general permutation on the 26 letters (so it need not be expressible as a product of disjoint two-cycles).

For each of the two modifications, indicate which, if any, of the two properties of Proposition 3.3 will continue to hold for such a modified machine and explain.

36. Suppose that a general Enigma machine as described in this chapter was allowed to have its reflector use any permutation of the 26 letters (so it need not be expressible as a product of disjoint two-cycles). Explain any resulting changes needed in the decryption process.

37. *Counting Plugboard Arrangements.*
    (a) For the six-socket plugboard of the scaled-down Enigma machine of Example 3.8, in how many ways could a single plugboard cable be installed? How many different ways could two plugboard cables be installed? How about three?
    (b) In the 26-socket plugboard of an Enigma machine, in how many ways could a single plugboard cable be installed? How many different ways could two plugboard cables be installed? How about three?
    (c) Suppose we have a plugboard with an even number n of labeled sockets (for an alphabet with n characters), and $k \leq n/2$ is a positive integer. Show that the number of ways that k plugboard cables can be installed is
    $$\frac{n!}{(n-2k)!\,k!\,2^k}.$$

    Suggestion: Use the combination numbers $C(n, j) = \frac{n!}{j!\cdot(n-j)!}$, which give the number of ways that j objects can be selected from a set of n objects. If we first think of the cables having different colors, we can count, for example, the number of ways to install a red, a green, and a blue plugboard cable into a 26-socket plugboard as follows: First there are $C(26, 2)$ ways to install the red cable, after this is done, 24 sockets remain so there are $C(24, 2)$ ways to install the green cable, and subsequently there will be $C(22, 2)$ ways to install the blue cable. By the multiplication principle, this gives a total of $C(26, 2)\cdot C(24, 2)\cdot C(22, 2)$ ways of installing these three different cables. This number simplifies to $\frac{26!}{20!\cdot 2^3}$. But since the cables are not colored (they are indistinguishable), this figure needs to be corrected by dividing by 3! (the number of permutations of the three cables).

38. *Every Permutation Can Be Decomposed into 2-Cycles.*
    (a) Express the three-cycle $(1, 2, 3)$ as a composition of two 2-cycles.

(b) Express a general 3-cycle (a, b, c) as a composition of two 2-cycles.

(c) Express a general 4-cycle (a, b, c, d) as a composition of three 2-cycles.

(d) Show how to express a general n-cycle $(a_1, a_2, \cdots, a_n)$ as a composition of $(n-1)$ 2-cycles.

(e) Prove that any permutation can be expressed as a composition of 2-cycles.

Chapter 3 Computer Implementations and Exercises

Note: As in Chapter 1, depending on which data structures are available or feasible on their computing platforms, readers may wish to accordingly modify the details of some of the data types in the program exercises below. For example, vectors may be replaced by strings, or vice versa, and alphabetic letters can be replaced with their integer equivalents (according to Table 3.1).

1. *Program for Affine Cipher.* Write a program StrOut = AffineCrypt(str,alpha,beta) that inputs a string str of plaintext in lowercase English letters, a multiplier alpha, which is relatively prime to 26, and a shift parameter beta, both integers mod 26. The output StrOut should be the corresponding ciphertext (in uppercase letters) after the affine operator with key (alpha, beta) is applied to the plaintext. Then use your program to redo the computations of (ordinary) Exercise 1.

   Note: In the decrypting parts, you will need to change your ciphertexts to lowercase (and choose the correct key).

   Suggestion: The programs of some of the computer exercises from Chapter 1 should be useful here.

2. *Ciphertext-Only Attack on the Affine Cipher.* It is known that the following ciphertexts were encrypted using (perhaps different) affine ciphers. Decrypt these messages, and determine the corresponding keys that were used.
   (a) VUJWLHLJLWHOPE
   (b) AGBCDHHFJOIAJAGBPJQAOJEOJJXZHSTNNBO QDTNGA
   (c) CTPDBCIXPFWTXSTRFWWOZPXMSZMXELOTWZRW MZZWFSUETMRWFGZSBZ
   (d) CNLQVAMVFLDFMMVAAFSNVWLNSNIEX

3. *Program for the Scytale.* Write a program that implements the scytale system according to the note that precedes ordinary Exercise 5. The syntax should be as follows: Ctext = Scytale Ptext, rows, cols, where the input variables are: Ptext, a string of lowercase letters representing the plaintext, and two positive integers rows and cols representing the number of rows and columns in a scytale grid (as in the explanatory note). The size of the grid (i.e., the number of rows times the number

of columns should be at least as large as the plaintext message. The output Ctext, should be the ciphertext obtained from the plaintext and grid as explained in the note. Run your program on the example done in the note, and then use it to redo part (a) of ordinary Exercises 5 and 6.

Note: Although the problems you are asked to solve with your program are such that the plaintext completely fills up the grid, you should design your program to accommodate cases where this does not occur (to be realistic with the original uses of scytales). For example, additional blank space in a grid could be filled with redundant letters, perhaps selected in some random fashion.

4. *Randomized Program for Affine Cipher Encryptions with Nulls.*
   (a) Write a program for an affine cipher encryption system with nulls that has the following syntax: Ctext = AffineNulls (Ptext, n, alpha, beta, Null FreqVector), where the input variables are Ptext, a string of lowercase letters representing the plaintext, n, a positive integer greater than 26, alpha, (the multiplier) a positive integer relatively prime to n, beta, a nonnegative integer less than n, and NullFreqVector, an optional final input, which is a vector of length $n - 26$ giving the desired frequencies of the nulls (these should be probabilities whose sum is less than 1). The output Ctext, will be a vector of integers mod n corresponding to the ciphertext (with nulls) using the affine map specified by the parameters. The encryption will proceed as suggested in the solution of part (d) of Example 3.4. Plaintext characters (after being converted to integers mod 26) get transformed by the affine map $\phi: \mathbb{Z}_n \to \mathbb{Z}_n :: \phi(x) = \alpha x + \beta$ into their ciphertext equivalents. Nulls $\phi(26)$, $\phi(27)$, $\cdots$, $\phi(n-1)$ get randomly inserted in the ciphertext according to the inputted NullFreqVector. In case this vector is not specified, the nulls should be randomly selected at each insertion, each with probability $(n - 26)/n$.
   (b) Run your program with n = 31, alpha = 10, beta = 5, and NullFreqVector as recommended in Example 3.4 to encrypt the plaintext message: "hostages have been taken, proceed with caution."
   (c) Use your program of part (a) with parameter settings: n = 37, alpha = 16, beta = 22, and no specified NullFreq Vector to encrypt the plaintext message: "sendnavalves-selsandsubmarines." Compare the number of nulls that were expected to be inserted with the number of nulls actually inserted.

5. *Decryption Program for Affine Cipher With Nulls.*
   (a) Write a program for decryption of an affine cipher system with nulls that has the following syntax: Ptext = AffineNullsDecrypt (Ctext, n, alpha, beta), where the input variables are: n, a positive integer greater than 26, Ctext, a vector of integers mod n, alpha, (the

multiplier) a positive integer relatively prime to n, beta, a nonnegative integer less than n, The output Ptext, will be a string of lowercase letters representing the corresponding plaintext. The decryption will proceed as explained in the section: Ciphertext characters get transformed by the inverse of affine map $\phi: \mathbb{Z}_n \to \mathbb{Z}_n :: \phi(x) = \alpha x + \beta$ into their plaintext integer equivalents (then converted into their corresponding letters). The nulls $\phi(26)$, $\phi(27)$, $\cdots$, $\phi(n-1)$ get ignored.

(b) Run your program with n = 31, alpha = 10, beta = 5, and NullFreqVector as recommended in Example 3.4 to decrypt the ciphertext message [27, 16, 20, 14, 22, 14, 27, 5, 30, 14, 0, 20, 23, 30, 21, 11, 14, 20, 30].

(c) Use the program to decrypt the ciphertext message obtained in part (c) of the preceeding computer exercise.

6. *Encryption/Decryption Programs for an Affine Cipher with 100 Homophones.*

(a) Write a program of an affine cipher encryption system with 100 homophones that has the following syntax: Ctext = AffineHomophones100 (Ptext, alpha, beta), where the input variables are: Ptext, a string of lower-case letters representing the plaintext; alpha (the multiplier), a positive integer relatively prime to 100; and beta, a nonnegative integer less than 100. The 100 homophones for the 26 lower case plaintext letters should be identified with elements of $\mathbb{Z}_{100}$ in a way that will make frequencies of each homophone to rather similar, as was done in Example 3.5, by referring to the English language frequency table (Table 1.1).[*] For each plaintext character, the program should randomly select one of the plaintext integer homophones for it, and then the affine map $\phi: \mathbb{Z}_{100} \to \mathbb{Z}_{100} :: \phi(x) = \alpha x + \beta$ is used to convert it to its corresponding ciphertext homophone. The output Ctext will be the vector of these ciphertext integers.

(b) Run your program with alpha = 21, and beta = 25, to encrypt the plaintext message: "hostages have been taken, proceed with caution."

(c) Write a program for decryption of ciphertexts produced by your program of part (a) that has the following syntax: Ptext = AffineHomophones100Decrypt (Ctext, alpha, beta), where the input and output variables have the same meaning as they did in the program of part (a).

(d) Run the decryption program of part (c) to decrypt the ciphertext that you produced in part (b).

Suggestions for part (a): As in Example 3.5, let the homophones for *a* be a set of consecutive integers starting at zero, say 0,1,2,3,4,5,6,7 (8 homophones since *a* has a frequency of 8.2%), let the homophones for *b* be a set of consecutive

---

[*] Of course, since Example 3.5 used 1000 homophones, we were able to adjust the homophones to each have the same frequency (according to Table 1.1). With only 100 homophones, such equal expected frequencies will not be possible, but we can nonetheless remove some of the strong distinctions of the original frequencies.

integers starting at 8, the last homophone should be 100 for z. Create a 26 element vector numHomophones giving the number of homophones for each of the letters in order; thus the first element of this vector should be 8 (the number of homophones for a, and the last should be 1 (the number of homophones for z). For each plaintext letter to be encrypted, randomly select an integer between 1 and the number of its homophones (accessed from the last vector) and encrypt this particular homophone.

7. *Encryption/Decryption Programs for Affine Cipher with 1001 Homophones*
   (a) Write a program for an affine cipher encryption system with 1001 homophones that has the following syntax: Ctext = AffineHomophones1001 (Ptext, alpha, beta), where the input variables are Ptext, a string of lowercase letters representing the plaintext; alpha (the multiplier), a positive integer relatively prime to 1001; and beta, a nonnegative integer less than 1001. The 1001 homophones for the 26 lowercase plaintext letters should be identified with elements of $\mathbb{Z}_{1001}$, as was done in Example 3.5, by referring to the English language frequency table (Table 1.1). For each plaintext character, the program should randomly select one of the plaintext integer homophones for it, and then the affine map $\phi: \mathbb{Z}_{1001} \to \mathbb{Z}_{1001} :: \phi(x) = \alpha x + \beta$ is used to convert it to its corresponding ciphertext homophone. The output Ctext, will be the vector of these ciphertex integers.
   (b) Run your program with alpha = 359, and beta = 207, to encrypt the plaintext message: "hostages have been taken, proceed with caution."
   (c) Write a program for decryption of ciphertexts produced by your program of part (a) that has the following syntax: Ptext = AffineHomophones1001Decrypt (Ctext, alpha, beta), where the input and output variables have the same meaning as they did in the program of part (a).
   (d) Run the decryption program of Part (c) to decrypt the ciphertext that you produced in part (b).

   Suggestion: See the suggestion for the previous computer exercise. This program here should be even easier to write since the frequencies can be directly read off Table 1.1.

8. *Encryption/Decryption Programs for Affine Cipher with 1001 Homophones and Nulls*
   (a) Write a program for an affine cipher encryption system with 1001 homophones that has the following syntax: Ctext = AffineHomophones1001WithNulls(n, Ptext, alpha, beta, NullFreqVector), where the input variables are: Ptext, a positive integer greater than 1001, Ptext, a string of lowercase letters representing the plaintext, alpha (the multiplier), a positive integer relatively prime to n; and beta, a nonnegative integer less than n; and NullFreqVector, an optional final input,

which is a vector of length $n - 1001$ giving the desired frequencies of the nulls (these should be probabilities whose sum is less than 1). The output Ctext, will be a vector of integers mod n corresponding to the ciphertext (with nulls) using the affine map specified by the parameters. The program will proceed by combining the ideas for the preceding computer exercises for affine ciphers with nulls and with homophones (see also Examples 3.4 and 3.5). Each character of ciphertext produced will either be a null (determined by generated random number) or a homophone for the next unprocessed plaintext character that will be randomly selected from all associated homophones. The output Ctext, will be the vector of the resulting ciphertex integers.

(b) Run your program with n = 1201 (so with 1001 homophones and 200 nulls), alpha = 359, beta = 207, to encrypt the plaintext message: "hostages have been taken, proceed with caution."

(c) Write a program for decryption of ciphertexts produced by your program of part (a) that has the following syntax: Ptext = AffineHomophones1001WithNulls(n, Ctext, alpha, beta), where the input and output variables have the same meaning as they did in the program of part (a).

(d) Run the decryption program of part (c) to decrypt the ciphertext that you produced in part (b).

Computer Representations of Permutations

Although cycles are convenient for theoretical purposes and hand calculations, for programming purposes it is often most convenient to represent permutations in tabular form, as two-row tables (matrices). This will facilitate easy access and computations with permutations. The next two exercises ask the reader to write programs that will compose a pair of permutations and compute the inverse of a permutation. Although both of these tasks can easily be programmed using FOR loops, more efficient approaches may be possible. For example, many computing platforms have a very useful sort utility that inputs an ordered list (vector) of numbers, and outputs the corresponding ordered list where the numbers have been sorted from smallest to largest, along with a same-sized ordered list of the indices where these sorted elements appeared in the original vector. If we were to apply such a sort utility to the ordered list [5, 3, 1, 2, 6, 4], the sorted list would be [1, 2, 3, 4, 5, 6] and the corresponding index vector would be [3, 4, 2, 6, 1, 5] (for example, the first element of the latter vector is 3 because the first element of the sorted vector 1 appeared as the third entry of the original vector). Such a utility could easily be applied to build the tabular form for the inverse of a permutation. For example, to invert the permutation given by

$$\begin{pmatrix} 1 & 2 & 3 & 4 & 5 & 6 \\ 5 & 3 & 1 & 2 & 6 & 4 \end{pmatrix}$$

we could simply apply the SORT utility to the second row of this table, [5, 3, 1, 2, 6, 4], and transplant the resulting index vector [3, 4, 2, 6, 1, 5] to be the second row of the table to arrive at the tabular form for the inverse permutation:

$$
\begin{pmatrix}
1 & 2 & 3 & 4 & 5 & 6 \\
3 & 4 & 2 & 6 & 1 & 5
\end{pmatrix}
$$

9. *Program for Computing Compositions of Permutations.*
   (a) Write a program $\mathtt{PermOut = PermComposer(Perm1, Perm2)}$ that inputs two permutations, $\mathtt{Perm1}$, $\mathtt{Perm2}$, on the same set $\{1, 2, \cdots, n\}$, that are stored in tabular form. The output $\mathtt{PermOut}$, will be the corresponding tabular form representation for the composition permutation $\mathtt{Perm1 \circ Perm2}$.
   (b) Use this program to redo ordinary Exercise 11.
   (c) Since 17 is prime, any affine mapping $\phi_{\alpha,\beta} : \mathbb{Z}_{17} \to \mathbb{Z}_{17} :: \phi(x) \equiv \alpha x + \beta \pmod{17}$ (with nonzero multiplier α) is invertible, and so defines a permutation on $\{1, 2, \cdots, 17\}$ (after we identify 0 with 17). Let σ be the permutation on $\{1, 2, \cdots, 17\}$ corresponding to $\phi_{12,5}$, and τ be the permutation on the same set corresponding to $\phi_{8,15}$. Use your program of part (a) to compute the tabular form for the composition permutation $\sigma \circ \tau$, and show that it corresponds to the composition of the associated affine mapping composition $\phi_{12,5} \circ \phi_{8,15}$. Compare with the result of ordinary Exercise 20.

10. *Program for Computing Inverses of Permutations.*
    (a) Write a program $\mathtt{InvPerm = PermInverter(Perm)}$ that inputs a permutation $\mathtt{Perm}$ on a set $\{1, 2, \cdots, n\}$, stored in tabular form. The output $\mathtt{InvPerm}$ will be the corresponding tabular form representation for the inverse permutation $\mathtt{Perm}^{-1}$.
    (b) Use this program to redo ordinary Exercise 11.
    (c) Since 17 is prime, any affine mapping $\phi_{\alpha,\beta} : \mathbb{Z}_{17} \to \mathbb{Z}_{17} :: \phi(x) \equiv \alpha x + \beta \pmod{17}$ (with nonzero multiplier α) is invertible and so defines a permutation on $\{1, 2, \cdots, 17\}$ (after we identify 0 with 17). Let σ be the permutation on $\{1, 2, \cdots, 17\}$ corresponding to $\phi_{12,5}$. Use your program of part (a) to compute the tabular form for the inverse permutation σ^{-1}, and show that it corresponds to the permutation of the associated inverse affine mapping $\phi_{12,5}^{-1}$. Compare with the result of ordinary Exercise 19.

11. *Program for the Scaled-Down Enigma Machine of Example 3.8.*
    (a) Write a program $\mathtt{StrOut = SmallEnigma(str,r1, r2,r3)}$ that inputs a string $\mathtt{str}$ of plaintext in lowercase English letters, and three integers $\mathtt{r1}$, $\mathtt{r2}$, $\mathtt{r3}$ each in the range $\{0,1,2,3,4,5\}$ corresponding to initial rotor settings of the scaled-down Enigma machine of Example 3.8. The output

StrOut should be the corresponding ciphertext (in uppercase letters) after the plaintext `str` is fed into this scaled-down Enigma machine with the indicated initial settings.

(b) Use your program to redo the computations of Example 3.8 and Exercise for the Reader 3.10.

(c) Use your program with initial rotor settings `r1` = 3, `r2` = 1, `r3` = 2 to encrypt the 50-character plaintext consisting of all a's: $aaa \cdots a$. Look at the whole answer, but write down every 10th ciphertext letter (so five letters total).

(d) The following ciphertext was created by the scaled-down Enigma machine of Example 3.8: BDECBFDABCEFC. It is known only that the third rotor setting was 0 when the transmission began. Use your program of part (a) to determine the plaintext, which consisted of four English words (put together in a nonsentence).

Note: In the decrypting parts, you will need to change your ciphertexts to lowercase.

Suggestion: The programs of some of the preceding computer exercises from this set and from Chapter 1 should be useful here; you may wish to invoke the programs of the preceding computer exercise.

12. *Program for the Scaled-Down Enigma Machine of Example 3.8 with Rotor Permutation and Plugboard Setting Capability.*

(a) Write a program StrOut = SmallEnigmaFullSettings (str,r1,r2,r3, PlugPerm,RotorPerm) whose first four inputs are the same as those of the preceding computer exercise, but that has two additional inputs, PlugPerm and RotorPerm, that are permutations (stored in tabular form) representing the plugboard settings and rotor permutation settings, respectively. Thus, PlugPerm will be a permutation of six objects, while RotorPerm will be a permutation of three objects. Although the plugboard was required to be equivalent to a product of disjoint two-cycles (so it used two plugboard cables), here we allow 0, 1, 2, or 3 cables to be used. This poses no additional programming effort and improves security. So as to remove ambiguity, we stipulate that the rotor setting variables `r1`, `r2`, `r3` correspond to the first, second, and third rotors that get applied *after* the rotors have been permuted. The output StrOut should be the corresponding ciphertext (in uppercase letters) after the plaintext `str` is fed into this scaled-down Enigma machine with the indicated rotor and plugboard settings.

(b) Use your program to redo the computations of Example 3.8 and Exercise for the Reader 3.10.

(c) Use your program with initial rotor settings `r1` = 3, `r2` = 1, `r3` = 2, with the rotors permuted according to the permutation (1 3 2), and with no plugboard cables attached to encrypt the 50-character plaintext consisting of all a's: $aaa \cdots a$. Look at the whole answer, but write down every 10th ciphertext letter (so five letters total).

(d) The following ciphertext was created by the scaled-down Enigma machine of Example 3.8 using the rotor settings of part (c): FBBBFDCFDACF. It is known that only one plugboard cable was used. Use your program of part (a) to determine the plaintext, which consisted of a certain English word repeated a certain number of times.

Suggestion: The program of the previous computer exercise can be made to be the core of this one; you simply need to remove the internal definition of the plugboard matrix and define (within the new program) the three rotors according to the given permutations. It is convenient to store the three original rotor permutations into a six-row array (so the first original rotor occupies the first two rows, the second original rotor the third and fourth rows, etc.). For a given rotor number r read off from the rotor permutation matrix, the starting row number in the corresponding six-row array will be $2(r-1)+1$.

13. *Program for the Original Enigma Machine with Rotor Permutation and Plugboard Setting Capability.*

   (a) Write a program `StrOut = OriginalEnigmaFull-Settings (str,r1,r2,r3, PlugPerm,RotorPerm)` with the same input variables, output variable, and functionality of the program of the previous computer exercise, except that the present program will simulate the original Enigma machine that was described in Chapter Exercise 30 (with the standard 26-letter plaintext and ciphertext alphabets, rather than the reduced-size six-letter alphabets of the scaled-down Enigma machine). The rotor and reflector details are as given above. Of course, `str` may now be taken from the full alphabet; `r1`, `r2`, `r3` may be any integers in the range $\{0, 1, 2, \cdots, 25\}$; and `PlugPerm` is now a permutation on the 26 plugs that is required to be equivalent to a product of disjoint two-cycles (so it corresponds to plugboard cables). We allow any number of plugboard cables to be used (from 0 to 13), since this poses no additional programming effort and improves security.

   (b) Use your program to encrypt the plaintext message: "generalrommelplanstosurrender," using the following settings: `r1 = 21`, `r2 = 11`, `r3 = 2`, with the rotors permuted according to the permutation (1 3 2), and with the following four plugboard cables attached: (c,f)(e,h)(g,m)(s,u).

   (c) Use your program with the settings of part (b) to encrypt the 50-character plaintext consisting of all *a*'s: *aaa···a*. Look at the whole answer, but write down every 10th ciphertext letter (so five letters total).

   Suggestion: If you have done the previous computer exercise, this one can be programmed with very little additional effort. Keying in the permutations will take most of the time.

Matrices and the Hill Cryptosystem

A very important data structure is that of a matrix, which is simply a rectangular spreadsheet of numbers. The numbers making up a matrix can lie in any of the number systems that we have thus far developed, as well as new ones that we will develop later. For example, entries can be integers or modular integers. Matrices are extremely convenient for computer applications since they can be easily stored and manipulated. Furthermore, when we restrict attention to matrices of a specific size, arithmetic operations can be defined on them so that the set of such matrices themselves become an abstract number system with a particularly rich structure. This chapter also discusses the Hill cryptosystem, which is based on matrix arithmetic and has the distinction of being one of the first cryptosystems that was based on advanced mathematics. Many modern computer-based cryptosystems rely on matrices and their arithmetic, as well as other advanced mathematics.

The Anatomy of a Matrix

We have already had several encounters with vectors, which are simply ordered lists of numbers. A more general and very natural data structure in computing is that of a **matrix** (plural: **matrices**), which is simply a rectangular array of numbers, see Figure 4.1. Matrices can be thought of simply as data spreadsheets, and these should certainly be familiar to anyone having basic computer experience. A matrix A with n rows and m columns is said to be an $n \times m$ matrix, and these two numbers determine the **dimensions** or the **size** of A. The **element** (or **entry**) a_{ij} of the matrix A is located in the ith row and the jth column (see Figure 4.1) and is called the (i, j) **entry** of A. By default, we will consider the entries of a matrix to be real numbers, but as needs arise, we will want to allow matrix elements to belong to other number systems.[*] For example, later in this chapter, we will be dealing with matrices of modular integers.

---

[*] In computer applications, the elements of a spreadsheet matrix may be any sorts of objects, not necessarily belonging to any number system (for example, elements could be English letters or words). In order to define the matrix operations of addition, subtraction, and multiplication, we need the matrix elements to belong to a number system in which such operations may be performed.

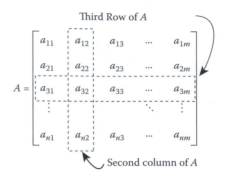

Figure 4.1 The anatomy of a matrix with n rows and m columns. The (3,2) entry that lies in the third row and the second column is denoted a_{32}.

For shorthand, a matrix A is often denoted by

$$A = [a_{ij}]$$

where the dimensions of A are either known from the context or are unimportant. An $n \times 1$ matrix has only one column and is called a **column matrix**. Similarly, a $1 \times m$ matrix is called a **row matrix**. Row and column matrices are simply ordered lists of numbers, otherwise known as **vectors**.[*] An $n \times n$ matrix has an equal number of rows and columns and is called a **square matrix**.

Matrix Addition, Subtraction, and Scalar Multiplication

Definition 4.1   Addition and Subtraction of Matrices

$A = [a_{ij}]$ and $B = [b_{ij}]$ of *the same size* are defined *componentwise*, that is, simply add or subtract corresponding entries. Similarly, if α is a number,[†] then the **scalar multiplication** αA is the matrix obtained from A by multiplying each of its entries by α. These matrix operations are summarized by the following formulas:

$$[a_{ij}] \pm [b_{ij}] = [a_{ij} \pm b_{ij}], \quad \alpha[a_{ij}] = [\alpha a_{ij}]$$

Multiplication of matrices, on the other hand, is not done componentwise. To define this concept, we first need to introduce the **dot product** $u \cdot v$ of two vectors, $u = [u_i]$ and $v = [v_i]$, of the same length.

---

[*] Often when we are dealing with vectors, we are only concerned that the list of numbers is ordered, and it is unimportant whether the vector is written as a row vector or a column vector. Also, vectors are sometimes written with commas inserted between elements; for example, [2 0 3] = [2, 0, 3].

[†] When dealing with vectors or matrices, numbers are sometimes referred to as *scalars*.

If the two vectors have length n, then this dot product $u \bullet v$ of the two vectors is the *number* (or scalar) defined by the formula

$$u \bullet v = [u_i] \bullet [v_i] = u_1 v_1 + u_2 v_2 + \cdots + u_n v_n = \sum_{i=1}^{n} u_i v_i$$

For this definition, we do not require that u and v have the same dimensions, only that they have the same length. For example, it still applies if u is a row matrix and v is a column matrix.

Example 4.1

Given the matrices and vectors

$$A = \begin{bmatrix} 2 & -4 \\ 1 & 6 \end{bmatrix}, \quad B = \begin{bmatrix} 1 & 0 & 9 \\ -4 & -2 & 4 \end{bmatrix}, \quad C = \begin{bmatrix} 3 & 7 \\ 5 & 5 \end{bmatrix},$$

$u = [6, -3, 12]$, $v = [4, 2, -2]$

compute (if possible) each of the following:

(a) $A + B$
(b) $A - 2C$
(c) $u \bullet v$

Solution: Part (a): Since A and B are different-sized matrices (A is 2×2, B is 2×3), their sum is undefined.

Part (b):

$$A - 2C = \begin{bmatrix} 2 & -4 \\ 1 & 6 \end{bmatrix} - 2 \cdot \begin{bmatrix} 3 & 7 \\ 5 & 5 \end{bmatrix} = \begin{bmatrix} 2 & -4 \\ 1 & 6 \end{bmatrix} - \begin{bmatrix} 6 & 14 \\ 10 & 10 \end{bmatrix}$$

$$= \begin{bmatrix} -4 & -18 \\ -9 & -4 \end{bmatrix}$$

Part (c):

$$u \bullet v = [6, -3, 12] \bullet [4, 2, -2] = 6 \bullet 4 - 3 \bullet 2 + 12 \bullet (-2)$$

$$= 24 - 6 - 24 = -6$$

Matrix Multiplication

Definition 4.2

Suppose that $A = [a_{ij}]$ is an $n \times m$ matrix, and $B = [b_{ij}]$ is an $m \times r$ matrix (that is, the number of columns of A is the number of rows

> of B). Then the **matrix product** $C = AB$ is defined as the matrix of
> dimensions $n \times r$, whose (i, j) entry c_{ij} is simply the dot product of the
> ith row of A with the jth column of B; that is,
>
> $$c_{ij} = a_{i1}b_{1j} + a_{i2}b_{2j} + a_{i3}b_{3j} + \cdots + a_{im}b_{mj} = \sum_{k=1}^{m} a_{ik}b_{kj}$$
>
> This procedure is illustrated in Figure 4.2.

At first glance, this definition of matrix multiplication probably seems
very unnatural, especially compared with the componentwise definitions
of matrix addition and subtraction. The reasons for this definition are that
it gives matrices a very rich arithmetic structure that results in a powerful
theory and manifold applications.

Example 4.2   Matrix Multiplication Is Not Commutative

Given the matrices

$$A = \begin{bmatrix} 2 & -4 \\ 1 & 6 \end{bmatrix}, \; B = \begin{bmatrix} 1 & 0 & 9 \\ -4 & -2 & 4 \end{bmatrix}, \; C = \begin{bmatrix} 3 & 7 \\ 5 & 5 \end{bmatrix},$$

compute (if possible) each of the following:

(a) AB and BA

(b) AC and CA

Solution: Part (a):

$$AB = \begin{bmatrix} 2 & -4 \\ 1 & 6 \end{bmatrix} \cdot \begin{bmatrix} 1 & 0 & 9 \\ -4 & -2 & 4 \end{bmatrix} = \begin{bmatrix} 17 & 8 & 2 \\ -23 & -12 & 33 \end{bmatrix}.$$

Each of the six entries of the 2×3 product matrix on the right
was computed as the appropriate dot product. For example, the
boxed (2,3) entry equals the dot product of the second row of
A, with the third column of B (both boxed), and was computed
as $1 \cdot 9 + 6 \cdot 4 = 9 + 24 = 33$.

The matrix BA is undefined since the inside dimensions of
the two matrices do not match.

Part (b): Since both A and C are square matrices of the same
size, both products may be computed:

$$AC = \begin{bmatrix} 2 & -4 \\ 1 & 6 \end{bmatrix} \begin{bmatrix} 3 & 7 \\ 5 & 5 \end{bmatrix} = \begin{bmatrix} -14 & -6 \\ 33 & 37 \end{bmatrix}.$$

$$CA = \begin{bmatrix} 3 & 7 \\ 5 & 5 \end{bmatrix} \begin{bmatrix} 2 & -4 \\ 1 & 6 \end{bmatrix} = \begin{bmatrix} 13 & 30 \\ 15 & 10 \end{bmatrix}.$$

$$\begin{bmatrix} a_{11} & a_{12} & a_{13} & \cdots & a_{1m} \\ a_{21} & a_{22} & a_{23} & \cdots & a_{2m} \\ a_{31} & a_{32} & a_{33} & \cdots & a_{3m} \\ \vdots & & \ddots & & \vdots \\ a_{n1} & a_{n2} & a_{n3} & \cdots & a_{nm} \end{bmatrix} \begin{bmatrix} b_{11} & b_{12} & b_{13} & \cdots & b_{1m} \\ b_{21} & b_{22} & b_{23} & \cdots & b_{2m} \\ b_{31} & b_{32} & b_{33} & \cdots & b_{3m} \\ \vdots & & \ddots & & \vdots \\ b_{m1} & b_{m2} & b_{m3} & \cdots & b_{mr} \end{bmatrix} = \begin{bmatrix} c_{11} & c_{12} & c_{13} & \cdots & c_{1r} \\ c_{21} & c_{22} & c_{23} & \cdots & c_{2r} \\ c_{31} & c_{32} & c_{33} & \cdots & c_{3r} \\ \vdots & & \ddots & & \vdots \\ c_{n1} & c_{n2} & c_{n3} & \cdots & c_{nr} \end{bmatrix}$$

$$n \times m \qquad\qquad m \times r \qquad\qquad n \times r$$

Figure 4.2 Illustration of the computation of an element of the product matrix $AB = C$: To get the (3,2) entry c_{32} of C, take the dot product of the third row of A (the first matrix) with the second column of B (the second matrix).

This example shows that *matrix multiplication is not commutative*. In part (a), we found that $AB \neq BA$, since the latter matrix was not defined. In part (b), even though both products are defined, they were not equal (in fact, each of the four corresponding entries was different). Despite this drawback, matrix arithmetic does possess many of the properties that we know are true for real numbers.

Exercise for the Reader 4.1

Preview of the Fact That Matrix Multiplication Is Associative

Given the matrices

$$A = \begin{bmatrix} 2 & -4 \\ 1 & 6 \end{bmatrix}, \; B = \begin{bmatrix} 8 & 0 \\ -4 & 1 \end{bmatrix}, \; C = \begin{bmatrix} 3 & 7 \\ 5 & 5 \end{bmatrix}$$

compute the two matrix products $(AB)C$ and $A(BC)$.

The associativity property $(AB)C = A(BC)$ that was witnessed in the preceding exercise for the reader turns out to be generally true, provided the sizes of the constituent matrices are such that both sides are defined. This and several other useful properties of matrix arithmetic are collected in the following.

Matrix Arithmetic

Proposition 4.1: Some Properties of Matrix Arithmetic

Suppose that A, B, and C are matrices and that α is a scalar. The following identities hold, where we assume in each part that the sizes of A, B, and C are such that the matrices on both sides of the identity are defined:

(a) *Commutativity of Addition.* $A + B = B + A$
(b) *Associativity.* $(A + B) + C = A + (B + C), \quad (AB)C = A(BC)$
(c) *Distributive Laws.* $A(B + C) = AB + AC, \quad (A + B)C = AB + AC, \alpha(A + B) = \alpha A + \alpha B$
(d) $\alpha(AB) = (\alpha A)B = A(\alpha B)$

Proof: Part (a) and the first identity of part (b) involve only matrix addition, and so these identities are direct consequences of the corresponding identities for real numbers (one entry at a time). To prove that matrix multiplication is associative [that is, the second identity of part (b)], we let $[f_{ij}] = (AB)C$ and $[g_{ij}] = A(BC)$. Since these two matrices have the same size, we need to show that each of the corresponding entries is equal; that is, we fix a pair of indices i and j, and aim to show that $f_{ij} = g_{ij}$. For definiteness, let the sizes of A, B, and C be $n \times m$, $m \times r$, and $r \times s$, respectively. By definition, f_{ij} is the dot product of the ith row of AB with the jth column of C. Since the kth entry in the ith row of AB is $\sum_{\ell=1}^{m} a_{i\ell}b_{\ell k}$, it follows that $f_{ij} = \sum_{k=1}^{r}\left(\sum_{\ell=1}^{m} a_{i\ell}b_{\ell k}\right)c_{kj} = \sum_{\ell=1}^{m} a_{i\ell}\left(\sum_{k=1}^{r} b_{\ell k}c_{kj}\right)$. (The commutative and associative properties for numbers allowed us to switch the order of summation.) The parenthesized sum on the right is simply the ℓth entry in the jth column of BC. Hence, $\sum_{\ell=1}^{n} a_{i\ell}\left(\sum_{k=1}^{r} b_{\ell k}c_{kr}\right) = g_{ij}$, so we have proved $f_{ij} = g_{ij}$. The proofs of the identities of parts (c) and (d) are easier than this one and are left to the next exercise for the reader and to the exercises. □

Exercise for the Reader 4.2

Prove the identities $A(B+C) = AB + AC$ and $\alpha(A+B) = \alpha A + \alpha B$ from part (c) of Proposition 4.1.

It quickly becomes impractical to multiply (or even add) large matrices without a computer; indeed, using Definition 4.2 to multiply an $n \times m$ matrix by an $m \times r$ matrix, each of the nr terms requires computing a dot product involving m multiplications and $m-1$ additions, giving a grand total of $nr(2m-1)$ mathematical operations. If all the dimensions are the same ($n = m = r$), this becomes $n^2(2n-1)$, which is asymptotic[*] to $2n^3$, or a constant times n^3. Since matrices have so many applications, the expense of multiplying them has been a bottleneck for many problems. Many mathematicians and computer scientists thought that this price of matrix multiplication could not be lowered until 1969, when a German mathematician Volker Strassen [Str-69] found a faster algorithm that multiplies a pair of $n \times n$ matrices using at most a constant times $n^{2.81}$ arithmetic operations. After this landmark discovery, scientists have been working hard to find ever more efficient algorithms for *fast matrix multiplication*. The fastest method known at present was discovered in 1990 by Don Coppersmith and Shmuel Winograd [CoWi-90] and brings the amount of arithmetic operations needed to multiply two $n \times n$ matrices down to a constant times $n^{2.376}$. This method relies on increasing complex generalizations of Strassen's idea and on ideas from the area of mathematics known as group theory. Researchers believe that it is likely that an algorithm can be found that will be able to multiply a pair of $n \times n$ matrices using only a constant times n^2.

---

[*] For two infinite sequences a_n and b_n to be *asymptotic*, it means that $a_n/b_n \to 1$, as $n \to \infty$.

This would, of course, be the holy grail of all efficiencies for this problem, since having to compute the n^2 entries will take an amount of operations work at least equal to n^2. The constants in these complexity estimates, as well as the overall intricacy of the algorithm, can make such algorithms practical only when the matrices involved are sufficiently large. The reader can find a good survey article on this very interesting topic in [Rob-05]. The exercises at the end of the chapter provide opportunities for the reader to examine Strassen's algorithm.

Definition of an Invertible (Square) Matrix

Thus far, we have learned how to add, subtract, and multiply matrices. It is natural to next consider the question of whether matrices can be divided. With real numbers, we are allowed to divide by any nonzero number. In modular arithmetic, we learned that division is possible as long as the divisor is relatively prime to the modulus. In each of these cases, the division by a number a can be viewed as multiplying by the inverse element a^{-1} that satisfies (if it exists) $a \cdot a^{-1} = a^{-1} \cdot a = 1$. For example, in the real numbers system, the division $7 \div 2$ can be viewed as the multiplication $7 \cdot 2^{-1} = 7 \cdot (1/2) = 3.5$. A similar situation exists for matrices. In order to define the inverse of a matrix, one needs to define an analogue for the number 1 for matrices; this will be the so-called *identity matrix* that is defined as follows.

Definition 4.3

For any positive integer n, the $\boldsymbol{n \times n}$ **identity matrix**, denoted as $\boldsymbol{I_n}$ or just I (when the size is clear from the context or unimportant) is the $n \times n$ matrix $[\delta_{ij}]$, where $\delta_{ij} = 1$, if $i = j$, and $\delta_{ij} = 0$, if $i \neq j$.

Below are the 1×1, 2×2, and 3×3 identity matrices.

$$I_1 = [1], \quad I_2 = \begin{bmatrix} 1 & 0 \\ 0 & 1 \end{bmatrix}, \quad I_3 = \begin{bmatrix} 1 & 0 & 0 \\ 0 & 1 & 0 \\ 0 & 0 & 1 \end{bmatrix}$$

The identity matrix is so named because it behaves like the number 1 does for numbers; that is, when it gets multiplied by a matrix, it does not change the matrix:

$$AI = A = IA \tag{4.1}$$

Exercise for the Reader 4.3

Prove Equation 4.1.

The definition of an *invertible matrix* is easily translated from that of an invertible real number by simply changing numbers to matrices and 1 to I.

Definition 4.4

A square matrix A is said to be **invertible** if there exists another matrix A^{-1} of the same size that, when multiplied by A on either side, gives the identity matrix:

$$AA^{-1} = A^{-1}A = I \qquad (4.2)$$

The matrix A^{-1} (if it exists) is unique [*Proof:* If B is another such matrix, then using Equation 4.1, Equation 4.2, and associativity, we obtain that $B = BI = B(AA^{-1}) = (BA)A^{-1} = IA^{-1} = A^{-1}$ □] and is called the **inverse of** A.

Example 4.3

Let $A = \begin{bmatrix} 3 & 5 \\ 4 & 7 \end{bmatrix}$, $B = \begin{bmatrix} 0 & 0 \\ 3 & 6 \end{bmatrix}$, $C = \begin{bmatrix} 7 & -5 \\ -4 & 3 \end{bmatrix}$

(a) Show that $A^{-1} = C$.
(b) Show that the matrix B is not invertible.

Solution: Part (a): The two multiplications

$$AC = \begin{bmatrix} 3 & 5 \\ 4 & 7 \end{bmatrix}\begin{bmatrix} 7 & -5 \\ -4 & 3 \end{bmatrix} = \begin{bmatrix} 1 & 0 \\ 0 & 1 \end{bmatrix} \text{ and }$$

$$CA = \begin{bmatrix} 7 & -5 \\ -4 & 3 \end{bmatrix}\begin{bmatrix} 3 & 5 \\ 4 & 7 \end{bmatrix} = \begin{bmatrix} 1 & 0 \\ 0 & 1 \end{bmatrix}$$

show that $AC = CA = I$, which means that $A^{-1} = C$ (and also that $C^{-1} = A$)

Part (b): If $M = \begin{bmatrix} a & b \\ c & d \end{bmatrix}$ is any 2×2 matrix and we compute the product BM, we notice that

$$BM = \begin{bmatrix} 0 & 0 \\ 3 & 6 \end{bmatrix}\begin{bmatrix} a & b \\ c & d \end{bmatrix} = \begin{bmatrix} 0 & 0 \\ * & * \end{bmatrix}$$

(where the asterisks denote numbers). The point is that no matter what the matrix M is, the product BM will inherit the property that the first row of B is all zeros. Thus it is impossible to have $BM = I$, so that B cannot be invertible.

Exercise for the Reader 4.4

(a) Generalize the idea given in the solution of part (b) of the preceding example to show that any square matrix A with a row of zeros cannot be invertible.

(b) Show, likewise, that if a square matrix A has a column of zeros, then it cannot be invertible.

The Determinant of a Square Matrix

As with real numbers and with modular arithmetic, it is sometimes important to know whether a given square matrix is invertible. Every square matrix A has a number called its *determinant*, denoted as $\det(A)$, associated with it. Whereas a real number is invertible if, and only if, it is nonzero, it turns out that a square matrix is invertible if, and only if, the determinant is nonzero (that is, if the determinant is an invertible real number). Many computing platforms have built-in functions that will compute determinants of square matrices (and inverses of invertible matrices). The following algorithm gives a method for computing determinants (*co-factor expansion*).[*]

Algorithm 4.1:   *Cofactor Expansion Algorithm for Computing Determinants*

Input: A square $n \times n$ matrix A.
Output: The determinant of A, $\det(A)$, *cofactor expansion along the first row.*

Case $n = 1$:   If $A = [a]$ (1×1 matrix), $\det(A) = a$.

Case $n = 2$:   If $A = \begin{bmatrix} a & b \\ c & d \end{bmatrix}$, $\det(A) = ad - bc$, i.e., just the product of the main diagonal entries (top left to bottom right) less the product of the off-diagonal entries (top right to bottom left).

Case $n > 2$:   For larger-sized matrices the algorithm will be recursive in that it will reduce the problem of computing the determinant into the computation of several related 2×2 determinants.

---

[*] Since we do not need it, we will not give the formal definition of determinants. The formal definition, as well as a proof that the result of the cofactor algorithm that we give is equivalent to it, can be found in books on linear algebra such as [HoKu-71]. We point out that cofactor algorithm is practical only for small-sized matrices, but it will be sufficient for our purposes.

For any entry a_{ij} of the $n \times n$ matrix A, we define the corresponding **submatrix** A_{ij} to be the $(n-1) \times (n-1)$ matrix obtained from A by deleting the row and column of A that contain the entry a_{ij}. Thus, for example, in the case of $n = 3$, the matrix A_{13} is the 2×2 matrix obtained from A by deleting the row and column determined by a_{13} (so the first row, and the third column). This operation is shown as

$$A_{13} = \begin{bmatrix} a_{11} & a_{12} & a_{13} \\ a_{21} & a_{22} & a_{23} \\ a_{31} & a_{32} & a_{33} \end{bmatrix} = \begin{bmatrix} a_{21} & a_{22} \\ a_{31} & a_{32} \end{bmatrix}$$

The recursive formula for the determinant of A is now given by

$$\det(A) = a_{11} \det(A_{11}) - a_{12} \det(A_{12}) + a_{13} \det(A_{13})$$
$$- \cdots + (-1)^{n+1} a_{1n} \det(A_{1n}) \tag{4.3}$$

Note that the signs alternate as we move across the entries of the top row of A.[*]

Thus the determinant of any $n \times n$ matrix (with $n > 2$) can be computed by repeatedly applying the recursion step (with Equation 4.3) until we have a combination that involves only 2×2 determinants, which can be handled directly with the $n = 2$ case. For example, we may compute

$$\det \begin{bmatrix} 8 & 2 & -2 \\ 4 & 0 & 6 \\ -2 & 6 & 5 \end{bmatrix} = 8 \det \begin{bmatrix} 0 & 6 \\ 6 & 5 \end{bmatrix} - 2 \det \begin{bmatrix} 4 & 6 \\ -2 & 5 \end{bmatrix} + (-2) \det \begin{bmatrix} 4 & 0 \\ -2 & 6 \end{bmatrix}$$

$$= 8(0 \cdot 5 - 6 \cdot 6) - 2(4 \cdot 5 - 6 \cdot (-2)) - 2(4 \cdot 6 - 0 \cdot (-2))$$

$$= -288 - 64 - 48$$

$$= -400$$

Exercise for the Reader 4.5

Use the cofactor expansion algorithm (Algorithm 4.1) to compute the determinant of the matrix $\begin{bmatrix} 5 & 6 & 9 \\ -12 & 2 & 7 \\ 2 & 3 & -7 \end{bmatrix}$

We now formally state the important relation that was mentioned above between the determinant and the invertibility of a matrix.

---

[*] It is proved in linear algebra books (for example, see [HoKu-71]) that one could instead take the corresponding (cofactor) expansion along any row or column of A, using the following rule to choose the alternating signs: The sign of $\det(d_{ij})$ is $(-1)^{i+j}$. See Proposition 4.6 in the Chapter Exercise set for a more complete statement of this result.

Theorem 4.2: On the Invertibility of Square Matrices

A square matrix A is invertible if, and only if, its determinant $\det(A)$ is nonzero.

We will not prove this result here (since it would require the formal definition of determinants), but the interested reader may refer to a decent book on linear algebra (for example, see [HoKu-71]).

Inverses of 2 × 2 Matrices

We have given an example of the inverse of a matrix but so far have given no clue as to how inverses of invertible matrices can be determined. The following result includes a simple formula for the inverse of a 2×2 matrix; later in this section we give a more general formula that will provide a way to find inverses of any invertible matrix.

Theorem 4.3: On the Invertibility of 2 × 2 Matrices

For a 2×2 matrix $A = \begin{bmatrix} a & b \\ c & d \end{bmatrix}$, with nonzero determinant $\det(A) = ad - bc$, the inverse matrix is given by

$$A^{-1} = \frac{1}{\det(A)} \begin{bmatrix} d & -b \\ -c & a \end{bmatrix}$$

Proof: We need only check that the matrix given by the formula satisfies Equation 4.2 (definition of an inverse matrix). We temporarily denote the matrix defined by the formula in the theorem as B. We will show that $BA = I$; the proof that $AB = I$ is similar.

$$BA = \frac{1}{\det(A)} \begin{bmatrix} d & -b \\ -c & a \end{bmatrix} \begin{bmatrix} a & b \\ c & d \end{bmatrix} = \frac{1}{\det(A)} \begin{bmatrix} ad - bc & db - bd \\ -ca + ac & -cb + ad \end{bmatrix}$$

$$= \frac{1}{\det(A)} \begin{bmatrix} \det(A) & 0 \\ 0 & \det(A) \end{bmatrix} = I \quad \square$$

For example, the matrix $A = \begin{bmatrix} 2 & 3 \\ 2 & 2 \end{bmatrix}$ has determinant $\det(A) = 2 \cdot 2 - 3 \cdot 2 = -2$, so this matrix has an inverse given by

$$A^{-1} = \frac{1}{-2} \begin{bmatrix} 2 & -3 \\ -2 & 2 \end{bmatrix} = \begin{bmatrix} 1 & -3/2 \\ -1 & 1 \end{bmatrix}$$

Exercise for the Reader 4.6

Use Theorem 4.3 to compute the inverses of the following matrices, if they exist:

$$M = \begin{bmatrix} 2 & 6 \\ 3 & -9 \end{bmatrix}, N = \begin{bmatrix} 2 & 6 \\ 3 & 9 \end{bmatrix}$$

Two final important yet basic concepts about matrices that we wish to mention are included in the following two definitions.

The Transpose of a Matrix

Definition 4.5

If $A = [a_{ij}]$ is an $n \times m$ matrix, then the **transpose** of A, denoted A', is the $m \times n$ matrix whose rows are the columns of A and whose columns are the rows of A (in order). Put differently, $A' = [b_{ij}]$, where $b_{ij} = a_{ji}$ ($1 \le i \le m$, $1 \le j \le n$). A square matrix A is called **symmetric** if $A = A'$.

Example 4.4

Find the transposes of the matrices $\begin{bmatrix} 1 & 2 & 1 \\ 4 & 6 & 9 \end{bmatrix}$ and $\begin{bmatrix} 6 & 5 \\ 5 & 1 \end{bmatrix}$.

Solution: Changing rows to columns, the transpose of the first

matrix is $\begin{bmatrix} 1 & 4 \\ 2 & 6 \\ 1 & 9 \end{bmatrix}$ while the transpose of the second matrix is

itself; thus the latter matrix is symmetric.

Modular Integer Matrices

In cryptographic applications, it is often useful to consider matrices whose entries are stipulated to lie in some other number system rather than the real numbers, which have been the basis of all examples considered up to this point. The next example illustrates how some of the preceding concepts play out when the elements of matrices are integers mod m [and the corresponding arithmetic is done (mod m)]. We refer to such matrices as **modular integer matrices**. We will compare and contrast some of the previously mentioned properties of matrices with real number arithmetic with modular integer matrices.

Definition 4.6

For a fixed integer $m > 1$, we consider matrices whose entries are integers mod m, that is, are elements in $\mathbb{Z}_m$. Such a matrix is called a **modular integer matrix**. All of the previously defined matrix operations are extended to such matrices with the proviso that all integer calculations are performed (mod m).

Example 4.5

If we are considering matrices mod 5, then here is how we would add and multiply the two matrices $A = \begin{bmatrix} 4 & 2 \\ 1 & 3 \end{bmatrix}$ and

$B = \begin{bmatrix} 3 & 0 \\ 4 & 3 \end{bmatrix}$:

$$A + B = \begin{bmatrix} 4+3 & 2+0 \\ 1+4 & 3+3 \end{bmatrix} = \begin{bmatrix} 7 & 2 \\ 5 & 6 \end{bmatrix} \equiv \begin{bmatrix} 2 & 2 \\ 0 & 1 \end{bmatrix} \text{ (mod 5)}$$

$$AB = \begin{bmatrix} 4 & 2 \\ 1 & 3 \end{bmatrix}\begin{bmatrix} 3 & 0 \\ 4 & 3 \end{bmatrix} = \begin{bmatrix} 20 & 6 \\ 15 & 9 \end{bmatrix} \equiv \begin{bmatrix} 0 & 1 \\ 0 & 4 \end{bmatrix} \text{ (mod 5)}$$

Almost all of the properties of matrix arithmetic that we have seen carry over very nicely to modular matrices. For example, each of the identities of Proposition 4.1 is valid for modular integer matrices. The proof is simple because the identities hold when real number arithmetic is used to compute the entries. Corresponding entries remain equal when we take their remainders mod m. Determinants of square modular integer matrices can be defined using the same definition as with ordinary matrices. But more importantly for our purposes, all of the alternative definitions, such as the cofactor expansion that was given in Algorithm 4.1, are also valid. It is delightfully surprising that both Theorems 4.2 and 4.3 on the invertibility of matrices have the following very natural analogue for modular matrices.

Theorem 4.4: On the Invertibility of Square Modular Integer Matrices

(1) A square modular integer matrix A is invertible (mod m) if, and only if, its determinant $\det(A)$ is relatively prime to the modulus m; i.e., $\gcd(\det(A), m) = 1$.

(2) For a 2×2 modular integer matrix $A = \begin{bmatrix} a & b \\ c & d \end{bmatrix}$, with

> determinant $\det(A) = ad - bc$ relatively prime to m, the inverse
>
> matrix is given by $A^{-1} = \det(A)^{-1} \begin{bmatrix} d & -b \\ -c & a \end{bmatrix}$ (mod m), where
>
> $\det(A)^{-1}$ is the inverse of $\det(A)$ (mod m).

Note that from Proposition 2.11, the condition that $\gcd(\det(A), m) = 1$ is equivalent to $\det(A)$ having an inverse (mod m). Since a real number is invertible precisely when it is nonzero, part (1) of Theorem 4.4 is really a direct translation of part (1) of Theorem 4.2, from the language of real number entries into the language of modular integer entries.

For the proof of part (1) of Theorem 4.4, we refer to the books by Koblitz [Kob-94] and Stinson [Sti-05]. The proof of part (2) is accomplished in the same way that we did it for ordinary matrices in Theorem 4.2.

Example 4.6

If it exists, find the inverse of the matrix $A = \begin{bmatrix} 3 & 2 \\ 2 & 3 \end{bmatrix}$ (a) (mod 10) and (b) (mod 12).

Solution: Part (a): $\det(A) = 3 \cdot 3 - 2 \cdot 2 = 5$ is not relatively prime to 10, therefore by Theorem 4.4, A has no inverse (mod 10).

Part (b): Since $\gcd(\det(A), 12) = \gcd(5, 12) = 1$, and since $5^{-1} = 5$ (mod 12) (either by trial and error or the extended division algorithm), it follows from Theorem 4.4(2) that

$$A^{-1} = 5^{-1} \begin{bmatrix} 3 & -2 \\ -2 & 3 \end{bmatrix} \equiv 5 \cdot \begin{bmatrix} 3 & 10 \\ 10 & 3 \end{bmatrix} \equiv \begin{bmatrix} 15 & 50 \\ 50 & 15 \end{bmatrix} \equiv \begin{bmatrix} 3 & 2 \\ 2 & 3 \end{bmatrix} \text{ (mod 12)}$$

The reader may wish to check that $AA^{-1} = I$. Coincidentally, here we have that $A^{-1} = A$. We also point out that this inverse is totally different from the real-number version (obtained from Theorem 4.3), which has fractions in all of its entries.

Exercise for the Reader 4.7

Consider the following matrices of integers:

$$A = \begin{bmatrix} 2 & 7 \\ 4 & 1 \end{bmatrix}, B = \begin{bmatrix} 1 & 2 \\ 9 & 8 \end{bmatrix}$$

(a)   Working mod 3, compute the following related matrices, if they are defined: $A + B$, AB, B^{-1}.

(b)   Repeat part (a), but this time working mod 10.

The Classical Adjoint (for Matrix Inversions)

The formulas of both Theorems 4.3 and 4.4 are special cases of the so-called *classical adjoint formula* for the inverse of an invertible square matrix. To state this formula, it will be convenient to first introduce the following definition.

Definition 4.7

Let $A = [a_{ij}]$ be an $n \times n$ matrix. As in Algorithm 4.1, for each pair of indices i, j, we let A_{ij} denote the $(n-1) \times (n-1)$ submatrix obtained from A by deleting the row and column containing a_{ij}. The **classical adjoint** of A is defined to be the matrix

$$\mathrm{adj}(A) = \begin{bmatrix} +\det(A_{11}) & -\det(A_{12}) & \cdots & (-1)^{1+n}\det(A_{1n}) \\ -\det(A_{21}) & +\det(A_{22}) & \cdots & (-1)^{2+n}\det(A_{2n}) \\ \vdots & & \ddots & \\ (-1)^{n+1}\det(A_{n1}) & (-1)^{n+2}\det(A_{n2}) & \cdots & +\det(A_{nn}) \end{bmatrix}'$$

$$= \left[(-1)^{i+j}\det(A_{ji}) \right]$$

Note the transpose operation in the upper right matrix.

Proposition 4.5: Classical Adjoint Formula for the Inverse of an $n \times n$ Matrix

If $A = [a_{ij}]$ is an invertible $n \times n$ (real-valued or modular integer) matrix, then $A^{-1} = \det(A)^{-1} \cdot \mathrm{adj}(A)$.

For an $n \times n$ matrix, this formula requires the computation of an $n \times n$ determinant and n^2 $(n-1) \times (n-1)$ determinants, so it is only practical for small-sized matrices.[*] Since we need to invert only matrices up to size 3×3 in this book, it will be sufficient for our purposes and can be used in conjunction with the cofactor expansion method for computing determinants of Algorithm 4.1. The proof of this formula easily follows from a slightly more general version of the cofactor expansion (see Algorithm 4.1) and is outlined in the chapter exercises. The proof and the formula are valid for modular integer matrices as well as ordinary matrices, provided (Theorem 4.4) that $\det(A)$ is relatively prime to the modulus.

---

[*] More efficient techniques for computing determinants of square matrices and inverses of invertible matrices can be found in any decent book on elementary linear algebra (such techniques are adaptations of Gaussian elimination); see, for example, [Poo-05] or [KoHi-99]; a more theoretically comprehensive text is the classical reference [HoKu-71].

Our next example involves the inversion of 3×3 integer modular matrices using Proposition 4.5.

Example 4.7

Find the inverse of the integer modular matrix $A = \begin{bmatrix} 7 & 5 & 2 \\ 0 & 6 & 4 \\ 8 & 2 & 5 \end{bmatrix}$ (mod 9), if it exists.

Solution: Using the cofactor expansion Equation 4.3, we find $\det(A) = 434 \equiv 2 \pmod 9$. [Alternatively, all intermediate calculations could have been performed (mod 9).] Since $\gcd(2, 9) = 1$, it follows that A^{-1} exists (mod 9). The classical adjoint matrix is[*]

$$\text{adj}(A) = \begin{bmatrix} \det\begin{pmatrix} 6 & 4 \\ 2 & 5 \end{pmatrix} & -\det\begin{pmatrix} 5 & 2 \\ 2 & 5 \end{pmatrix} & \det\begin{pmatrix} 5 & 2 \\ 6 & 4 \end{pmatrix} \\ -\det\begin{pmatrix} 0 & 4 \\ 8 & 5 \end{pmatrix} & \det\begin{pmatrix} 7 & 2 \\ 8 & 5 \end{pmatrix} & -\det\begin{pmatrix} 7 & 2 \\ 0 & 4 \end{pmatrix} \\ \det\begin{pmatrix} 0 & 6 \\ 8 & 2 \end{pmatrix} & -\det\begin{pmatrix} 7 & 5 \\ 8 & 2 \end{pmatrix} & \det\begin{pmatrix} 7 & 5 \\ 0 & 6 \end{pmatrix} \end{bmatrix}$$

$$= \begin{bmatrix} 22 & -21 & 8 \\ 32 & 19 & -28 \\ -48 & 26 & 42 \end{bmatrix}$$

so that $\text{adj}(A) \equiv \begin{bmatrix} 4 & 6 & 8 \\ 5 & 1 & 8 \\ 6 & 8 & 6 \end{bmatrix}$

Since $2 \cdot 5 = 10 \equiv 1 \pmod 9$, we have $2^{-1} \equiv 5 \pmod 9$, so (by Proposition 4.5)

$$A^{-1} = \det(A)^{-1} \cdot \text{adj}(A) = 5 \cdot \begin{bmatrix} 4 & 6 & 8 \\ 5 & 1 & 8 \\ 6 & 8 & 6 \end{bmatrix} \equiv \begin{bmatrix} 2 & 3 & 4 \\ 7 & 5 & 4 \\ 3 & 4 & 3 \end{bmatrix} \pmod 9$$

Note that in the final modular multiplication by 5, all but the two odd entries could have simply been divided by 2, rather than multiplied by its inverse. The reader may wish to check that $AA^{-1} = I = A^{-1}A$.

---

[*] Again, we could have been working (mod 9) throughout all intermediate calculations and would have obtained the same result. By the way, when viewed as an ordinary (real-valued) matrix, Proposition 4.5 tells us that the inverse matrix would be the prereduced (mod 9) adjoint divided by the pre-reduced determinant (434).

The above procedure is easily programmed, and this will be tasked in the computer exercises at the end of this chapter. But many computing platforms already have built-in programs for computing determinants (that are more efficient than Algorithm 4.1). The following note shows how to take advantage of such a feature to render an efficient and easy way of obtaining modular integer matrix inverses, whenever they exist.

Computing Note: As a consequence of Proposition 4.5, if working on a computing platform that already has programs for computing (ordinary) matrix inverses, any modular matrix inverse (if it exists) can easily be obtained from the corresponding (ordinary) matrix inverse by the following formula:

$$A^{-1} (\bmod n) \equiv (\det(A)^{-1} (\bmod n)) \cdot (\det(A) A^{-1}) \qquad (4.4)$$

where A^{-1} denotes the (ordinary) matrix inverse, $A^{-1} (\bmod n)$ and $\det(A)^{-1} (\bmod n)$ denote inverses (mod n), and the formula is valid whenever the modular inverses exist.

See Computer Exercise 11 at the end of this chapter for more on how to implement Equation 4.4. Moreover, if your system has a built-in determinant function (for computing determinants of real-valued matrices), it should be faster to compute determinants of modular matrices by directly using this function and then converting the answer to a mod n integer (by applying the mod function* as mod(det(A), n)).

Exercises for the Reader

Exercise 4.8

If it exists, find the inverse of the matrix $A = \begin{bmatrix} 1 & 1 & 3 \\ 8 & 4 & 5 \\ 5 & 0 & 1 \end{bmatrix}$

(a) (mod 15) and (b) (mod 16).

---

* Even on a fast computer, the cofactor expansion method starts to get impractically slow when the size of the matrices exceeds about 9×9. Faster determinant and matrix inverse methods that are usually built into computing platforms are based on Gaussian elimination (a technique in linear algebra) and can routinely invert and compute determinants of matrices exceeding sizes of 500×500. In any case, since we will not be needing to invert (or compute inverses of) matrices larger than 9×9 (even in the computer exercises), the methods and algorithms in this section are perfectly adequate for our purposes. There is one other issue concerning *round-off errors* that results in most standard (floating point arithmetic) computing systems, rendering the results unreliable if the number of significant digits is too large (usually about 15, which coincidentally occurs about when we try to compute determinants of mod 26 matrices of size larger than 9×9). We will discuss ways to circumvent this latter problem when we deal with large prime numbers in Chapter 8.

Exercise 4.9

Show that the formulas of Theorem 4.3 and of part 2(b) of Theorem 4.4 are both special cases of the classical adjoint inverse formula of Proposition 4.5.

The Hill Cryptosystem

Almost all of the cryptosystems that have been developed in the previous chapters are substitution ciphers and as such do not satisfy Shannon's requirement of diffusion. All of the contemporary ciphers of the computer age are block ciphers. We next describe an interesting block cryptosystem that was invented in the early 20th century known as the *Hill cryptosystem*.* The Hill cryptosystem has an important historical significance since it marked the beginning of a revolution in cryptography in which cryptosystems were reliant on mathematical sophistication. It is based on modular matrix multiplication by invertible matrices. The Hill system is no longer in widespread use today because of its vulnerability to attacks.

Definition 4.8   The Hill Cryptosystem

The key in a **Hill cryptosystem** is an invertible **encoding matrix** A of integers mod 26. The block length r of the cryptosystem is the number of rows (columns) of the matrix A. Given a plaintext message, we first translate it into a string of integers mod 26 using, for example, the natural scheme of Table 3.1 ($0 \leftrightarrow a$, $1 \leftrightarrow b$, ..., $25 \leftrightarrow z$). Thus, we may represent the plaintext message as a vector $[p_1 p_2 \cdots p_\ell]$, where each p_i is an element of $\mathbb{Z}_{26}$ representing the corresponding English letter, and ℓ is the length of the plaintext message. We then arrange the elements of this message into an r-rowed matrix P by vertically stacking the plaintext elements p_i in order:

$$P = \begin{bmatrix} p_1 & p_{r+1} & \cdots & \\ p_2 & p_{r+2} & & \\ \vdots & \vdots & \ddots & \vdots \\ p_r & p_{2r} & \cdots & \end{bmatrix}$$

Unless ℓ is a multiple of r, the final column of P will need to be padded with some additional elements; for the purposes of illustration and

---

* This system is named after its creator, Lester S. Hill (1891–1961), an American mathematician who developed his cryptosystem in a couple of published papers in 1929 and 1931. During this time, since computers had not yet been invented, the calculations involving matrix multiplications would have been extremely tedious. To make his idea more marketable, Hill went further and patented a purely mechanical machine to perform the encryptions and decryptions. Since Hill's cryptosystem depends on a particular invertible matrix (key) being used, and his machine only worked for a single matrix, he advocated use of a so-called *involutory matrix*, that is, a matrix A that is its own inverse, $A = A^{-1}$. (Example 4.6 exhibited such a matrix.) Hill's cryptosystem, along with his mechanical devices, were apparently used for radio call signals during World War II.

definiteness, we could use 13 (corresponding to a final string of n's). The encryption is performed by modular matrix multiplication,

$$C \equiv AP, \quad C \equiv \begin{bmatrix} c_1 & c_{r+1} & \cdots & \\ c_2 & c_{r+2} & & \\ \vdots & \vdots & \ddots & \vdots \\ c_r & c_{2r} & \cdots & \end{bmatrix} \pmod{26} \qquad (4.5)$$

where the cipher text vector $[c_1 \, c_2 \cdots c_\ell]$ is read off in the same order as P was formed, from the matrix product C of Equation 4.5. When translated back to English characters, the final string of n's (if any) would be ignored. Because the matrix A was chosen to be invertible, the decryption is performed simply by multiplying the encrypted matrix C by A^{-1}. (*Proof that this works:* $A^{-1}C = A^{-1}(AP) = (A^{-1}A)P = IP = P.$ □)

Example 4.8

Use the Hill cryptosystem with encoding matrix $A = \begin{bmatrix} 1 & 2 \\ 1 & 3 \end{bmatrix}$ to encrypt the message "code blue alert." Note that $\det(A) = 1 \equiv 1 \pmod{26}$ so the matrix A is invertible (mod 26) and is thus a legitimate encoding matrix.

Solution: The $\mathbb{Z}_{26}$ vector corresponding to the plaintext is (see Table 3.1)

$$[2 \; 14 \; 3 \; 4 \; 1 \; 11 \; 20 \; 4 \; 0 \; 11 \; 4 \; 17 \; 19]$$

Forming a two-rowed matrix (filling in the final entry with a 13) and multiplying by the encryption matrix gives the corresponding matrix of ciphertext elements:

$$C = AP \equiv \begin{bmatrix} 1 & 2 \\ 1 & 3 \end{bmatrix} \begin{bmatrix} 2 & 3 & 1 & 20 & 0 & 4 & 19 \\ 14 & 4 & 11 & 4 & 11 & 17 & 13 \end{bmatrix}$$

$$= \begin{bmatrix} 4 & 11 & 23 & 2 & 22 & 12 & 19 \\ 18 & 15 & 8 & 6 & 7 & 3 & 6 \end{bmatrix}$$

(mod 26). Translating the resulting $\mathbb{Z}_{26}$ ciphertext vector back into English letters gives the final ciphertext message:

plaintext:	c	o	d	e	b	l	u	e	a	l	e	r	t	n
plaintext in $\mathbb{Z}_{26}$:	2	14	3	4	1	11	20	4	0	11	4	17	19	13
ciphertext in $\mathbb{Z}_{26}$:	4	18	11	15	23	8	2	6	22	7	12	3	19	6
ciphertext:	E	S	L	P	X	I	C	G	W	H	M	D	T	G

Notice that the three instances of the plaintext letter *e* (shaded) were encrypted into three different letters in the ciphertext.

For the decryption process, the modular inverse of the encryption matrix needs to be computed. We presented methods for computing such modular matrix inverses (both by hand and by computer) earlier in this chapter. Recall that a $\mathbb{Z}_{26}$ modular matrix is invertible if, and only if, its determinant is relatively prime to 26.

Exercise for the Reader 4.10

The following ciphertext was encrypted using the Hill cryptosystem using the encoding matrix $A = \begin{bmatrix} 1 & 1 & 0 \\ 1 & 0 & 1 \\ 1 & 1 & 1 \end{bmatrix}$:

TARIDWXGXWNUANFHHU.
  Decode this message.

One natural question arises: How does one go about creating an encoding matrix of a desired size for the Hill cryptosystem? One effective method is to randomly fill in the entries of a square matrix of specified size with integers (mod 26) and check to see whether the determinant is relatively prime to 26. If it is, we have produced a suitable encoding matrix; otherwise, we repeat this process until we create one. The computer implementation material showcases this method.

Since changing one letter in the plaintext affects each of the letters in the corresponding ciphertext block, a ciphertext-only attack on the Hill cryptosystem would require a frequency analysis approach, by analyzing blocks governed by the size of the encryption matrix. For block sizes greater than 3, the number of four-letter sequences is at least 26^4, or about half a million, so that frequency attacks are rendered unfeasible. Although the Hill cryptosystem is quite difficult to crack with a ciphertext-only attack, it is easily compromised with any of the other three passive attacks: known plaintext, chosen plaintext, or chosen ciphertext. The common-thread scheme for such attacks is illustrated in the following example.

Example 4.9

Suppose that the Hill cryptosystem is being used and it is known that the plaintext message "stand by for further instructions" was encrypted into WTANVBOPGZRIJPTKZGBKTTQGZOARAH. Later on, the following ciphertext was intercepted: TAMDAQVAD EWPWDTKAUTCVEOMKMESXZQIWZIIRBOKTMADIKQNVSMM GCNZ. Decode this message.

Solution: The first step is to determine the size of the encryption matrix A. Since we know that the two ciphertext messages that were intercepted have lengths 30 and 56, it follows that the block length r (A is an $r \times r$ matrix) must divide gcd(30,56) = 2, so the key A must be a 2×2 matrix, which we write as $A = \begin{bmatrix} a & b \\ c & d \end{bmatrix}$.

We got a bit lucky here that the gcd was a prime number, so the block length was completely determined. In cases where it is not, we would proceed in the same fashion as we will be doing below, by trial and error on the possible block-length sizes.

The basic idea of how to compute A is simple. The first four letters of the known plaintext/ciphertext correspondence stan $\rightarrow$ WTAN are related by the matrix equation

$$A\begin{bmatrix} 18 & 0 \\ 19 & 13 \end{bmatrix} \equiv \begin{bmatrix} 22 & 0 \\ 19 & 13 \end{bmatrix} \pmod{26}$$

Now, if the matrix that is being multiplied by the unknown key A is invertible (mod 26), then we could simply right multiply this equation by its inverse to compute A. Since the determinant of this matrix is $18 \cdot 13 \equiv 0 \pmod{26}$, it is not invertible, so we cannot directly find A this way. Nonetheless, the above matrix equation does allow us to narrow down the possibilities for A. Rather than going this route, it is simpler to just try another two pairs of plaintext/ciphertext correspondence. For example, the next two pairs give us dbyf $\rightarrow$ VBOP, leading us to the matrix equation

$$A\begin{bmatrix} 3 & 24 \\ 1 & 5 \end{bmatrix} \equiv \begin{bmatrix} 21 & 14 \\ 1 & 15 \end{bmatrix} \pmod{26}$$

The matrix $M = \begin{bmatrix} 3 & 24 \\ 1 & 5 \end{bmatrix}$ has a determinant equal to 17 (mod 26), which is relatively prime to 26, so that we may now invert this matrix and use it to compute A:

$$M^{-1} \equiv 17^{-1}\begin{bmatrix} 5 & -24 \\ -1 & 3 \end{bmatrix} \equiv 23\begin{bmatrix} 5 & 2 \\ 25 & 3 \end{bmatrix} \equiv \begin{bmatrix} 11 & 20 \\ 3 & 17 \end{bmatrix} \pmod{26}$$

so

$$AM \equiv \begin{bmatrix} 21 & 14 \\ 1 & 15 \end{bmatrix} \Rightarrow A \equiv \begin{bmatrix} 21 & 14 \\ 1 & 15 \end{bmatrix} \cdot M^{-1} \equiv \begin{bmatrix} 21 & 14 \\ 1 & 15 \end{bmatrix} \cdot \begin{bmatrix} 11 & 20 \\ 3 & 17 \end{bmatrix}$$

$$\equiv \begin{bmatrix} 13 & 8 \\ 4 & 15 \end{bmatrix} \pmod{26}$$

Now that we have the key, the rest is straightforward: To perform the required decryption, we first compute the inverse of the key [since $\det(A) \equiv 7 \pmod{26}$]:

$$A^{-1} \equiv \begin{bmatrix} 13 & 8 \\ 4 & 15 \end{bmatrix}^{-1} \equiv 7^{-1} \begin{bmatrix} 15 & -8 \\ -4 & 13 \end{bmatrix} \equiv 15 \begin{bmatrix} 15 & -8 \\ -4 & 13 \end{bmatrix}$$

$$\equiv \begin{bmatrix} 17 & 10 \\ 18 & 13 \end{bmatrix} (\text{mod } 26)$$

Converting the ciphertext to a two-rowed matrix of integers using Table 3.1, left multiplying by the key inverse, and then converting back to text produces the plaintext message:

Leave at once to the safe house your location has been compromised.

A final n was discarded, and spaces have been inserted. We recommend that computations such as those carried out in this example be done on a computer.

One may wonder what would happen if we used the Hill cryptosystem in conjunction with a Vigenère cipher, or maybe added in a second Vigenère cipher. Indeed, such compositions of various ciphers would result in more secure and complicated systems, and they became the standard from the early 20th century into the computer age. Such systems were too complicated and inefficient to work with by hand, and before the computer age, mechanical and/or electronic devices were used to perform encryptions/decryptions for such complicated systems.

Chapter 4 Exercises

1. Given the matrices $A = \begin{bmatrix} 2 & -4 \\ 6 & 5 \end{bmatrix}$, $B = \begin{bmatrix} 8 & 4 \\ 3 & -2 \end{bmatrix}$, $C = \begin{bmatrix} 9 & -4 & 3 \\ -2 & 0 & 7 \\ -5 & -6 & 6 \end{bmatrix}$, $D = \begin{bmatrix} 2 & 6 \\ 7 & 4 \\ 5 & 6 \end{bmatrix}$, $E = \begin{bmatrix} 1 & 3 & 2 \\ 0 & 4 & 2 \\ 8 & 7 & 9 \end{bmatrix}$, do the following:

   (a) Determine the size of D.
   (b) Find the (1,3) entry, c_{13}, of the matrix C.
   (c) Find $A + B$.
   (d) Find $E - C$.
   (e) Find $C + D$.
   (f) Find $E + 2C$.

2. Given the matrices $A = \begin{bmatrix} 3 & 5 \\ 9 & 7 \end{bmatrix}$, $B = \begin{bmatrix} -4 & 7 \\ -7 & 5 \end{bmatrix}$, $C = \begin{bmatrix} 4 \\ 0 \\ 9 \end{bmatrix}$, $D = \begin{bmatrix} 8 & 0 & 1 \\ -5 & 5 & 2 \\ -5 & -9 & 9 \end{bmatrix}$, $E = \begin{bmatrix} 6 & 5 & -2 \\ -4 & 7 & 7 \\ -9 & 5 & -6 \end{bmatrix}$, do the following:

    (a) Determine the size of C.

    (b) Find the (2,1) entry, d_{21}, of the matrix D.

    (c) Find $A + B$.

    (d) Find $C + D$.

    (e) Find $E + D$.

    (f) Find $2E - 3D$.

3.   With the matrices A, B, C, D, and E, as in Chapter Exercise 1, find each of the following matrices (if it exists):

    (a) AB

    (b) BA

    (c) A^2

    (d) CD

    (e) DE

    (f) EC

4.   With the matrices A, B, C, D, and E, as in Chapter Exercise 2, find each of the following matrices (if it exists):

    (a) AB

    (b) BA

    (c) A^2

    (d) CD

    (e) DE

    (f) EC

5.   Find the transposes of the matrices A, C, and D of Chapter Exercise 1.

6.   Find the transposes of the matrices A, C, and E of Chapter Exercise 2.

7.   Find the inverses of the matrices A and B of Chapter Exercise 1 (if they exist).

8.   Find the inverses of the matrices A and B of Chapter Exercise 2 (if they exist).

9.   *Chain Matrix Multiplication.* One might expect, since matrix multiplication is associative [Proposition 4.1(b)], in computing a matrix product ABC, that it does not matter if we compute AB first or BC first. In practice, though, for large-sized matrices, it can make significant differences, as you will see in this exercise. Suppose that A has size 100×2, B has size 2×100, and C has size 100×2. Compare the total number of multiplications and additions required to compute $(AB)C$ and $A(BC)$.

10.   *Chain Matrix Multiplication.* (First read Chapter Exercise 9.) Suppose that A has size 100×1, B has size 1×1, C has size 1×100, and D has size 100×1. Compare the total number of

multiplications and additions required to compute $((AB)C)D$ and $A(B(CD))$.

11. Given the matrices $A = \begin{bmatrix} 2 & 1 \\ 3 & 5 \end{bmatrix}$ and $B = \begin{bmatrix} 7 & 4 \\ 0 & 3 \end{bmatrix}$, find each of the following matrices (if it is defined):

   (a) $A + B \pmod 8$
   (b) $A - B \pmod 8$
   (c) $A + 2B \pmod 8$
   (d) $3A - B \pmod 8$

12. Repeat each part of Chapter Exercise 11, but this time work in mod 9.

13. Redo Chapter Exercise 3, if the matrices are viewed as modular integer matrices (a) (mod 5) and (b) (mod 10). Then compare these answers to the answers obtained in Exercise 3 after the entries of each answer of Exercise 3 are converted into the appropriate modular integers.

14. Redo Chapter Exercise 3, if the matrices are viewed as modular integer matrices (a) (mod 7) and (b) (mod 12). Then compare these answers to the answers obtained in Exercise 3 after the entries of each answer of Exercise 3 are converted into the appropriate modular integers.

15. Find the determinants and the inverses of the matrices A and B of Chapter Exercise 11 (mod 9), if they exist.

16. Find the determinants and the inverses of the matrices A and B of Chapter Exercise 11 (mod 10), if they exist.

17. Letting $A = \begin{bmatrix} 9 & 1 & 5 \\ 9 & 5 & 6 \\ 8 & 6 & 7 \end{bmatrix}$, do the following:

   (a) Compute $\det(A)$.
   (b) Compute $\det(A) \pmod{21}$.
   (c) Compute $\det(A) \pmod{23}$.
   (d) Compute A^{-1}.
   (e) Compute $A^{-1} \pmod{21}$.
   (f) Compute $A^{-1} \pmod{23}$.

18. Letting $A = \begin{bmatrix} 2 & 5 & 0 \\ 6 & 0 & 7 \\ 5 & 11 & 7 \end{bmatrix}$, do the following:

   (a) Compute $\det(A)$.
   (b) Compute $\det(A) \pmod{21}$.
   (c) Compute $\det(A) \pmod{23}$.
   (d) Compute A^{-1}.

(e) Compute A^{-1} (mod 21).

(f) Compute A^{-1} (mod 23).

19. (a) Use the Hill cipher with encryption matrix $A = \begin{bmatrix} 5 & 2 \\ 11 & 5 \end{bmatrix}$

to encrypt each of the following plaintext strings:
   (i) the shipment will arrive at noon
   (ii) lay low until Friday
   (iii) always use the back door
   (iv) the phone is bugged

(b) Decrypt each of the following ciphertexts that were encrypted using the Hill cipher of part (a).
   (i) YFYCOUZEEPRTICADFJOCHS
   (ii) RBGAGUDSRTSQKNOXILANQVBBWWTCKE
   (iii) KMUKKHLLFKIODBRO
   (iv) NSOXEOTCLLQBIBEPQVMSQWFVMRQVQR

20. (a) Use the Hill cipher with encryption matrix $A = \begin{bmatrix} 5 & 6 \\ -2 & -3 \end{bmatrix}$

to encrypt each of the following plaintext strings:
   (i) two minutes until alarm sounds
   (ii) spread out your team
   (iii) reconnaissance is on schedule
   (iv) this hotel is safe

(b) Decrypt each of the following ciphertexts that were encrypted using the Hill cipher of part (a).
   (i) KIUFBWYBLOBJSIKVUNZUSYRB
   (ii) QGXVMUBJHAAEOIANOXHAHCOXUI
   (iii) KEHGHRHAFOVZQZCIFROG
   (iv) ICWUYFJOAUZGWUEYSOGYNI

21. (a) Use the Hill cipher with encryption matrix $A = \begin{bmatrix} 1 & 1 & 0 \\ 1 & 0 & 1 \\ 1 & 1 & 1 \end{bmatrix}$

to encrypt each of the strings of plaintext of Chapter Exercise 19, part (a).

(b) Decrypt each of the following ciphertexts that were encrypted using the Hill cipher of part (a).
   (i) MFFPZHZSM
   (ii) PSAALAXGKXWAPZHKRZLYC
   (iii) AXEGNOOOSHUMTMXGFTLLWTARIDWXGXWTAJEW
   (iv) GNRFVWBGZKZDVWZLJJHMPPEGNRGUFLPHWLLY

22. (a) Use the Hill cipher with encryption matrix

$A = \begin{bmatrix} 5 & 2 & 9 \\ -4 & 2 & 3 \\ 3 & -1 & -5 \end{bmatrix}$ to encrypt each of the strings of

plaintext of Chapter Exercise 20, part (a).

(b) Decrypt each of the following ciphertexts that were encrypted using the Hill cipher of part (a).

  (i) URMSCSWQKUBY

  (ii) MFKBLJTNDPFVHJMZVWWPINDW

  (iii) HRZCCLNWYHXMOOTBLLLQDRHKOWJPYZLEYWYAFOI

  (iv) RRVRRKBFBCSVDIFVCCPCEHBPUUPLZYQMH

23. In each part, a ciphertext is given, along with a portion of the plaintext. The Hill cryptosystem was used for each encryption, and the size of the encrypting matrix is also provided. Indicate whether the key (the encoding matrix) can be uniquely determined mathematically from the given information. Explain why this is or is not possible; in cases where it is possible, perform the decryption.

(a) A 2×2 encoding matrix was used.
Ciphertext: G R P Z B V L T Z C N V I R G C Q Z C I J Q M H Z N U C
Plaintext: * * * * * * * t w * * * * e o n b * * * * * * * * * *

(b) A 2×2 encoding matrix was used.
Ciphertext: I G H O E S Z W J F P I T W B M L L P O F R X J O R T J Z I
Plaintext: * * * r e o u * * * * * e r * * * * e l * * * * * * * e

(c) A 3×3 encoding matrix was used.
Ciphertext: S W M L Q L B U C I D V Q P J Y V Y J G K F K Y Z L S
Plaintext: * * * * * * r y l * * * * * * w i n * * * r o v * * *

24. Repeat the directions of Chapter Exercise 23 for each of the following parts:

(a) A 2×2 encoding matrix was used.
Ciphertext: M T A S W M T G S O J F D B N N A O K S K T U P L Z
Plaintext: * * s s * * * * * * * * a t * * * * * * * * r s *

(b) A 2×2 encoding matrix was used.
Ciphertext: E G S O I T T I W P M B H E I E Z F T S P C M O L W J M X W
Plaintext: a w * * * * * * * * * * * * * * n b e * * * * o v * * *

(c) A 3×3 encoding matrix was used.
Ciphertext: N T B E A H L H K Y N A A V C X R S A A R Y C J N K H
Plaintext: * * * m i l * * * n d o * * * r s i * * * * * * * *

25. *A Chosen Ciphertext Attack against the Hill Cipher.* Suppose that we know that a Hill cryptosystem is being used with an encoding matrix A of size $n \times n$. Suppose further that we have access to the encoder and are able to enter n plaintext strings, each having length n. What would be a good choice for these n plaintexts that would allow us to completely read off the encoding matrix A from the resulting ciphertexts?

26. *Problems with a Noninvertible Encoding Matrix for the Hill Cipher.* Give an example of a 2×2 matrix A of mod 26 modular integers that is not invertible (mod 26) and two different digraphs (plaintext pairs) that would encrypt under A to the same ciphertext under the Hill cryptosystem. (Of course, A would not be a legal encryption matrix since it is not invertible.)

27. Prove the following matrix distributive law from Proposition 4.1(c): $(A + B)C = AC + BC$.

28. Prove Proposition 4.1(d): $\alpha(AB) = (\alpha A)B = A(\alpha B)$.

29. *The Transpose of the Sum Is the Sum of the Transposes.* Prove that if A and B are matrices of the same size, then $(A + B)' = A' + B'$.

30. *The Transpose of the Product Is the Reverse Order Product of the Transposes.*
    (a) Prove that if A and B are matrices, then $(AB)' = B'A'$, whenever the left side is defined.
    (b) Extend the result in part (a) to general matrix products: $(A_1 A_2 \cdots A_t)' = A_t' A_{t-1}' \cdots A_1'$, whenever the left side is defined.

31. (a) Prove that if A and B are invertible matrices, then so is AB and $(AB)^{-1} = B^{-1}A^{-1}$.
    (b) Prove that if A is an invertible matrix and t is a positive integer, then A^t is also invertible and $(A^t)^{-1} = (A^{-1})^t$.

32. If A is an invertible matrix, prove that its transpose, A', is also invertible. What is $(A')^{-1}$?

Note: The following two exercises will provide a proof of the classical adjoint formula for the inverse of a matrix (Proposition 4.5) that will rely on a generalization of cofactor expansion formula for determinants that was given in Equation 4.3. That formula corresponds to a cofactor expansion along the first row; the general version here allows for an analogous expansion along any row or column of a square matrix A. As in the text,

---

Proposition 4.6: *General Cofactor Expansions for Determinants of Square Matrices*

Suppose that $A = [a_{ij}]$ is a square matrix. For any indices $i, j \, (1 \le i, j \le n)$, we have the following expansions for the determinant of A:

(a) Cofactor expansion along the ith row:

$$\det(A) = (-1)^{i+1} a_{i1} \det(A_{i1}) + (-1)^{i+2} a_{i2} \det(A_{i2}) + \cdots + (-1)^{i+n} a_{in} \det(A_{in})$$

(b) Cofactor expansion along the jth column:

$$\det(A) = (-1)^{1+j} a_{1j} \det(A_{1j}) + (-1)^{2+j} a_{2j} \det(A_{2j}) + \cdots + (-1)^{n+j} a_{nj} \det(A_{nj})$$

for an $n \times n$ matrix A and indices i and j, A_{ij} denotes the $(n-1) \times (n-1)$ submatrix obtained from A by deleting the ith row and jth column.
For a proof, the interested reader is referred to [HoKu-71].

33. Prove that if A is a square matrix that has either two identical columns or two identical rows, then $\det(A) = 0$.

Suggestion: Show that A cannot have an inverse as follows: if the i_1th row and the i_2th row of A were the same, and if B is any matrix (that can be multiplied by A), then the i_1th row and the i_2th row of AB must be the same. So there is no way that we could ever have $AB = I$. A similar argument works if A has two identical columns. Now apply Theorem 4.2.

34. Prove Proposition 4.5 (on the classical adjoint formula for the inverse of a matrix).

Suggestion: Use Proposition 4.6 along with the result of Chapter Exercise 33.

35. *Block Matrix Multiplication.* Suppose first that A and B are $n \times n$ matrices, where n is even.
    (a) Show that if $C = AB$, and if these three matrices are partitioned in the natural way as

$$\begin{bmatrix} C_{11} & C_{12} \\ C_{21} & C_{22} \end{bmatrix} = \begin{bmatrix} A_{11} & A_{12} \\ A_{21} & A_{22} \end{bmatrix} \begin{bmatrix} B_{11} & B_{12} \\ B_{21} & B_{22} \end{bmatrix}$$

where each of the submatrices has size $(n/2) \times (n/2)$, then we have

$$C_{11} = A_{11}B_{11} + A_{12}B_{21} \qquad C_{12} = A_{11}B_{12} + A_{12}B_{22}$$
$$C_{21} = A_{21}B_{11} + A_{22}B_{21} \qquad C_{22} = A_{21}B_{12} + A_{22}B_{22}$$

In other words, this block matrix multiplication behaves exactly like a 2×2 (ordinary) matrix multiplication. This technique is known as **block matrix multiplication**, and the smaller submatrices are known as the **blocks**.
    (b) Develop a similar version of the block matrix multiplication method of part (a) that uses a 3×3 array of block matrices in cases where n is a multiple of 3.
    (c) The general formulation of block matrix multiplication does not require square matrix blocks. This part examines what is needed for the most general formulation. Suppose that the matrices A and B are partitioned into blocks as follows:

$$A = \begin{bmatrix} A_{11} & A_{12} & \cdots & A_{1m} \\ A_{21} & A_{22} & \cdots & A_{2m} \\ \vdots & & \ddots & \vdots \\ A_{n1} & A_{n2} & \cdots & A_{nm} \end{bmatrix}, \quad B = \begin{bmatrix} B_{11} & B_{12} & \cdots & B_{1r} \\ B_{21} & B_{22} & \cdots & B_{2r} \\ \vdots & & \ddots & \vdots \\ B_{m1} & B_{m2} & \cdots & B_{mr} \end{bmatrix}$$

where the number of columns of each block A_{ik} equals the number of rows of each block B_{kj}. Find a corresponding decomposition into blocks of the product matrix $C = AB$, and write down a formula for each block of the product C_{ij} in terms of the blocks of A and B, so that the resulting method generalizes those of parts (a) and (b).

Note: Recall (as explained in this section) that the number of multiplications for an $n \times n$ matrix multiplication is n^3. The above block matrix multiplication scheme in part (a) involves eight $(n/2) \times (n/2)$ matrix multiplications, and thus a total of $8(n/2)^3 = n^3$ matrix multiplications; so we did not save any time with this scheme.

36. *Strassen's Method for Fast Matrix Multiplication.* [First read part (a) of Chapter Exercise 35.] Suppose first that A and B are $n \times n$ matrices, where n is even. In part (a) of Exercise 35, the block matrix multiplication method worked out to computing *eight* $(n/2) \times (n/2)$ matrix multiplications, and this worked out to exactly the same number of real-number multiplications as ordinary matrix multiplication. Strassen [Str-69] discovered a very clever way to reorganize the arithmetic so as to reduce the problem to computing only the following *seven* $(n/2) \times (n/2)$ matrix multiplications:

$$P_1 = (A_{11} + A_{22})(B_{11} + B_{22})$$

$$P_2 = (A_{21} + A_{22})B_{11}$$

$$P_3 = A_{11}(B_{12} - B_{22})$$

$$P_4 = A_{22}(B_{21} - B_{11})$$

$$P_5 = (A_{11} + A_{12})B_{22}$$

$$P_6 = (A_{21} - A_{11})(B_{11} + B_{12})$$

$$P_7 = (A_{12} - A_{22})(B_{21} + B_{22})$$

(a) Show that the blocks of the product matrix $C = AB$, as defined in the decomposition of part (a) of Chapter Exercise 31, are given in terms of the above products by

$$C_{11} = P_1 + P_4 - P_5 + P_7 \qquad C_{12} = P_3 + P_5$$

$$C_{21} = P_2 + P_4 \qquad\qquad\qquad C_{22} = P_1 - P_2 + P_3 + P_6$$

This form of block matrix multiplication is called **Strassen's method**.

(b) Show that if A and B are $n \times n$ matrices where $n = 2^K$ and if we recursively apply Strassen's method to perform the matrix multiplication AB until all of the block submatrices are of size 1×1 (i.e., after K recursions), then the total number of (single number) multiplications and additions required is less than 7^{K+1}. Explain how this fact can be translated into an upper bound for the total number of arithmetic operations being less than a constant times $n^{\log_2 7} \approx n^{2.81}$.

(c) Show more precisely that the number of single number arithmetic operations in recursively using Strassen's formula as indicated in part (b) is $7 \cdot 7^K - 6 \cdot 4^K$.

Suggestion: For part (a), verify that each of the asserted blocks agrees with the basic block formulas in part (a) of Chapter Exercise 31. Use the properties of matrix multiplication, and be careful to recall that it is not commutative. For part (c), use mathematical induction.

Chapter 4 Computer Implementations and Exercises

1. *Program for Matrix Multiplication.*
   (a) Even if your computing platform has a built-in function for this, write your own program with syntax C = Matrix Mult(A,B) that inputs two matrices A and B, and if the matrix product C = AB is defined, the program will output this product. If the matrix multiplication is not possible, the program will output an error message of the form "Error: the inside dimensions must agree in order for matrix multiplication to be possible."
   (b) Use your program to redo Chapter Exercise 3.
   (c) Use your program to redo Chapter Exercise 4.

2. Consider the matrices $A = \begin{bmatrix} 1 & 1 \\ 1 & 1 \end{bmatrix}$ and $B = \begin{bmatrix} 1 & 0 \\ 1 & 1 \end{bmatrix}$.

   (a) Compute A^2, A^3, A^4, A^{10}, A^{40}. Do you notice any patterns?
   (b) Compute B^2, B^3, B^4, B^{10}, B^{40}. Do you notice any patterns?

3. *Roots of Matrices.*
   (a) Can you find a 2×2 matrix A such that $A \neq I$ and $A^2 = I$?
   (b) Can you find a 2×2 matrix S such that $S^2 = A$, where A is the matrix of Computer Exercise 2?
   (c) Can you find a 2×2 matrix T such that $T^2 = B$, where B is the matrix of Computer Exercise 2?
   (d) Can you find a 3×3 matrix $U \neq I$ such that $U^3 = I$.

Suggestion: Try to do these by hand first; if you get stuck, do some computer searches.

4. *Program for Inverting a 2×2 Matrix.*
   (a) Write a program with syntax Ainv = MatInv2(A) that inputs a 2×2 matrix A and outputs its inverse (if it exists)

Ainv, using Theorem 4.3. If the determinant is 0, the program will output an error message of the form "Error: the matrix is not invertible."

(b)   Use your program to redo Chapter Exercise 7.

(c)   Use your program to redo Chapter Exercise 8.

5.   *Program for Modular Integer Matrix Addition and Scalar Multiplication.*

(a)   Write a program with syntax C = ModMatAdd(A,B,m) that inputs two same-sized modular integer matrices A and B, along with a modulus m (an integer > 1), and outputs the sum C = A + B (mod m).

(b)   Write a program with syntax, C = ModMatScalMult (a,A,m), whose inputs are a, a modular integer; A, a modular integer matrix; and m, a modulus (an integer > 1); and outputs the matrix C = aA (mod m).

(c)   Use your program to redo Chapter Exercise 11.

(d)   Use your program to redo Chapter Exercise 12.

6.   (a)   Write a program with syntax C = ModMatrixMult(A, B,m) that inputs two modular integer matrices A and B, along with a modulus m (an integer > 1), and outputs the matrix product C = AB (mod m). If the matrix multiplication is not possible, the program will output an error message of the form "Error: the inside dimensions must agree in order for matrix multiplication to be possible."

(b)   Use your program to redo Chapter Exercise 13.

(c)   Use your program to redo Chapter Exercise 14.

7.   *Program for Inverting a 2 × 2 Modular Integer Matrix.*

(a)   Write a program with syntax Ainv = ModMatInv2(A,m) that inputs a 2×2 matrix modular integer A and a modulus m (an integer > 1), and outputs its modular integer inverse (if it exists) Ainv, using Theorem 4.4. If the determinant is not relatively prime to the modulus, the program will output an error message of the form "Error: the matrix is not invertible mod m."

(b)   Use your program to redo the inverse questions of Chapter Exercise 15.

(c)   Use your program to redo the inverse questions of Chapter Exercise 16.

Suggestion: It will be helpful to call on one of the programs of the computer exercises from the previous section for computing inverses of modular integers.

8.   *Program for Inverting a 3 × 3 Modular Integer Matrix.*

(a)   Write a program with syntax Ainv = ModMatInv3(A,m) that inputs a 3×3 matrix of modular integers A and a modulus m (an integer > 1), and outputs its modular integer inverse (if it exists) Ainv, using Proposition 4.5. If the determinant is not relatively prime to the modulus, the program will output an error message of the form "Error: the matrix is not invertible mod m."

(b) Use your program to redo parts (e) and (f) of Chapter Exercises 17 and 18.

9. *Program for Computing Determinants of Modular Matrices Using Cofactor Expansion.*
   (a) Write a program with syntax detA = CofactorMod Det(A,m) that inputs an $n \times n$ matrix of modular integers A and a modulus m (an integer > 1), and outputs its modular integer determinant detA, using Algorithm 4.1 (cofactor expansion along the first row).
   (b) Use your program to redo the inverse questions of Chapter Exercises 15 and 16.
   (c) Use your program to redo parts (b) and (c) of Chapter Exercises 17 and 18.

10. *Program for Computing Inverses of Modular Matrices Using Cofactor Expansion and Classical Adjoints.*
    (a) Write a program with syntax invA = CofactorMod MatrixInv(A,m) that inputs an $n \times n$ matrix of modular integers A and a modulus m (an integer > 1), and outputs its modular integer inverse invA, using the classical adjoint formula for matrix inverses (Proposition 4.5) and cofactor expansion along the first row (Algorithm 4.1) to compute the needed determinants, or directly calling on the program of Computer Exercise 9.
    (b) Use your program to redo the inverse questions of Chapter Exercises 15 and 16.
    (c) Use your program to redo parts (e) and (f) of Chapter Exercises 17 and 18.

11. *Program for Computing Modular Matrix Inverses on Platforms with Ordinary Inverse Matrix Programs.* As mentioned in the text, computing classical adjoints and determinants using cofactor expansions turns out to be not very efficient. Many computing platforms have built-in programs for numerically computing matrix inverses and determinants. This exercise assumes the reader is working on such a platform.
    (a) Write a program with syntax Ainv=ModMatInv(A,m) that inputs an $m \times m$ matrix of modular integers A and a modulus m (an integer > 1), and outputs its modular integer inverse (if it exists) Ainv. The computation should be done using Equation 4.4, and your platform's numerical programs for (ordinary) determinants and (ordinary) matrix inverses. If the determinant is not relatively prime to the modulus, the program will output an error message of the form "Error: the matrix is not invertible mod m."
    (b) Use your program to redo the inverse questions of Chapter Exercises 15 and 16.
    (c) Use your program to redo parts (e) and (f) of Chapter Exercises 17 and 18.

Suggestion: Suppose that your platform's numerical programs for computing the determinant and inverse of a square matrix A have syntax det(A) and inv(A), respectively. To avoid any round-off errors that might come up in your platform's numerical functions, results that are known to be integer values should be rounded off; thus Equation 4.4 should be implemented in fashion such as:

```
ModMatInv(A, m)  = modinv(round(det(A)),
m)*round(det(A)* inv(A))
```

where the `modinv` program is that of Computer Exercise 10 of Chapter 2.

12. *Program for Hill Cipher.* Write a program `strOut = HillCrypt(str,A)` that inputs a string `str` of plaintext in lowercase English letters, and an invertible encryption matrix `A` (mod 26), representing the key. The output `strOut` should be the corresponding ciphertext (in uppercase letters) after the Hill cipher with encryption matrix `A` is applied to the plaintext.
    (a) Use your program to redo the encryptions of Chapter Exercise 19(a).
    (b) Use your program to redo the encryptions of Chapter Exercise 21(a).
    (c) Use your program to redo the decryptions of Chapter Exercise 19(b).

    Note: In part (c), you will need to change your ciphertexts to lowercase and compute the inverse of the encryption matrix (mod 26).

13. *Program for Decryption of the Hill Cipher.* Write a program `strOut = HillDeCrypt(STR,  A)` that inputs a string `STR` of ciphertext in uppercase English letters and an invertible encryption matrix `A` (mod 26), representing the key. The output `strOut` should be the corresponding plaintext (in lowercase letters) before the Hill cipher with encryption matrix `A` was applied.
    (a) Use your program to redo the decryptions of Chapter Exercise 19(b).
    (b) Use your program to redo the decryptions of Chapter Exercise 21(b).

    Note: This program can be written very simply if it uses the corresponding encryption program of the preceding exercise. The only thing needed will be to compute (inside the program) the inverse of the encryption matrix (mod 26), and this can be done by calling on one of the programs for computing modular inverses of Computer Exercise 10 or 11.

14. *Known Plaintext Attack on the Hill Cipher.* For each part, a corresponding pair of a known plaintext and ciphertext is given, along with an additional ciphertext. The encryption

was performed using a Hill cipher. Determine the encryption matrix and decrypt the additional ciphertext.

(a) plaintext: `gotothecourtyardatthreeoclock`, ciphertext: `EWEJBACEUONYSQVWXRBAOBWMJNQCBL`; additional ciphertext: `IUHRLIUGDFUFKRRFBAJZDHLTSVBAPRRFZC`

(b) plaintext: `proceedtothehamburgmaintrains tationplatformthree`, ciphertext: `BOGUWIHYALIQXVOVOVKIQWWJVHKT SLOJMCSJDITPMTOYVDRG`; additional ciphertext: `LIQNJIIDXKOKSHCWSCKUDIR MPJPILYTBGGKYFAZSQJ`

(c) plaintext: `threemillionisinthesuitcase`, ciphertext: `NACVRYPNKJNCPWRFXBWAALEXJCPL`; additional ciphertext: `TLOYDFGRFFWLETWCRKUXHS DDOYCYPWRFWXHTZVGNUNJDYWELXOWTZDAX`

(d) plaintext: `thediamondsareinasafedepositboxa tthecentralbankinkalamazoo`, ciphertext: `JGGIAGJJVAOXKKKNHBDQCALBFYFKS CYBAFPCFFHELPXYOGIMKYSHPBYANYJB`; additional ciphertext: `ACDQVKDJXCRCRWWHYJUYVT VUFHPBZDVFFBHHFSSTYBYMHATMXLBEXDEHEWWE`

15. *Program for Generating Encoding Matrices for the Hill Cipher.*
    (a) Write a program having the following syntax: `A = Hill EncodingMatrixGenerator(n)`, that inputs a positive integer $n > 1$, and outputs an $n \times n$ matrix A of integers that is invertible (mod 26), and thus may be used as a key for the Hill cryptosystem. The program should be based on the random generation trial-and-error method indicated in the text: Generate an $n \times n$ matrix of integers in the range 0 through 25, and check whether the determinant is relatively prime to 26; if it is, you have an invertible Hill matrix. Otherwise, repeat this process until you obtain one.
    (b) Use your program to generate a 3×3 Hill encryption matrix; check that the outputted matrix is indeed invertible.
    (c) Use your program to generate a 5×5 Hill encryption matrix; check that the outputted matrix is indeed invertible.
    (d) Use the encryption matrix that you obtained in part (c) in the Hill cryptosystem to encrypt the plaintext message: "gotothecourtyardatthreeoclock." Then decrypt the resulting ciphertext.

16. *Estimating the Probability That a Randomly Generated Matrix Is Invertible.*
    (a) How many modular integer 2×2 matrices are there (mod 26)?
    (b) Estimate the probability that a randomly generated modular integer 2×2 matrix will be invertible (mod 26) (and thus be a legal encoding matrix for the Hill cryptosystem) as follows: First initiate a counter: `count = 0`; then start a FOR loop to run through 10,000 trials of the following experiment: randomly generate a modular integer 2×2 matrix

of integers in the range 0 through 25, and check whether its determinant is relatively prime to 26; if it is (so the matrix is invertible), then add 1 to the counter: count = count + 1. After the FOR loop has executed, estimate the desired probability as the final value of the counter divided by the number of trials (i.e., count/10,000).

(c)   Exactly how many 2×2 modular integer matrices are invertible (mod 26)? Use this answer to compute the exact probability of the quantity that was estimated in part (b).

Suggestion: Part (a) is a simple counting problem; use the multiplication principle. For part (c), simply use a nested FOR loop to run through all of the matrices (mod 26) along with a counter to count the number that are invertible (mod 26). The exact probability will be the total number of invertible matrices divided by the total number of matrices (mod 26).

17.   *Estimating the Probability That a Randomly Generated Matrix Is Invertible.* Repeat parts (a) and (b) of Computer Exercise 16, but this time for 3×3 matrices (mod 26).

Note: There are too many matrices here to be able to do part (c) of Computer Exercise 16.

18.   *Programming Strassen's Algorithm for Fast Matrix Multiplication.*
(a)   Write a recursive program with syntax C = Strassen(A,B) that inputs two $n \times n$ matrices, A and B, where n is a power of 2. The output will be the product matrix C = AB, computed by recursively applying Strassen's algorithm as explained in part (b) of Chapter Exercise 36.
(b)   Randomly generate some pairs of $n \times n$ matrices A and B for each of the following values of n: 4, 8, 16, 64, 512.
(c)   Compare the runtimes of computing the matrix products AB for each of the pairs of matrices generated in part (b) with (i) the program you wrote in part (a), (ii) the MatrixMult program that you wrote in Computer Exercise 1, and (iii) your platform's built-in matrix multiplying program (if available). Comment on the results.

Note: Implementations and performance results of Strassen's algorithm of part (a) will vary and will depend on the particular platform that you are using.

5

The Evolution of Codebreaking until the Computer Era

The previous chapters have shown examples of some simple attacks that were tailored to some of the specific ciphers that have been introduced. In the present chapter, we provide details of some more sophisticated attacks. We begin with an example of how a frequency analysis can be used in a ciphertext-only attack on a substitution cipher. As mentioned in Chapter 1, frequency analysis was the first truly groundbreaking achievement in attacking cryptosystems. The techniques used here illustrate that, apart from the statistical information, several linguistic techniques are often employed in such attacks. We then move on to show the ciphertext-only attack on the Vigenère cipher that was discovered by Kasiski and Babbage. This is followed by the more mathematically elegant (and efficient) attack that was discovered by Friedman. Finally, we explain some conceptual and historical details relating to the famous attack on the Enigma machine that was initiated by Polish codebreakers.

Frequency Analysis Attacks

In Chapter 1 we explained how frequency analysis ciphertext-only attacks that were first discovered by the Arabs dethroned substitution ciphers from cutting-edge technology into obsolescence. Such attacks are possible in any language and are based on statistical frequency distributions of plaintext characters, as Table 1.1 provides for the English language. Of course, such distributions are long-term probabilities that are not going to agree exactly with distributions for a particular ciphertext sample, which typically might be of a small or moderate size. The longer the ciphertext, the more faithfully the long-term distributions of Table 1.1 should be represented by the actual frequencies of the ciphertext sample. If the two sets of frequencies were exact, the decryption would of course be very easy: we could simply match plaintext letters to ciphertext letters by matching their frequencies.[*]

---

[*] Only one pair and one triple of low-frequency letters share common frequencies in Table 1.1: g, y, and q, x, z. These could easily be distinguished if all other letter correspondences were known.

TABLE 5.1 Letters of the English Alphabet and Their Frequencies, Grouped by Frequency, and Separated into Vowels and Consonants

	Vowels	Consonants
High frequency: Greater than 4.5%	e .127	t .091
	a .082	n .067
	o .075	s .063
	i .070	h .061
		r .060
Moderate frequency: Greater than 1.5% to 4.5%	u .028	d .043
	y .020	l .040
		c .028
		m .024
		w .023
		f .022
		g .020
Low frequency: Up to 1.5%		b .015
		v .010
		k .008
		j .002
		q .001
		x .001
		z .001

In order to accommodate the discrepancies of actual plaintext frequencies with theoretical predictions, further linguistic properties of languages need to be employed and the frequency-based attack on a given ciphertext sample is accomplished in stages. First, identify a few of the more distinguished letters, substitute them into the plaintext, and examine and use the partially unmasked plaintext to aid in identifying additional letters. This process is repeated, sometimes with some trial and error. Linguistic properties are essential tools, which explains why in early cryptanalysis, linguistics was a more important skill than mathematics. Some useful linguistic properties have to do with how often certain letters appear adjacent to or in combination with other letters. Vowels tend to be very "social" letters in that they often appear adjacent to almost all other letters in words. Certain consonants are far less social and nearly all are almost always adjacent to a vowel (at least on one side) if they occur inside a word. High-frequency letters are usually easiest to identify; what is usually done is that a few high-frequency vowels are first identified, then some high-frequency consonants. Table 5.1 shows a reorganization of the frequency distribution of Table 1.1, where the letters are separated into vowels or consonants, and by high, moderate, or low frequency.

In order to avoid too many technicalities yet at the same time demonstrate the nature of a frequency analysis-based attack, we restrict ourselves to ciphertexts in which original spaces between words have been preserved from the corresponding plaintexts. This is, of course, valuable additional information that reduces the sizes of ciphertext samples needed for an effective frequency analysis. Readers looking for additional challenges may wish to reformat the ciphertext passages so that spaces are removed.[*] We caution the reader that even with spaces preserved, computers should be used to perform the tasks required to complete a frequency analysis ciphertext-only attack.

Example 5.1

Use a frequency analysis-based attack to decrypt the following ciphertext that was created using a substitution cipher in which the spaces between words have been preserved:

```
AHFB GXVUD PW P AXJ TKHYU XM WTU YHMBXYYUD YPMWP
FGPNP VPGGUE OHDJU ZXGGUNY IGPFU XW LPY FPGGUD XW
YWKKD APFB QNKZ WTU NKPD TPGQ TXDDUM PZKMJ WTU WNUUY
WTNKHJT LTXFT JGXZIYUY FKHGD AU FPHJTW KQ WTU LXDU
FKKG VUNPMDP WTPW NPM PNKHMD XWY QKHN YXDUY WTU TKHYU
LPY PIINKPFTUD AE JNPVUGGUD DNXVULPEY LTXFT LKHMD
PAKHW WTNKHJT LXDUYINUPDXMJ GPLMY PMD HMDUN WTU
XMWUNGPFXMJ AKHJTY KQ WPGG IKIGPNY PW WTU NUPN
WTXMJY LUNU KM UVUM P ZKNU YIPFXKHY YFPGU WTPM PW
WTU QNKMW WTUNU LUNU JNUPW YWPAGUY LTUNU P DKRUM
JNKKZY PMD AKEY TUGD QKNWT NKLY KQ VXMUFGPD YUNVPMWY
FKWWPJUY PM UMDGUYY PMD KNDUNGE PNNPE KQ KHWTKHYUY
GKMJ JNPIU PNAKNY JNUUM IPYWHNUY KNFTPNDY PMD AUNNE
IPWFTUY WTUM WTUNU LPY WTU IHZIXMJ IGPMW QKN WTU
PNWUYXPM LUGG PMD WTU AXJ FUZUMW WPMB LTUNU OHDJU
ZXGGUNY AKEY WKKB WTUXN ZKNMXMJ IGHMJU PMD BUIW FKKG
XM WTU TKW PQWUNMKKM
```

Solution: We first perform an initial frequency count for the most often occurring ciphertext letters by dividing the total number of occurrences of a given (often-appearing) ciphertext letter by the total number of ciphertext letters (less the spaces), of which there are 685. Table 5.2 shows the actual frequencies in the ciphertext sample of the most often-occurring ciphertext letters. Alongside each ciphertext letter we also give a count of the number of ciphertext letters that *do not* appear directly adjacent to it in any *cipherword*. Thus, low numbers for this second count mean that the letters are more "social."[†]

---

[*] We remind the reader that all ciphertext passages (both in the text proper, as well as those in the exercises) may be downloaded as text files from this book's Web site, the URL of which is given in the Preface of this book.

[†] Hand computations of this sort should be avoided. The applets created for this chapter will perform these and all other tasks needed to perform frequency analysis-based attacks on substitution ciphers. Alternatively, readers who wish to write their own programs for such tasks may consult the computer implementation material at the end of this chapter for some useful advice.

TABLE 5.2 Ciphertext Letters of Example 5.1 Having the Highest Frequencies, Along with the Number of Ciphertext Letters That Are Not Adjacent to Them in Any of the Ciphertext Words

Ciphertext Letter	Observed Frequency	Nonadjacent Count	Ciphertext Letter	Observed Frequency	Nonadjacent Count
U	.120	6	M	.064	14
P	.093	6	T	.061	16
W	.077	13	G	.050	12
N	.076	7	D	.048	15
K	.073	7	X	.042	8
Y	.067	10	J	.034	16

Comparing this with the predicted frequencies of Table 5.1, it seems quite likely that (the ciphertext letter) *U* represents (the plaintext letter) *e*. The next highest frequency vowels—*a*, *o*, and *i* (according to Table 5.1)—are a bit less obvious from the Table 5.2 statistics. Each of *P*, *N*, and *K* has a high frequency, and these are very social letters (due to their low nonadjacent count numbers of Table 5.2). The letter *X* is also very social but has a bit lower frequency. The most social consonants tend to be *r*, *s*, and *t* (in this order). Since the digraph *oo* is common in English, whereas *aa* and *ii* are rare, and since *KK* appears several times in the above ciphertext passage and not adjacent to any probable vowels, it seems quite plausible that *K* must represent *o*. Noting that *P* appears as a single ciphertext word, it must represent a vowel (either *a* or *i*), and because of its high frequency, it seems that *P* must represent *a*. Making these three vowel substitutions into the ciphertext produces the following (first stage):

```
AHFB GXVeD aW a AXJ ToHYe XM WTe YHMBXYYeD YaMWa
FGaNa VaGGeE OHDJe ZXGGeNY IGaFe XW LaY FaGGeD XW
YWooD AaFB QNoZ WTe NoaD TaGQ TXDDeM aZoMJ WTe WNeeY
WTNoHJT LTXFT JGXZIYeY FoHGD Ae FaHJTW oQ WTe LXDe
FooG VeNaMDa WTaW NaM aNoHMD XWY QoHN YXDeY WTe ToHYe
LaY aIINoaFTeD AE JNaVeGGeD DNXVeLaEY LTXFT LoHMD
aAoHW WTNoHJT LXDeYINeaDXMJ GaLMY aMD HMDeN WTe
XMWeNGaFXMJ AoHJTY oQ WaGG IoIGaNY aW WTe NeaN
WTXMJY LeNe oM eVeM a ZoNe YIaFXoHY YFaGe WTaM aW
WTe QNoMW WTeNe LeNe JNeaW YWaAGeY LTeNe a DoReM
JNooZY aMD AoEY TeGD QoNWT NoLY oQ VXMeFGaD YeNVaMWY
FoWWaJeY aM eMDGeYY aMD oNDeNGE aNNaE oQ oHWToHYeY
GoMJ JNaIe aNAoNY JNeeM IaYWHNeY oNFTaNDY aMD AeNNE
IaWFTeY WTeM WTeNe LaY WTe IHZIXMJ IGaMW QoN WTe
aNWeYXaM LeGG aMD WTe AXJ FeZeMW WaMB LTeNe OHDJe
ZXGGeNY AoEY WooB WTeXN ZoNMXMJ IGHMJe aMD BeIW FooG
XM WTe ToW aQWeNMooM
```

Although this is still quite cryptic, some probable words shed some important additional insights. The word *WTe* appears very often, and most probably represents *the*, leading to the correspondences: $W \leftrightarrow t$, and $T \leftrightarrow h$. Also, the high frequency of W makes t a very plausible plaintext correspondent.* The last three words now become: *the hot aQteNMooM*, which looks like *the hot afternoon*, and this, in turn, leads us to the correspondences: $Q \leftrightarrow f$, $N \leftrightarrow r$, and $M \leftrightarrow n$. The second of these implies, from what was said in the first stage, that the final frequent and social vowel i should be represented by X. (This correspondence is further supported by the presence of the word $XW \leftrightarrow it$.) Making these six additional substitutions into the Stage 1 partial ciphertext produces the following (second stage):

```
AHFB GiVeD at a AiJ hoHYe in the YHnBiYYeD Yanta
FGara VaGGeE OHDJe ZiGGerY IGaFe it LaY FaGGeD it
YtooD AaFB froZ the roaD haGf hiDDen aZonJ the treeY
throHJh LhiFh JGiZIYeY FoHGD Ae FaHJht of the LiDe
FooG VeranDa that ran aroHnD itY foHr YiDeY the hoHYe
LaY aIIroaFheD AE JraVeGGeD DriVeLaEY LhiFh LoHnD
aAoHt throHJh LiDeYIreaDinJ GaLnY anD HnDer the
interGaFinJ AoHJhY of taGG IoIGarY at the rear thinJY
Lere on eVen a Zore YIaFioHY YFaGe than at the front
there Lere Jreat YtaAGeY Lhere a DoRen JrooZY anD
AoEY heGD forth roLY of VineFGaD YerVantY FottaJeY
an enDGeYY anD orDerGE arraE of oHthoHYeY GonJ JraIe
arAorY Jreen IaYtHreY orFharDY anD AerrE IatFheY
then there LaY the IHZIinJ IGant for the arteYian
LeGG anD the AiJ FeZent tanB Lhere OHDJe ZiGGerY
AoEY tooB their ZorninJ IGHnJe anD BeIt FooG in the
hot afternoon
```

With the many key letters already deciphered, the remaining decryption is now very easy and can be accomplished by identifying some nearly complete passages (with probable words). For example, from the following segment:

```
it YtooD AaFB froZ the roaD haGf hiDDen aZonJ the
treeY throHJh
```

it seems the corresponding plaintext should be

it stood `AaFB` from the road half hidden among the trees through

This yields the correspondences $Y \leftrightarrow s$, $D \leftrightarrow d$, $Z \leftrightarrow m$, $G \leftrightarrow l$, $H \leftrightarrow u$, $J \leftrightarrow g$.

From these representations, the ciphertext passage: "tooB their ZorninJ IGHnJe" can be partially cleared up to read: "tooB their morning llunge" so it appears (from the first and last probable words) that $B \leftrightarrow k$, and $I \leftrightarrow p$. The subsequent ciphertext words "anD BeIt FooG in the hot afternoon" now clearly mean "and kept cool in the hot afternoon," providing

---

* There are other less probable possibilities (such as *she*), so if *the* did not seem to produce compatible results, we could go back and try another possibility (trial and error).

us with $F \leftrightarrow c$. Continuing in this fashion (perhaps with partial updates of the ciphertext) quickly leads us to the complete key as well as the original plaintext (shown here with original punctuation), which is a passage from the novel *The Call of the Wild*, written in 1917 by Jack London.[*]

Buck lived at a big house in the sun-kissed Santa Clara Valley. Judge Miller's place, it was called. It stood back from the road, half hidden among the trees, through which glimpses could be caught of the wide cool veranda that ran around its four sides. The house was approached by gravelled driveways which wound about through wide-spreading lawns and under the interlacing boughs of tall poplars. At the rear things were on even a more spacious scale than at the front. There were great stables, where a dozen grooms and boys held forth, rows of vine-clad servants' cottages, an endless and orderly array of outhouses, long grape arbors, green pastures, orchards, and berry patches. Then there was the pumping plant for the artesian well, and the big cement tank where Judge Miller's boys took their morning plunge and kept cool in the hot afternoon.

Exercise for the Reader 5.1

Use the applet tools for this chapter to perform a frequency analysis-based ciphertext-only attack to decrypt the following ciphertext that was created using a substitution cipher. (*Note*: All ciphertexts in the text and exercises of this book may be downloaded from the book's website.)

```
AUNDQHUDQA RWGEDN KWYN RUW YGMEDQRSK NWSYAWO YSS
UWN WMJWNEXWGAR YGO JYNAEYS RQTTWRRWR AD Y TYNWWN
OYESK DG AUW SELNYNK RAWJR DN EG AUW UYSS DP AUW
XYEG LQESOEGH AUW TDWOR AYSVWO DP BUYA RUYSS BW OD
BUWG BW PEGERU TDSSWHW WFWG AUW HENSR BUD VGWB AUYA
AUWK BWNW HDEGH AD LW XYNNEWO JNWAWGOWO AD LW
TDGREOWNEGH EXJDNAYGA LQREGWRR JDREAEDGR WFWG AUWK
BUD VGWB AUYA AUWK BDQSO UYFW AD BDNV UEGAWO YLDQA
PYLQSDQR RQEADNR YR PDN TYNDS RUW BYR YG DNJUYG UWN
DGSK GWYN NWSYAEFW BYR Y FYGESSY PSYFDNWO RERAWN
XYNNEWO AD YG DJAETEYG EG RA JYQS RUW UYO QRWO XDRA
DP AUW XDGWK PNDX UWN PYAUWNR WRAYAW RUW BYR GDA EG
SDFW AUYA ER GDA DPAWG GDN WFWN SDGH YA Y AEXW RUW
BDQSO WYNG UWN SEFEGH LQA UDB RUW BYR AD WYNG EA
UDB RUW BYR AD TDGZQWN AUW BDNSO YSXDRA WGAENWSK
PDN AUW BDNSOR DBG HDDO RUW OEO GDA RWW
```

As was demonstrated in the solution to this example, apart from the statistical data that aided some of the initial stages, the techniques were more linguistic than mathematical. In contrast, the remaining codebreaking methods that we present are entirely mathematical.

---

[*] The author wishes to acknowledge gratitude to the University of Virginia Library for permission to download this segment (as well as others that will give rise to other exercises in this chapter) from their online resource http://etext.lib.virginia.edu/ebooks/Llist.html that allows one to download entire novels from a vast collection that are in the public domain.

Figure 5.1 Charles Babbage (1791–1871), English mathematician and computer scientist.

The Demise of the Vigenère Cipher

The Vigenère cipher finally succumbed to a ciphertext-only attack in the mid-1850s, thanks to the efforts of the remarkable scientist Charles Babbage* (Figure 5.1). Babbage's breakthrough was not publicized. The British intelligence office needed to keep this valuable discovery confidential, as they were in the midst of the Crimean War, and knowing how to crack the Vigenère cipher would give the Brits a valuable strategic advantage over their rivals. The first published discovery of a successful ciphertext-only method for the Vigenère cipher appeared in 1863 by the German cryptographer and retired Prussian army officer Friedrich W. Kasiski (1805–1881).

Babbage became motivated to work on breaking the Vigenère cipher through a public challenge. An English dentist who was an amateur cryptographer published in 1854 what he thought was an ingenious new cryptosystem. Babbage read the article and found that this system had actually been discovered by Vigenère some 300 years prior. The dentist tried to defend himself by challenging Babbage to crack "his" cryptosystem. No one had accomplished this feat for three centuries, and despite the fact that cracking a system is not the same as inventing one, Babbage now became fixated on doing the latter. We briefly outline the elements of the Babbage/

---

* Although Babbage did some important work in mathematics and cryptography, as well as other diverse scientific fields, he is most famous for his work in designing mechanical calculators. He was motivated to do this from having to work with tedious tables of logarithms and from hearing stories of how the numerous errors in these tables arising from the hand calculations used to produce them had been responsible for several shipwrecks. His first machines were so well received that he was relieved from all of his teaching duties as a chaired professor of mathematics at Cambridge in order to pursue the design of his great "Analytical Engine." Despite this and the public financial support he was receiving, the sheer size of this project prevented its completion during his lifetime, although he did complete all of the specifications for it. The British government finally built his machine in 1991 to honor this part of its national heritage in the bicentennial year of Babbage's birth. It is on display in the London Science Museum. His designs bear a striking resemblance to the designs of modern electronic computers of this day. He was also the first to conceive of the contemporary science of what is known today as operations research. Babbage was remarkably ahead of his time with his vision and insights.

Figure 5.2 William F. Friedman (1891–1969), American cryptographer.

Kasiski method and then show a more efficient mathematical method that was subsequently discovered by American cryptographer William F. Friedman[*] (Figure 5.2).

The Babbage/Kasiski Attack

The main property that made the Vigenère cipher so recalcitrant against attacks for so many years was its ability to smooth out the frequencies of ciphertexts. This is nicely illustrated by comparing Figure 5.3 and Figure 5.4. Both figures are derived from two ciphertexts from the same plaintext of Example 5.1. Figure 5.3 shows the frequency distribution of the substitution cipher of Example 5.1, while Figure 5.4 shows the frequency distribution of the ciphertext obtained when the Vigenère cipher with keyword *friedman* is applied to the same plaintext.

The distribution of Figure 5.4 is starkly less informative than that of Figure 5.3; there are no distinguished peaks, and the bars also no longer get very low. Putting this into the context of Shannon's perfect secrecy theory (Chapter 1): As the plaintext passages get longer and longer, and as the Vigenère keyword also gets longer and longer (and is randomly constructed), the distribution of ciphertext letters would become *uniform* (that is, all letters would have the same height bars). As a first step, Babbage realized that if he could determine the length of the Vigenère keyword, and if the ciphertext passage was sufficiently long, then he could determine

---

[*] When he was an infant, Friedman's parents emigrated from Russia to the United States, where his father took work with the post office as an interpreter. After earning a BS degree in genetics, he went on to start a career as a geneticist at a private laboratory in Chicago in 1915. The company president had an interest in cryptography, and since Friedman was known for his photographic skills, Friedman was sent to England to photograph some historical documents. There Friedman met a cryptographer, Elizebeth Smith. Friedman became very interested both in this woman and in cryptography during this trip. They soon married, and he became a cryptographer for his company, in a newly formed Department of Codes and Ciphers. Friedman did extraordinary work in cryptography for the rest of his life. His company's services were contracted by many branches of the U.S. government for training and breaking codes. He published several seminal papers on the subject and was instrumental in cracking the Japanese "Purple" machine cipher (an analogue of the German Enigma machine) in 1939. He also invented the *index of coincidence*, which is a very useful mathematical tool for attacking an assortment of cryptosystems.

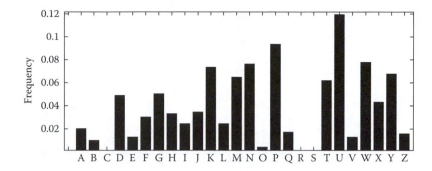

Figure 5.3 Frequency counts of the ciphertext letters for the ciphertext of the substitution cipher of Example 5.1.

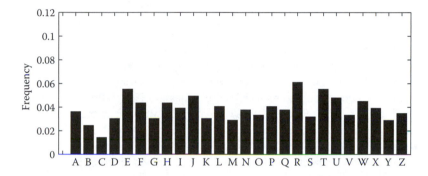

Figure 5.4 Frequency counts of the ciphertext letters for the ciphertext obtained when the Vigenère cipher with keyword "friedman" is applied to the plaintext of Example 5.1.

the keyword and thus break the cipher. Indeed, suppose it is known that the Vigenère keyword had length 8, but the exact keyword (*friedman*) was not known. By the periodic nature of the Vigenère cipher, any plaintext characters separated by a multiple of 8 would be subject to the same shift cipher. For example, the 1st, 9th, 17th, ... = $(8k + 1)$st characters would all be subjected to the shift determined by the first keyword character (f), the 2nd, 10th, 18th, ..., $(8k + 2)$nd characters would all be subjected to the shift determined by the second keyword character (r), and so on. Thus we could do 8 (or whatever keylength is being used) separate frequency analyses on these disjoint ciphertext character sets to determine the key characters one by one. Figure 5.5 shows the frequency distribution for the 1st, 9th, 17th, ... = $(8k + 1)$st ciphertext characters.

The resulting distribution is much more distinguished than the conglomerate distribution of Figure 5.4. Granted, the sample of characters is (1/8) the size of the complete sample, but nonetheless, it is much easier to infer the shift operator from this frequency distribution than it was to ascertain the exact permutation from a general substitution cipher.[*] By far, the highest bar belongs to the ciphertext letter *J*, so a natural first guess is

---

[*] A good analogy is this: A general permutation is like a perfectly shuffled deck of cards while a shift permutation is like a sorted deck of cards that has been "cut" just once (and subjected to no other shuffling).

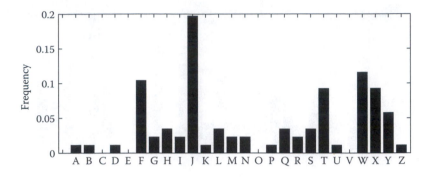

Figure 5.5 Frequency counts of the ciphertext letters for the 1st, 9th, 17th, ... = $(8k + 1)$st ciphertext characters obtained when the Vigenère cipher with keyword "friedman" is applied to the plaintext of Example 5.1.

that J represents e. This is corroborated by comparing other distinguishing characteristics of Figure 5.5 with the frequency distribution of Table 1.1. Indeed, F, which is four bars to the left of J, also has a relatively high bar, and a, which is four characters before e, has a relatively high frequency in Table 1.1. Also, E and C, which are one and three characters before F, never occurred in the ciphertext (no bars), and this corroborates with the extremely low frequencies of the plaintext characters z and x, which are one and three characters before a. So we have provided ample justification that $J \leftrightarrow e$ or that $F \leftrightarrow a$; that is, we have (correctly) determined the first character of the Vigenère keyword to be f.

The problem is thus reduced to determining the length of the Vigenère keyword, given only the ciphertext. The method will exploit the entropy of the Vigenère cryptosystem and make use of recurring *grams* in the ciphertext. A **gram** is just a short string that is embedded in either the plaintext or ciphertext, if a gram has n characters it is called an *n*-gram. Thus *the* is a common 3-gram (or *trigram*) in written English but typically will be quite rare in ciphertexts. We discussed how frequency analysis of ordinary substitution ciphers tends to break down when we consider plaintext grams of length greater than 2. The reason for this is that there are too many possibilities. For example there are $26^4 = 456,976$ possible 4-grams. In a plaintext passage of moderate length, certain grams of length at least 4 are bound to occur more than once; if a decent polyalphabet block cipher (like Vigenère's) is applied, the likelihood of a gram of length at least 4 appearing more than once becomes extremely low, unless the two occurrences are separated by a multiple of the keylength (so that they will be identical encryptions of the corresponding plaintext gram). Of course, it is still theoretically possible that a duplicated gram of length 4 or more could have occurred by pure accident, but such occurrences are rare. In general, the longer the gram, the more reliable the duplication data will be in determining the keylength. The keylength must be a divisor of all of these nonaccidental separation distances and, hence, must be a divisor of their GCD. This is the essence of how Babbage and Kasiski were able to determine keylengths from Vigenère ciphertexts; the details are illustrated in the following example.

Example 5.2

Identify all duplicated 4- and 5-grams in the following cipher-text that was produced by applying the Vigenère cipher with keyword *friedman* to the plaintext of Example 5.1. Deduce from the data the possible keylengths.

```
GLKOOUVRIRBEEUGUTLAILZTUJJCRNUSFJUAEQFAPQRZEYMLYJPR
YGSEZNCTIUEPYFTMMWIAFHRTPHPIGXKWSGNAPPWZSPFHRWFIHKM
LSMZLHHZAZTEOXKQTEJVAXKDOHLYELLOHTQZUTVQSPTLTHEQCNZ
XPXRRTUJNQHHOOBQMMVDZDNYYIXUMNNWFCRGUTFKFCVVUDRXKPI
KAUFJNIWDBPETRKLHPBLLIIZHXLRIUZMYQWNDJELLOHJTLVHDNO
HYKPVRGGUBZLIVBRRFUQRJXAJSJIRGGNQJIBLHUNGJITEFUNTGF
CKKEOSYRTPSAPYFIAEWFHRWVIVWTIALJEIUQOAJMMRDYOEJJXEF
UOHXJKEOQTUFEIXWTESWFVXWTEEJNMVHSRRFKAXDNLRXNPIUQAQ
TQMRJDOBRJIRGNOLXYMPGROEYYZSZEOSAZVIFXAQXVZZDZTFHFB
XDSEFFEMRGXEFXRVHRDDRWCGEUDALTWWYWTOHXVAPRZGTWRXIDD
BBWJOVHQNCFJBYUQSBWTPEUPSNSUJIUDYCFKKLHETUJEBLHDEJF
JBLHBUZUZVKSXAAYWWVWTENWKMWLMNJJCTEQPTUJSQKFQMRSKBE
QWWUJIMNXPGRRZTPHDSOTPAXRAKGMVQVPARANEOTOGNTJRVHNQP
GHFWPLZTUJYWXDRTRWEWSQ
```

Note: As with all ciphertexts in this book, this can be down-loaded from the book's Web page (ciphertext pages). Of course, it is not a good idea to do this by hand. The reader can use the applets for this chapter to find the needed duplication data for this and other ciphertexts. Readers who wish to write their own programs may consult the computer implementation material at the end of this chapter for some guidance.

Solution: Table 5.3 below is a summary of the relevant data for all duplicated 4- and 5-grams that appear in the ciphertext.

The length of the keyword should divide each of the numbers in the last column. In particular, it should also divide their gcd, which is 8. Interestingly, the gcd of any two of these numbers is also 8. This implies that the possible keylengths are 2, 4, or 8.

TABLE 5.3 Duplication Data for All Ciphergrams of Length 4 and 5 for the Ciphertext of Example 5.2

Ciphergram	First Occurrence Index	Number of Duplications	GCD of Separations
LZTU	21	2	648
ZTUJ	22	2	648
FHRW	94	2	232
ELLO	131	2	112
LLOH	132	2	112
JIRG	282	2	136
XWTE	372	2	8
LZTUJ	21	2	648
ELLOH	131	2	112

To determine the key, one would proceed with a frequency analysis of the ciphertext character sets whose separations are multiples of eight (as explained earlier). This would specify the key.

In principle, the Babbage/Kasiski attack will work to decrypt ciphertexts produced by any Vigenère cryptosystem. Of course, longer keywords would require larger cipherdata sets. For practical reasons (to keep the lengths of ciphertext a reasonable size), we will never use a Vigenère keyword of length greater than 8 in any such decryption exercises.

Exercise for the Reader 5.2

Use the Babbage/Kasiski attack to determine the keyword of Vigenère cryptosystem that produced the following ciphertext that comes from a passage of the novel *The Legend of Sleepy Hollow*, written by Washington Irving in 1886. This ciphertext can be downloaded from the book's Web page (ciphertext pages). The reader may use the applets for this chapter to find the needed duplication data for this and other ciphertexts; alternatively, readers interested in writing and using their own programs may consult the computer implementation material at the end of this chapter for some useful tips.

```
TCVTIGRZWPERMYIGDDLKWWISXJLJOHULMALLGXWLQWAZRIZLVRX
TXDKRTTBVTEHIIOETSACPGJAXBKTNSDLHWYMATVBUISQPEPGGMH
IZRVQQNGSCUHMYVDCJAKLTTHLJICQQZJXDZPBNEHAHBQISELMYE
EIQBTJTIUBKYGVHWWVPVWBTEATBMFXWMWBDIBIFAZRTIQWJXGWY
XYEGLWHIIPLMNJXXBDLZHXLVHKLTAKTWXHWIMYIHCQLDSIMWAIS
JOKMYIIPXGUIGAWHIQIPHZIINLRPETDCUPRWHEHIKEHQGXRRSDD
GZWWMGEZOTBKXKVPQOBEKVIUFVRIARYRKWWVMRFDDHFVMCBKXZ-
RIMQLVFACHHWXWMVNDQTZVDPWDUHYRMCBEKFACAKKVHHWIVCSJL
ZAZVAMGBEXDVRMYMCOQXJW
```

The Friedman Attack

Friedman published his attack on periodic substitution ciphers (like the Vigenère cipher) in 1925. Unlike the Babbage/Kasiski attack, Friedman's attack is purely mathematical and typically requires less ciphertext than the former (and no guesswork). The method uses some probabilistic concepts, for which the reader may wish to briefly review Appendix B. At the core of the Friedman attack is the so-called index of coincidence, which is defined for any pair of same-length strings by the following definition. The following notation will be convenient when dealing with such indices. *Notation:* For any two objects x and y, the **Kronecker delta** $\delta(x, y)$ is defined as follows:

$$\delta(x, y) = \begin{cases} 1, & \text{if } x = y \\ 0, & \text{if } x \neq y \end{cases}$$

The Index of Coincidence

Definition 5.1

Given two strings (in any alphabet) of the same length N:

$$\text{STR1} = \alpha_1\alpha_2\cdots\alpha_N, \quad \text{STR2} = \beta_1\beta_2\cdots\beta_N$$

their **index of coincidence** is given by

$$\mathscr{I}(\text{STR1}, \text{STR2}) = \frac{1}{N}\sum_{i=1}^{N}\delta(\alpha_i, \beta_i)$$

In words, the index of coincidence of two strings (of the same length) simply counts the number of exact matches in their corresponding characters and divides the total by the (common) length of the strings. Thus, $N \cdot \mathscr{I}(\text{STR1}, \text{STR2})$ is the total number of exact matches of corresponding characters in the strings.

Example 5.3

Find the indices of coincidence of the following pairs of strings:

(a) STR1 = ABCDEF, STR2 = BCDEFG

(b) STR1 = 01010, STR2 = 00000

Solution: Part (a):

$\mathscr{I}(\text{STR1}, \text{STR2}) = (1/6)[\delta(A,B) + \delta(B,C) + \delta(C,D) +$
$\delta(D,E) + \delta(E,G) + \delta(F,G)] = (1/6)[0 + 0 + 0 + 0 + 0 + 0] = 0$

Part (b):

$\mathscr{I}(\text{STR1}, \text{STR2}) = (1/5)[\delta(0,0) + \delta(1,0) + \delta(0,0) + \delta(1,0) + \delta(0,0)]$
$= (1/5)[1 + 0 + 1 + 0 + 1] = 3/5 = 0.6$

Expected Values of the Index of Coincidence

It is clear that $0 \le \mathscr{I}(\text{STR1}, \text{STR2}) \le 1$; the first equality is attained when the corresponding entries of each of the two strings is different [as in part (a) of the preceding example], whereas the second equality occurs when the strings are identical. What makes the index of coincidence such a useful tool is that its expected values are easily computed. The following proposition shows how this can be done if we know the underlying distributions of the characters in the strings.

Proposition 5.1: Expected Values of Indices of Coincidence

(1) Suppose that α, β are letters from the English alphabet (A, B, ..., Z), such that α came from a sampling distribution given

by a 26-component vector Vec1,[*] and β came from a sampling distribution given by a 26-component vector Vec2, then $E[\delta(\alpha, \beta)] = \text{Vec1} \cdot \text{Vec2}$.

(2) Suppose that $STR1 = \alpha_1 \alpha_2 \cdots \alpha_N$, $STR2 = \beta_1 \beta_2 \cdots \beta_N$ are strings taken from the English alphabet (A, B, ..., Z). If the characters (α_i) of STR1 have distribution given by a 26-component vector Vec1 and the characters (β_i) of STR2 have distribution given by a 26-component vector Vec2, then the expected value of the index of coincidence of the two strings is the dot product of their distribution vectors, that is,

$$E[\mathscr{I}(STR1, STR2)] = \text{Vec1} \cdot \text{Vec2}$$

Before proving this important result, we observe what will be its most important ramifications for the Friedman attack. These are collected in the following corollary.

Corollary 5.2

Suppose that $STR1 = \alpha_1 \alpha_2 \cdots \alpha_N$, $STR2 = \beta_1 \beta_2 \cdots \beta_N$ are strings taken from the English alphabet (A, B, ..., Z).

(a) If both string characters were randomly selected, then the expected value of their index of coincidence is $1/26 \approx 0.038$, i.e., $E[\mathscr{I}(STR1, STR2)] = 1/26$.

(b) If the string characters are sampled from "ordinary" written English, in accordance with the frequency distribution of Table 1.1, then the expected value of their index of coincidence is 0.067, i.e., $E[\mathscr{I}(STR1, STR2)] \approx 0.067$.

(c) If the strings are ciphertexts resulting from a substitution cipher applied to two plaintexts that came from "ordinary" written English, in accordance with the frequency distribution of Table 1.1, then the expected value of their index of coincidence is 0.067, i.e., $E[\mathscr{I}(STR1, STR2)] \approx 0.067$.

(d) If the characters in the two strings result from two different shift ciphers, then $0.032 \le E[\mathscr{I}(STR1, STR2)] \le 0.045$.

It is quite remarkable that if the strings result from different shifts, their indices of coincidence are significantly smaller than if they came from ordinary written English. We first show how the corollary follows from the proposition, and then give proof of the latter.

Proof of Corollary 5.2: Part (a): With random letter selection, the probability that any string character equals any particular letter (A through Z) is 1/26. So we have $\text{Vec1} = \text{Vec2} = (1/26, 1/26, \cdots, 1/26)$.

---

[*] This means that the ith component of Vec1 is the probability that α equals the ith letter of the alphabet. So $\text{Vec1} = (P[\alpha = A], P[\alpha = B], \cdots, P[\alpha = Z])$, and these 26 components must add up to $1 = 100\%$. For example, if α were randomly selected, then each of the components of Vec1 is 1/26.

There are 26 components in these vectors, so Proposition 5.1 yields $E[\mathscr{I}(STR1, STR2)] = Vec1 \cdot Vec2 = 26 \cdot (1/26)^2 = 1/26$.

Part (b): With letters being selected according to the distribution of Table 1.1, the probability that any string character equals any particular letter (A through Z) is specified by the corresponding entry in Table 1.1. Thus the dot product of Proposition 5.1 is simply the dot product of the vector of the 26 frequencies of Table 1.1 with itself:

$$Vec1 \cdot Vec2 = P(A)^2 + P(B)^2 + \cdots + P(Z)^2 \underset{\substack{From \\ Table\ 1.1}}{=} (.082)^2 + (.015)^2 +$$
$$\cdots + (.001)^2 \approx 0.067$$

The reader may check this dot product.

Part (c): Letting $STR1' = \mu_1\mu_2 \cdots \mu_N$, $STR2' = \lambda_1\lambda_2 \cdots \lambda_N$ denote the corresponding plaintext strings, we know from part (b) that $E[\mathscr{I}(STR1', STR2')] \approx 0.067$. We also know that for some permutation σ of the alphabet $\{A, B, \ldots, Z\}$ we have $\alpha_i = \sigma(\mu_i)$ and $\beta_i = \sigma(\lambda_i)$ for each index i. Thus, $P[\alpha_i = A] = P[\sigma(\mu_i) = A] = P[\mu_i = \sigma^{-1}(A)]$, and similarly $P[\beta_i = A] = P[\lambda_i = \sigma^{-1}(A)]$. Since the same relationship holds for the other 25 letters, and since σ^{-1} is a permutation, it follows that the 26 elements for the distribution vector for both α_i and β_i are the same as those for the distribution vector of part (b), but in a different order. From this and Proposition 5.1, we may infer that $E[\mathscr{I}(STR1, STR2)] = E[\mathscr{I}(STR1', STR2')] \approx 0.067$, as asserted.

Part (d): This part follows if we can show that the dot product of the distribution vector of Table 1.1, any (nonidentity) cyclic shift of this vector will result in a number that approximately falls in the range 0.032 to 0.045. As this task is readily completed with the aid of a computer, we leave it as Computer Exercise 8(c).□

Proof of Proposition 5.1: Part (1): Since $\delta(\alpha, \beta) = 0$ or 1, the expectation $E[\delta(\alpha, \beta)]$ is simply the probability $P[\delta(\alpha, \beta) = 1] = P[\alpha = \beta]$. This probability can be easily computed by conditioning on the outcome of α, which must be one of the 26 letters:

$$P[\alpha = \beta] = P[\alpha = \beta \mid \alpha = A] \cdot P[\alpha = A] + P[\alpha = \beta \mid \alpha = B] \cdot P[\alpha = B] +$$
$$\cdots + P[\alpha = \beta \mid \alpha = Z] \cdot P[\alpha = Z]$$
$$= P[\beta = A] \cdot P[\alpha = A] + P[\beta = B] \cdot P[\alpha = B] +$$
$$\cdots + P[\beta = Z] \cdot P[\alpha = Z]$$
$$= Vec1 \cdot Vec2$$

Part (2): By linearity of expectation (see Appendix B), we may write

$$E[\mathscr{I}(STR1, STR2)] = E\left[\frac{1}{N}\sum_{i=1}^{N} \delta(\alpha_i, \beta_i)\right] = \frac{1}{N}\sum_{i=1}^{N} E[\delta(\alpha_i, \beta_i)]$$

By part (1), each $E[\delta(\alpha_i, \beta_i)]$ in this sum equals $Vec1 \cdot Vec2$, and the result follows. Substituting this into the preceding formula produces

$$E[\mathscr{I}(STR1, STR2)] = \frac{1}{N} \sum_{i=1}^{N} E[\delta(\alpha_i, \beta_i)] = \frac{1}{N} N \cdot Vec1 \cdot Vec2$$

$$= Vec1 \cdot Vec2 \; \square$$

In the above proof, there was nothing special about the fact that the English alphabet was used. It can easily be modified into a proof of the following more general result (Exercise 41).

Theorem 5.3: Expected Values of Indices of Coincidence

Suppose that $STR1 = \alpha_1 \alpha_2 \cdots \alpha_N$, $STR2 = \beta_1 \beta_2 \cdots \beta_N$ are strings taken from any alphabet $(X_1, X_2, \cdots, X_\ell)$. If the characters (α_i) of STR1 have distribution given by an ℓ component vector Vec1, the characters (β_i) of STR2 have distribution given by an ℓ component vector Vec2, then the expected value of the index of coincidence of the two strings is the dot product of their distribution vectors, that is,

$$E[\mathscr{I}(STR1, STR2)] = Vec1 \cdot Vec2$$

Example 5.4

Suppose that $STR1 = \alpha_1 \alpha_2 \cdots \alpha_N$, $STR2 = \beta_1 \beta_2 \cdots \beta_N$ are strings whose characters come from a three-character alphabet {a, b, c} in a special written language where the letter a appears 50% of the time (as a character in the written language) and the other two characters each have frequencies of 25%.

(a) Compute the expected value of the index of coincidence $E[\mathscr{I}(STR1, STR2)]$.

(b) Use the index to estimate the expected number of matched characters in the two strings.

Solution: Part (a): The characters of both strings are governed by the same distribution vectors: $Vec1 = Vec2 = (1/2, 1/4, 1/4)$. By Theorem 5.3,

$$E[I(STR1, STR2)] = Vec1 \cdot Vec2$$
$$= (1/2, 1/4, 1/4) \cdot (1/2, 1/4, 1/4) = 1/2$$

Part (b): Since $N \cdot \mathscr{I}(STR1, STR2)$ equals the number of matched characters in the two strings, by using the linearity of expectation (Appendix B) along with the result of part (a), we may compute the expected number of matched letters as

$$E[N\mathscr{I}(STR1, STR2)] = N \cdot E[\mathscr{I}(STR1, STR2)] = N \cdot (1/2) = N/2$$

With the index of coincidence and Corollary 5.2 at our disposal, we are now ready to demonstrate the Friedman attack on the Vigenère cipher. This is the purpose of the following example.

Example 5.5

The following ciphertext came from the Vigenère cipher with an unknown keyword. (Such lengthy ciphertexts may be downloaded from the book's Web page.) We illustrate the Friedman attack by decrypting this ciphertext. The only hint we give here is that the keyword used contains at most eight characters.[*]

```
PDYTHXYNAWHMITHLLMGCIFHEWZPULSPEOICCLVNQZTHZPGZHVVDLH
WPNEBXGPXYMCZPHTPGZHVVDTMGCIXOVVHXUEZTCNASRCUYHECUGQT
OUGRGPUKSSLHWDAFHZSSWCDSPYUGWBLFTGTLWAKPGYVOBPUGRHSYK
CHPZHFXEMASHPYFSSWCDSPSIKGTXUDWCRUMOUPHVSDFNKOVPINGAJ
BBUWEBXAPYHXFDQBXFHNLTAQWYHJTCNASHPQTZADIYKPEYKWHLGRG
ITWMVXYATBSXIKSDGYKOIEBXHDAIYHWPGPSGPIKRXYUKWAJNASHPJ
KCQWYFGXYQAWIPQTHTCNASUZUFFPNCGUSZQGTGZGMVTDOFAXEIYSPNBP
OKPLXEJTLBBVLHXKAPUIOCOUESPAZKCBEBXOXC
```

Solution: As with the Babbage/Kasiski attack, there will be two steps.

> *Step 1.* Determine the length of the keyword.
> *Step 2.* Determine the keyword.

Once the keyword is known, the decryption can be readily performed as explained in Chapter 1. Although all of the tasks involved can be easily done on a computer (as will be further explained in the computer implementations material),[†] the reader should keep in mind that this method was originally done by hand.

Step 1: Determine the length of the keyword. We start by writing down two copies of the ciphertext, one directly above the other, but the bottom one horizontally shifted, k units to the right, where i is a positive integer. The idea of the method will be to find the best choice for i to represent the keyword length by looking to maximize the number of identical matches of a ciphertext character in one list with the one directly below it. Here is how the two lists would start off with $k = 4$:

```
P D Y T H X Y N A W H M I T H L L M G C I F H E W Z

    P D Y T H X Y N A W H M I T H L L M G C I F
    ▲           ▲
```

---

[*] This information is given only because of size limits. As would naturally be expected, longer keywords are more difficult to decrypt and would therefore require longer passages of ciphertext. We wanted to keep the size of the ciphertext passage in our example (reasonably) manageable for space considerations.

[†] Alternatively, the applets for this chapter contain all the necessary tools for readers to easily perform the dot products and other calculations needed to execute a Friedman attack.

```
P  U  L  S  P  E  O  I  C  C  L  V  N  Q  Z  T  H  Z  P  G  Z  H  V  ...
H  E  W  Z  P  U  L  S  P  E  O  I  C  C  L  V  N  Q  Z  T  H  Z  P  ...
            ▲
```

So thus far we have counted three matches. Table 5.4 shows the total number of matches when the ciphertext is horizontally shifted by each of the values $k = 1, 2, ..., 8$. The number of matches has a clear winner when the parameter $k = 5$ is used.

Let us now explain why this procedure should yield the key-length. Suppose N is the length of the ciphertext. When we shift one copy of the ciphertext i units to the right below another copy, we will be comparing the overlapping $N - i$ characters of these two superimposed strings. If we denote these two overlapping length $N - i$ string portions by STR1 (top) and STR2 (bottom), then we are counting the number of matches of STR1 and STR2, which is just $(N - i) \cdot \mathscr{I}(\text{STR1}, \text{STR2})$. (Recall that the index of coincidence of two strings of the same length is the number of corresponding matching characters divided by the common string length.) Since i is going to be small compared with N, this quantity is approximately equal to $N \cdot \mathscr{I}(\text{STR1}, \text{STR2})$. Now, if the horizontal shift parameter i is equal to (or is a multiple of) the Vigenère keylength, then corresponding characters of the strings STR1 and STR2 have the same distribution (governed by the Vigenère shift corresponding to the common keyword).[*] Since the plaintext came from the written English language, it follows from Corollary 5.2(c) that $N \cdot \mathscr{I}(\text{STR1}, \text{STR2})$ is expected to be $0.067N$. However, if the parameter i is not (a multiple of) the keyword length, then the Vigenère shifts for each pair of corresponding characters in STR1 and STR2 will be different (in fact, always different if the keyword has no repeated letters), and so by Corollary 5.2(d) $N \cdot \mathscr{I}(\text{STR1}, \text{STR2})$ is expected to be at most $0.045N$, or at least a whopping 33% smaller! (These are long-term averages, of course, which become more accurate with larger ciphertext samples.)

Step 2: Determine the keyword. Once we have established the keylength (5 in the running example), we show how to determine the (5-letter) Vigenère key, one letter at a time. In contrast with this step in the Babbage/Kasiski attack, the keyword letters can be mathematically determined by computing some dot products of the corresponding distribution vectors.

---

[*] In this explanation there are lurking two different sorts of shifts. First, there are the horizontal shifts corresponding to the shift parameter i that are used to shift the entire lower string of ciphertext to the right. Second, there are the Vigenère shifts that the cipher uses to compute individual ciphertext letters. If we horizontally shift the second ciphertext list to the right by an amount equal to (or a multiple of) the keylength, then any corresponding pair of letters (appearing directly above/below each other) in two lists of ciphertexts have resulted from Vigenère shift ciphers with the same Vigenère shift parameter (corresponding to the common key letter of the Vigenère cipher that was used).

TABLE 5.4 Total Number of Character Matches When One Copy of the Entire Ciphertext Is Horizontally Shifted a Number k Character to the Right below the Other, for $k = 1, 2, ..., 8$

Horizontal shift parameter k	1	2	3	4	5	6	7	8
Number of exact matches	11	12	11	12	30	15	15	18

As with the Babbage/Kasiski attack, we begin by computing the length-26 observed frequency vector of all ciphertext letters whose location is congruent to 1 (mod 5)—and thus all encrypted by the same shift corresponding to the first keyword letter. We denote this vector by F_1, and similarly for positive integers $i \leq 5$, F_i denotes the observed frequency vector for all ciphertext characters whose position is congruent to i (mod 5).

Using an applet for this chapter (or by writing and using a program for Computer Exercise 4 at the end of this chapter), we compute

$$F_1 = (.060, .036, .036, .072, .012, 0, .060, .108, .060,$$
$$.012, .012, .012, 0, .012, 0, .133, .024, .012, .060, .084,$$
$$.024, .024, .024, .084, 0, .036)$$

It will be convenient to introduce some relevant vectors. The length-26 vector whose entries are the frequencies of letters in standard plaintext (and specified by Table 1.1) will be denoted by V_0. Thus,

$$V_0 = (.082, .015, .028, \cdots, .020, .001)$$

We let V_i denote the vector that results from V_0 by shifting i places to the right (and cycling the last i components of V_0 at the beginning of V_i). Thus, $V_1 = (.001, .082, .015, .028,...,$ $.020)$, $V_2 = (.020, .001, .082, .015, .028,...)$, and so forth.

Now, theoretically, $F_1 = V_i$ for some index i. Determining i is the same as determining the first keyword character. To accomplish this, we need only look at the 26 dot products $F_1 \cdot V_0$, $F_1 \cdot V_1$, $F_1 \cdot V_2$, $\cdots$, $F_1 \cdot V_{25}$. One will be clearly larger than the rest (theoretically at least 49% larger, by Corollary 5.2), and this will correspond to the first key character.[*]

These dot products then are found to be

---

[*] A conceptual reason of why this happens comes from the following formula for a dot product of two vectors: $V \cdot W = \|V\|\|W\|\cos(\theta)$; that is, the dot product of two vectors equals the product of their magnitudes times the cosine of the angle between them. All of the vectors V_i have the same length (since they all have the same components, but are shifted different amounts). If $V = W$, then the angle between them is zero so the cosine is 1 (it is as large as possible), but in the dot product of two different V_i's, they will point in very different directions, because all of the components have been shifted, so the cosine of the angle between them gets smaller. Although this dot product formula is usually learned in two- and three-dimensional vector settings, it works for vectors in any dimension. This all works because of the different frequencies of English letters. If all English letters had the same frequency $(=1/26)$, then all of the vectors V_i would be the same $(1/26, 1/26, ..., 1/26)$, and the counts of number of matches for different shift amounts would all be close to one another.

.041, .040, .040, .042, .042, .033, .033, .041, .038, .032, .032, .047, .038, .033, .042, .062 , .037, .028, .033, .046, .032, .036, .043, .037, .031, .042

As anticipated, a single dot product (shaded) is not significantly larger than the rest. This component corresponds to V_{15}. This means that the first letter of the Vigenère key should be p (see Table 3.1). We continue in this fashion to obtain the remaining letters of the Vigenère key. For example, for the second key letter, the corresponding shifted dot products (after computing the frequency vector F_2) turn out to be

.040, .038, .038, .030, .041, .035, .032, .044, .036, .036, .036, .0687 , .038, .033, .027, .046, .030, .036, .037, .037, .036, .036, .048, .043, .047, .030

The clearly winning dot product (shaded) tells us the second Vigenère key element is l. Continuing in this fashion, we obtain the complete Vigenère key to be "pluto." This is easily checked by attempting to decrypt the given ciphertext using this key for the Vigenère cipher. Invoking the appropriate applet or program from Chapter 1, we obtain the following:

aseatinthisboatwasnotunlikeaseatuponabuckingbronchoandbythesameto kenabronchoisnotmuchsmallerthecraftprancedandrearedandplungedlike ananimalaseachwavecameandsheroseforitsheseemedlikeahorsemakingat afenceoutrageouslyhighthemannerofherscrambleoverthesewallsofwateris amysticthingandmoreoveratthetopofthemwereordinarilytheseproblemsin whitewaterthefoamracingdownfromthesummitofeachwaverequiringanew leapandaleapfromtheair

This makes sense, so we have successfully cracked the code! This is a passage from the short story by American author Stephen Crane, *The Open Boat*. It was written in 1898, and it shows that the much more recently published frequency distribution that we use (Table 3.1) also works very well for English written over a century ago.

Let us briefly summarize the Friedman attack that was derived in the course of the solution of the previous example.

Algorithm 5.1: *The Friedman Ciphertext-Only Attack on the Vigenère Cipher*

Input: A ciphertext produced by the Vigenère cipher
Output: The Vigenère keyword
Assumption: The algorithm assumes that the plaintext comes from ordinary English language and that the ciphertext is sufficiently lengthy to determine the keyword.[*]

---

[*] In principle the Friedman attack works to detect any finite-length keyword provided that the ciphertext is correspondingly lengthy. Basically, longer ciphertexts make the attack work more smoothly, while longer keywords make the attack work less smoothly.

Step 1: Determine the length of the keyword. We first count the number of character matches when one copy of the ciphertext (that is placed directly below another copy) is horizontally shifted different numbers of units to the right of the top copy. We make a table with the horizontal shift parameter k running from 1 to a number believed to be at least as large as the keyword length (8 will be sufficient for the examples in this text). The keylength will be the value of the shift parameter k that gives the largest number of matches.

Step 2: Determine the keyword. For $i = 1$ to k (length of keyword)
- We form the length frequency vector F_i for the ciphertext letters in all positions that are congruent to i (mod k).
- We compute the 26 dot products of F_i with the shifted English distribution vectors V_j (for $j = 0$ to 25), where V_0 is derived from Table 3.1. The index j for which the dot product is maximum corresponds to the ith keyword letter (by the basic integer/letter correspondence of Table 3.1).

Exercise for the Reader 5.3

The following ciphertext came from the Vigenère cipher with an unknown keyword applied to a passage of *The Legend of the Sleepy Hollow*, by Washington Irving. Download it from the book's Web page and perform the Friedman attack to decrypt it.

```
ZADOMOVQKLPPFTBKDEVCOPLJFATPFCBJRXWYXLAQEIDZUSWUVGG
OFWXQKICNBEMUFJIBFDKJFSAQIZRQPMCNIPAORVIIGEPGZVHOCU
MEKWDHUSMEFRILBCGJVESGJYQUKIGYEUCUKMRYXTBJUMHWSTUKE
EICPYZCKLTLUSIPJIKYSTBAKEZCORBJVFJLUSMPFJUNIPJCTOHI
GEPGNIPEBYLNRCXHHTBQEXWITPWHKLTMUCWPXCDOSXMTVTJHZDB
TZTACORBJRXLCONMFRXIBFWMCJXUFPFZKJLDZUSMTFHLUTAIUJI
SVZHQVYMCXVWOGEGTVVEBJVGAUJXAQWNJMUTKGNIGYTLBKJJXYE
MGKEJACDEQPXESIVMTGGSGNJZVQEWDGFWQVKPTNPFOJNVDHHSMC
UISVSZIFJOXLUPLFLXRBVCKJZRLBPDCNBISUOOAYVPAYELVFXVT
QEZOIVHPHEDCNCICVFYMCKLIBFMQTTLPFMEPKJLTWBWTGUHDCOR
PKJHJNZMGVYIXLQLZGEXHUOOPGEIKYSTVHCMRNFOIEYEHNJDMOV
RIQJEPQLXUIMWWYZRVCUMGVYIPMTFZCEGTMPNWPJSAUUZZAKSIB
FDUCIXXHHFZEYMCNILBJVADOMOZGDIBVFCQVRRSNILVMYMBZPCQ
VKLTFPYOGJXSUZSMJRHII MTDG
```

How Enigmas Were Attacked

Analyzing frequencies of a single ciphertext (or modular portions thereof) was at the core of the major cryptographic breakthroughs thus far discussed. For a given plugboard setting, the Enigma machines are periodic but with an inordinately large period of $26^3 = 17{,}576$, making an intercepted ciphertext (of finite length) essentially bulletproof against any frequency analysis. Although, if carefully implemented, Enigma machines were effectively unbreakable, there were certain flaws in their usage protocols by the Germans that allowed scientists to break them. This exceptionally remarkable feat was first achieved by a group of cryptographers in Poland, who were able to read Enigma ciphertexts for nearly a decade, despite the fact that the Germans continued to complicate matters by upgrading their machines and improving their usage protocols. The Poles, in turn, needed to be increasingly resourceful to continue their decryptions, until they eventually were unable to keep up with German countermeasures. At this point (around 1939), they shared their cryptographic breakthroughs with other Allied countries. The British, through their plentifully staffed Bletchley Park laboratory, were able to pick up where the Poles left off. The story is a truly remarkable one, and the mathematics is intriguing. We now move on to describe how some of the key individuals in this achievement were able to crack the Enigma. Although we cover almost all of the mathematical and cryptographic elements in this codebreaking, the implementations would require a tremendous amount of cataloging special types of permutations, making it infeasible to present a complete implementation in a reasonable amount of space.

We begin this discussion with reference to the basic (three-rotor, six-plugboard cable model) Enigma machine that was discussed in Chapter 3. In order to understand how these Enigmas were attacked, it is important to understand the protocols that the Germans used to send messages with them.

German Usage Protocols for Enigmas

All branches of the German military began using (the three-rotor, six-plugboard cable) Enigmas in 1928, and this continued throughout Hitler's *Third Reich* regime, which began in 1933 and lasted until its fall to the Allied forces in 1945. Each Enigma operator was given a codebook of daily rotor and plugboard settings (*Grundstellung*). Even with daily settings, however, with the enormous number of messages being sent (over radio waves) and intercepted, it would not have been sufficient for all plaintext messages to be encrypted with just these initial settings. Indeed, if this were to be done and, say, 1,000 of these messages were intercepted, then since the first character of each message would have been encrypted with the same substitution cipher, a frequency analysis could be performed to reveal it. Although the second plaintext characters would be encrypted by a different substitution cipher, a separate frequency analysis could be done just on the second characters of the 1,000 messages to break it. In this way, the separate substitution ciphers could be employed to break each character substitution for a given day. To circumvent this weakness, operators were instructed to select individual three-letter message keys for each transmission. In sending a message, the operators would first set their machines according to the daily setting, and use it to send encryptions of their key twice. So, for example, with a message key *kla*, the operator would begin the transmission by entering *klakla* into the Enigma machine, and would transmit the resulting ciphertext, say, "patoqk." The operator would then reset the rotors according to the established message key *kla* and encrypt the plaintext message. The intended recipient, who also would have the daily codebook, would intercept the ciphertext *patoqk* and decrypt it as *klakla*, reset his rotors to *kla*, and then be able to decrypt the ciphertext.

The key was sent twice to safeguard against transmission errors and noisy channels, but it was precisely this facet of the process that allowed their Enigma transmissions to succumb to attack.

The Polish Codebreakers

Once German Enigma transmissions began to hit the airways, other countries worked fastidiously to break them but were truly stymied for many years. The Poles were particularly well motivated in their efforts. After World War I, Germany had ceded some of their territory back to Poland, and the Poles were wary of losing it again to an increasingly powerful Germany. In the late 1920s, the Polish Cipher Bureau (Biuro Szyfrów) took the novel step of recruiting pure mathematicians into their staff. These individuals were initiated with an introductory course in cryptography. Three of them, who were all fluent in German and had impeccable mathematical skills, were Marian Rejewski (pronounced: Rey-EF-ski), Henryk Zygalski (pronounced: Zig-AHL-ski), and Jerzy Różycki (pronounced: Roozh-IT-ski) and became permanent members in 1932; see Figure 5.6.*

Rejewski's Attack

After the code of silence had been removed in 1973, Rejewski wrote some informative papers on how his team was able to crack Enigma.† When the team began working full-time for the Polish Cipher Bureau in 1932, they were in possession of a military Enigma machine and were aware of the transmission protocol of initially sending the three (ciphertext) for the three-character message key twice. Since a large number of messages

---

* It was Rejewski who had made the greatest contributions of the group, and in some cryptographic literature, his name appears as the sole innovator. All three were classmates studying mathematics at Poznan University. Before being recruited as full-time cryptographers in 1932, they did some part-time cryptography work and taught at Polish universities. The Polish Cipher Bureau obtained possession of an actual German military Enigma machine (indirectly from a German cipher employee turncoat), and this allowed them to break Enigma in the same year. The team's work allowed the Polish Cipher Bureau to read all German Enigma transmissions until about 1936, when the Germans made some modifications. Through arduous improvements in their work, the Polish team managed to adapt their attacks to continue to work, but at the cost of much greater time, even with substantial mechanical devices that they designed specifically for this purpose. Consequently, they were able to decrypt only about 75 percent of the intercepted Enigma transmissions. Then, in 1938, the Germans added further complications, which rendered the Polish team's methods no longer sufficient to break any codes. During their subsequent efforts to re-engineer their techniques, Poland fell to Germany in 1939. The Poles shared their Enigma breaking technology with the Allies at this point. Meanwhile, the three cryptographers fled first to Romania and then to France to avoid capture and internment. When France fell to the Third Reich in 1940, the three were sent to Algeria. Shortly thereafter, Rejewski and Zygalski returned to the unoccupied Vichy region in France to finally resume work on breaking Enigma codes. Różycki, along with over 200 others (including other prominent cryptographers), tragically perished in 1942 when their boat back to France sank—the circumstances of which remain a mystery. After the war, Rejewski returned to Poland to rejoin his family, while Zygalski went into exile in England. It was not until 1973, when Rejewski was able to tell his story, that he became quite a celebrity.

† Two of these papers appear as appendices in the book [Koz-84].

Figure 5.6 Marian Rejewski (1905–1980), Henryk Zygalski (1906–1978), and Jerzy Różycki (1909–1942), the Polish cryptography team credited with breaking Enigma.

were being intercepted daily, they focused on a corresponding collection of six-character first keys for large sets of ciphertext messages on a given day. For example, such a collection might begin with the following:

```
dmqvbn
vonpuw
pucfmq
```

For each line, the two pairs of three ciphertext characters were enciphered from the same three plaintext characters (the message key), and all were encrypted with the daily settings for the particular day that they were intercepted. Thus, in each line, the six characters were encrypted, in order, by the same list of permutations (on the 26 letters), which we denote by $\alpha, \beta, \gamma, \delta, \varepsilon, \varphi$. Although the corresponding plaintext (message keys) were not known, the collected data could nonetheless be used to gain insights about the permutations $\alpha, \beta, \gamma, \delta, \varepsilon, \varphi$ as follows: If we denote the unknown key for the first line as xyz, then since xyzxyz encrypts to dmqvbn, this tells us that $\alpha(x) = d, \beta(y) = m, \gamma(z) = q, \delta(x) = v, \varepsilon(y) = b, \varphi(z) = n$. But because of the self-encrypting property of the Enigma machine [Proposition 3.3(a)], we know also that $\alpha(d) = x, \beta(m) = y, \gamma(q) = z, \delta(v) = x, \varepsilon(b) = y, \varphi(n) = z$. By composing (and using the product notation for composing permutations), this allows us to eliminate the unknowns x, y, z by means of the following equations:

$$\delta\alpha(d) = v, \quad \varepsilon\beta(m) = b, \quad \varphi\gamma(q) = n$$

The same can be done for the first six characters of any intercepted ciphertext of that day; the second and third lines above similarly lead us to

$$\delta\alpha(v) = p, \quad \varepsilon\beta(o) = u, \quad \varphi\gamma(n) = w$$

and

$$\delta\alpha(p) = f, \quad \varepsilon\beta(u) = m, \quad \varphi\gamma(c) = q$$

After collecting a sufficient quantity of ciphertext transmissions for a given day, the extracting of the information as above eventually led to the complete determination of the three permutations $\delta\alpha$, $\varepsilon\beta$, $\varphi\gamma$. For example, this process might lead to the following representations:

$$\delta\alpha = \text{(dvpfkxgzyo)(eijmunqlht)(bc)(rw)(a)(s)}$$

$$\varepsilon\beta = \text{(blfqveoum)(hjpsyizrn)(axt)(cgw)(d)(k)}$$

and

$$\gamma\varphi = \text{(abviktjgfcqnw)(duzrehlxypsmo)}$$

It is no accident that in the above example each of the three permutations' disjoint cycle decomposition always consists of matched pairs of cycles of equal lengths. The Poles noted this and then mathematically proved why it always turned out like this (Exercise 42 contains an outline of Rejewski's proof).

One aspect of the Enigma machine that makes it so difficult to attack is that the rotor and plugboard settings combine to give an inordinately large number of settings to preclude any hope of a brute-force attack. A remarkable discovery that allowed Rejewski's team to advance was that they were able to decouple the plugboard and rotor settings in their analysis. They thought carefully about the effect of the plugboard settings, and they found that the cycle decomposition form of $\delta\alpha, \varepsilon\beta, \varphi\gamma$ is unaffected by the plugboard settings. That is, if, say, $\delta\alpha$ consisted of a product of two 10-cycles, two 2-cycles, and two 1-cycles (as above), and if the plugboard settings were to be modified in any way, then $\delta\alpha$ would still be of this same form. Instead of having to consider over 10,000,000,000,000,000 settings, they could thus focus attention only on the considerable smaller $3! \cdot 26^3 = 105,456$ possible rotor settings. It took a whole year, but the team went through the arduous task of using their actual Enigma machines to check for each rotor setting, the forms of the resulting three permutations $\delta\alpha$, $\varepsilon\beta$, $\varphi\gamma$. For each possible form, they found that only a handful of rotor settings could give rise to it. They put together a catalogue that could be used to look up all possible rotor settings from any intercepted forms for $\delta\alpha$, $\varepsilon\beta$, $\varphi\gamma$. Mathematically, this result follows from the following theorem about permutations. We leave the proof as the Exercise for the Reader 5.4, but will explain how the theorem applies to give this invariance of form property for Enigma transmissions.

Invariance of Cycle Decomposition Form

Theorem 5.4

Suppose that a permutation σ on any set $\{1, 2, 3, \ldots, n\}$ has disjoint cycle decomposition $\sigma = \sigma_1\sigma_2\cdots\sigma_k$. If τ is another permutation on the same set, then the disjoint cycle decomposition of $\tau\sigma\tau$ may be expressed as $\sigma_1'\sigma_2'\cdots\sigma_k'$, where the length of σ_i' equals the length of σ_i for each index i, $1 \leq i \leq k$.

In words, the theorem says that composing any permutation on both sides by any 2-cycle does not change its disjoint cycle form. The proof of the theorem can be discovered through an intriguing course of computations. The following exercise for the reader will guide the reader through this rewarding task, in the case that τ is a 2-cycle (the only case that was needed by Rejewski's team). The proof of the general theorem can be accomplished in a similar fashion and will be left as Exercise 45.

Exercise for the Reader 5.4

(a) If $\sigma = (a,b,c)$ is a 3-cycle and $\tau = (i,j)$ is a 2-cycle, show that $\tau\sigma\tau$ is always a 3-cycle.

(b) If $\sigma = (a_1,a_2,\cdots,a_k)$ is a k-cycle and $\tau = (i,j)$ is a 2-cycle, show that $\tau\sigma\tau$ is always a k-cycle.

(c) Prove Theorem 5.4, in the case that $\tau = (i,j)$ is a 2-cycle.

From the theorem, it is not hard to see why the plugboard settings do not change the cycle form of any of the permutations $\delta\alpha, \varepsilon\beta, \varphi\gamma$ (or any permutation, for that matter). This is because the plugboard corresponds to a disjoint product of 2-cycles $\tau_1\tau_2\cdots\tau_\ell$. Since disjoint cycles commute, we can apply Theorem 5.4 sequentially to see that each of the parenthesized permutations in the following equations have the same form:

$$\tau_1\tau_2\cdots\tau_{\ell-1}\tau_\ell(\sigma)\tau_{\ell-1}\tau_{\ell-2}\cdots\tau_2\tau_1 = \tau_1\tau_2\cdots\tau_{\ell-1}(\tau_\ell\sigma\tau_{\ell-1})\tau_{\ell-2}\cdots\tau_2\tau_1$$
$$= \tau_1\tau_2\cdots(\tau_{\ell-1}\tau_\ell\sigma\tau_{\ell-1}\tau_{\ell-2})\cdots\tau_2\tau_1$$
$$\cdots$$
$$= \tau_1(\tau_2\cdots\tau_{\ell-1}\tau_\ell\sigma\tau_{\ell-1}\tau_{\ell-2}\cdots\tau_2)\tau_1$$
$$= (\tau_1\tau_2\cdots\tau_{\ell-1}\tau_\ell\sigma\tau_{\ell-1}\tau_{\ell-2}\cdots\tau_2\tau_1)$$

This breakthrough allowed the Poles to determine the rotor settings for any day's transmission in about 15 minutes by using their catalogue. But what about the plugboard settings? Since unlike the rotors, the Enigma does not automatically change the plugboard settings, these could be determined by an old-fashioned frequency analysis on the ciphertexts. This system allowed the Poles to decrypt all German Enigma transmissions until 1936, when the Germans made some changes both to the machines (by adding rotors) and to the transmission protocols. At this point, the Poles needed to invent large electromechanical contraptions called "bombes" (in Polish: bomba krypto-logiczna), a name coined by Różycki, when the three were enjoying eating a Polish ice cream treat known as a bombe, which bore a striking resemblance to their new machine. With the aid of these bombes, the team continued to decrypt Enigma transmissions, but at a slower pace. In 1938, the Germans made additional enhancements to their military Enigma machines, and the Poles were no longer able to decrypt the new ciphertexts.

Alan Turing and Bletchley Park

The Allied forces were able to pick up Enigma decryptions from where the Poles left off, and this had a decisive impact on the timing of their

Figure 5.7 Alan M. Turing (1912–1954), English scientist of many disciplines.

defeat of Hitler. By some accounts, this intelligence breakthrough curtailed the war by about two years. This major intelligence breakthrough originated from a group of scientists stationed at the *British Government Code and Cypher School* at Bletchley Park (about 50 miles northwest of London), led by the illustrious Alan M. Turing*; see Figure 5.7.

With their knowledge of how to eavesdrop on Enigma communications, the British freely distributed some Enigma machines to governments in many of their colonies, allowing them easy access to first-rate intelligence. These innovative scientists who worked at Bletchley Park had to go for many years without any public recognition. Eventually, when the information became declassified, the surviving members of Bletchley Park team began to receive their due honor. A very interesting book ([HiSt-01]) published in 2001 recounts the events relating to the Bletchley Park years given by 27 individuals who were former team members.

---

* Turing and Babbage are two of the most remarkable computer scientists to have ever lived, the fact that both lived before the computer era notwithstanding. As with other notable geniuses such as Einstein, Turing did not find his school years sufficiently challenging to have demonstrated his genius, and so had a poor to mediocre school record. At one point, because of a strike, Turing had to bicycle 60 miles each day to commute to school. Throughout his life he remained an impressive athlete, continuing to run races and marathons. Rather than just not doing his schoolwork, however, Turing took to his own reading of some famous and high-level books. He flourished in college at Cambridge, and after graduating he wrote a paper that so impressed the leading logician of the time, Alonzo Church, that Church invited Turing to do his PhD with him at Princeton; Turing accepted. Turing designed his famous Turing machine, which is a theoretical device that has been successfully used for many applications in computer science. In full recognition of his contributions, the ACM (the main computer science society in the United States) offers an extremely prestigious *Turing Prize* for seminal achievements in computer science. Although Turing had many interests and made many important contributions in a wide range of mathematical fields, and also in biology and physics, his contributions and work in breaking Enigma at Bletchley Park perhaps saved more lives in World War II than the work of any other single individual. He personally considered this work period of his life to be the happiest and most fulfilling time of his career.

Chapter 5 Exercises

1. Suppose a certain language has only three letters: *X, Y, Z* (listed in alphabetical order), with respective frequencies: 20%, 75%, and 5%. For such three-character alphabets, there are three possible shifts corresponding to the shift parameters 0 (identity), 1 (shift one character to right), and 2 (shift two characters to right). Similarly, there are 3! = 6 permutation ciphers.

    (a) A shift was used to produce (in this language) the following ciphertext: XXXZZXXXXYZX. Perform a frequency analysis by hand to determine the likely shift parameter, and determine the corresponding plaintext.

    (b) A substitution cipher was used to produce (in this language) the following ciphertext: YYYZYYZYYYYY. Perform a frequency analysis by hand to determine the likely shift parameter, and determine the corresponding plaintext.

2. Suppose a certain language has only four letters: *A, B, C, D* (listed in alphabetical order), with respective frequencies: 65%, 8%, 2%, and 25%. For such four-character alphabets, there are four possible shifts corresponding to the shift parameters 0 (identity), 1 (shift one character to right), 2 (shift two characters to right), and 3. Similarly, there are 4! = 6 permutation ciphers.

    (a) A shift was used to produce (in this language) the following ciphertext: CCBCBCCCCBBC. Perform a frequency analysis by hand to determine the likely shift parameter, and determine the corresponding plaintext.

    (b) A substitution cipher was used to produce (in this language) the following ciphertext: BDBBDDDABDDA. Perform a frequency analysis by hand to determine the likely shift parameter, and determine the corresponding plaintext.

3. The following ciphertexts were obtained using shift ciphers. In each case, apply a frequency analysis to determine the shift parameter and decrypt.

    (a) ESCWAFALASDUGFLSULSKSTMKAFWKKESF

    (b) SFUVSOUPGJFMEPGGJDFOPX

4. The following ciphertexts were obtained using shift ciphers. In each case, apply a frequency analysis to determine the shift parameter and decrypt.

    (a) BTTFNCMFZPVSTUSJLFUJNFCZNJEOJHIU

    (b) OSALAFYXGJJAFKLJMULAGFK

    Exercises 5–12: The Web page for this book contains downloadable ciphertexts of each of these eight exercises that were obtained by applying eight different substitution ciphers to certain passages of some famous American novels. Download the ciphertexts and use the frequency analysis applet tools for this chapter to determine the key permutations, and decrypt.

    Exercises 13–20: The Web page for this book contains downloadable ciphertexts of each of these eight exercises that were obtained by applying the Vigenère cipher with eight different keywords to certain passages of some famous American novels. Download the ciphertexts and use the applets for this

chapter from the book's Web page to perform a Babbage/
Kasiski attack to determine the keywords. Then use the appropriate applets from Chapter 1 to decrypt.

21. For each pair of strings STR1 and STR2, compute their index
    of coincidence $\mathcal{I}(\mathrm{STR1,STR2})$.
    (a) STR1 = aaabbbccc, STR2 = cccaaabbb
    (b) STR1 = 123123123123, STR2 = 123412341234
    (c) STR1 = 123...123 (repeated 400 times), STR2 = 1234...
         1234 (repeated 300 times)

22. For each pair of strings STR1 and STR2, compute their index
    of coincidence $\mathcal{I}(\mathrm{STR1,STR2})$.
    (a) STR1 = ABCDEDCBA, STR2 = ABCABCABC
    (b) STR1 = 123445679, STR2 = 321321321
    (c) STR1 = 123445679...123445679 (repeated 100 times),
         STR2 = 321...321 (repeated 300 times)

23. Suppose a certain language has only three letters: X, Y, Z
    (listed in alphabetical order), with respective frequencies:
    20%, 75%, and 5%.
    (a) Suppose that STR1 and STR2 are a pair of strings of length
         $N > 0$ taken from written language in this alphabet. Com-
         pute the expected value of their index of coincidence,
         $E[\mathcal{I}(\mathrm{STR1,STR2})]$.
    (b) For such three-character alphabets, there are two nontriv-
         ial shifts corresponding to the shift parameters 1 (shift one
         character to right) and 2 (shift two characters to right). Sup-
         pose that STR1 and STR2 are a pair of strings of length
         $N > 0$, where STR1 is taken from written language, but
         STR2 is taken from a nontrivial shift of written text. Com-
         pute the range of expected values of their index of coinci-
         dence, $E[\mathcal{I}(\mathrm{STR1,STR2})]$.
         Suggestion: Use Theorem 5.3.

24. Suppose a certain language has only three letters: X, Y, Z (listed
    in alphabetical order), with respective frequencies: 10%, 89%,
    and 1%. For such three-character alphabets, there are $3! - 1 = 5$
    nontrivial permutation ciphers.
    (a) Suppose that STR1 and STR2 are a pair of strings of length
         $N > 0$ taken from written language in this alphabet. Com-
         pute the expected value of their index of coincidence,
         $E[\mathcal{I}(\mathrm{STR1,STR2})]$.
    (b) For such three-character alphabets, there are two nontriv-
         ial shifts corresponding to the shift parameters 1 (shift one
         character to right) and 2 (shift two characters to right). Sup-
         pose that STR1 and STR2 are a pair of strings of length $N >$
         0, where STR1 is taken from written language, but STR2 is
         taken from a nontrivial permutation cipher applied to some
         written text. Compute the range of expected values of their
         index of coincidence, $E[\mathcal{I}(\mathrm{STR1,STR2})]$.
         Suggestion: Use Theorem 5.3.

25. Suppose that they have an 11-letter alphabet, A, B, C, ..., I, J, K, in which the frequencies of the letters A, B, C, ..., K are all equal to 5%, but that the letter K has a 50% frequency.

   (a) Suppose that STR1 and STR2 are a pair of strings of length $N = 100$ taken from written language in this alphabet. Compute the expected value of their index of coincidence, $E[\mathcal{I}(STR1, STR2)]$.

   (b) Suppose that STR1 and STR2 are a pair of strings of length $N = 100$, where STR1 is taken from written language, but STR2 is taken from a nontrivial shift of written text. Compute the range of expected values of their index of coincidence, $E[\mathcal{I}(STR1, STR2)]$.

   (c) Explain how one could perform a ciphertext-only attack on a shift cipher by using indices of coincidence.
   Suggestion: Use Theorem 5.3.

26. Suppose that they have an 11-letter alphabet, *A, B, C, ..., I, J, K*, in which the frequencies of the letters *A, B, C, ..., I* all being equal, and with the remaining two letters *J, K* each having 10 times the frequency of the first letters.

   (a) Suppose that STR1 and STR2 are a pair of strings of length $N = 100$ taken from written language in this alphabet. Compute the expected value of their index of coincidence, $E[\mathcal{I}(STR1, STR2)]$.

   (b) Suppose that STR1 and STR2 are a pair of strings of length $N = 100$, where STR1 is taken from written language, but STR2 is taken from a nontrivial shift of written text. Compute the possible expected values of their index of coincidence, $E[\mathcal{I}(STR1, STR2)]$.

   (c) Explain how one could perform a ciphertext-only attack on a shift cipher by using indices of coincidence.
   Suggestion: Use Theorem 5.3.

27. The Hawaiian alphabet has 12 letters whose approximate frequencies (in written Hawaiian) are as follows:

Hawaiian Letter:	a	e	i	o	u	h	k	l	m	n	p	w
Frequency (%):	28	8	10	12	6	04	12	4	3	9	3	1

   (a) Suppose that STR1 and STR2 are randomly formed strings of length 30 in the Hawaiian alphabet. Determine the expected value of the incidence of coincidence: $E[\mathcal{I}(STR1, STR2)]$.

   (b) Suppose that STR1 and STR2 are strings of length 30 taken from written Hawaiian language. Determine the expected value of the incidence of coincidence: $E[\mathcal{I}(STR1, STR2)]$.

   (c) A shift cipher was used to encrypt the name of a famous Hawaiian king. If the ciphertext is NOPONOLO, use a frequency analysis to find the name of the king.

28. Suppose a certain language has only four letters: A,B,C,D (listed in alphabetical order), with respective frequencies: 65%, 8%, 2%, and 25%.

(a) Suppose that STR1 and STR2 are randomly formed strings of length 30 in this alphabet. Determine the expected value of the incidence of coincidence: $E[\mathscr{I}(\text{STR1}, \text{STR2})]$.

(b) Suppose that STR1 and STR2 are strings of length 30 taken from written documents in this language. Determine the expected value of the incidence of coincidence: $E[\mathscr{I}(\text{STR1}, \text{STR2})]$.

Exercises 29–36: The Web page for this book contains downloadable ciphertexts of each of these eight exercises that were obtained by applying the Vigenère cipher with eight different keywords to certain passages of some famous American novels. Download the ciphertexts and use the applets on the book's Web page to perform a Friedman attack to determine the keywords. Then use the appropriate applets from Chapter 1 to decrypt.

37. For each of the following statements, determine whether it is (always) true or false. For those that you decide are true, explain why, and for those that you decide are false, provide a counterexample.

(a) If $\mathscr{I}(\text{STR1}, \text{STR2}) = \mathscr{I}(\text{STR2}, \text{STR3})$ then $\mathscr{I}(\text{STR1}, \text{STR3}) = \mathscr{I}(\text{STR2}, \text{STR3})$.

(b) If STR1 and STR2 are randomly formed strings in the English alphabet each of length 100, and STR3 and STR4 are randomly formed strings in the English alphabet each of length 200, then $E[\mathscr{I}(\text{STR1}, \text{STR2})] < E[\mathscr{I}(\text{STR3}, \text{STR4})]$.

38. For each of the following statements, determine whether it is (always) true or false. For those that you decide are true, explain why, and for those that you decide are false, provide a counterexample.

(a) If $\mathscr{I}(\text{STR1}, \text{STR2}) = \mathscr{I}(\text{STR3}, \text{STR4})$, then $\mathscr{I}(\text{STR1STR1}, \text{STR2STR2}) = \mathscr{I}(\text{STR3STR3}, \text{STR4STR4})$.

(b) If STR1 and STR2 are two same-length strings that have been randomly selected from written text in letter alphabet (with a known frequency distribution) and STR3 comes from a shift cipher applied to such a text, then $E[\mathscr{I}(\text{STR1}, \text{STR3})] < E[\mathscr{I}(\text{STR1}, \text{STR2})]$.

39. Suppose that a certain company's systems administrator keeps an encrypted file of all employee passwords within the company's computer system (for access to e-mail, payroll information, etc.). If the encryption was done by a substitution cipher (which is unknown to the administrator) on each of the passwords—so the administrator can only view the cipherpasswords, for each scenario below, explain whether the administrator might be able to employ a frequency analysis to crack the substitution

cipher and thus have direct access to all employee passwords. Work with the following assumptions:

i This administrator is willing to dedicate one week's worth of time to the effort, and that he or she is able to test 1,000 passwords per second.

ii The company has 10,000 employees.

iii There are 150,000 English words, and that the average word length is 5.

(a) All Passwords that were selected consist of a single ordinary English word in lowercase (with no digits and no special characters).

(b) All passwords that were selected consist of a single ordinary English word that is mixed with upper- and lowercase letters (with no digits and no special characters).

(c) All passwords contain four digits and one ordinary English word in lowercase, with the digits occurring before and/or after (but not within) the English word. For example, *12lantern34* and *1234lantern* would be admissible passwords but *lan12tern34* would not.

(d) All passwords were nonsensical strings of eight characters of letters in lowercase.

40. Assuming that a hacker can use a computer to check 1 million passwords per hour, how well would each of the following password protocols safeguard against such a hacker who was prepared to dedicate his hacking efforts for 24 hours to break a single password?

(a) A password of length exactly 5 characters made up of lowercase letters that were randomly chosen.

(b) A password of length exactly 8 characters that is a word in lowercase letters taken from a standard English dictionary.

(c) A password of length exactly 8 characters that is a random string of lowercase letters.

(d) A password of length exactly 8 characters that is a random string of letters, each of which is either upper- or lowercase.

(e) A password of length exactly 8 characters that is a random string, each character of which is either upper- or lowercase or a single digit.

Suggestion: For each part, estimate the probability that the hacker will be able to discover the password in his or her 24 hours of attempts.

41. Prove Theorem 5.3.

42. Prove the following result about permutations that was discovered by Rejewski during his work on Enigma.

Proposition: If n is even, μ, σ are two permutations on $\{1, 2, \ldots, n\}$ whose disjoint cycle decompositions each consist of the maximum possible number ($n/2$) of 2-cycles, then the disjoint cycle decomposition of the product $\mu\sigma$ will consist of even numbers of cycles of each length that appears.

Suggestions: Separate the 2-cycles of μ, σ into three types: disjoint, identical, and those having one element in common. If you are having difficulty, try to follow the idea of Exercise for the Reader 5.4 by looking at some special smaller cases first.

43. (a) Using the vector V_0, which gives the frequencies (from Table 1.1) of the English alphabet, and with V_i denoting the vector obtained by shifting i units to the right (and recycling at the end)—just as in the development of the Friedman attack—prove that for any two indices i and j we have $V_i \bullet V_j = V_0 \bullet V_{|i-j|} = V_0 \bullet V_{|i-j|}$.
    (So all that matters here is the relative difference in the Vigenère shift parameters.)

    (b) Can the result of part (a) can be used to further streamline the Friedman attack as given in the text? Explain.

44. (a) Prove that the permutations generated by Algorithm 5.2 given in the computer implementation section of this chapter are indeed random.

    (b) Suppose that we modify Algorithm 5.2 by allowing index ℓ to decrease rather than increase, i.e., by replacing Step 2 with the following:

    Iterative Step 2′: FOR $\ell = k$ TO 2, repeat the following step: Use Algorithm 1.1 to generate a random integer j in the range $1 \le j \le \ell$, and interchange the jth and the ℓth element of VEC.

    Show that the resulting modified algorithm also produces random permutations.

    (c) Show the algorithm of part (b) that can be modified to create an efficient algorithm that will generate random subsets of a specified size of a given set of k objects.

    (d) Suppose that we modify Algorithm 5.2 by replacing step 2 with the following:
    Step 2″: Use Algorithm 1.1 to generate a random integer j in the range $1 \le j \le k$, and interchange the jth and the ℓ-element of VEC.
    Show that the resulting modified algorithm does not, in general, produce random permutation.

    Suggestions: For parts (a) and (b), let $\{\mu(i)\}_{i=1}^{k}$ denote a permutation generated by the algorithm, i.e., $\mu(i) = \text{VEC}(i)$, where VEC is the final value of VEC in the algorithm. Let σ be an arbitary permuation of $\{1, 2, ..., k\}$. It must be shown that $P(\mu = \sigma) = 1/k!$ Proceed using induction on k.

    For part (c), to generate a subset S of N objects from a set $T = \{a_1, a_2, ..., a_k\}$ $(N \le k)$, we separate into two cases. *Case 1:* $N \le k/2$: In *Iterative Step* 2′, we need only let ℓ run through N values (i.e., FOR $\ell = k$ TO $k - N + 1$). Output the following subset of N elements: $S = \{a_{\text{VEC}(k)}, a_{\text{VEC}(k-1)}, \cdots, a_{\text{VEC}(k-N+1)}\}$. *Case 2:* $N > k/2$: In this case, we can use the procedure in Case 1 to generate a random subset S_0 of $k - N$ elements. The complement of this set $S \triangleq T \sim S_0$ will then be a random subset of N objects. Supply a proof of these facts.

For part (b), set $n = 3$, and show that $P(\mu(3) = 3) = 2/9 \neq 1/3$, where $\{\mu(i)\}_{i=1}^3$ is a permutation generated by the (modified) algorithm.

For part (d), let $\{\mu(i)\}_{i=1}^3$ denote a permutation generated by the algorithm, i.e., $\mu(i) = VEC(i)$, where VEC is the final value of VEC in the algorithm. Show that $P(\mu(1) = 3) = 2/9 \neq 1/3$.

45. Prove Theorem 5.4.

Chapter 5 Computer Implementations and Exercises

Note: Many programs of this section involve dealing with strings of alphabetic characters as well as integer and real-number data types. Some of the programs from Chapter 1 will be useful to call on in the development of some of these new programs. As in Chapter 1, readers may wish to replace all strings with their integer counterparts, according to the standard correspondence of Table 3.1. Within such programs, most platforms will internally work with integer representatives of alphabetic characters, although in the external interface allowing inputs and outputs to be character strings is easier for users. As usual, we allow vectors to be interchangeably represented with or without commas; so, for example, [0 1 3] is the same object as [0, 1, 3].

Programs to Aid in Frequency Analysis

1. *Program for Frequency Counts.*
   (a) Write a program with syntax Vec = Frequency Counts(STR) with inputs STR, a string of uppercase letters (without spacing or punctuation). The output Vec is a 26-component vector that gives the observed frequencies of appearances of the 26 letters $A, B, C, \ldots$ (in order). Thus, for example, the call FrequencyCounts(AABAABBB) should produce the 26-component vector $(0.5, 0.5, 0, \ldots, 0)$, since exactly half of the observed characters are A, and the other half are B.
   (b) Run your program on the small example given above and check that it produces the indicated output.
   (c) Run your program on the ciphertext of Example 5.1 (which you should download from the book's Web page), and check that the output agrees with the observed frequency results presented in Table 5.2.

   Note: If possible, you should write your program so that it ignores spaces. Otherwise, before you do part (c) of this computer exercise, you will need to manually delete the spaces of the ciphertext.

2. *Program for Adjacent Letter Counts.*

(a) Write a program with syntax `Pre, Post = Adjac entLetter Counts(STR,k)`, with two inputs: `STR`, a string of uppercase ciphertext with spaces preserved between words, and `k`, a number between 0 and 25 corresponding to a ciphertext letter (via the standard correspondence of Table 3.1). The output consists of two length-26 vectors, `Pre` and `Post`, which give, for the ciphertext letter corresponding to `k`, the number of times each of the 26 letters occurs directly before (`Pre`) and directly after (`Post`) this letter corresponding to `k` in a word. If the letter occurs at the beginning/end of a word, the last letter of the previous word/first letter of the subsequent word should make no contributions to this letter.

Note: If we add the two vectors `Pre + Post` and count the number of zero components of this sum, this will be exactly the number of ciphertext letters that are not adjacent to the letter corresponding to `k` in any cipherword. This is the non-adjacent count quantity of Table 5.2.

(b) Run your program on the small example given above and check that it gives the results indicated above.

(c) Run your program on the ciphertext of Example 5.1 and check that the output agrees with the observed frequency results presented in Table 5.2.

3. *Program for Ciphertext Partial Substitutions.*

(a) Write a program with syntax `CtextSub = Cipher textPartialSub(Ctext,CLetters,PLetters)`, with inputs `Ctext`, a string of ciphertext in uppercase along with some possible lowercase letters with possible spaces between cipherwords, and two strings of the same-length `CLetters,PLetters`, the first a string of uppercase letters and the latter a string of lowercase letters. The output, `CtextSub`, is the string resulting from the inputted string `Ctext`, but with each occurrence of an uppercase letter appearing in the string `CLetters` being replaced by the corresponding lowercase letter of the string `PLetters`. For example, the call `CiphertextPartialSub(ABCDE, ACE, xyz)` would produce the output `xByDz`.

(b) Run your program on the small example given above and check that it gives the results indicated above.

(c) Run your program consecutively on the ciphertext of Example 5.1 to produce the stage 1 and stage 2 partial ciphertext substitutions, and check that the outputs agree with the results obtained in the section.

Programs to Aid in the Babbage/Kasiski Attack

4. *Program for Modular Frequency Counts.*

(a) Write a program with syntax `Vec = FrequencyCounts Modular(STR,i,n)` that has three inputs: `STR`, a string of uppercase letters (without spacing or punctuation), and two positive integers $i < n$. The output Vec is a 26-component vector that gives the tallies of the observed frequencies of appearances of the 26 letters $A, B, C, \ldots$ (in order) in all of the positions of the ciphertext whose places are congruent to i (mod n). Thus, for example, the call Frequency CountsModular(AACAACCA,2,3) would tally the frequencies of the 2nd, 5th, and 8th characters in the string AACAACCA [i.e., all character positions that are congruent to 2 (mod 3)] and would produce the 26-component vector $(1, 0, \ldots, 0)$, since all observed characters are A.

(b) Run your program on the small example given above and check that it gives the results indicated above.

(c) Run your program on the ciphertext of Example 5.1 and check that the output agrees with the observed frequency results presented in Table 5.2.

Extracting Ciphergram Data from a Ciphertext String. The Babbage/Kasiski attack on the Vigenère cipher required locating all duplicated grams of length at least 4 within the ciphertext, and recording the number of occurrences and the gcd of the separation distances. As was pointed out in the text, if there are no "accidents," then the keylength must divide the gcd of these separation distances. We indicate some ideas that will help the reader to write programs for such a task, which would be extremely tedious by hand. We explain how to record data for 4-grams, the method for grams of larger lengths is similar. The following algorithm will produce two data sets: a matrix Record that has three columns and the number of rows equal to the number of duplicated 4-grams. For each row of Record, the first element will be the position in the ciphertext where the duplicated 4-gram first occurs, the second will be the number of times it is duplicated in the ciphertext (at least 2), and the last entry is the gcd of the separation distances of these duplicated 4-grams. The second output DuplicatedGrams will be the corresponding list of duplicated 4-grams (in the same order as they appear in Record. Note that Record is an integer valued matrix, whereas DuplicatedGrams is a vector of (4-gram) strings. This is really just a brute-force search and record algorithm; it might possibly be simplified if the user chooses to use some of the sorting utilities available on his or her platform. We assume that STR is a string of uppercase ciphertext characters (as would be produced by the Vigenère cipher) without spaces between words. The data structures involved in this program include integers, strings, vectors, matrices, and sets.

Step 1. Let N be the number of characters in STR.

Step 2. Form a vector Fourgrams of length $N - 3$ whose elements consist of all imbedded 4-grams in the ciphertext. [The ith element of this vector will be the substring of STR from the ith to the $(i + 3)$rd character.] A simple FOR loop can be used to form this vector.

Step 3. Initialize a DuplicatedIndexSet as the empty set, and row = 1 (row counter).

Step 4. The following nested FOR loop is the core of the algorithm:

```
FOR i = 1 TO N - 3
   IF (i does not belong to DuplicatedIndexSet)
      set fourgram = ith element of Fourgrams vector
      set dupCounter = 0, DupIndSet = {i}
      FOR j = i+1 TO N - 3
         IF (fourgram equals jth element of Fourgrams
         vector)
            DupIndSet = {j} union DupIndSet
            dupCounter = dupCounter + 1;
            sepdist = j-i;
            IF (dupCounter equals 1)
               set sepdistGCD = sepdist;
            ELSE
               set sepdistGCD = gcd(sepdistGCD,
               sepdist);
            END IF
         END IF
      END j FOR
   END IF
IF dupCounter > 0
   set ith row of Record = [i dupCounter+1 sepdistGCD]
   set ith row of Record DuplicatedGrams = fourgram
   update row = row + 1;
   update
      DupFourgramIndexSet = unionDupFourgramIndex
      Set union DupIndSet
   END IF
END i FOR
```

As an example, if the inputted string STR were "ABCEFABCA
BCEABCEF," the outputs would be

$$
\text{Record} = \begin{bmatrix} 1 & 3 & 4 \\ 2 & 2 & 12 \end{bmatrix}
\qquad
\text{DuplicatedGrams} = \begin{bmatrix} \text{ABCE} \\ \text{BCEF} \end{bmatrix}
$$

5. *Program for Extracting 4-Gram Data.*
   (a) Write a program with syntax Record, Duplicat
       edGrams = FourGramLocator(STR), a string of
       uppercase letters (without spacing or punctuation) STR,
       and whose outputs Record, DuplicatedGrams are
       as described in the previous algorithm, giving data on
       all duplicated 4-grams in STR.
   (b) Run your program twice using the input ABCEFABCABCE
       ABCEF, and check that it gives the results indicated above.
   (c) Run your program on the ciphertext of Example 5.2 and
       check that it gives the results indicated in the section.

6. *Program for Extracting 5-Gram Data.*
   (a) Write a program with syntax Record, Duplicated
       Grams = FiveGramLocator(STR), a string of uppercase

letters (without spacing or punctuation) STR, and whose outputs Record, DuplicatedGrams describe all duplicated 5-grams appearing in STR, just as these outputs described duplicated 4-grams in the preceding algorithm.

(b) Run your program on the ciphertext of Example 5.2 and check that it gives the results indicated in the section.

Programs Related to the Friedman Attack

7. *Program for Horizontal Shifted Match Counts.*

   (a) Write a program Count = ShiftedMatches(STR,i) that inputs a string STR of ciphertext in uppercase English letters (which does not need to have been produced by a Vigenère cipher), and a positive integer i, the shift parameter. The output Count will be a nonnegative integer that gives the number of exact matches, when two copies of the ciphertext are written out, the first on top of the second, with the second copy shifted horizontally i characters to the right. By an exact match, we mean that a certain ciphertext letter of the first list has the same ciphertext character directly below it in the second list.

   (b) Use this program to confirm the counts shown in Table 5.4 for the ciphertext of Example 5.5.

   (c) What should the program produce (for a given inputted string) if we set the parameter i to be 0?

8. *Program for Horizontally Shifted Dot Products.*

   (a) Write a program Vec = ShiftedDotProducts(F) that inputs a length-26 vector F for a portion of ciphertext characters in uppercase English letters (which in a Friedman attack would typically have been produced by the program in Computer Exercise 4 after the keyword length has been determined). The output Vec will be a 26-component vector recording the dot products of F with each of the shifted vectors V_0, V_1, $\cdots$, V_{25}, where V_0 gives the frequencies (from Table 1.1) of the English alphabet and V_i is obtained by shifting i characters to the right (and recycling at the end).

   (b) Use this program to confirm the dot products $F_1 \cdot V_0, F_1 \cdot V_1, F_1 \cdot V_2, \cdots, F_1 \cdot V_{25}$ that were computed in the solution of Exercise 5.5.

   (c) Use this program to supply the missing details in the proof of part (d) of Corollary 5.2, by computing the 26 dot products $V_0 \cdot V_0, V_0 \cdot V_1, V_0 \cdot V_2, \cdots, V_0 \cdot V_{25}$.

9. *Program for Computing Indices of Coincidence.*

   (a) Write a program Ind = CoincidenceIndex-(STR2,STR2) that inputs two strings STR2, STR2 of the same length, and outputs Ind, their index of coincidence.

   (b) Use your program to compute the indices of coincidence for each part of Chapter Exercise 21.

Generating Random Permutations. Generating a random permutation of some vector (or ordered list) of objects can be thought of as shuffling a deck of cards. The ability to generate a random permutation is useful for numerous applications, but in cryptography a principal application will be the creation of random substitution ciphers. The following algorithm, which calls on the random integer generation algorithm of Chapter 1 (Algorithm 1.1 in the computer implementation section of that chapter), easily accomplishes this task. For simplicity, we assume that the list of objects to be permuted is a list of consecutive integers of the form $[1, 2, 3, \cdots, k]$, where k is a positive integer.

Algorithm 5.2 *Generating Random Permutations*
***of a Vector* $[1, 2, 3, \cdots, k]$**

Given a positive integer k, the following steps will produce a random permutation VEC of the vector $[1, 2, 3, \cdots, k]$:

Step 1: Initialize VEC $= [1, 2, 3, \cdots, k]$

Iterative Step 2: While $\ell = 2$ TO k, repeat the following step:

Use Algorithm 1.1 to generate a random integer j in the range $1 \le j \le \ell$, and interchange the jth and the ℓth element of VEC.

For example, if with $k = 4$, we start with VEC $= [1, 2, 3, 4]$ and $\ell = 2$. In the first iteration of Step 2, if $j = 1$ were generated, we would interchange the first and second elements of VEC to result in the update: VEC $= [2, 1, 3, 4]$. In the next iteration with $\ell = 3$, if we were to generate $j = 1$, we would interchange the first and third elements of (the current) VEC $= [2, 1, 3, 4]$ to produce VEC $= [3, 1, 2, 4]$. In the final iteration with $\ell = 3$, if we were to generate $j = 4$, then we would interchange the fourth and the fourth elements of VEC, which of course leaves it unchanged as the final output of the algorithm: VEC $= [3, 1, 2, 4]$. Ordinary Exercise 44 outlines a proof that its algorithm indeed produces random permutations.

10. *Program for Generating Random Permutations.*
    (a) Write a program with syntax OutVec = RandPerm Generator(k) that inputs a positive integer k, and whose output OutVec is a vector of length k that is a random permutation of $[1, 2, 3, \ldots, k]$, generated by Algorithm 5.1.
    (b) Run your program twice using the following input values: $k = 2, k = 4, k = 10$, and record the results.

        With $k = 26$, the program of the previous exercise will generate a random permutation of the integers mod 26 that can be put into tabular form using some of the programs from Chapter 3 (the Computer Implementation Section). To ensure the numbers are in the range $\{0, 1, \ldots, 25\}$, we could simply subtract 1 from (each component of) the output vector. A slight modification allows us to produce a permutation of the integers {0, 1, ..., 26}, where 26 is left fixed. Such a construction corresponds to generating random permutations for a substitution cipher on a plaintext where it is desired to preserve spaces (we simply let 26 correspond to "space"). The next computer exercise guides the reader through such an implementation.

11. *Random Substitution Ciphers.*
    (a) Write a program with syntax `RandPerm = RandPerm Generator1to27fix27`, that has no input, and whose output `RandPerm` is a 2×27 matrix representing the tabular form of permutation of the set {1, 2, ..., 27}, where 27 is left fixed, and the remaining elements are permuted randomly using Algorithm 5.1.
    (b) Run your program to produce a random permutation then subtract 1 (from each entry) of the output matrix `RandPerm` to produce a random permutation of [0, 1, ..., 26]. Then, with the aid of some programs from Chapter 1, use the resulting substitution cipher (with the usual correspondence of Table 3.1) to encrypt the message: "the money is under the doghouse."
    (c) Use the PermInverter program from Computer Exercise 10 of Chapter 3 to compute the inverse of the permutation generated in part (b), and use this inverse to decrypt the ciphertext that you produced in part (b) to recover the original plaintext.

12. *Randomly Generated Permutations in Tabular Form.*
    (a) Write a program with syntax `OutVec = RandPerm Generator(k)` that inputs a positive integer k, and whose output `OutVec` is a vector of length k that is a random permutation of [1, 2, 3, ..., k], generated by Algorithm 5.1.
    (b) Run your program twice using the following input values: $k = 2$, $k = 4$, $k = 10$, and record the results.

Representation and Arithmetic of Integers in Different Bases

Up to this point we have seen an assortment of precomputer era crypto-systems. Although different number systems have been implemented (for example, modular integers and matrices), almost all the arithmetic that we have used has been boiled down to standard decimal (base 10) arithmetic of integers. It turns out that integers can be represented in terms of any positive integer base $b > 1$. Just as the familiar base 10 (decimal) system uses digits from 0 to 9 ($= 10 - 1$), a base b system will use digits from 0 to $b - 1$. Computers often work internally with *binary* (base 2) expansions, or expansions with respect to some other power of 2, such as *hexadecimal* (base 16), because electric circuitry is based on components being on (1) or off (0). In order to take greatest advantage of the speed of computers, it is best to design algorithms to work in such power of 2 bases. In this chapter, we discuss how to represent integers with respect to an arbitrary base b, and then move on to discuss how arithmetic of integers can be done directly in terms of such representations. These arithmetic algorithms are extensions of the ones we all learned in grade school for elementary arithmetic. The chapter ends with a very fast algorithm for modular exponentiation that is based on binary expansions, which was promised in Chapter 2.

Representation of Integers in Different Bases

We begin this section on some very common ground. Across contemporary society, languages vary, but all have adopted base 10 arithmetic with Arabic numerals. For example, if you take a taxi ride in any country in which you cannot speak the language, and afterwards give the driver a pad of paper and pull out your wallet (or purse), he or she will understand your request and write down a number that you will understand is the fee. We can view any positive integer as a series in powers of 10. Here is a simple example of how this works:

$$12,307 = 1 \times 10^4 + 2 \times 10^3 + 3 \times 10^2 + (0 \times 10^1) + 7 \times 10^0$$

Notice that the value of a digit in the decimal expansion depends on the position that it occupies. This seems so natural to us since we have been working in base 10 arithmetic our whole lives.[*] It is interesting to note that we could have also built our number system using *any* integer $b > 1$ for the base. This fact is summarized in the following theorem.

Theorem 6.1: Base b Representations of Integers

Let $b \geq 2$ (the **base**) be an integer. Any positive integer n can be uniquely represented in the form

$$n = c_K b^K + c_{K-1} b^{K-1} + \cdots + c_2 b^2 + c_1 b + c_0 = \sum_{k=0}^{K} c_k b^k \qquad (6.1)$$

where K and c_i $(0 \leq i \leq K)$ are nonnegative integers with $c_i \in \{0, 1, \cdots, b-1\}$ and $c_K > 0$. The expansion in Equation 6.1 is called the **base b expansion of n** and is abbreviated as

$$n \sim [c_K \, c_{K-1} \cdots c_2 \, c_1 \, c_0] \ (\text{base } b)$$

Thus, in the base b expansion, each digit corresponds to (and gets multiplied by) a power of b. In the notation of this theorem, we may write $12307 \sim [12307]$ (base 10). Since $11 = 2^3 + 2^1 + 2^0 = 1 \times 2^3 + 0 \times 2^2 + 1 \times 2^1 + 1 \times 2^0$, we may write $11 \sim [1011]$ (base 2). Note that exactly b digits are available to use in the base b expansion of an integer. We will prove Theorem 6.1 a bit later; for now, we give a general algorithm that explains how to obtain such expansions. (Then, later when we prove the existence portion of this theorem, we will need only to explain why this algorithm works.) The reverse process is quite simple: to convert a base b expansion to its integer equivalent, we simply sum the corresponding series (Equation 6.1).

Algorithm 6.1: Construction of the Base b Expansion of a Positive Integer n

Step 1. Put $R = n$ (initialize remainder).

Step 2. Determine the largest nonnegative integer k such that $b^k \leq R$. Next, determine and set c_k to be the largest positive integer such that $c_k b^k \leq R$. Update $R \to R - c_k b^k$.

Step 3. If $R = 0$, go to Step 4, otherwise, return to Step 2.

Step 4. Let K be the first (i.e., the largest) value of k found in Step 2. The base b expansion of n will be $n \sim [c_K \, c_{K-1} \cdots c_2 \, c_1 \, c_0]$ (base b), where any unassigned coefficients (i.e., skipped in Step 2) are taken as zeros.

---

[*] The decimal system perhaps gained popularity since human beings have 10 fingers. Other ancient societies had used other bases for their arithmetic. For example, over five centuries ago the Babylonians developed a number system with a base of 60, called a *sexagesimal positional system*, where they had separate symbols for each of their 60 digits from 0 to 59.

The next example should help to familiarize the reader with the use of Algorithm 6.1 to perform conversions among different bases and integer equivalents. We point out that in each instance of Step 2, once k is determined, the corresponding coefficient c_k is simply the quotient (in the integer division algorithm from Chapter 2) when R is divided by b^k, which is just floor$(c_k/b^k) = \lfloor c_k/b^k \rfloor$.

Example 6.1

(a) Find the integer equivalents for each of the following expansions: [10011011] (base 2), [1234567] (base 8), and [22222] (base 3).
(b) Use Algorithm 6.1 to find the base 2 expansion of 69, the base 8 expansion of 225, and the base 16 expansion of 729.

Solution: Part (a): To translate any base b expansion into its integer equivalent, we simply substitute the expansion coefficients into Equation 6.1, and then evaluate the sum. The expansion [10011011] (base 2) has eight place holders (digits), so the leading (highest) power of the base 2 will be 2^7.[*] The corresponding series expansion Equation 6.1 of this number is thus

$$1 \times 2^7 + 0 \times 2^6 + 0 \times 2^5 + 1 \times 2^4 + 1 \times 2^3 + 0 \times 2^2 + 1 \times 2^1 +$$
$$1 \times 2^0 = 128 + 32 + 16 + 8 + 2 + 1 = 187$$

Similarly, working in base 8, we obtain $[1234567] \sim 1 \cdot 8^6 + 2 \cdot 8^5 + 3 \cdot 8^4 + 4 \cdot 8^3 + 5 \cdot 8^2 + 6 \cdot 8 + 7 \cdot 1 = 342{,}391$ while working in base 3, we have $[22222] \sim 2 \cdot (3^4 + 3^3 + 3^2 + 3^1 + 1) = 242$.

Part (b): In each case, we use Algorithm 6.1.

(i) In Step 1, we simply initialize the remainder as the number to be converted: $R = 69$. In the first application of Step 2, we find the largest power of 2 not exceeding $R = 69$ is $2^6 = 64$, so we put $(K = 6)$ and $c_6 = 1$. R is now updated to $69 - 64 = 5$, and Step 2 is repeated. Now, $k = 2$ is the largest exponent such that $2^k \le 5 (= R)$, so we set $(c_5 = c_4 = c_3 = 0)$ and $c_2 = 1$, and update R to be $5 - 4 = 1$. In the next (and last) iteration of Step 2, $k = 0$ is the largest exponent for which $2^k \le 1 (= R)$, so we determine the final digits of the expansion $c_1 = 0$, $c_0 = 1$. We now arrive at Step 4, to produce the resulting binary expansion: $69 \sim [1000101]$ (base 2).

(ii) In Step 1, we initialize the remainder $R = 225$. Since $K = 2$ is the largest power of 8 not exceeding $R = 225$, we determine the coefficient $c_2 = \lfloor 225/64 \rfloor = 3$, and we update $R = 225 - 3 \cdot 8^2 = 33$. In the next iteration of Step 2, we get $k = 1$, and $c_1 = \lfloor 33/8 \rfloor = 4$, so we update $R = 33 - 4 \cdot 8 = 1$.

---

[*] Always one less than the number of digits, since the powers of b descend to the 0th power.

> One final application of Step 2 gives the remaining coefficient
> $c_0 = 1$, and we then move on to Step 4 of Algorithm 6.1 to
> obtain the base 8 expansion $225 \sim [341]$ (base 8).
> (iii) In Step 1, we initialize the remainder $R = 729$. The largest
> power of 16 not exceeding 729 is $16^2 = 256$. Hence $K = 2$,
> $c_2 = \lfloor 729 / 256 \rfloor = 2$, and we update R to be $729 - 2 \cdot 256 = 217$. In the next application of Step 2, $k = 1$, $c_1 = \lfloor 217 / 16 \rfloor = 13$,
> and R becomes $217 - 13 \cdot 16 = 9$. The final iteration of Step 2
> gives $c_0 = 9$, and hence $729 \sim [2\ 13\ 9]$ (base 16).

This algorithm is nicely amenable to be coded into computer programs, and this is done in the computer implementation material at the end of the chapter. We give one useful fact here that will often be helpful in implementing the above algorithm with the aid of either a scientific calculator or computer.

Computing Note: Step 2 of the above algorithm involves finding the largest integer with the property that $b^k \le R$. A brute-force approach would be to simply start off with $k = 0$ and continue computing increasing powers of 2 until the result equals or exceeds R: $b^0 = 1$, $b^1 = b$, $b^2, b^3, \cdots.$* Although this would work, it is not very efficient. Since most calculators (and computers) have built-in logarithm functions, it is much faster to use one of these when we compute this largest integer k. Indeed, for each base b, the associated logarithm function, denoted $y = \log_b(x)$ associates for each positive real number x (in the domain) the real number y with the property that $b^y = x$. The output of this function will generally not be an integer even when the input is, but since we are seeking the largest integer k such that $b^k \le x$, we can simply take $k = \text{floor}(\log_b(x))$. Finally, since any calculator or computing platform will have only a handful of built-in log functions, the good news is that we need only one. Suppose that we have some logarithm function, denoted as $\text{LOG}(x)$ on our computing tool. The so-called *change of base formula* allows us to express any log in terms of our log as follows:

$$\log_b(x) = \text{LOG}(x) / \text{LOG}(b)$$

In summary, the determination of k in Step 2 of Algorithm 6.1 can be accomplished in a single shot with the formula

$$k = \text{floor}(\text{LOG}(R) / \text{LOG}(b)) \tag{6.2}$$

For example, with $b = 2$ and $R = 136,825$, the largest integer k such that $2^k \le 136,825$ is given by $k = \text{floor}(\text{LOG}(136,825)/\text{LOG}(2)) = \text{floor}(17.062...) = 17$.

Hex(adecimal) and Binary Expansions

The most important, often-used bases are given special terminology and notation; some of these will be given in the following definition.

---

* On a computer, this would be implemented with a WHILE loop.

Definition 6.1

As we have mentioned, base 2 expansions are also known as **binary expansions** (and are made up of *bit strings*). Expansions in base 3, 8, and 16 are known, respectively, as **ternary**, **octal**, and **hexadecimal**. In hexadecimal expansions, the double digits 10, 11, 12, 13, 14, and 15 are customarily replaced with the letters, A, B, C, D, E, and F, respectively, so that each element in the representing expansion string will be a single character. Thus, from the above example, we may write 729 ~ [2D9] (base 16), which can also be written as 729 ~ [2D9] (hex). This convention allows hexadecimal expansions to be viewed as strings rather than vectors (without it, the string [125] could mean either [12 5] = [C5] or [1 2 5]).

In cryptography in general, and in particular for the remaining chapters of this book, the most important bases are decimal, binary, and hex(adecimal). It is also sometimes convenient to consider expansions of a fixed length (these will be used in the DES block cryptosystem in the next chapter, for example), but with this convention, expansions are not going to be unique. For reference, Table 6.1 converts the hex digits into these other bases. The reader is advised to verify the entries of this table with Equation 6.1. In the binary expansions, we have made all of the strings be of equal length 4 so, for example, the binary expansion for 2, which is 10 $(= 1 \cdot 2^1 + 0 \cdot 2^0)$, is expressed as 0010 $(= 0 \cdot 2^3 + 0 \cdot 2^2 + 1 \cdot 2^1 + 0 \cdot 2^0)$. The hex format is thus the least cumbersome. In the DES systems of Chapter 7, a key would consist of a dizzying 64-bit string (that is, a binary string of length 64), whereas in hex notation the same string would have a much smaller length of 16.

Exercise for the Reader 6.1

(a) Find the integer equivalents for each of the following expansions: [110100111] (binary), [777] (octal), and [123ABC] (hexadecimal).

(b) Use Algorithm 6.1 to find
   (i) The binary expansion of 122
   (ii) The base 32 expansion of 9675
   (iii) The hexadecimal expansion of 52,396

Unlike in the previous introductory and historical chapters, in subsequent cryptographic developments, we will customarily avoid the basic translation steps of changing English-language plaintexts into their numerical equivalents. Typically, the plaintext and ciphertext alphabets with which we will be working will be digits in some base (or in some other number system on which the underpinnings of the system will be based). Interested readers can easily go through the additional steps needed to transform English plaintexts into strings in the alphabet being used. The following example provides some typical schemes for such conversions.

TABLE 6.1 Conversions between Hexadecimal, Decimal, and Binary Digits

Hex	Decimal	Binary
0	0	0000
1	1	0001
2	2	0010
3	3	0011
4	4	0100
5	5	0101
6	6	0110
7	7	0111
8	8	1000
9	9	1001
A	10	1010
B	11	1011
C	12	1100
D	13	1101
E	14	1110
F	15	1111

Example 6.2: Conversions between English Plaintexts and Strings of Digits

(a) Using Table 3.1 as a reference, develop a corresponding natural scheme for translating English letters into bit strings (binary strings) and use this scheme to translate the English plaintext *retreat* into its corresponding bit string.

(b) Develop a similar scheme for translating English letters into hexadecimal strings and use this scheme to translate the English plaintext *retreat* into its corresponding bit string.

Solution: Part (a): Since the (non-case-sensitive) English alphabet has 26 characters, we would need to use length-5 bit strings to be able to represent all of them. There are $2^5 = 32$ such strings, so we would have six left unused. Length-4 bit strings would not be adequate since there are only $2^4 = 16$ of them; see Table 6.1. If we simply use the length-5 bit string equivalents of the integer representatives in Table 3.1, we obtain the results shown in Table 6.2.

Using this table, the English plaintext *retreat* would be represented by the following bit string:

```
(100001 00100 10011 10001 00100 00000 10011)
10000100100100111000100100000010011
```

Part (b): Since there are only 16 single-digit hexadecimal strings, but $16^2 = 256$-digit hexadecimal schemes, we would need to use length-2 hexadecimal strings to accommodate all 26 English letters. The natural correspondence is shown in Table 6.2, which leads us to the following representation for the English plaintext *retreat*: (11 04 13 11 04 00 13) 11041311040013.

Computing Note: Most computing platforms have built-in conversion utilities for all of the 256 ASCII (*American Standard Code for Information Interchange*) standard English letters, and digits, as well as some common foreign letters, special symbols, and control characters. We caution the reader that the correspondence that we have set up in Table 6.2 (and Table 3.1) for a single set of 26 (non-case-sensitive) English letters is different from the ASCII correspondence. For example, in ASCII, the digits 0–9 have (decimal) integer equivalents 48–57, the lowercase English letters $a-z$ have integer equivalents 97–122, and the uppercase English letters $A-Z$ have integer equivalents 65–90. Thus the complete ASCII alphabet of 256 symbols has a 2-digit hexadecimal representation and an 8-digit binary representation.

Next we give a proof of Theorem 6.1 in which we will explain why Algorithm 6.1 always works.

TABLE 6.2 Correspondence of the English Alphabet Using the Integers mod 26 ($\mathbb{Z}_{26}$), Length-5 Binary Strings, and Length-2 Hexadecimal Strings

English	A	B	C	D	E	F	G	H	I	J	K	L	M
Integer	0	1	2	3	4	5	6	7	8	9	10	11	12
Binary	00000	00001	00010	00011	00100	00101	00110	00111	01000	01001	01010	01011	01100
Hex	00	01	02	03	04	05	06	07	08	09	0A	0B	0C

English	N	O	P	Q	R	S	T	U	V	W	X	Y	Z
Integer	13	14	15	16	17	18	19	20	21	22	23	24	25
Binary	01101	01110	01111	10000	10001	10010	10011	10100	10101	10110	10111	11000	11001
Hex	0D	0E	0F	10	11	12	13	14	15	16	17	18	19

Proof of Theorem 6.1.

Fix a base b (an integer greater than 1), and let n be a positive integer. We must show two things:

(a) n has a base b expansion (existence)
(b) any two base b expansions for n must be the same (uniqueness)

Part (a): *Existence.* Existence of a base b expansion of n will be accomplished by showing that Algorithm 6.1 performs correctly. The iterated Step 2 will clearly construct a decreasing sequence of integers $K = k_1 > k_2 > \cdots > k_t \geq 0$, along with a corresponding sequence of coefficients $c_{k_i} = c(k_i)$, each of which is a nonnegative integer less than b. Keeping track of all of the terms that get subtracted from n, we see that each k_i is the largest exponent such that

$$b^{k_i} \leq n - c(k_1)b^{k_1} - c(k_2)b^{k_2} - \cdots - c(k_{i-1})b^{k_{i-1}}$$

The right sides are the updated values of R in the algorithm. We need to check that when all of the coefficients multiplied by their respective powers of b are subtracted from n, we are left with zero; that is,

$$n - c(k_1)b^{k_1} - c(k_2)b^{k_2} - \cdots - c(k_t)b^{k_t} = 0$$

This would prove that R is eventually zero, so Step 3 eventually moves to Step 4 [also, this equation is equivalent to Equation (6.1) and corresponds to the base b expansion produced by the algorithm]. Let us temporarily denote the left side of the above equation by S. By the way in which the coefficient $c(k_t)$ is chosen, we must have $0 \leq S < b^{k_t}$. Since the process has ended, there can be no other nonnegative power of b that is less than or equal to S. This means that $S < b^0 = 1$, so indeed $S = 0$, as we needed to show.

Part (b): *Uniqueness*. Suppose that we have two base b expansions of n:

$$c_K b^K + c_{K-1} b^{K-1} + \cdots + c_2 b^2 + c_1 b + c_0$$

$$= n = d_L b^L + d_{L-1} b^{L-1} + \cdots + d_2 b^2 + d_1 b + d_0$$

where $c_i, d_j \in \{0, 1, 2, \cdots, b-1\}$ and $c_K, d_L > 0$. We accomplish the proof by method of contradiction: We assume that the above expansions are different, and from this assumption we will reach a contradiction.

With the assumption that the above two expansions are different, there must be a smallest index s for which $c_s \neq d_s$. Since we can cancel out all (common) terms in the above equation with smaller indices, we are led to the equation

$$c_K b^K + c_{K-1} b^{K-1} + \cdots + c_{s+1} b^{s+1} + c_s b^s = d_L b^L + d_{L-1} b^{L-1}$$

$$+ \cdots + d_{s+1} b^{s+1} + d_s b^s$$

If we divide both sides of this equation by b^s, and then solve for $c_s - d_s$, we are led to

$$c_s - d_s = c_K b^{K-s} + c_{K-1} b^{K-s-1} + \cdots + c_{s+1} b^1 - d_L b^{L-s} -$$

$$d_{L-1} b^{L-s-1} - \cdots - d_{s+1} b^1$$

Now since b certainly is a factor of the right side of this equation, it must also be a factor of the left side, that is, $b|(c_s - d_s)$. But since $c_s, d_s \in \{0, 1, 2, \cdots, b-1\}$, it follows that $c_s - d_s \in \{-(b-1), -(b-2), \cdots, -2, -1, 0, 1, 2, \cdots, b-1\}$, so the only way we could have $b|(c_s - d_s)$ is if $c_s - d_s = 0$, that is, $c_s = d_s$ —a contradiction! Uniqueness of base b expansions is thus established. □

Addition Algorithm with Base b Expansions

The way computer and software systems work to perform arithmetic operations is as follows. First, they input the numbers in the usual decimal format (the human interface), then the numbers get converted to computer numbers (in some base b that is a power of 2), next the computer performs the arithmetic operations directly with these computer numbers, and finally the answer gets converted back to decimal format. We have already discussed how such translations work. As it turns out, the decimal arithmetic algorithms that one learns about in grade school easily generalize to base b algorithms. For example, after having first learned how to add single digits (such as $6 + 8$), we then learned how to add two multidigit positive integers in grade school by stacking the numbers vertically (so corresponding digits were directly above/below each other), and then, starting from the right, we added successive digits, moving any *carries* (when two single-digit additions are larger than 9) over to the next digits on the left. When adding two numbers, any carry must equal 1 (since the biggest that two single decimal digits could add up to is 18). In order to avoid too much formality, we will assume that we are able to add, subtract, and multiply single digits in any base (perhaps with carries or borrows), just as in grade-school arithmetic. This can be accomplished by converting to ordinary integer arithmetic. For example, to perform the hexadecimal subtraction E − B, we convert to the corresponding integer subtraction (see Table 6.1) $14 - 11 = 3$, whose decimal answer is the same as its hexadecimal form. To perform the single-digit hexadecimal multiplication E · B, we first perform the corresponding integer multiplication $14 \cdot 11 = 154$. Since the result exceeds 15 (the largest for a single hexadecimal digit), we apply the division algorithm to this result with divisor 16: $154 = 9 \cdot 16 + 10 = 9 \cdot 16 + A$, to find that the hexadecimal result is [9A], or A with a carry of 9.

We next give an analogous algorithm that shows how to add two integers $c = [c_{K-1} \cdots c_2\ c_1\ c_0]$ and $d = [d_{K-1} \cdots d_2\ d_1\ d_0]$ of the same base b.[*] We begin by adding the rightmost digits in base b: $[c_0] + [d_0] = [\text{Car}_0\ s_0]$, where Car_0 (the "carry") can be either 0 or 1. In ordinary integer language, this corresponds to the equation $c_0 + d_0 = b \cdot \text{Car}_0 + s_0$.[†] We then continue this process, each time moving one digit to the left and adding the existing carry to the sum. The next step would be to add second-to-right digits with Car_0 : $[c_1] + [d_1] + [\text{Car}_0] = [\text{Car}_1\ s_1]$. Note that since in ordinary integer arithmetic this corresponds to the equation $c_1 + d_1 + \text{Car}_0 = b \cdot \text{Car}_1 + s_1$, it follows that $\text{Car}_1 = \text{floor}([c_1 + d_1 + \text{Car}_0]/b)$. We also point out the new carry, Car_1, is either 0 or 1, because $0 \le c_1 + d_1 + \text{Car}_0 \le (b-1) + (b-1) + 1 \le 2b - 1$. This process continues as we move through the digits, with the carries always being either 0 or 1. After we have moved through the K digits, to obtain $s_0, s_1, \cdots, s_{K-1}$, we take s_K to be the value of the last carry, Car_{K-1}. It thus follows that $[c_{K-1} \cdots c_2\ c_1\ c_0] + [d_{K-1} \cdots d_2\ d_1\ d_0] = [s_K s_{K-1} \cdots s_2\ s_1\ s_0]$, in base b. This process is summarized in the following algorithm in which

---

[*] We may always arrange things so that the two base b numbers being added have the same number of (base b) digits by padding shorter numbers with additional zeros on the left.

[†] In other words, if we apply the division algorithm to the division of $c_0 + d_0$ by b, will be the quotient and s_0 will be the remainder.

we do not bother storing the individual carries, since they may be discarded (overwritten) once they have been used.

Algorithm 6.2: Addition of Two Base b Integers

Assume that we have two K-digit base b integers[*] of the form

$$c = [c_{K-1} \cdots c_2 \ c_1 \ c_0] \text{ and } d = [d_{K-1} \cdots d_2 \ d_1 \ d_0]$$

This algorithm will compute the base b expansion $[s_K s_{K-1} \cdots s_2 \ s_1 \ s_0]$ of the sum $c + d$.

Step 1.   Put Car $= 0$ <Initialize carry>
Step 2.   FOR $i = 0$ TO $K - 1$

   NewCar $= \text{floor}([c_i + d_i + \text{Car}]/b)$ <Compute new carry>
   Set $s_i = c_i + d_i + \text{Car} - b \cdot \text{NewCar}$
   Update Car $=$ NewCar
   END <i FOR>

Step 3.   Set $s_K = $ Car
Step 4.   Form $[s_K s_{K-1} \cdots s_2 \ s_1 \ s_0]$, the base b expansion of $c + d$.

This algorithm may be carried out with a similar notation to what is taught in elementary school for addition of numbers by hand. We use this notation in the following example.

Example 6.3

Use Algorithm 6.2 to perform each of the indicated additions and then translate the operation to an ordinary (base 10) integer addition.

(a) The binary (base 2) addition: [11011] + [10110]
(b) The octal (base 8) addition: [744] + [552]

Solution: Part (a): The process proceeds as follows: In Step 2, the first value of NewCar is floor([1 + 0 + 0]/2) = 0, so $s_0 = 1 + 0 + 0 - 0 = 1$. Moving into the $i = 1$ iteration of Step 2, the carry (Car) is 0, and so NewCar becomes floor ([1 + 1 + 0]/2) = 1 and $s_1 = 1 + 1 + 0 - 2 = 0$. Moving into the $i = 2$ iteration of Step 2, the carry (Car) is now 1, and we have NewCar = floor ([0 + 1 + 1]/2) = 1, and so $s_2 = 0 + 1 + 1 - 2 = 0$. In the $i = 3$ iteration, we have Car = 1, NewCar = floor([1 + 0 + 1]/2) = 1 and $s_3 = 1 + 0 + 1 - 2 = 0$. In the final $i = 4$ iteration, we have Car = 1, NewCar = floor([1 + 1 + 1]/2) = 1 and $s_4 = 1 + 1 + 1 - 2 = 1$.

---

[*] This algorithm subsumes additions of single-digit base b numbers along with some very restricted floor operations involving divisions of small integers by b. Of course, in the design of any computer arithmetic system, such simple operations would need to be programmed in before Algorithm 6.2 could function correctly.

In Step 3, the final carry Car = 1 gets transferred to $s_5 = 1$. Thus we obtain [11011] + [10110] = [110001] (base 2). This corresponds to the integer addition: 27 + 22 = 49.

```
  1   1   1   1
      1   1   0   1   1
      1   0   1   1   0
  _____
  1   1   0   0   0   1
```

Part (b): The process proceeds as follows: In Step 2, the first value of NewCar is floor([4 + 2 + 0]/8) = 0, so $s_0 = 4 + 2 + 0 - 0 = 6$. Moving into the $i = 1$ iteration of Step 2, the carry (Car) is 0, and so NewCar becomes floor([4 + 5 + 0]/8) = 1 and $s_1 = 4 + 5 + 0 - 8 = 1$. Moving to the (final) $i = 2$ iteration of Step 2, the carry (Car) is now 1, and we have NewCar = floor([7 + 5 + 1]/8) = 1, and so $s_2 = 7 + 5 + 1 - 8 = 5$. In Step 3, the final carry (Car) = 1 gets transferred to $s_3 = 1$. Thus we obtain [744] + [552] = [1516] (base 8). This corresponds to the integer addition: 484 + 362 = 846.

```
   1   1
       7   4   4
       5   5   2
   _____
   1   5   1   6
```

Use Algorithm 6.2 to perform each of the indicated additions and then translate the operation to an ordinary (base 10) integer addition.

(a)   The binary addition: [101111] + [001111]

(b)   The hexadecimal addition: [7D4E] + [1AA2]

Subtraction Algorithm with Base b Expansions

Just as was done for addition, an algorithm for subtraction in terms of base b expansions will be fashioned after the usual method that is taught in grade school for subtracting integers by hand. We first explain how the algorithm works and then formally summarize the steps.

Suppose that we wish to perform a subtraction $c - d$ of two integers $c = [c_{K-1} \cdots c_2\ c_1\ c_0]$ and $d = [d_{K-1} \cdots d_2\ d_1\ d_0]$ expanded in the same base b, with $c > d$. We begin by subtracting the rightmost digits in base b, $c_0 - d_0 = \text{Borr}_0 \cdot b + s_0$ where Borr_0 (the "borrow") can be either 0, if $c_0 - d_0 \geq 0$, or -1. The borrow is what is needed to take from the next digit of c to ensure that the current digit subtraction will have a nonnegative result. In ordinary integer language, this corresponds to the equation $c_0 - d_0 = b \cdot \text{Borr}_0 + s_0$.* We then continue this process, each time moving one digit to the left and adding the existing borrow to the difference of the new digits. The next step would be to subtract the second-to-right digits: $c_1 - d_1 + \text{Borr}_0 = b \cdot \text{Borr}_1 + s_1$. Note that since in ordinary integer arithmetic this corresponds to the equation $c_1 - d_1 + \text{Borr}_0 = b \cdot \text{Borr}_1 + s_1$, it follows that $\text{Borr}_1 = \text{floor}([c_1 - d_1 + \text{Borr}_0]/b)$. We also point out that the new borrow, Borr_1, is again either 0 or -1 because $-b \leq c_1 - d_1 + \text{Borr}_0 < b$. This process continues as we move through the digits, with the borrows always being either 0 or -1. After we have moved through the K digits to obtain $s_0, s_1, \cdots, s_{K-1}$, we note that because of the assumption that $c > d$, the last borrow, Borr_{K-1}, will always be 0. It thus follows that $[c_{K-1} \cdots c_2\ c_1\ c_0] - [d_{K-1} \cdots d_2\ d_1\ d_0] = [s_{K-1} \cdots s_2\ s_1\ s_0]$, in base b. This process is summarized in the following algorithm in which we do not bother storing the individual borrows, since they may be discarded (overwritten) once they have been used.

Algorithm 6.3: Subtraction of Two Base b Integers

Assume that we have two K-digit base b integers[†] of the form $c = [c_{K-1} \cdots c_2\ c_1\ c_0]$ and $d = [d_{K-1} \cdots d_2\ d_1\ d_0]$ with $c > d$. This algorithm will compute the base b expansion $[s_{K-1} \cdots s_2\ s_1\ s_0]$ of the difference $c - d$.

Step 1.   Put Borr = 0 <Initialize borrow>

Step 2.   FOR $i = 0$ TO $K - 1$

   NewBorr = floor($[c_i - d_i + \text{Borr}]/b$) <Compute new borrow>
   Set $s_i = c_i - d_i + \text{Borr} - b \cdot \text{NewBorr}$
   Update Borr = NewBorr
   END <i FOR>

Step 3.   Form $[s_{K-1} \cdots s_2\ s_1\ s_0]$, the base b expansion of $c - d$.

This algorithm may be carried out with a similar notation to what is taught in elementary school for subtraction of numbers by hand. We use this notation in the following example.

---

* If we apply the division algorithm to the division of $c_0 - d_0$ by b, Borr_0 will be the quotient and s_0 will be the remainder. In the case $c_0 - d_0 > 0$ since $c_0 - d_0 < b$, we will have $\text{Borr}_0 = 0$; while in the case $c_0 - d_0 < 0$, we will have $\text{Borr}_0 = -1$.

† This algorithm subsumes subtractions of single-digit base b numbers, along with some very restricted floor operations involving divisions of small integers by b. Of course, in the design of any computer arithmetic system, such simple operations would need to be programmed in before Algorithm 6.3 could function correctly.

Example 6.4

Use Algorithm 6.3 to perform the subtraction [6FAA] − [4FED] in hexadecimal arithmetic.

Solution: The process proceeds as follows:

$$
\begin{array}{ccccc}
 & -1 & -1 & -1 & \\
6 & F & A & A \\
4 & F & E & D \\
\hline
1 & F & B & D \\
\end{array}
$$

Step 1 always initializes the borrow to be 0: Borr = 0. In Step 2, the first value of NewBorr is floor([A − D + 0]/16) = floor([10 − 13 + 0]/16) = −1, so s_0 = 10 − 13 + 0 − 16 · (− 1) = 26 − 13 = 13 = D. Moving into the $i = 1$ iteration of Step 2, the borrow (Borr) is −1, and so NewBorr becomes floor([A − E − 1]/16) = floor([10 − 14 − 1]/16) = −1, and so

$$s_1 = c_1 - d_1 + \text{Borr} - 16 \cdot \text{NewBorr}$$

$$= 10 - 14 - 1 - 16 \cdot (-1) = 26 - 15 = 11 = B$$

Moving into the $i = 2$ iteration of Step 2, the borrow (Borr) is now −1, and we have NewBorr = floor([F − F − 1]/16) = floor([15 − 15 − 1]/16) = −1, and so

$$s_1 = c_2 - d_2 + \text{Borr} - 16 \cdot \text{NewBorr}$$

$$= 15 - 15 - 1 - 16 \cdot (-1) = 16 - 1 = 15 = F$$

In the final $i = 3$ iteration of Step 2, the borrow (Borr) is still −1, NewBorr = floor([6 − 4 − 1]/16) = 0, and

$$s_3 = c_3 - d_3 + \text{Borr} - 16 \cdot \text{NewBorr}$$

$$= 6 - 4 - 1 - 16 \cdot (0) = 1$$

We thus obtain [6FAA] − [4FED] = [1FBD] (base 16). This corresponds to the integer subtraction

$$[6 \cdot 16^3 + 15 \cdot 16^2 + 10 \cdot 16^1 + 10 \cdot 16^0]$$
$$- [4 \cdot 16^3 + 15 \cdot 16^2 + 14 \cdot 16^1 + 13 \cdot 16^0]$$
$$= 1 \cdot 16^3 + 15 \cdot 16^2 + 11 \cdot 16^1 + 13 \cdot 16^0$$

or 28,586 − 20,461 = 8125.

Exercise for the Reader 6.3

Use Algorithm 6.3 to perform each of the indicated subtractions and then translate the operation to an ordinary (base 10) integer subtraction.

(a)   The binary subtraction: $[101101] - [001111]$

(b)   The hexadecimal subtraction: $[7D4E] - [1AA2]$

Multiplication Algorithm in Base b Expansions

The general algorithm for multiplying two numbers using their base b expansions follows easily from the following two simple observations:

1. If k is a positive integer, the base b expansion of b^k is $[1\ 0\ 0 \cdots 0]$, where there are k zeros following the 1. $\underbrace{\qquad}_{k \text{ zeros}}$

2. If we multiply a base b expansion $[c_{K-1} \cdots c_2\ c_1\ c_0]$ by $b^k \sim [1\ 0\ 0 \cdots 0]$, the former expansion gets shifted to the left k places, with a string of k zeros appended on the right:
$$b^k \cdot [c_{K-1} \cdots c_2\ c_1\ c_0] = [c_{K-1} \cdots c_2\ c_1\ c_0 \underbrace{0\ 0 \cdots 0}_{k \text{ zeros}}].$$

Since it is a direct generalization of the grade-school multiplication algorithm (in base 10), we motivate the algorithm of multiplying two integers $c = [c_{K-1} \cdots c_2\ c_1\ c_0]$ and $d = [d_{K-1} \cdots d_2\ d_1\ d_0]$ of the same base b, with a simple grade-school hand multiplication of 126×63.

The algorithm is usually performed schematically:

```
          1     3 / 1
          1     2     6
        ×       6     3
      ─────────────────
          3     7     8
    7     5     6
  ─────────────────
    7     9     3     8
```

We line up the two numbers at the right margin and begin by multiplying the last digit of the lower number, by each of the digits of the upper number (starting from the right and going left). Each time we get a product of digits larger than 9 ($= b - 1$), we have a carry that we move to the next digit (that will be added to the product). When we get through multiplying all of the top number's digits by the last digit of the bottom, we move to the second-to-last digit of the bottom number and repeat this, but we shift the answers that we get one digit to the left and put them in a second row. The number of such shifted rows that we get will equal the number of digits of the lower number. Note that each row can give rise to different carries; previous carries can be crossed off when we are done using them. We simply included the single pair of multiple carries with a slash symbol. We then need to add these rows using base b addition.[*]

---

[*] That is, using Algorithm 6.2. Technically, when there are more than two rows, we would need to use Algorithm 6.2 repeatedly, adding two numbers at a time.

The role of the carries is quite similar to their role in base b addition (Algorithm 6.2); the reason for the shifting of rows can be seen by the distributive law and the shifting property 2 above as follows:

$$cd = c[d_0 \cdot b^0 + d_1 \cdot b^1 + \cdots d_{K-1} \cdot b^{K-1}]$$
$$= (cd_0) \cdot b^0 + (cd_1) \cdot b^1 + \cdots (cd_{K-1}) \cdot b^{K-1}$$

Each term in the latter sum is thus a single digit of the latter number, multiplied by (all the digits of) the former number, multiplied by a power of the base. This latter multiplication has the effect of shifting all of the digits of the multiplication to the left (by the shifting property 2), with the number of slots shifted equaling the power of the base. The general procedure is summarized in the following algorithm.

Algorithm 6.4: Multiplication of Two Base b Integers

Assume that we have two base b integers[*] of the form

$$c = [c_{L-1} \cdots c_2 \ c_1 \ c_0] \text{ and } d = [d_{K-1} \cdots d_2 \ d_1 \ d_0]$$

This algorithm will compute the base b expansion of the product $c \cdot d$.

Step 1. Set $P = 0$ <Initialize product; will consist of a sum of terms>

Step 2. FOR $i = 0$ TO $K - 1$ <i will be the index of a digit of d>

Set $p_0 = p_1 = \cdots = p_{i-1} = 0$ <this is the shifting corresponding to d_i>
Set Car $= 0$ <Initialize carry>
FOR $j = 0$ TO $L - 1$ <j will be the index of a digit of c>
NewCar $= \text{floor}([c_j \cdot d_i + \text{Car}]/b)$ <Create new carry>
Set $p_{i+j} = c_j \cdot d_i + \text{Car} - b \cdot \text{NewCar}$
Update Car $= $ NewCar <Update carry>
END <j FOR>
Set $p_{i+L} = $ Car
Update $P \rightarrow P + [p_{i+L} \cdots p_2 \ p_1 \ p_0]$. <Using Algorithm 6.2>
END <i FOR>

Step 3. Form $P = [p_{K+L} \ p_{K+L-1} \cdots p_2 \ p_1 \ p_0]$, the base b expansion of $c \cdot d$.

We point out that (in the case $b = 10$) this algorithm deviates only slightly from the elementary school algorithm in that the updating of the cumulative sum (P) is done after each d-digit multiplication, rather than all at once at the end.[†]

---

[*] In contrast with Algorithm 6.2, no advantage is gained by assuming the base b expansions of the numbers being multiplied have the same length.

[†] This more properly fits with the use of Algorithm 6.2, since the latter was developed for adding two (base b) numbers.

Example 6.5

Use Algorithm 6.4 to perform each of the indicated multiplications and then translate each to an ordinary (base 10) integer multiplication.

(a) The binary multiplication: $[1101] \times [110]$
(b) The hexadecimal multiplication: $[2A4] \times [12E]$

Solution: Part (a): In binary arithmetic, carries never arise in the digit multiplication process since the largest that a product of digits can be is 1 (a single binary digit). The process is outlined in the diagram.* In Step 2 of Algorithm 6.4, all of the values of Car and NewCar are zero, and the formula $s_{i+j} = c_j \cdot d_i + \text{Car} - b \cdot \text{NewCar}$ reduces to $s_{i+j} = c_j \cdot d_i$. It follows that in Step 2, for each digit multiplication d_j, the result will be either a string of zeros (which we can ignore) in the case $d_j = 0$ (like d_0), or simply a shifted copy of the binary string for c (as happens for d_1, d_2). Thus, the result will be $[0000] + [11010] + [110100] = [1001110]$. In the language of ordinary (base 10) integers, this result corresponds to the multiplication $13 \times 6 = 78$.

$$
\begin{array}{ccccccc}
 & & & 1 & 1 & 0 & 1 \\
 & & \times & & 1 & 1 & 0 \\
\hline
 & & & 0 & 0 & 0 & 0 \\
 & & 1 & 1 & 0 & 1 & \\
 & 1 & 1 & 0 & 1 & & \\
\hline
1 & 0 & 0 & 1 & 1 & 1 & 0 \\
\end{array}
$$

Part (b): The process is outlined in the diagram, and the reader might wish to simply verify this diagram rather than read through the following details. In Step 2, starting with $i = 0$ (corresponding to the digit E of the latter number), we need to multiply each of the digits of the top number by $E = 14$.

$$
\begin{array}{ccccc}
2/ & 8/1 & 3/ & & \\
 & & 2 & A & 4 \\
 & \times & 1 & 2 & E \\
\hline
 & 2 & 4 & F & 8 \\
 & 5 & 4 & 8 & \\
2 & A & 4 & & \\
\hline
3 & 1 & D & 7 & 8 \\
\end{array}
$$

---

* To become more comfortable with such multiplications, readers are encouraged to construct their own single digit multiplication tables, say, for binary and hex multiplication. For binary, the table would be two by two, while for hex it would be sixteen by sixteen and include all single digit hex products such as $F \times B = A5$ [which corresponds to the integer multiplication $15 \times 11 = 165 = 10 \cdot 16 + 5 (= A5)$.]

The first (rightmost) digit multiplication is $4 \cdot E = 4 \cdot 14 = 56 = 3 \cdot 16 + 8$, so $p_0 = 8$, and the first carry Car $= 3$. Next, with $j = 1$, we multiply $A \cdot E + \text{Car} = 10 \cdot 14 + 3 = 143 = 8 \cdot 16 + 15$, so $p_1 = 15 = F$, and the next carry is Car $= 8$. Moving to the last digit with $j = 2$, the multiplication (and carry) is $2 \cdot E + \text{Car} = 2 \cdot 14 + 8 = 36 = 2 \cdot 16 + 4$, so $p_2 = 4$, and the last carry (2) is the value of p_3. The cumulative sum P is now updated from its previous value (0) to the new value [24F8] (being added to it). Next we move to $i = 1$, corresponding to the middle digit 2 of the latter number. Here we pad ($i = 1$ digit) $p_0 = 0$. The first multiplication $4 \cdot 2 = 8$ has no carry (Car $= 0$), and gives $p_1 = 8$. Next, with $j = 1$, we multiply $A \cdot 2 + \text{Car} = 10 \cdot 2 + 0 = 20 = 1 \cdot 16 + 4$, to obtain $p_2 = 4$ and Car $= 1$. Moving to the last digit with $j = 2$, the multiplication (and carry) is $2 \cdot 2 + \text{Car} = 2 \cdot 2 + 1 = 5$, so $p_3 = 5$, and there is no final carry to import to p_4. The cumulative sum is now updated to $P = [24F8] + [5480] = [7978]$ (base 16; using Algorithm 4.2). The last $i = 2$ iteration is simple, since it corresponds to multiplying by the digit one, and the resulting product is simply the top number with two zeros padded at the right. Hence, the final answer of the multiplication is obtained by updating $P = [7978] + [2A400] = [31D78]$. The resulting multiplication [2A4] $\times$ [12E] $= [31D78]$ (base 16) translates to $676 \times 302 = = 204{,}152$ (base 10) in ordinary integer language.

Exercise for the Reader 6.4

Use Algorithm 6.4 to perform each of the indicated multiplications and then translate the operation to an ordinary (base 10) integer multiplication.

(a)   The binary multiplication: [1111] $\times$ [1111]

(b)   The base 7 multiplication: [262] $\times$ [520]

The preceding algorithms provide prototypes for more complete collections of algorithms for general arithmetic operations in base b arithmetic. We end this chapter with an application of base b arithmetic to obtain an algorithm for fast modular exponentiation that was promised in Chapter 2. But first we say a few words about doing computations with large numbers (which is, as we will see in later chapters, a very relevant topic to public key cryptography).

Arithmetic with Large Integers

By default, most computing platforms perform their computations using a so-called floating point arithmetic system. In terms of integers (the real numbers most used in cryptosystems), this means that any inputted integer is stored as a bit string, but the number of bits used to store the significant digits is limited. This number, called **word size**, is usually a power of 2. For example, in IEEE[*] double precision arithmetic, the word size is 52, which means that numbers larger than $2^{52} - 1 = 4.5036... \times 10^{15}$ cannot be accurately stored or computed in the system. Some computing platforms have special features to accommodate computations with large integers, often known as *symbolic functionality*. The basic idea on how such accommodations are made is to devote several words to represent larger integers. For example, if we needed to multiply two 250-bit numbers, the answer would typically be of size 501 bits, so we would need to use 10 words to represent the numbers in the system (which would accommodate numbers up to 520 bits). Algorithm 6.3 could be suitably modified to multiply the numbers, with any carries of the leftmost character of one word being transferred to the rightmost character of the next word. The computer implementation material at the end of this chapter provides some more details on this. The following example demonstrates the main ideas in an artificially smaller scale.

Example 6.6

Suppose that we are working in a very small computing platform whose (binary) word length is just 3. Outline a scheme by which this system could be used to perform the addition [10111] + [11101] (base 2). Then check the result using ordinary integer arithmetic.

Solution: The answer to an addition of five-bit numbers will require at most six bits, so we will need to work with two words. The addition is thus recast as: [010][111] + [011][101]. We abbreviate these four words as: $[w_1^{\text{left}}][w_1^{\text{right}}] + [w_2^{\text{left}}][w_2^{\text{right}}]$. Now, the addition algorithm (Algorithm 6.2) is modified to accommodate such two-word additions as follows: The addition proceeds just as before to create the first three bits of the sum of the "right" words. If the final carry Car is 1, this gets carried to initialize Car for the "left" words' addition (ordinarily the Car is initialized to be 0). Adding the right words gives [111] + [101] = [100] with the final Car = 1. This initializes the carry of the "left" word addition: [010] + [011], which we show in steps. The last digit is 0 (since the sum of the last two digits + Car is 0 + 1 + 1) and gives a carry Car = 1. The middle digit is 1 (since the sum of the middle digits + Car = 1 + 1 + 1) and gives a carry Car = 1. The first digit is 1 (since the sum of the first digits + Car = 0 + 0 + 1). Thus the left and right words of the answer are

---

[*] The IEEE (Institute of Electrical and Electronics Engineers) is a nonprofit, technical professional association of more than 350,000 individual members in 150 countries.

[110][100], and the result of this addition can be summarized as [10111] + [11101] = [110100] (base 2), which, in ordinary integer arithmetic, corresponds to 23 + 29 = 52.

Fast Modular Exponentiation

With the thorough development of binary expansions behind us, we are now well poised to formalize an algorithm for modular exponentiation that was promised in Chapter 2. The following algorithm is extremely efficient and is very useful in our work relating to public key and elliptic curve cryptography.

Algorithm 6.5: Fast Modular Exponentiation

Input: An integer base c, an integer exponent x, and an integer modulus $m > 1$.

Output: A nonnegative integer $a < m$ that satisfies $a \equiv c^x \pmod{m}$.

Step 1. Use Algorithm 6.1 to create the binary expansion of the exponent x:

$$x \sim [d_K \ d_{K-1} \ \cdots d_1 \ d_0] \quad \text{(base 2)}$$

Step 2. Repeatedly square the number $c \pmod{m}$ as we run through the binary digits d_k of x, including the result in the cumulative product only when $d_k = 1$.

    Set $a = 1$ <Initialize cumulative product a>
    Set $s = c \pmod{m}$ <Initialize squaring>
    FOR $k = 0$ TO K
        IF $d_k = 1$
            Update $a \rightarrow a \times s \pmod{m}$
        END <IF>
        Update $s \rightarrow s^2 \pmod{m}$ <Squaring need not be done
                           when $k = K$>
    END <k FOR>

Step 3. Output: a

The proof that this algorithm works follows simply from writing out the binary expansion of the exponent and using the laws of exponents:

$$c^x \equiv c^{\sum_{k=0}^{K} d_k 2^k} \equiv c^{d_0 2^0} c^{d_1 2^1} \cdots c^{d_K 2^K} \pmod{m}$$

and since $c^{2^{k+1}} = (c^{2^k})^2$. All of the successive squares need to be computed $\pmod{m}$, but only those for which the corresponding $d_k = 1$ make an appearance in the above (cumulative) expression for b^x. Algorithm 6.5 turns out to be a very fast and efficient scheme for modular exponentiation, just requiring an amount of work roughly equal to performing K successive squarings mod m, where K is the number of binary digits in the exponent x.

Example 6.7

Use the fast exponentiation algorithm to compute the following modular power: $2^{825} (\mod 173)$.

Solution: We first apply Algorithm 6.1 to obtain the binary expansion $825 \sim [1100111001]$ (base 2). After initializing $a = 1$ and $s = 2$, the iterations of Step 2 produce the following updates:

$k = 0$:   Since $d_k = 1$, we update $a \to a \times s = 2$ and then update the square $s \to s^2 = 2^2 = 4$.

$k = 1$:   Since $d_k = 0$, we only update the square $s \to s^2 = 4^2 = 16$.

$k = 2$:   Since $d_k = 0$, we only update the square $s \to s^2 = 16^2 = 256 \equiv 83 (\mod 173)$.

$k = 3$:   Since $d_k = 1$, we update $a \to a \times s = 2 \cdot 83 = 166$ and then update the square $s \to s^2 = 83^2 = 6889 \equiv 142 (\mod 173)$.

$k = 4$:   Since $d_k = 1$, we update $a \to a \times s = 166 \cdot 142 = 23572 \equiv 44$ and then update the square $s \to s^2 = 142^2 = 20164 \equiv 96 (\mod 173)$.

$k = 5$:   Since $d_k = 1$, we update $a \to a \times s = 44 \cdot 96 = 4224 \equiv 72$ and then update the square $s \to s^2 = 96^2 = 9216 \equiv 47 (\mod 173)$.

$k = 6$:   Since $d_k = 0$, we only update the square $s \to s^2 = 47^2 = 2209 \equiv 133 (\mod 173)$.

$k = 7$:   Since $d_k = 0$, we only update the square $s \to s^2 = 133^2 = 17689 \equiv 43 (\mod 173)$.

$k = 8$:   Since $d_k = 1$, we update $a \to a \times s = 72 \cdot 43 = 3096 \equiv 155$ and then update the square $s \to s^2 = 43^2 = 1849 \equiv 119 (\mod 173)$.

$k = 9$:   Since $d_k = 1$, we update $a \to a \times s = 155 \cdot 119 = 18445 \equiv 107$.

The algorithm thus tells us that $2^{825} \equiv 107 (\mod 173)$.

Computing Note: On any computing platform that operates in standard floating point arithmetic with binary word length of about 50, the simple approach of evaluating 2^{825} and taking its remainder (mod 173) would lead to inaccurate results since the number is too large to be accurately stored. The algorithm works around this issue since at each iteration of Step 2, the numbers that arise are less than 172^2 and are then immediately converted to an integer (mod 173). Even if one is using a computing platform with symbolic capabilities, the above algorithm is a much more efficient method.

Exercise for the Reader 6.5

Use the fast exponentiation algorithm to compute the following modular power: $289^{225} (\mod 311)$

Chapter 6 Exercises

1. Count from 0 to 25 in each of the following number systems:
   (a) binary (base 2)
   (b) octal (base 8)
   (c) hexadecimal (base 16)

2. Count from 0 to 25 in each of the following number systems:
   (a) base 3
   (b) base 5
   (c) base 11

3. Count from 100 to 125 in each of the following number systems:
   (a) binary (base 2)
   (b) octal (base 8)
   (c) hexadecimal (base 16)

4. Count from 100 to 125 in each of the following number systems:
   (a) base 3
   (b) base 5
   (c) base 11

5. Convert each of the following expansions to an (ordinary) decimal integer:
   (a) [101010] (binary)
   (b) [3123123] (base 4)
   (c) [ABCDEF] (hex)
   (d) [22333] (base 4)
   (e) [12345AAAA] (hex)
   (f) [9876412] (base 11)

6. Convert each of the following expansions to an (ordinary) decimal integer:
   (a) [111000111] (base 2)
   (b) [70073] (base 8)
   (c) [123ABC] (hex)
   (d) [244343] (base 5)
   (e) [CAB12FF] (hex)
   (f) [1122334455] (base 6)

7. Convert each of the following decimal (base 10) integers to (i) binary (base 2) form, (ii) octal (base 8) form, and (iii) hexadecimal (base 16) form:
   (a)   66
   (b)   237
   (c)   1925
   (d)   12,587
   (e)   28,000
   (f) 150,269

8. Convert each of the following decimal (base 10) integers to (i) binary (base 2) form, (ii) octal (base 8) form, and (iii) hexadecimal (base 16) form:
    (a)   87
    (b)   126
    (c)   8000
    (d)   12,347
    (e)   77,895
    (f) 186,000

9. Convert each of the following decimal (base 10) integers to (i) base 3 form, (ii) base 9 form, and (iii) base 27 form:
    (a)   66
    (b)   237
    (c)   1925
    (d)   12,587
    (e)   28,000
    (f) 150,269

10. Convert each of the following decimal (base 10) integers to (i) base 3 form, (ii) base 9 form, and (iii) base 27 form:
    (a)   87
    (b)   126
    (c)   8000
    (d)   12,347
    (e)   77,895
    (f) 186,000

11. (a) Using Table 6.2, convert the following English plaintexts into binary notation and into hexadecimal notation:
        (i)   agent
        (ii)   met
        (iii) liaison
    (b) Using Table 6.2, convert the following strings (which are either binary or hexadecimal) into their corresponding English plaintext:
        (i)   00111001000101101111
        (ii)   0A040B0B0411
        (iii) 0D0E16

12. (a) Using Table 6.2, convert the following English plaintexts into binary notation and into hexadecimal notation:
        (i)   take
        (ii)   cover
        (iii) intown
    (b) Using Table 6.2, convert the following strings (which are either binary or hexadecimal) into their corresponding English plaintext:
        (i)   13070418
        (ii)   00111000001010100100
        (iii) 1604000F0E0D12

13. Perform the following additions in the indicated bases; then translate each into decimal (ordinary) integer language as a check.
   (a) [1100] + [1111] in binary arithmetic
   (b) [5551] + [3333] in octal arithmetic
   (c) [AACC] + [9998] in hexadecimal arithmetic
   (d) [22 8] + [2 10] in base 25

14. Perform the following additions in the indicated bases; then translate each into decimal (ordinary) integer language as a check.
   (a) [1111] + [1010] in binary arithmetic
   (b) [6767] + [3277] in octal arithmetic
   (c) [FACE] + [1AA2] in hexadecimal arithmetic
   (d) [22 18] + [22 13] in base 23

15. Perform the following additions in the indicated bases; then translate each into decimal (ordinary) integer language as a check.
   (a) [110101100] + [10111011] in binary arithmetic
   (b) [55544471] + [333322] in octal arithmetic
   (c) [AABBCC] + [99988FF] in hexadecimal arithmetic
   (d) [22 18 9 8] + [2 10 22 13] in base 25

16. Perform the following additions in the indicated bases; then translate each into decimal (ordinary) integer language as a check.
   (a) [101100111000] + [10011101100] in binary arithmetic
   (b) [5678765] + [1357531] in base 9 arithmetic
   (c) [11AAFF] + [92929292] in hexadecimal arithmetic
   (d) [22 18 13 8] + [2 20 29 13] in base 30

17. Perform the following subtractions in the indicated bases; then translate each into decimal (ordinary) integer language as a check.
   (a) [1100] − [1011] in binary arithmetic
   (b) [7211] − [1127] in octal arithmetic
   (c) [AA22CC] − [9988FF] in hexadecimal arithmetic
   (d) [22 18 9 8] − [2 10 22 13] in base 25

18. Perform the following subtractions in the indicated bases; then translate each into decimal (ordinary) integer language as a check.
   (a) [1101] − [11] in binary arithmetic
   (b) [5674235] − [1357538] in base 9 arithmetic
   (c) [92929292] − [11AAFF] in hexadecimal arithmetic
   (d) [22 18 13 8] − [2 20 29 13] in base 30

19. Perform the following multiplications in the indicated bases; then translate each into decimal (ordinary) integer language as a check.
    (a) [110] × [111] in binary arithmetic
    (b) [555] × [33] in octal arithmetic
    (c) [ACC] × [999] in hexadecimal arithmetic
    (d) [22 8] × [2 10] in base 25

20. Perform the following multiplications in the indicated bases; then translate each into decimal (ordinary) integer language as a check.
    (a) [1111] × [100] in binary arithmetic
    (b) [676] × [377] in octal arithmetic
    (c) [CAB] × [1A9] in hexadecimal arithmetic
    (d) [22 18] × [22 13] in base 23

21. Perform the following multiplications in the indicated bases; then translate each into decimal (ordinary) integer language as a check.
    (a) [11001] × [11111] in binary arithmetic
    (b) [5544] × [3333] in octal arithmetic
    (c) [AACC] × [99FF] in hexadecimal arithmetic
    (d) [22 18 9] × [2 22 13] in base 25

22. Perform the following multiplications in the indicated bases; then translate each into decimal (ordinary) integer language as a check.
    (a) [10110] × [101] in binary arithmetic
    (b) [56787] × [31] in base 9 arithmetic
    (c) [AAFF2] × [92] in hexadecimal arithmetic
    (d) [22 18 13 8] × [2 20] in base 30

23. Use the fast modular exponentiation algorithm (Algorithm 6.5) to perform the following modular exponentiations:
    (a) 2^{58} (mod 5)
    (b) 7^{394} (mod 17)
    (c) 24^{1422} (mod 29)
    (d) 177^{998} (mod 223)

24. Use the fast modular exponentiation algorithm (Algorithm 6.5) to perform the following modular exponentiations:
    (a) 3^{97} (mod 5)
    (b) 12^{117} (mod 17)
    (c) 28^{213} (mod 43)
    (d) 275^{884} (mod 307)

25. Develop a (simple) algorithm for directly converting a binary expansion into its corresponding hexadecimal expansion (without converting to decimal expansions as an intermediate step).

26. Develop a (simple) algorithm for directly converting a binary expansion into its corresponding octal expansion (without converting to decimal expansions as an intermediate step).

27. Show that when Algorithm 6.2 is used to add two base b expansions, each with a total of n digits, it will always require between $2n$ and $3n$ single-digit (or carry-digit) additions/subtractions.

28. Formulate a simple criterion to determine which of two base b expansions (represents an integer that) is greater than or equal to the other.

29. Explain why any integer weight between 0 and $2^n - 1$ (inclusive) can be determined exactly using a balance scale if we have at our disposal exactly one each of the following weights: $\{1, 2, 2^2, \cdots, 2^{n-1}\}$.

30. (a) Show that it would not be possible to determine all integer weights in the range $0 \le x \le 2^n - 1$ on a balance scale if we had at our disposal any set of $n - 1$ weights and are allowed to place them on/off the right side of the scale if the weight x (to be determined) is placed on the left side.

    (b) Show that if $n \ge 4$, then there is a set of $n - 1$ weights so that any weight in the range $0 \le x \le 2^n - 1$ can be balanced on the scale by placing some (or all) of the given weights on either the left or the right side of the scale. Can you find a different algorithm that would actually determine the weight?

    Suggestion for part (b): Use weights that are powers of 3.

The Two's Complement Representation Scheme

A common scheme that computers use to internally store integers is the so-called **two's complement representation** scheme, which works as follows. Given a positive integer n, any integer a in the range $-2^{n-1} \le a < 2^{n-1}$ is represented by a length-n bit string $a \sim [b_{n-1} \; b_{n-2} \; \cdots b_1 \; b_0]$, where the leftmost bit stores the sign of a: $b_{n-1} = 0$ if a is a nonnegative integer, and $b_{n-1} = 1$ if a is negative. In case a is nonnegative, the remaining bits $[b_{n-2} \cdots b_1 \; b_0]$ are just those of the binary expansion of a, whereas if a is negative, $[b_{n-2} \cdots b_1 \; b_0]$ is the binary expansion of $2^{n-1} - |x|$. Thus, for example, if we use length $n = 5$ bit strings, the two's complement representation of 8 would be [01000], and the two's complement representation of -5 would be (since $2^{5-1} - |-5| = 16 - 5 = 11$) [11011].

31. (a) Using $n = 6$ bits, find the two's complement representation of each of the following integers:
    (i)   17
    (ii) -22
    (iii) -32

    (b) Determine the integers that have the following 6-bit two's complement representations:
    (i)   [110011]
    (ii) [001100]
    (iii) [011111]

32. (a) Using $n = 6$ bits, find the two's complement representation of each of the following integers:
      (i) -2
      (ii) -17
      (iii) 25
    (b) Determine the integers that have the following 6-bit two's complement representations:
      (i)   [011011]
      (ii) [101100]
      (iii) [110001]

33. (a) Find a simple relationship between the two's complement representation of an integer in the range $-2^{n-1} \le a < 2^{n-1}$, and the remainder when a is divided by 2^{n-1}.
    (b) Write out a simple algorithm that inputs an integer n and an integer a within the range $-2^{n-1} \le a < 2^{n-1}$, and outputs the two's complement representation of a. Apply your algorithm to Exercise 31(a).
    (c) Write out a simple algorithm that inputs a two's complement representation vector $[b_{n-1}\ b_{n-2}\ \cdots b_1\ b_0]$, and outputs the integer that it represents. Apply your algorithm to Exercise 31(b).

34. (a) Write out a simple algorithm that inputs two length-n vectors, $[c_{n-1}\ c_{n-2}\ \cdots c_1\ c_0]$ and $[d_{n-1}\ d_{n-2}\ \cdots d_1\ d_0]$, that are two's complement representations of a pair of integers, c and d, in the range $-2^{n-1} \le c, d < 2^{n-1}$, and outputs a length $n + 1$ vector $[s_n\ s_{n-1}\ \cdots s_1\ s_0]$ that gives the two's complement representation of sum $c + d$.
    (b) Apply your algorithm to evaluate the following integer sums: $2 + 7$, $(-22) + 16$, and $(-22) + (-20)$.

35. Suppose that we are working in a very small computing platform whose (hexadecimal) word length is just 3; outline a scheme by which this system could be used to perform the addition [8FF8] + [9BA2] (hexadecimal). Then check the result using ordinary integer arithmetic.

36. Suppose that we are working in a very small computing platform whose (hexadecimal) word length is just 4; outline a scheme by which this system could be used to perform the addition [A8FF8] + [29BA2] (hexadecimal). Then check the result using ordinary integer arithmetic.

Note: **Assessing the Work Required to Execute an Algorithm-Complexity Analysis**

Since computers often work in binary arithmetic (or in some other power of two base), it is common to estimate the amount of work required to execute an algorithm in terms of bit operations, where a bit operation consists of either adding, subtracting, or multiplying two single bits (binary integers), including processing any carry or borrow that might arise. Since different

computing platforms take different amounts of time for bit operations, it is common to make such estimates in terms of an invis-
ible (unimportant) constant times some (important) function of
the input sizes n (the total length of bit operations). This provides a common platform-independent method for understanding
and comparing the efficiency of algorithms. The **big-O** notation
that is used for such estimates works as follows: If $f(n)$ and $g(n)$
both represent positive-valued functions of positive integers n,
we write $f(n) = O(g(n))$ [and say $f(n)$ **is big-O of** $g(n)$] if there
is a positive constant C such that $f(n) \leq C \cdot g(n)$ for all integers
n past a certain point. For example, $6n^3 + 10000n^2 + 28n = O(n^3)$,
since $6n^3 + 10000n^2 + 28n \leq 6n^3 + 10000n^3 + 28n^3 = 10034n^3$ for
all positive integers n. Note also that although $n^2 = O(n^3)$, it is
false that $n^3 = O(n^2)$, since n^3 is not bounded by any constant
times n^2 for arbitrarily large values of n. The complexity analysis of algorithms is concerned with the determination of a most
suitable (and simple) function $f(n)$ such that the algorithm takes
$O(f(n))$ bit operations, and that no essentially smaller function
would work. For example, if the number of bit operations of a
certain algorithm with input size n were determined to be exactly
$6n^3 + 10000n^2 + 28n$, we would estimate this number as $O(n^3)$,
but we could not use $O(n^a)$ for any exponent $a < 3$.

37. For each pair of positive functions $f(n)$ and $g(n)$ with domain
    taken to be the set of positive integers, determine whether
    $f(n) = O(g(n))$, whether $g(n) = O(f(n))$, or neither.
    (a) $f(n) = n$, $g(n) = n^2$
    (b) $f(n) = n \log_2(n), g(n) = n$
    (c) $f(n) = 2^n$, $g(n) = n!$

38. For each pair of positive functions $f(n)$ and $g(n)$ with domain
    taken to be the set of positive integers, determine whether
    $f(n) = O(g(n))$, whether $g(n) = O(f(n))$, or neither.
    (a) $f(n) = (3n^3 + 9)^2$, $g(n) = n^5$
    (b) $f(n) = n + \log_2(n), g(n) = n$
    (c) $f(n) = n^n$, $g(n) = n!$

39. *Complexity of Addition of Binary Sequences.* Suppose that
    $[c_{n-1} \, c_{n-2} \cdots c_1 \, c_0]$ and $[d_{n-1} \, d_{n-2} \cdots d_1 \, d_0]$ are two length-n
    bit strings. Show that the number of bit operations to use
    either Algorithm 6.2 to perform the corresponding addition
    or Algorithm 6.3 to perform the corresponding subtraction (if
    possible) is $O(n)$.

40. *Complexity of Multiplication of Binary Sequences.* Suppose
    that $[c_{n-1} \, c_{n-2} \cdots c_1 \, c_0]$ and $[d_{n-1} \, d_{n-2} \cdots d_1 \, d_0]$ are two length-n
    bit strings. Show that the number of bit operations to use
    Algorithm 6.4 to perform the corresponding multiplication
    is $O(n^2)$.

Chapter 6 Computer Implementations and Exercises

Note: There are two natural *data structures* for storing base *b* expansions: either as vectors or as strings. Strings are only appropriate in case the base *b* "digits" are single characters: if *b* < 10 or if *b* = 16 (due to our special hexadecimal notation). Vectors tend to be more amenable to writing programs, but strings display more efficiently. You should contemplate both possibilities on your particular computing platform and decide which of these options (or perhaps another option) would be most suitable for the exercises and applications of this section.

1. Write a program n = bin2int(v) that will take as input a vector v for a binary expansion (zeros and/or ones) and will output its corresponding equivalent decimal integer n. Perform, by hand, the corresponding conversions for the binary strings [1011], [11111], and [1011110], and run your program on these inputs (debug, as necessary).

2. Write a program n = oct2int(v) that will take as input a vector v for an octal (base 8) expansion and will output its corresponding equivalent decimal integer n. Perform, by hand, the corresponding conversions for the octal expansions [5027], [23456], and [7031410], and run your program on these inputs (debug, as necessary).

3. Write a program n = hex2int(v) that will take as input a vector v for a hexadecimal (base 16) expansion and will output its corresponding equivalent decimal integer n. Perform, by hand, the corresponding conversions for the hexadecimal expansions [8B7], [1AEE], and [AAAA6], and run your program on these inputs (debug, as necessary).

   Note: The reader may wish to instead use vectors of integers in the range 0–15 for hexadecimal sequences.

4. Write a program n = base92int(v) that will take as input a vector v for a base 9 expansion and will output its corresponding equivalent decimal integer n. Perform, by hand, the corresponding conversions for the base 9 expansions [857], [1444], and [65626], and run your program on these inputs (debug, as necessary).

5. Write a program n = baseb2int(v,b) that will take two inputs: a vector v for a base b expansion, and b, an integer greater than 1 (for the base of the expansion). The output will be the corresponding equivalent decimal integer n. Run your program on each of the expansions of Chapter Exercise 5 (debug, as necessary).

6. Write a program v = int2bin(n) that will take as input a nonnegative integer n and will output its binary expansion vector v using Algorithm 6.1. Perform, by hand, the corresponding conversions to binary expansions for *n* = 8, 107, 327, and 12,557, and run your program on these inputs (debug, as necessary).

7. Write a program v = int2oct(n) that will take as input a nonnegative integer n and will output its octal (base 8) expansion vector v using Algorithm 6.1. Perform, by hand, the

corresponding conversions to binary expansions for $n = 8$, 107, 327, and 12,557, and run your program on these inputs (debug, as necessary).

8. Write a program $v = \text{int2hex}(n)$ that will take as input a nonnegative integer n and will output its hexadecimal (base 16) expansion vector v using Algorithm 6.1. Perform, by hand, the corresponding conversions to binary expansions for $n = 8$, 107, 327, and 12,557, and run your program on these inputs (debug, as necessary).

    Note: Read Computer Exercise 3.

9. Write a program $v = \text{int2baseb}(n,b)$ that will take as inputs a nonnegative integer n and an integer b (the base) greater than 1. The output will be the base b expansion vector v of the integer n, determined by using Algorithm 6.1. Perform, by hand, the corresponding conversions for $n = 8$, 107, 327, and 12,557, with each of the bases 2, 8, 16, and run your program on these (14) inputs (debug, as necessary).

10. Write a program $w = \text{bin\_add}(u,v)$ that will take as inputs two vectors u and v representing binary (base 2) expansions and will output the vector w representing the binary expansion of the sum $u + v$, computed using Algorithm 6.2. Perform, by hand, the binary additions [101] + [111], [110110] + [1010111], and run your program for these additions (debug, as necessary).

11. Write a program $w = \text{hex\_add}(u,v)$ that will take as inputs two vectors u and v representing hexadecimal (base 16) expansions (using digits from 0 to 15) and will output the vector w representing the hexadecimal expansion of the sum $u + v$, computed using Algorithm 6.2. Perform, by hand, the hexadecimal additions [C42] + [A1A], [86B4D] + [76A0C], and run your program for these additions (debug, as necessary).

    Note: The reader may wish to use instead vectors of integers in the range 0–15 for hexadecimal sequences.

12. Write a program $w = \text{baseb\_add}(u,v,b)$ that will take as inputs two vectors u and v representing base b expansions, and a third input b (the base) being an integer greater than 1. The output will be the vector w representing the base b expansion of the sum $u + v$, computed using Algorithm 6.2. Run your program on the base b subtractions of Chapter Exercise 15 (debug, as necessary).

13. Write a program $w = \text{bin\_sub}(u,v)$ that will take as inputs two vectors u and v representing binary (base 2) expansions where $u \geq v$, and will output the vector w representing the binary expansion of the difference $u - v$, computed using Algorithm 6.3. Perform, by hand, the binary subtractions [101] − [011], [110110] − [101011], and run your program on them (debug, as necessary).

14. Write a program $w = \text{hex\_sub}(u,v)$ that will take as inputs two vectors u and v representing hexadecimal (base 16) expansions where $u \geq v$, and will output the vector w representing

the hexadecimal expansion of the difference u − v, computed using Algorithm 6.3. Perform, by hand, the hexadecimal subtractions [A42] − [A1A], [86A4D] + [76C8B], and run your program on them (debug, as necessary).

Note: The reader may wish to use instead vectors of integers in the range 0–15 for hexadecimal sequences.

15. Write a program w = baseb _ sub(u,v,b) that will take as inputs two vectors u and v representing base b expansions where u ≥ v, and a third input b (the base) being an integer greater than 1. The output will be the vector w representing the base b expansion of the difference u − v, computed using Algorithm 6.3. Run your program on the base *b* subtractions of Chapter Exercise 17 (debug, as necessary).

16. Write a program w = bin _ mult(u,v) that will take as inputs two vectors u and v representing binary (base 2) expansions and will output the vector w representing the binary expansion of the product u × v, computed using Algorithm 6.4. Perform, by hand, the binary multiplications [101] × [111], [110110] × [1010111], and run your program for these multiplications (debug, as necessary).

17. Write a program w = hex _ mult(u,v) that will take as inputs two vectors u and v representing hexadecimal (base 16) expansions (using digits from 0–15) and will output the vector w representing the hexadecimal expansion of the product u × v, computed using Algorithm 6.4. Perform, by hand, the hexadecimal multiplications [C42] × [A1A], [86B4D] × [76A0C], and run your program for these multiplications (debug, as necessary).

Note: The reader may wish to use instead vectors of integers in the range 0–15 for hexadecimal sequences.

18. Write a program w = baseb _ mult(u,v,b) that will take as inputs two vectors u and v representing base b expansions, and a third input b (the base) being an integer greater than 1. The output will be the vector w representing the base b expansion of the product u × v, computed using Algorithm 6.4. Run your program on the base *b* multiplications of Chapter Exercises 19 and 21 (debug, as necessary).

19. *The Squaring Algorithm for Modular Exponentiation.*
    (a) Write a program with syntax a = FastExp(b,x,m) that will take as inputs a modular integer b ≢ 0 (mod m), a positive integer exponent x, and a corresponding modulus m > 1. The output a will be the unique nonnegative integer < m that satisfies $a \equiv b^x \pmod{m}$. The mechanics of the program should follow Algorithm 6.5
    (b) Use this program to redo Chapter Exercise 23.
    (c) Use this program to compute the following:
        (i) $2^{1234567890} \pmod{169}$
        (ii) $12^{123456789012} \pmod{1865}$

Block Cryptosystems and the Data Encryption Standard (DES)

With the exception of the one-time pad cryptosystem, which was perfectly secure but impractical because of the unwieldy key requirements, all of the cryptosystems introduced in the precomputer era would not be sufficiently secure in today's environment of high-speed computers. To answer the increasing needs of many U.S. industries, the National Bureau of Standards (now the National Institute of Standards and Technology, NIST) set up a competition in 1973 to create a national standard for a secure cryptosystem that could be used industry-wide. The winning system was submitted by IBM, but before it was adopted, the system was sent to the National Security Agency (NSA) for some final modifications. The completed cryptosystem then became freely available and was named the **Data Encryption Standard** (**DES**). DES was used all the way up to the late 1990s by an assortment of industries in the United States and the rest of the world. This chapter discusses the rise and fall of DES, and provides a complete description of its functionality. After some related history, we next discuss some preliminary concepts of the more general block cryptosystems on which DES is based. We then give the description of a scaled-down version of DES, which will make our subsequent description of the real DES easier to follow. After we have completed the description of DES, we discuss some interesting technological breakthroughs spurred by efforts to crack DES, along with how DES eventually was broken. After its fall, DES was reengineered in a simple way (triple-DES) that has made it once again an extremely effective cryptosystem. The chapter ends with descriptions of some developments of various modes of operation for general block ciphers.

The Evolution of Computers into Cryptosystems

The first programmable electronic computer dates back to the pre–World War II era and was created by German scientist and civil engineer Konrad Zuse (1910–1995). He completed his Z1 programmable computer in 1938. The Z1 did not work very well because of inferior parts. This and the state of German relations with the United States and Britain meant that news of his work was initially not very well disseminated and was confined primarily to Germany. Zuse continued to improve his designs in the 1940s and actually started one of the world's first computer companies. Throughout the war, British cryptographers at Bletchley Park were effectively using their massive bombes at full capacity to just barely break German Enigma codes. When the Germans came out with the more intricate Lorenz encryption machines, the British were stymied at first since the bombes would need a very long series of different configurations to have any hope of decrypting the Lorenz system. The Lorenz system was performed by a mechanical machine that was much more complicated than the Enigmas; it was the system that Hitler used for direct communications with his generals. The top brass at Bletchley Park were convinced that achieving the design of a programmable computer that would be capable of breaking the Lorenz cipher would be a hopelessly impractical goal and decided to abandon further effort on this approach. Unfettered by this decision, one British scientist, Tommy Flowers (1905–1998), began to build such a machine using existing blueprints from another Bletchley scientist. It took him 10 months, but he eventually succeeded in producing such a machine in 1943. Known as *Colossus*, it was made up of 1,500 vacuum tubes, was considerably faster than the bombes, and, most importantly, could be programmed. The British built 10 of these machines during the war, and they helped to decrypt some vital intelligence regarding Hitler's plans during the D-day invasion.

As with other important achievements that were discovered under the cloak of the secrecy of government-sponsored cryptographic research, Flowers was not recognized for the creation of his computer. After the war, his machines and the blueprints were destroyed by the British government, and credit for the invention of the first fully programmable computer was given in 1945 to American scientists J. Presper Eckert and John W. Mauchly, who created the *ENIAC* (Electronic Numerical Integrator And Calculator). The construction of this machine was funded by the U.S. Army; it measured 8.5 feet in height and 80 feet in length, contained 18,000 vacuum tubes, and weighed 17 tons! It had a clock speed of 100 kHz. By comparison, clock speeds of modern personal computers are measured in GHz (1 GHz = 1 million kHz). Due to their size and cost, only a select few large companies and government industries were initially able to use programmable computers. Less visible were the state-of-the-art computers used by big governments for encryption and decryption. The invention of the transistor in 1947 led to great reductions in the size of computers. IBM began producing computers in 1953 and created the first universal programming language, FORTRAN, in 1957. In 1952, President Harry S. Truman created the NSA (National Security Agency), the cryptographic branch of the U.S. Department of Defense responsible for the analysis of foreign intelligence. The NSA's duties have since grown to include the protection of all federal computer networks against attacks.

DES Is Adopted to Fulfill an Important Need

As computers evolved into efficient and affordable designs, numerous large companies began using them throughout the 1950s and 1960s, and cryptographic systems were employed to keep the internal data records

Figure 7.1 Horst Feistel (1915–1990), German American cryptographer.

and transmissions safe from compromise. There were natural efficiency problems, however, since companies wishing to exchange commerce needed to first agree on a symmetric key cryptographic protocol and to exchange keys. In order to address the first hurdle, the U.S. government initiated a competition to create an efficient cryptographic protocol that would be suitably secure for financial transactions and business communications to serve as the national standard. The public announcement came in 1973 and was made by the *National Bureau of Standards* (*NBS*), which has since become the *National Institute of Standards and Technology* (*NIST*). The top choice was a program called *Lucifer* that was developed by a team of IBM scientists. One of this team's leaders was Horst Feistel* (see Figure 7.1), who had previously developed the more general cryptosystem (now called a *Feistel system*) on which DES is based.

Lucifer was a complicated block cipher that worked with 64-bit blocks and processed them through a series of 16 rounds of complicated substitutions and permutations, all determined by a 112-bit key. The NBS forwarded Lucifer to the NSA for further evaluation, and the latter group (after thoroughly analyzing and evaluating it) made some changes to it to form the **Data Encryption Standard**, or **DES cryptosystem**. One of the most significant changes was that the (binary string) keylength was decreased from 112 to 56. Keylength is a significant security feature of any cryptosystem, and the number of possible keys of DES is $2^{56} \approx 10^{17}$ whereas for Lucifer it was $2^{112} \approx 10^{34}$. Some experts believe that NSA made changes to safeguard against the possibility that IBM had built a trapdoor into their system, thus allowing them to tap into any DES transmissions at will. Additionally, it has been speculated that NSA made the

---

* Feistel was born in Berlin in 1905 and immigrated to the United States in 1934. He studied physics and earned a bachelor's degree at MIT and then a master's degree at Harvard. His initial attempt to become a U.S. citizen failed as America was preparing to join the Allied forces against Germany in the Second World War, and for national security reasons he was placed under house arrest. Finally in 1944 he became a U.S. citizen and obtained his security clearance, and he immediately began working for the U.S. Air Force Cambridge Research Center on spy detection devices. He eventually took a cryptography position at IBM in the 1950s. His work on block ciphers was pioneering and quite general, and his work at IBM led to several internal awards and some important patents. The Lucifer system was originally developed to fulfill a request to IBM from a large bank to develop a secure cryptosystem for ATM transactions.

key size more manageable in order that it might, with its most powerful computing power and technical experts, be able to decrypt any DES transmissions in cases where national security was threatened. By the late 1990s, it was becoming evident that the DES was susceptible to brute-force key searches and that its days were coming to an end. So in 1997, the NIST put out another call for a new more secure system. The resulting system, known as Advanced Encryption Standard (AES), will be the subject of Chapter 11. We have already developed all of the mathematical prerequisites for DES; AES, however, relies on a new number system that we have not yet encountered. This number system, denoted $GF(2^8)$, is what is known as the *Galois field of order* 2^8, and it will be discussed in Chapter 10. It contains $2^8 = 256$ elements and has multiplication and inverse operations that are highly nonlinear and help to give the AES a very strong level of security.

Convention: Since computers work internally primarily with binary strings, throughout this chapter all cryptosystems will be described and implemented using the binary alphabet for both plaintexts and ciphertexts: $P = C = \{0, 1\}$, so the plaintext/ciphertext strings will simply be binary (or bit) strings. *Throughout this chapter, all plaintexts/ciphertexts are given as bit strings or vectors, or for brevity, using their hexadecimal equivalents* (by means of Table 6.1). Such strings can be easily used to represent the ordinary (or expanded ASCII) English alphabet (or any alphabet) using a basic substitution scheme, as was seen in Chapter 6.

The XOR Operation

The XOR operation is a simple scheme for combining two bit strings of the same length to create a new bit string. The process, which is easily reversed, is described in the following definition.

Definition 7.1

The **XOR operation**, denoted by $\oplus$, is defined on pairs of bit strings of the same length, by simply adding corresponding components mod 2. Thus, for a single bit, we have $0 \oplus 0 = 1 \oplus 1 = 0$, $1 \oplus 0 = 0 \oplus 1 = 1$. Here is an example with a length-3 bit string: $110 \oplus 100 = 010$.*

Since it is based on mod 2 addition, the XOR operation inherits some very convenient properties of modular addition that are used often in cryptography. We summarize some of these in the following.

---

* The reason for the name "XOR" of this operator has to do with logic. XOR is an abbreviation for "exclusive or." If P and Q are two logical statements (that are either true or false), then "P exclusive or Q" is true if exactly one of P or Q is true, and false if P and Q are both true or both false. In computer science/Boolean logic, 0 corresponds to "false" and 1 corresponds to "true," so the XOR operator on single bits just corresponds to the logical exclusive or, if the bits represent truth values of the corresponding logical statements.

Proposition 7.1: Properties of the XOR Operator

Suppose that A, B, and C represent bit strings of the same length N. We let $0_N = 0$ denote the length-N bit string of zeros. The following properties hold:

1. *Commutativity.* $A \oplus B = B \oplus A$
2. *Associativity.* $(A \oplus B) \oplus C = A \oplus (B \oplus C)$
3. **0** *is the XOR Identity.* $A \oplus \mathbf{0} = A$
4. *Self-canceling property.* $A \oplus B \oplus B = A$

Each of these properties directly follows bit-by-bit from the corresponding properties of addition in $\mathbb{Z}_2$ (which in turn are inherited, as explained in Chapter 2, from the corresponding properties of addition of integers); see Exercise 28 in this chapter. Note that because of associativity, we never need to use parentheses when XORing several bit strings. The self-canceling property provides a very simple cryptosystem on bit strings of a certain length N: For any fixed bit string B (the key), the function $A \mapsto A \oplus B$ takes a plaintext bit string to its ciphertext $C = A \oplus B$. If the recipient knows the key B, he or she can decrypt by XORing the ciphertext with the key $C \oplus B = A \oplus B \oplus B = A$. This concept is used on several occasions throughout this chapter.

Feistel Cryptosystems

The DES (and the scaled-down DES) are block cryptosystems that fall under the general category of what are called *Feistel cryptosystems*. The underlying concept is explained in the following definition.

The plaintext and ciphertext spaces both consist of the length-$2t$ bit strings.

Definition 7.2

A **Feistel cryptosystem** (also known as a **Feistel network** or **system**) is a block cryptosystem determined by the following components:

- The **block size** $2t$ (an even number) and a key size N
- The **number of rounds** NR (a positive integer)
- A **key schedule**: A mechanism for generating NR **round keys** $\kappa^1, \kappa^2, \cdots, \kappa^{NR}$ from the single cryptosystem key κ.
- A **round key function**, f_{κ^i}, for each round key κ^i, which inputs any t-bit string R, and outputs another t-bit string $f_{\kappa^i}(R)$.

Encryption: Given a plaintext P, a bit string of length $2t$, we first split P into two bit strings of length t: $P = (L_0, R_0)$, with L_0 being the left half, and R_0 being the right half. We then proceed through NR rounds for the following transformations:

FOR round $i = 1$ TO NR: $L_i = R_{i-1}$, $R_i = L_{i-1} \oplus f_{\kappa^i}(R_{i-1})$

The ciphertext will then be $C = (R_{NR}, L_{NR})$. Notice that the ciphertext permuted the left and right parts of the result from the final round.

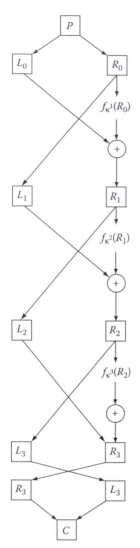

Figure 7.2 Encryption process in a three-round Feistel system.

Decryption: We feed the ciphertext C back into the above encryption process, the only change being that the keys are used in the reverse order.

It may not be so obvious that the decryption process indeed works, and perhaps quite surprising that the round functions need not be invertible. The self-decrypting property of a Feistel cryptosystem stems from its careful construction and the properties of XOR; an outline of the proof is provided in Exercises 30 and 31, and a special case is previewed in Exercise for the Reader 7.2. It is the round functions and the round key generation scheme that make a Feistel system so powerful.* Figure 7.2 shows a schematic diagram of a general Feistel cipher with three rounds. We next give an example of a very small (and minimally secure) Feistel system.

Example 7.1: A Simple Three-Round Feistel Cryptosystem

The block size will be 8, the keylength will be 12, and there will be $NR = 3$ rounds. For each length-12 bit string $\kappa = k_1 k_2 \cdots k_{12}$, representing a system key, the key scheduling algorithm will take the ith keys ($i = 1, 2, 3$) to be the following 4-bit strings:

$$\kappa^1 = k_1 \cdots k_4 \oplus k_5 \cdots k_8$$

$$\kappa^2 = k_5 \cdots k_8 \oplus k_9 \cdots k_{12}$$

$$\kappa^3 = k_9 \cdots k_{12} \oplus k_1 \cdots k_4$$

The round key function $f_{\kappa^j}(R)$ is simply obtained by XORing an inputted 4-bit string R with the round key κ^i.

(a) With a 12-bit system key $\kappa = \text{ABC}$ (represented in hex form), use this Feistel cryptosystem to encrypt the 8-bit plaintext $P = \text{DF}$ (represented in hex form).

(b) Perform the corresponding decryption to the ciphertext that resulted from P in part (a).

Solution: Part (a): According to the indicated key scheduling rule, the three round keys are given by (see Table 6.1):

$$\kappa^1 = \text{A} \oplus \text{B} = 1010 \oplus 1011 = 0001$$

$$\kappa^2 = \text{B} \oplus \text{C} = 1011 \oplus 1100 = 0111$$

$$\kappa^3 = \text{C} \oplus \text{A} = 1100 \oplus 1010 = 0110$$

We must go through the three rounds to encrypt $P = \text{DF}$; the left and right halves of P are just D and F: $L_0 = D = 1101$, $R_0 = F = 1111$.

---

* Many well-known cryptosystems, such as DES, employ a Feistel system; examples include Blowfish, FEAL, Camellia, and KASUMI.

Round 1:

$L_1 = R_0 = 1111$

$R_1 = L_0 \oplus f_{\kappa^1}(R_0) = L_0 \oplus [\kappa^1 \oplus R_0]$

$$= 1101 \oplus [0001 \oplus 1111] = 0011$$

Round 2:

$L_2 = R_1 = 0011$

$R_2 = L_1 \oplus f_{\kappa^2}(R_1) = L_1 \oplus [\kappa^2 \oplus R_1]$

$$= 1111 \oplus [0111 \oplus 0011] = 1011$$

Round 3:

$L_3 = R_2 = 1011$

$R_3 = L_2 \oplus f_{\kappa^3}(R_2) = L_2 \oplus [\kappa^3 \oplus R_2] = 0011 \oplus [0110 \oplus 1011]$

$$= 1110$$

The ciphertext is now obtained by switching the left and right halves of the final (third) round: $C = (R_3, L_3) = 11101011 = $ EB (in hex format).

Part (b): To decrypt, we proceed just as above, but we reverse the key sequence. For greater clarity in this first example, we will denote all left/right intermediate strings using primes. The left and right halves of C are just E and B: $L_0' = $ E $= 1110$, $R_0' = $ B $= 1011$.

Round 1:

$L_1' = R_0' = 1011$

$R_1' = L_0' \oplus f_{\kappa^3}(R_0') = L_0' \oplus [\kappa^3 \oplus R_0'] = 1110 \oplus [0110 \oplus 1011]$

$$= 0011$$

Round 2:

$L_2' = R_1' = 0011$

$R_2' = L_1' \oplus f_{\kappa^2}(R_1') = L_1' \oplus [\kappa^2 \oplus R_1'] = 1011 \oplus [0111 \oplus 0011]$

$$= 1111$$

Round 3:

$L_3' = R_2' = 1111$

$R_3' = L_2' \oplus f_{\kappa^1}(R_2') = L_2' \oplus [\kappa^1 \oplus R_2'] = 0011 \oplus [0001 \oplus 1111]$

$$= 1101$$

The plaintext is now obtained by switching the left and right halves of the final (third) round: $C = (R_3', L_3') = 11011111 = \mathtt{DF}$ (in hex format), which is, as expected, the original plaintext.

Exercises for the Reader

Exercise 7.1

(a) Use the Feistel system of Example 7.1 to encrypt the plaintext $P = \mathtt{DE}$ with the key $\kappa = \mathtt{0E6}$.

(b) Perform the corresponding decryption process on the ciphertext resulting from part (a).

Exercise 7.2

Verify that the decryption process specified in Definition 7.2 for (a) a general one-round and (b) a general two-round Feistel cryptosystem indeed works.

Suggestion: Use notation similar to what was introduced in the solution of Example 7.1. For example in the two round system, the left/right strings introduced in the encryption process are $P = (L_0, R_0), (L_1, R_1), (L_2, R_2)$. The left/right strings introduced in the decryption are $C = (R_2, L_2) \triangleq (L_0', R_0'), (L_1', R_1'), (L_2', R_2')$. Since the output of the decryption process will be (R_2', L_2'), we need to verify that $R_2' = L_0$ and $L_2' = R_0$.

A Scaled-Down Version of DES

From our brief description of DES given above, the sheer size of its blocks and keys and the number of rounds make it completely infeasible to completely work through any specific examples without a computer. As we did with the scaled-down version of Enigma in Chapter 3, we will circumvent this difficulty by first introducing a scaled-down DES cryptosystem that has the same features as the actual DES system but with reduced sizes and numbers. A basic comparison of the size characteristics of the two cryptosystems is shown in Table 7.1.

The (scaled-down) DES algorithm is a Feistel cryptosystem that will be further enhanced with an initial bit permutation applied before the Feistel system, and its inverse will be applied after the action to the Feistel system. We stress that such bit permutations are *not* substitution ciphers, as

TABLE 7.1 Size Characteristics Comparison of the DES Cryptosystem with the Scaled-Down DES Cryptosystem That We Will Use to Motivate the Former

	Scaled-Down DES	DES
Block size	8 bit	64 bit
Key size	12 bit	64 bit
Number of rounds	2	16

were most of the permutations that were applied in earlier cryptosystems. Previously, we used permutations on the actual alphabet to change each character; here, we use them as transposition ciphers: to interchange the locations of the bits (but not their values).[*]

Algorithm 7.1: Scaled-Down DES Encryption. Part I: Outline

Plaintext, Ciphertext Spaces: $\mathscr{P} = \mathscr{C} = \{$8-bit strings$\}$

Keyspace: $\mathscr{K} = \{$12-bit strings $b_1 b_2 \cdots b_{10} b_{11} b_{12}$, with the property that the 6th and 12th bits, b_6, b_{12}, are parity check bits chosen so that the sums of the consecutive 6-bit blocks are odd$\}$. Thus, because of the redundancy of these two bits in any key, $\mathscr{K}$ should be viewed as a 10-bit keyspace.

Encryption Scheme: Let $P = b_1 b_2 \cdots b_8$ be the plaintext.

Step 1. *Initial Permutation.* First we perform the following initial permutation IP(P).

$$P \mapsto b_2 b_5 b_1 b_3 b_8 b_4 b_7 b_6 \tag{7.1}$$

Step 2. *Two-Round Feistel Cipher.* Each of the two rounds will use an 8-bit **round key** κ^i that is generated (as explained later) from the 12-bit system key κ. The details of the round keys and the round key functions f_{κ^i} will be explained shortly.

Step 3. *Perform Inverse of Initial Permutation.* We apply the inverse of the initial permutation of Step 1 to the result of Step 2: $c = IP^{-1}(R_2 L_2)$. This will be the ciphertext. We point out that this inverse initial permutation (see Figure 7.3) is specified by:

$$c_1 c_2 c_3 c_4 c_5 c_6 c_7 c_8 \mapsto c_3 c_1 c_4 c_6 c_2 c_8 c_7 c_5 \tag{7.2}$$

Since the details of the round key schedule and functions are somewhat technical, we first outline the components of the scalded-down DES and then elaborate on the details.

Figure 7.3 gives a schematic of Algorithm 7.1. Before we can give an example, we first need to complete all of the details in the above outline.

Algorithm 7.1: Scaled-Down DES Encryption. Part II: Details

1. Instructions on how to generate the two 8-bit round keys κ^1, κ^1 from the 12-bit DES key κ:

   First we extract a permutation of the 10 nonparity check bits of the key κ (so we discard the bits in slots 6 and 12). The permutation that is used is specified in Table 7.2 and is separated into the left-half 5 bits C_0 (top) and the right-half 5 bits D_0 (bottom).

---

[*] Note that the binary alphabet $\{0,1\}$ is so small that there is only one nontrivial permutation. Thus, it would not have been a suitable alphabet in the precomputer era, when most cryptosystems were on (character) substitution ciphers.

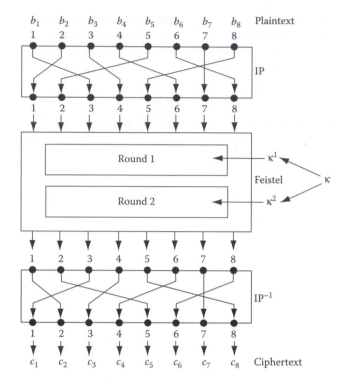

Figure 7.3 Schematic diagram of the scaled-down DES (Algorithm 7.1).

We let C_1 and D_1 be the bit strings resulting from cyclically shifting the bits of the 5-bit strings C_0 and D_0, respectively, one unit to the left, and let C_2 and D_2 be the bit strings resulting from cyclically shifting C_0 and D_0, respectively, two units to the left. (For example, if $C_0 = 01011$, then $C_1 = 10110$ and $C_2 = 01101$.) Finally, from the resulting 10-bit juxtaposition $C_i D_i$ we extract a permutation of 8 of its bits that is specified in Table 7.3, and this will be the ith round key κ^i.

2. Instructions on how to evaluate each of the two round key functions $f_{\kappa^i} : \{4\text{-bit strings}\} \to \{4\text{-bit strings}\}$ $(i = 1, 2)$ from the round keys κ^i:

Let R be a 4-bit string. We must show how to compute $f_{\kappa^i}(R)$. We first apply an expansion function $E : \{4\text{-bit strings}\} \to \{8\text{-bit strings}\}$ that uses (and reuses some) of the bits of a 4-bit string to form an 8-bit string. Table 7.4 gives a tabular description of this expansion function, and Figure 7.4 gives a schematic diagram. (The

TABLE 7.2 Initial Permutation of the Nonredundant Bits of the Key of the Scaled-Down DES

10	4	1	5	11	$\to C_0$
8	9	3	7	2	$\to D_0$

TABLE 7.3 Final Round Key Extraction Permutation

8	2	4	9	7	10	6	3

Note: Only 8 of the 10 bits are selected.

reader should convince himself or herself that the table and the figure specify the same function.)

TABLE 7.4 Ordered List of the 8-Bit Expansion Function for the Scaled-Down DES

3	1	2	4	2	4	3	1

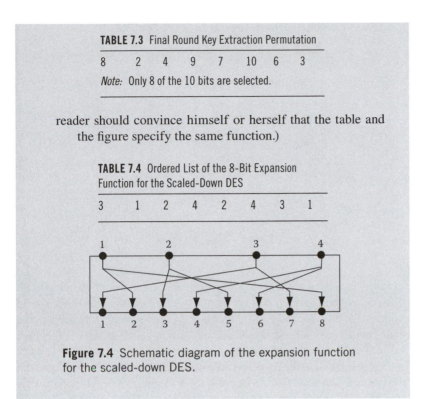

Figure 7.4 Schematic diagram of the expansion function for the scaled-down DES.

We first XOR $E(R)$ with the 8-bit round key to produce an 8-bit string

$$B = B_1 B_2 \triangleq E(R) \oplus \kappa^i$$

The two strings B_1, B_2 are the 4-bit strings that make up the left half and right half of the above 8-bit string. (This will be the only use of the round key.)

This 8-bit string B will be converted into the output $f_{\kappa^i}(R)$ 4-bit string by use of a couple of special functions that are called **S-boxes**. The first S-box will be used to convert the first 4-bit string, B_1, into a 2-bit string, C_1, and the second S-box will be used to convert the second 4-bit string, B_2, into a second 2-bit string, C_2. We need to explain what these S-boxes are and how they work.[*]

Each S-box, also called a **substitution box**, is a function {4-bit strings} $\rightarrow$ {2-bit strings}. The two S-boxes are specified in Table 7.5. Letting S_1, S_2 denote the two S-boxes, we show how to use Table 7.5 so that they will both be functions {4-bit strings} $\rightarrow$ {2-bit strings}. Let $B = b_1 b_2 b_3 b_4$ be a 4-bit string. To compute $S_j(B)$, we use the jth 4-by-4 S-box array of Table 7.5 as follows: The integer represented by the outer bits $b_1 b_4$ of B tells which row of the jth S-box to use, while the two inside bits $b_1 b_3$ specifies the binary expansion of the integer telling which column of the jth S-box to use. We set $S_j(B)$ to be the binary expansion of the integer showing in this row and column of the jth S-box, padded with a zero on the left, if

---

[*] These bit strings C_1, C_2 are not the same as those with the same name in the round key generation procedure.

TABLE 7.5 Two S-Boxes for the Scaled-Down DES

Rows	Columns			
	0	1	2	3
S-Box 1:				
0	3	1	0	2
1	2	0	1	3
2	2	3	1	0
3	1	2	3	0
S-Box 2:				
0	1	3	2	0
1	3	1	2	0
2	2	0	3	1
3	0	3	2	1

necessary, so that it has length 2. For example, if we needed to compute S_2 (0111), since the two outer bits are 01 (the binary expansion of 1), we would use row 1 of the second S-box, and since the two inner bits are 11 (the binary expansion of 3), we would use column 3; the table entry is 0, which is 0 in binary expansion. Padding one zero on the left gives $S_2(0111) = 00$. In the same fashion, the reader should check that $S_1(0101) = 01$. Since the S-boxes are new concepts, we digress for a moment to provide the reader with a small opportunity to practice with them.

Exercise for the Reader 7.3

Determine the following S-box values: $S_1(1110)$, $S_2(1110)$.

The final output of the round key function, $f_{K^i}(R)$, will be a permutation of the 4-bit string C_1C_2, and this final permutation is shown in Table 7.6.

This completes the description of the round key functions, and hence also the encryption algorithm for the scaled-down DES. Figure 7.5 shows a schematic diagram of the steps needed to evaluate a round key function.

We next give an example of a scaled-down DES encryption that we will be able to work through by hand. Although we will not be able to do such examples of the actual DES algorithm without the aid of a computer, all of the concepts of DES are nicely realized with this scaled-down version.[*]

TABLE 7.6 Final Permutation in the Round Key Function for the Scaled-Down DES

2	4	1	3

---

[*] At a first glance, with all of its intricate details, the scaled-down DES may appear to be a reasonably secure cryptosystem. But the (effective) key size of 10 bits is too small— there are only $2^{10} = 1024$ keys! Such a small number of keys could easily be tested with an exhaustive search on a computer.

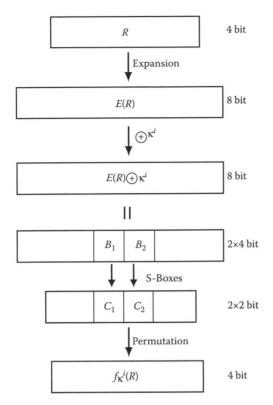

Figure 7.5 Schematic diagram of the round key functions for the scaled-down DES (Algorithm 7.1).

Example 7.2

Use the scaled-down DES to encrypt (with Algorithm 7.1) the 8-bit plaintext $P = $ DF (represented in hex form). Use the 12-bit system key $\kappa = $ 2A8 (represented in hex form).

Solution: With the aid of Table 6.1, we write out the binary expansion: $P = $ DF $= 11011111$.

Step 1. We apply initial permutation from formula 7.1 ($p \mapsto b_2 b_5 b_1 b_3 b_8 b_4 b_7 b_6$):

$$P = 11011111 \mapsto 11101111$$

Step 2. We apply the two-round Feistel system:

We first prepare the two keys κ^1, κ^2. To do this, we write out the 12-bit system key as its binary expansion $\kappa = $ 2A8 $= 001011011001$. We select the bits of this key indicated by Table 7.2 to form the two 5-bit vectors: the "top" $C_0 = 00010$ and the "bottom" $D_0 = 11100$. We then cyclically shift each of these vectors to the left one and then two units to obtain the resulting shifted vectors:

$$C_1 = 00100, \ D_1 = 11001; \ C_2 = 01000, \ D_2 = 10011$$

The first round key is the 8-bit extracted permutation (indicated in Table 7.3) of the 10-bit string $C_1D_1 = 0010011001 \to \kappa^1 = 00001111$.

The second round key is the 8-bit extracted permutation (indicated in Table 7.3) of the 10-bit string $C_2D_2 = 0100010011 \to \kappa^2 = 01010110$.

We enter the Feistel system with the left and right halves extracted from the result of Step 1: $11101111 \mapsto L_0 = 1110, R_0 = 1111$.

Feistel Round #1:

We first evaluate $f_{\kappa^1}(R_0)$, by following the outline of Figure 7.5:

- Expansion: $E(R_0) = 11111111$ (cf. Table 7.4)
- XOR with the round key: $\kappa^1 \oplus E(R_0) = 11110000 \triangleq B_1B_2$
- Apply S-boxes: $C_1 = S_1(B_1) = S_1(1111) = 00, C_2 = S_2(B_2) = S_2(0000) = 01$
- Apply final permutation (Table 7.6): $C_1C_2 = 0001 \to 0100$

So we have $f_{\kappa^1}(R_0) = 0100$.

The results of Round #1 are thus
$L_1 = R_0 = 1111, R_1 = L_0 \oplus f_{\kappa^1}(R_0) = 1110 \oplus 0100 = 1010$

Feistel Round #2:

We first evaluate $f_{\kappa^2}(R_1)$, by following the outline of Figure 7.5:

- Expansion: $E(R_1) = 11000011$ (cf. Table 7.4)
- XOR with the round key: $\kappa^2 \oplus E(R_1) = 10010101 \triangleq B_1B_2$
- Apply S-boxes: $C_1 = S_1(B_1) = S_1(1001) = 01, C_2 = S_2(B_2) = S_2(0101) = 10$
- Apply final permutation (Table 7.6): $C_1C_2 = 0110 \to 1001$

So we have $f_{\kappa^2}(R_1) = 1001$.

The results of Round #2 are thus
$L_2 = R_1 = 1010, R_2 = L_1 \oplus f_{\kappa^2}(R_1) = 1111 \oplus 1001 = 0110$

The result of the Feistel system is $R_2L_2 = 01101010$

Step 3. Apply the inverse initial permutation from Equation 7.2 $(c_1c_2c_3c_4c_5c_6c_7c_8 \mapsto c_3c_1c_4c_6c_2c_8c_7c_5)$ to the result of Step 2 to obtain the ciphertext

$01101010 \mapsto 10001011$

In hex format this ciphertext is 8B.

The decryption process is the same as the scaled-down DES algorithm, except that the order of the key sequence in the Feistel cipher (Step 2) is reversed; this follows from the corresponding self-decrypting property of general Feistel ciphers.

Algorithm 7.2: Scaled-Down DES Decryption

To decrypt an 8-bit ciphertext block that was encrypted with the scaled-down DES (Algorithm 7.2), we simply apply the encryption Algorithm 7.1 but with the two round keys used in the opposite order.

Exercise for the Reader 7.4

Use Algorithm 7.2 to decrypt the ciphertext that was produced in Example 7.2.

DES

With our detailed development of general Feistel systems and the scaled-down DES behind us, the general DES cryptosystem will be very easy to understand. Although we provide an example of encrypting with DES, it is not at all feasible to verify the details by hand. Readers may either use the applets that accompany this chapter for DES-related computations (including complete DES encryption/decryption) or refer to the Computer Implementation material at the end of this chapter to design their own programs.[*]

Algorithm 7.3: Data Encryption Standard (DES). Part I: Outline

Plaintext, Ciphertext Spaces: $\mathscr{P} = \mathscr{C} = \{$64-bit strings$\}$.

Keyspace: $\mathscr{K} = \{$64-bit strings $b_1 b_2 \cdots b_{64}$, with the property that every eighth bit $b_8, b_{16}, \cdots, b_{64}$ is a parity check bit chosen so that the sums of the consecutive 8-bit blocks are odd$\}$. Thus, because of the redundancy of some of the bits in any key, $\mathscr{K}$ should be viewed as a 56-bit keyspace.

Encryption Scheme: Let $P = b_1 b_2 \cdots b_{64}$ be the plaintext.

Step 1. *Initial Permutation.* First we perform the following initial permutation IP(P):

$$p \mapsto \begin{matrix} b_{58}b_{50}b_{42}b_{34}b_{26}b_{18}b_{10}b_{2}b_{60}b_{52}b_{44}b_{36}b_{28}b_{20}b_{12}b_{4}b_{62}b_{54}b_{46}b_{38}b_{30}b_{22}b_{14}b_{6} \\ b_{64}b_{56}b_{48}b_{40}b_{32}b_{24}b_{16}b_{8}b_{57}b_{49}b_{41}b_{33}b_{25}b_{17}b_{9}b_{1}b_{59}b_{51}b_{43}b_{35}b_{27}b_{19}b_{11}b_{3}b_{61} \\ b_{53}b_{45}b_{37}b_{29}b_{21}b_{13}b_{5}b_{63}b_{55}b_{47}b_{39}b_{31}b_{23}b_{15}b_{7} \end{matrix} \quad (7.3)$$

Step 2. *16-Round Feistel Cipher.* Each of the 16 rounds will use a 48-bit **round key** κ^i that is generated (as explained later) from the 64-bit system key κ. The details of the round keys and the round key functions f_{κ^i} will be explained shortly.

Step 3. *Perform Inverse of Initial Permutation.* We apply the inverse of the initial permutation of Step 1: $c = IP^{-1}(R_{16}L_{16})$. This will be the ciphertext.

We will provide some concrete examples, but we first need to complete all of the details in the above outline.

---

[*] We have also included a text file on the book's Web page that includes the lengthy permutation and S-box data that is associated with the DES algorithm.

Algorithm 7.3: Data Encryption Standard (DES). Part II: Details

1. Instructions on how to generate each of the 48-bit round keys $\kappa^i (1 \le i \le 16)$ from the 64-bit DES key κ:

   First we extract a permutation of the 56 nonparity check bits of the key κ (so we discard the bits in slots 8, 16, 24, etc.). The permutation that is used is specified in Table 7.7 and is separated into the left-half 28 bits C_0 (top) and the right-half 28 bits D_0 (bottom).

   Next, we proceed through the following iterative process to generate the 16 round keys. We will need the following left shift parameter sequence for the iterations:

   $$\ell_i = \begin{cases} 1, & \text{if } i = 1, 2, 9, 16 \\ 2, & \text{otherwise} \end{cases} \tag{7.4}$$

   In the ith round of the key generation process $(1 \le i \le 16)$, we let C_i and D_i be the bit strings resulting from cyclically shifting the bits of C_{i-1} and D_{i-1}, respectively, ℓ_i units to the left. Finally, from the resulting 56-bit juxtaposition $C_i D_i$, we extract a permutation of 48 of its bits that is specified in Table 7.8, and this will be the ith round key κ^i.

2. Instructions on how to evaluate each of the round key functions $f_{\kappa^i} : \{32\text{-bit strings}\} \rightarrow \{32\text{-bit strings}\}$ $(1 \le i \le 16)$ using the round keys κ^i:

   Let R be a 32-bit string. We must show how to compute $f_{\kappa^i}(R)$. We first apply an **expansion function** $E : \{32\text{-bit strings}\} \rightarrow \{48\text{-bit strings}\}$ that uses (and reuses some) of the bits of a 32-bit string to form a 48-bit string. This function is described in Table 7.9.

TABLE 7.7 Initial Permutation of the Key

57	49	41	33	25	17	9	1	58	50	42	34	26	18
10	2	59	51	43	35	27	19	11	3	60	52	44	36
63	55	47	39	31	23	15	7	62	54	46	38	30	22
14	6	61	53	45	37	29	21	13	5	28	20	12	4

TABLE 7.8 Final Round Key Selection Permutation

14	17	11	24	1	5	3	28	15	6	21	10
23	19	12	4	26	8	16	7	27	20	13	2
41	52	31	37	47	55	30	40	51	45	33	48
44	49	39	56	34	53	46	42	50	36	29	32

Note: Only 48 of the 56 bits are selected.

TABLE 7.9 Expansion Function E:{32-Bit Strings} → {48-Bit Strings}

32	1	2	3	4	5	4	5	6	7	8	9
8	9	10	11	12	13	12	13	14	15	16	17
16	17	18	19	20	21	20	21	22	23	24	25
24	25	26	27	28	29	28	29	30	31	32	1

Thus, except for the first and last bits, the expansion function recopies the bits of the inputted permutation, in order, but with a slight yet regular "stutter." We next partition the expanded 48-bit string $E(R) \oplus \kappa^i$ into eight blocks of 6-bit strings:

$$E(R) \oplus \kappa^i \triangleq B_1 B_2 \cdots B_8 \qquad (7.5)$$

where each B_j is a 6-bit string.

From the 48-bit string (Equation 7.5), we next form a 32-bit string $C_1 C_2 \cdots C_8$, where each C_j is a 4-bit string computed from the corresponding 6-bit string B_j using the jth **S-box**. The eight S-boxes of the DES are specified in Table 7.10, and we now explain how to use this table.

Each S-box is a function S_j : {6-bit strings} → {4-bit strings}. Suppose that $B = b_1 b_2 \cdots b_6$ is a 6-bit string. To compute $S_j(B)$, we use the jth 4-by-16 S-box array of Table 7.10 as follows: The integer represented by the outer bits $b_1 b_6$ of B tells which row of the jth S-box to use, while the remaining four inside bits $b_2 b_3 b_4 b_5$ specify the binary expansion of the integer telling which column of the jth S-box to use. Now $S_j(B)$ is the binary expansion of the integer showing in this row and column of the jth S-box, padded with zeros on the left, as necessary, so that it has length 4. For example, if we needed to compute $S_2(011011)$, since the two outer bits are 01 (the binary expansion of 1), we would use row 1 of the second S-box, and since the four inner bits are 1101 (the binary expansion of $8 + 4 + 1 = 13$), we would use column 13. The S-box entry is 9, which is 1001 in binary expansion, giving us $S_2(011011) = 1001$.

The final output $f_{\kappa^i}(R)$ is now obtained by applying the permutation shown in Table 7.11 to the 32-bit string $C_1 C_2 \cdots C_8$. The construction of these round key functions is schematically illustrated in Figure 7.6.

This completes the description of the DES encryption algorithm. *The decryption process is just the same as the DES algorithm, except that the order of the key sequence in the Feistel cipher (Step 2) is reversed*; the justification of this fact is left as Exercise 32.

The S-boxes might seem rather peculiar, but their design was by no means arbitrary. IBM created them to be highly nonlinear; in particular, they have the following interesting properties:

(a) If two inputs of an S-box differ by only one bit, then the outputs differ by two bits.
(b) If two inputs of an S-box differ in their first two bits, but have the same last two bits, then the outputs will not be equal.
(c) For any given 6-bit string, there are 32 pairs whose XOR will equal this string. If for each of these pairs we compute the XOR of their output 4-bit strings under any S-box, no more than eight of them should be the same.

TABLE 7.10 Eight S-Boxes for the DES System

Rows	Columns															
	0	1	2	3	4	5	6	7	8	9	10	11	12	13	14	15
Box 1:																
0	14	4	13	1	2	15	11	8	3	10	6	12	5	9	0	7
1	0	15	7	4	14	2	13	1	10	6	12	11	9	5	3	8
2	4	1	14	8	13	6	2	11	15	12	9	7	3	10	5	0
3	15	12	8	2	4	9	1	7	5	11	3	14	10	0	6	13
Box 2:																
0	15	1	8	14	6	11	3	4	9	7	2	13	12	0	5	10
1	3	13	4	7	15	2	8	14	12	0	1	10	6	9	11	5
2	0	14	7	11	10	4	13	1	5	8	12	6	9	3	2	15
3	13	8	10	1	3	15	4	2	11	6	7	12	0	5	14	9
Box 3:																
0	10	0	9	14	6	3	15	5	1	13	12	7	11	4	2	8
1	13	7	0	9	3	4	6	10	2	8	5	14	12	11	15	1
2	13	6	4	9	8	15	3	0	11	1	2	12	5	10	14	7
3	1	10	13	0	6	9	8	7	4	15	14	3	11	5	2	12
Box 4:																
0	7	13	14	3	0	6	9	10	1	2	8	5	11	12	4	15
1	13	8	11	5	6	15	0	3	4	7	2	12	1	10	14	9
2	10	6	9	0	12	11	7	13	15	1	3	14	5	2	8	4
3	3	15	0	6	10	1	13	8	9	4	5	11	12	7	2	14
Box 5:																
0	2	12	4	1	7	10	11	6	8	5	3	15	13	0	14	9
1	14	11	2	12	4	7	13	1	5	0	15	10	3	9	8	6
2	4	2	1	11	10	13	7	8	15	9	12	5	6	3	0	14
3	11	8	12	7	1	14	2	13	6	15	0	9	10	4	5	3
Box 6:																
0	12	1	10	15	9	2	6	8	0	13	3	4	14	7	5	11
1	10	15	4	2	7	12	9	5	6	1	13	14	0	11	3	8
2	9	14	15	5	2	8	12	3	7	0	4	10	1	13	11	6
3	4	3	2	12	9	5	15	10	11	14	1	7	6	0	8	13
Box 7:																
0	4	11	2	14	15	0	8	13	3	12	9	7	5	10	6	1
1	13	0	11	7	4	9	1	10	14	3	5	12	2	15	8	6
2	1	4	11	13	12	3	7	14	10	15	6	8	0	5	9	2
3	6	11	13	8	1	4	10	7	9	5	0	15	14	2	3	12
Box 8:																
0	13	2	8	4	6	15	11	1	10	9	3	14	5	0	12	7
1	1	15	13	8	10	3	7	4	12	5	6	11	0	14	9	2
2	7	11	4	1	9	12	14	2	0	6	10	13	15	3	5	8
3	2	1	14	7	4	10	8	13	15	12	9	0	3	5	6	11

TABLE 7.11 Final Permutation for the Round
Key Function of DES

16	7	10	21
29	12	28	17
1	15	23	26
5	18	31	20
2	8	24	14
32	27	3	9
19	13	30	6
22	11	4	25

Such properties were carefully designed to make the system resilient to an assortment of anticipated attacks, and it required several months of computer searching for IBM researchers to create the S-boxes.

A complete example illustrating the DES would necessarily be quite involved and would best be performed with the aid of a computer (the platform for which DES was designed). The computer exercises of this chapter will guide the reader through the development of complete programs for DES encryption/decryption. The following example illustrates some of the core concepts of DES encryption.

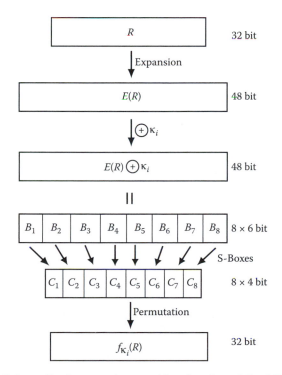

Figure 7.6 Schematic diagram of a round key function of the DES cryptosystem.

TABLE 7.12 Plaintext for Example 7.3

1	0	1	1	1	0	1	1	0	1	0	1	1	1	0	1
1	1	1	0	1	1	1	0	0	0	0	1	1	0	0	1
0	1	0	1	1	0	0	0	0	1	1	0	0	1	1	1
0	0	1	1	1	0	0	1	1	0	1	0	1	0	0	1

Note: To be read in reading order.

TABLE 7.13 DES System Key for Example 7.3

1	0	0	1	0	0	1	0	1	0	1	0	0	1	1	1
0	1	1	0	0	0	1	0	0	0	1	1	1	1	0	1
1	1	1	1	1	1	1	0	0	1	1	0	1	0	1	1
1	0	0	1	1	0	0	0	1	0	0	1	1	0	1	1

Example 7.3

We consider the problem of encrypting the 64-bit plaintext, given in hex form as $P = $ BB5DEE19586739A9 (which was randomly generated) and whose binary form is shown in Table 7.12, using DES with the key given in hex form as $\kappa = $ 92A7623DFE6B989B (also randomly generated, except for the parity check bits), and whose binary form is shown in Table 7.13.

(a) Apply the initial permutation IP (from Equation 7.3) to the plaintext.
(b) Determine the first round key κ^1.
(c) Determine the resulting 32-bit (left and right) strings L_1 and R_1 resulting from the first round of the DES encryption algorithm.

Solution: Part (a): The result of the initial permutation applied to the plaintext of Table 7.12 is shown in Table 7.14.
We digress with a brief Exercise for the Reader.

TABLE 7.14 Initial Permutation of the Plaintext in Example 7.3

0	0	1	1	0	1	1	0	0	1	0	1	1	0	1	1
0	0	1	0	0	1	1	0	1	1	1	0	1	0	1	1
1	0	0	0	0	1	0	1	1	1	1	0	0	1	0	1
1	1	0	1	1	1	1	1	0	0	1	0	0	1	0	1

Note: The first 32 bits make up L_0 and the last 32 bits make up R_0.

Exercise for the Reader 7.5

Will the length-16 hex string corresponding to the initial permutation 64-bit string necessarily be a permutation of the elements of the hex string representing the plaintext? Explain your answer.

TABLE 7.15 Initial Permutation of the Nonparity Bits for the DES Key of Table 7.13

1	1	0	1	0	0	1	1	0	0	1	1	0	1
0	0	0	0	1	1	1	1	1	0	1	1	0	1
1	0	1	1	0	1	1	1	0	0	0	1	1	0
1	0	1	1	1	1	1	0	0	0	1	0	0	1

Note: The bits in the upper half of the table constitute C_0, while those in the lower half are D_0.

Part (b): We outline the details. The initial permutation of the non-parity check bits of the key given in Table 7.13 is shown in Table 7.15.

For the first round key, we perform a left shift of one bit in each of the two vectors C_0, D_0 (from the upper half and lower half, respectively, of Table 7.15) to create C_1, D_1 and then take the final round key selection permutation (Table 7.8) from the resulting juxtaposed vector. The result is the first round key, κ^1, shown in Table 7.16.

Part (c): We start with the 32-bit vectors L_0 and R_0 that are the top and bottom halves, respectively, of the initially permuted string of Table 7.14. From the Feistel encryption process (Definition 7.2), we know that $L_1 = R_0$; the work will be to compute $R_1 = L_0 \oplus f_{\kappa^1}(R_0)$.

Using the round key κ^1 found in part (b) (Table 7.16), and XORing with the expansion (using Table 7.9) $E(R_0)$, we obtain the expression for $E(R_0) \oplus \kappa^1 = B_1 B_2 \cdots B_8$ that is shown in Table 7.17. Now we need to apply the S-boxes: the ith row of six bits in Table 7.15 is fed into the ith S-box to give the corresponding four bits.

The final permutation (Table 7.10) is then done to the result to produce $f_{\kappa^1}(R_0)$, which is shown in Table 7.18. Last, we need to XOR this 32-bit vector with L_0 to obtain R_1. This final result is shown in Table 7.19.

The remaining 15 rounds would be performed in the same fashion. The computer exercises at the end of this chapter will develop the complete program.

TABLE 7.16 First Round Key κ^1 of the DES System of Example 7.3

0	0	1	1	1	0	1	1	0	1	1	1
0	1	0	0	0	0	0	1	1	1	1	0
0	1	1	0	1	1	1	1	0	1	1	1
1	0	1	1	1	0	1	1	0	0	0	0

TABLE 7.17 Resulting 32-Bit String when $E(R_0)$ is XORed with κ^1

1	1	1	1	1	0
1	1	1	1	0	0
1	0	1	1	0	0
0	1	0	1	0	1
1	0	0	0	0	0
0	0	1	0	0	1
0	0	1	0	1	0
1	1	1	0	1	1

TABLE 7.18 $f_{\kappa^1}(R_0)$

0	1	0	0
0	1	0	0
0	1	1	0
0	1	0	0
0	0	1	0
1	0	0	0
0	0	1	0
1	1	0	0

TABLE 7.19 R_1 in Example 7.3

0	1	1	1	0	0	1	0	0	0	1	1	1	1	1	1
0	0	0	0	1	1	1	0	1	1	0	0	0	1	1	1

Our next example partially illustrates the fact that the DES cryptosystem indeed possesses Shannon's properties of diffusion and confusion for a cryptosystem.

Example 7.4

If the DES cryptosystem with the key of Example 7.3 is used to encrypt the following two plaintext 64-bit strings—one of 64 zeros and one of 63 zeros with a single 1 at the end—the resulting ciphertexts will differ in 26 bits! Such diffusion and confusion are rather typical of the DES; it is so strong that cryptographers sometimes refer to such a system as having the *avalanche effect*. It also makes it practical to use DES to encode plaintext messages of any length, since the last block corresponding to the tail of the message need not contain a specified number of plaintext characters. (For example, the end of the message may be padded with zeros, to fill the block, without compromising security.) The facts presented in this example will be verified in Computer Exercise 20.

The Fall of DES

We next give some indications on how DES finally met its demise and how it could be easily modified to render it much more resilient to present-day attacks. The security of DES had been extensively studied throughout its 20-year tenure, and such studies have spurred the invention of new areas of cryptanalysis for block cryptosystems. Two such areas are *linear cryptanalysis* and *differential cryptanalysis*; detailed information about these methods can be found in the treatise by Stinson [Sti-06]; see also [Sch-96] or [WaTr-06].

Differential cryptanalysis is a chosen plaintext attack that basically traces differences in certain chosen plaintext pairs through the various rounds of encryption (differences in bit strings are measured by their XOR). It was first published in a 1991 paper by Eli Biham and Adi Shamir [BiSh-91]; it works much more effectively if there are a smaller number of rounds (as the complexity of the algorithm grows exponentially with the number of Feistel rounds). Indeed, as was indicated in a 1994 paper by Don Coppersmith, an IBM scientist who was part of the DES design team, the technique was known to the IBM team in the 1970s during the design phase (the team originally called it the *T-attack*), and measures were taken to safeguard DES against it. IBM decided not to publish their "T-attack," since it was a general and powerful method that might have fallen into the wrong hands if it were to have been made public. Linear cryptanalysis is also a chosen plaintext attack but has a very different approach of approximating the DES encryption function by a linear function on the input bits. This method, discovered by Japanese cryptographer Mitsuru Matsui [Mat-94], did not seem to be anticipated by the IBM team.

It has turned out that the first successful attacks on DES have involved exhaustive key searches through the space of $2^{56} \approx 7.2 \times 10^{16}$ possible keys. Although this number exceeds the number of 64-bit plaintext/ciphertext pairs estimated to be needed to achieve an attack with differential cryptanalysis (approximately $2^{47.2}$) and linear cryptanalysis (approximately 2^{43}), these latter attacks are more complicated to implement and are not easily distributed on multiple machines.[*] The first well-known attack on DES was in response to a 1997 $10,000 challenge by the RSA encryption security company. The prize was claimed by an independent computer programmer and consultant, Rocke Verser. Verser recruited a large network of PCs (by offering to give the owner of the PC that found the winning key $4,000 of his winnings) and found the correct key in about five months. A year later, a similar challenge by RSA was answered in only 39 days. With networks of PCs, a DES system can now be broken in a few hours.

Triple DES

DES has an interesting property that is not shared by most of the cryptosystems considered in the preceding chapters (or by the AES system considered in the Chapter 11): It turns out that if we perform two (or more) DES encryptions using different keys, the result will not, in general, be equivalent to a single DES encryption. More precisely, if κ and κ' are two DES keys, with DES_κ and $\text{DES}_{\kappa'}$ denoting the corresponding encryption functions, then the composition encryption function $\text{DES}_{\kappa'} \circ \text{DES}_\kappa$ will in general not equal $\text{DES}_{\kappa''}$ for any other DES key κ''.[†] From this fact, it may seem plausible that using an encryption scheme of one DES followed by another (with different keys)—the so-called **double DES**—may effectively double the key size of DES from 56 bits to 112 bits. The hope for this idea was dashed early on by a 1981 discovery by Merkle and Hellman [MeHe-81] that the effective key size increase in security would be only 1 bit (from 56 to 57 bits), thereby only doubling the time needed to search through the keyspace. Despite the minimal gain experienced through an effective doubling of the key size through double DES, a **triple DES**

---

[*] The fact that brute-force key searches were the first and remain the most efficient attacks on DES is further justification that the DES was an extremely successful cryptosystem. Any cryptographer can only aspire to designing a system that will be as difficult to crack as it would be to do a brute-force trial-and-error search on the entire keyspace.

[†] In mathematical terms, this fact can be expressed by saying that the set of all DES encryption functions does not form a *group*, i.e., it is not closed under composition. In fact, it has been shown that the smallest set of encryption functions that contains all DES encryption functions and is closed under composition (*the group generated by* DES) has size at least 10^{2499}. The interested reader will find more details about this interesting fact in the original reference [CaWi-93] or in the handbook by Menezes, van Oorschot, and Vanstone [MeOoVa-96].

cryptosystem using three different keys—κ, κ', and κ''—has thus far appeared to have extended the effective security level to 112 bits.[*]

Another way to increase the security of DES would be to use the same system with a larger key size (say, using a 128- or 256-length key). Of course, the core parts of the algorithm, including the key generation process, the number of rounds to use, and the round key functions (along with the S-boxes) would all need to be redone in a way that would help to reap the benefits of this larger key size.

Modes of Operation for Block Cryptosystems

We end this chapter with some practical implementation concepts for general block cryptosystems (as is any Feistel-based cryptosystem). Such algorithms are set up to encrypt only single blocks of a specified size (for example, for DES the blocks have to be 64-bit strings). Oftentimes, however, it is necessary to encrypt numerous plaintexts occupying several blocks, some of which have sizes that typically are not a multiple of the block size. By a **mode of operation** for a block cipher, we simply mean a system by which such encryptions can be carried out. We present the four most widely used modes of operation; the first two are **block modes** since they transmit only whole blocks at a time, while the latter two are **stream modes** since they may transmit smaller subblocks, or even one character or bit at a time.

Notation: Throughout this development, we denote a sequence of plaintext blocks by $P_1, P_2, P_3, \cdots$; each will be a block of ℓ bits (ℓ = the block size). The encryption mapping of the particular block cipher being employed are denoted as E_κ, and its inverse (the decryption mapping) are denoted as $D_\kappa \triangleq E_\kappa^{-1}$. The corresponding blocks of ciphertexts that the modes of operation will send are denoted as $C_1, C_2, C_3, \cdots$, each also size ℓ.

Electronic Codebook (ECB) Mode

The simplest mode of operation for a block cryptosystem is to simply encrypt consecutive blocks (with the same key) and to transmit the corresponding ciphertext blocks. This is called the **electronic codebook (ECB) mode** and is represented by the following equation:

$$C_i = E_\kappa(P_i), \qquad i = 1, 2, 3, \cdots \tag{7.6}$$

To decrypt, the intended recipient need only apply the inverse encryption mapping to the cipherblocks: $P_i = D_\kappa(C_i)$. The electronic codebook mode is susceptible to what is called a **codebook attack**, when large

---

[*] In practical terms, this means that unless there is a novel cryptographic breakthrough for DES (other than those that have been discovered over the heavy activity period of the past three decades), the triple DES system should take 7.2×10^{16} times however much time it will take to crack a single DES system. Even if the time needed to crack a single DES goes down from its current requirement of a few hours to, say, a millionth of a second, triple DES would still take thousands of years to crack.

sets of data are being sent. In a code-book attack, an adversary collects and analyzes large sets of intercepted cipherblocks with whatever tools and additional information might be available, such as frequency analysis, cribs (known plaintext/ciphertext pairs), and so forth. Once certain cipherblocks are ascertained, they are entered into a codebook that can be used to check subsequent messages for partial decryptions without actually having the key. To avoid this sort of attack, it is best to use a mode that adds some additional "noise" in the processing of blocks so that identical plaintext blocks need not be processed into identical cipherblocks. Each of the following three modes of operation accomplishes this goal.

Cipherblock Chaining (CBC) Mode

Rather than directly encrypting each plaintext block, the **cipherblock chaining (CBC) mode** first XORs each plaintext block P_i with the previously produced ciphertext block C_{i-1}. The operation is summarized by the following equation:

$$C_i = E_\kappa(P_i \oplus C_{i-1}), \qquad i = 1, 2, 3, \cdots \tag{7.7}$$

There is one piece of information that is missing: To get started with this mode of operation, we do not yet have a "previously produced" ciphertext block; that is, when we transmit the first cipherblock, we use $i = 1$ into Equation 7.7 to obtain $C_1 = E_\kappa(P_1 \oplus C_0)$. It is clear that we need to artificially define the zeroth ciphertext block C_0 (which does not actually correspond to any plaintext block but is the "seed" needed to get this mode of operation started). Everything will work if we take C_0 to be any ℓ-bit string; but if we were to use the same C_0 for repeated transmissions, then the corresponding first ciphertext blocks C_1 would be susceptible to a codebook attack. The best policy would be to randomly produce the zeroth cipherblock C_0 each time a new transmission is started. These seed cipherblocks could simply be transmitted unencrypted, since without knowing the encryption key, C_0 would be of no additional help to a potential intruder. To decrypt, the intended recipient would use the following procedure:

$$P_i = D_\kappa(C_i) \oplus C_{i-1}, \qquad i = 1, 2, 3, \cdots \tag{7.8}$$

Proof: Using Equation 7.7, and the properties of XOR (Proposition 7.1): $D_\kappa(C_i) \oplus C_{i-1} = D_\kappa(E_\kappa(P_i \oplus C_{i-1})) \oplus C_{i-1} = E_\kappa^{-1}(E_\kappa(P_i \oplus C_{i-1})) \oplus C_{i-1} = (P_i \oplus C_{i-1}) \oplus C_{i-1} = P_i.$ □

Exercise for the Reader 7.6

We consider the following (very simple) block encryption function $E_\kappa = E$ on 2-bit blocks (so the block size is $\ell = 2$) that is defined in Table 7.20. The following sequence of plaintext is to be transmitted: 1010100011.

TABLE 7.20 Block Encryption Function of Exercise for the Reader 7.5

P	00	01	10	11
E(P)	10	00	11	01

(a)   Determine the corresponding ciphertext sequence that gets transmitted if electronic codebook mode is used.

(b)   Determine the corresponding ciphertext sequence that gets transmitted if cipherblock chaining mode is used with seed $C_0 = 10$.

Cipher Feedback (CFB) Mode

While the preceding two modes process entire ℓ-bit blocks at a time, the **cipher feedback (CFB) mode** works on smaller **subblocks** of any size k, where $k \mid \ell$. Thus, in the extreme case $k = 1$, CFB would process one bit at a time. A much more typical case is when $k = 8$, so that each subblock would represent one of the 256 ASCII characters. In fact, this is such a common representation scheme that a string of eight bits is given a special name.

Definition 7.3

A **byte** is an 8-bit string.

Thus, a byte is equivalent also to a hexadecimal string of length 2.

Since the CFB mode operates on subblocks of size k, each plaintext block P_i is decomposed into its $n = \ell / k$ subblocks:

$$P_i = p_i^1 p_i^2 \cdots p_i^n, \qquad i = 1, 2, 3, \cdots \qquad (7.9)$$

Thus each p_j^m represents a single k-bit subblock. Since the CFB mode is somewhat more complicated than the previous two modes, we enunciate it with the following algorithm.

Algorithm 7.4: Cipher Feedback (CFB) Mode Encryption

We assume that we have a string of a certain number of plaintext subblocks, $p_1, p_2, p_3, \cdots$, that we need to encrypt and send.[*] The resulting stream of cipher subblocks that the algorithm produces will be denoted as $c_1, c_2, c_3, \cdots$. Note that in light of the above notation, the encryption mapping E_κ works with strings of n k-bit subblocks.

Step 1.   Generate an ℓ-bit **shift register** S_1. Random generation is recommended. Initialize subblock counter: $i = 1$.

---

[*] If the blocks are $P_1, P_2, \cdots$, then, in the notation of (9), the corresponding subblock stream $p_1, p_2, p_3, \cdots$ would be: $p_1^1, p_1^2, \cdots, p_1^n, p_2^1, p_2^2, \cdots, p_2^n, \cdots$.

> *Step 2.* Encrypt $S_i \rightarrow E_\kappa(S_i)$, let T_i be the leftmost k-bit subblock of $E_\kappa(S_i)$, and let R_i be the string of the rightmost $n-1$ subblocks of the shift register S_i.
>
> *Step 3.* Define $c_i = p_i \oplus T_i$, update the next shift register as $S_{i+1} = R_i c_i$, and update $i \rightarrow i+1$.
>
> *Step 4.* Return to Step 2 until all plaintext subblocks have been encrypted.

Notice that at each iteration of Step 3, the new shift register is obtained by deleting the previous shift register's leftmost subblock and appending the current cipher subblock on the right. Algorithm 7.4 is illustrated in Figure 7.7.

Example 7.5

Suppose that we are given the plaintext `101110`. Determine the ciphertext that gets transmitted if the encryption function of Table 7.20 is used in cipher feedback mode with parameters $k = 1$ and initial shift register $S_1 = 10$.

Solution: Since $k = 1$, the plaintext gets processed one bit at a time:

$$101110 \rightarrow p_1 = 1, p_2 = 0, p_3 = 1, p_4 = 1, p_5 = 1, p_6 = 0$$

Thus, we will need to run through Steps 2 and 3 of Algorithm 7.4 six times:

$i = 1$: $E(S_1) = E(10) = 11$. So $T_1 = 1, R_1 = 0$,
$c_1 = p_1 \oplus T_1 = 1 \oplus 1 = 0$, and $S_2 = R_1 c_1 = 00$.

$i = 2$: $E(S_2) = E(00) = 10$. So $T_2 = 1, R_2 = 0$,
$c_2 = p_2 \oplus T_2 = 0 \oplus 1 = 1$, and $S_3 = R_2 c_2 = 01$.

$i = 3$: $E(S_3) = E(01) = 00$. So $T_3 = 0, R_3 = 1$,
$c_3 = p_3 \oplus T_3 = 1 \oplus 0 = 1$, and $S_4 = R_3 c_3 = 11$.

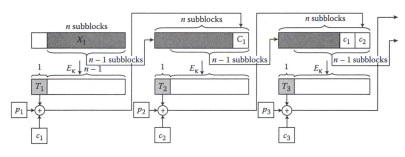

Figure 7.7 The cipher feedback (CFB) mode of encryption for block cryptosystems.

$i = 4$: $E(S_4) = E(11) = 01$. So $T_4 = 0, R_4 = 1$,
$\qquad c_4 = p_4 \oplus T_4 = 1 \oplus 0 = 1$, and $S_5 = R_4 c_4 = 11$.

$i = 5$: $E(S_5) = E(11) = 01$. So $T_5 = 0, R_5 = 1$,
$\qquad c_5 = p_5 \oplus T_5 = 1 \oplus 0 = 1$, and $S_6 = R_5 c_5 = 11$.

$i = 6$: $E(S_6) = E(11) = 01$. So $T_6 = 0$, and
$\qquad c_6 = p_6 \oplus T_6 = 0 \oplus 0 = 0$.

Thus, the ciphertext that gets transmitted is $c_1 c_2 c_3 c_4 c_5 c_6 = $ 011110.

The decryption procedure of CFB mode is given in the following.

Algorithm 7.5: Cipher Feedback (CFB) Mode Decryption

Step 1. Initialize subblock counter $i = 1$.

Step 2. Encrypt $S_i \rightarrow E_\kappa(S_i)$, let T_i be the leftmost subblock of $E_\kappa(S_i)$, and R_i be the string of the rightmost $n - 1$ subblocks of S_i.

Step 3. Set $p_i = c_i \oplus T_i$ and $S_{i+1} = R_i c_i$, and update $i \rightarrow i + 1$.

Step 4. Return to Step 2 until all plaintext subblocks have been encrypted.

Exercises for the Reader

Exercise 7.7

Use Algorithm 7.5 to decrypt the ciphertext that was obtained in Example 7.5.

Exercise 7.8

Show that Algorithm 7.5 correctly performs CFB mode decryptions.

Output Feedback (OFB) Mode

The cipher feedback mode is a stream cipher protocol that can be used with any block cipher. If used correctly, it is very secure and effective, but if any error were to be introduced in the entering of any plaintext subblock, the error would propagate through the stream to corrupt the next $n - 1$ cipher characters, after which the corrupted stream would be flushed from the system; see Exercise 29. A more robust mode can be created with a very slight modification of the cipher feedback mode. The **output feedback (OFB) mode** encrypts as in Algorithm 7.4, but with the shift register update $S_{i+1} = R_i c_i$ of Step 3 being changed to $S_{i+1} = R_i T_i$. In other words, in creating new shift registers, the first subblock of the encryption of the previous shift register gets appended on the right, rather than the previous cipher subblock. A schematic diagram for this mode is shown in Figure 7.8. With this modification, an error in entering a single plaintext subblock will corrupt only the corresponding cipher subblock.

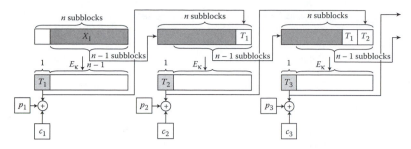

Figure 7.8 The output feedback (OFB) mode of encryption for block cryptosystems.

Exercises for the Reader

Exercise 7.9

Determine the ciphertext that gets transmitted if the encryption function of Table 7.20 is used in output feedback mode with the plaintext and parameters of Example 7.5.

Exercise 7.10

Explain the decryption algorithm for the output feedback mode.

Chapter 7 Exercises

1. Let $A = 010101$, $B = 110011$, $C = 111000$. Compute the following:
    (a) $A \oplus B$
    (b) $A \oplus C$
    (c) $A \oplus B \oplus A$
    (d) $C \oplus C \oplus C$

2. Let $A = 1100000$, $B = 1100111$, $C = 1111000$. Compute the following:
    (a) $A \oplus B$
    (b) $A \oplus C$
    (c) $A \oplus B \oplus A$
    (d) $C \oplus C \oplus C$

3. Let $\mathbf{1}_N = \mathbf{1}$ denote the N-bit string of 1's. With $A = 010101$, $B = 110011$ (and $N = 6$), compute:
    (a) $A \oplus \mathbf{1}$
    (b) $B \oplus \mathbf{1}$
    (c) Explain the general relationship between a bit string S and $S \oplus \mathbf{1}$.

4. *Complements of Bit Strings.* For a bit string A, define its *complement* bit string $\overline{A}$ to be the same length bit string whose bits are the opposite of the corresponding bits of A. For example, $0001 = 1110$.
    (a) Let $A = 1100000$, $B = 1100111$. Compute both $A \oplus B$ and $\overline{A} \oplus \overline{B}$.

(b) If A and B are any bit strings of the same length, what can you say about the relationship between their XOR $A \oplus B$ and the XOR of their complements $\overline{A} \oplus \overline{B}$?

(c) Suppose that A and B are length-N strings that satisfy $A \oplus B = \mathbf{1}_N$, where $\mathbf{1}_N$ denotes the N-bit string of 1's. What can you say about the relationship between the bits of A and B?

5. *Two-Round Feistel System.* Consider the two-round Feistel cryptosystem as explained in Example 7.1, but with the third round being removed.

(a) Using this system of a 12-bit system key $\kappa = 4\text{D}7$ (represented in hex form), use this Feistel cryptosystem to encrypt the 8-bit plaintext $P = 12$ (represented in hex form).

(b) Using this system of a 12-bit system key $\kappa = \text{F}88$ (represented in hex form), use this Feistel cryptosystem to encrypt the 8-bit plaintext $P = 7\text{D}$ (represented in hex form).

(c) Perform the corresponding decryption to the ciphertext that you obtained in part (a).

(d) Perform the corresponding decryption to the ciphertext that you obtained in part (b).

6. *Two-Round Feistel System.* Consider the two-round Feistel cryptosystem as explained in Example 7.1, but with the first round being removed (so the second round in Example 7.1 becomes the first round here, etc.).

(a) With a 12-bit system key $\kappa = 4\text{D}7$ (represented in hex form), use this Feistel cryptosystem to encrypt the 8-bit plaintext $P = 12$ (represented in hex form).

(b) With a 12-bit system key $\kappa = \text{F}88$ (represented in hex form), use this Feistel cryptosystem to encrypt the 8-bit plaintext $P = 7\text{D}$ (represented in hex form).

(c) Perform the corresponding decryption to the ciphertext that you obtained in part (a).

(d) Perform the corresponding decryption to the ciphertext that you obtained in part (b).

7. *Three-Round Feistel System.* Using the three-round Feistel cryptosystem of Example 7.1, do the following:

(a) With a 12-bit system key $\kappa = 48\text{C}$ (represented in hex form), encrypt the 8-bit plaintext $P = \text{AB}$ (represented in hex form).

(b) With a 12-bit system key $\kappa = \text{A}62$ (represented in hex form), encrypt the 8-bit plaintext $P = 22$ (represented in hex form).

(c) Perform the corresponding decryption to the ciphertext that you obtained in part (a).

(d) Perform the corresponding decryption to the ciphertext that you obtained in part (b).

8. *Three-Round Feistel System.* Using the three-round Feistel cryptosystem of Example 7.1, do the following:

(a) With a 12-bit system key $\kappa = \text{FAA}$ (represented in hex form), encrypt the 8-bit plaintext $P = \text{D}2$ (represented in hex form).

(b)   With a 12-bit system key $\kappa = $ C02 (represented in hex form), encrypt the 8-bit plaintext $P = $ 30 (represented in hex form).

(c)   Perform the corresponding decryption to the ciphertext that you obtained in part (a).

(d)   Perform the corresponding decryption to the ciphertext that you obtained in part (b).

9.   *Scaled-Down DES S-Boxes.* Perform the following S-box computations using the S-boxes for the scaled-down DES given in Table 7.5:

(a) $S_1(0000)$

(b) $S_1(1010)$

(c) $S_2(1111)$

(d) $S_2(0011)$

10.   *Scaled-Down DES S-Boxes.* Perform the following S-box computations using the S-boxes for the scaled-down DES given in Table 7.5:

(a) $S_1(1111)$

(b) $S_1(1011)$

(c) $S_2(0000)$

(d) $S_2(1011)$

11.   *Scaled-Down DES Round Key Functions.* Perform the following round key evaluations computations using the round key functions for the scaled-down DES cryptosystem that were developed in Algorithm 7.1, and using the round key $\kappa^i = $ 10110111:

(a) $f_{\kappa^i}(0000)$

(b) $f_{\kappa^i}(1010)$

(c) $f_{\kappa^i}(1111)$

(d) $f_{\kappa^i}(0011)$

12.   *Scaled-Down DES Round Key Functions.* Perform the following round key evaluations computations using the round key functions for the scaled-down DES cryptosystem that were developed in Algorithm 7.1, and using the round key $\kappa^i = $ 00111100:

(a) $f_{\kappa^i}(0110)$

(b) $f_{\kappa^i}(1001)$

(c) $f_{\kappa^i}(1111)$

(d) $f_{\kappa^i}(1011)$

13.   *Scaled-Down DES.*

(a)   Use scaled-down DES to encrypt (Algorithm 7.1) the 8-bit plaintext $P = $ D2 (represented in hex form) with 12-bit system key $\kappa = $ FAA (represented in hex form).

(b)   Use scaled-down DES to encrypt (Algorithm 7.1) the 8-bit plaintext $P = $ 30 (represented in hex form) with 12-bit system key $\kappa = $ C42 (represented in hex form).

(c) Perform the corresponding decryption (with Algorithm 7.2) to the ciphertext that you obtained in part (a).

(d) Perform the corresponding decryption (with Algorithm 7.2) to the ciphertext that you obtained in part (b).

14. *Scaled-Down DES.*

(a) Use scaled-down DES to encrypt (Algorithm 7.1) the 8-bit plaintext $P = $ AB (represented in hex form) with 12-bit system key $\kappa = $ 59C (represented in hex form).

(b) Use scaled-down DES to encrypt (Algorithm 7.1) the 8-bit plaintext $P = $ 22 (represented in hex form) with 12-bit system key $\kappa = $ A63 (represented in hex form).

(c) Perform the corresponding decryption (with Algorithm 7.2) to the ciphertext that you obtained in part (a).

(d) Perform the corresponding decryption (with Algorithm 7.2) to the ciphertext that you obtained in part (b).

15. *DES S-Boxes.* Perform the following DES S-box computations using Table 7.10:

(a) $S_7(110010)$

(b) $S_3(110010)$

(c) $S_7(110011)$

(d) $S_4(111110)$

16. *DES S-Boxes.* Perform the following DES S-box computations using Table 7.10:

(a) $S_3(101010)$

(b) $S_6(011010)$

(c) $S_3(111010)$

(d) $S_1(111111)$

17. *DES Round Keys.* Determine (in hex form) the second round key κ_2 for the DES key given in Example 7.3.

18. *DES Round Keys.* Determine (in hex form) the third round key κ_3 for the DES key given in Example 7.3.

19. *DES Calculation.* Determine (in hex form) the resulting 32-bit (left and right) strings L_2 and R_2 resulting from the second round of the DES encryption algorithm of Example 7.3.

20. *DES Calculation.* Determine (in hex form) the resulting 32-bit (left and right) strings L_3 and R_3 resulting from the third round of the DES encryption algorithm of Example 7.3.

21. *Block Cryptosystem Encryptions.* Consider the following encryption mapping: $E_\kappa = E$: {4-bit strings} $\rightarrow$ {4-bit strings} defined by first cyclically shifting a 4-bit string one unit to the left, and then XORing the result with the fixed string 1011. For example, to compute $E(0011)$, we first shift the inputted string one unit to the left, $0011 \rightarrow 0110$, and then we XOR the result with 1011 to produce $E(0011) = 0110 \oplus 1011 = 1101$. Suppose that we need to transmit the plaintext $P = 111000$, and we will implement one of the block cryptosystem modes of operation. Determine the resulting transmitted ciphertext if:

(a) The electronic codebook mode is used, padding any uncompleted blocks with 0s.

(b) The cipherblock chaining mode is used, padding any uncompleted blocks with 0s. Take $C_0 = 1110$.

(c) The cipher feedback mode is used, with subblock size $k = 2$, and initial shift register $S_1 = 1110$. Use 0's to pad, as needed.

(d) The output feedback mode is used, with subblock size $k = 2$, and initial shift register $S_1 = 1110$. Use 0's to pad, as needed.

22. *Block Cryptosystem Encryptions.* Consider the following encryption mapping: $E_\kappa = E$: {6-bit strings} $\rightarrow$ {6-bit strings} defined by first cyclically shifting a 6-bit string two units to the left, and then XORing the result with the fixed string 101100. For example, to compute $E(001111)$, we first shift the inputted string two units to the left, $001111 \rightarrow 111100$, and then we XOR the result with 101100 to produce $E(001100) = 111100 \oplus 101100 = 010000$. Suppose that we need to transmit the plaintext $P = 111000101$, and we will implement one of the block cryptosystem modes of operation. Determine the resulting transmitted ciphertext if:

(a) The electronic codebook mode is used, padding any uncompleted blocks with 0s.

(b) The cipherblock chaining mode is used, padding any uncompleted blocks with 0's. Take $C_0 = 111001$.

(c) The cipher feedback mode is used, with subblock size $k = 3$, and initial shift register $S_1 = 111001$. Use 0's to pad, as needed.

(d) The output feedback mode is used, with subblock size $k = 3$, and initial shift register $S_1 = 111001$. Use 0's to pad, as needed.

23. *Block Cryptosystem Decryptions.*

(a) Show that the inverse of the encryption mapping $E_\kappa = E : \{4\text{-bit strings}\} \rightarrow \{4\text{-bit strings}\}$ of Exercise 21 is the mapping $D_\kappa = D : \{4\text{-bit strings}\} \rightarrow \{4\text{-bit strings}\}$ defined by first XORing the inputted 4-bit string with the fixed string 1011, and then cyclically shifting the resulting 4-bit string one unit to the right. For example, to compute $D(1101)$, we first XOR the input with 1011 to produce $1101 \oplus 1011 = 0110$, and then shift the resulting string one unit to the right $0110 \rightarrow 0011$, to give that $D(1101) = 0011$. (This is the inverse calculation of the one given in Exercise 21.)

Next, using the appropriate decryption algorithm for the different modes of operation, perform the following decryptions:

(b) Decrypt the ciphertext that was obtained in part (a) of Exercise 21.

(c) Decrypt the ciphertext that was obtained in part (b) of Exercise 21.

(d) Decrypt the ciphertext that was obtained in part (c) of Exercise 21.

(e) Decrypt the ciphertext that was obtained in part (d) of Exercise 21.

Suggestion: For part (a), use the result of Exercise for the Reader 3.7.

24.   *Block Cryptosystem Decryptions.*
   (a) Show that the inverse of the encryption mapping $E_\kappa = E : \{6\text{-bit strings}\} \rightarrow \{6\text{-bit strings}\}$ of Exercise 22 is the mapping $D_\kappa = D : \{6\text{-bit strings}\} \rightarrow \{6\text{-bit strings}\}$ defined by first XORing the inputted 6-bit string with the fixed string 101100, and then cyclically shifting the resulting 6-bit string two units to the right. For example, to compute $D(010000)$, we first XOR the input with 101100 to produce $010000 \oplus 101100 = 111100$, and then shift the resulting string two units to the right, $111100 \rightarrow 001111$, to give that $D(010000) = 001111$. (This is the inverse calculation of the one given in Exercise 22.)

   Next, using the appropriate decryption algorithm for the different modes of operation, perform the following decryptions:
   (b) Decrypt the ciphertext that was obtained in part (a) of Exercise 22.
   (c) Decrypt the ciphertext that was obtained in part (b) of Exercise 22.
   (d) Decrypt the ciphertext that was obtained in part (c) of Exercise 22.
   (e) Decrypt the ciphertext that was obtained in part (d) of Exercise 22.
   Suggestion: For part (a), use the result of Exercise for the Reader 3.7.

25.   *Complementary Keys and Plaintext Yield Complementary Plaintexts in DES.* The *complement* of a bit vector (or string) P is the bit vector $\bar{P}$ of the same length whose bits are the opposites of those of P. Put differently, $\bar{P} = P \oplus 11111\cdots$ (see Exercise 4). Prove that the complement $\bar{C}$ of a ciphertext message produced by DES from a plaintext P and using a key κ is the same as the ciphertext message produced (directly) by DES using plaintext $\bar{P}$ and key $\bar{\kappa}$. This result may be symbolized as $\overline{\text{DES}_\kappa(P)} = \text{DES}_{\bar{\kappa}}(\bar{P})$.

Suggestion: The fine details of DES (like the S-boxes) are not needed for this justification.

Note: For a brute-force key search attack on DES, the result of this exercise only reduces the number of keys to search by a factor of 2, from 2^{56} to 2^{55}.

26.   *Weak Keys for DES.* A key κ for DES is called a *weak key* if it turns out that all of the round keys are the same, i.e., $\kappa^1 = \kappa^2 = \cdots = \kappa^{15} = \kappa^{16}$. Prove that there are exactly four weak keys for DES, and determine these weak keys.

Note: Because of the DES decryption scheme, a weak key would render the DES decryption algorithm exactly the same as the DES decryption algorithm.

27.   *Peculiar Property of One of the S-Boxes.*
   (a) Show that each entry in the second row of the S-box S_4 can be obtained by the corresponding entry in the first row of S_4 by XORing the latter entry with the 4-bit string 0110 (the former entry should also be written as a 4-bit string).
   (b) Show that any row of S_4 can be obtained from any other row of S_4 by a similar procedure.

28.   *Properties of XOR.* Prove Proposition 7.1.

29.   *Propagation of Errors in Block Cryptosystem Modes of Operation.* Throughout this exercise, we assume (as in the development of the modes of operation) that we have an underlying block cryptosystem with block size ℓ. We denote the encryption mapping by E, and the corresponding decryption mapping by D. In cases of a stream mode of operation, we let k denote the subblock size, so that $k \mid \ell$. Suppose that a single plaintext bit has been entered incorrectly.
   (a)   How many ciphertext bits could be possibly corrupted if the electronic code-book mode is used?
   (b)   How many ciphertext bits could be possibly corrupted if the cipherblock chaining mode is used?
   (c)   How many ciphertext bits could be possibly corrupted if the cipher feedback mode is used?
   (d)   How many ciphertext bits could be possibly corrupted if the output feedback mode is used?

30.   *Self-Decryption Proof of Three-Round Feistel Systems.* Verify that any three-round Feistel cryptosystem is self-decrypting (if the order of the keys is reversed).

      Suggestion: Examine the corresponding verification for two-round Feistel cryptosystems in the solution of Exercise for the Reader 7.2. Use the same notation and take it one more step.

31.   *Self-Decryption Proof of General Feistel Systems.* Verify that any Feistel cryptosystem is self-decrypting (if the order of the keys is reversed).

      Suggestion: Examine the corresponding verification for two-round Feistel cryptosystems in the solution of Exercise for the Reader 7.2. With notation similar to what was used there, use mathematical induction to establish the identities $L_i' = R_{NR-i}$, $R_i' = L_{NR-i}$, for each $i = 0, 1, \ldots, NR$.

32.   *Self-Decryption Proof of DES.* Verify that the DES is self-decrypting (if the order of the keys is reversed).

33.   *An Active Attack on the OFB Mode.*
   (a)   Suppose that Mallory has obtained a corresponding pair of plaintext and ciphertext from an OFB transmission. Show how she could subsequently send a legitimate ciphertext encryption in the same mode that corresponds to whatever malicious plaintext message she wishes to send.
   (b)   Provide a simple safeguard that users of the OFB mode could employ that would render ineffective the attack in part (a).

34.   *The Counter Mode of Operation.* The **counter (CTR) mode of operation** is another stream mode similar to the OFB mode, but with an advantage that the shift registers may be computed separately (rather than recursively), thus making it amenable to parallel processing (i.e., distributed computation). The encryption algorithm for the CTR mode is similar to that for the OFB mode, the only change is in how the shift registers S_i are updated. Recall that each S_i is an ℓ-bit string. Such a string can be thought of as a binary expansion for a nonnegative integer in the range $0 \le x < 2^\ell$. The shift register updating will be by

modular addition of these integers, adding 1 at each iteration: $S_{i+1} = S_i + 1 (\mod 2^\ell)$.

(a) Show that $S_k = S_1 + k - 1 (\mod 2^\ell)$, for and positive integer k.

(b) Consider the following encryption mapping: $E_K = E$: {4-bit strings} $\rightarrow$ {4-bit strings} defined by first cyclically shifting a 4-bit string one unit to the left, and then XOR-ing the result with the fixed string 1011. For example, to compute $E(0011)$, we first shift the inputted string one unit to the left, $0011 \rightarrow 0110$, and then we XOR the result with 1011 to produce $E(0011) = 0110 \oplus 1011 = 1101$. Suppose that we need to transmit the plaintext $P = 111000$, and we will implement one of the block cryptosystem modes of operation. Determine the resulting transmitted ciphertext if the counter mode is used with subblock size $k = 2$ and initial shift register $S_1 = 1110$.

(c) Describe the general decryption algorithm for the counter mode.

(d) Apply the decryption algorithm that you obtained in part (c) to decrypt the ciphertext that you obtained in part (b).

Chapter 7 Computer Implementations and Exercises

Note: As in previous chapters, while we sometimes distinguish between vectors and strings (both in displays and words), the reader is free either to follow our conventions or to use whichever data structures are more suitable for his or her computing platforms.

1. *Bit String to Vector Conversion.* Write a program $BinVec = BitStr2BinVec(BitStr)$ that inputs a bit string $BitStr$ and outputs the corresponding binary vector $BinVec$. Thus, for example, the command $BitStr2BinVec(01110)$ should produce the output [0 1 1 1 0]. Run your program with the following inputs and record the outputs:
   (a) 11001111
   (b) 111110000101
   (c) 001001000111

2. *Vector to Bit String Conversion.* Write a program $BitStr = Vec2BitStr(BinVec)$ that inputs a binary vector $BinVec$ and outputs the corresponding bit string $BitStr$. Thus, for example, the command $Vec2BitStr([0 1 1 1 0])$ should produce the output 01110. Run your program with the following inputs and record the outputs:
   (a) [0 1 1 1 0 1]
   (b) [1 1 1 1 1 0 1]
   (c) [1 0 1 1 1 0 1]

   Check that the programs produced here and in Computer Exercise 1 are inverses of one another by computing $BinVec2BitStr(BitStr2Vec(01110))$ and also $BitStr2Vec(BinVec2BitStr([0 1 1 1 0]))$.

3. *XOR Program for Vectors.*
   (a) Write a program `BinOutVec = XORVec(BinVec1, BinVec2)` that inputs a pair of binary vectors of the same length `BinVec1`, `BinVec2` and outputs the binary vector of their XOR, `BinOutVec = BinVec1 ⊕ BinVec2`.
   (b) Use your program to perform the calculations of Chapter Exercise 1.

4. *XOR Program for Strings.*
   (a) Write a program `BinOutStr = XORStr(BinStr1, BinStr2)` that inputs a pair of bit strings of the same length `BinStr1`, `BinStr2` and outputs the bit string of their XOR, `BinOutStr = BinStr1 ⊕ BinStr2`.
   (b) Use your program to perform the calculations of Chapter Exercise 1.

5. *Program for Three-Round Feistel Cryptosystem Encryption.*
   (a) Write a program with syntax `Ctext = FeistelSystem3 RoundsEx7_1_Encrypt(Ptext,Key)` that will perform the three-round Feistel system encryption process of Example 7.1. The inputs are `Ptext`, an 8-bit string representing the plaintext, and `Key`, a 12-bit string representing the system key. The output `Ctext` will be the corresponding ciphertext.
   (b) Use your program to perform the encryptions of parts (a) and (b) of Chapter Exercise 7.

6. *Program for Three-Round Feistel Cryptosystem Decryption.*
   (a) Write a program with syntax `Ptext = FeistelSystem3 RoundsEx7_1_Deccrypt(Ctext,Key)` that will perform the three-round Feistel system decryption process of Example 7.1. The inputs are `Ctext`, an 8-bit string representing the ciphertext, and `Key`, a 12-bit string representing the system key. The output `Ptext` will be the corresponding plaintext.
   (b) Use your program to perform the decryptions of parts (c) and (d) of Chapter Exercise 7.

Writing Programs for DES

The next four exercises will build up a full program for the scaled-down DES encryption. Taken one at a time, the programming tasks are not difficult. In later computer exercises, a similar outline will be employed to write encryption programs for the full DES cryptosystem. Once the former task has been completed, the latter will be very easy. Most of the programs for the former will be easy to modify for the larger DES system. The large data sets of needed permutations and S-box data may be downloaded from the book's Web page.

7. *Program for Scaled-Down DES Round Key Generation.*
   (a) Write a program with syntax `RoundKeys = Scaled Down DESRoundKeys(Key)` that will generate the two round keys of the scaled-down DES cryptosystem of Algorithm 7.1. The input is `Key`, a 12-bit string representing the system key. If the key is not admissible (i.e., if the parity

check bits do not satisfy the required properties), the program will produce an error message. The output Round Keys will be a 2×8 matrix, whose two rows will be (in order) the two round keys for DES.

(b) Use your program to obtain the two round keys that were obtained in part (a) of the solution of Example 7.2.

(c) Use your program to obtain the two round keys for each of parts (a) and (b) of Chapter Exercise 13.

8. *Program for Scaled-Down DES S-Boxes.*
   (a) Write a program with syntax Out2 = ScaledDownDESS Box(In4, n) that will perform S-box evaluations for scaled-down DES cryptosystem (according to the S-boxes specified by Table 7.5). The inputs are In4, a 4-bit string, and n, which will specify the S-box and so will either be 1 or 2. The output Out2 will be the 2-bit string representing $S_n($In4$)$.
   (b) Use your program to perform the S-box evaluations of Exercise for the Reader 7.3.
   (c) Use your program to perform the S-box evaluations of Chapter Exercise 9.

9. *Program for Scaled-Down DES Round Key Functions.*
   (a) Write a program with syntax ROut = ScaledDownDES RoundKeyFunction(RIn,RoundKey) that will perform round key evaluations for scaled-down DES cryptosystem (according to Algorithm 7.1). The inputs are RIn, a 4-bit string, and RoundKey, an 8-bit string representing the round key κ^i. The output ROut will be the 4-bit string representing $f_{\kappa^i}($RIn$)$.
   (b) Use your program to perform the two round key function evaluations that were done in the solution of Example 7.2
   (c) Use your program to perform the two round key function evaluations of Chapter Exercise 11.

10. *Program for Scaled-Down DES Encryption.*
    (a) Write a program with syntax Ctext = ScaledDown DES(Ptext,Key) that will perform encryptions with the scaled-down DES cryptosystem (according to Algorithm 7.1). The inputs are Ptext, an 8-bit string representing the plaintext, and Key, a 12-bit string representing the system key. The output Ctext will be the corresponding ciphertext.
    (b) Use your program to perform the encryption that was done in the solution of Example 7.2.
    (c) Use your program to perform encryptions of parts (a) and (b) of Chapter Exercise 13.

11. *Program for Scaled-Down DES Decryption.*
    (a) Write a program with syntax Ptext = ScaledDown DESDecrypt(Ctext,Key) that will perform encryptions with the scaled-down DES cryptosystem (according to Algorithm 7.2). The inputs are Ctext, an 8-bit string

representing the ciphertext, and Key, a 12-bit string representing the system key. The output Ptext will be the corresponding plaintext.

(b) Use your program to perform the decryption that was done in the solution of Exercise for the Reader 7.4.

(c) Use your program to perform decryptions of parts (c) and (d) of Chapter Exercise 13.

12. *Initial Permutation of DES.* Write a program OutVec = DESInitPerm(InVec) that inputs a length-64 binary (plaintext) vector InVec and outputs another length-64 binary vector OutVec that results from the first vector by applying the initial permutation of DES to it. Check this program with the result obtained in Example 7.3(a).

13. *Inverse Initial Permutation of DES.* Create a table giving the inverse of the initial permutation IP^{-1} that is needed in the last step of the DES.

14. *Inverse of the Initial Permutation of DES.*

(a) Write a program OutVec = DESInvInitPerm(InVec) that inputs a length-64 binary vector InVec and another length-64 binary vector OutVec that results from the first vector by applying the inverse of the initial permutation of DES to it.

(b) Apply this program to the initially permuted 64-bit plaintext string of Table 7.14, and check that the output is the original plaintext string of Table 7.12

15. *Creation of DES Round Keys.* Write a program RoundKeys = DESRoundKeys(Key) that inputs a length-64 binary vector Key that serves as an admissible DES key. If the key is not admissible (i.e., if the parity check bits do not satisfy the required properties), the program should produce an error message. The output, RoundKeys, will be a 16-by-48 matrix whose *i*th row will be the corresponding *i*th round key for DES. Check this program with the results obtained in part (b) of Example 7.6 (for the first round key) and in Chapter Exercise 3 (for the second round key).

16. *S-Box Evaluation Program.*

(a) Write a program Out4 = DESSBox(In6, n) that inputs a 6-bit string In6 and an S-box number n (between 1 and 8). The output, Out4, will be the 4-bit string resulting from applying the *n*th S-box from Table 7.10 to the 6-bit string In6.

(b) Use your program to perform the S-box evaluations of Chapter Exercise 15.

17. *Evaluation of the DES Round Key Functions.*

(a) Write a program ROut=DESRoundKeyFunction(RIn, RoundKey) that inputs a length-32 binary vector RIn, along with a length-48 binary vector, RoundKey. The output, ROut, will be the length-32 binary vector that is the output of the corresponding DES round key function.

(b) Check this program with the results obtained in part (c) of Example 7.3 (for the first round key) and in Chapter Exercise 19 (for the second round).

Note: It is intended that the RoundKey input be taken from the output of the program in Computer Exercise 15. In a complete DES implementation, the latter program will thus only be executed once.

18. *Complete Program for the DES Algorithm.* Write a program Ctext = DES(Ptext,Key) that inputs a 64-bit string (plaintext) Ptext, along with a 64-bit string Key that serves as an admissible DES key. If the key is not admissible (i.e., if the parity check bits do not satisfy the required properties), the program should produce an error message. The output, Ctext, will be the length-64 binary (ciphertext) vector that corresponds to the output of the DES encryption algorithm. Use your program to compute the ciphertext when the DES is applied with the key and plaintext of Example 7.3, and check that it produces the following ciphertext (written as a string):

01000101001100100001011001011011001101011000100000011110100010000

Note: This program should be a simple one to write if one calls on some of the programs that were developed in the preceding exercises.

19. *Complete Program for the DES Algorithm-Hex Format.* Write a program Ctext = DESHex(Ptext,Key) that inputs a hex string of length-16 (plaintext) Ptext, along with another length-16 hex string Key that serves as an admissible DES key. If the key is not admissible (i.e., if the binary parity check bits do not satisfy the required properties), the program should produce an error message. The output, Ctext, will be a hex string of length-16 vector that corresponds to the ciphertext output of the DES encryption algorithm. Use your program to compute the ciphertext when the DES is applied with the key and plaintext of Example 7.3, and check that it produces the following ciphertext (written as a string):

45322CB66B103D10

Note: This program should be a simple one to write if one calls on some of the programs that were developed in the preceding exercises, or a minor modification of the one of Computer Exercise 18.

20. *Avalanche Effect of the DES Cryptosystem.* Use your program of Computer Exercise 18, with the DES key of Example 7.3, to encode the plaintext message consisting of 64 zero bits. Check that your ciphertext is

111000101001000010111010000000011000001010101100001010000000011010

In the same fashion, encode the plaintext message that is the same as the first, except that the last bit is one rather than zero. Check that your ciphertext is

```
0110000101111010101101101010101001101011010101101
00111100010001111
```

Thus, the ciphertexts of the nearly identical plaintexts differ by 26 bits!

21. *Program for Decrypting DES.*
    (a) Write a program `Ptext = DESDecrypt(Ctext,Key)` that inputs a 64-bit string (ciphertext) vector `Ctext` and another 64-bit string `Key` that serves as an admissible DES key. If the key is not admissible (i.e., if the parity check bits do not satisfy the required properties), the program should produce an error message. The output, `Ptext`, will be the 64-bit string representing the plaintext that DES would have encrypted to the given ciphertext.
    (b) Use your program to compute the plaintext when the DES is applied with the key of Example 7.3 to produce the ciphertext string shown in Computer Exercise 18, and check that it produces the corresponding plaintext (shown in Table 7.12).

22. *Program for Decrypting DES-Hex Format.*
    (a) Write a program `Ptext = DESDecryptHex(Ctext, Key)` that inputs a hex string of length-16 (ciphertext) `Ctext` and another length-16 hex string `Key` that serves as an admissible DES key. If the key is not admissible (i.e., if the binary parity check bits do not satisfy the required properties), the program should produce an error message. The output, `Ptext`, will be the hex string of length 16 representing the plaintext that DES would have encrypted to the given ciphertext.
    (b) Use your program to compute the plaintext when the DES is applied with the key of Example 7.3 to produce the ciphertext string shown in Computer Exercise 18, and check that it produces the corresponding plaintext (as given in the example).

23. *Programs for Triple DES.*
    (a) Write a program `C=TripleDES(P,Key1,Key2,Key3)` that inputs a length-64 binary (plaintext) vector `P`, along with three length-64 binary vectors `Key1`, `Key2`, `Key3` that serve as three admissible DES keys. If any of these keys is not admissible (i.e., if the parity check bits do not satisfy the required properties), the program should produce an error message. The output, `C`, will be the length-64 binary (ciphertext) vector that corresponds to the result when three applications of DES are applied to the plaintext, first using `Key1`, then `Key2`, and finally `Key3`.

(b) Write a program P = TripleDESDecrypt(C,Key1, Key2, Key3) that inputs a length-64 binary (ciphertext) vector C, along with three length-64 binary vectors Key1, Key2, Key3 that serve as three admissible DES keys (in order of successive encryption). If any of these keys is not admissible (i.e., if the parity check bits do not satisfy the required properties), the program should produce an error message. The output, P, will be the length-64 binary (plaintext) vector that corresponds to the input of the triple DES encryption algorithm of part (a).

(c) Use your program of part (a) to triple DES encrypt the plaintext message of Example 7.3 (shown in Table 7.10) using the following encryption keys: for Key1, use the key of Example 7.3 (shown in Table 7.12); for Key2, use the key obtained from Key1 by left shifting by 23 bits, followed by complementing all odd-numbered bits; and Key3 will be the result of performing the initial permutation (Computer Exercise 12) on Key1. Next, apply your program of part (b) to the result, and verify if this brings you back to the original plaintext.

Some Number Theory and Algorithms

Number theory has been a subject of extensive research that has continuously occupied a central position of pure mathematics since mathematics became a subject of study. Until the 1970s it remained a pure science in that its connections and applications were primarily to other branches of mathematics. When public key cryptography was discovered in the 1970s, however, number theory suddenly extended itself into a major applied subject, and a renewed interest has reenergized the subject to a new significance. Much of the security in contemporary public key cryptosystems relies on certain very difficult number theoretic problems that continue to remain recalcitrant. This chapter develops some number theoretic concepts and algorithms that, although fascinating on their own, have turned out particularly relevant and useful for public key cryptography. We begin with a discussion of the prime number theorem, which gives a quantitative description of how densely the primes are distributed among the integers. We also reveal an important periodicity in powers of (invertible) integers modulo an integer. This naturally leads us to the concepts of *orders* and *primitive roots*, which, when they exist, are integers of the highest possible order in a given modulus. Large primes and primitive roots are important ingredients for many cryptosystems. We develop some algorithms for determining whether an integer is prime, for factoring, and for the determination of primitive roots.

The Prime Number Theorem

Euclid's proof that there are infinitely many primes, although very elegant, does not tell us much about how the primes are distributed among the positive integers. A more informative but much deeper theorem is known as *the prime number theorem*; it gives a very sharp estimate of the number of primes less than any given number x.

Theorem 8.1: Prime Number Theorem

If $\pi(x)$ denotes the number of prime numbers p satisfying $p < x$, then we have

$$\pi(x) \sim \frac{x}{\ln x}$$

In other words, $\pi(x)$ is asymptotic to the ratio $x/\ln x$, meaning that the ratio $\pi(x)/(x/\ln x) \to 1$ as $x \to \infty$.

The prime number theorem was first proved in 1896 independently by Jacques Hadamard (1865–1963, a French mathematician) and Charles Jean Gustave Nicolas Baron de la Vallée Poussin (1866–1962, a Belgian mathematician). Their proofs were quite sophisticated and used complex analysis. Elementary (but difficult) proofs have later been found and the interested reader may refer to the books by Hardy and Wright [HaWr-80] or Nagell [Nag-01]. We will forgo giving a proof here. The prime number theorem shows that the integers are quite densely populated by primes. For example, if one were to attempt to factor a certain 304-digit number[*] n_{364} by checking for prime factors up to $\sqrt{n_{364}}$, since $\sqrt{n_{364}} > 10^{152}$, this would possibly entail checking up to $\pi(10^{152}) \, 10^{152} / \ln(10^{152}) \approx 2.86 \times 10^{149}$ divisions. Even if all the computers in the world could be programmed to work together on this task and if they each could check one trillion divisions per second, it would take many millions of life spans of our universe. There are better algorithms for factoring into primes, but the prime factorization problem has remained one of the hardest problems in computational number theory, and it is overwhelmingly believed among specialists that a truly efficient (polynomial time) algorithm probably does not exist. These guaranteed difficulties are at the heart of many effective cryptographic applications of prime numbers. The prime number theorem has many practical ramifications.

Example 8.1

    (a) Use the prime number theorem to estimate the number of 50-digit prime numbers.

    (b) Suppose that we were to randomly select a 50-digit odd integer (so the first digit is randomly selected from 1 to 9, the last digit from 1, 3, 5, 7, and 9, and all other 48 digits are randomly selected from 0 to 9). Estimate the probability that we select a prime number.

---

[*] RSA Security (a high-tech cryptographic security company) had offered a number of public challenges on its company Web site. One of these offered a $100,000 prize to the first person to factor a certain 304-digit number (larger prizes were available for factoring larger integers). This particular challenge remained open for several years. Such challenges actually benefit the company by helping to test the security of some of their secret codes (that rest on the infeasibility of being able to factor such large or even larger integers) against potential hackers.

Solution: Part (a): There are $\pi(10^{50}) - \pi(10^{49})$ 50-digit prime numbers. The prime number theorem estimates this number to be $10^{50}/\ln(10^{50}) - \pi(10^{49})/\ln(10^{49}) \approx 7.7996 \times 10^{47}$.

Part (b): By part (a), the proportion of 50-digit numbers that are prime is approximately $7.7996 \times 10^{47}/10^{50} \approx 1/128$. This means that if we were to randomly select odd 50-digit numbers, the chances of selecting a prime number would be about twice this number, or 1/64.

Exercise for the Reader 8.1

(a)   Use the prime number theorem to estimate the number of 300-bit prime numbers (where a 300-bit string represents the binary representation of an integer as in Chapter 6, and we assume that the first bit is 1 so the number really requires specification of 299 bits).

(b)   Suppose that we were to randomly select a 300-bit odd integer (so the first and last bits are set to 1, and all other 298 bits are randomly selected).[*] Estimate the probability that we select a prime number.

Fermat's Little Theorem

The probabilistic primality tests that we develop at the end of this chapter will help us to astoundingly improve odds that were seen in Example 8.1 of randomly generating primes of a specified size. The prototypical primality test of this sort is based on a small but important theorem of Pierre de Fermat; see Figure 8.1.[†] We motivate this theorem with an example

---

[*] The last bit being 1 makes the integer odd, and the first bit being 1 makes the effective length of the string 300 (since any zero bits on the left are redundant).

[†] After completing (the equivalent of) a bachelor's degree in mathematics, Fermat earned a law degree and followed an impressive career path as a government lawyer and council member in the city of Toulouse. He maintained his ardent interest in mathematics throughout his life and was involved in cutting-edge mathematics among the leaders in the field. His published works are relatively small in number, however, but this was due to his preference for solving new problems rather than taking the time to formally write up his work on problems that he had finished, the fact that he held another very demanding full-time job notwithstanding. Fermat often made bold scientific claims, such as finding faults or simpler approaches to the famous development of optics by René Descartes (the founder of analytic geometry). This sometimes put him at the ire of his subjects and caused him difficulties in getting his work accepted, but in most cases, Fermat's assertions later proved to be correct. In one of Fermat's notebooks he wrote that the equation $a^n + b^n = c^n$ can never have any positive integer solutions a, b, c, whenever n is an integer greater than 2 (when $n = 2$, this equation has many such integer solutions, such as $a, b, c = 3, 4, 5$). He wrote further that he had a truly remarkable fact of this result, which the margin of the notebook was too small to contain. This result, known as Fermat's last theorem, had captivated number theorists for three centuries. It was not until over three centuries after Fermat's death when a proof of this result was finally discovered and published by a Princeton mathematician, Andrew Wiles. Wiles claimed to have worked over seven years in his attic on his proof, which had been a fascination of his since childhood. His proof led to his being awarded the *Field's Medal,* which is the most coveted prize in mathematics (often called the Nobel Prize of mathematics). Apart from his prestige in number theory, Fermat is also credited as being (with Blaise Pascal) one of the founders of the subject of probability.

Figure 8.1 Pierre de Fermat (1601–1665), French mathematician.

involving modular exponentiation with a prime base and perform the exponentiation in two ways: the first method reviews the fast exponentiation algorithm (Algorithm 6.5), while the second is a much more efficient method that will initially appear to be quite mysterious. We then clear up the mystery by presenting Fermat's little theorem.

Example 8.2

Compute $2^{1452} \bmod 19$.

Solution: As was mentioned in Chapter 6, a horribly inefficient way to do this would be to first compute 2^{1452} directly, using integer arithmetic, and then convert it to its representative, modulo 19, in the set $\{0, 1, 2, \cdots, 18\}$. The first number would be so large that it would overflow on many computer systems.[*] We will instead perform this computation with two much more efficient schemes; the first is simply the fast modular exponentiation method (Algorithm 6.5), but with an informal notation.

Method 1: Fast Modular Exponentiation. We begin with $2^2 \equiv 4 \,(\bmod\, 19)$, and continue to square both sides until the exponents exceed at least half of the desired exponent:

$$2^4 \equiv 4^2 \equiv 16$$

$$2^8 \equiv 16^2 \equiv 256 \equiv 9$$

$$2^{16} \equiv 9^2 \equiv 81 \equiv 5$$

$$2^{32} \equiv 6$$

---

[*] *Computing Note*: This should serve as a caution to students not to take for granted or to rely too much on the capabilities of computing platforms, depending on what type of system you are working on, floating point or symbolic. A floating point system only has accuracy to about 16 digits. For example, the integer 13^{20} has 23 digits, so its computation on a floating point system would only be accurate to the first 15 or so of these digits. Thus, if we took the remainder of this computation, say mod 29, we would probably get the wrong answer. Symbolic systems have much greater accuracy but usually work more slowly, but even these have limitations in the sizes of the numbers they can deal with.

$$2^{64} \equiv 17$$

$$2^{128} \equiv 4$$

$$2^{256} \equiv 16$$

$$2^{512} \equiv 9$$

$$2^{1024} \equiv 5$$

We will now be able to use the above powers to compute the desired power of 2 (mod 19). This is because of the binary expansion $1452 \sim [10110101100]$ (base 2), which is equivalent to $1452 = 1024 + 256 + 128 + 32 + 8 + 4$, as the reader can easily check. It follows that we may compute

$$2^{1452} = 2^{1024} \cdot 2^{256} \cdot 2^{128} \cdot 2^{32} \cdot 2^8 \cdot 2^4$$
$$\equiv 5 \cdot 16 \cdot 4 \cdot 6 \cdot 9 \cdot 16 \equiv 11 (\mathrm{mod}\, 19).$$

Method 2: At first glance, this method will appear to use a lucky trick. But we will soon give Fermat's little theorem, which will show that this trick can be easily replicated in general situations. We note that $2^{18} \equiv 1(\mathrm{mod}\, 19)$. If we apply the division algorithm to the integer division of 1452 by 18, we obtain $1452 = 80 \cdot 18 + 12$. It follows that $2^{1452} \equiv (2^{18})^{80} \cdot 2^{12} \equiv 1^{80} \cdot 11 \equiv 11(\mathrm{mod}\, 19)$. This was quite a bit less work than Method 1.

In general, the same trick used in the second method of the above example can be used to compute any power a^e (mod m), provided that we can find a special exponent s (less than m) such that $a^s \equiv 1$ (mod m). We will show that such an exponent always exists (the same will work for any a), as long as a and m are relatively prime, and show how to find it. We first deal with the case in which the modulus m is prime (as in the above example when m was 19). The following classical theorem of Pierre de Fermat shows that such a "magical" exponent is very easily found for any prime modulus.

Theorem 8.2: Fermat's Little Theorem

Suppose that p is a prime and a is an integer that is not a multiple of p, then $a^{p-1} \equiv 1$ (mod p).

Proof: For greater clarity in this proof, we will denote elements of $\mathbb{Z}_p$ (the integers mod p) using square brackets: $[k]$ represents the set of all integers that are congruent to k (mod p). We will construct a function with domain and codomain both being the set of nonzero integers mod p: $A = \{[1], [2], \cdots, [p-1]\}$, $f : A \to A$, defined by $f([x]) = [ax]$. By Proposition 2.10, this definition gives the same output no matter which representative we use of $[x]$, so it is a well-defined function on elements of A. But we still need to check that the images will never be $[0]$ (that is, so the codomain of the function can be taken to be A). Indeed, if $[ax] = [0]$, this would mean that $ax \equiv 0(\mathrm{mod}\, p)$, so that $p \mid ax$. But then Euclid's lemma (Proposition 2.7) would imply that either $p \mid a$ or $p \mid x$. Both of these options are not possible since $[a] \neq [0]$ and since $[x] \neq [0]$.

Next we will show that f is one-to-one. Suppose that $f([x]) = f([y])$. This means that $[ax] = [ay]$ or $ax \equiv ay \pmod{p}$. By definition, this means that $p \mid (ax - ay)$ or $p \mid a(x - y)$. Euclid's lemma (Proposition 2.7) then tells us that either $p \mid a$ or $p \mid (x - y)$. Since we know the former is false, the latter must hold, which means that $x \equiv y \pmod{p}$ or $[x] = [y]$, so f is one-to-one.

Since f is a one-to-one function of the set A to itself, it follows that the images of f: $f([1]), f([2]), \cdots, f([p-1]) = [a], [2a], \cdots, [(p-1)a]$ are simply a relisting of the elements of A: $[1], [2], \cdots, [p-1]$, in perhaps a different order. It follows that if we multiply representatives from these to listings of the nonzero elements of $\mathbb{Z}_p$, we will get the same result (mod p):

$$1 \cdot 2 \cdot \ldots \cdot (p-1) \equiv a \cdot 2a \cdot \ldots \cdot (p-1)a \equiv a^{p-1}(1 \cdot 2 \cdot \ldots \cdot (p-1)) \pmod{p}$$

This equation implies that $p \mid [a^{p-1}(1 \cdot 2 \cdot \cdots \cdot (p-1)) - 1 \cdot 2 \cdot \cdots \cdot (p-1)]$, or $p \mid [(a^{p-1} - 1)(1 \cdot 2 \cdot \cdots \cdot (p-1))]$, and Euclid's lemma tells us that p must divide one of the factors on the right. The only possibility is that $p \mid (a^{p-1} - 1)$, so that $a^{p-1} \equiv 1 \pmod{p}$, as we wished to prove. $\square$

In light of Example 8.1, it is now easy to see how Fermat's little theorem can help us to easily raise any integer a to any power e modulo a prime p; we simply make use of the "magic" exponent $p - 1$ and use the division algorithm to write $e = q(p-1) + r$, where $0 \leq r < p - 1$. It then follows that $a^e \equiv a^r \pmod{p}$.

Exercise for the Reader 8.2

Compute $18^{802} \pmod{29}$, using each of the two methods shown in Example 8.2. (In Method 2, use Fermat's little theorem.)

The Euler Phi Function

Our next theorem, which is due to Leonhard Euler,[*] will generalize Fermat's little theorem to work for any modulus m. The analogue of the "magic" exponent p is determined by the following very useful integer function:

---

[*] Leonhard Euler (pronounced "Oiler") entered into mathematics during one of its most exciting eras; calculus had recently been invented and the field was transforming with numerous consequences and applications. Euler's life was nothing short of phenomenal. Educated in Switzerland, he was first appointed as a professor at age 19 at the renowned St. Petersburg University in Russia, and six years later he was appointed to the Berlin Academy and became its leader. His published works were significant and touched on practically all of the fields of mathematics. He was the most prolific mathematician ever, even during the last 17 years of his life when he was completely blind (in fact, this was perhaps his most productive period). His papers were assembled into a collected works compendium that filled over 100 encyclopedia-sized tomes! His mental skills remained remarkably acute throughout his life. At age 70, for example, he could recite an entire novel, as well as stating the first and last sentences on each page, and he once settled an argument between two students whose answers differed in the 15th decimal place, by a fast computation in his head. Euler had 13 children, and he told stories about having made some of his most seminal mathematical discoveries as he was holding one child on his lap while others were playing at his feet.

Figure 8.2 Leonhard Euler (1707–1783), Swiss mathematician.

Definition 8.1

Euler's phi function is a function defined on the set of positive integers: $\phi: \mathbb{Z}_+ \to \mathbb{Z}_+$, by $\phi(n) =$ the number of integers in the set $\{1, 2, \cdots, n\}$ that are relatively prime to n.

In the case $n = p$ is a prime number, then each of $1, 2, \ldots, p-1$, is relatively prime to p, so that $\phi(p) = p - 1$. The numbers in the set $\{1, 2, \cdots, 10\}$ that are relatively prime to 10 are 1, 3, 7, 9, so $\phi(10) = 4$. The following result makes it easy to compute $\phi(n)$ for any positive integer n, provided that we have the prime factorization of n.

Proposition 8.3

If $n = p_1^{\alpha_1} p_2^{\alpha_2} \cdots p_n^{\alpha_n}$, where $p_1, p_2, \cdots, p_n$ are distinct primes, and $\alpha_1, \alpha_2, \cdots, \alpha_n \in \mathbb{Z}_+$ then

$$\phi(n) = (p_1 - 1) p_1^{\alpha_1 - 1} (p_2 - 1) p_2^{\alpha_2 - 1} \cdots (p_n - 1) p_n^{\alpha_n - 1} \qquad (8.1)$$

For example, since the prime factorization of 10 is $2 \cdot 5$, Proposition 8.3 tells us that $\phi(10) = (2 - 1) \cdot 2^0 \cdot (5 - 1) \cdot 5^0 = 4$, as we had shown earlier. For another example, since the prime factorization of 378 is $3^3 \cdot 7^2$, the proposition tells us that $\phi(378) = \phi(3^3 \cdot 7^2) = (3 - 1) \cdot 3^2 \cdot (7 - 1) \cdot 7^1 = 108$. A proof of Proposition 8.3 is outlined in the exercises; see Chapter Exercises 41–43.

Exercise for the Reader 8.3

Compute the following values of Euler's phi function: $\phi(15)$, $\phi(20)$, $\phi(208)$, $\phi(2208)$.

Euler's Theorem

Notice that when $n = p$ is prime, the formula in Proposition 8.3 gives $\phi(p) = p - 1$, which was the "magic" exponent in Fermat's little theorem. Euler generalized Fermat's little theorem to work for any modulus, with his phi function continuing to play the role of the "magic" exponent.

Theorem 8.4: Euler's Theorem

Suppose that a and m are relatively prime positive integers with $m > 1$, then $a^{\phi(m)} \equiv 1 \pmod{m}$.

Exercise for the Reader 8.4

Prove Euler's theorem.

Suggestion: Mimic the proof of Fermat's little theorem, but now take the set A to be the set of all positive integers less than m that are relatively prime to m.

Example 8.3

Make use of Euler's theorem to perform each of the following tasks:

(a) Compute $18^{2551} \pmod{25}$.

(b) Find the last (one's) digit of the integer 13^{2017}.

Note: The integer in part (b) has 2246 digits.

Solution: Part (a): Since $\gcd(18, 25) = 1$ and $\phi(25) = \phi(5^2) = (5 - 1) \cdot 5^1 = 20$, Euler's theorem tells us that $18^{20} \equiv 1 \pmod{25}$. Using the division algorithm for the integer division of 2551 by 20 gives $2551 = 127 \cdot 20 + 11$, and consequently $18^{2551} \equiv (18^{20})^{127} \cdot 18^{11} \equiv 1 \cdot 7 \equiv 7 \pmod{25}$.

Part (b): Finding the one's digit of any number is the same as the answer we would get by converting it to an integer modulo 10. Thus, we wish to find $13^{2017} \pmod{10}$. Since $\gcd(13, 10) = 1$ and $\phi(10) = 4$, Euler's theorem tells us that $13^4 \equiv 1 \pmod{10}$. Applying the division algorithm to the given exponent divided by 4 gives $2017 = 504 \cdot 4 + 1$, hence $13^{2017} \equiv (13^4)^{504} \cdot 13^1 \equiv 1 \cdot 13 \equiv 3 \pmod{10}$.

Exercise for the Reader 8.5

(a) Compute $7^{8486} \pmod{58}$.

(b) Find the last three digits of the integer 13^{2017}.

Despite the speed and apparent magic with which Euler's theorem helps us to perform such modular exponentiations, there is one serious drawback: Evaluating $\phi(n)$ requires the prime factorization of n. As we had mentioned, this is a very hard problem for which no efficient algorithm exists (and according to the general consensus) never will. Consequently, the fast modular exponentiation algorithm generally works more quickly. Modular exponentiation has many applications; for example, it is at the core of the so-called ElGamal cryptosystem that will be studied in Chapter 9.

Two intimately related number theoretic concepts that will be useful for an assortment of public key cryptosystems are *orders* and *primitive roots*.

Modular Orders of Invertible Modular Integers

Definition 8.2

For integers $1 \le a < n$, with a and n relatively prime, we define the **order of a relative to n** (or **order of a mod n**) to be the smallest positive exponent k for which $a^k \equiv 1 \pmod{n}$. We write this as $k = \operatorname{ord}_n(a)$.[*]

The following example illustrates the different orders of elements relative to two small moduli. Note that from Euler's theorem, we know that $\operatorname{ord}_n(a) \le \phi(n)$.

Example 8.4

(a) Compute the orders mod n of all positive integers less than (and relatively prime to) $n = 7$.
(b) Do the same for $n = 8$.

Solution: Table 8.1 and Table 8.2 illustrate all of the powers [up to the $\phi(n)$th].

TABLE 8.1 Powers (mod 7) of Integers Relatively Prime to $n = 7$

	$a^k \pmod{7}$					
	$k = 1$	$k = 2$	$k = 3$	$k = 4$	$k = 5$	$k = 6$
$a = 1$	1	1	1	1	1	1
$a = 2$	2	4	1	2	4	1
$a = 3$	3	2	6	4	5	1
$a = 4$	4	2	1	4	2	1
$a = 5$	5	4	6	2	3	1
$a = 6$	6	1	6	1	6	1

Note: In each row, the order is at the top of the column containing the highlighted box.

---

[*] This notation for orders is due to C. F. Gauss.

TABLE 8.2 Powers (mod 8) of Integers Relatively Prime to $n = 8$

	$a^k \pmod 8$			
	$k = 1$	$k = 2$	$k = 3$	$k = 4$
$a = 1$	1	1	1	1
$a = 3$	3	1	3	1
$a = 5$	5	1	5	1
$a = 7$	7	1	7	1

Note: In each row, the order is at the top of the column containing the highlighted box.

The next proposition collects some useful facts about orders and modular exponentiation.

> ### Proposition 8.5
>
> Suppose that a and $n > 1$ are relatively prime positive integers.
>
> (a) If k is a positive integer with $a^k \equiv 1 \pmod n$, then $\mathrm{ord}_n(a) \mid k$.
> (b) $\mathrm{ord}_n(a) \mid \phi(n)$.
> (c) If i and j are nonnegative integers, then $a^i \equiv a^j \pmod n$ if, and only if, $i \equiv j \pmod{\mathrm{ord}_n(a)}$.

Proof: Part (a): Use the division algorithm to write $k = q \cdot \mathrm{ord}_n(a) + r$, where q and r are integers with $0 \le r < \mathrm{ord}_n(a)$. By the assumption that $a^k \equiv 1 \pmod n$, and then using the definition of order, we may write

$$1 \equiv a^k \equiv a^{q \cdot \mathrm{ord}_n(a) + r} \equiv (a^{\mathrm{ord}_n(a)})^q a^r \equiv 1^q a^r \equiv a^r \pmod n$$

So we have found that $a^r \equiv 1 \pmod n$. But since $\mathrm{ord}_n(a)$ is the smallest positive exponent with this property, we must have $r = 0$, which means that $\mathrm{ord}_n(a) \mid k$.

Part (b): By Euler's theorem we know that $a^{\phi(m)} \equiv 1 \pmod m$. Thus, part (b) follows from part (a) with $k = \phi(n)$.

Part (c): We first assume that $a^i \equiv a^j \pmod n$ and (switching the roles of i and j if necessary) that $i \ge j$. Since $\gcd(a, n) = 1$, we know from Proposition 2.11 that the multiplicative inverse of a, a^{-1}, exists mod n. If we multiply both sides of the congruence $a^i \equiv a^j \pmod n$ by $(a^{-1})^j$, we obtain that $a^{i-j} \equiv a^{j-j} \equiv 1 \pmod n$, so the result of part (a) (with $k = i - j$) tells us that $\mathrm{ord}_n(a) \mid i - j$; that is, $i \equiv j \pmod{\mathrm{ord}_n(a)}$. Conversely, if $i \equiv j \pmod{\mathrm{ord}_n(a)}$, then, assuming as before that $i \ge j$, we may write $i - j = q \cdot \mathrm{ord}_n(a)$, or $i = j + q \cdot \mathrm{ord}_n(a)$, for some nonnegative integer q. From this equation, we obtain

$$a^i \equiv a^{q \cdot \mathrm{ord}_n(a) + j} \equiv (a^{\mathrm{ord}_n(a)})^q a^j \equiv 1^q a^j \equiv a^j \pmod n \quad \square$$

Primitive Roots

Note that with regard to Table 8.1, Euler's theorem tells us [since $\phi(7) = 6$] that $a^6 \equiv 1 \pmod 7$ for each a relatively prime to 7, while in the context of Table 8.2, it says that $a^4 \equiv 1 \pmod 8$, for each a relatively prime to 8 [since $\phi(8) = 4$]. But the tables are more revealing. Table 8.1 shows that only two elements have (maximum) order equal to Euler's exponent 6, while Table 8.2 shows that none of the elements have order equal to Euler's exponent 4. This motivates the following definition.

Definition 8.3

We say that a positive integer g that is less than n and relatively prime to n is a **primitive root** mod n if the order of g is $\phi(n)$.

From Tables 8.1 and 8.2, we see that there are two primitive roots mod 7, namely 3 and 5, but that there are no primitive roots mod 8. Notice also that the powers of the primitive roots cycle through all of the $\phi(n)$ positive integers that are less than and relatively prime to n. The following proposition shows that this is true in general.

Proposition 8.6

If g is a primitive root modulo a positive integer n, then the powers $g, g^2, g^3, \cdots, g^{\phi(n)}$ are all different (mod n), and hence this list of powers is congruent to the set of all positive integers less than n that are relatively prime to n.

Proof: If there were ever a repeated term in this sequence of powers $g, g^2, g^3, \cdots, g^{\phi(n)}$, say $g^i \equiv g^j \pmod{n}$, then part (c) of Proposition 8.5 would tell us that $i \equiv j \pmod{\mathrm{ord}_n(g)}$. Since g is a primitive root, $\mathrm{ord}_n(g) = \phi(n)$, this congruence is impossible if i and j are different integers in the range $\{1, 2, \cdots, \phi(n)\}$. It follows that all of the modular powers $g, g^2, g^3, \cdots, g^{\phi(n)}$ are distinct mod n. $\square$

Exercise for the Reader 8.6

(a)   As in the solution of Example 8.4, construct tables for the modular powers of all positive integers less than (and relatively prime) to $n = 4$. Use the table to find all primitive roots.

(b)   Repeat the instruction of part (a) for $n = 9$.

Since some major public key cryptosystems (that will be studied in the next chapter) depend on primitive roots and orders, we would like to now address the following natural questions:

1. *Existence of Primitive Roots.* For which moduli n do primitive roots exist, and when they exist, how many will there be?
2. *Finding Primitive Roots.* When primitive roots exist mod n, how can we find them?
3. *Computing Orders.* How can we compute the order $\mathrm{ord}_n(a)$ of an element a that is relatively prime to n?
4. *The Discrete Logarithm Problem.* If g is a primitive root mod n, and a is an integer relatively prime to n, how can we find the exponent j [unique mod $\phi(n)$ by Proposition 8.6] such that $g^j \equiv a \pmod{n}$?

We are able to give some reasonable answers to the first three of these questions. The discrete logarithm problem of the fourth question is very difficult, and no known efficient algorithms for it are known. Like the prime factorization problem, its difficulty is the basis for the security of some widely used public key systems (the ElGamal cryptosystem and digital signature scheme, and the Diffie–Hellman key exchange protocol) that will be introduced in the next chapter.

Existence of Primitive Roots

The following theorem nicely and completely answers the first question.

Theorem 8.7: Existence and Number of Primitive Roots

Suppose that n is a positive integer greater than 1. Primitive roots exist mod n if, and only if, n is of the form 2, 4, p^s or $2p^s$, where p is an odd prime and t is a positive integer. In this case, there are exactly $\phi(\phi(n))$ primitive roots mod n.

Theorem 8.7 can be proved in tandem with our next theorem on the determination of primitive roots. The proofs are rather lengthy, although not particularly difficult, so we will omit them. Interested readers may refer to any good book on number theory, such as [Ros-05].

Example 8.5

For which of the following moduli n do primitive roots exist? In cases where primitive roots exist, how many will there be (mod n)?

(a) $n = 20$
(b) $n = 59$
(c) $n = 30$
(d) $n = 1250$

Solution: The prime factorizations of these integers are $20 = 2^2 \cdot 5$, 59 (prime), $30 = 2 \cdot 3 \cdot 5$, and $1250 = 2 \cdot 5^4$. Only those of parts (b) and (d) satisfy the conditions of Theorem 8.7 and thus have primitive roots. The corresponding number of primitive roots are given by Euler's phi function (by Theorem 8.7), so that (using Proposition 8.3) 59 has $\phi(\phi(59)) = \phi(58) = \phi(2 \cdot 29) = 28$ primitive roots, and 1250 has $\phi(\phi(1250)) = \phi(\phi(2 \cdot 5^4)) = \phi(500) = \phi(2^2 \cdot 5^3) = 200$ primitive roots.

Determination of Primitive Roots

There is unfortunately no simple algorithm for determining a primitive root modulo an integer for which they are known to exist. A brute-force search of checking orders of numbers relatively prime to n can be used, but that is slow. [It can be significantly speeded up using Proposition 8.5(b), however, if the prime factorization of n is known, since the only possible orders are divisors of $\phi(n)$.]

Gauss developed a more efficient implementation of this brute-force approach for determining primitive roots modulo a prime. We will discuss Gauss's algorithm shortly; his algorithm still seems to be the best general tool for this task. We first present a theorem that reduces the problem of finding primitive roots (when they exist) to the determination of primitive roots modulo a prime.

Theorem 8.8: Determination of Primitive Roots

In each of the cases for a modulus n of Theorem 8.7 in which primitive roots exist, the following rules show how primitive roots can be found:

(a) If g is a primitive root modulo an odd prime power p^k, then g or $g + p^k$ (whichever is odd) will be a primitive root mod $2p^k$.

(b) If g is a primitive root modulo an odd prime p, then g or $g + p$ (whichever is odd) will be a primitive root mod p^2.

(c) If p is an odd prime and g is a primitive root mod p^2, then g will also be a primitive root mod any higher power p^k of p.

Before we state Gauss's algorithm for computing primitive roots modulo a prime, we will give a principle that is sometimes useful in computing orders of integers in a certain modulus. The following result gives a formula for the order of any modular power of an integer in terms of the order of the integer (mod m).

Order of Powers Formula

Proposition 8.9: Order of Powers Formula

If a, j, and $n > 1$ are positive integers, with a relatively prime to n, then

$$\mathrm{ord}_n(a^j) = \frac{\mathrm{ord}_n(a)}{\gcd(j, \mathrm{ord}_n(a))} \tag{8.2}$$

We point out one immediate consequence of this proposition shows that once we have one primitive root mod n, we can easily get them all.

Corollary 8.10

If g is a primitive root (mod n), then the other primitive roots (mod n) are the same as the modular powers g^j (mod n), as j runs through all of the positive integers less than and relatively prime to $\phi(n) = \mathrm{ord}_n(g)$.

The corollary easily follows from the proposition since with $a = g$, a primitive root used in Equation 8.2, the right side will equal $\mathrm{ord}_n(a)$ exactly when the denominator $\gcd(j, \mathrm{ord}_n(a))$ is 1. To illustrate, in Example 8.4, we found that $g = 3$ is a primitive root of 7. By the above corollary, all the primitive roots (mod 7) will be given by the modular powers g^j whose exponents j are relatively prime to $\phi(7) = 6$, which are $3^1 \equiv 3, 3^5 \equiv 5$. (This is confirmed by the rest of Table 8.1.)

Proof of Proposition 8.9: Let $m = \text{ord}_n(a)$. From Proposition 8.5(a), it follows that $a^{j\ell} \equiv 1 \pmod{n}$ if, and only if, $j\ell$ is a multiple of m. But by definition of order, ℓ is the smallest positive integer such that $a^{j\ell} \equiv 1 \pmod{n}$. It thus follows that $j \cdot \text{ord}_n(a^j) = \text{lcm}(j, m)$. But [see Exercise for the Reader 2.2(c)] $\text{lcm}(j, m) = \dfrac{jm}{\gcd(j, m)}$, which, when combined with the previous equation, gives

$$\cancel{j} \cdot \text{ord}_n(a^j) = \frac{\cancel{j}m}{\gcd(j, m)} \implies \text{ord}_n(a^j) = \frac{m}{\gcd(j, m)} = \frac{\text{ord}_n(a)}{\gcd(j, \text{ord}_n(a))} \quad \square$$

Example 8.6

If possible, do each of the following:

(a) Compute $\text{ord}_{59}(7)$.
(b) Find an integer between 1 and 59 whose mod 59 order is 22.
(c) Find a primitive root of 59.
(d) Find an exponent j such that $g^j \equiv 7 \pmod{59}$, where g is the primitive root that was found in part (c).

Solution: Part (a): By Proposition 8.5(b), the order of any integer (relatively prime to 59) must divide $\phi(59) = 58 = 2 \cdot 29$, so the only possible orders (apart from 1 for integers congruent to 1 mod 59) are 2, 29, and 58 (in which case we have a primitive root). So to check $\text{ord}_{59}(7)$, we need only compute at most two modular powers $7^2, 7^{29}$; if one of these is congruent to 1, the exponent will be the $\text{ord}_{59}(7)$; otherwise, this order will be 58 (and 7 would be a primitive root). Indeed, $7^2 \equiv 49$, and using fast modular exponentiation (Algorithm 6.5), $7^{29} \equiv 1 \pmod{59}$, so $\text{ord}_{59}(7) = 29$.

Part (b): Since $22 \nmid 58 = \phi(59)$, we know from Proposition 8.5(a) that 22 cannot be an order mod 59.

Part (c): By Theorem 8.7, since 59 is prime, it will have $\phi(\phi(59)) = \phi(58) = 28$ primitive roots. This is nearly half of the nonzero mod 59 integers. We could do a simple brute-force search starting with $a = 2$. We would compute the powers a^2, a^{29} if neither is congruent to 1, then (as explained above) a would be a primitive root. If $a = 2$ fails to be a primitive root, we increase a by 1: $a \to a + 1 = 3$. We continue this process until we find a primitive root. Since $2^2 \equiv 4$, and (using Algorithm 6.5) $2^{29} \equiv 58 \pmod{29}$, we can stop in our first attempt and declare 2 as a primitive root.

Part (d): This is a discrete logarithm question. As mentioned earlier, there are no efficient algorithms to do this. By Proposition 8.9, since $g = 2$ is a primitive root, we have

$$\operatorname{ord}_{59}(2^j) = \frac{\operatorname{ord}_2(59)}{\gcd(j, \operatorname{ord}_2(59))} = \frac{58}{\gcd(j, 58)}$$

Since we know from part (a) that $\operatorname{ord}_{59}(7) = 29$, the above equation tells us that in order to have $2^j \equiv 7 \pmod{59}$, j must be even. The brute-force approach would be to continue computing modular even powers of 2: 2^2, 2^4, 2^6, $\cdots$ (mod 59) until we obtain 7. Since each term is just four times the previous term (mod 59), the numbers are small enough for the computation to be feasible:

$$2^2 \equiv 4,\ 2^4 \equiv 16,\ 2^6 \equiv 5,\ 2^8 \equiv 20,\ 2^{10} \equiv 21,$$

$$2^{12} \equiv 25,\ 2^{14} \equiv 41,\ 2^{16} \equiv 46,\ 2^{18} \equiv 7$$

Thus we have found the desired discrete logarithm to be $j = 18$.

Exercise for the Reader 8.7

(a)   How many primitive roots does $n = 334$ have?
(b)   What is the smallest primitive root?
(c)   How many integers mod 334 have order equal to 2?

Algorithm 8.1: Gauss's Algorithm for Finding a Primitive Root Modulo a Prime p

The input will be an odd prime p, and output will be a primitive root g (mod p).

Step 1. Initialize the set of noncandidates for a primitive root mod p: $N = \varnothing$. (So initially the candidates can be any integer in the range $1 < a < p$.) Select a candidate a (preferably randomly), and successively compute the powers of a (mod p) until we first obtain 1: a^2, a^3, $\cdots$, $a^\ell \equiv 1$ [thus $\operatorname{ord}_n(a) = \ell$]. If $\ell = p - 1$, then a is a primitive root, so output $g = a$, and exit the algorithm. Otherwise go to Step 2.

Step 2. Update $N \to N \cup \{a, a^2, \cdots, a^{\ell-1}\}$. Select a modular integer candidate outside of N: b (preferably randomly), and compute $m = \operatorname{ord}_n(b)$. If $m = p - 1$, then b is a primitive root, so output $g = b$, and exit the algorithm. Otherwise, let $u = \operatorname{lcm}(m, \ell)$, and write $u = xy$, where $x \mid m$, $y \mid \ell$, and $\gcd(x, y) = 1$. Compute the modular powers $a' \triangleq a^{\ell/y}, b' \triangleq b^{m/x}$ (mod p), and set $G = a'b'$ (mod p). G has order $xy = u$. If $u = p - 1$, then G is a primitive root, so output $g = G$, and exit the algorithm. Otherwise go to Step 3.

Step 3. Return to Step 2 with the following updates: $\ell \to u, a \to G$.

This algorithm will always terminate since in each application of Step 2, it will always be the case that $u > \ell$ (see Chapter Exercise 44 for an outline of the proof of this fact) so that the orders of the elements found continue to increase. Let us illustrate with a small example. Of course, for primes greater than 200 or so, this algorithm should be implemented on a computing platform.

Example 8.7

Use Gauss's algorithm (Algorithm 8.1) to find a primitive root for the prime $p = 101$.

Solution: Let us begin Step 1 by choosing $a = 36$. Computing successive powers mod 101, we have $a^i \equiv 36, 84, 95, 87, 1$, so we have $\ell = \mathrm{ord}_n(a) = 5$. Since a is not a primitive root, we move on to Step 2: We select a (nonzero) modular integer b different from the powers of a, say $b = 88$. We compute $m = \mathrm{ord}_n(b) = 25$. Since $u = lcm(m, \ell)$, we can just take $G = m$. But since G is still not a primitive root, Step 3 tells us to go back again to Step 2, now with $a = 88$, and ℓ taken to be its order 25. We need to select b to be a nonzero modular integer different from the powers of a: $a^i \equiv 88, 68, 25, 79, 84, 19, 56, 80, 71, 87, 81, 58, 54, 5, 36, 37, 24, 92, 16, 95, 78, 97, 52, 31, 1$. We take $b = 51$ and compute $m = \mathrm{ord}_n(b) = 100$. We have thus found a primitive root, and output $G = 51$.

Prime Number Generation

The most fundamental need in public key cryptography is the generation of large prime numbers that can serve to create new and secure keys. Here is a typical outline of a scheme for constructing a prime of a prescribed size: Say that we need a random prime of size 1000 bits. We would set the first and last bits equal to 1 (the last bit being set to one makes the number odd), and then randomly assign the remaining 998 bits. A 1000-bit number will have size approximately $\log_{10}(2^{1000}) \approx 10^{300}$, and so by the prime number theorem (Theorem 8.1), the density of primes among integers of such size is approximately $1/\ln(10^{300}) \approx 1/691$. Since we are only checking odd integers, this doubles the density of primes to be approximately $1/345$. Thus, on average, it will take about 345 randomly generated odd 1000-bit numbers before we hit on a prime. On each attempt, we use a *probabilistic primality test* that will screen out probable primes. The probability that the number produced is not prime can be prescribed to be as small as we would like with such a test. Although these probabilistic probability tests cannot guarantee a prime is produced, some will allow the user to prescribe the probability that a prime is not produced to be as small as is needed for most practical purposes, say $1/10^{15}$. If it is absolutely necessary to confirm that the probable prime is actually prime, there are more expensive prime certification tests that can be applied. The analogy is similar to the public health problem of screening people for certain chronic illnesses. Usually a quick inexpensive test is given to the mass population, and for those who test positive, a more accurate (and expensive and slower) test is given to see if they really are infected. We will focus on only the first type of probabilistic primality tests. Readers interested in learning about the prime certification tests may refer to one of the following books on algorithmic and computational number theory: [BaSh-96], [Coh-93]; see also [BrWa-99].

Initially it may seem hard to believe, but the problem of checking whether a positive integer is prime is much easier than the problem of factoring a positive integer. In fact, as mentioned in Chapter 1, it is strongly believed among number theorists that an efficient algorithm for factoring integers will never exist. Checking primality, on the other hand, can be done with deterministic algorithms that run in polynomial time.[*]

Fermat's Primality Test

One basic reason that primality checks can be done faster than factoring is that integers can be proved to be composite without actually producing a factorization. This phenomenon will be seen in our first simple primality test, which is based on the contrapositive of Fermat's little theorem (Theorem 8.2): If p is a prime and $1 < a < p - 1$, then $a^{p-1} \equiv 1 \pmod{p}$.[†] The contrapositive[‡] can be formulated as:

> **Contrapositive of Fermat's Little Theorem**
>
> If n is a positive integer and $a^{n-1} \not\equiv 1 \pmod{n}$, for some number a, $1 < a < n - 1$, then n is composite.

As a simple example, we can compute (with the fast exponentiation Algorithm 6.5) that $2^{1002} \equiv 990 \pmod{1003}$, so by the contrapositive of Fermat's little theorem (with $a = 2$), this proves that 1003 is composite. This is a nice example to demonstrate how it is possible to determine that an integer is composite without actually factoring it. Since it is not clear how to choose an appropriate a to attempt to use this criterion to prove a given integer n is composite, it is best to simply make random choices and to apply the test a certain number k times. If one of these k trials results in $a^{n-1} \not\equiv 1 \pmod{n}$, then the test has proved n to be composite, and a corresponding base a for which $a^{n-1} \not\equiv 1 \pmod{n}$, is called a **witness** to the

---

[*] It has been well known by specialists since the time of Gauss that the primality check problem is easier to solve than the factoring problem and that a polynomial time deterministic algorithm for the former should exist. But it was not until 2002 when such an algorithm was discovered by a group of three computer scientists—Manindra Agrawal, Neeraj Kayal, and Nitin Saxena from the Indian Institute of Technology in Kangpur, India. The algorithm is quite simple and elegant and is named after the three or, more briefly, the *AKS test*; see [AgKaSa-04]. These three scientists have received several prestigious awards for this very important discovery that had eluded many great minds for a long period of time.

[†] We intentionally omitted the values $a = 1$ and $a = p - 1$, since, even if p were composite (but an odd number greater than 1), these values would always equal 1 when risen to the (even) power $p - 1$.

[‡] The *contrapositive* of any logical implication of the form "If P, then Q." is the statement "If not Q, then not P." Any implication is logically equivalent to its contrapositive.

fact that n is composite. If each of the k trials results in $a^{n-1} \equiv 1 \pmod{n}$, then n is declared as probably prime. Here is a formal summary of this randomized algorithm.

Algorithm 8.2: Randomized Fermat Primality Test

Inputs: An integer $n > 3$ and a positive integer k.
Output: Either a declaration that n is composite, along with a witness integer a that satisfies $a^{n-1} \not\equiv 1 \pmod{n}$ (and thus proves that n is composite), or a declaration that n is probably prime.

Step 1. Initialize iteration counter: $i = 0$:
Step 2. Randomly* choose an integer a, $1 < a < n-1$, compute (with Algorithm 6.5) $a^{n-1} \pmod{n}$, and update the iteration counter $i \rightarrow i+1$.
Step 3. If $a^{n-1} \not\equiv 1 \pmod{n}$, then declare n is composite, and a as a witness to this fact, and exit the algorithm. Otherwise, go back to Step 2, unless $i = k$, in which case we declare n is probably prime and exit the algorithm.

We point out that even though the above algorithm and the others that follow are called primality tests, they are not capable of proving that a number is prime; they can only prove compositeness. One drawback of the Fermat primality test is that it does not come with any guarantee on the probability that any declared probable prime really is prime. (Our next primality test will come with such a guarantee, however.)

Example 8.8

Apply the Fermat primality test (Algorithm 8.2) with $k = 4$ to the following odd integers n: (a) $n = 409$, (b) $n = 721$
Solution: Part (a): $n = 409$

Step 1. Initialize the trial counter $i = 0$.
Step 2. Randomly generate a base: $a = 238$. We use Algorithm 6.5 to compute the modular power $a^{n-1} \equiv 238^{408} \equiv 1 \pmod{409}$. So 409 has passed Fermat's primality test with this value of a; since i is now 1, we repeat Step 2.
Step 2. (Second repetition) Randomly generate a base: $a = 222$. We use Algorithm 6.5 to compute the modular power $a^{n-1} \equiv 222^{408} \equiv 1 \pmod{409}$. So 409 has passed Fermat's primality test with this value of a; since i is now 2, we repeat Step 2.
Step 2. (Third repetition) Randomly generate a base: $a = 356$. We use Algorithm 6.5 to compute the modular power $a^{n-1} \equiv 356^{408} \equiv 1 \pmod{409}$. So 409 has passed Fermat's primality test with this value of a; since i is now 3, we repeat Step 2.

---

* We remind the reader that the computer implementation material of Chapter 1 provides schemes for computer generations of random integers in specified ranges.

Step 2. (Fourth repetition) Randomly generate a base: $a = 109$. We use Algorithm 6.5 to compute the modular power $a^{n-1} \equiv 109^{408} \equiv 1 (\text{mod } 409)$. So 409 has passed Fermat's primality test with this value of a, and since i is now 4, this is the final iteration of Step 2.

Step 3. Declare 409 as probably prime. (The reader may check that 409 is indeed prime; the test worked.)

Part (b): $n = 721$

Step 1. Initialize the trial counter $i = 0$.

Step 2. Randomly generate a base: $a = 230$. We use Algorithm 6.5 to compute the modular power $a^{n-1} \equiv 230^{720} \equiv 484 (\text{mod } 721)$.

Step 3. Since $a^{n-1} \not\equiv 1 (\text{mod } n)$, Fermat's test has proved that $n = 721$ is composite with witness $a = 230$. The reader may check this by factoring $721 = 7 \cdot 103$.

Fermat's test worked very well in the above example. For part (b), compositeness was detected in just one (random) try. Indeed, a computer calculation can quickly verify that 684 out of the 718 possible bases in the range $1 < a < 721 - 1$ would have worked as witnesses to show that 721 is composite.

Exercise for the Reader 8.8

Apply the Fermat primality test (Algorithm 8.2) with $k = 4$ to the following odd integers n:

(a) $n = 2581$
(b) $n = 1889$

Carmichael Numbers

Usually if a number n is not prime, it will succumb to Fermat's primality test for almost all values of a. Although very rare, there are extreme exceptions for which Fermat's test will detect compositeness only if the base a that is selected is an actual factor of n. (Thus the Fermat test for detecting compositeness for one of these numbers would be about as effective as trying to factor the number by randomly guessing at factors.)

Definition 8.4

A composite number $n > 1$ is called a **Carmichael number** if $a^{n-1} \equiv 1 (\text{mod } n)$, for each integer a that is relatively prime to n.

These numbers are named after the American mathematician Robert D. Carmichael (1879–1967), who introduced these numbers and some of their interesting properties. Carmichael conjectured in 1912 that there were infinitely many Carmichael numbers, but this fact was not proved until 80 years later (see [AlGrPo-92]). The first three Carmichael numbers are 561, 1105, and 1729, and there are just 2163 Carmichael numbers that are less than 25 billion.

The Miller–Rabin Test

Apart from the existence of Carmichael numbers, another drawback of the Fermat primality test is that no matter how many iterations are used, it comes with no guaranteed confidence level of the probability that any probable primes produced will actually be prime. With a bit more work, a much more effective probabilistic primality test, known as the *Miller–Rabin primality test*,[*] can be developed that transcends both of these weaknesses. There will be a very quantitative performance guarantee, and there will be no analogue of Carmichael numbers for the Miller–Rabin test. As with Fermat's test, this one will depend on Fermat's little theorem, but it will hinge on the following two additional results, the second of which is a refined version of the contrapositive of Fermat's little theorem.

> ### Lemma 8.11: Square Roots of One mod p
>
> If p is an odd prime, then $\sqrt{1} \equiv \pm 1 \pmod{p}$; that is, modulo an odd prime, 1 has exactly two square roots, namely ± 1.

Proof: Modulo any integer n, $(\pm 1)^2 \equiv 1 \pmod{n}$, so ± 1 are always (modular) square roots of 1. We need to show that modulo an odd prime p, there are no others. Indeed, if x is a square root of 1 (mod p), then $x^2 \equiv 1 \Rightarrow x^2 - 1 \equiv 0 \Rightarrow (x+1)(x-1) \equiv 0 \pmod{p} \Rightarrow p \mid (x+1)(x-1)$. By Euclid's lemma (Lemma 2.7) this, in turn, implies that either $p \mid (x+1)$, or $p \mid (x-1)$; that is, $x \equiv -1$ or $x \equiv 1 \pmod{p}$, as asserted. $\square$

The test derives from the contrapositive of the following result, just as the Fermat primality test came from the contrapositive of Fermat's little theorem.

> ### Proposition 8.12
>
> Suppose that p is an odd prime and $1 < a < p - 1$. Write $p - 1 = 2^f m$, where m is an odd integer. Then either $a^m \equiv 1 \pmod{p}$ or $a^{2^j m} \equiv -1 \pmod{p}$, for some j, $0 \le j < f$.

---

[*] The nonprobabilistic version of this test was discovered by Gary L. Miller in 1976 [Mil-76]. It came with an effectiveness guarantee, but the drawback was that this guarantee rested upon an unproved conjecture in number theory (the generalized Riemann hypothesis). In 1980, Michael O. Rabin ([Rab-80]) converted this algorithm into a probabilistic one that (importantly) came with the following unconditional guarantee: If the test finds a number to be composite, it is guaranteed to be composite; if the test finds the number to be "probably prime," it has at least a 75% chance of being prime. This guarantee can be improved to attain as high a percentage as one wishes, simply by (independently) iterating the test. Moreover, it has been found that the Miller–Rabin algorithm does much better, on average, than the conservative 75% guarantee level; see [DaLaPo-93].

Proof: Fermat's little theorem (Theorem 8.2) tells us that $a^{p-1} \equiv 1 \pmod{p}$. Since $a^{(p-1)/2} \equiv a^{2^{f-1}m}$ is a square root of $a^{p-1} \pmod{p}$, Lemma 8.11 allows us to conclude that $a^{2^{f-1}m} \equiv \pm 1 \pmod{p}$. If $a^{2^{f-1}m} \equiv -1 \pmod{p}$, then the second assertion holds; otherwise, we can continue to take square roots in this fashion, either verifying the second assertion of the proposition, or else getting all the way to $a^m \equiv 1 \pmod{p}$, which is the first assertion of the proposition. □

The contrapositive of Proposition 8.12 can be formulated as follows:

Contrapositive of Proposition 8.12

If $n > 1$ is an odd integer, with $n - 1 = 2^f m$, where m is an odd integer, and we can find an integer a, with $1 < a < n - 1$ such that $a^m \not\equiv 1 \pmod{n}$ and $a^{2^j m} \not\equiv -1 \pmod{n}$, for all j, $0 \le j < f$, then n must be composite.

As with the Fermat test, there does not seem to be a good (deterministic) method for choosing such an a to prove compositeness, so a random choice is most effective. The following implementation of the Miller–Rabin test is set up to minimize its complexity. It relies on the simple fact that if $a^{2^j m} \equiv 1 \pmod{n}$ for some nonnegative integer j, then $a^{2^\ell m} \equiv 1 \pmod{n}$, for each $\ell \ge j$.

Algorithm 8.3: The Miller–Rabin Primality Test

Inputs: An odd integer $n > 3$, suspected to be prime, and a positive integer k.
Output: Either a declaration that n is composite, along with a witness integer a that violates the contrapositive of Proposition 8.12 (and thus proves that n is composite), or a declaration that n is probably prime.

Step 1. First express $n - 1$ as $2^f m$, where m is an odd integer. Initialize the main iteration counter: $i = 0$.

Step 2. Randomly choose an integer a, $1 < a < n - 1$. Calculate (with Algorithm 6.5) $A_0 \equiv a^m \pmod{n}$. If $A_0 \equiv \pm 1 \pmod{n}$, update main iteration counter $i \to i + 1$, and move on to Step 3, otherwise enter into the following FOR loop:

FOR $j = 1$ TO $f - 2$

Compute $A_j \equiv A_{j-1}^2 \pmod{n}$ <this is $a^{2^j m} \pmod{n}$>

IF $A_j \equiv 1 \pmod{n}$
   Declare n as composite, output the witness a, and EXIT algorithm.

ELSE IF $A_j \equiv -1 \pmod{n}$

   Update main iteration counter $i \to i + 1$, and move on to Step 3.
END <IF>
END <FOR>

Compute $A_{f-1} \equiv A_{f-2}^2 \pmod{n}$ <this is $a^{(n-1)/2} \pmod{n}$>

IF $A_{f-1} \not\equiv -1 \pmod{n}$
    Declare n as composite, output the witness a, and EXIT program.
ELSE Update the main iteration counter $i \to i + 1$.
END <IF>

Step 3.   If i equals k, declare n is probably prime, and exit the algorithm. Otherwise, go back to Step 2.

An integer a that proves n is composite in the Miller–Rabin primality test is called a **witness** for the compositeness of n. Observe that if the Miller–Rabin test declares n to be probably prime by using a certain integer a, then $a^{2^j m} \equiv \pm 1 \pmod{n}$ for some j, $0 \le j < k$, and from this it follows (by repeated squaring) that $a^{n-1} \equiv a^{2^k m} \equiv 1 \pmod{n}$, so that the Fermat test would have also declared n to be probably prime. Thus the Miller–Rabin test is at least as effective as the Fermat test; in fact, it is more effective—the Computer Implementation and Exercises material at the end of this chapter will examine this phenomenon. But what is even more important and useful is that the Miller–Rabin algorithm comes with the following performance guarantee.

Theorem 8.13: Performance Guarantee for the Miller–Rabin Primality Test

If $n \ge 3$ is an odd composite number, then at most $(n-1)/4$ of the numbers in the set $\{1, 2, \cdots, n-1\}$ that are relatively prime to n will not serve as witnesses to the compositeness of n in the Miller–Rabin primality test. Thus, the probability that the Miller–Rabin test of Algorithm 8.3 with k independent iterations declares n to be probably prime is at most $(1/4)^k$.

So, for example, if we perform $k = 20$ iterations, and the Miller–Rabin test has declared that n is probably prime, the probability that this is incorrect is smaller than $(1/4)^{20} = 9.0949... \times 10^{-13}$, or less than 1 in 1 trillion! Of course, just like with the Fermat test, if a compositeness conclusion is produced, the result is 100 percent correct. Such a test can be performed very quickly on most computing platforms to produce primes of several hundred digits; the failure rate is so low that the results are reliable for most practical purposes.[*] The contemporary French number theorist Henri Cohen has referred to such probable primes as **industrial-grade primes**.

---

[*] The reader will have the opportunity to experiment with such examples in the Computer Implementation and Exercises material at the end of the chapter. One important fact to point out is that unless one is working on a symbolic computing platform, the default floating point arithmetic systems on most computing platforms will only be able to deal accurately with integers up to 15 digits or so. Thus, in order to effectively implement the algorithms in this chapter with larger integers, one will need to make sure that the computing platform has symbolic functionality. (Most computing platforms have this capability but not necessarily as a default mode, so some modifications in syntax might be required.) All of the applets for this book are designed to have symbolic functionality.

The Miller–Rabin Test with a Factoring Enhancement

With an additional minor observation, it is sometimes possible to get factors of n, in cases where the Miller–Rabin test finds n to be composite. Indeed, within (and using the notation of) Step 2 of Algorithm 8.3, if it is found that $A_j \equiv 1 \pmod{n}$ (and so n is declared composite) at a particular value of j, $1 \le j \le f - 1$, this means that $a^{2^j m} \equiv 1 \pmod{n}$, which means that $n \mid a^{2^j m} - 1$ and $a^{2^{j-1} m} \not\equiv \pm 1 \pmod{n}$ (the latter being true since the FOR loop would have exited previously if this were not the case). But since $n \mid a^{2^j m} - 1 = (a^{2^{j-1} m} - 1)(a^{2^{j-1} m} + 1)$, we may conclude that n must share a nontrivial common factor with $a^{2^{j-1} m} - 1$ (because $a^{2^{j-1} m} + 1 \not\equiv n$, so n could not divide completely into this second factor). This means that gcd $(a^{2^{j-1} m} - 1, n)$ is a nontrivial factor of n. We restate this modified Miller–Rabin algorithm.

Algorithm 8.4: The Miller–Rabin Primality Test with a Factoring Enhancement

Inputs: An odd integer $n > 3$, suspected to be prime, and a positive integer k.

Output: Either a declaration that n is composite, along with either (i) a nontrivial factor of n, or (ii) a witness integer a that violates the contrapositive of Proposition 8.11 (and thus proves that n is composite), or a declaration that n is probably prime.

The algorithm is the same as Algorithm 8.3, with the following two changes:

(1) The IF statement within the FOR loop of Step 2 needs the following modification:

...

IF $A_j \equiv 1 \pmod{n}$
Declare n as composite, output the nontrivial factor gcd(A_{j-1}, n), of n
(computed with the Euclidean Algorithm 2.1), and EXIT algorithm

ELSE ...

(2) The IF statement immediately following the FOR loop of Step 2 needs the following modification:

IF $A_{f-1} \not\equiv \pm 1 \pmod{n}$
Declare n as composite, output the witness a, and EXIT program
ELSE IF $A_{f-1} \equiv 1 \pmod{n}$
Declare n as composite, output the nontrivial factor gcd(A_{f-2}, n), of n
(computed with the Euclidean Algorithm 2.1), and EXIT algorithm
END <IF>

Example 8.9

Apply the enhanced Miller–Rabin primality test (Algorithm 8.4) with $k = 2$ to the following odd integers n:

(a) $n = 409$
(b) $n = 721$

Solution: Part (a): $n = 409$

Step 1. First we express $n - 1 = 408$ as $2^f \cdot m$, where $f = 3$ and $m = 51$. Initialize counter $i = 0$.

Step 2. Randomly generate a base: $a = 216$. We compute $A_0 = a^m \equiv 216^{51} \equiv 1 \pmod{409}$. Since $A_0 \equiv \pm 1 \pmod n$, this trial finds n to be probably prime. Since i is now 1, we repeat Step 2.

Step 2. (Second repetition) Randomly generate a base: $a = 196$. We compute $A_0 = a^m \equiv 196^{51} \equiv 143 \pmod{409}$. Since $A_0 \not\equiv \pm 1 \pmod n$, we enter into the FOR loop. For $j = 1$, we compute $A_1 \equiv A_0^2 \equiv 143^2 \equiv 408 \pmod{409}$. Since $A_1 \equiv -1 \pmod n$, this trial finds n to be probably prime. Since i is now 2, this was the final iteration of Step 2.

Step 3. Declare 409 as probably prime. (The reader may check that 409 is indeed prime so the test worked.)

Part (b): $n = 721$

Step 1. First we express $n - 1 = 720$ as $2^f \cdot m$, where $f = 4$ and $m = 45$. Initialize counter $i = 0$.

Step 2. Randomly generate a base: $a = 641$. We compute $A_0 = a^m \equiv 641^{45} \equiv 64 \pmod{721}$. Since $A_0 \not\equiv \pm 1 \pmod n$, we enter into the FOR loop. For $j = 1$, we compute $A_1 \equiv A_0^2 \equiv 64^2 \equiv 491 \pmod{721}$. Since $A_1 \not\equiv \pm 1 \pmod n$, we next (for $j = 2$) compute $A_2 \equiv A_1^2 \equiv 491^2 \equiv 267 \pmod{721}$. Since the FOR loop has completed (without finding n composite or probably prime), we do one last squaring: $A_3 \equiv A_2^2 \equiv 267^2 \equiv 631 \pmod{721}$. Since $A_3 \not\equiv -1 \pmod n$, this proves that n is composite, with witness $a = 641$.

Exercise for the Reader 8.9

Apply the enhanced Miller–Rabin primality test (Algorithm 8.4) with $k = 2$ to the following odd integers n:

(a) $n = 2581$
(b) $n = 1889$

The Pollard $p - 1$ Factoring Algorithm

We close this chapter with a factoring algorithm. Although the enhanced Miller–Rabin algorithm can sometimes help with factoring, this is not its primary goal, and it is not guaranteed to do anything more than test for primality. Although factoring is a much more difficult problem than primality testing and there are no known efficient algorithms for this problem, there are sometimes better approaches than a brute-force approach (of attempting to factor n by checking divisibility of all prime numbers $p \le \sqrt{n}$). Factoring algorithms have been designed to perform well if the composite number being factored has certain properties (that the algorithm is designed to take advantage of). We present such a method due to John Pollard [Pol-74], known now as *Pollard's $p-1$ method*. This method tends to work well if n has a prime factor p for which $p-1$ has only small prime factors. In Chapter 12, we will give another factoring algorithm based on elliptic curves that is more powerful since it requires only that n has a prime factor p for which some numbers close to p have only small prime factors. As a countermeasure to Pollard's method, the notion of *strong primes* has evolved, initially defined to be a prime p for which $p-1$ has at least one large prime factor. Since Pollard's method was announced, primes used in cryptosystems should be strong primes. The notion has been extended to include primes that are resistant to other specialized factoring algorithms. The chapter Exercises and Computer Implementation material examine some methods for generating strong primes.

As with the primality tests developed earlier, Pollard's $p-1$ factorization algorithm is based on Fermat's little theorem. We explain how and why the algorithm works and then summarize it. We assume that a composite number n (that we wish to factor) has a prime factor p for which $p-1$ has only small prime factors. It follows that we will have $p-1 | B!$ for a not-too-large positive integer B. For example, if $p = 26{,}951$, then $p-1 = 2 \cdot 5^2 \cdot 7^2 \cdot 11$, and the reader may check that $p-1 | 14!$ and that any larger but no smaller factorial will do. If we choose any base $a > 1$ ($a = 2$ is most often used), then Fermat's little theorem tells us that

$$a^{B!} \equiv a^{(p-1)q} \equiv (a^{(p-1)})^q \equiv (1)^q \equiv 1 (\bmod\, p) \;\Rightarrow\; p\,|\,a^{B!} - 1$$

We consider $d = \gcd(a^{B!} - 1, n)$. Now, if n has another prime factor q, it is unlikely that $q\,|\,a^{B!} - 1$, unless $q-1$ also has only small prime factors. This means that d will be a nontrivial factor of n.

Algorithm 8.5: Pollard's $p-1$ Factorization Algorithm

Inputs: An odd composite integer $n > 3$ and a positive integer B. An optional third input is a positive integer base a, with default value $a = 2$.
Output: Either a nontrivial factor of n or no output in case the algorithm does not find one.

Step 1.   Compute $a^{B!} (\bmod\, n)$ by using fast modular exponentiation B times in the following chain:

$$a^{1!} \equiv a,\; a^{2!} \equiv (a^{1!})^2,\; a^{3!} \equiv (a^{2!})^3, \cdots a^{B!} \equiv (a^{[B-1]!})^B (\bmod\, n).$$

Step 2.   Use the Euclidean algorithm (Algorithm 2.1) to compute $d = \gcd(a^{B!} - 1, n)$ (using the representative for $a^{B!}$ that was found in Step 1). If $d > 1$, output d as a nontrivial factor of n.

We now give a "small" example, although the sizes of the numbers are about the limit on what can be done with any common floating point arithmetic computing platform. Examples with larger integers would require symbolic computing platforms and will be considered in the Computer Implementation material at the end of the chapter.

Example 8.10

Apply Pollard's $p - 1$ factorization algorithm to the integer 58,932,967.

Solution: We will apply the algorithm using $B = 12$.

Step 1. We compute $a^{B!}(\bmod n)$ by using fast modular exponentiation 12 times:

$2^{1!} \equiv 2$
$2^{2!} \equiv (2^{1!})^2 \equiv 4$
$2^{3!} \equiv (2^{2!})^3 \equiv 64$
$2^{4!} \equiv (2^{3!})^4 \equiv 16,777,216$
$2^{5!} \equiv (2^{4!})^5 \equiv 6,054,079$
$2^{6!} \equiv (2^{5!})^6 \equiv 56,169,321$
$2^{7!} \equiv (2^{6!})^7 \equiv 55,888,294$
$2^{8!} \equiv (2^{7!})^8 \equiv 47,597,184$
$2^{9!} \equiv (2^{8!})^9 \equiv 9,175,828$
$2^{10!} \equiv (2^{9!})^{10} \equiv 35,101,026$
$2^{11!} \equiv (2^{10!})^{11} \equiv 41,033,283$
$2^{12!} \equiv (2^{11!})^{12} \equiv 37,504,803$

Step 2. We use the Euclidean algorithm (Algorithm 2.1) to compute $d = \gcd(a^{B!} - 1, n) = \gcd(37,504,802; 58,932,967) = 7351$. This is the outputted nontrivial factor of n. Dividing n by this gives 8017, thereby making significant progress in factoring n.

Note that the factor that was found (7351) in the above example is a prime p (as the reader may check) and $p - 1$ factors as $2 \cdot 3 \cdot 5^2 \cdot 7^2$. This shows that the condition needed for Pollard's algorithm is indeed satisfied. Interestingly, although our theoretical explanation of the method would have suggested that we use a value of B to be at least 14, the example showed that the smaller value 12 worked. It turns out that for this example, any value of B greater than 9 would work to produce the above prime factor.

Exercise for the Reader 8.10

Apply Pollard's $p - 1$ factorization algorithm to the integer 12,637,211 using $B = 15$.

Although it is not guaranteed to produce prime factors of n, Pollard's $p - 1$ factorization algorithm is an often-used tool. In practice, one first tries a few trial divisions with smaller primes (on a computer, primes of up to 1 billion or so could be quickly checked), and then a primality test should be applied to what is left. If any composite factors remain,

Pollard's $p - 1$ factorization algorithm is repeatedly applied (perhaps by increasing the values of B). Anything remaining should be tested with a primality test, and if found to be composite, more sophisticated methods (such as the elliptic curve method that we will present in Chapter 12) can be applied. One nice feature of Pollard's algorithm is that if we try it and it does not produce a factor, the modular powers that were computed can still be used if we run the algorithm again with a larger value of B.

Chapter 8 Exercises

1. (a) Use the prime number theorem to estimate the number of primes that are less than 1 billion.
    (b) Use the prime number theorem to estimate the number of primes that lie between 1 billion and 10 billion.
    (c) Use the prime number theorem to estimate the number of primes that lie between 1 billion and 1 trillion.

2. (a) Use the prime number theorem to estimate the number of primes that are less than 1 thousand.
    (b) Use the prime number theorem to estimate the number of primes that lie between 1 thousand and 10 thousand.
    (c) Use the prime number theorem to estimate the number of primes that lie between 1 thousand and 1 million.

3. (a) Use the prime number theorem to estimate the number of 100-bit primes. This will be the number of primes between 2^{99} and 2^{100}.
    (b) If we randomly pick a 100-bit odd integer, use the result of part (a) to estimate the probability that it will be prime.
    (c) Use the prime number theorem to estimate the number of 1000-bit primes. This will be the number of primes between 2^{999} and 2^{1000}.
    (d) If we randomly pick a 1000-bit odd integer, use the result of part (c) to estimate the probability that it will be prime.

4. (a) Use the prime number theorem to estimate the number of 50-bit primes. This will be the number of primes between 2^{49} and 2^{50}.
    (b) If we randomly pick a 50-bit odd integer, use the result of part (a) to estimate the probability that it will be prime.
    (c) Use the prime number theorem to estimate the number of 5000-bit primes. This will be the number of primes between 2^{4999} and 2^{5000}.
    (d) If we randomly pick a 5000-bit odd integer, use the result of part (c) to estimate the probability that it will be prime.

5. Perform each of the following modular exponentiations, first using (i) fast modular exponentiation (Algorithm 6.5) and then (ii) Fermat's little theorem.
    (a) $2^{58} \pmod{11}$
    (b) $9^{102} \pmod{13}$
    (c) $12^{207} \pmod{23}$
    (d) $17^{1236} \pmod{47}$

6. Perform each of the following modular exponentiations, first using (i) fast modular exponentiation (Algorithm 6.5) and then (ii) Fermat's little theorem.
    (a) $3^{45} \pmod{13}$
    (b) $6^{101} \pmod{17}$
    (c) $11^{1977} \pmod{29}$
    (d) $22^{1437} \pmod{53}$

7. Compute each of the indicated values of Euler's phi function:
    (a) $\phi(60)$
    (b) $\phi(248)$
    (c) $\phi(1224)$
    (d) $\phi(9900)$

8. Compute each of the indicated values of Euler's phi function:
    (a) $\phi(50)$
    (b) $\phi(360)$
    (c) $\phi(987)$
    (d) $\phi(10,000)$

9. (a) Show that if n is an even positive integer, then $\phi(2n) = 2\phi(n)$.
    (b) Show that if n is an odd positive integer, then $\phi(2n) = \phi(n)$.

10. (a) Show that if n is a positive integer with $n \equiv 0 \pmod{3}$, then $\phi(3n) = 3\phi(n)$.
    (b) Show that if n is a positive integer with $n \not\equiv 0 \pmod{3}$, then $\phi(3n) = 2\phi(n)$.

11. Find all positive integer solutions (if any) of the following equations:
    (a) $\phi(n) = 1$
    (b) $\phi(n) = 4$
    (c) $\phi(n) = 5$
    (d) $\phi(n) = 12$

12. Find all positive integer solutions (if any) of the following equations:
    (a) $\phi(n) = 2$
    (b) $\phi(n) = 3$
    (c) $\phi(n) = 6$
    (d) $\phi(n) = 14$

13. Use Euler's theorem to compute each of the following modular exponentiations. Write each answer as an integer in $\{1, 2, \cdots, m-1\}$, if you are working mod m.
    (a) $2^{1256} \pmod{15}$
    (b) $7^{3945} \pmod{20}$
    (c) $2^{22,970} \pmod{25}$
    (d) $8^{32,149} \pmod{35}$

14. Compute each of the indicated powers, working in modular arithmetic that is specified. Write each answer as an integer in $\{1, 2, \cdots, m-1\}$, if you are working mod m.
    (a) $3^{1256} (\text{mod } 8)$
    (b) $12^{3945} (\text{mod } 25)$
    (c) $3^{22,970} (\text{mod } 40)$
    (d) $13^{32,149} (\text{mod } 15)$

15. (a) As in the solution of Example 8.4, create a table of all modular powers of the modular integers a that are relatively prime with $n = 6$ (up to the $\phi(n)th$). Use the table to identify the orders of each of these modular integers as well as any primitive roots.
    (b) Repeat part (a) for $n = 12$.

16. (a) As in the solution of Example 8.4, create a table of all modular powers of the modular integers a that are relatively prime with $n = 10$ (up to the $\phi(n)th$). Use the table to identify the orders of each of these modular integers as well as any primitive roots.
    (b) Repeat part (a) for $n = 11$.

17. Compute each of the following orders, if they exist:
    (a) $\text{ord}_{10}(3)$
    (b) $\text{ord}_{21}(6)$
    (c) $\text{ord}_{304}(21)$

18. Compute each of the following orders, if they exist:
    (a) $\text{ord}_{11}(5)$
    (b) $\text{ord}_{17}(2)$
    (c) $\text{ord}_{427}(21)$

19. For each of the following integers n, do the following:
    (i) Use Theorem 8.7 to determine whether there are any primitive roots mod n; if so, how many will there be?
    (ii) If there are primitive roots, find one.
        (a) $n = 12$
        (b) $n = 13$
        (c) $n = 14$

20. For each of the following integers n, do the following:
    (i) Use Theorem 8.7 to determine whether there are any primitive roots mod n; if so, how many will there be?
    (ii) If there are primitive roots, find one.
        (a) $n = 16$
        (b) $n = 17$
        (c) $n = 18$

21. For each of the following integers n, do the following:
    (i) Determine whether there are any primitive roots mod n; if so, how many will there be?
    (ii) If there are primitive roots, find one. For parts (a) and (b) give the smallest primitive root; for parts (c) and (d), give any primitive root.

(iii) If there are primitive roots, use the one you found in (ii)
to construct another.
(a) $n = 25$
(b) $n = 39$
(c) $n = 31$
(d) $n = 50$
(e) $n = 52$
(f) $n = 29{,}791$

22. For each of the following integers n, do the following:
(i) Determine whether there are any primitive roots mod n; if
so, how many will there be?
(ii) If there are primitive roots, find one. For parts (a) and (b)
give the smallest primitive root, for parts (c) and (d), give
any primitive root.
(iii) If there are primitive roots, use the one you found in (ii)
to construct another.
(a) $n = 17$
(b) $n = 81$
(c) $n = 323$
(d) $n = 289$
(e) $n = 4913$
(f) $n = 162$

23. (a) Verify that $g = 3$ is a primitive root of 223.
(b) How many integers mod 223 have order 6? If such elements
exist, find one.
(c) How many integers mod 223 have order 74? If such elements exist, find one.
(d) How many integers mod 223 have order 10? If such elements exist, find one.

24. (a) Verify that $g = 3$ is a primitive root of 566.
(b) How many integers mod 556 have order 12? If such elements exist, find one.
(c) How many integers mod 556 have order 6? If such elements
exist, find one.
(d) How many integers mod 223 have order 94? If such elements exist, find one.

25. Use Gauss's algorithm (Algorithm 8.1) to find a primitive root
of the following primes:
(a) $p = 107$
(b) $p = 211$
(c) $p = 653$

26. Use Gauss's algorithm (Algorithm 8.1) to find a primitive root
of the following primes:
(a) $p = 127$
(b) $p = 233$
(c) $p = 733$

27. Apply the Fermat primality test (Algorithm 8.2) with $k = 4$ to the following odd integers:
   (a) $n = 527$
   (b) $n = 523$
   (c) $n = 943$
   (d) $n = 5963$
   (e) $n = 11{,}303$
   (f) $n = 1811$

28. Apply the Fermat primality test (Algorithm 8.2) with $k = 4$ to the following odd integers:
   (a) $n = 449$
   (b) $n = 629$
   (c) $n = 1147$
   (d) $n = 4559$
   (e) $n = 8893$
   (f) $n = 9727$

29. Apply the Miller–Rabin test (Algorithm 8.3) with $k = 4$ to each of the odd integers given in Exercise 27.

30. Apply the Miller–Rabin test (Algorithm 8.3) with $k = 4$ to each of the odd integers given in Exercise 28.

31. Apply the enhanced Miller–Rabin test (Algorithm 8.4) with $k = 4$ to each of the odd integers given in Exercise 27.

32. Apply the enhanced Miller–Rabin test (Algorithm 8.4) with $k = 4$ to each of the odd integers given in Exercise 28.

33. Apply Pollard's $p - 1$ factorization algorithm (Algorithm 8.5) to each of the following integers:
   (a) $n = 7{,}781{,}707$
   (b) $n = 12{,}418{,}223$
   (c) $n = 47{,}486{,}269$

34. Apply Pollard's $p - 1$ factorization algorithm (Algorithm 8.5) to each of the following integers:
   (a) $n = 7{,}427{,}207$
   (b) $n = 8{,}468{,}039$
   (c) $n = 16{,}701{,}131$

35. Determine the ones digit of $3^{100!}$.

36. Determine the ones digit of $17^{100!}$.

37. *True or False*. Indicate whether the following statement is (always) true or false. Then either give an explanation of why it is true (i.e., a proof) or a counterexample of a case where it can be false:

   If p is a prime, g is a primitive root mod p, and q is a prime factor of $p - 1$, then $g^{(p-1)/q} \not\equiv 1 \pmod{p}$.

38. *True or False.* Indicate whether the following statement is (always) true or false. Then either give an explanation of why it is true (i.e., a proof) or a counterexample of a case where it can be false:

   If a and b are relatively prime positive integers less than a prime p, then $\mathrm{ord}_p(ab) = \mathrm{ord}_p(a) \cdot \mathrm{ord}_p(b)$.

39. *True or False.* Indicate whether the following statement is (always) true or false. Then either give an explanation of why it is true (i.e., a proof) or a counterexample of a case where it can be false:

   If a and b are different integers greater than 1, $\phi(ab) = \phi(a)\phi(b)$.

40. Prove that if $n > 2$ is any integer, then $\phi(n)$ is even.

   Note: The next three exercises together will build up a proof of Proposition 8.3.

41. Prove that if p is a prime and k is a nonnegative integer, then $\phi(p^k) = p^{k-1}(p-1)$.

   Suggestion: By definition, $\phi(p^k)$ is the number of integers from the list $1, 2, \cdots, p^k$ that are relatively prime to p^k. Being relatively prime to p^k is equivalent to being relatively prime to p. Thus, the elements of this list that are not counted are precisely every pth element (the multiples of p).

42. Suppose that n and m are relatively prime positive integers.
   (a) Show that an integer a is relatively prime to nm if, and only if, it is relatively prime to both n and m.
   (b) Prove that if n and m are relatively prime positive integers, then $\phi(nm) = \phi(n)\phi(m)$.
   (c) Prove that if $n_1, n_2, \cdots, n_\ell$ are pairwise relatively prime integers, then $\phi(n_1 n_2 \cdots n_\ell) = \phi(n_1)\phi(n_2)\cdots\phi(n_\ell)$.

   Note: In number theory, functions on the positive integers that satisfy the condition of part (b) are called *multiplicative functions*.

   Suggestion: For part (b), arrange the integers $1, 2, \cdots, mn$ into the following array:

1	2	$\cdots$	k	$\cdots$	m
$m+1$	$m+2$	$\cdots$	$m+k$	$\cdots$	$2m$
$2m+1$	$2m+2$	$\cdots$	$2m+k$	$\cdots$	$3m$
$\vdots$	$\vdots$		$\vdots$		$\vdots$
$(n-1)m+1$	$(n-1)m+2$	$\cdots$	$(n-1)m+k$	$\cdots$	nm

   By part (a), if we remove the numbers from this array that are relatively prime to either n or m, the number of integers left will be $\phi(nm)$. Use the fact that $\gcd(qm+k,m) = \gcd(k,m)$ to conclude that either *all* of the numbers in the kth column are relatively prime to m or none will be, and conclude that the columns of relatively prime to m numbers will be those $\phi(m)$ columns corresponding to the values of k that are relatively prime to m. It suffices to show that in each of these $\phi(m)$ columns, exactly $\phi(n)$ of the n numbers will be relatively prime to n. Accomplish this by first showing that the n numbers in any such column are pairwise incongruent (mod n).

43. Prove Proposition 8.3 by using the results of the previous two exercises.

44. Suppose that $n > 1$ is a positive integer that has a primitive root.
    (a) If k is a positive integer and $k|\phi(n)$, show that the equation $x^k \equiv 1(\bmod n)$ has exactly k different solutions (mod n).
    (b) *Convergence of Gauss's Primitive Root Finding Algorithm 8.1.* Show that in Step 2 of Algorithm 8.1, we must have $u > \ell$, and hence (as pointed out after the statement of the algorithm) Algorithm 8.1 will always terminate by finding a primitive root.

Suggestions: For part (a), let g be a primitive root. If x is any modular integer that satisfies $x^k \equiv 1(\bmod n)$ then x must be relatively prime to n, so by Proposition 8.6, we must have $x \equiv g^j$, for some exponent j. Thus, $1 \equiv x^k \equiv (g^j)^k \equiv g^{jk}(\bmod n)$, so by Proposition 8.5(a), we may conclude $\phi(n)|jk$, from which we obtain that j must be a multiple of $\phi(n)|k$. Since there are only k such multiples that clearly correspond to different roots of $x^k \equiv 1(\bmod n)$ the result follows.

For part (b), use the result of part (a).

Chapter 8 Computer Implementations and Exercises

Note: As mentioned in the chapter text, if your computing platform is a floating point arithmetic system, it may allow you up to only 15 or so significant digits of accuracy. Symbolic systems allow for much greater precision, being able to handle hundreds of significant digits. Some platforms allow users to choose if they wish to work in floating point or symbolic arithmetic. If you are working on such a dual-capability platform, you may wish to create two separate programs for those that might work with large integers: an ordinary version and a symbolic version (perhaps attaching a Sym suffix to the names of those of the latter type). In case you do not have access to a symbolic system, some particular questions may need to be skipped or modified so the numbers are of a manageable size.

1. *Program for the Euler Phi Function.*
   (a) Write a program $y = \texttt{EulerPhi(n)}$ that inputs an integer $n > 1$ and outputs $y = \phi(n)$.
   (b) Check your program with the results of Chapter Exercise 7 and then use it to compute $\phi(n)$ for each of the following values of n: 18,365, 222,651, 1,847,773, 22,991,877.

   Suggestion: Call on the program of Computer Exercise 2 for factoring integers and use the formula for $\phi(n)$ given in Proposition 8.3. Since this program relies on factorization, it is not intended to be used with large integers (of, say, 10 digits or more).

2. *Program for Computing Orders.*
   (a) Write a program $k = \texttt{Order(a, n)}$ that inputs two relatively prime positive integers a and n, with $a < n$, and outputs $k = \text{ord}_n(a)$.
   (b) Check your program with the results of Chapter Exercise 17 and then use it to compute these two orders: $\text{ord}_{1807}(3)$, $\text{ord}_{10543}(54)$.

   Suggestion: This can either be done with a brute-force approach (of computing successive powers of a until we reach 1) or using Proposition 8.5(b) (and calling on the program of Computer Exercise 1). In either case, the programs are not suitable for large integers.

3. *Program for Finding Primitive Roots.*
   (a) Write a program $\texttt{pRoot = SmallestPrimitive}$ $\texttt{Root(n)}$ that inputs a positive integer n that is of the form listed in Theorem 8.8, and that outputs the smallest primitive root of n (guaranteed to exist by Theorem 8.8). It is fine to call on the program Order of the preceding exercise.
   (b) Check your program with the results of Chapter Exercise 21(a)–21(c) and then run it on the following values of n: 8893, 17786, 123457, 246914.
   (c) Write a related program $\texttt{PRoots = AllPrimitive}$ $\texttt{Roots(n)}$ that inputs a positive integer n that is of the form listed in Theorem 8.8, and that outputs a vector *PRoots* of all of the $\phi(\phi(n))$ primitive roots of n (see Theorem 8.8).
   (d) Run your program of part (c) on each of the inputs of part (b).

   Suggestion: For part (c), make use of Theorem 8.8.

4. *Gauss's Algorithm for Finding a Primitive Root Modulo a Prime*
   (a) Write a program with syntax $g = \texttt{GaussPrimRoot(p)}$ that inputs an odd prime p and will apply Gauss's Algorithm 8.1 to output a primitive root g mod p.
   (b) Run your program on each for each of the following prime inputs: $p = 17$, $p = 67$, $p = 257$, $p = 1747$, $p = 76801$, $p = 354{,}037$, $p = 4{,}245{,}679$, $p = 82{,}243{,}801$. Check (if feasible) that the results are indeed primitive roots by using the Order program of Computer Exercise 2.

5. *Fermat's Primality Test.*

   (a) Write a program with syntax `y = FermatTest(n,k)`, that inputs an integer n > 3 and will apply the Fermat primality test k times. The second input variable k is optional (default value is 1). The output y will be 1 if at least one of the k tests has found n to be composite, and 0 in case all of the tests were inconclusive (meaning, informally, they all found n to be probably prime). Set your program up so that in cases that it proves n is composite, it should also output the witness a that resulted in the composite conclusion.

   (b) Apply your program to the following integers n, using $k = 10$: $n = 215, 841, 1931, 3973, 22879$. Check the outputs by factoring each of these integers (in the usual way or using a built-in utility); also, for each corresponding witness a, compute $a^{n-1} \pmod{n}$.

   (c) Use your program to attempt to prove compositeness of each of the following (composite) integers: 3668963, 154915253, 6271549451, 6732432725687, 52322940983667496651.

   (d) The following two large prime numbers give rise to a famous factoring challenge by the RSA cryptosystems company:

   $p = 16347336458092538484431338838865090859841783$
   $67003309231218111085238933310010450815121211817$
   67511579
   $q = 19008712816648221131268515739354139754718967$
   $89968515493666638539088027103802104498957191261$
   61465571

   Their product $n = pq$, is known as RSA-640.* Apply your program of part (a) to the RSA-640 (174 digit) number using $k = 1$, and repeat this 10 times.

   Note: As with any probabilistic algorithm, results will vary. When the author did part (d), in all 10 trials, the (single application of) Fermat's primality test proved that RSA-640 was prime very quickly (in a couple of seconds). By contrast, recall that it took over four years before the international challenge to factor RSA-640 was finally completed. This fact alone gives good evidence that primality checking is a much easier problem than factoring.

6. *The Miller–Rabin Primality Test.*

   (a) Write a program with syntax `MillerRabinTest(n, k)`, that inputs an odd integer $n > 3$, and will apply the Miller–Rabin primality test k times. The second input variable k is optional (default value is 1). The program will indicate that n was found to be composite, if at least one of the k tests has found n to be composite, and that n is probably prime, in case all of the tests were inconclusive (meaning, informally, they all found n to be probably prime). Set

---

* The name RSA-640 refers to size of the binary representation. RSA-640 has 640 binary digits, or 174 decimal digits.

your program up so that in cases that it proves n is composite, it should also output the witness a, and corresponding exponent parameters [j,m] that resulted in the composite conclusion.

(b) through (d): Apply your program to redo each of the corresponding parts of Computer Exercise 5.

7. *Comparison of the Miller–Rabin and the Fermat Primality Tests.* As was explained above, the Miller–Rabin primality test is at least as effective as the Fermat primality test in detecting compositeness. Construct an example that will show the Miller–Rabin test to be stronger than the Fermat test. More specifically, find an example of a composite integer n and a corresponding Miller–Rabin witness a, for which the Fermat primality test would find n to be probably prime (with this same witness a).

    Suggestion: Apply both tests to some randomly generated numbers with a large number of digits.

8. *Comparison of the Miller–Rabin and the Fermat Primality Tests.* For this exercise, it will be convenient to modify the programs of Computer Exercises 5 and 6 into basic versions with syntax y = FermatTestBasic(n, k), and y = MillerRabinTestBasic(n, k) that work exactly in the same fashion as the corresponding programs in those exercises, but which output only the value of y (1 if n is found composite, and 0 if n is found probably prime by the test); so all other output is suppressed.

    (a) There are exactly 560 primes p in the range $5000 < p < 10,000$. Use the Fermat primality test with k = 1 to run through all odd integers in this range. How many possible primes were encountered? Repeat this using k = 4, and then 10.

    (b) Repeat part (a) using the Miller–Rabin primality test.

    (c) There are exactly 9590 primes in the range $3 < p < 100,000$. Use the Fermat primality test with k = 1 to run through all odd integers in this range. How many possible primes were encountered? Repeat this using k = 4, and then 10.

    (d) Repeat part (a) using the Miller–Rabin primality test.

    Note: As with any probabilistic algorithm, results will vary.

9. *The Miller–Rabin Primality Test with Factoring Enhancement.*

    (a) Write a program with syntax MillerRabinTest Premium(n, T), that functions exactly like the program of Computer Exercise 6 (or the basic version of Computer Exercise 9) except that if a factor of n can be found as in Algorithm 8.4, then this factor should also be outputted.

    (b) Apply this program in a FOR loop to all odd numbers from n = 5 to n = 9999, and using k = 1. For how many of these odd numbers that were found to be composite was a factor actually found?

The next exercise will develop a prime number generator using the Miller–Rabin algorithm in conjunction with
the technique mentioned earlier. The user will be able to
input the desired confidence level (for the guarantee that
the number found by the program is actually prime).

10. *Pseudoprime Generating Program.*
   (a) Write a program with syntax pProb = MillerRabin
        PrimeGenerator(nbits, tol) that inputs an integer
        nbits > 3, as well as a real number tol, $0 < $ tol $ < 1$.
        The program proceeds as follows: A binary string of
        length nbits is generated by setting the first and last
        bits = 1 (first bit equaling 1 defines the bit string's length,
        the last bit equaling 1 ensures it is an odd integer), and
        randomly generating the remaining (inside) bits. This
        string is then converted into an integer n, and the pro-
        gram MillerRabinTest(n, k) of Computer Exer-
        cise 6 is applied, where the second input k is chosen to be
        $\lfloor -(1/2)\log_2(\text{tol}) \rfloor$.[*] This is repeated until such randomly
        generated integer n is encountered for which this primality
        test declares as probably prime. This n will be the ouput
        pProb.
   (b) Run this program with nbits = 25, and tol = .01, and
        then with tol = .001.
   (c) If you have a factoring utility on your computing platform,
        run this program 200 times with nbits = 25 and tol = .1,
        and check the veracity of the 200 probable primes that
        were produced and the proportion of composites that were
        declared probable primes with the number tol.
   (d) Use your program to produce a 100-bit and a 200-bit prob-
        able prime using tol = .001.
   (e) Use your program to produce a 100-bit and a 200-bit prob-
        able prime using tol $= 10^{-12}$.

11. *Program for Pollard's p – 1 Factorization Algorithm.*
   (a) Write a program with syntax Pollardpminus1(n,B)
        that inputs an odd composite number $n > 3$, and a positive
        integer B. The program will apply Pollard's $p-1$ factor-
        ization algorithm (Algorithm 8.5). When used, a primal-
        ity test (such as Miller–Rabin) should be applied first to
        ensure that n is composite. If no factor is found, the algo-
        rithm may be applied with a larger value of B.
   (b) Use the program of part (a) to redo Example 8.10.
   (c) Use the program of part (a) to redo Chapter Exercise 33.
   (d) Apply the program of part (a) to try and factor following
        integers (perhaps experimenting with different values of
        the parameter B):
        (i) $n = 1{,}575{,}409{,}999$
        (ii) $n = 180{,}049{,}737{,}691$
        (iii) $n = 2{,}045{,}607{,}156{,}689$

---

[*] From Theorem 8.13, this value for k is the smallest number of iterations of the Miller–
Rabin test to ensure that the probability that a composite number is declared probably
prime is less than tol.

Public Key Cryptography

One of the main drawbacks in all of the symmetric key cryptosystems discussed up to this point is the fact that in order for such a system to be employed, the keys must be distributed to all participating parties before any secure communication can take place. This task by itself is often difficult or impractical. This drawback can now be circumvented thanks to a remarkable revolution in cryptography known as **public key cryptography** or **asymmetric key cryptography** that occurred in the 1970s. The discovery was first published in a groundbreaking 1976 paper by American cryptographers Whit Diffie (1944–) and Martin Hellman (1945–) [DiHe-76]. Although Diffie and Hellman did not provide a complete practical implementation of a public key cryptosystem, they provided an important key exchange protocol (the *Diffie–Hellman key exchange*) by which two remote parties can establish a secure key using public (insecure) channels. Inspired by the Diffie–Hellman paper and the need for a practical implementation of their concept cryptosystem, MIT scientists Ronald Rivest, Adi Shamir, and Leonard Adleman introduced their RSA cryptosystem.[*] This has turned out to be one of the most important and widely used public key cryptosystems. We have already developed the necessary mathematical ingredients necessary to completely describe some of the most important contemporary public key cryptosystems. Apart from the Diffie–Hellman protocol and RSA, this chapter describes the ElGamal cryptosystem and the knapsack cryptosystems of Hellman and Merkle. Applications to provide digital signatures and nonrepudiation are also developed. As usual, we provide information on some known weaknesses and potential attacks, as well as a brief account of some of the political ramifications of these powerful cryptosystems, which are considered weapons by many national governments (including that of the United States). The chapter ends with an example of a theoretical result that shows the problem of being able to break RSA is at least as difficult as factoring.

An Informal Analogy for a Public Key Cryptosystem

A good analogy between a symmetric key cryptosystem (like all that we have so far developed) and a public key cryptosystem is the following: Imagine a mathematically primitive world where the only way to encrypt a message would be to lock it in a case. If Alice, who resides

---

[*] It was in their RSA paper [RiShAd-78] that the characters "Alice" and "Bob" were introduced as permanent fixtures into the cryptography saga.

in California, wants to send Bob, who lives in Maryland, a confidential message using a symmetric key cryptosystem, she would need to arrange ahead of time with Bob to have a matching set of keys to a certain padlock. She could then place her message in a steel case, lock the case with her padlock, and mail it to Bob. Since Bob, the intended recipient, has Alice's key, he (and only he) should be able to unlock the case and read the message. The main drawback is the need to exchange keys before the communication can take place. The need for a common key was thought for many years to be an essential ingredient in cryptography, and many cryptographers simply took this as an axiom. Now consider the following change of protocol: Alice and Bob have their own padlocks and keys but have not exchanged any common keys. Alice puts her confidential message to Bob in a steel case and locks the case with her padlock. She then mails the locked case to Bob. When Bob receives the case, he cannot open it because he does not have the key to Alice's lock. But he locks it up a second time with his own padlock and mails the double-locked case back to Alice. Alice then removes her lock, and mails the case back to Bob who is now the only one able to open the case and read Alice's message. Although the second protocol was more time consuming (and this would not be relevant if the mailings were done electronically), its ramifications are quite striking.

Note: With one minor exception, throughout this chapter, plaintexts and ciphertexts will be modular integers in some base. As explained in Chapter 6, it is a routine matter to convert such integers into strings of English letters (or more general ASCII characters), so we will not devote further attention to such matters in this chapter.

Notation: Since most of the cryptosystems in this chapter will involve modular arithmetic, it will be convenient for us to adopt the following notation for the set of **invertible elements** in $\mathbb{Z}_n$:

$$\mathbb{Z}_n^\times = \{\text{all invertible elements of } \mathbb{Z}_n\}. \tag{9.1}$$

By Proposition 2.11 (and the definition of the Euler phi function), $\mathbb{Z}_n^\times$ consists of the $\phi(n)$ positive integers that are less than n and relatively prime to n. In particular, if p is prime, then $\mathbb{Z}_p^\times = \{1, 2, \cdots, p-1\}$.

The Quest for Secure Electronic Key Exchange

The above analogy kept the early innovators Diffie and Hellman captivated with the problem of developing a feasible implementation. The reason that this method cannot be readily applied stems from a simple observation about composition of functions. In mathematical terms, Alice would have her own cryptosystem (corresponding to her padlock in the above analogy), along with a corresponding encryption and decryption functions: $E_A, D_A = E_A^{-1}$. Likewise, Bob would have his own encryption and decryption functions: $E_B, D_B = E_B^{-1}$. Now Alice locking her plaintext P and sending it to Bob would correspond to Alice sending Bob $E_A(P)$. Bob applying his padlock to this would, in turn, produce $E_B \circ E_A(P)$. Now, when he sends this back to Alice, Alice will not be able to remove her padlock by applying her decryption function; she would only get $D_A \circ E_B \circ E_A(P)$. The general rule is that the inverse of a composition of functions is the composition of the inverses in the reverse order (see Exercise for the Reader 3.7)—informally, "first on, first off." Despite these seemingly insurmountable mathematical difficulties and the discouragement from many colleagues and other cryptographers, Diffie and Hellman thought to the future of a networked society and the absolute need for such communications. They continued their efforts and eventually were able to achieve the important public key distribution algorithm. It was not a complete cryptosystem (although one would be developed shortly thereafter, based on their innovation); it basically allowed secure cryptographic communications to take place without the old-fashioned hindrance of having to manually agree upon and exchange a common key. For example, the protocol could be used to securely establish a DES (or triple DES) key between two parties that could subsequently communicate with the latter symmetric key cryptosystem.

One-Way Functions

The following definition of a one-way function, although somewhat informal, lies at the core of any symmetric key system. The basic idea is that a one-way function can be given out publicly so that anyone can use it, but it would be an intractable problem for anyone to compute the inverse function.

Definition 9.1

A **one-way function** (also known as a **trapdoor function**) is any bijective function $f : D \to R$, such that any of its values, $f(x)$, is easy to compute but whose inverse function values, $f^{-1}(y)$, are computationally intractable to compute.

The trapdoor terminology stems from the fact that with some additional information (the private key), the inverse function will be easy to compute. For example, if $P_N = \{p \mid p \geq N,$ and p is prime$\}$, then the function f with domain being the set of finite-length vectors of primes in P_N, with codomain being all finite products of such primes, and which is defined by $f([p_1, p_2, \cdots, p_k]) = p_1 \cdot p_2 \cdots \cdot p_k$, is a one-way function if N is sufficiently large. Indeed, as we pointed out earlier, the problem of factoring integers with more than 300 digits (according to present standards) is intractable, whereas the inverse problem of multiplying numbers even with millions of digits is easy (on a computer). A trapdoor into factoring a certain output $f([p_1, p_2, \cdots, p_k])$ of the above function might be knowing some of the factors, thereby reducing the factoring problem to one for an integer of tractable size (say, 150). As computer technology increases, the value of N needed to ensure that the above function is one-way will increase, but it is generally believed that with such modifications, this function should remain a one-way function in perpetuity.

We next explain the Diffie–Hellman key exchange system. Its elegance and simplicity are striking, but the reader should keep in mind that generations of cryptographers had been convinced that such a system could not possibly exist. They published their discovery in a landmark 1976 paper [DiHe-76] that completely revolutionized the field. As pointed out earlier, this is not a complete cryptosystem but can be used for two parties (Alice and Bob) to create a mutual symmetric key over a public channel (to which Eve has access). Its security rests on the difficulty of computing discrete logarithms, as does the related ElGamal complete cryptosystem that we will introduce subsequently. The algorithm requires a large prime p and a corresponding primitive root g. Algorithms for generating such parameters were provided in Chapter 8.[*]

Review of the Discrete Logarithm Problem

The security of the Diffie–Hellman key exchange, as well as that of the ElGamal cryptosystem that will be introduced later in this chapter, both rely on the difficulty of computing *discrete logarithms*. This topic was discussed in the previous chapter, and we review it here in a slightly less general sense, as will be described in the following definition. Recall that a primitive root modulo a prime number p is a positive integer $g < p$, whose order (mod p) equals $p - 1$ (that is, is as large as possible). This means that $k = p - 1$ is the smallest positive integer for which $g^k \equiv 1$ (mod p). There always exist primitive roots modulo any prime, and the powers of any primitive root: $g, g^2, g^3, \cdots, g^{p-1} \equiv 1$ are all distinct and consist precisely of the integers mod p that are relatively prime to p, that is, $\mathbb{Z}_p^\times$.

Definition 9.2   Discrete Logarithms

Given a prime number, p, and a primitive root g (mod p), the **discrete logarithm (mod p) with base g** is the inverse function of the modular exponentiation function $E_g : \mathbb{Z}_p^\times \to \mathbb{Z}_p^\times :: E_g(\ell) \equiv g^\ell \pmod{p}$.

Since g is a primitive root, this modular exponential function is both one-to-one and onto, and so has an inverse function $L_g : \mathbb{Z}_p^\times \to \mathbb{Z}_p^\times$, which is the corresponding discrete logarithm function. Although it would be more proper to include the modulus p in these notations (for example, to use $E_{g,p}$ and $L_{g,p}$ instead), the prime p will usually be clear from the context and this allows us to avoid such cumbersome symbols. Computing any value $L_g(a)$, where we are given $a \in \mathbb{Z}_p^\times$, is called an instance of the **discrete logarithm problem**.

---

[*] As in Chapter 8, our illustrative examples will use primes of smaller size so the concepts can be more easily understood. The computer implementation material at the end of this chapter provides opportunities for readers to use this and other public key algorithms from this chapter with realistically large primes (provided that they have access to a computing platform with symbolic functionality).

Unlike its inverse function E_g, the discrete logarithm function L_g has no explicit formula. More importantly for cryptographic applications, although there are fast algorithms for evaluating E_g (using fast modular exponentiation—Algorithm 6.5), evaluating discrete logarithms is extremely difficult, and it is believed (like the factoring problem on which RSA's security rests) that there cannot exist an efficient algorithm for evaluating discrete logarithms.

Example 9.1

The number $p = 53$ is a prime number, so it has $\phi(\phi(53)) = \phi(52) = 26$ different primitive roots (mod p) (Theorem 8.7). Thus, half of all the integers (mod 53) serve as primitive roots. Very often $g = 2$ and $g = 3$ are primitive roots, but in any case, we can find one by computing orders of successive integers (which is easily done on a computer, using Algorithm 6.5). To determine the smallest primitive roots, we could simply run through all integers greater than 1 and compute their orders until we find one whose order equals $\phi(p) = p - 1$, that is, $\mathrm{ord}_{53}(g) = 52$. We begin testing with the integer 2. By Proposition 8.5(b) $\mathrm{ord}_{52}(2) \,|\, 52 = 2^2 \cdot 13$, so the only possibilities for $\mathrm{ord}_{52}(2)$ are 2, 4, 13, 26, and 52, and we need only show that the first four options are not possible by showing that $2^2, 2^4, 2^{13}, 2^{26} \not\equiv 1 (\mathrm{mod}\ 53)$. Indeed:

$2^2 \equiv 4$

$2^4 \equiv (2^2)^2 \equiv 16$

$2^{13} \equiv (2^4)^2 \cdot 2^4 \cdot 2 \equiv 16^2 \cdot 16 \cdot 2 \equiv 44 \cdot 32 \equiv 30$

$2^{26} \equiv (2^{13})^2 \equiv 30^2 \equiv 52(\equiv -1)\ (\mathrm{mod}\ 53)$

Thus $g = 2$ is the smallest primitive root of 52.
Using fast exponentiation (Algorithm 6.5), we can quickly compute any power of $g(\mathrm{mod}\ 53)$, but the inverse problem of computing a discrete logarithm is much more tedious. For example, if we wanted to find $\ell = L_g(5)$—that is, to solve the equation $g^\ell \equiv 5(\mathrm{mod}\ 53)$—the only obvious way would be a brute-force approach of computing successive powers of g until we get 5 (mod 53). Of course, with such a relatively small example, a computer could quickly get us the answer $\ell = L_g(5) = 47$, but for a very large value of p, say of size 10^{200}, this brute-force scheme would require, on average, $(p-1)/2$ fast exponential evaluations, or more than 10^{199} of them, way too many to make this approach feasible. By contrast, the inverse function would involve only a single fast exponentiation.

The Diffie–Hellman Key Exchange

Algorithm 9.1: The Diffie–Hellman Key Exchange Protocol

Purpose: Alice and Bob need to create a secret key between them using a public (insecure) communication channel.

Requirements: A large prime number p and a primitive root g (mod p); both g and p can be made public.

Note: All modular powers in this algorithm should be computed with fast modular exponentiation (Algorithm 6.5).

Step 1.   Alice and Bob select integers a and b, respectively (preferably randomly), with $1 \leq a, b \leq p - 2$. Each keeps his or her exponent number secret.

Step 2.   Alice computes $A \equiv g^a \pmod{p}$ and sends this to Bob, while Bob computes $B \equiv g^b \pmod{p}$ and sends this to Alice. These numbers A and B may be sent over public channels.

Step 3.   Alice and Bob obtain the common key K by raising the number they received from the other to their secret exponent (mod p). Indeed,

$$B^a \equiv (g^b)^a \equiv g^{ba} \equiv g^{ab} \equiv (g^a)^b \equiv A^b \pmod{p}$$

Example 9.2

Suppose we use the primitive root, $g = 2$ of the prime $p = 53$ (see Example 9.1). If Alice chooses her secret exponent to be $a = 22$ and Bob chooses his to be $b = 47$, determine the resulting Diffie–Hellman key.

Solution: Bob (computes and) sends Alice the number $B = g^b \equiv 2^{47} \equiv 5 \pmod{53}$, and Alice sends Bob the number $A = g^a \equiv 2^{22} \equiv 43 \pmod{53}$. On her end, Alice computes the common (secret) key as $K = B^a \equiv 5^{22} \equiv 29 \pmod{53}$ and Bob computes it as $K = A^b \equiv 43^{47} \equiv 29 \pmod{53}$. Notice that the resulting common key $K = 29$ is now a shared secret between Alice and Bob, and it was never sent over the (insecure) channel.

Computing Note: In practice, a large-sized prime p would be needed to ensure a sufficiently large keyspace (in order to meet the security needs of the particular symmetric key cipher for which the key is going to be used). As we learned in Chapter 8, finding a primitive root can be a difficult mathematical problem in general, particularly if a factorization of $p - 1$ is not available. The task is much more manageable if p is of a special form such as $p = 2q + 1$, where q is another prime.* Another approach is to simply search for an integer g whose order (mod p) is sufficiently large. A common key will be created by the Diffie–Hellman algorithm, regardless of what integer is used for g (primitive root or not), but higher order g's result in added security.

Exercise for the Reader 9.1

(a) Determine the smallest primitive root g of the prime $p = 79$.

(b) Using the prime p and the primitive root g of part (a), suppose that Alice chooses her secret exponent to be $a = 51$, and Bob chooses his to be $b = 33$. Determine the resulting Diffie–Hellman key.

Suppose that Eve intercepts the agreed-upon parameters p and g, as well as the numbers A and B that were sent. How could she determine the resulting Diffie–Hellman key K of Algorithm 9.1? If she knew either Bob's secret exponent b or Alice's a, she could easily obtain K with a singular modular exponentiation, but there is no other known way to do this. So Eve would be faced with the problem of either determining a from the knowledge that $g^a \equiv A \pmod{p}$ or determining b from the equation $g^b \equiv B \pmod{p}$. But this is just the discrete logarithm problem that was discussed in Chapter 8. It is a very difficult problem for which it is believed that there can exist no efficient algorithm. Up to now, this is the only known attack on the Diffie–Hellman key exchange. It is an unproved conjecture that a successful attack on the Diffie–Hellman key exchange would translate into an efficient scheme for computing discrete logarithms. Of course, in the small-scale setting of Example 9.2, this could easily be done by a brute-force check of all possible modular exponents. But if instead p had 500 digits, this procedure would probably take about 10^{500} attempts— much too large even if all the computers in the world were to be able work together on it for billions of billions of years!

The Quest for a Complete Public Key Cryptosystem

Soon after the revelation of Diffie and Hellman's groundbreaking discovery, cryptographers worked hard to create a full public key cryptosystem, for which we adopt the following definition.

---

* The reason is that $\phi(p) = p - 1 = 2q$ and thus the order of any element in $\mathbb{Z}_p^\times$, can only be 1, 2, q, or $2q$ (in which case it is a primitive root). Prime numbers q for which $2q + 1$ is also prime are called *Sophie Germain primes*. Many Sophie Germain primes have been discovered, so there is no shortage of them for cryptographic applications. In 2007, the largest known Sophie Germain prime had 51,910 digits. It is conjectured, but has not yet been proved, that there are infinitely many Sophie Germain primes; they are named after the French mathematician Marie-Sophie Germain (1776–1831).

Definition 9.3

A **public key cryptosystem** requires that any participating party have two keys: a **public key**, which is available to everyone, inside the system or not, and corresponds to a one-way encryption function, and a **private key**, which is kept secret and corresponds to the decryption function. If Alice needs to send Bob a private message, she simply encrypts the plaintext P to Bob using his (readily available) public key E_B. When Bob receives the resulting ciphertext $C = E_B(P)$, only he has the needed decryption function from his private key $D_B = E_B^{-1}$, so he applies it to the ciphertext to recover the plaintext, that is, $D_B(C) = D_B(E_B(P)) = E_B^{-1}(E_B(P)) = P$.

The security of such a system relies on the security of the one-way functions corresponding to the public keys; that is, it should be essentially impossible to determine the corresponding private key from any public key. In order for this to be done, the problem of this determination should boil down to an intractable mathematical problem, just as breaking the Diffie–Hellman key exchange depended on the discrete logarithm problem. Table 9.1 shows the intractable mathematical problems on which each of the public key cryptosystems that we will present are based.

TABLE 9.1 Intractable Mathematical Problems That Form the Basis of the Security of the Public Key Cryptosystems

Public Key Cryptosystem	Mathematical Problem That Provides Security
RSA	Factoring Integers
ElGamal	Discrete Logarithm Problem
Knapsack	Knapsack Problem*

The RSA Cryptosystem

The first publicly announced discovery of a complete and secure public key cryptosystem was made by a team of three researchers at MIT: Ronald Rivest, Adi Shamir, and Leonard Adleman[†]; see Figure 9.1. Rivest was

---

* The knapsack problem is discussed later in this chapter.
[†] Shamir was born in Tel Aviv. He obtained his computer science PhD at the Weizmann Institute where he later returned as a faculty member (joint in mathematics and computer science—the most prevalent subject areas for cryptographers). He was a visiting professor at MIT from 1977–1980, when he helped to create RSA. He has made many other important contributions to cryptography and has been awarded several very prestigious prizes for this work. Rivest was born in New York. He earned his PhD in computer science from Stanford in 1974 and subsequently earned a professorship at MIT, which he still retains. He founded the company RSA Data Security. He has authored a major textbook on algorithms and created several important cryptographic tools (apart from RSA). Adleman was born in California. In 1976 he earned his PhD in electrical engineering and computer science from Berkeley. His first position was as a mathematics professor at MIT. His research has grown to include several areas of biology, and he is now a chaired professor at USC.

Figure 9.1 Adi Shamir (1952–), Israeli cryptographer; Ronald Rivest (1947–), American cryptographer; and Leonard Adleman (1945–).

extremely intrigued by the 1976 Diffie–Hellman paper, which suggested that such a cryptosystem could be constructed. He recruited the other two to join him in a quest for building the first public key cryptosystem. The difficulty was in designing a true one-way function that could be incorporated into such a system, and the team was perfectly suited for the task. Rivest and Shamir would come up with new ideas for implementation schemes with candidate one-way functions and pass them on to Adleman, who had the strongest mathematical background of the three. Adleman would routinely find flaws in the systems or one-way functions, which sent Rivest and Shamir back to look for new ideas. The ideas being presented by Rivest and Shamir were primarily based on number theoretic problems, Adleman's area of expertise.

This back-and-forth went on for about a year, with Rivest and Shamir never losing hope, while gradually hovering in toward their cherished holy grail, and Adleman helping to keep them from wasting time going down the wrong roads. One April evening in 1977, after the three had celebrated a holiday party given by an MIT student, Rivest was returning to his home when a new idea for a public key cryptosystem sparked into his mind. He was so excited and confident about this idea that he did not sleep that night; rather, he worked straight until dawn ironing out the details that seemed to be coming together very nicely. By morning, Rivest had a draft paper written up that he apprehensively gave to Adleman. This time, however, Adleman could find no flaws and was able to certify the system as a legitimate public key cryptosystem. Adleman's only request was that Rivest change the ordering of the names that he had listed on the draft joint paper, which were in alphabetical order, to put his name last.[*] This change was made, and their system was coined the **RSA cryptosystem**. Its security is based on the difficulty of the factorization problem, and it remains one of the most successful and powerful public key cryptosystems. A year

---

[*] In mathematical publications, it is customary to always list the author's names in alphabetical order. This is most certainly not the case in some of the other sciences, such as biology, where being the "first author" of a paper carries significantly more weight. Practices in the other academic fields vary between these two extremes.

> ### Algorithm 9.2 The RSA Public Key Cryptosystem
>
> *Purpose:* Alice needs to send Bob a private message, but they have not yet met to exchange any cryptosystem keys.
>
> *Requirements:* Bob will first need two different prime numbers, $p \neq q$, that he should keep secret. He multiplies these primes to obtain an integer $n = pq$; n will be made public. He selects (preferably randomly) an integer d that is relatively prime to $\phi(n) = (p-1)(q-1)$.[*] He then computes (using the extended Euclidean Algorithm 2.2) the inverse e of $d \pmod{\phi(n)}$.
>
> *Plaintext, Ciphertext Spaces:* $\mathscr{P} = \mathscr{C} = \mathbb{Z}_n$, where $n = pq$, with p and q prime numbers.
>
> *Keyspace:* $\mathscr{K} = \{(p,q,d,e) : de \equiv 1 \pmod{\phi(n)}, \text{ where } n = pq\}$. From a system key $\kappa = (p,q,d,e)$, Bob's private key is (n, d) and the corresponding public key is (n, e). The parameter e is called the **encryption exponent** and the parameter d is called the **decryption exponent**.
>
> *Encryption Scheme:* $e_\kappa : \mathbb{Z}_n \to \mathbb{Z}_n$, defined by $e_\kappa(P) \equiv P^e \pmod{n}$.
> *Decryption Scheme:* $d_\kappa : \mathbb{Z}_n \to \mathbb{Z}_n$, defined by $d_\kappa(C) \equiv C^d \pmod{n}$.

later, they showed how to incorporate *digital signatures* that allow receivers to *authenticate* that a certain message or document did actually come from the alleged person who sent it. It also makes it impossible for the sender to deny having sent a message that has his or her digital signature (*nonrepudiation*).

We now describe the RSA cryptosystem.

Both encryption and decryption can be performed efficiently using the squaring algorithm for fast modular exponentiation (Algorithm 6.5). Although the RSA algorithm was set up to accept only single integers as plaintexts, there are natural and effective ways to expand the algorithm into a block cipher (see Chapter Exercise 27). Before showing that the decryption functions above are the inverses of the corresponding encryption functions, we give a small example of RSA. As explained earlier, the primes used in the chapter examples are too small to be secure, but they are chosen so that the concepts are more easily illustrated. The computer implementation material at the end of the chapter provides the reader with opportunities to work with much larger keys.

Example 9.3

Suppose that Bob adopts the RSA cryptosystem with primes $p = 37$ and $q = 41$. He chooses the (public key) encryption exponent to be $e = 49$.

(a) Show that Bob's choice of encryption exponent is legitimate, and find his corresponding (private key) decryption exponent d.

---

[*] This evaluation of $\phi(pq)$ easily follows from Proposition 8.3.

(b) Suppose that Alice encrypts the plaintext message $P = 44$ using the RSA cryptosystem with Bob's public key $(n, e) = (1517, 49)$. What is the resulting ciphertext that would be sent to Bob?

(c) Go through the decryption process that would need to be done at Bob's end, using his private key (n, d) with the decryption exponent that was determined in part (a).

Solution: Part (a): Since $\phi(n) = (p - 1)(q - 1) = 36 \cdot 40 = 1440$, we may compute that $\gcd(e, \phi(n)) = \gcd(49, 1440) = 1$, so $e = 49$ is indeed a legitimate encryption exponent. We use the extended Euclidean Algorithm 2.2 to compute $d \equiv e^{-1} \equiv 529$, which is the decryption exponent.

Part (b): The ciphertext is computed as $C \equiv 44^{49} \equiv 1069$ (mod 1517)..[*]

Part (c): Bob's decryption system would need to raise the ciphertext $c = 1069$ to the decryption exponent $d = 529$ (mod $n = 1517$): $1069^{529} \equiv 44 \pmod{1517}$—the original plaintext message.

Proposition 9.1

The decryption functions listed in Algorithm 9.2 work; that is, $d_\kappa(e_\kappa(m)) \equiv m \pmod{n}$ for any $m \in \mathbb{Z}_n$.

Proof: Since $de \equiv 1 \pmod{\phi(n)}$, we can write $de = \ell\phi(n) + 1$, for some $\ell \in \mathbb{Z}_+$. By Proposition 8.3, we have $\phi(n) = \phi(pq) = (p - 1)(q - 1)$. Also, by Fermat's little theorem (Theorem 8.2), it follows that $d_\kappa(e_\kappa(m)) \equiv (m^e)^d \equiv m^{\ell\phi(n)+1} \equiv (m^{(p-1)})^{(q-1)\ell} m \equiv 1^{(q-1)\ell} m \equiv m \pmod{p}$, provided that $\gcd(m, p) = 1$. But if this latter condition does not hold, this would mean that $p \mid m$, so that $d_\kappa(e_\kappa(m)) \equiv (m^e)^d \equiv 0 \equiv m \pmod{p}$. We have thus shown that $d_\kappa(e_\kappa(m)) \equiv m \pmod{p}$ for any $m \in \mathbb{Z}_n$. The same argument shows that $d_\kappa(e_\kappa(m)) \equiv m \pmod{q}$ for any $m \in \mathbb{Z}_n$. It follows from the Chinese remainder theorem (Theorem 2.13) [or more directly since p and q are different primes that both divide $d_\kappa(e_\kappa(m)) - m$], that $d_\kappa(e_\kappa(m)) \equiv m \pmod{n}$ for any $m \in \mathbb{Z}_n$, as was needed to show. $\square$

The above proof breaks down if $p = q$ (why?), and thus shows one reason why the RSA cryptosystem requires $p \neq q$. Besides, if n were chosen

---

[*] We remind the reader of the problems of working with large integers on computing platforms (as we have discussed in Chapters 6 and 8). The number 44^{49} has 81 digits and could be easily computed (exactly) on a symbolic computing platform. However, for users of a floating point system, it should not be computed directly since the answer would not be accurate. Using Algorithm 6.5 would circumvent this problem. The same comments apply to the calculation needed in part (b), where the power, if computed first without modular arithmetic, would have over 1,600 digits. Working with integers having more than 16 digits or so would require a symbolic computing platform for all calculations. Thus, with the exception of a few (clearly marked) exercises, all of the material in this section involves values of n with at most eight digits (so the product of two numbers would have at most 16 digits, before reducing mod n). This is sufficient for presenting and understanding the concepts of RSA.

to be a perfect square, it would be easy to factor just by taking its square root. Care should also be taken to ensure that p and q are not too close, since in this case also n can be easily factored (see Exercise 28). From the proof of Proposition 9.1, we extract the following corollary that will have important consequences regarding the security of RSA systems.

Corollary 9.2

If d' is any positive integer that satisfies $d'e \equiv 1 \pmod{\mathrm{lcm}(p-1, q-1)}$, then d' could also be used as a decryption exponent for the RSA algorithm.

As an extreme case where this corollary can show a weakness in an RSA system, consider the two Mersenne primes $p = 2^{521} + 1$ and $q = 2^{607} + 1$. Ostensibly, the RSA decryption exponent d is found modulo $(p-1)(q-1) = 2^{1128} \approx 10^{340}$. Corollary 9.2, however, says that the search can be restricted modulo $\mathrm{lcm}(p-1, q-1) = 2^{607} \approx 10^{183}$, a significantly less complex problem. One way to avoid this sort of weakness in RSA algorithms is to use primes that do not have a lot of small factors. There are some other recommendations that can be made regarding effective choices for the two primes p and q, in order to enhance the security of RSA against factorization of the modulus attacks; for more details, see [Mol-03] or [MeOoVa-96].

We discuss the security of the RSA system and how it relies on the difficulty of the factorization problem, together with the security of the other methods we present, later in this chapter. We next summarize a few implementation issues.

In order to use RSA, one first needs to generate or obtain two sufficiently large primes $p \neq q$. Chapter 8 dealt with this problem and presented some practical algorithms. Note that the primes used to construct $n = pq$ will be much smaller and, thus, much more tractable in size than n. The size of the modulus n for an RSA system is determined by the current state-of-the-art methods for factoring. For example, if a 300-digit modulus is required, then we will need two primes with at least 150 digits each. Once n is determined, an encrypting exponent e can be randomly generated from the integers less than $\phi(n)$ and relatively prime to it (that is, from $\mathbb{Z}_{\phi(n)}^{\times}$). The decryption exponent is then the inverse of $e \pmod{\phi(n)}$ and can be found using the extended Euclidean algorithm (Algorithm 2.2).

Exercise for the Reader 9.2

Suppose that it is desired to create an RSA cryptosystem using the two primes $p = 67$ and $q = 37$.

(a)   Determine the smallest encryption exponent e that is greater than 1000, and could be used for such an encryption exponent, and use it to encrypt the plaintext message $P = 2012$.

(b)   What is the decryption exponent d for the RSA of part (a)?
Decrypt the ciphertext message $C = 1999$.

Digital Signatures and Authentication

One problem that needs special attention for public key cryptosystems is that since public keys are freely available, Mallory could send Bob a message saying that it was from Alice. With just a basic public key algorithm (such as RSA), Bob would have no way of knowing whom it came from (perhaps after responding to Alice about the message using her public key, they might eventually detect the forgery, but this might be too late). Written documents are usually certified using an actual signature of the sender or authorizing party. The recipient then uses the signature to authenticate that the document came from the indicated sender. Moreover, once the recipient has a signed document, it is not possible for the signatory to repudiate the signature (or renege the document or contract). Of course, this traditional system does have its weaknesses because of susceptibility to forgeries, but countermeasures (such as notarizations) have allowed the system to continue to be used for thousands of years in all modern societies. Public key communication therefore also needs an effective signature and authentication system. The forgery risks mentioned above would become unacceptably high if we were to simply scan our (handwritten) signature and paste it onto electronic documents, since Mallory might be able to do the same (if she got hold of any signed document, such as a used check or credit card slip, or even an electronic file with the scanned signature).

A **digital signature** system generally modifies the document to be signed in a way that the recipient will be able to unequivocally ascertain that it was indeed sent from the designated sender (signatory). The following algorithm shows a method for creating digital signatures in any public key cryptosystem.

Algorithm 9.3: A General Digital Signature
for Public Key Cryptosystems

We assume that a public key cryptosystem has been established that includes Bob and Alice as participating parties. Suppose that Bob needs Alice to sign a certain document P. Alice uses her private key D to *encrypt* P and sends Bob the resulting **digital signature** $s = D(P)$. Bob then uses Alice's corresponding public key $E = D^{-1}$ and applies it to s to obtain $E(s) = D^{-1}(D(P)) = P$, that is, the original document, thereby verifying Alice's signature. If $E(s) \neq P$, then Bob would reject the signature as invalid.

As with all public key cryptosystems, it is not feasible to determine Alice's private key D from public information (including Alice's public key E). Thus, if Mallory wanted to attach Alice's digital signature to a different message p' she would need to transmit the corresponding signature $s' = D(p')$, and this would not be feasible to compute with access to only Alice's public key E.

Example 9.4: An RSA Digital Signature

(a) Suppose that Bob needs Alice to sign the document message $P = 1776$. Compute Alice's corresponding RSA digital signature s (that she will send Bob over public channels) given that Alice's RSA system key is $\kappa = (p, q, d, e) = (23, 79, 103, 883)$.

(b) Go through the authentication process that would need to be done by Bob when he receives Alice's digital signature.

Solution: Part (a): From Algorithm 9.2, Alice's (private key) decryption mapping is $D(P) \equiv P^d (\bmod n)$, where $n = pq = 1817$. Thus, by Algorithm 9.3, her digital signature will be $s = D(P) \equiv P^d \equiv 1776^{103} \equiv 1790 (\bmod 1817)$.

Part (b): To verify Alice's signature, according to Algorithm 9.3, Bob must apply Alice's (public key) encryption mapping, $D(m) \equiv m^e (\bmod n)$, to the purported signature $s = 1790$: $s^e \equiv 1790^{883} \equiv 1776 (\bmod 1817)$. Since this is the original document P, Alice's signature has been verified.

We point out that the digital signature scheme of Algorithm 9.3 is not designed to encrypt the plaintext document. Since Eve has access to Alice's public key, just like Bob, she could apply it to the digital signature to reconstruct the plaintext document. With a small amount of additional work, however, Alice could encrypt her digitally signed document using Bob's public key E_B. Thus, she would send Bob $E_B(s) = E_B(D_A(P))$, where D_A denotes Alice's private key decryption mapping. Bob would both decrypt the plaintext and verify Alice's signature by first applying his private key and then Alice's public key. [*Proof:* $E_A(D_B(E_B(s))) \equiv E_A(s) \equiv E_A(D_A(P)) \equiv P$. □] We point out one important observation: In sending a single RSA communication or digital signature, the communicating parties need not share the same modulus (or two primes) since only one party's encryption/decryption mappings are used for both creating the ciphertext (or digital signature) and decrypting it (or authenticating it). In this latter protocol of encrypting a document with a digital signature, both parties' encryption/decryption mappings are used, so they should choose their keyspaces so that the plaintext/ciphertext spaces will be the same.[*]

---

[*] Here is an extremely simple example to show what can go wrong. Suppose that Bob's n was $n_B = 11 \cdot 7 = 77$ and Alice's is 6887 (as in the example) and that Alice's digital signature turned out to be 770. Then, in Bob's system, $s \equiv 0$, so when Alice applies E_B to it, she will obtain $0^{e_B} \equiv 0 (\bmod 77)$ and when Bob applies E_A to the result he would likewise obtain 0—not the original plaintext. Actually, all that is needed is that $n_B > s$.

Exercises for the Reader

Exercise 9.3

(a) Suppose that Bob needs Alice to sign the document $P = 5$. Compute Alice's corresponding RSA digital signature s (that she will send Bob over public channels), given that Alice's RSA private key is $\kappa = (p,q,d,e) = (71, 97, 139, 2569)$.

(b) Go through the authentication process that would need to be done on Bob's end, using Alice's public key $(n,e) = (6887, 2659)$.

(c) If Alice wanted to send an encrypted version of her digitally signed document of part (a), what would she send to Bob, given that Bob's public key is $(n,e_B) = (6887, 1007)$?

(d) Compute Bob's private key exponent d_B and perform the decryption and digital signature verification that Bob would need to do with the ciphertext of part (c).

Exercise 9.4

(a) Show how, in an RSA cryptosystem, if one knows n (always public) and $\phi(n)$, then the factorization $n = pq$ can be readily obtained.

(b) Demonstrate your method to factor $n = 628883$, given that $\phi(n) = 627288$.

The ElGamal Cryptosystem

We are now nicely prepared to discuss the *ElGamal cryptosystem.*[*] It is a bit more complicated than the RSA system, but it is still quite elementary and serves as a nice extension of the original Diffie–Hellman key exchange into a full cryptosystem. Moreover, unlike the RSA system, ElGamal has natural extensions to more sophisticated cryptosystems (for example, elliptic curve cryptosystems).

---

[*] This system was developed in 1985, along with an associated digital signature scheme, by Taher Elgamal (1955–), an Egyptian-American cryptographer. Elgamal did his undergraduate work in Cairo, and then studied at Stanford where he earned a PhD in computer science. His digital signature algorithm was adopted by the NIST as the *Digital Signature Standard (DSS)*. Elgamal founded his own security company (*Securify*), and before that he worked as a senior scientist at *Netscape* and at *RSA Security*.

Algorithm 9.4: The ElGamal Public Key Cryptosystem

Purpose: Alice needs to send Bob a private message, but they have not yet met to exchange any cryptosystem keys.

Requirements: Bob will first need a (large) prime number p and a corresponding primitive root $g \pmod p$. These parameters will be made public. He chooses a secret exponent b (preferably randomly) in the range $0 < b < p-1$. Alice will need to also choose her own private exponent a in the range $0 < a < p-1$.

Plaintext, Ciphertext Spaces: $\mathscr{P} = \mathbb{Z}_p$, $\mathscr{C} = \{(x,y) : x, y \in \mathbb{Z}_p\}$, where p is a prime number. Thus the ciphertexts consist of ordered pairs of mod p integers (that is, length-2 vectors of mod p integers).

Keyspace: $\mathscr{K} = \{(p, g, b) : g \text{ is a primitive root } (\bmod\, p), 0 < b < p - 1\}$. An element $\kappa = (p, g, b)$ is (Bob's) **private key**; it will contain all the information needed to encrypt or decrypt a message. The corresponding (Bob's) **public key** is (p, g, B), where $B \equiv g^b \pmod p$.

Encryption Scheme: (Alice) randomly selects an integer a (her private key), with $0 < a < p-1$, and she computes $A \equiv g^a \pmod p$. She then computes $C \equiv B^a P \pmod p$,* where P is the plaintext message. The ciphertext will be the ordered pair (A,C). In summary, $e_\kappa : \mathbb{Z}_p \to \mathbb{Z}_p^2 :: e_\kappa(m) \equiv (A, C) \pmod p$.

Decryption Scheme: $d_\kappa : e_\kappa(\mathbb{Z}_p) \to \mathbb{Z}_p :: d_\kappa((A, C)) \equiv A^{p-1-b}C \pmod p$. Note that the c, the second component of the ciphertext, is simply the plaintext message multiplied by B^a, which is the common Diffie–Hellman key between Alice and Bob, so that $B^a \equiv A^b \pmod p$. Thus, Bob can decrypt by simply multiplying C by the inverse of the Diffie–Hellman key. The formula that was given amounts to the same operation, but avoids having to compute a modular inverse (so is a bit quicker). Here is a proof of this fact:

$$A^{p-1-b}C \equiv g^{a(p-1-b)}A^b P \equiv (g^{(p-1)})^a (g^a)^{-b} A^b P \equiv 1^a A^{-b} A^b P \equiv P \pmod p$$

Notice that since $0 < b < p-1$, the **decryption exponent** $p - 1 - b$ lies also in the same (nonnegative) range.

As with RSA, both ElGamal encryption and decryption can be performed efficiently using the squaring algorithm for fast modular exponentiation (Algorithm 6.5).

Example 9.5

Using the primitive root $g = 2$ of the prime $p = 53$ from Example 9.1, suppose that Alice chooses her secret exponent to be $a = 22$ and Bob chooses his to be $b = 47$.

(a) Using the ElGamal system, what would be the ciphertext when Alice encrypts the message $P = 44$ to send to Bob?

(b) Perform the ElGamal decryption process that would need to be done at Bob's end to decrypt Alice's message.

---

* Note that a, b, A, and B are exactly as in the Diffie–Hellman key exchange, so a and b are Alice's and Bob's private keys, and A and B are Alice and Bob's public keys, respectively. Furthermore, $B^a \equiv A^b$ is the Diffie–Hellman key.

Solution: Part (a): As in Example 9.2, $A = 43$, $B = 5$, so that Alice can compute $C \equiv B^a P \equiv 29 \cdot 44 \equiv 4 \pmod{53}$ Thus, the entire ciphertext would be $(A, C) = (43, 4)$.

Part (b): The decryption exponent is $p - 1 - b = 5$ and $A^5 \equiv 43^5 \equiv 11 \pmod{53}$, and so $d_\kappa((A,C)) \equiv A^{p-1-b} C \equiv 11 \cdot 4 \equiv 44 \pmod{53}$, as expected.

Exercise for the Reader 9.5

(a) Determine the smallest primitive root g greater than 700 for the prime $p = 1231$.

(b) Suppose that Alice and Bob will be using the ElGamal crypto-system with these parameters and that Alice chooses her secret exponent to be $a = 212$. What would be the ciphertext when Alice encrypts the message $P = 44$ to send to Bob?

(c) Perform the ElGamal decryption process that would need to be done at Bob's end to decrypt Alice's message.

Many of the guidelines for choosing "safe" primes that make $p - 1$ resistant to factorizations for use in RSA cryptosystems apply equally well to ElGamal cryptosystems. In particular, it is important to use primes p for which $p - 1$ does not have a lot of small prime factors. See our previous comments and the references cited for more details on this.

Digital Signatures with ElGamal

The general digital signature algorithm (Algorithm 9.3) can be used with the ElGamal cryptosystem to digitally sign documents, but there is an improved digital signature algorithm for the ElGamal setting. This algorithm allows the possibility of several different legitimate signatures per document, per individual—a foreign concept in the traditional signature setting, as well as in the RSA digital signature setting.

Algorithm 9.5: The ElGamal Digital Signature Protocol

We assume that an ElGamal cryptosystem has been established that includes Bob and Alice as participating parties. Suppose that Alice needs to sign a certain document P and send it to Bob. As in Algorithm 9.4, we let a denote Alice's private key, and $A \equiv g^a$ her corresponding public key.

Step 1: Attaching a Signature. Alice randomly selects a **signature exponent** d, $1 \le d \le p - 2$, which is relatively prime to $p - 1$. She uses it to compute $r \equiv g^d \pmod{p}$, and then $s \equiv d^{-1}(P - ar) \pmod{p - 1}$.[*]

---

[*] As usual, Alice could use the extended Euclidean algorithm (Algorithm 2.2) to compute $d^{-1} \pmod{p - 1}$.

We emphasize that whereas r is computed using mod p arithmetic, s is computed (using r) but with mod p − 1 arithmetic. The **digitally signed document** that Alice sends to Bob will be the ordered triple (P, r, s). (Note that the first two components of this vector are mod p integers, while the third is a mod $p − 1$ integer.)

  Step 2: Authentication of Signature. Upon receipt of the signed document (P, r, s), to verify that it was digitally signed by Alice, Bob uses Alice's public key A and checks to see whether $A^r r^s \equiv g^P \pmod{p}$. The signature is authentic if this congruence is valid. We caution the reader not to confuse the uppercase P (plaintext document) that is the exponent on g with the lowercase p (modulus).

Unlike the RSA digital signature scheme, the document P that has been signed is included (unencrypted) in the ElGamal digital signature scheme.[*] This is because it is not feasible for Bob to compute it directly from the quantities r and s. We need to explain why the authentication process (Step 2) works. Indeed, the signature scheme of Step 1 is equivalent to $ds \equiv P − ar$ or $P \equiv ds + ar \pmod{p − 1}$. Since [by Proposition 8.5(c)] powers with the same base and exponents that are congruent mod $p − 1$ are congruent mod p, we thus obtain (in the case of a valid signature) $g^P \equiv g^{ds+ar} \equiv (g^a)^r (g^d)^s \equiv A^r r^s \pmod{p}$, as asserted.

Example 9.6

  (a) Using the ElGamal digital parameters of Example 6.5, what is the digital signature that Alice would send to Bob if she picks her signature exponent to be $d = 15$ to sign the document $P = 44$?

  (b) Perform the digital signature verification process that would need to be done at Bob's end to authenticate Alice's signature.

Solution: Part (a): We first use the extended Euclidean algorithm (Algorithm 2.2) to compute $d^{-1} \equiv 7 \pmod{52}$. Now we may compute both $r \equiv g^d \equiv 2^{15} \equiv 14 \pmod{53}$ and $s \equiv d^{-1}(P − ar) \equiv 7(44 − 22 \cdot 14) \equiv 24 \pmod{52}$. Thus, the digitally signed document that Alice sends to Bob is $(P, r, s) = (44, 14, 24)$.

  Part (b): To authenticate that the document 44 (the first component of the vector received from Alice) was really signed by her, Bob uses the second two components of the vector received along with Alice's public ElGamal key $A = 43$, computes $A^r r^s \equiv 10 \cdot 10 \equiv 47 \pmod{53}$, and compares this with $g^P \equiv 2^{44} \equiv 47 \pmod{53}$. Since the results agree, Alice's signature has been authenticated.

---

[*] But recall that in the general digital signature protocol of Algorithm 9.3, the plaintext, although not explicitly sent, is readily computed by the public keys and parameters.

Exercise for the Reader 9.6

(a)   Using the ElGamal digital parameters of Exercise for the Reader 9.5(b), what is the digital signature that Alice would send to Bob if she picks her signature exponent to be $d = 337$ to sign the document $P = 44$?

(b)   Perform the digital signature verification process that would need to be done at Bob's end to authenticate Alice's signature.

Caution: It is very important in the ElGamal digital signature scheme that a new signature exponent be used for each new document to be signed. This is because if Mallory were to intercept two different digitally signed documents of the form (P, r, s), and (P', r, s'), where both used the same signature exponent and hence had the same value of r, then she might be able to use this information to obtain both Alice's secret digital signature exponent d and her private key a. Thus, Mallory would subsequently be able to digitally forge Alice's signature and read her encrypted messages. The following exercise for the reader asks to verify this assertion.

Exercise for the Reader 9.7

Show how Mallory might be able deduce Alice's private key a if she intercepted two different digitally signed documents, (P, r, s) and (P', r, s'), from Alice in which the same signature exponents were used.

The last public key cryptosystem that we present has its security based on the following difficult number theoretic problem, which does not involve prime numbers.

Knapsack Problems

Definition 9.4

Given a vector of distinct positive integers $[a_1\ a_2\ a_3\ \cdots\ a_n]$, the **object weights** and a subcollection of these numbers, $a_{i1}, a_{i2}, \cdots, a_{ik}$ (thought of as the weights of objects selected to be put into a knapsack), it is easy to compute the sum $s = a_{i1} + a_{i2} + \cdots + a_{ik}$ (the total weight of the objects put into the knapsack). The **knapsack problem** is the opposite: given a positive integer s, determine, if possible, a subcollection $a_{i1}, a_{i2}, \cdots, a_{ik}$ having total weight s. (In other words, if we know the total weight of the objects that were put into the knapsack, the problem is to determine the objects.) Such a subcollection of objects is called a **solution** to the given instance of the knapsack problem.

Instances of the knapsack problem need not have any solution, and they may have multiple solutions.

Example 9.7

Find all solutions to the knapsack problem with the following parameters: $[a_1, a_2, a_3, a_4, a_5, a_6] = [3, 4, 6, 8, 10, 12]$, and $s = 28$.

Solution: By inspection we find that there are two different solutions to this knapsack problem:

$a_3 + a_5 + a_6 = 6 + 10 + 12 = 28$, and $a_2 + a_3 + a_4 + a_5 = 4 + 6 + 8 + 10 = 28$

Although the above example was small enough to do by hand, note that the number of possible knapsack collections that would have needed to be considered using a brute-force approach would be $2^6 = 64$ (by the multiplication principle, since for each of the six objects we have two choices: include it or do not include it). In general, the knapsack problem is an *NP complete problem*, which puts it in the same class as some of the most difficult discrete problems for which polynomial time solutions are strongly believed not to exist. Since both problems are *NP* complete, if a polynomial time algorithm for the knapsack problem is ever found, then it could be translated into a polynomial time algorithm for prime factorization (and vice versa); see [GaJo-79] for more details.

It will be convenient to reformulate the knapsack problem as follows: We introduce a length-n binary vector $[x_1 \ x_2 \ x_3 \ \cdots \ x_n]$. If we use the interpretation that object i (with weight a_i) is included in the knapsack (solution) if, and only if, $x_i = 1$, then the total weight of the included objects is simply the dot product $[x_1 \ x_2 \ x_3 \ \cdots \ x_n] \cdot [a_1 \ a_2 \ a_3 \ \cdots \ a_n] = x_1 a_1 + x_2 a_2 + x_3 a_3 + \cdots + x_n a_n$. So a solution of the knapsack problem is a bit vector $[x_1 \ x_2 \ x_3 \ \cdots \ x_n]$ for which the above dot product equals s. In the previous example, the two solutions correspond to the binary vectors $[x_1 \ x_2 \ x_3 \ x_4 \ x_5 \ x_6] = [0, 0, 1, 0, 1, 1]$ and $[0, 1, 1, 1, 1, 0]$.

We will need the vector of object weights to have some additional properties when we create knapsack-based cryptosystems. Although general knapsack problems are intractable, certain classes can be solved very quickly. One such class is described in the following definition.

Definition 9.5

An object weight vector $[a_1 \ a_2 \ a_3 \ \cdots \ a_n]$ in a knapsack problem is said to be **superincreasing** if each object weight is greater than the sum of all of the preceding object weights: $a_i > a_1 + a_2 + \cdots + a_{i-1}$, for $i = 2, 3, \ldots, n$.

Any knapsack problem with superincreasing weights will have at most one solution, and it can be solved very quickly. We give an example that will motivate the general algorithm, and then present the algorithm.

Example 9.8

(a) Check that the sequence $[a_1, a_2, a_3, a_4, a_5, a_6] = [1, 2, 4, 9, 20, 48]$ is superincreasing.

(b) Find all solutions to the knapsack problem with object weights specified by the sequence of part (a), and $s = 27$.

Solution: Part (a): $a_2 = 2 > 1 = a_1$, $a_3 = 4 > 3 = a_1 + a_2$, $a_4 = 9 > 7 = a_1 + a_2 + a_3$, $a_5 = 20 > 16 = a_1 + a_2 + a_3 + a_4$, $a_6 = 48 > 46 = a_1 + a_2 + a_3 + a_4 + a_5$. This shows that $[a_1, a_2, a_3, a_4, a_5, a_6]$ is superincreasing.

Part (b): We seek a binary vector $[x_1\ x_2\ x_3\ x_4\ x_5\ x_6]$ whose dot product with $[a_1, a_2, a_3, a_4, a_5, a_6]$ is $s = 27$; that is, $x_1 + 2x_2 + 4x_3 + 9x_4 + 20x_5 + 48x_6 = 27$. First, it is clear that we must have $x_6 = 0$ (the last weight is too heavy). Next, it is easy to see that $x_5 = 1$. The reason is that if we left out the fifth weight of 20 (which fits), our total weight could only add up to at most $a_1 + a_2 + a_3 + a_4$, which is less than the fifth weight. Now the problem is reduced to solving the smaller binary vector equation $x_1 + 2x_2 + 4x_3 + 9x_4 = 7$. By the same token, we see that $x_4 = 0$ and $x_3 = 1$, and the problem is again reduced to solving $x_1 + 2x_2 = 3$, which forces $x_1 = x_2 = 1$ We have thus found the unique solution: $1 + 2 + 4 + 20 = 27$.

In general, the same approach that was used in part (b) of the above example can be used to prove (see Exercise 31) that any knapsack problem with superincreasing weights can be solved with the following simple approach: Go through all weights, starting with the heaviest and going down. Each time a weight is considered, add it to the knapsack if it will fit, otherwise move on. After we have finished with the last weight, if the total weight of the knapsack equals s, the unique solution has been found; otherwise, there is no solution. We record this general result and follow it with a formal statement of this algorithm.

Proposition 9.3

A knapsack problem with superincreasing weights can have at most one solution, and the following is a fast algorithm for solving any such problem.

Algorithm 9.6: Solving a Knapsack Problem with Superincreasing Weights

Input: A superincreasing weight vector $[a_1\ a_2\ a_3\ \cdots\ a_n]$ and a positive integer s.

Output: Either a binary vector $[x_1\ x_2\ x_3\ \cdots\ x_n]$ that specifies the unique solution of the knapsack problem with inputted parameters, or a message that there is no solution.

Step 1: Initialize $S = s$, Index $= n$

Step 2: IF $S \geq a_{\text{Index}}$
    Set $x_{\text{Index}} = 1$, and Update $S = S - a_{\text{Index}}$
ELSE
    Set $x_{\text{Index}} = 0$
END <IF>
Update Index = Index − 1

Step 3: If Index > 0, go back to Step 2; otherwise, go to Step 4.
Step 4: IF S equals 0
    Output solution vector x
ELSE
    Output message: No Solution
END <IF>

Exercises for the Reader

Exercise 9.8

(a)   Check that the sequence $[a_1, a_2, a_3, a_4, a_5, a_6] = [3, 5, 9, 18, 36, 100]$ is superincreasing.

(b)   Find all solutions to the knapsack problem with object weight specified by the sequence of part (a) and $s = 27$.

Exercise 9.9

(a)   Suppose that $[a_1 \ a_2 \ a_3 \ \cdots \ a_n] = [1 \ 2 \ 4 \ \cdots \ 2^n]$. Show that this sequence is superincreasing and describe the set of all positive integers s for which the corresponding knapsack problem will have a solution.

(b)   Show that any superincreasing sequence $[a_1 \ a_2 \ a_3 \ \cdots \ a_n]$ must satisfy $a_i \geq 2^{i-1}$ for each i. Thus, the superincreasing sequences of part (a) are the smallest superincreasing sequences.

The Merkle–Hellman Knapsack Cryptosystem

In the same year that RSA was published, another public key cryptosystem was introduced by Ralph Merkle and Martin Hellman [MeHe-78]. Such systems are known as *knapsack cryptosystems*, and their security rests on the difficulty of corresponding knapsack problems. There are many different special sorts of knapsack problems that give rise to an assortment of knapsack cryptosystems. In this development, we restrict our attention to a single prototypical knapsack problem and its associated cryptosystem. This knapsack cryptosystem is the original one in the Merkle–Hellman paper. For a more complete treatment of knapsack problems, along with their complexity and applications, we refer the reader to [KePfPi-04].

The Merkle–Hellman cryptosystem is based on the clever idea that the secret decryption key will convert an intractable general knapsack problem into one with superincreasing object weights that can be easily solved. As with any public key cryptosystem, the essential ingredient is an appropriate one-way function. The following proposition provides the one that will be used.

Proposition 9.4: One-Way Functions for Knapsack Cryptosystems

Given a superincreasing object weight vector $[a_1\ a_2\ a_3\ \cdots\ a_n]$ for a knapsack problem, we define an associated function $f_a : \{\text{length-}n\text{ by binary vectors}\} \to \mathbb{Z}_+$ defined by $f_a(X) = x_1 a_1 + x_2 a_2 + x_3 a_3 + \cdots + x_n a_n$, where $X = [x_1\ x_2\ x_3\ \cdots\ x_n]$. Choose a modulus $m > a_1 + a_2 + a_3 + \cdots + a_n$. Next, randomly choose a multiplier w, $1 < w < m$, that is relatively prime to m. We use these parameters to define a new vector of object weights $[b_1\ b_2\ b_3\ \cdots\ b_n]$, where b_i is the smallest nonnegative integer that is congruent to $wa_i \pmod{m}$, for each index i.

(a) Both f_a and f_b are injective.[*] Also, if their codomains are restricted to their ranges (so both become bijections), the function f_a is never a one-way function, but f_b will typically be a one-way function.

(b) *Trapdoor for* f_b. Given a knapsack problem with object weights $[b_1\ b_2\ b_3\ \cdots\ b_n]$ and parameter $s > 0$, a solution that corresponds to a solution of $f_b(X) = s$ can be found (using the number w—the **trapdoor**) as the corresponding solution of $f_a(X) = s'$, where w^{-1} is the inverse of $w \pmod{m}$, and s' is the least positive integer congruent to $w^{-1} \cdot s \pmod{m}$.

Proof: Part (a): Since $f_a(X) = s$, if, and only if, $X = [x_1\ x_2\ x_3\ \cdots x_n]$ solves the knapsack problem with parameters $[a_1\ a_2\ a_3\ \cdots a_n]$ and s, the fact that f is injective follows from the uniqueness of solutions to knapsack problems with superincreasing weight sequences. If we have $f_b(X) = f_b(X')$, for two binary vectors $X = [x_1\ x_2\ x_3\ \cdots\ x_n]$ and $X' = [x_1'\ x_2'\ x_3'\ \cdots\ x_n']$, this means that $\sum_{i=1}^{n} b_i x_i = \sum_{i=1}^{n} b_i x_i'$, which in turn implies $\sum_{i=1}^{n} w^{-1} b_i x_i \equiv \sum_{i=1}^{n} w^{-1} b_i x_i' \pmod{m}$. But since $w^{-1} b_i \equiv a_i$, the latter congruence can be rewritten as $\sum_{i=1}^{n} a_i x_i \equiv \sum_{i=1}^{n} a_i x_i' \pmod{m}$. By choice of m, each of these sums is a nonnegative integer less than m; thus, the sums must be equal so that $f_a(X) = f_a(X')$. But since f_a is one-to-one, we conclude that $X = X'$, and the proof that f_b is one-to-one is thus complete.

---

[*] The function f_b is defined in the same fashion as f_a. (The sequences need not be superincreasing for this definition.)

Since solving $f_a(X) = s$ is equivalent to a superincreasing knapsack problem, which is easy to solve using Algorithm 9.6, the function f_a is not a one-way function. Since solving $f_b(X) = s$ is equivalent to a knapsack problem with a weight sequence $[b_1 \ b_2 \ b_3 \ \cdots \ b_n]$ where each of the weights is obtained from the a_i's by modular multiplication by $w \pmod m$. The sequence typically will no longer be superincreasing, so that the knapsack problem will be a general one that typically will be difficult to solve.

Part (b): If we need to solve $f_b(X) = s$ and we know w, we can proceed as follows: Since $f_b(X) = s$ is equivalent to $\sum_{i=1}^{n} b_i x_i = s$, which implies $\sum_{i=1}^{n} b_i x_i \equiv s$ (mod m), we can (first compute w^{-1}) multiply both sides of this congruence by w^{-1} to obtain $\sum_{i=1}^{n} w^{-1} b_i x_i \equiv w^{-1} s \Rightarrow \sum_{i=1}^{n} a_i x_i \equiv w^{-1} s \triangleq s' \pmod m$. Since the sum on the left is a nonnegative integer less than m (by choice of m), the last congruence is actually an equality $\sum_{i=1}^{n} a_i x_i = s'$. This equation can be uniquely solved using Algorithm 9.6, thus determining the vector X. □

We are now nicely poised to explain the Merkle–Hellman knapsack cryptosystem.

Algorithm 9.7: Merkle–Hellman Knapsack Cryptosystem

Purpose: Alice needs to send Bob a private message, but they have not yet met to exchange any cryptosystem keys.

Requirements: Bob will first need a (large) integer n, a superincreasing sequence $[a_1 \ a_2 \ a_3 \ \cdots \ a_n]$, and an integer m larger than $a_1 + a_2 + a_3 + \cdots + a_n$. He then chooses (preferably randomly) a positive integer w in the range $1 < w < m$ that is relatively prime to m.

Plaintext Space: $\mathscr{P} = \{$length-n binary vectors$\}$
Ciphertext Space: $\mathscr{C} = \mathbb{Z}_{\geq 0}$ (the nonnegative integers)

Keyspace: $\mathscr{K} = \{w, [a_1 \ a_2 \ a_3 \ \cdots \ a_n], m : [a_1 \ a_2 \ a_3 \ \cdots \ a_n]$ is superincreasing, $m > \sum_{i=1}^{n} a_i$, and $1 < w < m$ with gcd$(w,m) = 1\}$. From a system key $\{w, [a_1 \ a_2 \ a_3 \ \cdots \ a_n], m\}$, which is Bob's **private key**, the corresponding **public key** is $\{[b_1 \ b_2 \ b_3 \ \cdots \ b_n]\}$, where b_i is the smallest nonnegative integer that is congruent to $w a_i \pmod m$ for each index i.

Encryption Scheme: $e_\kappa : \{$length-n binary vectors$\} \to \mathbb{Z}_{\geq 0}$, defined by $e_\kappa([x_1 \ x_2 \ x_3 \ \cdots \ x_n]) = x_1 b_1 + x_2 b_2 + x_3 b_3 + \cdots + x_n b_n = s$.

Decryption Scheme: $d_\kappa : \text{Range}(e_\kappa) \to \{$length-$n$ binary vectors$\}$, defined by $d_\kappa(s) = $ the (unique) solution of the superincreasing knapsack problem with weight vector $[a_1 \ a_2 \ a_3 \ \cdots \ a_n]$, and knapsack weight s', the least positive integer congruent to $w^{-1} \cdot s \pmod m$. This can be quickly found using Algorithm 9.6.

Notice that the encryption function is the function f_b of Proposition 9.4. As with the rest of the section examples, we use an artificially small keyspace for ease of illustration. The computer exercises allow readers to experiment with larger parameters.

Example 9.9

Suppose that Bob uses the Merkle–Hellman knapsack cryptosystem with superincreasing sequence $[a_1\ a_2\ a_3\ a_4\ a_5\ a_6] = [1, 2, 4, 9, 20, 48]$, $m = 101$ (which is greater than Σa_i) and $w = 38$ (which is relatively prime to m).

(a) What is Bob's public key?
(b) If Alice uses Bob's public key to encrypt the plaintext 011101, determine the resulting ciphertext.
(c) Perform the decryption process that would need to be done when Bob receives the ciphertext of part (b).

Solution: Part (a): Working mod 101, we have $w \cdot [a_1\ a_2\ a_3\ a_4\ a_5\ a_6] \equiv 38 \cdot [1, 2, 4, 9, 20, 48] \equiv [38, 76, 51, 39, 53, 6]$. Thus, this latter vector is Bob's public key $[b_1\ b_2\ b_3 \cdots b_n]$.

Part (b): The ciphertext is $f_b([0, 1, 1, 1, 0, 1]) = x_1b_1 + x_2b_2 + x_3b_3 + x_4b_4 + x_5b_5 + x_6b_6 = 0 \cdot 38 + 1 \cdot 76 + 1 \cdot 51 + 1 \cdot 39 + 0 \cdot 53 + 1 \cdot 6 = 172 = s$.

Part (c): Using the extended Euclidean algorithm (Algorithm 2.2), we compute $w^{-1} \equiv 8$. (This need only be computed once and could be supplied with the rest of Bob's private key.) Since $w^{-1} \cdot s \equiv 8 \cdot 172 \equiv 63$, the plaintext will be the solution of the superincreasing knapsack problem with weight vector $[a_1\ a_2\ a_3\ a_4\ a_5\ a_6]$ and knapsack weight $s' = 63$. Algorithm 9.6 quickly produces the original plaintext.

Exercise for the Reader 9.10

Suppose that Bob uses the Merkle–Hellman knapsack cryptosystem with superincreasing sequence $[a_1\ a_2\ a_3\ a_4\ a_5\ a_6] = [3, 5, 9, 18, 36, 100]$, $m = 201$ (which is greater than Σa_i), and $w = 77$ (which is relatively prime to m).

(a) What is Bob's public key?

(b) If Alice uses Bob's public key to encrypt the plaintext 111000, determine the resulting ciphertext.

(c) Perform the decryption process that would need to be done when Bob receives the ciphertext of part (b).

In 1982, Adi Shamir [Sha-82] discovered a polynomial time algorithm that could effectively crack the Merkle–Hellman cryptosystem. His attack was based on the fact that an adversary need not exactly determine the private key parameters m and w, but any other pair m', w' that will render $(w')^{-1} \cdot [b_1 \; b_2 \; b_3 \; \cdots \; b_n] \pmod{m'}$ a superincreasing sequence will work to decrypt ciphertexts. A stronger knapsack cryptosystem (the *Chor–Rivest cryptosystem*) was subsequently developed; see [Mol-03], but this system met its demise in 2001 by an attack discovered by Serge Vaudenay [Vau-01]. Although promising new knapsack cryptosystems have been developed, much confidence in them has been lost, due to the relative short life of the above two much-heralded versions, so they are presently not so often used in practice.

Government Controls on Cryptography

With the emergence of simple yet highly secure public key cryptosystems, RSA in particular, in tandem with the development of personal wireless and electronic communications, national governments have been struggling to adapt their national security policies. Governments have always strived to remain one step ahead of cryptographers, and to a great extent this has been possible until the advent of public key cryptography. The U.S. government, for example, has strict regulations on what sorts of cryptographic technologies can be sold or given, and to which countries, as well as the permissible key sizes. Violators of these rules (even people who freely distribute cryptosystems on Web sites) are considered as committing acts of treason and are subject to imprisonment. While on one hand, many people feel that it is their right to be able to use cryptography to protect their secrets, the flip side of this coin is that cryptography may be used as a weapon. Terrorist groups have taken advantage of simple public key cryptosystems for secure communications, many of which have remained unbreakable.

With sufficiently large key sizes, for example, a properly implemented RSA cryptosystem appears to be essentially unbreakable. This state of affairs could change in only two ways:

1. If a polynomial time algorithm for factoring primes gets discovered, or
2. If a new revolution occurs in computing hardware

Both of these possibilities are extremely remote, as most experts believe. We have already discussed that if it were ever found, a polynomial time algorithm for factoring primes would translate into polynomial time algorithms for almost all of the intractable problems in discrete structures. After hundreds of years of research on developing the best algorithms for such problems, researchers have all but conceded total defeat to the problem of finding such an algorithm.

According to *Moore's law*, computing speeds on the latest computers double approximately every 18 months. This was first observed in the 1960s, by Gordon Moore (1929–), one of the founders of Intel Corp. At present, it takes about an hour (on a network of computers) for an exhaustive search attack on a 56-bit key. It would require $512 - 56 = 456$ powers of two more time for such an attack on a 512-bit key. Moore's law's estimated waiting time for computers to be able to achieve the same performance on a 512-bit key search as they do today with a 56-bit key search would therefore be $456 \cdot 1.5 = 684$ years! One idea that originated in the early 1980s was a suggestion by Nobel Physics Prize Laureate Richard Feynman that it might be possible to build a new type of **quantum computer**, designed on quantum mechanical principles, which would operate under different physical axioms than present computers. In 1994, it was shown by AT&T Labs researcher Peter Shor that if such a computer could be built, then a polynomial time prime factorization algorithm could be implemented on it. Interested readers may refer to [KaLaMo-07] for more information on quantum computing.

Although the general public is not privy to the latest cryptographic technologies in use by large government organizations (such as the NSA), much evidence suggests that governments are unable to decrypt strong RSA encryptions. For example, so-called **tempest devices** are extremely sensitive electromagnetic detectors that can be used to intercept keystrokes on a computer. This would allow a government to park a tempest-enabled van outside a suspect's house or office to pick up any plaintexts before they are encrypted. Buildings can be fitted with a special insulation procedure that protects against tempest devices, but any company or individual in the United States who would have this insulation must first obtain a license from the federal government.

A Security Guarantee for RSA

We close this chapter with a sampling from the area of *provable security* in public key cryptography. We want to have some justification to contentions that are made about a public key cryptosystem, to the effect that being able to determine the private key from knowledge of the public key is as difficult as the mathematical problem on which the system rests. We prove such a result for the RSA cryptosystem by showing that being able to determine the secret key (d) from the public key (n, e) of an RSA cryptosystem (Algorithm 9.2) is at least as difficult as factoring the integer $n = pq$. The precise result is contained in the following.

Theorem 9.5: Security Guarantee for the RSA Cryptosystem

Suppose that the decryption exponent has been determined for a certain RSA cryptosystem having public key (n, e). Then there is an algorithm with polynomial complexity (more technically: that operates with time proportional to $O(\log^2 n)$, which means that in a time less than a constant time the square of the number of digits of n) that will obtain the factorization $n = pq$ with probability at least 1/2. This algorithm can be independently iterated any number K of times to result in another algorithm with polynomial complexity that will find the factorization of n with probability at least $1 - (1/2)^K$, that is, as close to 1 as one desires.

Although the theorem does not guarantee the factorization, it can be informally translated into the statement: If we can figure out the RSA decryption key d, then we can probably factor the RSA modulus n. For example, if we employ 20 iterations of the algorithm, then the chance that the algorithm we provide would not factor n would be less than one in a million. The proof is rather involved and some readers may wish to skip all of the details, but the probabilistic algorithm that is included in the proof should at least be looked at.

Proof: In order to describe the algorithm, we need to introduce two numbers s and f that depend on n, e, and d. Let s denote the largest exponent for which $2^s \mid ed - 1$. We know that s is a positive integer, since $\phi(n)$ is even and $ed \equiv 1 \pmod{\phi(n)}$. We let $f = (ed - 1) / 2^s$. We may now describe the algorithm.

Algorithm 9.8: Probabilistic Factoring Algorithm for an RSA Modulus n Given the Decryption Exponent d

Input: n, d (the private exponent) and e (the public exponent).
Output: Either one of the factors of n (that is, either p or q if $n = pq$) or the message that a factor of n was not found.

Step 1. Define s to be the largest integer such that $2^s \mid ed - 1$. Randomly choose a seed a, $1 \le a \le n - 1$, and compute $D = \gcd(a, n)$. Initialize $t = s - 1$.

Step 2. *Iterative Step.* If $D \notin \{1, n\}$, go to Step 3; otherwise, compute $b \equiv a^{2^t f} - 1 \pmod{n}$, $\gcd(b, n)$ and Update $D = \gcd(b, n)$ and $t = t - 1$.

Step 3. If $D \notin \{1, n\}$, then Output that a factor of D will be p or q (factorization was found) and exit; otherwise, either go back to Step 2 if $t \ge 0$, or Output that the algorithm did not succeed with the random seed a, and exit.

It suffices to show that this algorithm is successful at least 50 percent of the time. Unless a is relatively prime to n, the algorithm will immediately determine the factorization (in Step 1), so we may assume that $\gcd(a, n) = 1$. Thus, for the remainder of the proof, we assume that $a \in \mathbb{Z}_n^\times$. The 50-percent success rate of Algorithm 9.8 easily follows from the next two claims:

Claim 1. If $\operatorname{ord}_p(a^f) \ne \operatorname{ord}_q(a^f)$, Algorithm 6.8 will succeed with seed a.

Claim 2. The number of integers $z \in \mathbb{Z}_n^\times$ for which $\operatorname{ord}_p(z^f) \ne \operatorname{ord}_q(z^f)$ is at least $(p-1)(q-1) / 2$.

Indeed, since $\mid \mathbb{Z}_n^\times \mid = \phi(n) = (p-1)(q-1)$, we know from the two claims that the algorithm will succeed with at least half of the seeds a that are relatively prime with n, but we already know it (immediately) succeeds for all of the seeds a that are not relatively prime with n.

Proof of Claim 1: Since $ed \equiv 1 \pmod{\phi(n)}$, we may write $ed - 1 = \ell\phi(n)$ for some $\ell \in \mathbb{Z}$. From this we obtain $(a^f)^{2^s} \equiv a^{ed-1} \equiv a^{\ell\phi(n)} \equiv (a^{\phi(n)})^\ell \equiv 1$ (the last congruence by Euler's theorem) $\pmod{n}$. From this it follows from Proposition 8.5(a) that $\operatorname{ord}_n(a^f) \mid 2^s$. Also, since $p \mid n$, and $(a^f)^{2^s} \equiv 1$ $\pmod{n}$, we also have that $(a^f)^{2^s} \equiv 1 \pmod{p}$, and for the same reasons as before $\operatorname{ord}_p(a^f) \mid 2^s$. By the same token, $(a^f)^{2^s} \equiv 1 \pmod{q}$. Since $\operatorname{ord}_p(a^f) \neq \operatorname{ord}_q(a^f)$, and the roles of p and q are interchangeable, we may assume that $\operatorname{ord}_p(a^f) = 2^t < \operatorname{ord}_q(a^f)$. In terms of congruences, this implies $a^{f2^t} \equiv 1 \pmod{p}$, but $a^{f2^t} \not\equiv 1 \pmod{q}$; put slightly differently $p \mid a^{f2^t} - 1$, but $q \nmid a^{f2^t} - 1$. Since $n = pq$, the last two items tell us that $\gcd(a^{f2^t} - 1, n) = p$, which means that Algorithm 9.8 will succeed in factoring n if the seed a is used.

Proof of Claim 2: Since prime numbers always have primitive roots (see Theorem 3.21), it follows that there exists a primitive root $g_p \bmod p$ and a primitive root $g_q \bmod q$. By the Chinese remainder theorem (Theorem 2.13), it follows that there is a single positive integer g that can serve as a primitive root both mod p and mod q.[*]

Case 1: $\operatorname{ord}_p(g^f) > \operatorname{ord}_q(g^f)$. The proof of Claim 1 showed that both of these orders are powers of 2. Take x to be any odd integer in the set $\{1, 2, \cdots, p - 1\}$, and y to be any integer in the set $\{1, \cdots, q - 1\}$. By the Chinese remainder theorem, there exists an integer z in $\{1, 2, \cdots, n - 1\}$ that satisfies both $z \equiv g^x \pmod{p}$ and $z \equiv g^y \pmod{q}$. Notice that these $(p - 1)/2 \cdot (q - 1)$ numbers z are all different, because g is a primitive root both mod p and mod q. Since x is odd and $\operatorname{ord}_p(g^f)$ is a power of 2, it follows that $\operatorname{ord}_p(z^f) = \operatorname{ord}_p(g^f)$. But since $z^f \equiv g^{yf} \pmod{q}$, it follows that $\operatorname{ord}_q(z^f) \le \operatorname{ord}_q(g^f)$, and hence $\operatorname{ord}_q(z^f) < \operatorname{ord}_p(z^f)$. In summary, the assertion of Claim 2 has been established in the present Case 1.

Case 2: $\operatorname{ord}_p(g^f) < \operatorname{ord}_q(g^f)$. The proof is identical to the one we just gave in Case 1, simply reversing the roles of p and q.

Case 3: $\operatorname{ord}_p(g^f) = \operatorname{ord}_q(g^f)$. Since $p - 1$ is even, g is a primitive root mod p, and f is odd, it follows that these orders are at least 2. By analogy with what we did in Case 1, we take x to be any integer in the set $\{1, 2, \cdots, p - 1\}$ and y to be any integer in the set $\{1, 2, \cdots, q - 1\}$, but this time either x or y is even, and the other is odd. By the Chinese remainder theorem, there exists an integer z in $\{1, 2, \cdots, n - 1\}$ that satisfies both $z \equiv g^x \pmod{p}$ and $z \equiv g^y \pmod{q}$. As in Case 1, these $(p-1)(q-1)/2$ integers z are distinct. We just need to show that for each such z, z^f will have different orders mod p and mod q. But we already know that $\operatorname{ord}_p(g^f) = \operatorname{ord}_q(g^f) = 2^t$ for some positive integer t. If x is even and y is odd, then it follows that $\operatorname{ord}_p(z^f) = \operatorname{ord}_p(g^{xf}) \le t - 1$ and $\operatorname{ord}_q(z^f) = \operatorname{ord}_q(g^{yf}) = t$, so that $\operatorname{ord}_p(z^f) < \operatorname{ord}_q(z^f)$. Similarly, if x is odd and y is even, then $\operatorname{ord}_p(z^f) > \operatorname{ord}_q(z^f)$. This completes the proof of Claim 2, and hence also of Theorem 9.5. □

We illustrate Algorithm 9.8 with an example.

---

[*] This does not say that g can serve as a primitive root mod n; indeed, by Theorem 3.21, $n = pq$ has no primitive roots if $p, q > 2$.

Example 9.10

Apply Algorithm 9.8 with the RSA parameters $n = 562013$, $e = 18223$, $d = 15439$, using the random seed $a = 87643$.

Solution: We first compute $s = 7$ and $f = (ed - 1) / 2^s = 2198007$.

Step 1. Using the Euclidean algorithm, we compute $D = \gcd(87643, n) = 1$.

Step 2. We put $t = 6$, and then compute (with the aid of Algorithm 6.5) $b \equiv a^{2^t} - 1 \equiv 0 \pmod{n}$, so D gets updated to $\gcd(b,n) = n$, and $t = 5$. In the second iteration, we have $b \equiv a^{2^t} - 1 \equiv 0 \pmod{n}$, again, so D stays equal to n, and t becomes 4. The same thing happens in the next two iterations, after which t goes down to 2. In the $t = 2$ iteration, we get that $b \equiv a^{2^t} - 1 \equiv 246194 \pmod{n}$, and that $D = \gcd(b, n) = 557$. The WHILE loop stops.

Step 3. The algorithm has successfully found a factor $p = 557$ of n. The other is therefore $q = n/p = 1009$.

Exercise for the Reader 9.11

Apply Algorithm 9.8 with the RSA parameters of Example 9.10, but now using the seed $a = 234167$.

Chapter 9 Exercises

1. Suppose that Alice and Bob wish to create a secret key between them using the Diffie–Hellman protocol. They select the prime $p = 773$ and corresponding primitive root $g = 2$. Alice takes $a = 333$ as her secret key, and Bob takes his to be $b = 603$.
    (a) Verify that g is indeed a primitive root mod p.
    (b) Compute the number A that Alice (publicly) sends Bob, and the number B that Bob sends Alice.
    (c) Compute the shared secret Diffie–Hellman key for Alice and Bob in two different ways, as would be done on Alice's end and on Bob's end.

2. Suppose that Alice and Bob wish to create a secret key between them using the Diffie–Hellman protocol. They select the prime $p = 821$ and corresponding primitive root $g = 2$. Alice takes $a = 404$ as her secret key, and Bob takes his to be $b = 769$.
    (a) Verify that g is indeed a primitive root mod p.
    (b) Compute the number A that Alice (publicly) sends Bob, and the number B that Bob sends Alice.
    (c) Compute the shared secret Diffie–Hellman key for Alice and Bob in two different ways, as would be done on Alice's end and on Bob's end.

3. Suppose that Alice and Bob wish to create a secret key between them using the Diffie–Hellman protocol. They select the prime $p = 1553$, and use the smallest primitive root $g > 300$. Alice takes $a = 1333$ as her secret key, and Bob takes his to be $b = 807$.

   (a) Determine the primitive root g that they use.

   (b) Compute the number A that Alice (publicly) sends Bob, and the number B that Bob sends Alice.

   (c) Compute the shared secret Diffie–Hellman key for Alice and Bob in two different ways, as would be done on Alice's end and on Bob's end.

4. Suppose that Alice and Bob wish to create a secret key between them using the Diffie–Hellman protocol. They select the prime $p = 2267$ and use the smallest primitive root $g > 2000$. Alice takes $a = 1197$ as her secret key, and Bob takes his to be $b = 62$.

   (a) Determine the primitive root g that they use.

   (b) Compute the number A that Alice (publicly) sends Bob, and the number B that Bob sends Alice.

   (c) Compute the shared secret Diffie–Hellman key for Alice and Bob in two different ways, as would be done on Alice's end and on Bob's end.

5. (a) Use the RSA algorithm with (Bob's) public key $(n, e) = (6887, 143)$ to encrypt the plaintext message $P = 1234$.

   (b) Go through the resulting decryption process that would need to be done on the recipient's (Bob's) end to decode the ciphertext that was produced in part (a) using the private key $(n, d) = (6887, 47)$.

   (c) If Alice were to apply her digital signature to the document $P = 1234$ of part (a), compute the resulting digitally signed document, given that her private key is $d_A = 367$ and she uses the RSA modulus of part (a).

   (d) Use Alice's public key $e_A = 4303$ to perform the corresponding authentication on the digital signature produced in part (c) to authenticate Alice's signature.

6. (a) Use the RSA algorithm with (Bob's) public key $(n, e) = (7493, 229)$ to encrypt the plaintext message $P = 125$.

   (b) Go through the resulting decryption process that would need to be done on the recipient's (Bob's) end to decode the ciphertext that was produced in part (a) using the private key $(n, d) = (7493, 4021)$.

   (c) If Alice were to apply her digital signature to the document $P = 125$ of part (a), compute the resulting digitally signed document, given that her private key is $d_A = 1805$ and she uses the RSA modulus of part (a).

   (d) Use Alice's public key $e_A = 3737$ to perform the corresponding authentication on the digital signature produced in part (c) to authenticate Alice's signature.

7. (a) Use the RSA algorithm with (Bob's) public key $(n, e) = (69353, 4321)$ to encrypt the plaintext message $P = 12345$.

(b) Go through the resulting decryption process that would need to be done on the recipient's (Bob's) end to decode the ciphertext that was produced in part (a) using the private key $\kappa = (n,d) = (69353,\ 29401)$.

(c) If Alice were to apply her digital signature to the document $P = 12345$ of part (a), compute the resulting digitally signed document, given that her private key is $d_A = 367$ and she uses the RSA modulus of part (a).

(d) Use Alice's public key $e_A = 50443$ to perform the corresponding authentication on the digital signature produced in part (c) to authenticate Alice's signature.

8. (a) Use the RSA algorithm with (Bob's) public key $(n,\ e) = (66277,\ 4321)$ to encrypt the plaintext message $P = 12345$.

(b) Go through the resulting decryption process that would need to be done on the recipient's (Bob's) end to decode the ciphertext that was produced in part (a) using the private key $\kappa = (n,d) = (66277,\ 25301)$.

(c) If Alice were to apply her digital signature to the document $P = 12345$ of part (a), compute the resulting digitally signed document, given that her private key is $d_A = 1807$ and she uses the RSA modulus of part (a).

(d) Use Alice's public key $e_A = 41183$ to perform the corresponding authentication on the digital signature produced in part (c) to authenticate Alice's signature.

9. Suppose that Bob adopts the RSA cryptosystem with primes $p = 37$ and $q = 67$. He chooses the (public key) encryption exponent to be $e = 169$.

(a) Show that Bob's choice of encryption exponent is legitimate, and find his corresponding (private key) decryption exponent d.

(b) Suppose that Alice encrypts the plaintext message $P = 1234$ using the RSA cryptosystem with Bob's public key $(n,\ e) = (2479,\ 169)$. What is the resulting ciphertext that would be sent to Bob?

(c) Go through the decryption process that would need to be done at Bob's end, using his private key (n,d) with the decryption exponent that was determined in part (a).

(d) If Alice were to apply her digital signature to the document $P = 1234$ of part (b), compute the resulting digitally signed document, given that her private key is $d_A = 367$ and she uses the RSA modulus of part (a).

(e) Compute Alice's public key e_A and then perform the corresponding authentication on the digital signature produced in part (d) to authenticate Alice's signature.

10. Suppose that Bob adopts the RSA cryptosystem with primes $p = 43$ and $q = 73$. He chooses the (public key) encryption exponent to be $e = 1195$.

(a) Show that Bob's choice of encryption exponent is legitimate and find his corresponding (private key) decryption exponent d.

(b)   Suppose that Alice encrypts the plaintext message $P =$ 1234 using the RSA cryptosystem with Bob's public key $(n,e) = (3139, 1195)$. What is the resulting ciphertext that would be sent to Bob?

(c)   Go through the decryption process that would need to be done at Bob's end, using his private key (n,d) with the decryption exponent that was determined in part (a).

(d)   If Alice were to apply her digital signature to the document $P = 1234$ of part (b), compute the resulting digitally signed document, given that her private key is $d_A = 1975$ and she uses the RSA modulus of part (a).

(e)   Compute Alice's public key e_A and then perform the corresponding authentication on the digital signature produced in part (d) to authenticate Alice's signature.

11.   Suppose that Alice and Bob decide to communicate with an ElGamal cryptosystem using the prime p, corresponding primitive root g, and individual keys a and b as given in Exercise 1.

(a)   Compute the ciphertext in this system if Alice sends Bob the message $P = 321$.

(b)   Perform the ElGamal decryption process that would need to be done at Bob's end to decrypt Alice's message.

(c)   What is the digital signature that Alice would send to Bob, if she picks her signature exponent to be $d = 215$ to sign the document $P = 321$?

(d)   Perform the corresponding authentication on the answer to part (c) to authenticate Alice's signature.

12.   Suppose that Alice and Bob decide to communicate with an ElGamal cryptosystem using the prime p, corresponding primitive root g, and individual keys a and b as given in Exercise 2.

(a)   Compute the ciphertext in this system if Alice sends Bob the message $P = 321$.

(b)   Perform the ElGamal decryption process that would need to be done at Bob's end to decrypt Alice's message.

(c)   What is the digital signature that Alice would send to Bob, if she picks her signature exponent to be $d = 239$ to sign the document $P = 321$?

(d)   Perform the corresponding authentication on the answer to part (c) to authenticate Alice's signature.

13.   Suppose that Alice and Bob decide to communicate with an ElGamal cryptosystem using the prime $p = 6469$, and individual keys $a = 2256$ and $b = 4127$, and using the smallest primitive root g of p that satisfies $g > 5050$.

(a)   Determine the primitive root g.

(b)   Compute the ciphertext in this system if Alice sends Bob the message $P = 4321$.

(c)   Perform the ElGamal decryption process that would need to get done at Bob's end to decrypt Alice's message.

(d)   What is the digital signature that Alice would send to Bob, if she picks her signature exponent to be $d = 2011$ to sign the document $P = 4321$?

(e) Perform the corresponding authentication on the answer to part (d) to authenticate Alice's signature.

14. Suppose that Alice and Bob decide to communicate with an ElGamal cryptosystem using the prime $p = 8263$, and individual keys $a = 856$ and $b = 3127$, and using the smallest primitive root g of p that satisfies $g > 1700$.
    (a) Determine the primitive root g.
    (b) Compute the ciphertext in this system if Alice sends Bob the message $P = 4321$.
    (c) Perform the ElGamal decryption process that would need to get done at Bob's end to decrypt Alice's message.
    (d) What is the digital signature that Alice would send to Bob, if she picks her signature exponent to be $d = 3127$ to sign the document $P = 4321$?
    (e) Perform the corresponding authentication on the answer to part (d) to authenticate Alice's signature.

15. For each of these Merkle–Hellman knapsack cryptosystem keys:
    (i) [3, 5, 9, 18, 36, 100], $m = 175$, $w = 88$
    (ii) [18, 36, 100, 184, 360, 750], $m = 1450$, $w = 371$
    (ii) [5, 9, 18, 34, 72, 144], $m = 286$, $w = 205$

    do the following:
    (a) Verify that it is a legitimate Merkle–Hellman knapsack cryptosystem key.
    (b) Determine the corresponding public key.
    (c) Use the public key of part (a) to encrypt the plaintext message $P = 101010$.
    (d) Perform the decryption process that would need to be done on the receiving end for each of the ciphertexts of part (c).

16. For each of these Merkle–Hellman knapsack cryptosystem keys:
    (i) [1, 2, 6, 10, 25, 55], $m = 101$, $w = 77$
    (ii) [2, 6, 10, 25, 55, 205], $m = 310$, $w = 161$
    (iii) [3, 13, 23, 43, 83, 173], $m = 339$, $w = 220$

    do the following:
    (a) Verify that it is a legitimate Merkle–Hellman knapsack cryptosystem key.
    (b) Determine the corresponding public key.
    (c) Use the public key of part (a) to encrypt the plaintext message $P = 101010$.
    (d) Perform the decryption process that would need to be done on the receiving end for each of the ciphertexts of part (c).

17. Suppose that Bob uses the Merkle–Hellman knapsack cryptosystem with superincreasing sequence $[a_1 \ a_2 \ a_3 \ a_4 \ a_5 \ a_6 \ a_7 \ a_8 \ a_9] = [3, 5, 9, 18, 36, 100, 184, 360, 750]$, $m = 1499$ (which is greater than Σa_i), and $w = 365$ (which is relatively prime to m).
    (a) What is Bob's public key?

   (b) If Alice uses Bob's public key to encrypt each of the following plaintexts:
      (i)   111000111
     (ii)   101010101
    (iii)   110011001
       determine the resulting ciphertexts.
   (c) Perform the decryption process that would need to be done
when Bob receives each of the ciphertexts of part (b).

18. Suppose that Bob uses the Merkle–Hellman knapsack cryptosystem with superincreasing sequence $[a_1\ a_2\ a_3\ a_4\ a_5\ a_6\ a_7$
$a_8\ a_9] = [1, 2, 6, 10, 25, 55, 105, 205, 505]$, $m = 999$ (which is
greater than Σa_i), and $w = 334$ (which is relatively prime to m).
   (a) What is Bob's public key?
   (b) If Alice uses Bob's public key to encrypt each of the following plaintexts:
      (i)   111000111
     (ii)   101010101
    (iii)   110011001
       determine the resulting ciphertexts.
   (c) Perform the decryption process that would need to be done
when Bob receives each of the ciphertexts of part (b).

19. *A Common Modulus Attack on RSA.*
   (a) Suppose that in a certain RSA cryptosystem, Alice needs
to send the same message P to two individuals, Bob and
Ben, whose public keys are e and $\tilde{e}$, respectively. Suppose
that these two encryption exponents are relatively prime.
If Eve intercepts both of the corresponding ciphertexts
$C \equiv P^e$ and $\tilde{C} \equiv P^{\tilde{e}}$ (mod n), show how she will be able to
decrypt the plaintext message P.
   (b) Demonstrate your technique by decrypting the two common modulus ciphertexts (of the same plaintext message)
$C = 2254$ and $\tilde{C} = 1902$ that were created using the public
keys $e = 143$ and $\tilde{e} = 2209$, respectively, with the common
RSA modulus $n = 6887$.
   (c) Demonstrate your technique by decrypting the two common modulus ciphertexts (of the same plaintext message)
$C = 747126$ and $\tilde{C} = 189255$ that were created using the
public keys $e = 31547$ and $\tilde{e} = 6251$, respectively, with the
common RSA modulus $n = 976901$.

20. Explain whether it would add any additional security to an RSA
system if a certain user required that any incoming messages be
encrypted twice using different encryption exponents, $e, \tilde{e}$, both
relatively prime to $\phi(n)$. More precisely, this user's public key
would consist of the RSA modulus n along with the two RSA
encryption exponents $e, \tilde{e}$, with the instructions to first encrypt
with exponent e and then encrypt the result with the exponent $\tilde{e}$.

21. Consider the following cryptosystem: Fix a large prime number p. The plaintext and ciphertext spaces are both $\mathbb{Z}_p$. For a
given element $e \in \mathbb{Z}_{p-1}^{\times}$, the associated encryption function is
defined by $E_e : \mathbb{Z}_p \to \mathbb{Z}_p :: E_e(P) \equiv P^e (\text{mod } p)$.

(a) Describe the corresponding decryption function and find it explicitly in the case $p = 1009$ and $e = 275$. In the latter specific setting, decrypt the ciphertext $C = 777$.

(b) How does the security of such a cryptosystem compare with that of RSA with a comparably sized modulus? Could this system be considered a public key cryptosystem?

22. *Broadcast Attack on RSA.* Suppose that Alice broadcasts the same message P to a group of users. Each user is free to choose his or her own RSA modulus and encryption exponent. If in the group, there are e users who share the same small encryption exponent e and whose RSA moduli are pairwise relatively prime, then Eve will be able to use the corresponding e ciphertexts along with the Chinese remainder theorem to recover the plaintext message m. For example, if three recipients with RSA pairwise relatively prime moduli n_1, n_2, and n_3 all use the encryption exponent $e = 3$, this means that the corresponding ciphertexts satisfy $C \equiv P^3 \pmod{n_1}$, $C' \equiv P^3 \pmod{n_2}$, and $\tilde{C} \equiv P^3 \pmod{n_3}$. Eve can use the (constructive proof of the) Chinese remainder theorem (Theorem 2.13) to find an integer D that simultaneously solves all three congruences, and from this it follows that $D \equiv P^3 \pmod{n_1 n_2 n_3}$. Since $P^3 < n_1 n_2 n_3$, this last congruence is actually valid in regular arithmetic, i.e., $D = P^3$, so the plaintext message can be decoded by taking the cube root: $P = \sqrt[3]{D}$.

(a) Apply the broadcast attack to decrypt the plaintext message m, if it was broadcast to a network of users, including three who share the same encryption exponent $e = 3$ and with moduli $n_1 = 1207$, $n_2 = 2407$, and $n_3 = 3649$. The corresponding three ciphertexts were $C_1 = 494$, $C_2 = 113$, and $C_3 = 2372$.

(b) Give an example of a broadcast attack on RSA with $e = 4$, if possible.

23. *A Chosen Ciphertext Attack on RSA.* Assume that Alice has used RSA to send Bob a certain plaintext message P and that Eve has intercepted the ciphertext C. Eve really wants to know the plaintext message P. Assume that Bob, playing a risky game with Eve, offers to decrypt for her any other single ciphertext $\tilde{C} \neq C$. Show how by judiciously choosing $\tilde{C}$, Eve will be able to deduce the original plaintext P. We assume, as usual, that Eve knows the public key (n, e).

Note: This does not suggest Eve will be able to determine Bob's secret key d.

Suggestion: Let x be any element of $\mathbb{Z}_n^\times$ such that $Cx^e \not\equiv C \pmod{n}$ and take $\tilde{C} \equiv Cx^e \pmod{n}$.

24. *Cracking the ElGamal Cryptosystem Is Equivalent to Cracking the Diffie–Hellman Key Exchange.* Prove the following two assertions:

(a) If one is able to determine the plaintext P from any given ciphertext C of a given ElGamal cryptosystem with knowledge only of the public key (without knowing the private key), then one will also be able to determine Diffie–Hellman keys from knowing only the transmitted information (A or B

in Algorithms 9.1 and 9.4) and the public key. It is assumed that the publicly known prime p and corresponding primitive root g are the same for both algorithms.

(b) Prove the converse of the statement in part (a).

Sketch of Proof for Part (a): Assume that we know Bob's transmission $B (\equiv g^b \pmod p)$ and Alice's transmission $A (\equiv g^a \pmod p)$. Take $C = 1$ in Algorithm 9.5 and decrypt the ElGamal ciphertext (A, C) to get $P \equiv 1 \cdot A^{-b} \equiv g^{-ab} \pmod p$, so the Diffie–Hellman key $(= g^{ab} \pmod p)$ will be $P^{-1} \pmod p$.

25. *Properties of Discrete Logarithm Functions.* Establish each of the following "laws of discrete logarithms" that are direct analogues to the laws for ordinary logarithms that one learns about in (pre-)calculus courses. Throughout, g is a primitive root for some prime integer p and $L_g : \mathbb{Z}_p^\times \to \mathbb{Z}_p^\times$ is the corresponding discrete logarithm function (as in Definition 9.2).

(a) $L_g(ab) \equiv L_g(a) + L_g(b) \pmod{p-1}$.

(b) $L_g(a^{-1}) = -L_g(a) \pmod{p-1}$.

(c) $L_g(a^k) = kL_g(a) \pmod{p-1}$.

(d) If h is another primitive root mod p, then $L_h(a) \equiv L_h(g) \cdot L_g(a) \pmod{p-1}$.

26. *A Known Plaintext Attack on ElGamal.* Assume that Alice has sent Bob two messages, P_1 and P_2, resulting in ciphertext (second components) C_1 and C_2, respectively, using the ElGamal system, and that she used the same secret key a to encrypt both messages. Show how, if Eve finds out P_1 and has learned that it is the plaintext for C_1, then she will be able to determine P_2.

Note: Such an attack can be prevented by (randomly) generating a new secret key for each message sent.

27. *Formulation of RSA into a Block Cryptosystem.* Suppose that we wish to send messages in a certain alphabet that has K symbols in it. We first use $\mathbb{Z}_K$ to represent this alphabet. For example, with the ordinary (lowercase) English alphabet, Table 6.2 shows a natural representation using $\mathbb{Z}_{26}$. We assume that the RSA modulus n being used is several times larger than K. (This will usually be automatic in any real-life RSA system, since n will have to be very large to ensure security.) We let $r = \text{floor}(\log_K(n))$; this will be the *block length* of plaintext messages in the alphabet $\mathbb{Z}_K$ that will be transmitted. An r block of letters, represented by a length-r vector $V = [m_1\ m_2 \cdots m_r]$, with elements $m_i \in \mathbb{Z}_K$ is transmitted by one viewing V as a base K representation of an integer and converting V to an ordinary integer as in Chapter 6:

$$M = [m_1\ m_2 \cdots m_r] \ \to \ m = \sum_{j=1}^{r} K^{r-j} m_j$$

[Recall that this is a one-to-one correspondence, and the representing integer m will always lie in the range $0 \le m \le \sum_{j=1}^{r} K^{r-j}(K-1) = K^r - 1 < n.$] We then use the (ordinary) RSA system (with modulus n) to encrypt the integer m into an integer c. This integer c is converted into its corresponding base K representation, which will have length at most $r + 1$.

$$c = \sum_{j=0}^{r} K^{r-j} c_j \to C = [c_0 \ c_1 \ c_2 \cdots c_r]$$

The ciphertext is this length-$r + 1$ vector C (padded with zeros on the left, if necessary) of elements of $\mathbb{Z}_K$.

(a) Use the RSA system of Exercise 1 to transmit the length-8 plaintext block [0 0 1 1 1 1 0 1] (in the binary alphabet) into a corresponding length-9 ciphertext vector.

(b) Use the RSA system of Exercise 3 to transmit the message "clearout" (in a 26 lowercase English-letter alphabet), using the correspondence of Table 6.2, into corresponding ciphertext blocks.

(c) With the block version of RSA of part (a), decrypt the ciphertext block: [0 1 1 0 1 0 1 0].

(d) With the block version of RSA of part (b), decrypt the ciphertext blocks: [FQYL] [FGOU] [CXET].

28. *An Elementary Factoring Method.* It was mentioned in the text that the RSA prime factors p and q should not be chosen too close together since then $n = pq$ would be easy to factor. The following simple factoring method shows why this is so.

(a) If $n = pq$ and $p > q$, show that $q < \sqrt{n} < p$.

(b) (*For readers who have studied calculus*) Show that the quantity $\frac{1}{2}(p+q)$, thought of as a function of the real variable p ($q = n/p$), strictly increases from $\sqrt{n}$ to $(n + 1)/2$ as p runs from $\sqrt{n}$ to n.

(c) Show that $n = \left(\dfrac{p+q}{2}\right)^2 - \left(\dfrac{p-q}{2}\right)^2$.

(d) If p and q are close together, then by part (b), the integer $x = \frac{1}{2}(p+q)$ will be (greater than but) close to $\sqrt{n}$, and the integer $y = \frac{1}{2}(p-q)$ will be small. Use part (c) to obtain the factorization $n = (x + y)(x - y)$.

Note: In practice, since $x^2 - n = y^2$, if p and q are close together, then the factorization of part (d) can be found by checking integer values of x larger than $\sqrt{n}$ until $x^2 - n$ is a perfect square. Then this perfect square will be y^2 and we will have $n = (x + y)(x - y)$.

(e) Use the above factoring method to factor the RSA modulus $n = 3301453$.

29. Apply Algorithm 9.8 with RSA parameters $n = 529651$, $e = 1241$, and $d = 61289$ to search for a factor of n using the following values for the seed:
    (a) $a = 431519$
    (b) $a = 483770$
    (c) $a = 40236$

30. Apply Algorithm 9.8 with RSA parameters $n = 1044541$, $e = 6781$, and $d = 754081$ to search for a factor of n using the following values for the seed:
    (a) $a = 268978$
    (b) $a = 878163$
    (c) $a = 265608$

31. Prove Proposition 9.3.

Chapter 9 Computer Implementations and Exercises

Note: Although some of the cryptographic systems in this chapter are very simple to program using algorithms from previous sections (like the fast exponentiation Algorithm 6.5), we nonetheless aim for the creation of a complete set of programs. As mentioned in chapter text, if your computing platform is a floating point arithmetic system, it may allow you only up to 15 or so significant digits of accuracy. Symbolic systems allow for much greater precision, being able to handle hundreds of significant digits. Some platforms allow users to choose if they wish to work in floating point or symbolic arithmetic. If you are working on such a dual-capability system, you may wish to create two separate programs (for those that might work with large integers): an ordinary version and a symbolic version (perhaps attaching a `Sym` suffix to the names of those of the latter type). In case you do not have access to a symbolic system, some particular questions below may need to be skipped or modified so the numbers are of a manageable size.

1. *Program for the Diffie–Hellman Key Exchange.*
    (a) Write a program with syntax `K = DiffieHellmanKey (p,g,B,a)` that will produce the Diffie–Hellman key (using Algorithm 9.1) given the (public) inputs: `p` = the prime number, `g` = the primitive root (mod p), `B` = Bob's public key, and `a` = Alice's private key. The output, `K`, is a mod p integer representing the resulting Diffie–Hellman key.
    (b) Run your program with the parameters of Example 9.2, and confirm the results.
    (c) Run your program on the parameters of Example 9.2 but with Bob's and Alice's parameters switched [so you should run `DiffieHellmanKey(p,g,A,b)`], and check to see that the result is the same as was obtained in part (b).

2. Suppose that Alice and Bob wish to create a secret key between them using the Diffie–Hellman protocol. They select the prime $p = 24259676230523707727576331569769824969681$ and corresponding primitive root $g = 3$. If Alice takes $a = 8866446688$ as her secret key, and Bob takes his to be $b = 196819691970$:

    (a) Apply the Miller–Rabin test to certify the primality of p so that the probability of an incorrect certification is less than 1 in 1 trillion. Then check that g is the smallest primitive root mod p.

    (b) Compute the number A that Alice (publicly) sends Bob, and the number B that Bob sends Alice.

    (c) Compute the shared secret Diffie–Hellman key for Alice and Bob in two different ways, as would be done on Alice's end and on Bob's end.

Answers:

    (a) $\text{ord}_p(2) = (p-1)/2$.

    (b) $A = 1749037741156461472834806859126142492373$, and $B = 1519836956974222839962608060419998836078$.

    (c) DH Key = $836565563359907232787680145037578098830$

3. *Program for the RSA Encryption.*

    (a) Write a program with syntax C = RSAEncrypt(P,e,n) that will perform the RSA encryption (using Algorithm 9.2) given the inputs P = the plaintext (an integer mod n), n = the RSA modulus, and e = the (public) encryption exponent. The output, C, is a mod n integer representing the ciphertext.

    (b) Run your program with the parameters of Example 9.3(b), and confirm the results.

    (c) Suppose that Bob adopts the RSA cryptosystem with primes $p = 153,817$ and $q = 1,542,689$, and public key encryption exponent $e = 202,404,606$. If Alice uses this system to send Bob the plaintext message $P = 888,999,000$, apply your program of part (a) to determine the ciphertext.

4. *Program for the RSA Decryption.*

    (a) Write a program with syntax P = RSADecrypt(C,d,n) that will perform the RSA decryption (using Algorithm 9.2) given the inputs C = the ciphertext (an integer mod n), n = the RSA modulus, and d = the (private) decryption exponent. The output, P, is a mod n integer representing the plaintext.

    (b) Run your program with the parameters of Example 9.3(c), and confirm the results.

    (c) For the RSA system described in Computer Exercise 3(c), compute Bob's corresponding decryption exponent d. Apply your program of part (a) to decrypt Alice's ciphertext (produced in Computer Exercise 3(c)).

5. *Program for the RSA Digital Signatures.*

    (a) Write a program with syntax s = RSADigitalSig(P,d,n) that will create RSA signatures (using Algorithm 9.3) given the inputs P = the plaintext (an integer mod n), n = the RSA modulus, and d = the (public) decryption exponent of the sender. The output, s, is a mod n integer representing the digital signature.

(b) Run your program with the parameters of Example 9.4(a), and confirm the results.

(c) Suppose that Bob needs Alice to digitally sign her plaintext message P of Computer Exercise 3(c), where the primes are as before, and Alice's public key parameter is $e_A = 727{,}641{,}838{,}100$. Determine Alice's corresponding private key parameter d_A, and then use your program of part (a) to compute Alice's digital signature.

6. *Program for the RSA Digital Signature Authentication.*

(a) Write a program with syntax P = RSADigitalSig Authenticate(s,e,n) that will authenticate RSA signatures (using Algorithm 9.3) given the inputs s = the digital signature (an integer mod n), n = the RSA modulus, and e = the (public) encryption exponent of the sender. The output, P, is a mod n integer that should coincide with the plaintext (if the signature is authentic).

(b) Run your program with the parameters of Example 9.4(b), and confirm the results.

(c) Use your program of part (a) to perform Bob's authentication on the digital signature produced by Alice in part (c) of the previous computer exercise.

7. (a) Factor the integer $n = 21463366383055728841$.

(b) Verify that (Bob's) exponent $e = 22347$ is relatively prime to $\phi(n)$, and so is an admissible RSA encryption exponent. Use the RSA algorithm with modulus n as in part (a), and encryption exponent e to encrypt the plaintext message $P = 123456789$.

(c) Determine the (Bob's) corresponding decryption exponent d, and then go through the resulting decryption process that would need to be done on the recipient's (Bob's) end to decode the ciphertext that was produced in part (b).

(d) If Alice were to apply her digital signature to the document $P = 123456789$, compute the resulting transmitted digitally signed document, given that Alice's private key is $d_A = 3651$.

(e) Perform the corresponding authentication on the answer to part (c) to authenticate Alice's signature.

8. (a) Factor the integer $n = 34523666179842135318488756939 6 0104104481$.

(b) Verify that the (Bob's) exponent $e = 8877665544332211$ is relatively prime to $\phi(n)$, and so is an admissible RSA encryption exponent. Use the RSA algorithm with modulus n as in part (a), and encryption exponent e to encrypt the plaintext message $P = 1234567890987654$.

(c) Determine the (Bob's) corresponding decryption exponent d, and then go through the resulting decryption process that would need to be done on the recipient's (Bob's) end to decode the ciphertext that was produced in part (b).

(d) If Alice were to apply her digital signature to the document $P = 1234567890987654$, compute the resulting transmitted digitally signed document, given that Alice's private key is $d_A = 76123655$.

(e) Perform the corresponding authentication on the answer to part (d) to authenticate Alice's signature.

Answers:

(a) $n = pq$, with $p = 71755440315342536873$, and $q = 48112959837082048697$.

(b) $c = 17892360866680854316334267016010129257134$

(c) $d = 30149247924591350080675379792995583280635$

(d) $s = 26133341095182669938621283116955149302 85$

(e) Need to compute Alice's public key: $e_A = 2962912700774$
$94314799277079946100 8445751$

9. *An Industrial-Grade RSA System.* The following two prime numbers give rise to a rather special RSA cryptosystem:
$p = 16347336458092538484 43133883865$
$09085984178367003309 2312181110852$
$38933310010450815121 2118167511579$
$q = 19008712816648221131 26851573935$
$41397547189678996851 5493666638539$
$08802710380210449895 71912614 65571$

Their product, $n = pq$, is known as RSA-640 that was introduced in the last chapter. The RSA corporation put out a number of public challenges (with monetary awards) to factor some very large RSA moduli. RSA no longer offers such challenges, but RSA-640 was the last of these challenges that they put out in 2001 to be factored. The factorization was announced in 2005 by a team of scientists and it took 30 2.2GHz-Opteron-CPU years (in over five months of calendar time).

Note: Rather than entering p and q by hand into your computer, it would be much quicker (and more pleasant) to navigate to the appropriate RSA Web page, and copy and paste these integers into your computer.

(a) If you have a built-in factoring utility on your platform, factor p and q (to check that they are prime); then factor $p - 1$. You will find that "factoring" p and q should go quite a bit quicker (since these really amounted to primality checks). **Caution:** Do not try to factor n or even $q - 1$ (using your built-in factoring utility).

(b) Verify that the (Bob's) exponent $e = 12345678910111213$
141516171819 is relatively prime to $\phi(n)$, and so is an admissible RSA encryption exponent. Use the RSA algorithm with modulus $n = pq$, and encryption exponent e to encrypt the plaintext message $P = 11223344556677889900 9988776655$
44332211.

(c) Determine the (Bob's) corresponding decryption exponent d, and then go through the resulting decryption process that would need to get done on the recipient's (Bob's) end to decode the ciphertext that was produced in part (b).

Answers:

(a) $p - 1 = 2^2 \cdot 3 \cdot 53 \cdot 47041965497216811220810358707 \cdot$ 29596457774818880277216 7 $\cdot$ 1554503367019 $\cdot$ 6706111 $\cdot$ 10987 $\cdot$ 55057 $\cdot$ 7129.

(b) $c = 13153732987882471249579502704073862802800243$ 97121229495725892550921089990223014363655486752 53735490833445879911996290866476311987326677746 9874179432552037896130218401997625 9

(c) $d = 13400166866899423912075336250535256532775694$ 9565457343285686874309095239732015041352042615 9 36237040934760876266406890681692923024853062591 93823620928165198761750561703990061 9

10. *Program for the ElGamal Encryption.*

(a) Write a program with syntax `A, C = ElGamalEncrypt (P,a,B,p,g)` that will perform the ElGamal encryption (using Algorithm 9.4) given the inputs `P` = the plaintext (an integer mod p), `p` = the ElGamal modulus (a prime), `g` = a corresponding primitive root that has been adopted, `a` = the (private) ElGamal encryption exponent of the sender, and `B` = the (public) ElGamal key of the receiving party. The outputs `A`, `C` are mod p integers representing the ciphertext.

(b) Run your program with the parameters of Example 9.5(a), and confirm the results.

(c) Determine the smallest primitive root g of the prime p of Computer Exercise 9. If Alice and Bob choose their corresponding ElGamal private keys to be $d_A = 111,222,333$ and $d_B = 444,666,888$, use your program of part (a) to determine the ciphertext that Alice would send to Bob as the encryption of the plaintext message $P = 333,555,777$.

11. *Program for the ElGamal Decryption.*

(a) Write a program with syntax `P = ElGamalDeccrypt (A,C,b,p)` that will perform the ElGamal decryption (using Algorithm 9.4) given the inputs `A`, `C`, mod p integers representing the ciphertext, `p` = the ElGamal modulus (a prime), `b` = the (private) ElGamal encryption exponent of the receiver. The outputs `A`, `C` are mod p integers representing the ciphertext.

(b) Run your program with the parameters of Example 9.5(a), and confirm the results.

(c) Use your program of part (a) to decrypt the ciphertext that was produced in Computer Exercise 9(c).

12. *Program for the ElGamal Digital Signatures.*

(a) Write a program with syntax `P,r,s = ElGamalDigital Sig(P,a,d,p,g)` that will create ElGamal signatures (using Algorithm 9.5) given the inputs: `P` = the plaintext (an integer mod p), `p` = the ElGamal modulus (a prime), `g` = a corresponding primitive root that has been adopted, `a` = the (private) ElGamal cryptosystem exponent of the sender, and `d` = the (private) digital signature exponent of the sender. The outputs `P,r,s` are three integers representing the digital signature; P and r are mod p integers, while r is a mod p − 1 integer.

(b) Run your program with the parameters of Example 9.6(a), and confirm the results.

(c) Use your program of part (a) to create Alice's digitally signed copy of her plaintext from Computer Exercise 9(c).

13. *Program for the ElGamal Digital Signature Authentication.*

(a) Write a program with syntax Q = ElGamalDigitalSig Authenticate(P,r,s,A,p,g) that will authenticate ElGamal signatures (using Algorithm 9.5) given the inputs P,r,s the three integers representing the digital signature; P and r are mod p integers, while r is a mod p − 1 integer, A = sender's public key, p = the ElGamal modulus (a prime), and g = a corresponding primitive root that has been adopted. The outputs, Q should be the plaintext P if the digital signature is a valid one, if it is not, the program should output the message: "Digital signature does not authenticate."

(b) Run your program with the parameters of Example 9.6(b), and confirm the results.

(c) Use your program of part (a) to authenticate Alice's digital signature produced in Computer Exercise 11(c).

14. *Brute-Force Program for General Knapsack Problem.*

(a) Write a program with syntax Bestx BestValue = Kn apsackBruteFindAllBest(Weights, s) that will perform a brute-force search for a knapsack problem, and collect all solution vectors corresponding to the best solution. The inputs are Weights, the vector of object weights (positive integers), and s = the knapsack capacity. The outputs will be a matrix Bestx and a scalar BestValue that have the following meanings: BestValue will be the greatest possible knapsack weight that is less than or equal to s, and Bestx will be a matrix of all corresponding binary vectors that describe the different knapsack configurations. Each row of this matrix is a binary vector of the same size as Weights and corresponds to such a configuration.

(b) Run your program with the parameters of Example 9.7, and confirm the results.

(c) Randomly generate some knapsack problems with increasing numbers of weights, and run your program of part (a) on them until it takes more than five seconds to execute. What is the size of the problem for which this first occurs (results will vary)?

Suggestion for Part (c): For the weight vectors, randomly generate individual weights in a specified range (say, between 1 and 40); it is fine if some weights are the same. For the value of s, randomly generate some values between SumWgts/2 and SumWgts, where SumWgts is the sum of all the weights. Run two or three randomly generated data sets for each size vector, starting with $n = 8$ weights.

15. *Efficient Program for Superincreasing Knapsack Problem.*

(a) Write a program with syntax $x = $ KnapsackSuper Increasing(Weights, s) that will use Algorithm 9.6 to solve a superincreasing knapsack problem. The inputs are Weights, the vector of superincreasing object weights (positive integers), and s = the knapsack capacity. The output will be a binary vector x that solves the problem (if a solution exists). If there is no solution, the program should print an error message to this effect.

(b) Run your program with the parameters of Exercise for the Reader 9.8, and confirm the results.

(c) Randomly generate some *superincreasing* knapsack problems with increasing numbers of weights, and run your program of part (a) on them until it takes more than five seconds to execute. What is the size of the problem for which this first occurs (results will vary)?

16. *Program for Merkle–Hellman Encryption.*

(a) Write a program with syntax $C = $ MerkleHellman Encrypt(PublicWeights, xPlaintext) that will use Algorithm 9.7 to perform encryptions using the Merkle–Hellman knapsack cryptosystem. The inputs are: PublicWeights, the public key vector object weights (positive integers), and xPlaintext = a binary vector representing the plaintext. The output is a nonnegative integer C that is the corresponding ciphertext.

(b) Run your program with the parameters of Example 9.9(b), and confirm the results.

(c) Run your program on the encryptions of Chapter Exercise 17.

17. *Program for Merkle–Hellman Decryption.*

(a) Write a program with syntax xPlaintext = Merkle HellmanDecrypt(PrivateWeights, m, w, C) that will use Algorithm 9.7 to perform decryptions using the Merkle–Hellman knapsack cryptosystem. The inputs are PrivateWeights, the private key vector superincreasing object weights (positive integers), m, w, the private key modulus and multiplier, and C = an integer representing the ciphertext. The output will be the binary vector xPlaintext representing the corresponding plaintext.

(b) Run your program with the parameters of Example 9.9(c), and confirm the results.

(c) Run your program on the decryptions of Chapter Exercise 17.

Finite Fields in General, and *GF*(2⁸) in Particular

Although many readers may not have heard these words before (in a mathematical context), *rings* and *fields* are names for generic number systems that possess a number of important arithmetic axioms. Both require that there be two operations on an underlying set of elements: addition and multiplication. Some familiar axioms like associativity and the distributive law are required, and there must be an additive identity (0) and a multiplicative identity (1). Fields are rings with an additional requirement that it is always possible to divide by any nonzero element; thus rings are more general objects. Examples of fields include the real numbers, the rational numbers (all fractions), and $\mathbb{Z}_p$, whenever p is prime. Examples of rings that are not fields include the integers $\mathbb{Z}$ and $\mathbb{Z}_n$, whenever $n > 1$ is composite. A finite field is a field whose underlying set is a finite set, such as $\mathbb{Z}_p$, whenever p is prime, which contains p elements. It turns out that any finite field must contain p^n elements, where p is a prime and n is a positive integer, and that for any such p and n, there is always a unique finite field with p^n elements. This unique finite field is denoted $GF(p^n)$. We show how to construct any $GF(p^n)$ using a concrete procedure involving polynomials with coefficients in $\mathbb{Z}_p$. The process is entirely analogous to the way that $\mathbb{Z}_p$ was constructed from the integers. Finite fields are useful number systems in constructing cryptosystems. We also make some specialized comments about $GF(2^4)$ and $GF(2^8)$, since both of these finite fields play important roles in the next chapter.

Binary Operations

Definition 10.1

A **binary operation** on any nonempty set S is simply a function whose domain is the set of all ordered pairs of elements of S and whose codomain is S.

Rather than using a notation such as $f(s_1, s_2)$ for a specific binary operation to be applied to an ordered pair of elements s_1, s_2, it is more common to adopt a symbol for the binary operation and simply place the symbol between the elements. For example, if we adopt the "triangle" symbol $\triangle$ for a certain binary operation, the image on an ordered pair s_1, s_2 would be denoted as $s_1 \triangle s_2$.

Example 10.1

(a) Here are some very familiar examples of binary operations: the addition operation (+) that is defined on many sets of numbers, such as the integers $\mathbb{Z}$, the mod n integers $\mathbb{Z}_n$, and also on the set of all matrices of a fixed size (whose entries are in a fixed set, such as the real numbers or the integers).

(b) The following equation defines a binary operation, which we denote by the "triangle" symbol $\triangle$ on the set of positive integers $\mathbb{Z}_+ : a \triangle b = a + a \cdot b^2$. Thus, for example, $2 \triangle 3 = 2 + 2 \cdot 3^2 = 20$.

As can be imagined from part (b) of this example, there is no limit to the number of binary operations that can be conceived. Any nonempty set can be endowed with binary operations. If these binary operations satisfy some useful axioms, this can lead to some new and interesting number systems.

Rings

A ring is a nonempty set with two binary operations that are called addition (+) and multiplication ($\cdot$) which satisfy many of the common arithmetic properties of the integers $\mathbb{Z}$, which we think of as a prototypical ring. The following definition specifies the required properties. Although the definition is long, it is really not so complicated since all of the axioms are familiar ones that we deal with in typical mathematical calculations. It should be helpful to the reader to think about the integers while reading this definition.

Definition 10.2

A **ring** is a set R that is endowed with two binary operations: addition (+) and multiplication ($\cdot$), and for which the following axioms hold, where a, b, c denote generic elements of R.

1. *Commutativity of Addition.* $a + b = b + a$.
2. *Associativity of Addition.* $(a + b) + c = a + (b + c)$.
3. *Additive Identity.* There exists in R an **additive identity** element, denoted as 0 and called **zero**, that satisfies $a + 0 = a$.

> 4. *Additive Inverses.* For each ring element a, there exists a corresponding **additive inverse**, denoted as $-a$, that satisfies $a + (-a) = 0$.
> 5. *Commutativity of Multiplication.* $a \cdot b = b \cdot a$.
> 6. *Associativity of Multiplication.* $(a \cdot b) \cdot c = a \cdot (b \cdot c)$.
> 7. *Multiplicative Identity.* There exists in R a **multiplicative identity** element, denoted as 1 and called **one**, that satisfies $a \cdot 1 = a$.
> 8. *Distributive Law.* $a \cdot (b + c) = a \cdot b + a \cdot c$.

A ring is sometimes formally denoted as a triple $(R, +, \cdot)$, or simply as R, if its two binary operations are clear from the context.

Note: In some books, the definition of a ring does not include Axiom 5 (commutativity of multiplication) and/or Axiom 7 (multiplicative identity). In such treatments, what we call a ring might be called a commutative ring and/or a commutative ring with identity. Since we will only be needing to work with rings that satisfy all of the above axioms, we avoid this more threadbare definition.*

Example 10.2

The integers $\mathbb{Z}$ and the modular integers $\mathbb{Z}_n$ (with $n > 1$ a positive integer) when endowed with their usual addition and multiplication are rings. From our early school days, we know that the integers satisfy all of the above ring axioms. As was pointed out in Chapter 2, the modular integers $\mathbb{Z}_n$ inherit each of these axioms from the integers. Other familiar examples of rings include the real numbers $\mathbb{R}$ and the rational numbers $\mathbb{Q}$, which is the set of all real numbers that are expressible as fractions of integers (with the denominator being nonzero).

Definition 10.3

Let R be a ring, and a, b be elements of R. Just as in ordinary arithmetic, we define **subtraction** in a ring by the following equation:

$$a - b \triangleq a + (-b)$$

In other words, subtracting b from a is the same as adding the additive inverse of b to a. Also, a nonzero element a has a (**multiplicative**) **inverse**

---

* With our definition, we could use phrases such as "noncommutative ring" and/or "ring without identity" to describe some of these more general systems. An example of a noncommutative ring is the set of all square matrices (with real number, integer, or modular integer entries) of a certain size. We show in Chapter 4 that all of the ring axioms are satisfied, with the exception of commutativity of matrix multiplication. An example of a commutative ring without identity is the set of all even integers $2\mathbb{Z} = \{0, \pm 2, \pm 4, \cdots\}$. All of the other ring axioms are inherited from the ring $\mathbb{Z}$, but $2\mathbb{Z}$ does not contain the multiplicative identity 1.

(also we say a is **invertible**) if there exists another element $a^{-1} \in R$ (called the **inverse** of a) with the property that

$$a \cdot a^{-1} = 1$$

If a is invertible, we define **division** by a as simply multiplying by its inverse.

The set of all **invertible elements** in a ring R is denoted as $R^{\times}$.

Exercise for the Reader 10.1

Let R be a ring. Show that the set of invertible elements $R^{\times}$ satisfies the following axioms (with multiplication as a binary operation inherited from R):

1. *Closure under Multiplication.* If $a, b \in R^{\times}$, then $a \cdot b \in R^{\times}$.
2. *Closure under Inverses.* If $a \in R^{\times}$, then $a^{-1} \in R^{\times}$.
3. *Multiplicative Identity.* $1 \in R^{\times}$.

Many other familiar algebra rules are consequences of the ring axioms. The following proposition gives a sampling of this phenomenon; other examples will appear in the exercises. When working with rings, it is customary to use the axioms without always specifying them. We do this often with commutativity and associativity. For example, commutativity of multiplication allows us to restate the distributive law $a \cdot (b + c) = a \cdot b + a \cdot c$ as $(a + b) \cdot c = a \cdot c + b \cdot c$.

Proposition 10.1

Let R be a ring. The following identities are valid for any $a, b, c \in R$:

(1) $0 \cdot a = 0$.
(2) Additive inverses and multiplicative inverses are unique.
(3) $-(-a) = a$.
(4) $a \cdot (-b) = (-a) \cdot b = -(a \cdot b)$ and $(-a)(-b) = ab$.
(5) $a \cdot (b - c) = a \cdot b - a \cdot c$.

Proof: We will prove parts (1) and (4); the proofs of parts (2), (3), and (5) will be left as exercises at the end of the chapter (Exercise 23).

(1) Using the distributive law, we may write $0 \cdot a = (0 + 0) \cdot a = 0 \cdot a + 0 \cdot a$. If we subtract $0 \cdot a$ from both sides we are left with $0 \cdot a = 0$.

(2) Using the distributive law, we have $a \cdot b + a \cdot (-b) = a \cdot (b + (-b)) = a \cdot 0 = 0$, where the last equation is true from part (1). It follows that $a \cdot (-b)$ is an additive inverse of $a \cdot b$, and since additive inverses are unique by part (2), we must have $a \cdot (-b) = -(a \cdot b)$. An analogous argument shows that $(-a) \cdot b = -(a \cdot b)$. The last identity follows from two applications of the first, and by using part (3):

$$(-a)(-b) = -a(-b) = -(-ab) = ab. \quad \square$$

Fields

We were brought up (in our work with the real numbers) with the notion that we cannot divide by zero, but it is possible to divide by any nonzero number. A field is a ring with this additional axiom.

Definition 10.4

A **field** F is a ring in which every nonzero element is invertible. A **finite field** is a field having a finite number of elements.

Examples

Example 10.3

With the usual binary operations of addition and multiplication, the ring of real numbers $\mathbb{R}$ and the ring of rational numbers $\mathbb{Q}$ are fields, because every nonzero element is invertible. (For a nonzero rational number a/b, the inverse is just the reciprocal fraction b/a.)

Example 10.4

The ring of integers $\mathbb{Z}$ is not a field. For example, the integer 2 does not have an inverse in $\mathbb{Z}$; that is, there is no *integer k* such that $2 \cdot k = 1$.

Example 10.5

We learned in Chapter 2 that whenever p is prime, every nonzero element of $\mathbb{Z}_p$ is invertible (since it is relatively prime to p); thus, $\mathbb{Z}_p$ is a finite field. If $n > 1$ is composite, however, there will always be nonzero elements of $\mathbb{Z}_n$ that are not relatively prime to n and, hence, (from Proposition 2.11) will not be invertible. Thus, whenever n is composite, the ring $\mathbb{Z}_n$ is not a field.

Exercise for the Reader 10.2

Let the set $F = \{0,1,a\}$ of three elements have addition defined using the ring axioms and $1+1 = a$, $a+a = 1$, $1+a = 0$, and multiplication defined using the ring axioms and $a \cdot a = 1$.

(a) Fill in complete addition and multiplication tables for F.

(b) Is F a field? Explain your answer.

The fields in Example 10.5 are already familiar. The main purpose of this chapter is to learn about some new finite fields. The following theorem tells us all of the possible sizes for a finite field, and that finite fields are uniquely determined by their size. Since a proof of this theorem would require quite a diversion, we refer the interested reader to any decent book on abstract algebra, such as [Hun-96] or [Her-96]. We will soon provide a procedure that will, in principle, allow us to construct any finite field.

> ### Theorem 10.2: Inventory of Finite Fields
>
> If F is a finite field with at least two elements, then there exists a prime p such that the number of elements of F is a power of p; that is, $|F| = p^n$, for some positive integer n.[*] Conversely, for any such prime power p^n, there exists a finite field having p^n elements, and this field is unique.[†] The unique field with p^n elements is denoted as **GF(p^n)** and is called the **Galois field with p^n elements**, in honor of the youth prodigy mathematician Évariste Galois[‡] (see Figure 10.1), who made significant contributions to the theory of fields.

---

[*] Technically, there is a field with just one element, but it is not a very interesting one and will henceforth be ignored in our developments.

[†] By saying a field with a certain number of elements is unique, the technical mathematical concept is that of *isomorphism*. More precisely, if F and F' are both fields with the same number p^n (by Theorem 10.2) elements, then there exists a bijective function $\Phi : F \to F'$ that preserves addition and multiplication and takes the 0/1 element of F to the 0/1 element of F'. Preserving addition means that $\Phi(a+b) = \Phi(a) + \Phi(b)$, for any two elements $a, b \in F$. Preserving multiplication means that $\Phi(a \cdot b) = \Phi(a) \cdot \Phi(b)$, for any two elements $a, b \in F$ Such a function Φ is called a *field isomorphism*, and F and F' are said to be *isomorphic* if an isomorphism exists between them. Informally, all this means is that the elements of F can be relabeled as the elements of F', in such a way that when this is done, the addition and multiplication tables for F become those for F'. In cases where two fields F and F' are isomorphic, we will sometimes abuse notation a bit and simply say (as common in practice) that they are equal: $F = F'$.

[‡] Evariste Galois grew up in a small town outside of Paris, where his father was the mayor. Starting from his school days, Galois was definitely a nonconformist. He rarely kept up with his assignments, and his extraordinary mathematical aptitude and intelligence were not noticed until he began his college studies. His literature professor did not believe that Galois had much intelligence at all until he heard from colleagues about Galois's superior mathematical talents. Galois tried to get into the top French mathematics university (École Polytechnique), but after failing the entrance exam on several attempts, he opted for the lesser École Normale. As a student, he published his first mathematical paper at the age of 17, and that same year he wrote another paper on a very difficult topic that was sent to the illustrious French mathematician Augustus Cauchy to referee. Galois did not take the time to carefully write up his work, and although it impressed many in the mathematical community, even the top mathematicians were often not able to follow his reasoning and needed him to revise his papers before they could be further considered. As he continued to flourish as a professional mathematician while he was a still student, he was also quite politically active during a tumultuous era in the history of France. He was imprisoned twice for periods of several months for his political acts, which included making a public threat to the king. After his second prison release in 1832, he fell in love with a daughter of a prison physician (Stephanie-Felice du Motel), and circumstances surrounding this relationship led to his being challenged to a duel. He was compelled to accept the challenge. In the duel he was badly injured by a hit to the stomach and died the following day. Shortly before his death, Galois gathered his unpublished work and instructed a friend to pass them to Gauss and Carl Gustav Jacobi (1804–1851, another topnotch German mathematician). With his papers, he wrote a note that said: "*il se trouvera, j'espère, des gens qui trouveront leur profit à défchiffrer tout ce gâchis*" (I hope that later some people will find it to their advantage to decipher all this mess). His work was indeed not only published but was soon after recognized to contain some most significant mathematical breakthroughs of the century. Despite his tragically brief life, he is considered among the greatest French mathematicians to have ever lived.

Figure 10.1 Evariste Galois (1811–1832), French mathematician.

We have had quite a bit of experience with the field $GF(p)$, which is just $\mathbb{Z}_p$. We give now an abstract formulation of the smallest new finite field $GF(2^2)$, which has only four elements. After this we give a more general and intuitive construction that will allow us to construct any of the finite fields $GF(p^n)$.

Example 10.6

Consider the set F of all 2-bit strings: $F = \{00, 01, 10, 11\}$. We define addition on F as simply XORing a pair of 2-bit strings (and will write addition using the XOR symbol $\oplus$ rather than the ordinary addition symbol $+$). Thus, for example, $10 \oplus 11 = 01$. We define multiplication of elements of F (denoted by the usual symbol $\cdot$) using the ring axioms with 01 being the multiplicative identity, and by the rules $10 \cdot 10 = 11$, $11 \cdot 11 = 10$, and $10 \cdot 11 = 01$.

(a) Fill in complete addition and multiplication tables for F.
(b) Show that F is a field and, thus, (by Theorem 10.2) must be $GF(4)$.

Solution: Part (a): Note that 00 is the additive identity since $00 \oplus ab = ab$. This, together with the given multiplication information, allows us to completely fill out the addition and multiplication tables, which are shown in Table 10.1 and Table 10.2.

Part (b): Looking through the eight ring axioms of Definition 10.2 that need verification, from the tables of part (a), it is clear that both addition and multiplication are commutative [(1) and (5)], that 00 is the additive identity (3), and that 01 is the multiplicative identity (7). The associativity of addition axiom (2) follows from the associativity of XOR [Proposition 7.1(2)]. To verify that multiplication is associative (Axiom 7), note that the identity $(a \cdot b) \cdot c = a \cdot (b \cdot c)$ is clearly true if any of a, b, or c is either zero (00) or one (01). So we need only check the remaining eight instances of this identity (where a, b, and c each runs

TABLE 10.1 Addition Table for Field F of Example 10.6

$\oplus$	00	01	10	11
00	00	01	10	11
01	01	00	11	10
10	10	11	00	01
11	11	10	01	00

TABLE 10.2 Multiplication Table for Field F of Example 10.6

$\cdot$	00	01	10	11
00	00	00	00	00
01	00	01	10	11
10	00	10	11	01
11	00	11	01	10

through 10 and 11), less the two where $a = b = c$, which are also clearly true. Here are the verifications of these six:

$(10 \cdot 10) \cdot 11 = 10 = 10 \cdot (10 \cdot 11),$
$(10 \cdot 11) \cdot 10 = 10 = 10 \cdot (11 \cdot 10),$

$(11 \cdot 10) \cdot 10 = 10 = 11 \cdot (10 \cdot 10),$
$(10 \cdot 11) \cdot 11 = 11 = 10 \cdot (11 \cdot 11),$
$(11 \cdot 10) \cdot 11 = 11 = 11 \cdot (10 \cdot 11),$ and
$(11 \cdot 11) \cdot 10 = 11 = 11 \cdot (11 \cdot 10).$

The last ring axiom to check is the distributive law: $a \cdot (b + c) = a \cdot b + a \cdot c$. Of the $4^3 = 64$ possible instances of this equation, some are clearly valid: in cases where any of a, b, or c is zero (00), or if a is 1 (01). Also, since we already know addition is commutative, if we know the distributive law in the case, say, $b = 10$ and $c = 11$, there is no need to check it when the values are reversed to $b = 11$ and $c = 10$. This reduces the number of equations that need to be checked to only $2 \cdot 6 = 12$. This routine verification is left to the next exercise for the reader.[*]

Finally, to show that the ring F is a (finite) field, we need to show that every nonzero element has a multiplicative inverse. This is clear from the multiplication table, which tells us that $01^{-1} = 01$, $10^{-1} = 11$ and $11^{-1} = 10$. By Theorem 10.2, it follows that $F = GF(4)$.

Exercise for the Reader 10.3

Verify the 12 remaining cases of the distributive law verification in the solution of Example 10.6

---

[*] With a bit more thought, it is easy to eliminate half of these 12 in which $b = c$, since both sides of the distributive law will involve an XOR of identical strings and so will be 00.

$\mathbb{Z}_p[X]$ = the Polynomials with Coefficients in $\mathbb{Z}_p$

At this juncture, the creation of the finite field of Example 10.6 may have appeared quite contrived and mysterious, but we will now enter into a general construction procedure from which we will be able to produce this and all other finite fields. The relevant field operations will be nicely modeled with the addition and multiplication of polynomials that one learns about in high school mathematics courses. The only difference is that we now assume that the coefficients of the polynomials lie in $\mathbb{Z}_p$, for some prime number p.

Definition 10.5

Given a prime number p, a **polynomial** with coefficients in $\mathbb{Z}_p$ is any expression of the form

$$f = a_n X^n + a_{n-1} X^{n-1} + \cdots a_1 X + a_0$$

where $n \geq 0$ is an integer, X is an **indeterminate** (a formal symbol), and the **coefficients** $a_n, a_{n-1}, \cdots, a_1, a_0$ are elements of $\mathbb{Z}_p$. Any term with a zero coefficient may be omitted, $1 \cdot X^k$ may be written simply as X^k, and X^0 may be written as 1. Thus, we may express a polynomial using the compact sigma notation as follows:

$$f = \sum_{i=0}^{n} a_i X^i$$

The **zero polynomial** arises when all coefficients above are zero, and it can be written simply as 0. For a nonzero polynomial f, with the above representation, if $a_n \not\equiv 0 \pmod{p}$, then we say that the polynomial has **degree** n, and we write this as $\deg(f) = n$. So the degree is the highest exponent of the indeterminant appearing in the expression of the polynomial that has a nonzero coefficient.[*] A polynomial of degree zero is also called a **constant polynomial**. The set of all polynomials with coefficients in $\mathbb{Z}_p$ in the indeterminant X is denoted by $\mathbf{Z}_p[X]$. Two polynomials in $\mathbb{Z}_p[X]$ are **equal** if all corresponding coefficients are equal $\pmod p$.

Example 10.7

The following are polynomials in $\mathbb{Z}_2[X]$: 0, X, $X + 1$, $X^2 + X + 1$. The polynomials X and $X + 1$ have degree 1, while the polynomial $X^2 + X + 1$ has degree 2. We point out that $X^2 + X + 1 = X^2 - X + 1 = X^2 - X - 1$ in $\mathbb{Z}_2[X]$ because $-1 \equiv 1 \pmod 2$.

---

[*] It will be convenient to define the degree of the zero polynomial as $-\infty$, as is customarily done in the literature.

Addition and Multiplication of Polynomials in $\mathbb{Z}_p[X]$

The following definition shows how to add and multiply polynomials in $\mathbb{Z}_p[X]$. The procedure is exactly like the one that is taught in high school, except that the coefficient arithmetic needs to be done in $\mathbb{Z}_p$. Here is a simple motivating example involving polynomials in $\mathbb{Z}_2[X]$ [recall $2 \equiv 0(\mod 2)$]:

$$(X^2 + X + 1) + (X + 1) = X^2 + 2X + 2 = X^2$$

$$(X^2 + X + 1) \cdot (X + 1) = (X^2 + X + 1) \cdot X + (X^2 + X + 1) \cdot 1$$

$$= X^3 + X^2 + X + X^2 + X + 1$$

$$= X^3 + 1$$

Definition 10.6   Addition and Multiplication of Polynomials in $\mathbb{Z}_p[X]$

Suppose that $f = \sum_{i=0}^{n} a_i X^i$ and $g = \sum_{i=0}^{m} b_i X^i$ are polynomials in $\mathbb{Z}_p[X]$.

(1) We define the **sum** $f + g$ to be the polynomial $\sum_{i=0}^{N} c_i X^i$ with $N = \max(n,m)$, and $c_i \equiv a_i + b_i(\mod p)$.

(2) We define the **product** $f \cdot g$ to be the polynomial $\sum_{i=0}^{n+m} d_i X^i$ with $d_i \equiv \sum_{j=0}^{i} a_j \cdot b_{i-j}(\mod p)$. (As usual, we take coefficients that do not appear in the expressions of f and g to be zero.)

Note that the coefficients d_i in (2) come from multiplying all combinations of terms in f with terms in g such that the total degree adds up to i; this is based on the algebraic identity $X^j \cdot X^k = X^{j+k}$. Indeed:

$$a_j X^j \cdot b_{i-j} X^{i-j} = a_j b_{i-j} X^{j+(i-j)} = a_j b_{i-j} X^i$$

To see why the coefficient formula in (2) really gives the same result as when we use the standard polynomial multiplication rules that were learned in high school, let us redo the example $(X^2 + X + 1) \cdot (X + 1)$ that was done above. The coefficients of the first polynomial are $a_2 = a_1 = a_0 = 1$, and those for the second are $b_1 = b_0 = 1$. The degrees of the two polynomials are $n = 2$ and $m = 1$, respectively. The coefficients of the product $\sum_{i=0}^{n+m} d_i X^i$, as given in (2), are as follows:

$$d_3 \equiv \sum_{j=0}^{3} a_j \cdot b_{i-j} \equiv a_0 b_3 + a_1 b_2 + a_2 b_1 + a_3 b_0 \equiv 1 \cdot 0 + 1 \cdot 0 + 1 \cdot 1 + 0 \cdot 1 \equiv 1(\mod 2)$$

$$d_2 \equiv \sum_{j=0}^{2} a_j \cdot b_{i-j} \equiv a_0 b_2 + a_1 b_1 + a_2 b_0 \equiv 1 \cdot 0 + 1 \cdot 1 + 1 \cdot 1 \equiv 0(\mod 2)$$

$$d_1 \equiv \sum_{j=0}^{1} a_j \cdot b_{i-j} \equiv a_0 b_1 + a_1 b_0 \equiv 1 \cdot 1 + 1 \cdot 1 \equiv 0 \pmod{2}$$

$$d_0 \equiv \sum_{j=0}^{0} a_j \cdot b_{i-j} \equiv a_0 b_0 \equiv 1 \cdot 1 \equiv 1 \pmod{2}$$

Thus, $\sum_{i=0}^{n+m} d_i X^i = 1 + X^3$, as we obtained above.

In performing polynomial multiplications by hand, it is often convenient to distribute the multiplications among the terms of one of the factors (as one does in high school polynomial multiplications and as we did in the motivating examples), but doing the coefficient arithmetic mod p. We will soon give a theorem that shows such distributive laws remain valid in $\mathbb{Z}_p[X]$ arithmetic.

Exercise for the Reader 10.4

Perform the following polynomial computations:

(a) $(X^5 + 4X^3 + 2) + (3X^3 + 2X)$ and $(X^5 + 4X^3 + 2) \cdot (3X^3 + 2X)$ in $\mathbb{Z}_5[X]$

(b) $(X^n + X^{n-1} + X^{n-2} + \cdots + X + 1) \cdot (X - 1)$ in $\mathbb{Z}_p[X]$, where n is any positive integer and p is any prime

The **leading term** of any nonzero polynomial is the term of highest degree (= the degree of the polynomial) that has a nonzero coefficient. It is clear that when we multiply two nonzero polynomials, the leading term of the product is the product of the leading terms of the factors.[*] Since the coefficients are nonzero elements in a field $\mathbb{Z}_p$, their product must be nonzero (see Exercise for the Reader 10.1). Thus the degree of the product of two nonzero polynomials is the sum of the degree of the two polynomials. We enunciate this important observation as the following.

Proposition 10.3

If p is a prime, and f and g are polynomials in $\mathbb{Z}_p[X]$, then $\deg(fg) = \deg(f) + \deg(g)$.[†]

Vector Representation of Polynomials

Although we discuss this topic more thoroughly in the computer implementation material at the end of the chapter, we briefly indicate here how polynomials can be easily stored and manipulated on a computer. It is often

---

[*] In terms of the coefficient formula in (2) of Definition 10.6, this means that the formula $d_{n+m} \equiv \sum_{j=0}^{n+m} a_j \cdot b_{n+m-j} \pmod{p}$ really has only a single term in the sum (corresponding to $j = n$), since if $j > n$, $a_j = 0$ whereas if $j < n$, then $n + m - j > m$, so $b_{n+m-j} = 0$. Thus, $d_{n+m} \equiv a_n \cdot b_m \pmod{p}$.

[†] Note that this identity is true even if one of the polynomials is the zero polynomial, since in this case their product will also be the zero polynomial, and $-\infty + A = -\infty$, for any real number A.

convenient to store a polynomial $f = a_n X^n + a_{n-1} X^{n-1} + \cdots a_1 X + a_0 \in \mathbb{Z}_p[X]$ by the $\mathbb{Z}_p$ vector of its coefficients: $f = \sum_{i=0}^{n} a_i X^i \sim [a_n, a_{n-1}, \cdots, a_1, a_0]$. The addition and multiplication operations can be converted into corresponding operations on such vectors. Suppose that $g = \sum_{i=0}^{m} b_i X^i \sim [b_m, b_{m-1}, \cdots, b_1, b_0]$ is another polynomial in $\mathbb{Z}_p[X]$. From the definition of addition of polynomials, we may write:

$$f + g \sim [c_N, c_{N-1}, \cdots, c_1, c_0] \tag{10.1}$$

where $N = \max(n, m)$, and $c_i \equiv a_i + b_i \pmod{p}$ for $1 \leq i \leq N$.

To understand the vector version of polynomial multiplication, we first see how it will work if we multiply $f \neq 0$ by a **monomial**, which is a nonzero polynomial consisting of a single term, $b_k X^k$:

$$f \cdot X^k = (a_n X^n + a_{n-1} X^{n-1} + \cdots a_1 X + a_0) \cdot X^k =$$
$$a_n X^{n+k} + a_{n-1} X^{n+k-1} + \cdots a_1 X^{1+k} + a_0 X^k$$

In vector notation, this multiplication becomes

$$[a_n, a_{n-1}, \cdots, a_1, a_0] \cdot [b_k, \underbrace{0, 0, \cdots, 0}_{k \text{ zeros}}] = b_k [a_n, a_{n-1}, \cdots, a_1, a_0, \underbrace{0, 0, \cdots, 0}_{k \text{ zeros}}] \tag{10.2}$$

where the latter scalar/vector multiplication is done mod p. By (repeatedly) using the distributive law, a general polynomial multiplication can be broken down into a sum of multiplications of a polynomial by a monomial:

$$f \cdot g = f \cdot \sum_{i=0}^{m} b_i X^i = \sum_{i=0}^{m} f \cdot b_i X^i \Rightarrow f \cdot g$$

$$\sim \sum_{i=0}^{m} b_i [a_n, a_{n-1}, \cdots, a_1, a_0, \underbrace{0, 0, \cdots, 0}_{i \text{ zeros}}] \tag{10.3}$$

Thus, with this method of storing polynomials along with the associated Equation 10.1 and Equation 10.3 for their addition and multiplication, we have an efficient means for manipulating polynomials on computing platforms. This will serve as a basis for the computer implementation of soon-to-be developed arithmetic of finite fields.

$\mathbb{Z}_p[X]$ Is a Ring

With the addition and multiplication operations defined above, it is not hard to show that $\mathbb{Z}_p[X]$ is a ring. The proof that the ring axioms are satisfied will be left to Exercise 25.

> ### Theorem 10.4
>
> If p is any prime number, then $\mathbb{Z}_p[X]$ is a ring. The zero polynomial is the zero element, and the constant polynomial 1 serves as the multiplicative identity.

Divisibility in $\mathbb{Z}_p[X]$

We are now ready to discuss some aspects of the rich divisibility theory for the ring $\mathbb{Z}_p[X]$ that very nicely parallels that which was developed in Chapter 2 for the ring $\mathbb{Z}$ of integers. Before reading on, the reader would do well to take a few moments to glance back over the first several pages of Chapter 2.

Recall that earlier in this chapter, we computed $(X^2 + X + 1) \cdot (X + 1) = X^3 + 1$ in $\mathbb{Z}_2[X]$. We say that either of the polynomials on the left is a *factor* or *divides* the polynomial on the right, and use the integer notation for divisibility. So we could write, for example, $(X^2 + X + 1) \mid X^3 + 1$. Here is the general definition (cf. Definition 2.1).

Definition 10.7

Suppose that f and g are polynomials in $\mathbb{Z}_p[X]$ with $f \neq 0$. We say that f **divides** g (written as $f \mid g$) if there is another polynomial $h \in \mathbb{Z}_p[X]$ such that $g = fh$. This can also be expressed by saying f is a **factor** of g, or g is a **multiple** of f. If f does not divide g, we write $f \nmid g$.

The next result records some simple facts about polynomial divisibility.

Theorem 10.5

Suppose that f, g, and h are polynomials in $\mathbb{Z}_p[X]$, for some prime p.

  (a) If $f \mid g$ and $g \neq 0$, then $\deg(f) \leq \deg(g)$.
  (b) *Divisibility is transitive.* If $f \mid g$ and $g \mid h$, then $f \mid h$.
  (c) If $f \mid g$ and $f \mid h$, then $f \mid sg + th$ for any polynomials $s, t \in \mathbb{Z}_p[X]$.

Proof: Part (a): If $f \mid g$, then $g = fH$ for some $H \in \mathbb{Z}_p[X]$. By Proposition 10.3, it follows that $\deg(g) = \deg(f) + \deg(H)$. Since $\deg(H) \geq 0$ (because H cannot be the zero polynomial), the previous equation implies that $\deg(g) \geq \deg(f)$.

Parts (b) and (c): The proofs of the corresponding parts of Theorem 2.1 for integers work equally well in the setting of polynomials. □

Example 10.8

  (a) For which prime numbers p (if any) is it true that $X + 1 \mid (X^2 - 1)$ in $\mathbb{Z}_p[X]$?
  (b) For which prime numbers $p < 10$ is it true that $X + 1 \mid (X^2 + 1)$ in $\mathbb{Z}_p[X]$?

Solution: Part (a): Some readers might remember the factorization identity $X^2 - 1 = (X + 1)(X - 1)$. (If not, it is easily checked by multiplying out the right product.) Any such polynomial identity that is valid for ordinary polynomials (with integer arithmetic) will automatically be valid in any $\mathbb{Z}_p[X]$ (because if the coefficients are equal as integers, then they will certainly be equal mod p). Thus, $X + 1 \mid (X^2 - 1)$ in $\mathbb{Z}_p[X]$ for any prime number p.

Part (b): If $X + 1 \mid (X^2 + 1)$ $\mathbb{Z}_p[X]$, then there would be a polynomial $f \in \mathbb{Z}_p[X]$ such that $X^2 + 1 = f \cdot (X + 1)$. By Proposition 10.3, we must have deg(f) = 1, so $f = aX + b$ for some coefficients $a, b \in \mathbb{Z}_p$, $a \not\equiv 0$. Since

$$X^2 + 1 = f \cdot (X + 1) = (aX + b)(X + 1) = aX^2 + (a + b)X + b$$

we can compare coefficients to conclude that $a \equiv 1$, $a + b \equiv 0$, and $b \equiv 1$. Substituting the outer two equations into the middle equation shows that this is possible if, and only if, $2 \equiv 0 \pmod{p}$. Thus, $X + 1 \mid (X^2 + 1)$ in $\mathbb{Z}_p[X]$, only in the case $p = 2$. Also, we have shown that the polynomial identity $X^2 + 1 = (X + 1)(X + 1) = (X + 1)^2$ (which we know is false in ordinary arithmetic) is true in $\mathbb{Z}_2[X]$.

We next would like to define *irreducible* polynomials, which play the same role in $\mathbb{Z}_p[X]$ as primes play in the ring of integers. Since all nonzero elements of $\mathbb{Z}_p$ have inverses, it follows that the leading coefficient of any nonzero polynomial in $\mathbb{Z}_p[X]$ always can be factored out, so we may always assume that the leading coefficient of any polynomial to be factored is 1. Moreover, by Proposition 10.3, no polynomial of positive degree can have an inverse in $\mathbb{Z}_p[X]$. Thus, the units in $\mathbb{Z}_p[X]$, $\mathbb{Z}_p[X]^\times$ are just the set of (nonzero) constant polynomials.

Definition 10.8

A polynomial f in $\mathbb{Z}_p[X]$ of positive degree is said to be **irreducible** in $\mathbb{Z}_p[X]$ if f has no polynomial factor $g \in \mathbb{Z}_p[X]$ with $0 < \deg(g) < \deg(f)$.

Exercise for the Reader 10.5

(a) For which prime numbers p (if any) is it true that $X^2 - 1$ is irreducible in $\mathbb{Z}_p[X]$?

(b) For which prime numbers p (if any) is it true that $X^2 + 1$ is irreducible in $\mathbb{Z}_p[X]$?

We next develop some more systematic mechanisms for testing divisibility relations that can be used to determine whether a given polynomial is irreducible.

The Division Algorithm for $\mathbb{Z}_p[X]$

We need an efficient way to determine whether a polynomial g divides into another polynomial f, whenever $0 < \deg(g) < \deg(f)$. This is provided by a following division algorithm that very nicely corresponds to the one we developed for integers in Chapter 2, except that the "sizes" of polynomials are gauged by their degrees. The mechanics of this polynomial division algorithm are just as those for polynomial long division that are taught in high school, so before formally stating the algorithm, we motivate it by performing the division $(X^4 + X^3 + 1) \div (X^2 + 1)$ in $\mathbb{Z}_2[X]$. The process is displayed with the usual schematic diagram that one learns about in high school; see Figure 10.2. With reference to this diagram, the steps taken are as follows (the only difference with the high school algorithm is that the coefficient arithmetic is performed mod 2): We first put the **divisor** $X^2 + 1$ on the left side of the long division symbol $)$ and, after inserting zero coefficients as placeholders in the **dividend** $X^4 + X^3 + 1 \rightarrow X^4 + X^3 + 0 \cdot X^2 + 0 \cdot X + 1$, we put this on the right and under the top of the division symbol. We next divide the leading term of the dividend ($= X^4$) by the leading term of the divisor ($= X^2$) to obtain the *first quotient* X^2 (because $X^2 \cdot X^2 = X^4$) and put this quotient X^2 directly over X^4 on top of the long division symbol $)$.

Next, we multiply this first quotient with the divisor to obtain $X^2(X^2 + 1) = X^4 + X^2$, and subtract this result from the dividend to obtain the current remainder: $X^3 + X^2 + 1$.[*] The current remainder is lined up two rows below the dividend in the diagram. Since the degree of this current remainder ($= 3$) is at least as large as that of the divisor ($= 2$), we repeat this process and divide the leading term of the current remainder ($= X^3$) by the leading term of the divisor ($= X^2$) to obtain the *second quotient* X (because $X \cdot X^2 = X^3$) and put this quotient X directly over X^3 on top of the long division symbol $)$. We then multiply this second quotient with the divisor to obtain $X(X^2 + 1) = X^3 + X$, and subtract this result from the previous remainder to obtain the current remainder: $X^2 + X + 1$. This current remainder is lined up two rows below the previous one in the diagram. Since the degree of the current remainder ($= 2$) is the same

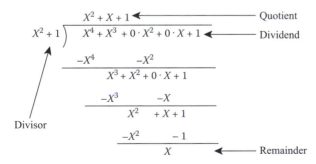

Figure 10.2 Schematic diagram of the polynomial division $(X^4 + X^3 + 1) \div (X^2 + 1)$ in the ring $\mathbb{Z}_2[X]$.

---

[*] We remind the reader that we are working in mod 2 arithmetic with the coefficients, so that $-X^2 = X^2$.

as that of the divisor, we need to repeat the process once more. We obtain the third and last quotient of 1 (since the leading coefficient of the current remainder and the divisor are the same), which is placed directly over $0 \cdot X^2$ on top of the long division symbol $\big)$. We then multiply this third quotient with the divisor to obtain $1 \cdot (X^2+1) = X^2+1$, and subtract this result from the previous remainder to obtain the current remainder, X. This current remainder is lined up two rows below the previous one in the diagram. Since the degree of this current remainder is less than that of the divisor, the algorithm terminates, and produces the result that $(X^4+X^3+1) \div (X^2+1) = X^2+X+1$ Remainder X, which corresponds to the equation $X^4+X^3+1 = (X^2+X+1)\cdot(X^2+1) + X$. This example had the luxury of working in $\mathbb{Z}_2$, where the only nonzero number (1) is its own inverse, and adding it is the same as subtracting it. The following note explains how to deal with finding monomial quotients in the general ring $\mathbb{Z}_p[X]$.[*]

Note: If p is any prime, and aX^ℓ, bX^k are two nonzero monomials in $\mathbb{Z}_p[X]$ with $\ell \geq k$, then the quotient $aX^\ell \div bX^k$ is $ab^{-1}X^{\ell-k}$. In the case $p = 2$, this simply becomes $X^\ell \div X^k = X^{\ell-k}$. In most of our divisions, the divisor will have a leading coefficient of 1, so that there is no need to find $b^{-1} (\bmod\ p)$.

Algorithm 10.1: Division Algorithm in $\mathbb{Z}_p[X]$

Input: Two polynomials $f, g \in \mathbb{Z}_p[X]$ with $g \neq 0$.
   Output: Two polynomials $q, r \in \mathbb{Z}_p[X]$ that satisfy $f = q \cdot g + r$, with $\deg(r) < \deg(g)$.
   As in the analogous Proposition 2.4 for integers, f is called the **divisor**, g is called the **dividend**, q the **quotient**, and r the **remainder**.

Step 1.   Initialize the current remainder CurrRem = f, and the Quotient = 0.
Step 2.   Divide the leading coefficient of CurrRem by that of g, call the result as QuotTemp. Update Quotient $\rightarrow$ Quotient + QuotTemp.
Step 3.   Update CurrRem $\rightarrow$ CurrRem – QuotTemp $\cdot g$.
Step 4.   If $\deg(\text{CurrRem}) \geq \deg(g)$, go back to Step 2, otherwise assign outputs: $q =$ Quotient, $r =$ CurrRem, and exit algorithm.

Note that in Step 3, since the leading coefficients of CurrRem, and QuotTemp $\cdot g$ have been designed to match, the degrees of the remainders go down in each iteration, so it follows that the algorithm will terminate. It is also not hard to show using the distributive law that the outputted quotient and remainder do indeed satisfy $f = q \cdot g + r$, and furthermore that the quotient and remainder are the unique polynomials that satisfy this

---

[*] Readers who are solely interested in the aspects of polynomials for AES may skip over comments regarding polynomial arithmetic in $\mathbb{Z}_p[X]$, for specific values of $p > 2$.

equation under the condition that $\deg(r) < \deg(g)$; see Exercise 32. This result will be used often, so we enunciate it as the following proposition.[*]

> **Proposition 10.6:** **The Division Algorithm**
>
> Suppose that p is a prime, and that $f, g \in \mathbb{Z}_p[X]$, with $g \neq 0$. Then there exist unique polynomials $q, r \in \mathbb{Z}_p[X]$ that satisfy $f = q \cdot g + r$, with $\deg(r) < \deg(g)$. These polynomials can be determined by Algorithm 10.1.

Example 10.9

Find the quotient and remainder when the division algorithm is applied to the following polynomial division in $\mathbb{Z}_5[X]$: $(X^4 + 3X + 2) \div (X^2 + 2)$.

Solution: The division algorithm is summarized by the schematic diagram shown here. A more detailed outline of the steps follows.

$$
\begin{array}{r}
X^2 \qquad\quad\ +3 \\
\hline
X^2 + 2 \,\big)\ \overline{X^4 + 0X^3 + 0X^2 + 3X + 2} \\
\underline{-X^4 \qquad\quad -2X^2} \\
3X^2 + 3X + 2 \\
\underline{-3X^2 \qquad -1} \\
3X + 1
\end{array}
$$

Step 1. Initialize CurrRem $= X^4 + 3X + 1$, Quotient $= 0$.

Step 2. First we divide the leading coefficient of the divisor $(= X^2)$ into that of the current remainder $(= X^4)$ to obtain the temporary quotient $(= X^2)$. We add this to update the quotient, which is now Quotient $= X^2$.

Step 3. CurrRem gets updated to $X^4 + 3X + 2 - X^2(X^2 + 2) = 3X^2 + 3X + 2$ in $\mathbb{Z}_5[X]$. Since this has degree $2 \geq 2 = \deg(X^2 + 2)$, we return to Step 2 (for the second and final iteration).

Step 2. *Second iteration.* We divide the leading coefficient of the divisor $(= X^2)$ into that of the current remainder $(= 3X^2)$ to obtain the temporary quotient $(= 3)$. We add this to update the quotient, which is now Quotient $= X^2 + 3$.

Step 3. *Second iteration.* CurrRem gets updated to $3X^2 + 3X + 2 - 3(X^2 + 2) = 3X + 1$.

Step 4. Since CurrRem has degree $1 < 2 = \deg(X^2 + 2)$, the algorithm terminates with the outputs: $q = $ Quotient $= X^2 + 3$, and $r = $ Remainder $= 3X + 1$.

---

[*] Although the statement of this proposition is not an algorithm, it is common practice to refer to this statement as "the division algorithm." Compare with Proposition 2.4.

Exercise for the Reader 10.6

Use the division algorithm (Algorithm 10.1) to perform the indicated polynomial divisions:

(a) $X^5 + 1 \div X^3 + 1$ in $\mathbb{Z}_2[X]$

(b) $X^5 + 4X^2 \div X^2 + 5$ in $\mathbb{Z}_7[X]$

Computing Note: Algorithm 10.1 can easily get quite complicated and time consuming. But since each step is easily translated into a corresponding vector operation based on additions and multiplications of polynomials whose vector counterparts were described earlier, Algorithm 10.1 is easily implemented on a computing platform. Details are provided in the computer implementation material at the end of the chapter.

The next Exercise for the Reader will allow the reader to discover that $g \mid f$ if, and only if, the remainder r in the above division is zero. (Just as was the case for the integers.)

Exercise for the Reader 10.7

Show that if we have two polynomials $f, g \in \mathbb{Z}_p[X]$, with $0 < \deg(g) \leq \deg(f)$, then $g \mid f$ if, and only if, the remainder r of the division algorithm (Algorithm 10.1) is the zero polynomial.

The result of this exercise for the reader provides one brute-force algorithm for determining whether a given polynomial is irreducible, which we summarize as an algorithm.

Algorithm 10.2: Brute-Force Irreducibility Test for Polynomials in $\mathbb{Z}_p[X]$

Input: A polynomial $f \in \mathbb{Z}_p[X]$ of degree at least 2 which we suspect might be irreducible.[*]

Output: Either a nontrivial factor of f (showing that it is not irreducible) or a statement that f is irreducible.

Step 1.   Initialize the factor degree deg = 1 (the smallest possible degree of a nontrivial factor).

Step 2.   Run through all polynomials g having degree deg and leading coefficient 1: $g = X^{\deg} + a_{\deg-1} X^{\deg-1} + \cdots + a_1 X + a_0$, where the deg coefficients $a_{\deg-1}, \cdots, a_1, a_0$ range through all possible mod p integers. Apply the division algorithm to $f \div g$. If the remainder r of such a division ever turns out to be zero, then (by Exercise for the Reader 10.7) the quotient q will be a nontrivial factor of f. So output this quotient and exit the algorithm.

Step 3.   Update deg $\rightarrow$ deg + 1. If this new value of deg is greater than floor($\deg(f)/2$), declare f as irreducible and exit the algorithm; otherwise, return to Step 2.

---

[*] Any polynomial of degree 1 in $\mathbb{Z}_p[X]$ is irreducible. (Why?)

The reason that (in Step 3) f can be declared irreducible if no non-trivial factor has been found of degree $\leq \text{floor}(\deg(f)/2)$ is because of Proposition 10.3. For polynomials of small degree (say, up to 5 or maybe 7) and $p = 2$ (our main modulus for our use of finite fields), this algorithm is feasible for hand computations. When testing a polynomial of degree 5 with $p = 2$, for example, there could be up to three division algorithms to perform.[*] Using computer implementations, the algorithm continues to be effective for testing polynomials of moderate degree (for example, for polynomials up to degree 20 or so, with $p = 2$, a computer can perform Algorithm 10.2 in less than a second). Exercise 25 presents a quick method to check whether a given first-degree polynomial is a factor of a given polynomial, and this method could be embellished into Algorithm 10.1.

Exercise for the Reader 10.8

(a) Use Algorithm 10.2 to determine whether the polynomial $X^3 + X + 1$ is irreducible in $\mathbb{Z}_2[X]$.

(b) Use Algorithm 10.2 to determine whether the polynomial $X^4 + X^3 + X + 1$ is irreducible in $\mathbb{Z}_2[X]$.

Congruences in $\mathbb{Z}_p[X]$ Modulo a Fixed Polynomial

We may define congruences in $\mathbb{Z}_p[X]$, modulo any nonzero polynomial $m \in \mathbb{Z}_p[X]$, just as we defined congruences in $\mathbb{Z}$ modulo any nonzero integer n.

Definition 10.9

Two polynomials $f, g \in \mathbb{Z}_p[X]$ are said to be **congruent** mod(ulo) a third (nonzero) polynomial $m \in \mathbb{Z}_p[X]$ if $m \mid (f - g)$. In this case, we write $f \equiv g \pmod{m}$.

It follows from the fact that $\mathbb{Z}_p[X]$ is a ring that all of the favorable properties of the congruence relations for integers (see, for example, Propositions 2.8–2.10) also carry over to hold for congruences modulo a polynomial. The proofs are essentially identical to those given in Chapter 2 in the setting of integers.

---

[*] Any factor with no constant term, such as $X^2 + X$, need not be checked with the division algorithm, since it factors into polynomials of smaller degree that have already been checked.

Example 10.10

Find a polynomial in $\mathbb{Z}_2[X]$ of degree less than 3 that is congruent to $X^5 + X$ (mod $X^3 + X + 1$).

Solution: Do this problem using two different methods:

Method 1: Reduction Method. In the notation of Definition 10.8, we have $m = X^3 + X + 1$. Since $m \mid (m - 0)$, we have $m \equiv 0 \pmod{m}$, or $X^3 + X + 1 \equiv 0$, which implies (subtracting $X + 1$ from both sides) that $X^3 \equiv -X - 1 \equiv X + 1$ (since the coefficient work is done mod 2). Using this congruence twice, we can perform the following reductions of X^5 (mod $X^3 + X + 1$):

$$X^5 + X \equiv X^3 \cdot X^2 + X \equiv (X + 1)X^2 + X \equiv X^3 + X^2 + X$$
$$X^3 + X^2 + X \equiv (X + 1) + X^2 + X \equiv X^2 + 1$$

We have thus shown that $X^5 \equiv X^2 + 1$ (mod $X^3 + X + 1$).

Method 2: Division Algorithm. By the division algorithm, any polynomial $f \in \mathbb{Z}_p[X]$ can be expressed as $f = q \cdot m + r$, where $q, r \in \mathbb{Z}_p[X]$ with $\deg(r) < \deg(m)$. It follows that $f \equiv r$, its remainder on division by m. Thus, the answer to the question of the example will be the remainder in the $\mathbb{Z}_2[X]$ polynomial division $(X^5 + X) \div (X^3 + X + 1)$, shown in schematic.

$$
\begin{array}{r}
X^2 \qquad\qquad +1 \\
X^3 + X + 1 \overline{\smash{)}\ X^5 + 0X^4 + 0X^3 + 0X^2 + X + 0} \\
\underline{-X^5 \qquad\quad -X^3 - X^2 \qquad\qquad} \\
X^3 + X^2 + X + 0 \\
\underline{-X^3 \qquad\quad -X - 1} \\
X^2 \qquad +1
\end{array}
$$

The remainder is the same as the answer obtained by Method 1, as expected.

Exercise for the Reader 10.9

Find a polynomial in $\mathbb{Z}_2[X]$ of degree less than 4 that is congruent to $X^7 + X^4 + X^2 + X$ (mod $X^4 + X + 1$).

Building Finite Fields from $\mathbb{Z}_p[X]$

We finally have all of the machinery needed to construct all of the finite fields $GF(p^n)$ of Theorem 10.2. The process nicely parallels how the fields $GP(p) = \mathbb{Z}_p$ were constructed from the integers in Chapter 2. Table 10.3 indicates the correspondences between the two processes. Just as we were able to build the finite ring $\mathbb{Z}_m$ from the infinite ring

TABLE 10.3 Parallels between Constructions of Finite Fields $GF(p)$ from $\mathbb{Z}$ and $GF(p^n)$ from $\mathbb{Z}_p[X]$

	Integers	**Polynomials**
Infinite ring	$\mathbb{Z}$	$\mathbb{Z}_p[X]$
Fixed ring element	m (integer > 1)	m (polynomial of degree > 0)
Finite modular ring	$\mathbb{Z}_m$ (has m elements)	$\mathbb{Z}_p[X](\mathrm{mod}\,m)$ (has $p^{\deg(m)}$ elements)
When is finite modular ring a field?	$\mathbb{Z}_m$ is a field if, and only if, $m = a$ prime p	$\mathbb{Z}_p[X](\mathrm{mod}\,m)$ is a field if, and only if, m is an irreducible polynomial

$\mathbb{Z}$, using congruence modulo a positive integer $m > 1$, we will be able to build a finite ring from $\mathbb{Z}_p[X]$ for each nonconstant polynomial $m \in \mathbb{Z}_p[X]$.

The finite polynomial rings are described in the following.

Definition 10.10:   *Finite Polynomial Rings from $\mathbb{Z}_p[X]$*

If p is a prime number and m is a nonconstant polynomial in $\mathbb{Z}_p[X]$, then the set $\mathbb{Z}_p[X](\mathrm{mod}\,m)$ is defined to be the set of all polynomials in $\mathbb{Z}_p[X]$ having degree less than $\deg(m)$. Thus, $\mathbb{Z}_p[X](\mathrm{mod}\,m)$ consists of all possible remainders that can arise when any polynomial in $\mathbb{Z}_p[X]$ is divided by m. The addition and multiplication binary operations in $\mathbb{Z}_p[X](\mathrm{mod}\,m)$ (which we denote by their usual symbols "+" and "·") are performed by first doing the corresponding usual polynomial operations, and then converting the answer to a polynomial in $\mathbb{Z}_p[X](\mathrm{mod}\,m)$.

The most important facts about $\mathbb{Z}_p[X](\mathrm{mod}\,m)$ are provided in the following theorem, whose proof is similar to the corresponding proofs in Chapter 2, and will be left to the exercises.

Theorem 10.7

If p is a prime number and m is nonconstant polynomial in $\mathbb{Z}_p[X]$, then $\mathbb{Z}_p[X](\mathrm{mod}\,m)$ is a ring consisting of the $p^{\deg(m)}$ polynomials in $\mathbb{Z}_p[X]$ that have degree less than $\deg(m)$. This ring will be a field precisely when m is irreducible, in which case (by Theorem 10.2) we will have $\mathbb{Z}_p[X](\mathrm{mod}\,m) = GF(p^{\deg(m)})$.

As a set,

$$\mathbb{Z}_p[X](\mathrm{mod}\,m) = \{a_{\deg(m)-1}X^{\deg(m)-1} + a_{\deg(m)-2}X^{\deg(m)-2}$$
$$+ \cdots + a_1 X + a_0 : a_i \in \mathbb{Z}_p\} \tag{10.4}$$

also has the vector representation

$$\mathbb{Z}_p[X](\operatorname{mod} m) = \{[a_{\deg(m)-1}, a_{\deg(m)-2}, \cdots, a_1, a_0] : a_i \in \mathbb{Z}_p\} \quad (10.5)$$

The latter representation leads to an efficient way to code addition and multiplication in $\mathbb{Z}_p[X](\operatorname{mod} m)$, and this concept is expounded upon in the computer implementation material at the end of this chapter. Note that addition in $\mathbb{Z}_p[X](\operatorname{mod} m)$ is just ordinary polynomial addition (since the degree of a sum of two polynomials is no greater than the maximum degree of the two polynomials); it is the product of $\mathbb{Z}_p[X](\operatorname{mod} m)$ that is new.

The above development shows that we can construct $GF(p^n)$ provided that we can find an irreducible polynomial m in $\mathbb{Z}_p[X]$ with $\deg(m) = n$. It can be shown that irreducible polynomials of any degree exist in $\mathbb{Z}_p[X]$, and Algorithm 10.2 can be used to find them. Let us first show how this method can be used to construct the field with four elements $GF(2^2)$ that was constructed in Example 10.6 using a more ad hoc scheme.

Example 10.11

(a) Show that there is only one irreducible polynomial $m \in \mathbb{Z}_2[X]$ that has degree 2.
(b) Create addition and multiplication tables for the finite field $GF(4) = \mathbb{Z}_2[X](\operatorname{mod} m)$, where m is the irreducible polynomial of part (a).

Solution: Part (a): Of the four degree-2 polynomials in $\mathbb{Z}_2[X]$: $X^2, X^2 + 1, X^2 + X, X^2 + X + 1$, the first three factor as $X^2 = X \cdot X$, $X^2 + X = X(X + 1)$, and (as was seen in Example 10.8) $X^2 + 1 = (X + 1)(X + 1)$ in $\mathbb{Z}_2[X]$. This leaves only one possibility: $m = X^2 + X + 1$. To show that m is irreducible, we can verify that there will be a nonzero remainder when m is divided by either of the two degree-1 polynomials in $\mathbb{Z}_2[X]$. Applying the division algorithm (Algorithm 10.1), to either of the divisions of m by X or $X + 1$ produces the same result: $m = X^2 + X + 1 = X \cdot (X + 1) + 1$. Since the remainder ($= 1$) is nonzero, it follows that m is irreducible in $\mathbb{Z}_2[X]$.

Part (b): Tables 10.4 and 10.5 show the addition and multiplication tables for $\mathbb{Z}_2[X](\operatorname{mod} X^2 + X + 1) = \{0, 1, X, X + 1\}$.

TABLE 10.4 Addition Table for Field $GF(4) = \mathbb{Z}_2[X](\operatorname{mod} X^2 + X + 1) = \{0, 1, X, X + 1\}$ of Example 10.11

+	0	1	X	X + 1
0	0	1	X	X + 1
1	1	0	X + 1	X
X	X	X + 1	0	1
X + 1	X + 1	X	1	0

TABLE 10.5 Multiplication Table for Field $GF(4) = \mathbb{Z}_2[X](\text{mod } X^2 + X + 1) = \{0,1,X,X+1\}$ of Example 10.11

·	0	1	X	$X + 1$
0	0	0	0	0
1	0	1	X	$X + 1$
X	0	X	$X + 1$	1
$X + 1$	0	$X + 1$	1	X

Notice that these tables are identical with Tables 10.1 and 10.2 for the field F of Example 10.6, with the identifications $00 \leftrightarrow 0, 01 \leftrightarrow 1, 10 \leftrightarrow X, 11 \leftrightarrow X + 1$, and since these identifications correspond exactly to the vector representation of polynomials, Tables 10.1 and 10.2 simply give the vectorized version of these tables.

Exercise for the Reader 10.10

Create addition and multiplication tables for the finite ring $\mathbb{Z}_2[X](\text{mod } X^2)$, and use one of these tables to provide one piece of evidence that demonstrates that $\mathbb{Z}_2[X](\text{mod } X^2)$ is not a field.

The Fields $GF(2^4)$ and $GF(2^8)$

We now present constructions of the 16-element finite field $GF(2^4)$ and of the 256-element finite field $GF(2^8)$. These fields will play crucial roles in the scaled-down AES and the full AES cryptosystems of the next chapter. Thus, from a cryptographic perspective, the construction and understanding of these two fields is the main purpose of this chapter.

Example 10.12: *The Field GF(16)*

(a) Show that the degree-4 polynomial $m = X^4 + X + 1$ is irreducible in $\mathbb{Z}_2[X]$. Thus, $\mathbb{Z}_2[X]$ (mod $X^4 + X + 1$) must be (by Theorems 10.2 and 10.7) $GF(2^4)$, the unique finite field of 16 elements.

(b) Perform the following multiplications in $\mathbb{Z}_2[X]$ (mod $X^4 + X + 1$): $(X + 1) \cdot X$, $(X^3 + X + 1) \cdot (X^2 + X)$, $(X^2 + X + 1) \cdot (X^5 + 1)$.

(c) Using the hexadecimal representation for the polynomials in $\mathbb{Z}_2[X]$ (mod $X^4 + X + 1$) that derives from the standard binary/hex correspondence of Table 6.1, create a multiplication table for the finite field $GF(2^4) = \mathbb{Z}_2[X]$ (mod $X^4 + X + 1$).

Solution: Part (a): We use Algorithm 10.2. The only degree-1 polynomials (in $\mathbb{Z}_2[X]$) are X and $X + 1$. Certainly X is not a factor of m. Since the division algorithm (Algorithm 10.1) applied to the division $m \div (X + 1)$ produces a remainder of 1, we know $X + 1$ is not a factor. We now move on to check for degree-2 factors. The possibilities are $X^2, X^2 + X, X^2 + 1, X^2 + X + 1$. We need check only the last two of these. Since the division algorithm produces respective remainders of X and 1, it follows that none of these are factors. Algorithm 10.2 now terminates with the conclusion that m is irreducible.

Part (b): Since $(X + 1) \cdot X = X^2 + X$ has degree less than 3, no further reductions are needed.

The other two products will need to be reduced after performing the usual multiplication process in $\mathbb{Z}_2[X]$:

$$(X^3 + X + 1) \cdot (X^2 + X) \equiv (X^3 + X + 1) \cdot X^2 + (X^3 + X + 1) \cdot X$$

$$\equiv X^5 + X^3 + \cancel{X^2} + X^4 + \cancel{X^2} + X$$

$$\equiv X^4 \cdot X + X^4 + X^3 + X$$

$$\equiv (X + 1) \cdot X + (X + 1) + X^3 + X$$

$$(\text{since } X^4 \equiv X + 1)$$

$$\equiv X^2 + X + (X + 1) + X^3 + X$$

$$\equiv X^3 + X^2 + X + 1 \quad \text{in } \mathbb{Z}_2[X] \ (\text{mod } X^4 + X + 1)$$

$$(X^2 + X + 1) \cdot (X^3 + 1) \equiv (X^2 + X + 1) \cdot X^3 + (X^2 + X + 1) \cdot 1$$

$$\equiv X^5 + X^4 + X^3 + X^2 + X + 1$$

$$\equiv (X + 1) \cdot X + (X + 1) + X^3 + X^2 + X + 1$$

$$(\text{since } X^4 \equiv X + 1)$$

$$\equiv \cancel{X^2} + \cancel{X} + \cancel{X} + \cancel{1} + X^3 + \cancel{X^2} + X + \cancel{1}$$

$$\equiv X^3 + X \quad \text{in } \mathbb{Z}_2[X] \ (\text{mod } X^4 + X + 1)$$

Part (c): Table 10.6 shows a natural representation of the 16 $\mathbb{Z}_2[X]$ polynomials (of degree less than 4) using the 16 hexadecimal characters that is based on the binary correspondence (see Table 6.1) and the vector representation of polynomials that was developed in this chapter. Such efficient representations will be used in the coding of AES cryptosystems in the next chapter. They also help to make the multiplication table for $GF(2^4)$ that is shown in Table 10.5 much more concise than if we had used polynomials instead. For example,

TABLE 10.6 Hexadecimal and Binary Representation for the Polynomials in *GF*(16)

Polynomial	Binary	Hex
0	0000	0
1	0001	1
X	0010	2
$X+1$	0011	3
X^2	0100	4
X^2+1	0101	5
X^2+X	0110	6
X^2+X+1	0111	7
X^3	1000	8
X^3+1	1001	9
X^3+X	1010	A
X^3+X+1	1011	B
X^3+X^2	1100	C
X^3+X^2+1	1101	D
X^3+X^2+X	1110	E
X^3+X^2+X+1	1111	F

in the hex notation of Table 10.6, the three $GF(2^4)$ multiplications that were computed in part (b) can be economically represented as:

$$(X+1)\cdot X = X^2+6 \quad \rightarrow \quad 3\cdot 2 \;=\; 6$$
$$(X^3+X+1)\cdot(X^2+X) \equiv X^3+X^2+X+1 \quad \rightarrow \quad B\cdot 6 \;=\; F$$
$$(X^2+X+1)\cdot(X^3+1) \equiv X^3+X \quad \rightarrow 7\cdot 9 \;=\; A$$

In Table 10.7, the complete multiplication table for $GF(2^4)$ is given, in which one can find the three multiplications of part (b).

Exercise for the Reader 10.11

Using the hex notation of Table 10.6, perform the following computations in *GF*(16):

(a) $6+D$

(b) $D\cdot(4+A^2)$

(c) $A\cdot B\cdot C\cdot D$

We point out that there are two other degree-4 irreducible polynomials in $\mathbb{Z}_2[X]$: X^4+X^3+1 and $X^4+X^3+X^2+X+1$. Each of these would, by

TABLE 10.7 Multiplication Table $GF(16)$ Using the Hexadecimal Notation of Table 10.6.

.	0	1	2	3	4	5	6	7	8	9	A	B	C	D	E	F
0	0	0	0	0	0	0	0	0	0	0	0	0	0	0	0	0
1	0	1	2	3	4	5	6	7	8	9	A	B	C	D	E	F
2	0	2	4	6	8	A	C	E	3	1	7	5	B	9	F	D
3	0	3	6	5	C	F	A	9	B	8	D	E	7	4	1	2
4	0	4	8	C	3	7	B	F	6	2	E	A	5	1	D	9
5	0	5	A	F	7	2	D	8	E	B	4	1	9	C	3	6
6	0	6	C	A	B	D	7	1	5	3	9	F	E	8	2	4
7	0	7	E	9	F	8	1	6	D	A	3	4	2	5	C	B
8	0	8	3	B	6	E	5	D	C	4	F	7	A	2	9	1
9	0	9	1	8	2	B	3	A	4	D	5	C	6	F	7	E
A	0	A	7	D	E	4	9	3	F	5	8	2	1	B	6	C
B	0	B	5	E	A	1	F	4	7	C	2	9	D	6	8	3
C	0	C	B	7	5	9	E	2	A	6	1	D	F	3	4	8
D	0	D	9	4	1	C	8	5	2	F	B	6	3	E	A	7
E	0	E	F	1	D	3	2	C	9	7	6	8	4	A	B	5
F	0	F	D	2	9	6	4	B	1	E	C	3	8	7	5	A

Theorem 10.7, give rise to the "same" finite field $GF(16)$. The polynomial $X^4 + X + 1$ was chosen because the reduction operation (in modular polynomial multiplication) proceeds more efficiently than with the other two. Similar comments apply to the choice of degree-8 polynomial that is made in the following construction of $GF(256)$.

Example 10.13

(*The Field GF(256)*).

It can be shown that the degree-8 polynomial $m = X^8 + X^4 + X^3 + X + 1$ is irreducible in $\mathbb{Z}_2[X]$ (see Computer Exercise 4). Thus, $\mathbb{Z}_2[X] \pmod{X^8 + X^4 + X^3 + X + 1}$ must be (by Theorems 10.2 and 10.7) $GF(2^8)$, the unique finite field of 256 elements.

(a)   Perform the following multiplication in $\mathbb{Z}_2[X] \pmod{X^8 + X^4 + X^3 + X + 1}$: $(X^6 + X + 1) \cdot (X^4 + X)$.

(b)   Develop binary/hex correspondence for elements of $GF(2^8)$ similar to what was done in Table 6.1 for the finite field $GF(2^4) = \mathbb{Z}_2[X] \pmod{X^4 + X + 1}$, where each element of $GF(2^8)$ is represented by a string of two hex characters. Represent the computation done in part (b) using this notation, and then compute $79 + A4$ in $GF(2^8)$.

Solution: Part (a): As with any modular polynomial multiplication, we first multiply the polynomials in $\mathbb{Z}_2[X]$, and then reduce powers that the result will have degree at most $7 = \deg(m) - 1$.

$$(X^6 + X + 1) \cdot (X^4 + X) \equiv (X^6 + X + 1) \cdot X^4 + (X^6 + X + 1) \cdot X$$

$$\equiv X^{10} + X^5 + X^4 + X^7 + X^2 + X$$

$$\equiv (X^4 + X^3 + X + 1) \cdot X^2 + X^7 + X^5 + X^4 + X^2 + X$$

$$(\text{mod } X^8 + X^4 + X^3 + X + 1)$$

$$\equiv (X^6 + X^5 + X^3 + X^2) + X^7 + X^5 + X^4 + X^2 + X$$

$$\equiv (X^6 + X^5 + X^3 + X^2) + X^7 + X^5 + X^4 + X^2 + X$$

$$\equiv X^7 + X^6 + X^4 + X^3 + X$$

Part (b): As explained earlier, any element of $GF(2^8)$ naturally corresponds to its binary string of coefficients:

$$a_7 X^7 + a_6 X^7 + \cdots + a_1 X + a_0 \sim [a_7, a_{7-1}, \cdots, a_1, a_0] \sim a_7 a_6 a_5 a_4 a_3 a_2 a_1 a_0$$

We can then use the natural binary/hex correspondence (shown in Table 10.6) to convert the half strings of the first and last four bits into their hex equivalents

$$a_7 a_6 a_5 a_4 a_3 a_2 a_1 a_0 \rightarrow [a_7 a_6 a_5 a_4] [a_3 a_2 a_1 a_0] \rightarrow H_1 H_2$$

where H_1, H_2 is each a single hexadecimal character.

Making this conversion in the computation of part (b), the steps are as follows:

$$(X^6 + X + 1) \cdot (X^4 + X) \equiv X^7 + X^6 + X^4 + X^3 + X$$
$$\rightarrow ([0100][0011]) \cdot ([0001][0010]) \equiv [1101][1010]$$
$$\rightarrow 43 \cdot 12 \equiv \text{DA}$$

Since adding polynomials in $GF(2^8)$ corresponds to XORing their bit strings, we perform the addition 79 + A4 by working directly on the corresponding bit strings:

```
79 + A4 → ([0111][1001]) ⊕ ([1010][0100]) = ([1101]
[1101]) → DD
```

Exercise for the Reader 10.12

Using the hex notation, perform the following computations in $GF(256)$:

(a) 64+CB

(b) AA^2

(c) A4·(B9+12)

The Euclidean Algorithm for Polynomials

We have discussed efficient algorithms for computing sums and products in the rings $\mathbb{Z}_p[X](\bmod m)$, where m is any nonconstant polynomial in $\mathbb{Z}_p[X]$, but the question naturally arises about how to find inverses (when they exist). Of course, for such important and relatively small fields such as $GF(16)$ and $GF(256)$, look-up tables for the inverses could be easily constructed. But at this point, we do not yet have a method for computing such inverses, except by trial and error or if we have a multiplication table at our hands. For example, from Table 10.7, we see that $\mathrm{D} \cdot 4 = 1$, from which we conclude that $\mathrm{D}^{-1} = 4$ in $GF(16)$.

We recall from Chapter 2 that to compute inverses of an element $a \in \mathbb{Z}_n$, we could apply the Euclidean algorithm (Algorithm 2.1) to compute $\gcd(a,n) = r$, and that a^{-1} if, and only if, this gcd is 1. Furthermore, in case $r = 1$, a^{-1} could be found by working backwards through the steps of the Euclidean algorithm to eventually compute a^{-1}. This latter process was formalized in the extended Euclidean algorithm (Algorithm 2.2). Since the Euclidean algorithm is based on the division algorithm for integers, and we have a polynomial analogue for this, this whole program turns out to extend very nicely to the setting of polynomials. The sizes of polynomials are measured by their degrees. The following example will illustrate this algorithm.

Example 10.14

Compute the inverse of the element F4 in $GF(256)$.

Solution: We need to work with polynomials, so we convert as explained in Example 10.13:

$$\mathrm{F4} \to [1111][0100] \to X^7 + X^6 + X^5 + X^4 + X^2$$

The polynomial version of the Euclidean algorithm, which we use here for the computation of $\gcd(X^8 + X^4 + X^3 + X + 1, X^7 + X^6 + X^5 + X^4 + X^2)$, follows the same strategy of repeatedly applying the division algorithm until we get a zero remainder.

$$\text{remainder} \to \text{divisor} \to \text{dividend} \to \text{not used}$$

$$X^8 + X^4 + X^3 + X + 1 = (X + 1)(X^7 + X^6 + X^5 + X^4 + X^2) + (X^2 + X + 1) \ X^7 + X^6 + X^5 + X^4 + X^2 = (X^5 + X^2 + X + 1)(X^2 + X + 1) + 1$$

Since we have reached a remainder of 1, the Euclidean algorithm terminates to tell us the gcd is 1. We already knew this (Why?), but we next work backwards with the above equations to express the remainder as a polynomial combination of the original two polynomials. (Throughout the computations below, please remember that all coefficient work is mod 2, so $-X^k \equiv X^k$.)

$$X^7 + X^6 + X^5 + X^4 + X^2 = (X^5 + X^2 + X + 1)(X^2 + X + 1) + 1 \Rightarrow$$
$$1 = (X^7 + X^6 + X^5 + X^4 + X^2) + (X^5 + X^2 + X + 1)(X^2 + X + 1)$$

Next, we use the first division algorithm equation:

$X^8 + X^4 + X^3 + X + 1 = (X + 1)(X^7 + X^6 + X^5 + X^4 + X^2) +$
$(X^2 + X + 1) \implies X^2 + X + 1 = (X^8 + X^4 + X^3 + X + 1) + (X + 1)$
$(X^7 + X^6 + X^5 + X^4 + X^2)$

to substitute into the previous equation to produce:

$$
\begin{aligned}
1 \; = \; & (X^7 + X^6 + X^5 + X^4 + X^2) \\
& + (X^5 + X^2 + X + 1)[(X^8 + X^4 + X^3 + X + 1) \\
& + (X + 1)(X^7 + X^6 + X^5 + X^4 + X^2)] \\
= \; & (X^5 + X^2 + X + 1)(X^8 + X^4 + X^3 + X + 1) \\
& + [1 + (X^5 + X^2 + X + 1)(X + 1)](X^7 + X^6 + X^5 + X^4 + X^2) \\
= \; & (X^5 + X^2 + X + 1)(X^8 + X^4 + X^3 + X + 1) \\
& + [X^6 + X^5 + X^3](X^7 + X^6 + X^5 + X^4 + X^2)
\end{aligned}
$$

From this (and the definition of polynomial congruence), it follows that

$[X^6 + X^5 + X^3](X^7 + X^6 + X^5 + X^4 + X^2)$

$\equiv 1 (\mathrm{mod}\, X^8 + X^4 + X^3 + X + 1)$

Thus we have found that $(X^7 + X^6 + X^5 + X^4 + X^2)^{-1} \equiv$ $X^6 + X^5 + X^3 \to$ [0110] [1000]. In hex notation: F4^{-1} = 68 in *GF*(256).

Exercise for the Reader 10.13

Use the Euclidean algorithm (polynomial version) to compute the following inverses:

(a)   A^{-1} in *GF*(16)

(b)   1A^{-1} in *GF*(256)

Check your answer for part (a) with Table 10.7.

The analogies—with many of the properties of the ring integers $\mathbb{Z}$ and those for the rings of polynomials $\mathbb{Z}_p[X]$—go further than those we have developed in this chapter. For example, there is a corresponding theory of unique factorization in $\mathbb{Z}_p[X]$ in terms of irreducible polynomials. Interested readers may consult any good book on abstract algebra for more on these topics; see, for example, [Hun-96] or [Her-96]. Exercise 31 will give an irreduciblity testing algorithm based on the Euclidean algorithm for polynomials that is much more efficient than Algorithm 10.1 for testing irreducibility of modular polynomials of high degree. Computer Exercise 10 will make some efficiency comparisons.

Chapter 10 Exercises

1. In each part, a set S along with an addition and multiplication operation is specified. Determine whether an S is a ring. If S is not a ring, identify at least one of the ring axioms that is violated. If a ring is formed, identify the additive and multiplicative identities.

   (a) The set $S = \{0\}$, with the binary operations $0 + 0 = 0$, $0 \cdot 0 = 0$.

   (b) The subset $S = \{0, 3\}$ of $\mathbb{Z}_6$ with addition and multiplication taken as those operations in $\mathbb{Z}_6$.

   (c) The subset $S = \{0, 2, 4, 6, 8\}$ of even integers in $\mathbb{Z}_{10}$ with addition and multiplication taken as those operations in $\mathbb{Z}_{10}$. (That these are binary operations on S comes from the fact that adding or multiplying even numbers in $\mathbb{Z}_{10}$ always produces an even number in $\mathbb{Z}_{10}$.)

   Note: In parts (b) and (c), the addition and multiplication binary operators come from known rings, being restricted to some subset. Thus, many of the ring axioms will automatically be inherited to hold true (when restricted to a smaller set of elements), and so need not be checked.

2. In each part, a set S along with an addition and multiplication operation is specified. Determine whether an S is a ring. If S is not a ring, identify at least one of the ring axioms that is violated. If a ring is formed, identify the additive and multiplicative identities.

   (a) The subset $S = \{0, 2\}$ of $\mathbb{Z}_4$, with addition and multiplication taken as those operations in $\mathbb{Z}_4$.

   (b) The set of all nonnegative integers $\mathbb{Z}_{\geq 0}$, with addition and multiplication taken as those operations in $\mathbb{Z}$.

   (c) The set S of all real numbers of the form $a + b\sqrt{2}$, where $a, b \in \mathbb{Q}$, with addition and multiplication taken as those operations in $\mathbb{R}$. (Recall that $\mathbb{Q}$ is the field of rational numbers, i.e., numbers expressible as fractions of integers.)

   Note: In each part, the addition and multiplication binary operators come from known rings, being restricted to some subset. Thus, many of the ring axioms will automatically be inherited to hold true (when restricted to a smaller set of elements), and so need not be checked.

3. For each of the number systems of Exercise 1 that was found to be a ring, determine whether it is a field.

4. For each of the number systems of Exercise 2 that was found to be a ring, determine whether it is a field.

5. (a) Let $\mathbb{Z}_2^3$ be the set of all length-3 binary vectors, and let addition and multiplication operations be defined by using the corresponding $\mathbb{Z}_2$ operations on the separate components:

$$[a,b,c]+[a',b',c']=[a+a'(\text{mod } 2),\, b+b'(\text{mod } 2),c+c'(\text{mod } 2)]$$
$$[a,b,c]\cdot[a',b',c']=[a\cdot a'(\text{mod } 2),\, b\cdot b'(\text{mod } 2),c\cdot c'(\text{mod } 2)]$$

Notice that addition simply XORs the vectors. For example, $[1,0,1]+[1,1,0]=[0,1,1]$ and $[1,0,1]\cdot[1,1,0]=[1,0,0]$.

   (b)   Show that $\mathbb{Z}_2{}^3$ is a ring.

   (c)   Extend the result of part (a) to $\mathbb{Z}_2{}^k$, the set of all length-k binary vectors, where $k>1$ is an integer.

   (d)   Does the result of part (a) continue to hold true for $\mathbb{Z}_n{}^k$, the set of all length-k vectors of entries in $\mathbb{Z}_n$, where $k,\, n>1$ are integers (and addition and multiplication of vectors is defined using the corresponding $\mathbb{Z}_n$ operations on the separate components)?

6.   Refer back to the definitions and notation of Exercise 5.

   (a)   Construct a multiplication table for $\mathbb{Z}_2{}^3$. Identify the set of invertible elements $(\mathbb{Z}_2{}^3)^\times$, and explain why $\mathbb{Z}_2{}^3$ is not a field.

   (b)   Identify the set of invertible elements $(\mathbb{Z}_2{}^k)^\times$ when $k>1$ is an integer.

   (c)   In the case $k,\, n>1$ are integers and $\mathbb{Z}_n{}^k$, determine the corresponding set of invertible elements $(\mathbb{Z}_n{}^k)^\times$.

7.   Perform the indicated polynomial additions:

   (a) $(X^5+X^3+1)+(X^3+X+1)$ in $\mathbb{Z}_2[X]$

   (b) $(4X^3+3X^2+9)+(8X^3+5X+5)$ in $\mathbb{Z}_{11}[X]$

   (c) $(4X^3+3X^2+2X+1)+(3X^3+5X+5)$ in $\mathbb{Z}_7[X]$

8.   Perform the indicated polynomial additions:

   (a) $(X^5+X^3+X^2+X+1)+(X^3+X^2+X+1)$ in $\mathbb{Z}_2[X]$

   (b) $(4X^3+3X^2+6)+(5X^3+5X+5)$ in $\mathbb{Z}_7[X]$

   (c) $(2X^3+X^2+2)+(2X^3+2X+1)$ in $\mathbb{Z}_3[X]$

9.   Perform the indicated polynomial multiplications:

   (a) $(X^5+X^3+1)\cdot(X+1)$ in $\mathbb{Z}_2[X]$

   (b) $(4X^3+3X^2+9)\cdot(8X^3+5X)$ in $\mathbb{Z}_{11}[X]$

   (c) $(4X^3+3X^2+1)\cdot(3X^3+5X+5)$ in $\mathbb{Z}_7[X]$

10.   Perform the indicated polynomial multiplications:

   (a) $(X^5+X^3+X^2+X+1)\cdot(X^2+1)$ in $\mathbb{Z}_2[X]$

   (b) $(4X^3+6)\cdot(5X^3+5X+5)$ in $\mathbb{Z}_7[X]$

   (c) $(2X^3+X^2+2)\cdot(2X^3+2X+1)$ in $\mathbb{Z}_3[X]$

11.   Use the division algorithm (Algorithm 10.1) to perform the indicated polynomial divisions:

   (a) $X^5+X^3+X^2+1\div X^2+1$ in $\mathbb{Z}_2[X]$

   (b) $X^4+X^3+X^2+1\div X+1$ in $\mathbb{Z}_2[X]$

   (c) $X^5+4X^2+7X\div X^2+2X$ in $\mathbb{Z}_{11}[X]$

   (d) $X^5+2X^2+1\div 2X^2+1$ in $\mathbb{Z}_3[X]$

12. Use the division algorithm (Algorithm 10.1) to perform the indicated polynomial divisions:
   (a) $X^5 + X^3 + X^2 + 1 \div X + 1$ in $\mathbb{Z}_2[X]$
   (b) $X^6 + X^3 + X^2 + 1 \div X^3 + 1$ in $\mathbb{Z}_2[X]$
   (c) $X^5 + 5X^2 + 6 \div X^2 + 4X$ in $\mathbb{Z}_5[X]$
   (d) $X^5 + 3X^2 + 2 \div 2X^2 + 1$ in $\mathbb{Z}_5[X]$

13. (a) Use Algorithm 10.2 to determine whether the polynomial $X^3 + X^2 + 1$ is irreducible in $\mathbb{Z}_2[X]$.
   (b) Use Algorithm 10.2 to determine whether the polynomial $X^5 + X^4 + X^2 + X + 1$ is irreducible in $\mathbb{Z}_2[X]$.

14. (a) Use Algorithm 10.2 to determine whether the polynomial $X^3 + X^2 + 1$ is irreducible in $\mathbb{Z}_2[X]$.
   (b) Use Algorithm 10.2 to determine whether the polynomial $X^4 + X^3 + 1$ is irreducible in $\mathbb{Z}_2[X]$.

15. (a) Find a polynomial in $\mathbb{Z}_2[X]$ of degree less than 6 that is congruent to $X^{10} + X^9 + 1$ (mod $X^6 + X^3 + X^2 + 1$).
   (b) Find a polynomial in $\mathbb{Z}_7[X]$ of degree less than 4 that is congruent to $X^6 + 2X^4 + 3X$ (mod $X^4 + 5X + 1$).

16. (a) Find a polynomial in $\mathbb{Z}_2[X]$ of degree less than 4 that is congruent to $X^8 + X^4 + X$ (mod $X^4 + X^2 + 1$).
   (b) Find a polynomial in $\mathbb{Z}_5[X]$ of degree less than 4 that is congruent to $X^6 + 2X^4 + 3X$ (mod $X^4 + 3X^3 + 1$).

17. (a) Is the ring $\mathbb{Z}_3[X] \,(\mathrm{mod}\, X^3 + X + 1)$ a field? Explain your answer.
   (b) Compute $(2X^2 + X + 2) + (2X + 1)$ and $(2X^2 + X + 2) \cdot (2X + 1)$ in $\mathbb{Z}_3[X] \,(\mathrm{mod}\, X^3 + X + 1)$.

18. (a) Is the ring $\mathbb{Z}_3[X] \,(\mathrm{mod}\, X^3 + X^2 + X + 1)$ a field? Explain your answer.
   (b) Compute $(2X^2 + X + 2) + (2X + 1)$ and $(2X^2 + X + 2) \cdot (2X + 1)$ in $\mathbb{Z}_3[X] \,(\mathrm{mod}\, X^3 + X^2 + X + 1)$.

19. Using hex notation, perform the following computations in $GF(256)$:
   (a) E1 + 24
   (b) 12^2
   (c) 4D·(C7 + 1F)

20. Using hex notation, perform the following computations in $GF(256)$:
   (a) 74 + AE
   (b) $0F^2$
   (c) CD·(CE + F5)

21. Use the polynomial Euclidean algorithm to determine whether the following inverses exist. In cases where an inverse exists, find it.
    (a) The element $X + 1$ in $\mathbb{Z}_2[X] \, (\mathrm{mod}\, X^2)$
    (b) The element $X + 1$ in $\mathbb{Z}_3[X] \, (\mathrm{mod}\, X^3 + X + 1)$
    (c) The element 9 in $GF(16)$
    (d) The element 1D in $GF(256)$

22. Use the polynomial Euclidean algorithm to determine whether the following inverses exist. In cases where an inverse exists, find it.
    (a) The element $X + 1$ in $\mathbb{Z}_3[X] \, (\mathrm{mod}\, X^2)$
    (b) The element $X + 1$ in $\mathbb{Z}_3[X] \, (\mathrm{mod}\, X^3 + X^2 + X + 1)$
    (c) The element C in $GF(16)$
    (d) The element 2E in $GF(256)$

23. Prove parts (2), (3), and (5) of Proposition 10.1.

24. Prove Theorem 10.4.

25. *Roots of Polynomials in* $\mathbb{Z}_p[X]$. We say that a modular integer $r \in \mathbb{Z}_p$ is a *root* (or a *zero*) of a polynomial $f = \sum_{i=0}^{n} a_i X^i \in \mathbb{Z}_p[X]$ if $f(r) \triangleq \sum_{i=0}^{n} a_i r^i \equiv 0 \,(\mathrm{mod}\, m)$.

    (a) Find all roots of the following polynomials:
        (i) $X^3 + X^2 + X + 1$ in $\mathbb{Z}_2[X]$
        (ii) $X^3 + 2X^2 + 2X + 1$ in $\mathbb{Z}_3[X]$
        (iii) $X^3 + 2X^2 + 2X + 1$ in $\mathbb{Z}_5[X]$
        (iv) $X^5 + 2X^2 + 4X + 1$ in $\mathbb{Z}_7[X]$

    (b) Show that $r \in \mathbb{Z}_p$ is a *root* (or a *zero*) of a polynomial $f \in \mathbb{Z}_p[X]$ if, and only if, $X - r$ is a factor of f. Thus, a polynomial $f \in \mathbb{Z}_p[X]$ of degree at least 2 that has a root $r \in \mathbb{Z}_p$ cannot be irreducible.
    (c) Use the result of part (b) to obtain any partial factorizations of the polynomials given in part (a) that are obtainable from roots.

26. (a) Suppose that $m \in \mathbb{Z}_p[X]$ is a polynomial of positive degree and that $f_1, f_2, g_1, g_2 \in \mathbb{Z}_p[X]$ are polynomials that satisfy $f_1 \equiv f_2, g_1 \equiv g_2 \,(\mathrm{mod}\, m)$. Show that (i) $f_1 + f_2 \equiv g_1 + g_2 \,(\mathrm{mod}\, m)$, and that (ii) $f_1 \cdot f_2 \equiv g_1 \cdot g_2 \,(\mathrm{mod}\, m)$,
    (b) Prove Theorem 10.7.

27. *Integral Domains.* A ring R is called an *integral domain* if the product of two nonzero elements is always nonzero, i.e., $a, b \in R, a \neq 0, b \neq 0 \Rightarrow ab \neq 0$. Note that the ring $\mathbb{Z}$ of integers is an example of an integral domain.
    (a) Show that a field is always an integral domain.
    (b) For which positive integers $n > 1$ is the ring of modular integers $\mathbb{Z}_n$ an integral domain?
    (c) If p is a prime, is $\mathbb{Z}_p[X]$ an integral domain?
    (d) Show that if m is a nonconstant polynomial in $\mathbb{Z}_p[X]$, where p is a prime, then the ring $\mathbb{Z}_p[X](\mathrm{mod}\, m)$ is an integral domain if, and only if, m is irreducible in $\mathbb{Z}_p[X]$.

28. *Products of Rings.*

(a) Suppose that $R = (R, +_R, \cdot_R)$ and $S = (S, +_S, \cdot_S)$ are rings. We let $R \times S$ be the set of all vectors $[r, s]$, where $r \in R$ and $s \in S$, and define addition and multiplication on $R \times S$ componentwise: $[r, s] + [r', s'] = [r +_R r', s +_S s']$, $[r, s] \cdot [r', s'] = [r \cdot_R r', s \cdot_S s']$. Show that $R \times S$ is a ring; it is called the product ring of the rings R and S.

(b) If, in part (a), both R and S are fields, is their product ring $R \times S$ also a field? Explain your answer.

29. Suppose that R is a ring with at least two elements. Is it possible to have $1 = 0$ (i.e., for the additive and multiplicative identity to coincide)? Either provide an example showing this is possible or carefully use the ring axioms to prove that such an example is not possible.

Note: Exercise 1(a) shows it to be possible if R has just one element.

30. Show that -1 has a square root modulo a prime p if, and only if, $p \equiv 1 \pmod 4$.

Suggestion: Let g be a primitive root mod p. By Euler's theorem and Proposition 8.6, it follows that $g^{(p-1)/2} \equiv -1$. If a is a square root of -1, write $a \equiv g^j$, and obtain $g^{2j} \equiv g^{(p-1)/2} \pmod p$. Use Proposition 8.5(c).

31. *Ben-Or's Irreducibility Determination Algorithm for Polynomials in $\mathbb{Z}_p[X]$.* The following is an efficient algorithm for checking irreducibility of polynomials in $\mathbb{Z}_p[X]$. It can be used in place of Algorithm 10.2 for checking irreducibility of higher degree polynomials. See Theorem 3.20 of [LiNi-86] for a proof of its correctness.

(a) Use Ben Or's algorithm to redo Exercise 13.

(b) Use Ben Or's algorithm to show that the polynomial $m = X^8 + X^4 + X^3 + X + 1$, that was used in the construction of $GF(256)$, is irreducible in $\mathbb{Z}_2[X]$.

Algorithm 10.3:   Ben Or's Irreducibility Test

Input: A polynomial $f \in \mathbb{Z}_p[X]$, of degree at least 2 that we suspect might be irreducible.

Output: Either a nontrivial factor of f (showing that it is not irreducible) or a statement that f is irreducible.

Step 1.   Initialize the index $i = 1$, and the test polynomial $h = X^p$.

Step 2.   Compute the polynomial $H \triangleq h - X \pmod f$, and then use the polynomial Euclidean algorithm to compute $g \triangleq \gcd(f, H)$. If g is a nonconstant polynomial, output g as a nontrivial factor of f and exit the algorithm.

Step 3. Update $i \to i + 1$. If this new value of i is greater than floor$(\deg(f)/2)$, declare f as irreducible and exit the algorithm; otherwise, update $h \to h^p (= X^{p^i})$ and return to Step 2.

31. $\mathbb{Z}_n[X]$. Suppose that, in Definition 10.5 of modular polynomials, we allow the modulus to be any integer $n > 1$, rather than a prime. The resulting systems $\mathbb{Z}_n[X]$ that arise satisfy some but not all of the properties of $\mathbb{Z}_p[X]$. This exercise elaborates on some similarities and differences.

(a) Show that Proposition 10.3 will never hold for $\mathbb{Z}_n[X]$, if $n > 1$ is composite.

(b) Show that $\mathbb{Z}_n[X]$ is a ring (i.e., Theorem 10.4 continues to hold).

(c) Does the division algorithm (Proposition 10.6) hold in $\mathbb{Z}_n[X]$?

(d) If $n > 1$ is composite, show that $\mathbb{Z}_n[X]$ is never an integral domain (see Exercise 27).

32. Prove Proposition 10.6.

Chapter 10 Computer Implementations and Exercises

Computer Storage and Arithmetical Operations on Polynomials. As pointed out in the text, vectors of coefficients are very efficient means for storing and manipulating polynomials on computers. The basic storage strategy is as follows:

Polynomial in $\mathbb{Z}_p[X]$	Vector of integers mod p
$a_n X^n + a_{n-1} X^{n-1} + \cdots a_1 X + a_0$	$[a_n, a_{n-1}, \cdots, a_1, a_0]$

1. *Program for Polynomial Addition in $\mathbb{Z}_p[X]$.*

(a) Write a program with syntax `Answer= ZpPolyAdd(px, qx, p)` that will add two polynomials mod p (i.e., in $\mathbb{Z}_p[X]$). The first two inputs, `px` and `qx`, are vectors representing the polynomials to be added, and the third input variable is the modulus `p`. The output, `Answer`, is a vector representing the sum of the inputted polynomials. If the sum is the zero polynomial, the output should be [0]; otherwise, the output should have a nonzero first component (so that the degree of the sum is one less than the length of the output vector).

(b) Run your program on the polynomial additions of Chapter Exercise 7.

2. *Program for Polynomial Multiplication in $\mathbb{Z}_p[X]$.*

(a) Write a program with syntax `Answer= ZpPolyMult(px, qx, p)` that will multiply two polynomials mod p (i.e., in

$\mathbb{Z}_p[X]$) The first two inputs, px and qx, are vectors representing the polynomials to be multiplied, and the third input variable is the modulus p. The output, Answer, is a vector representing the product of the inputted polynomials. If the sum is the zero polynomial, the output should be [0]; otherwise, the output should have a nonzero first component (so that the degree of the sum is one less than the length of the output vector).

(b) Run your program on the polynomial additions of Chapter Exercise 9.

Suggestion: Use the scheme discussed in the text that is based on Equation 10.2 and Equation 10.3.

3. *Program for the Polynomial Division Algorithm in $\mathbb{Z}_p[X]$.*
   (a) Write a program with syntax qx, rx = ZpDivAlg(fx, gx,p) that will compute the quotient and remainder when the division algorithm (Algorithm 10.1) is applied in the division of the polynomial represented by the first inputted vector, fx, by the polynomial represented by the second inputted vector, gx. The third input is the (prime) modulus p. The output variables, qx and rx, are vectors representing the quotient and remainder of the division.
   (b) Run your program on the polynomial additions of Chapter Exercise 11.

   Suggestion: Follow the algorithm, and make use of the programs for adding and multiplying polynomials that were created in the previous two computer exercises.

4. *Program for the Checking Irreducibility of a Polynomial in $\mathbb{Z}_p[X]$.*
   (a) Write a program with syntax Answer = ZpIrredCheck (fx, p) that will determine whether the polynomial represented by the first inputted vector, fx, is irreducible in $\mathbb{Z}_p[X]$. The second input is the (prime) modulus p. The output variable, Answer, will be either a vector representing a nontrivial factor or a message that the polynomial is irreducible. Algorithm 10.2 should be used.
   (b) Run your program on the polynomials of Exercise for the Reader 10.8 and then those of Chapter Exercise 13.
   (c) Use your program to show that the degree-8 polynomial $m = X^8 + X^4 + X^3 + X + 1$ that is used in the construction of $GF(2^8)$ is indeed irreducible in $\mathbb{Z}_2[X]$. Also, find all other degree-8 irreducible polynomials in $\mathbb{Z}_2[X]$.
   (d) Use your program to check whether the polynomial $f = X^{13} + X^9 + X^2 + X + 1$ is irreducible in $\mathbb{Z}_p[X]$ when $p = 2, 3, 5$, and 7.
   (e) Use your program to check whether the polynomial $f = X^{11} + 2X^8 + 2X^2 + 3X + 1$ is irreducible in $\mathbb{Z}_p[X]$ when $p = 5, 7$, and 11.

5. *Program for the Multiplication in $\mathbb{Z}_p[X](\bmod m)$.*
   (a) Write a program with syntax fgx = ZpModmMult(fx, gx, mx, p) that will compute the product of two polynomials

represented by the first two inputted vectors, fx, gx, is computed in $\mathbb{Z}_p[X](\bmod m)$, where m is a nonconstant polynomial represented by the third input, mx. The last input is the (prime) modulus p. The output variables, fgx, is a vector representing the product and should have degree less than that of m.

(b) Run your program on the polynomial multiplications of Example 10.12(b) and in Example 10.13.

(c) Use the program of part (a), in conjunction with that of Computer Exercise 1 to perform the *GF*(256) calculations of Exercise for the Reader 10.12 and those of Chapter Exercise 19.

Suggestion: This will be an easy program to write if you use those of Computer Exercises 2 and 3.

6. *Programs for GF(16) Addition/Multiplication.*

(a) Write a program with syntax aPlusbHex = GF16Add (aHex, bHex) that inputs two elements of GF(16), aHex, bHex, as single-character hex strings, and outputs their sum, aPlusbHex, in *GF*(16), also as a single hex character.

(b) Write a program with syntax aTimesbHex = GF16Mult (aHex, bHex) that inputs two elements of *GF*(16), aHex, bHex, as single-character hex strings, and outputs their product, aTimesbHex, in *GF*(16), also as a single hex character.

(c) Run your programs on the computations of Exercise for the Reader 10.11.

Suggestion: Make use of the programs of Computer Exercises 1, 2, and 5. The only thing to be careful with is the internal hex to binary vector conversions.

7. *Programs for GF(256) Addition/Multiplication.*

(a) Write a program with syntax aPlusbHex = GF256Add (aHex, bHex) that inputs two elements of *GF*(256), aHex, bHex, as two-character hex strings, and outputs their sum, aPlusbHex, in *GF*(256), also as a double hex character.

(b) Write a program with syntax aTimesbHex = GF256 Mult(aHex, bHex) that inputs two elements of *GF*(256), aHex, bHex as two character hex strings, and outputs their product, aTimesbHex, in *GF*(256), also as a double hex character.

(c) Run your programs on the computations of Exercise for the Reader 10.11 (and check the results with the answers in the back of the book).

Suggestion: Make use of the programs of Computer Exercises 1, 2, and 5. The only thing to be careful with is the internal hex to binary vector conversions.

8. *Program for the Extended and Regular Euclidean Algorithm for Polynomials in* $\mathbb{Z}_p[X]$.
    (a) Write a program with syntax gcd = ZpEuclidAlg(ax, bx, p) that will compute the $\mathbb{Z}_p[X]$ gcd of two polynomials, not both zero, that are represented by the first two inputs. The third input is the (prime) modulus p. The output variable, gcd, is a vector representing the gcd, as computed by the polynomial version of the Euclidean algorithm.
    (b) Write a program with syntax gcd, ex, fx = ZpExt EuclidAlg(ax, bx, p) that will compute the $\mathbb{Z}_p[X]$ gcd of two polynomials, not both zero, that are represented by the first two inputs. The third input is the (prime) modulus p. The output variables are vectors representing the gcd, and two polynomials *e* and *f*, such that gcd = *ae* + *bf* The analogue for the extended Euclidean algorithm (Algorithm 2.2) should be followed.
    (c) Run your programs for parts (a) and (b) on Example 10.14, and check their correctness.

    Suggestion: Make use of Computer Exercise 3; see the formal statements and computer implementation material for the integer versions of these algorithms in Chapter 2.

9. *Programs for Computing Inverses in GF(16) and in GF(256).*
    (a) Write a program with syntax aInvHex=GF16Inv(aHex) that inputs a nonzero element of *GF*(16), aHex, as a single-character hex string, and outputs its inverse, aInvHex, in *GF*(16), also as a single hex character.
    (b) Write a program with syntax aInvHex = GF256Inv (aHex) that inputs a nonzero element of *GF*(256), aHex, as a single-character hex string, and outputs its inverse, aInvHex, in *GF*(256), also as a single hex character.
    (c) Run your program of part (a) to compute the inverses of all 15 nonzero elements of *GF*(16), and check their correctness using Table 10.7, or using the program of Computer Exercise 6(b).
    (d) Use your program of part (b) to compute the inverses of the following elements of *GF*(256): A1, B2, C3, D4, E5, F6, FB, and check their correctness by using the program of Computer Exercise 7(b).

    Suggestion: Make use of the programs of Computer Exercises 8. The only thing to be careful with is the internal hex to binary vector conversions.

10. *Implementation of Ben Or's Irreducibility Test.*
    (a) Write a program with syntax Answer=ZpIrredBenOr (fx, p) that will determine whether the polynomial represented by the first inputted vector, fx, is irreducible in $\mathbb{Z}_p[X]$. The second input is the (prime) modulus p. The output variable, Answer, will either be a vector representing a nontrivial factor or a message that the polynomial is irreducible. Algorithm 10.3 (of Chapter Exercise 31) should be used.

(b) Run your program on the polynomials of Exercise for the Reader 10.8 and then those of Chapter Exercise 13.

(c) Use your program to check whether the polynomial $f = X^{11} + 2X^8 + 2X^2 + 3X + 1$ is irreducible in $\mathbb{Z}_p[X]$ when $p = 5, 7,$ and 11.

(d) Use your program to check that the following high-degree polynomials are irreducible over $\mathbb{Z}_2[X]$: $f = X^{45} + X^4 + X^3 + X$, $g = X^{140} + X^{15}$, $h = X^{208} + X^9 + X^3 + X$.

Caution: Do not attempt to apply the program of Computer Exercise 4 to these polynomials! You have been warned.

The Advanced Encryption Standard (AES) Protocol

We begin by describing some of the history and basic facts about the AES cryptosystem. Then we present a scaled-down version of AES, similarly to what was done in Chapter 7 for the DES. This scaled-down version exhibits almost all of the salient features of AES, but its size allows complete examples to be worked out by hand. The actual AES cryptosystem is designed to accept keylengths of size 128, 192, or 256 bits. We focus our treatment on the 128-bit version, since there are essentially no new ideas in the extension to larger key sizes. The underlying number systems for the scaled-down AES and for the full AES are the finite fields $GF(2^4)$ and $GF(2^8)$, respectively, that were described completely in the previous chapter. The notation and concepts of Chapter 10 regarding these finite fields are used freely throughout this chapter. The AES system breaks 128-bit strings into 16 bytes (recall a byte is 8 bits), so that each of these bytes can be viewed as a single element of $GF(2^8)$. Thus, bytes can be added and multiplied using the field operations. The highly nonlinear multiplication operations in $GF(2^8)$ are vital to the security of AES.

An Open Call for a Replacement to DES

In 1997, as DES was becoming increasingly vulnerable to attacks, the U.S. *National Institute of Standards and Technology (NIST)* put out an open international call for a new encryption standard to replace DES. They gave some specific guidelines, including that the system should be a block cipher that operates on 128-bit blocks, and that the system should accommodate keys of sizes 128, 192, or 256 bits. The result of the ensuing arduous selection process is truly a *David versus Goliath story*, with a couple of young independent cryptographers outclassing huge corporate competitors, such as IBM (the winner of the last competition when DES had been selected over 25 years earlier). A total of 21 systems were submitted by the June 1998 due date. From these, the NIST selected a first round of 15 that were presented

Figure 11.1 Joan Daemen (1965–), Belgian cryptographer.

Figure 11.2 Vincent Rijmen (1970–), Belgian cryptographer.

to the cryptographic community through two international conferences that it organized in August 1998 and in March 1999. The international flavor of the competition was nicely demonstrated by the list of countries of origin of these 15 cryptosystems: Australia, Belgium, Canada, Costa Rica, France, Germany, Israel, Japan, Korea, Norway, the United Kingdom, and the United States. Based on feedback from the community and its own further research, five finalists were announced in August 1999: *MARS* (from IBM), *RC6* (from RSA Laboratories), and the following three that were submitted by groups of individuals: *Rijndael*, *Serpent*, and *Twofish*. The NIST sponsored a third conference in April 2000 for final feedback from the cryptographic community. At last, on October 2, 2000, after this very thorough screening process, the NIST announced that the new **Advanced Encryption Standard (AES)** would be **Rijndael**.

Rijndael was created by two young Belgian cryptographers, Joan Daemen and Vincent Rijmen*; see Figures 11.1 and 11.2. Security was, above all, the most important property for the new AES, but computational efficiency and feasibility for implementation on a variety of platforms were important considerations as well. The five finalists were all found to be extremely secure, but Rijndael distinguished itself because of its computational speed on a variety of hardware configurations. Although

---

* Daemen studied electromechanical engineering, and subsequently was employed as cryptographic researcher at the COSIC Laboratory within the Katholieke Universiteit Leuven. While there, he completed his Ph.D. in cryptography in 1995. Rijmen first studied electrical engineering and then went on to work on his doctorate at the COSIC Laboratory as a post doc. Rijmen and Daemen became friends and worked on several joint projects, including Rijndael. At the time of this writing, Rijmen is a professor at the Katholieke Universiteit Leuven, and Daemen works as a professional cryptographer for STMicroelectronics in Brussels. Rijndael is an artificial name obtained using portions of the last names of its creators. They originally chose the name as a "teaser" that would be difficult to pronounce for non-Flemish speakers, but they have subsequently acquiesced to either of the following pronunciations: <Rain Doll> or <Rhine Dahl>.

it uses multiple rounds, Rijndael is not a Feistel system but an example of what is called a *substitution permutation network*. The AES satisfies the strong avalanche condition: changing one bit of plaintext results in a ciphertext with the property that each bit is different from that of the original ciphertext with probability approximately 1/2. The system thus very nicely exhibits Shannon's confusion and diffusion properties. Rijndael is actually more general than the AES: it supports any block-length and keylength between 128 and 256 bits that are multiples of 32 (making 25 combinations in total), whereas AES has a block length of 128 bits, and three choices for the keylength, 128, 192, and 256 bits (three combinations in total).

Nibbles

Our scaled-down AES system will operate at the level of 4-bit strings. The following definition will thus help to streamline our development.

Definition 11.1

A **nibble** (also called a **nybble**) is a string of 4 bits.

We recall that a byte is a string of 8 bits. Thus, 1 byte = 2 nibbles = 8 bits. Using the standard correspondence of Table 6.1, a nibble can be represented by a single hexadecimal digit. Similarly, a nibble can be viewed as an element of the 16-element field $GF(2^4)$ that was described in Chapter 10 (see, in particular, Example 10.12). Using the vector representation of any polynomial $aX^3 + bX^2 + cX + d \leftrightarrow [a, b, c, d]$ (where $a, b, c, d \in \mathbb{Z}_2$), we are led to a natural scheme to multiply nibbles.

Definition 11.2

Addition and Multiplication of Nibbles

Given two nibbles N and N', the **product** $N \cdot N'$ is defined as follows:

Step 1. Convert the nibbles to binary vectors: $N = [a, b, c, d]$, $N' = [a', b', c', d']$

Step 2. Convert the binary vectors of Step 1 into their corresponding polynomials in $GF(2^4)$:

$$N \to aX^3 + bX^2 + cX + d, \ N' \to a'X^3 + b'X^2 + c'X + d'$$

Step 3. Multiply the polynomials of Step 2 in $GF(2^4)$: (mod $X^4 + X + 1$)

$$(aX^3 + bX^2 + cX + d)(a'X^3 + b'X^2 + c'X + d') \equiv a''X^3 + b''X^2 + c''X + d$$

Step 4. From the coefficients of the product of Step 3, read off the product nibble $N \cdot N'$.

The nibble **sum** $N + N'$ is defined in the same fashion, except in Step 3, the multiplication changes to an addition.

Note that nibble addition is just XORing the bit strings. This is a consequence of the similar property for polynomial addition in $\mathbb{Z}_2[X]$, as was discussed in Chapter 10.[*] In what follows, we will not go through the above four steps in working with nibble arithmetic. For example, we will freely make nibble associations such as 0100 (binary) = 4 (hex) = X^2 (polynomial). The following example provides a brief review of the sorts of manipulations that were taught in Chapter 10. Readers may wish to review portions of Chapter 10 in order to make themselves proficient with such computations.

Example 11.1

Perform the following nibble operations (where the nibbles are represented as single hexadecimal digits):

(a) $A + 8$
(b) $F \cdot A$

Solution: Part (a): *Method 1:* XOR the corresponding bit strings, and then convert the result back into hex: $A + 8 = 1010 \oplus 1000 = 0010 = 2(\text{hex})$.

Method 2: Convert to polynomials in $GF(16)$, and add using field addition (polynomial addition):

$$A + 8 = 1010 + 1000 = (X^3 + X) + (X^3) = X = 0010 \text{ (bin)} = 2 \text{ (hex)}$$

Part (b): We convert the hex characters to their corresponding polynomials in $\mathbb{Z}_2[X]$, and then use $GF(16)$ multiplication:

$$F \cdot A = 1111 \cdot 1010 = (X^3 + X^2 + X + 1) \cdot (X^3 + X)$$
$$= (X^3 + X^2 + X + 1) \cdot X^3 + (X^3 + X^2 + X + 1) \cdot X$$
$$= (X^6 + X^5 + X^4 + X^3) + (X^4 + X^3 + X^2 + X)$$
$$\equiv X^6 + X^5 + X^2 + X$$
$$\equiv X^2 \cdot (X + 1) + X \cdot (X + 1) + X^2 + X$$
$$(\text{Since } X^4 \equiv X + 1)$$
$$\equiv (X^3 + X^2) + (X^2 + X) + X^2 + X$$
$$\equiv X^3 + X^2 \ (\text{mod } X^4 + X + 1) = 1100 \text{ (bin)} = C \text{ (hex)}$$

---

[*] The reader should take care to remember that nibble addition is different from the ordinary (decimal) addition that was done with bit or hex string notation in Chapter 6. For example, in Chapter 6, the addition $A + A = 14$ (hex) is the hexadecimal notation for the decimal addition $10 + 10 = 20$. The corresponding nibble addition, however, can be thought of as either XORing the bit string 1010 (which corresponds to A) with itself to get 0000, or as adding the corresponding $GF(16)$ polynomial $X^3 + X$ to itself to get zero. The result is written in hex notation as $A + A = 0$.

As pointed out in Chapter 10, the same polynomial could have been computed as the remainder polynomial when the division algorithm (Algorithm 10.1) is applied to the polynomial division $(X^6 + X^5 + X^2 + X) \div (X^4 + X + 1)$.

Exercise for the Reader 11.1

Perform the following nibble operations (where the nibbles are represented as single hexadecimal digits):

(a) $9 + 9$
(b) $9 \cdot 9$

Note that since nibble addition and multiplication are really just the operations in $GF(2^4)$, all of the field axioms will be satisfied, including commutativity and associativity, and any nonzero nibble has an inverse nibble.

A Scaled-Down Version of AES

The scaled-down version of AES that we present was developed by Edward Schaefer, a Santa Clara University professor, who was motivated to create it while trying to teach AES to his cryptography students. The creation was published in [MuScWe-03]. First we present a table (Table 11.1) that contrasts the main size differences between the scaled-down AES and the 128-bit key size version that we will present later.

TABLE 11.1 Size Characteristics Comparison of the 128-Bit Key AES Cryptosystem with the Scaled-Down AES Cryptosystem

	Scaled-Down AES	128-Bit Key AES
Block size	16 bit	128 bit
Key size	16 bit	128 bit
Number of rounds	3	11

We break the scaled-down AES encryption algorithm into two parts: outline and details.

Algorithm 11.1: Scaled-Down AES Encryption. Part I: Outline

Plaintext, Ciphertext, Keyspaces: $\mathcal{P} = \mathcal{C} = \mathcal{K} = \{16\text{-bit strings}\}$.
   Encryption Scheme: Let $P = b_1 b_2 \ldots b_{12}$ be the plaintext.
   We express P in terms of four nibbles with the following subscripting notation:

$$P = [b_{0,0} \ b_{1,0} \ b_{0,1} \ b_{1,1}]$$

The encryption consists of several steps; in each one, the plaintext will be transformed through various so-called **states**, the final state corresponding to the ciphertext. It will be convenient to express the plaintext and all subsequent states as two-by-two matrices of nibbles (by stacking the bytes into the columns, from left to right):

$$P = \begin{bmatrix} b_{0,0} & b_{0,1} \\ b_{1,0} & b_{1,1} \end{bmatrix}$$

The double indices now essentially correspond to the usual matrix subscript notation, where the first index gives the row and the second index gives the column of the entry. But the rows and columns are thus numbered from 0 to 1. To facilitate our explanations of the various state transformations (mappings), we make the following notational conventions:

$$\text{Current State} = \begin{bmatrix} c_{0,0} & c_{0,1} \\ c_{1,0} & c_{1,1} \end{bmatrix}, \quad \text{Next State} = \begin{bmatrix} d_{0,0} & d_{0,1} \\ d_{1,0} & d_{1,1} \end{bmatrix}$$

There will be three **rounds**, each one requiring its own **round key**. We will explain later how to obtain the round keys from the scaled-down AES key κ; they will be denoted as $\kappa^0, \kappa^1, \kappa^2$ (the rounds are labeled as the zeroth through second), and each will be the same size as κ.

Step 1. *Round Zero.* To pass to the next state, we XOR the plaintext P with the round key κ^0. We refer to this operation as the **Add Round Key Operation**. It will be used in each of the rounds.

Step 2. *Round One.* This round will consist of the following sequence of four state transformations:

(i) We first apply the **Nibble Sub Transformation**: This is just a nibble-by-nibble substitution scheme (for each of the four nibbles) that will be done by means of a scaled-down AES S-box table that will be explained later:

$$\begin{bmatrix} c_{0,0} & c_{0,1} \\ c_{1,0} & c_{1,1} \end{bmatrix} \rightarrow \begin{bmatrix} S(c_{0,0}) & S(c_{0,1}) \\ S(c_{1,0}) & S(c_{1,1}) \end{bmatrix} \triangleq \begin{bmatrix} d_{0,0} & d_{0,1} \\ d_{1,0} & d_{1,1} \end{bmatrix}$$

Here S denotes the S-box function.

(ii) Next we apply the **Shift Row Transformation**: This is a very simple row transformation of the matrix of nibbles; row i gets cyclically shifted to the left by i places (recall that the rows are labeled as 0 through 1):

$$\begin{bmatrix} c_{0,0} & c_{0,1} \\ c_{1,0} & c_{1,1} \end{bmatrix} \rightarrow \begin{bmatrix} c_{0,0} & c_{0,1} \\ c_{1,1} & c_{1,0} \end{bmatrix} \triangleq \begin{bmatrix} d_{0,0} & d_{0,1} \\ d_{1,0} & d_{1,1} \end{bmatrix}$$

(iii) Apply the **Mix Column Transformation**: The columns of the state will be transformed by a nonlinear scheme that uses the nibble multiplication operation [via $GF(2^4)$ multiplication] that was defined earlier (Definition 11.2). The complete transformation can be easily expressed as the following $GF(2^4)$ matrix multiplication:

$$\begin{bmatrix} 1 & X^2 \\ X^2 & 1 \end{bmatrix} \cdot \begin{bmatrix} c_{0,0} & c_{0,1} \\ c_{1,0} & c_{1,1} \end{bmatrix} \triangleq \begin{bmatrix} d_{0,0} & d_{0,1} \\ d_{1,0} & d_{1,1} \end{bmatrix}$$

We will give an illustrative example shortly of this transformation; for now, we simply point out that each entry of the new state is obtained by the dot product of the corresponding row and column of the matrices on the left. For example, to get $d_{0,1}$, we need to take the dot product of row 0 of the first matrix with column 1 of the second matrix: $d_{0,1} = 1 \cdot c_{0,1} + X \cdot c_{1,1}$. [This is just like ordinary matrix multiplication, except the additions and multiplications are done with the arithmetic of $GF(2^4)$.]

(iv) The round is completed by an application of the Add Round Key Operation that was introduced in Step 1.

Step 3. *Round Two.* This round is like Round One, except that the Mix Column transformation is omitted: Nibble Sub, Shift Row, and then Add Round Key.

Figure 11.3 shows a schematic diagram of Algorithm 11.1. Next, we complete all of the details in the above outline.

Algorithm 11.1: Scaled-Down AES Encryption. Part II: Details

1. Instructions on how to substitute nibbles in the Nibble Sub Transformation:

The nibble-by-nibble substitution is accomplished using the scaled-down AES S-box (or substitution box) given in Table 11.1, which is used as follows: The first two bits of an input nibble determine the row of the S-box, and the last two bits determine the column of the S-box table to use. The entry of the S-box in the indicated row and column is the transformed (output) bit. For example, if we wish to use the S-box to transform the nibble $1001 = 9$ (hex), we look up the entry in row 2 (the decimal equivalent of the bit string 10) and column 1 (the decimal equivalent of the bit string 01) of Table 11.2 to obtain the nibble 8 (hex) = 1000 (binary). Notice that this S-box functions in a very similar fashion as those for DES. Although this S-box (and the corresponding larger one we give later on for the full AES) might appear to be mysterious, it actually comes from a simple mathematical formulation in $GF(2^4)$; see

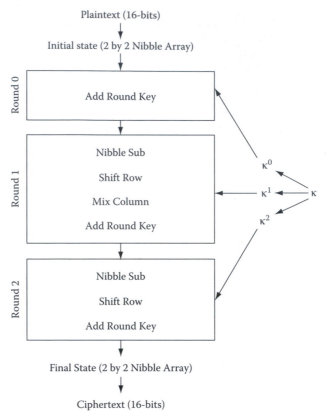

Figure 11.3 Schematic diagram of the scaled-down AES encryption algorithm (Algorithm 11.1).

TABLE 11.2 Scaled-Down AES S-Box Used in the Byte Sub Transformation

	0	1	2	3
0	9	D	6	C
1	4	1	2	E
2	A	8	0	F
3	B	5	3	7

Exercises 13 and 14. This is quite a different situation than for DES, where the S-boxes were constructed element by element, with a much more intricate procedure.

2. Instructions on how to generate the three 16-bit round keys $\kappa^0, \kappa^1, \kappa^2$ from 16-bit scaled-down AES system key κ:

The three round keys will be read off from successive columns of a two-by-six nibble matrix W that is constructed recursively as follows: We denote the jth column of the matrix W by $W(j)$. Thus, each column $W(j)$ consists of two nibbles (or 8 bits). We take the first two columns of W, $W(0)$, $W(1)$, to be the two columns of the AES key κ, when the four nibbles of the latter are expressed as a two-by-two matrix, as was done when the plaintext was converted into a state matrix of nibbles:

$$\kappa = [k_{0,0}\ k_{1,0}\ k_{0,1}\ k_{1,1}] \rightarrow \begin{bmatrix} k_{0,0} & k_{0,1} \\ k_{1,0} & k_{1,1} \end{bmatrix}$$

Thus, the round zero key κ^0 is simply the AES key κ. The remaining four columns $W(j)$ of W will be defined using a recursive procedure that we will soon specify. The procedure will rely on the following:

- **Round Constants**: $RC(j)$ that are the elements of $GF(2^4)$ (nibbles) defined by $RC(j) \equiv X^{j+2} \pmod{X^4 + X + 1}$. The round constants that we will need are

$$RC(1) \equiv X^{1+2} \equiv X^3 = 1000(\text{binary}) = 8 \text{ (hex)}$$

$$RC(2) \equiv X^{2+2} \equiv X + 1 = 0011(\text{binary}) = 3 \text{ (hex)}$$

- From these, we then define a corresponding byte by suffixing the nibble 0000: $RCON(j) = RC(j)0000$.

- **RotNib = Rotate Nibble operator**: For a pair of consecutive nibbles $N_1 N_2$ (which together form a byte), we define as $RotNib(N_1 N_2) = N_2 N_1$.

- **SubNib = Sub Nibble operator**: For a pair of consecutive nibbles $N_1 N_2$ (which together form a byte), each nibble is transformed using the scaled-down S-box of Table 11.1, so $N_1 N_2$ transforms to

$$SubNib(N_1 N_2) = \text{S-Box}(N_1)\text{S-Box}(N_2)$$

We are now ready to give the recursive procedure for defining the remaining columns $W(j)$ of W. The treatment will depend on whether j is even or odd.

Case 1: j is even:

$$W(j) = W(j-2) \oplus RCON(j/2) \oplus SubNib(RotNib(W(j-1))) \qquad (11.1)$$

Case 2: j is odd:

$$W(j) = W(j-2) \oplus W(j-1)$$

Figure 11.4 shows schematic diagram of how the $W(2)$ and $W(3)$ are constructed from $W(0)$ and $W(1)$. $W(4)$ and $W(5)$ are then constructed from $W(2)$ and $W(3)$ in the same fashion.

With the details of the scaled-down AES encryption algorithm now having been completely specified, we are ready to present a complete example. Although we will not be able to do such examples of the actual

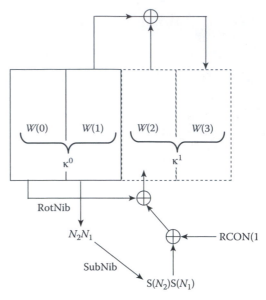

Figure 11.4 Schematic diagram recursive construction of two columns for the key generation matrix W for the scaled-down AES algorithm.

AES algorithm without the aid of a computer, all of the concepts of AES are nicely realized with this scaled-down version.[*]

Example 11.2

Use the scaled-down AES (with Algorithm 11.1) to encrypt the 16-bit plaintext $P = $ D2A6 (represented in hex form). Use the 16-bit system key $\kappa = $ 2A74 (represented in hex form).

Solution: We first prepare the three round keys $\kappa^0, \kappa^1, \kappa^2$. The zeroth round key is simply the system key: $\kappa^0 = \kappa = $ 2A74. The left and right two nibbles are used to form the first two columns of the matrix W: $W(0) = \begin{bmatrix} 2 \\ A \end{bmatrix}$, $W(1) = \begin{bmatrix} 7 \\ 4 \end{bmatrix}$. The next column, $W(2)$, falls under Case 1 in the construction, and is given by Equation 11.1 with $j = 2$:

$$W(2) = W(0) \oplus \text{RCON}(1) \oplus \text{SubNib}(\text{RotNib}(W(1)))$$

$$= \begin{bmatrix} 2 \\ A \end{bmatrix} \oplus \begin{bmatrix} 8 \\ 0 \end{bmatrix} \oplus \text{SubNib}\left(\text{RotNib}\left(\begin{bmatrix} 7 \\ 4 \end{bmatrix}\right)\right)$$

---

[*] We remind the reader that, despite all of its intricate details, the scaled-down AES is not a secure cryptosystem because of its small keyspace—there are only $2^{16} = 65{,}536$ keys! Such a small number of keys could be tested with an exhaustive search on a computer in a fraction of a second.

$$= \begin{bmatrix} 0010 \\ 1010 \end{bmatrix} \oplus \begin{bmatrix} 1000 \\ 0000 \end{bmatrix} \oplus \text{SubNib}\left(\begin{bmatrix} 4 \\ 7 \end{bmatrix}\right)$$

$$= \begin{bmatrix} 1010 \\ 1010 \end{bmatrix} \oplus \text{SubNib}\left(\begin{bmatrix} 4 = 0100 \\ 7 = 0111 \end{bmatrix}\right)$$

$$= \begin{bmatrix} 1010 \\ 1010 \end{bmatrix} \oplus \begin{bmatrix} 4 = 0100 \\ E = 1110 \end{bmatrix} \qquad \text{(Using Table 11.1)}$$

$$= \begin{bmatrix} 1110 = E \\ 0100 = 4 \end{bmatrix}$$

The next column, $W(3)$, falls under Case 2 in the construction, so is the XOR of the previous two:

$$W(3) = W(1) \oplus W(2) = \begin{bmatrix} 7 \\ 4 \end{bmatrix} \oplus \begin{bmatrix} E \\ 4 \end{bmatrix} = \begin{bmatrix} 0111 \\ 0100 \end{bmatrix} \oplus \begin{bmatrix} 1110 \\ 0100 \end{bmatrix} = \begin{bmatrix} 1001 \\ 0000 \end{bmatrix} = \begin{bmatrix} 9 \\ 0 \end{bmatrix}$$

The last two columns of W are constructed in the same fashion, but using these two most recently constructed columns in place of the first two:

$$W(4) = W(2) \oplus \text{RCON}(2) \oplus \text{SubNib}(\text{RotNib}(W(3)))$$

$$= \begin{bmatrix} E \\ 4 \end{bmatrix} \oplus \begin{bmatrix} 3 \\ 0 \end{bmatrix} \oplus \text{SubNib}\left(\text{RotNib}\left(\begin{bmatrix} 9 \\ 0 \end{bmatrix}\right)\right)$$

$$= \begin{bmatrix} 1110 \\ 0100 \end{bmatrix} \oplus \begin{bmatrix} 0011 \\ 0000 \end{bmatrix} \oplus \text{SubNib}\left(\begin{bmatrix} 0 \\ 9 \end{bmatrix}\right)$$

$$= \begin{bmatrix} 1101 \\ 0100 \end{bmatrix} \oplus \text{SubNib}\left(\begin{bmatrix} 0 = 0000 \\ 9 = 1001 \end{bmatrix}\right)$$

$$= \begin{bmatrix} 1101 \\ 0100 \end{bmatrix} \oplus \begin{bmatrix} 9 = 1001 \\ 8 = 1000 \end{bmatrix} \qquad \text{(Using Table 11.1)}$$

$$= \begin{bmatrix} 0100 = 4 \\ 1100 = C \end{bmatrix}$$

$$W(5) = W(3) \oplus W(4) = \begin{bmatrix} 9 \\ 0 \end{bmatrix} \oplus \begin{bmatrix} 4 \\ C \end{bmatrix} = \begin{bmatrix} 1001 \\ 0000 \end{bmatrix} \oplus \begin{bmatrix} 0100 \\ 1100 \end{bmatrix} = \begin{bmatrix} 1101 \\ 1100 \end{bmatrix} = \begin{bmatrix} D \\ C \end{bmatrix}$$

This completes the construction of the matrix of round key nibbles:

$$W = \begin{bmatrix} 2 & 7 & E & 9 & 4 & D \\ A & 4 & 4 & 0 & C & C \end{bmatrix}$$

from which we read the first and second round keys, $\kappa^1 = $ E490, $\kappa^2 = $ 4CDC. We now proceed through the rounds of the encryption algorithm. We convert the plaintext into the corresponding initial state array: $P \rightarrow \begin{bmatrix} D & A \\ 2 & 6 \end{bmatrix}$.

Step 1. Round Zero. To pass to the next state, we XOR the plaintext P with the round key κ^0: $P \oplus \kappa^1 \rightarrow$

$$\begin{bmatrix} D & A \\ 2 & 6 \end{bmatrix} \oplus \begin{bmatrix} 2 & 7 \\ A & 4 \end{bmatrix} = \begin{bmatrix} 1101 & 1010 \\ 0010 & 0110 \end{bmatrix} \oplus \begin{bmatrix} 0010 & 0111 \\ 1010 & 0100 \end{bmatrix}$$

$$= \begin{bmatrix} 1111 & 1101 \\ 1000 & 0010 \end{bmatrix} = \begin{bmatrix} F & D \\ 8 & 2 \end{bmatrix}$$

Step 2. Round One.
    (i)   Nibble Sub: Using the S-box of Table 11.1:

$$\begin{bmatrix} F & D \\ 8 & 2 \end{bmatrix} = \begin{bmatrix} 1111 & 1101 \\ 1000 & 0010 \end{bmatrix} \rightarrow \begin{bmatrix} 7 & 5 \\ A & 6 \end{bmatrix}$$

    (ii)   Shift Row:

$$\begin{bmatrix} 7 & 5 \\ A & 6 \end{bmatrix} \rightarrow \begin{bmatrix} 7 & 5 \\ 6 & A \end{bmatrix}$$

    (iii)   Mix Column:

$$\begin{bmatrix} 7 & 5 \\ 6 & A \end{bmatrix} \rightarrow \begin{bmatrix} 1 & X^2 \\ X^2 & 1 \end{bmatrix} \cdot \begin{bmatrix} X^2+X+1 & X^2+1 \\ X^2+X & X+1 \end{bmatrix} \triangleq \begin{bmatrix} d_{0,0} & d_{0,1} \\ d_{1,0} & d_{1,1} \end{bmatrix}$$

where

$$d_{0,0} \equiv 1 \cdot (X^2+X+1) + X^2 \cdot (X^2+X) \equiv X^4 + X^3 + X^2 + X + 1$$

$$\equiv (X+1) + X^3 + X^2 + X + 1 \equiv X^3 + X^2 (\bmod X^4 + X + 1) = 1100 = C$$

$$d_{0,1} \equiv 1 \cdot (X^2+1) + X^2 \cdot (X^3+X) \equiv X^5 + X^3 + X^2 + 1$$

$$\equiv X(X+1) + X^3 + X^2 + 1 \equiv X^3 + X + 1 (\bmod X^4 + X + 1) = 1011 = B$$

$$d_{1,0} \equiv X^2 \cdot (X^2+X+1) + 1 \cdot (X^2+X) \equiv X^4 + X^3 + X^2 + X^2 + X$$

$$\equiv (X+1) + X^3 + X \equiv X^3 + 1 (\bmod X^4 + X + 1) = 1001 = 9$$

$$d_{1,1} \equiv X^2 \cdot (X^2+1) + 1 \cdot (X^3+X) \equiv X^4 + X^2 + X^3 + X$$

$$\equiv (X+1) + X^3 + X^2 + X \equiv X^3 + X^2 + 1 (\bmod X^4 + X + 1) = 1101 = D$$

(iv)   Add Round Key: $\begin{bmatrix} C & B \\ 9 & D \end{bmatrix} \oplus \kappa^1 \rightarrow$

$$\begin{bmatrix} C & B \\ 9 & D \end{bmatrix} \oplus \begin{bmatrix} E & 9 \\ 4 & 0 \end{bmatrix} = \begin{bmatrix} 1100 & 1011 \\ 1001 & 1101 \end{bmatrix} \oplus \begin{bmatrix} 1110 & 1001 \\ 0100 & 0000 \end{bmatrix}$$

$$= \begin{bmatrix} 0010 & 0010 \\ 1101 & 1101 \end{bmatrix} = \begin{bmatrix} 2 & 2 \\ D & D \end{bmatrix}$$

Step 3. Round Two.
   (i)   Nibble Sub: Using the S-box of Table 11.1,

$$\begin{bmatrix} 2 & 2 \\ D & D \end{bmatrix} = \begin{bmatrix} 0010 & 0010 \\ 1101 & 1101 \end{bmatrix} \rightarrow \begin{bmatrix} 6 & 6 \\ 5 & 5 \end{bmatrix}$$

(ii)   Shift Row:

$$\begin{bmatrix} 6 & 6 \\ 5 & 5 \end{bmatrix} \rightarrow \begin{bmatrix} 6 & 6 \\ 5 & 5 \end{bmatrix}$$

(iii)   Add Round Key: $\begin{bmatrix} 6 & 6 \\ 5 & 5 \end{bmatrix} \oplus \kappa^2 \rightarrow$

$$\begin{bmatrix} 6 & 6 \\ 5 & 5 \end{bmatrix} \oplus \begin{bmatrix} 4 & D \\ C & C \end{bmatrix} = \begin{bmatrix} 0110 & 0110 \\ 0101 & 0101 \end{bmatrix} \oplus \begin{bmatrix} 0100 & 1101 \\ 1100 & 1100 \end{bmatrix}$$

$$= \begin{bmatrix} 0010 & 1011 \\ 1001 & 1001 \end{bmatrix} = \begin{bmatrix} 2 & B \\ 9 & 9 \end{bmatrix}$$

From the final state $\begin{bmatrix} 2 & B \\ 9 & 9 \end{bmatrix}$, we read off the ciphertext $C = 29B9$.

Exercise for the Reader 11.2

Use the scaled-down AES (with Algorithm 11.1) to encrypt the 16-bit plaintext $P = 47B0$ (represented in hex form). Use the 16-bit system key $\kappa = 468C$ (represented in hex form).

Decryption in the Scaled-Down Version of AES

The general principle that the inverse of a composition of functions is the reverse order composition of the inverses allows us to easily derive the decryption algorithm for Algorithm 11.1. Roughly speaking, we need only reverse the order of the rounds, and within the rounds, the inverses of the individual operations should be used in the reverse order. The **Shift Row** mapping is (clearly) its own inverse. Also, by the self-cancelling property of XOR [Proposition 7.1(4)], the **Add Round Key** mapping is its own inverse. We just need to find the inverses of the Nibble Sub and the Mix Column transformations.

1. The inverse of the **Nibble Sub** mapping is the **Inv Nibble Sub** mapping defined by

$$
\begin{bmatrix} c_{0,0} & c_{0,1} \\ c_{1,0} & c_{1,1} \end{bmatrix} \rightarrow \begin{bmatrix} S^{-1}(c_{0,0}) & S^{-1}(c_{0,1}) \\ S^{-1}(c_{1,0}) & S^{-1}(c_{1,1}) \end{bmatrix} \triangleq \begin{bmatrix} d_{0,0} & d_{0,1} \\ d_{1,0} & d_{1,1} \end{bmatrix}
$$

where S^{-1} denotes the inverse S-box function. This inverse S-box function can be described by its own box with which it can be evaluated in the same fashion as the S-box itself. The resulting inverse S-box's table is shown in Table 11.2; the following exercise for the reader asks for verification of these facts.

As a simple example, we will show that $S^{-1}(S(\mathbf{A})) = \mathbf{A}$: First, we use the S-box Table 11.1 to evaluate $S(\mathbf{A}) = S(\mathbf{1010}) = 0$. (We used the $10 = 2$nd row and the $10 = 2$nd column.) Next, since we use the $00 = 0$th row, the $00 = 0$th column of Table 11.3 is used to evaluate $S^{-1}(0) = S^{-1}(\mathbf{0000}) = \mathbf{A}$.

Exercise for the Reader 11.3

Verify that the function from the set of all 16 nibbles to itself described by Table 11.3 (and with the same usage as the S-box function defined by Table 11.2) really is the inverse function of the S-box for the scaled-down AES.

2. The **Mix Column** mapping is defined by a $GF(16)$ a matrix multiplication of the state matrix by the matrix $M = \begin{bmatrix} 1 & X^2 \\ X^2 & 1 \end{bmatrix}$.

Our experience with matrices whose entries lie in the ring of modular integers would lead us to think that the inverse of this Mix Column mapping should multiply a state matrix by the inverse of this matrix M (on the left), if it exists. Since $\det(M) = 1 \cdot 1 - X^2 \cdot X^2 \equiv 1 - X^4 \equiv 1 - (X+1) \equiv X \not\equiv 0 (\mathrm{mod}\, X^4 + X + 1)$, and the proof of Theorem 4.3 used only the ring axioms on the entries in the matrices, we can use the formula of that theorem to compute

$$
M^{-1} = \det(M)^{-1} \begin{bmatrix} 1 & -X^2 \\ -X^2 & 1 \end{bmatrix} \equiv X^{-1} \cdot \begin{bmatrix} 1 & X^2 \\ X^2 & 1 \end{bmatrix} \equiv \begin{bmatrix} X^{-1} & X \\ X & X^{-1} \end{bmatrix}
$$

TABLE 11.3 Table for Scaled-Down AES S-Box Inverse Function

	0	1	2	3
0	A	5	6	E
1	4	D	2	F
2	9	0	8	C
3	3	1	7	B

Furthermore, since $X^{-1} \equiv X^3 + 1$ in $GF(16)$,[*] we can summarize the **Inv Mix Column Transformation** as

$$\begin{bmatrix} X^3+1 & X \\ X & X^3+1 \end{bmatrix} \cdot \begin{bmatrix} c_{0,0} & c_{0,1} \\ c_{1,0} & c_{1,1} \end{bmatrix} \triangleq \begin{bmatrix} d_{0,0} & d_{0,1} \\ d_{1,0} & d_{1,1} \end{bmatrix}$$

The scaled-down AES decryption algorithm is schematically illustrated in Figure 11.5, which was simply obtained from the corresponding schematic for encryption (Figure 11.3), but with the arrows reversed and the inverses of the component transformations replacing the component transformations. To make the two processes easier to compare, we kept the same names for the rounds; the price to be paid is that, in decryption, we start with the second round and end with the zeroth round. Since the decryption algorithm is so readily understood from Figure 11.5 (along with our corresponding developments of the component transformations), we do not bother making a formal statement of it.

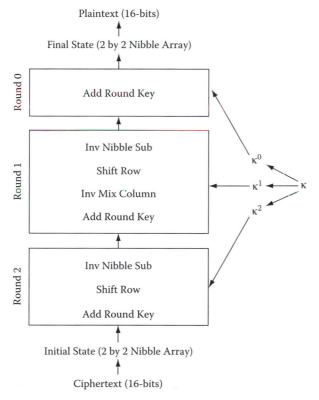

Figure 11.5 Schematic diagram of the scaled-down AES decryption algorithm.

---

[*] This can be computed either by trial and error or by using the extended Euclidean algorithm, as was shown in Chapter 10. It is easy to check: $X \cdot (X^3 + 1) \equiv X^4 + X \equiv (X+1) + X \equiv 1 \pmod{X^4 + X + 1}$.

Exercise for the Reader 11.4

Perform the scaled-down AES decryption algorithm on the cipher-text that was produced in Example 11.2.

AES

Having completely developed the scaled-down AES cryptosystem, it will now be a conceptually simple task to explain the full AES cryptosystem. Although AES was designed to work with key sizes of 128, 192, or 256 bits, we will restrict our treatment to 128-bit keys for simplicity. The modifications needed for the larger key sizes involve no significant new ideas. Nonetheless, the sheer sizes involved even with this smallest possible key size are too large to perform encryptions by hand, so complete examples with AES will be delegated to the computer exercises at the end of this chapter.

Byte Representation and Arithmetic

Since the scaled-down AES was based on the arithmetic of the finite field $GF(2^4)$, it was most efficient to break down key and plaintext data into nibbles that could then be identified with elements of $GF(2^4)$, so that all of the associated field operations and arithmetic could then be carried out on the nibbles. By the same token, since the full AES is based on the arithmetic of the 256 element field $GF(2^8)$, we will be breaking down the data into bytes (groups of 8 bits), and for ease of illustration we will use hex notation to represent bytes. Bytes are thus the "atoms" of AES—all of the components of the AES algorithm are operations on bytes rather than on individual bits. To describe the various parts of the AES algorithm, it will be convenient to view a byte b alternatively in one of the following three forms:

1. *Bitwise Representation.* As an 8-bit string: $b = b_0 b_1 \cdots b_6 b_7$.
2. *Hexadecimal Representation.* As a length-2 hexadecimal string: $b = h_1 h_2$; here, h_1 is the hexadecimal representation of the first four bits $b_0 b_1 b_2 b_3$, and h_2 is the hexadecimal representation of the last four bits $b_4 b_5 b_6 b_7$.
3. $GF(2^8)$ *Polynomial Representation.* As an element of $GF(2^8)$, represented as a polynomial in the abstract variable X: $b = $
$$\sum_{i=0}^{7} b_i X^{7-i} = b_0 X^7 + b_1 X^6 + \cdots + b_6 X + b_7.$$ These $2^8 = 256$ polynomials in X of degree at most 7 and coefficients in $\mathbb{Z}_2$ make up the number system $GF(2^8)$.

The first and the third notations are functional in the sense that they will help us to naturally define and compute certain byte operations (addition and multiplication of bytes). The hexadecimal notation is introduced solely for its economy of notation. It is much less cumbersome to express bytes in hexadecimal notation.

Bytes can be naturally added if they are either in bitwise or $GF(2^8)$ notation; if they are presented in hexadecimal form, they should be converted first to one of these other forms before adding them (and then converted back to hexadecimal form). In order to understand and work with AES, the reader should become proficient with shifting gears between these three representations and performing additions and multiplications of bytes. We provide two brief examples here, but readers needing more practice are advised to spend some time studying the relevant parts of Chapter 10 before moving on past these two examples.

Example 11.3

We compare the mechanics of the byte addition: 10110001 + 00110011 in the three notational forms.

1. *Bitwise Representation.* Performing each of the eight component additions (mod 2) is the same as XORing the strings and gives the result 10000010.

2. *Hexadecimal Representation.* To put the first byte in hexadecimal notation (see Table 6.1), 1011 → B, and 0001 → 1 = h_2, so the first byte is B1; similarly, the second byte is 33, and the representation of their sum 10000010 is 82. *Caution:* This is not the same as the Chapter 6 addition algorithm in hex notation (which is just a different way of writing integer additions): B1 + 33 = (B + 3) (1 + 3) = E4—an incorrect result in byte addition context.

3. $GF(2^8)$ *Representation.* In polynomial form, the byte addition looks like

$$(X^7 + X^5 + X^4 + 1) + (X^5 + X^4 + X + 1) = X^7 + X$$

where the coefficients of like powers have been added (mod 2).

So far, the finite field $GF(2^8)$ has not brought anything new (apart from giving an alternative way to add bytes). What makes this $GF(2^8)$ useful is that it has a very interesting multiplication operation that we use to define the multiplication of bytes.

We recall that multiplication of polynomials in $GF(2^8)$, which is $\mathbb{Z}_2[X]$ modulo the irreducible polynomial $g = X^8 + X^4 + X^3 + X + 1$, can be broken down into two steps:

Step 1. Multiply the polynomials using standard polynomial multiplication (Definition 10.5), using mod 2 arithmetic on the coefficients.

Step 2. The result needs to be converted to an equivalent polynomial with degree less than 8 (mod g). This latter conversion can be done in one of two ways:

- Starting with the term of highest degree (if at least 8), substitute with the congruence $X^8 \equiv X^4 + X^3 + X + 1 \pmod{g}$ to convert the term into terms of lower degree. Simplify the resulting expression, and repeat until there are no terms of degree greater than 7.

- Apply the polynomial division algorithm (Algorithm 10.1) to the product polynomial (if it has degree at least 8) divided by $g = X^8 + X^4 + X^3 + X + 1$. The remainder of this division will be congruent to the product.

Example 11.4

Perform the following multiplication of bytes: B1 · 0A (where the bytes are written in hexadecimal notation).

Solution: In binary string notation, B1 = 10110001 and 0A = 00001010, so translating to polynomials in $GF(2^8)$, the problem is to perform the following multiplication:

$$(X^7 + X^5 + X^4 + 1) \cdot (X^3 + X)$$

Step 1. Ordinary Polynomial Multiplication mod 2.

$$(X^7 + X^5 + X^4 + 1) \cdot (X^3 + X) \equiv (X^7 + X^5 + X^4 + 1) \cdot X^3$$

$$+ (X^7 + X^5 + X^4 + 1) \cdot X$$

$$\equiv X^{10} + \cancel{X^8} + X^7 + X^3 + \cancel{X^8} + X^6 + X^5 + X \equiv X^{10} + X^7$$

$$+ X^6 + X^5 + X^3 + X$$

Step 2. Reduction of Terms Whose Degree Is > 7. We use only the first method; the reader may check that the second method (with the division algorithm) produces the same result. Only the X^{10} term needs to be reduced. Since

$$X^{10} \equiv X^8 \cdot X^2 \equiv (X^4 + X^3 + X + 1) \cdot X^2 \equiv X^6 + X^5 + X^3 + X^2 \pmod{g}$$

the preceding equation is congruent to

$$(X^6 + X^5 + X^3 + X^2) + X^7 + X^6 + X^5 + X^3 + X \equiv X^7 + X^2 + X$$

The final result has binary expansion 10000110 and hexadecimal expansion 86. In summary, we have computed B1 · 0A = 86.

Exercise for the Reader 11.5

Perform the following byte operations: 77 + 99, D7 · 2F (where the bytes are written in hexadecimal notation).

Now being able to add and multiply bytes, we are ready to explain the AES algorithm. As we did with the scaled-down version, we first outline the general encryption algorithm for AES, and this is followed by the details needed to completely understand it.

The AES Encryption Algorithm

Algorithm 11.2: Scaled-Down AES Encryption. Part I: Outline
Plaintext, Ciphertext, and Keyspaces: $\mathscr{P} = \mathscr{C} = \mathscr{K} = \{128\text{-bit strings}\}$.
All three elements will be expressed as length-16 hex strings.
   Encryption Scheme: Let the plaintext be expressed in terms of
16 bytes with the following subscripting notation:

$$P = [b_{0,0} \; b_{1,0} \; b_{2,0} \; b_{3,0} \; b_{0,1} \; b_{1,1} \; b_{2,1} \; b_{3,1} \; b_{0,2} \; b_{1,2} \; b_{2,2} \; b_{3,2} \; b_{0,3} \; b_{1,3} \; b_{2,3} \; b_{3,3}]$$

The encryption consists of several steps; in each one, the plaintext
will be transformed through various so-called **states**, the final state
corresponding to the ciphertext. It will be convenient to express the
plaintext and all subsequent states as four-by-four matrices of bytes (by
stacking the bytes into the columns, from left to right):

$$p = \begin{bmatrix} b_{0,0} & b_{0,1} & b_{0,2} & b_{0,3} \\ b_{1,0} & b_{1,1} & b_{1,2} & b_{1,3} \\ b_{2,0} & b_{2,1} & b_{2,2} & b_{2,3} \\ b_{3,0} & b_{3,1} & b_{3,2} & b_{3,3} \end{bmatrix}$$

The double indices now essentially correspond to the usual
matrix subscript notation, where the first index gives the row and
the second index gives the column of the entry. The rows and columns are thus numbered from 0 to 3. To facilitate our explanations of
the various state transformations, we make the following notational
conventions:

$$\text{Current State} = \begin{bmatrix} c_{0,0} & c_{0,1} & c_{0,2} & c_{0,3} \\ c_{1,0} & c_{1,1} & c_{1,2} & c_{1,3} \\ c_{2,0} & c_{2,1} & c_{2,2} & c_{2,3} \\ c_{3,0} & c_{3,1} & c_{3,2} & c_{3,3} \end{bmatrix}$$

$$\text{Next State} = \begin{bmatrix} d_{0,0} & d_{0,1} & d_{0,2} & d_{0,3} \\ d_{1,0} & d_{1,1} & d_{1,2} & d_{1,3} \\ d_{2,0} & d_{2,1} & d_{2,2} & d_{2,3} \\ d_{3,0} & d_{3,1} & d_{3,2} & d_{3,3} \end{bmatrix}$$

The AES algorithm goes through 11 **rounds**, each one requiring
its own **round key**. We will explain later how to obtain the round keys
from the AES key κ; they will be denoted as κ^0, κ^1, $\cdots$, κ^{10} (the rounds
labeled as zeroth through tenth), and each will be the same size as
κ. Rounds 1 through 9 are identical while the 0th and 10th rounds are
slightly different.

Step 1. *Round Zero.* To pass to the next state, we XOR the plaintext state P with the round key κ^0. We refer to this operation, XORing the current state with the current round key, as the Add Round Key operation. It will be used again.

Step 2. *Rounds One through Nine.* Each of these nine identical rounds applies the following four state transformations:

(i) We first apply the Byte Sub Transformation: This is just a (complicated) byte-by-byte substitution scheme (for each of the 16 bytes) that will be done using an AES S-box table that will be given later.

(ii) Next we apply the Shift Row Transformation: This is a very simple row transformation of the matrix of bytes; row i gets cyclically shifted to the left by i places (recall that the rows are labeled as 0 through 3):

$$
\begin{bmatrix}
c_{0,0} & c_{0,1} & c_{0,2} & c_{0,3} \\
c_{1,0} & c_{1,1} & c_{1,2} & c_{1,3} \\
c_{2,0} & c_{2,1} & c_{2,2} & c_{2,3} \\
c_{3,0} & c_{3,1} & c_{3,2} & c_{3,3}
\end{bmatrix}
\rightarrow
\begin{bmatrix}
c_{0,0} & c_{0,1} & c_{0,2} & c_{0,3} \\
c_{1,1} & c_{1,2} & c_{1,3} & c_{1,0} \\
c_{2,2} & c_{2,3} & c_{2,0} & c_{2,1} \\
c_{3,3} & c_{3,0} & c_{3,1} & c_{3,2}
\end{bmatrix}
$$

$$
\triangleq
\begin{bmatrix}
d_{0,0} & d_{0,1} & d_{0,2} & d_{0,3} \\
d_{1,0} & d_{1,1} & d_{1,2} & d_{1,3} \\
d_{2,0} & d_{2,1} & d_{2,2} & d_{2,3} \\
d_{3,0} & d_{3,1} & d_{3,2} & d_{3,3}
\end{bmatrix}
$$

(iii) We apply the Mix Column Transformation: The columns of the state will be transformed by a nonlinear scheme that uses the byte multiplication operation [via $GF(2^8)$ multiplication] that was explained earlier. The complete transformation can be easily expressed as the following $GF(2^8)$ matrix multiplication:

$$
\begin{bmatrix}
x & x+1 & 1 & 1 \\
1 & x & x+1 & 1 \\
1 & 1 & x & x+1 \\
x+1 & 1 & 1 & x
\end{bmatrix}
\cdot
\begin{bmatrix}
c_{0,0} & c_{0,1} & c_{0,2} & c_{0,3} \\
c_{1,0} & c_{1,1} & c_{1,2} & c_{1,3} \\
c_{2,0} & c_{2,1} & c_{2,2} & c_{2,3} \\
c_{3,0} & c_{3,1} & c_{3,2} & c_{3,3}
\end{bmatrix}
$$

$$
\equiv
\begin{bmatrix}
d_{0,0} & d_{0,1} & d_{0,2} & d_{0,3} \\
d_{1,0} & d_{1,1} & d_{1,2} & d_{1,3} \\
d_{2,0} & d_{2,1} & d_{2,2} & d_{2,3} \\
d_{3,0} & d_{3,1} & d_{3,2} & d_{3,3}
\end{bmatrix}
$$

We will give an illustrative example of this transformation shortly; for now, we simply point out that each entry of the new state is obtained by the dot product of the corresponding row and column of the matrices on the left. For example, to get $d_{1,1}$, we need to take the dot product of row 1 of the first matrix with column 1 of the second matrix: $d_{1,1} = 1 \cdot c_{0,1} + x \cdot c_{1,1} + (x+1) \cdot c_{2,1} + c_{3,1}$. [This is just like ordinary matrix multiplication, except the additions and multiplications are done with the arithmetic of $GF(2^8)$.]

(iv)   The round is completed by an application of the Add Round Key operation that was introduced in Step 1.

Step 3.   Round Ten. This round is like Rounds One through Nine, except that the Mix Column Transformation is omitted: Byte Sub, Shift Row, and then Add Round Key.

Figure 11.6 shows a schematic diagram of this algorithm. We now explain the remaining details.

Algorithm 11.2:   *Advanced Encryption Standard (AES) with 128 Bit Keys. Part II: Details*

1. Instructions on how to substitute bytes in the Byte Sub Transformation:

   The byte-by-byte substitution is accomplished using the AES S-box given in Table 11.4, which is used as follows: The first four bits (or first hexadecimal symbol) of a byte determines the row of the S-box, and the last four bits (or the second hexadecimal symbol) determine the column of the S-box table to use. The entry of the S-box in the indicated row and column is the transformed bit. For example, if we wish to use the S-box to transform the byte 10011010 = 9A, we look up the entry in row 9 and column A in Table 11.4 to get B8 = 10111000.

2. Instructions on how to generate the 11 round keys κ^i $(0 \le i \le 10)$ from the 16-byte AES key κ:

   The 11 round keys will be read off from successive columns of a 4-by-44 byte matrix W that is constructed recursively as follows: We denote the jth column of the matrix W by $W(j)$. We take the first four columns of W: $W(1)$, $W(2)$, $W(3)$, $W(4)$, to be the four columns of the AES key κ, when the 16 bytes of the latter are expressed as a four-by-four matrix, as shown above. The remaining 40 columns of W are then defined recursively: Each successive group of four columns is computed from the previous four by using the same procedure.

Case 1.   If the column index j is not a multiple of 4, we use the formula:

$$W(j) = W(j-4) \oplus W(j-1) \qquad (11.2)$$

For example, we would first use Equation 11.2 (with $j = 5$) to compute $W(5) = W(1) \oplus W(4)$. After this is done, we could use Equation 11.2 again (with $j = 6$) to compute $W(6) = W(2) \oplus W(5)$, and finally compute $W(7)$. The computation of $W(8)$ is more involved, and handled in the next case.

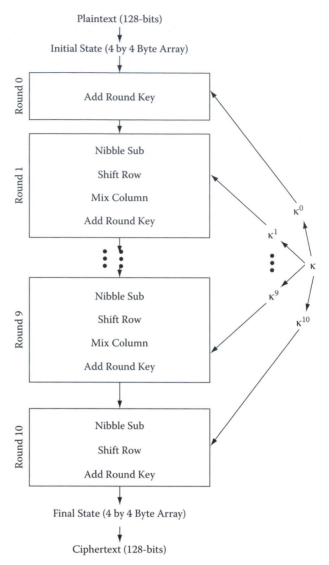

Figure 11.6 Schematic diagram of the AES encryption algorithm (Algorithm 11.2).

Case 2. If the column index $j = 4k$, for some integer k:

    Step 1. Write $W(j-1) = [a\ b\ c\ d]'$. [That is, a, b, c, d are the bytes making up the column $W(j-1)$, which is assumed to have been computed in this recursive procedure.]*

    Step 2. Cyclically shift the bytes one unit to the left: $[a\ b\ c\ d]' \rightarrow [b\ c\ d\ a]'$.

---

* The prime is the matrix transpose symbol.

Step 3. Apply the S-box of Table 11.4 to the four bytes:
$$[b \ c \ d \ a]' \rightarrow [S(b) \ S(c) \ S(d) \ S(a)]'.$$

Step 4. Compute the round constant $RC = X^{k-1} (\text{mod } X^8 + X^4 + X^3 + X + 1)$ as a byte, and use it to form $U \triangleq [(x^J \oplus S(b)) \ S(c) \ S(d) \ S(a)]'.$

Step 5. Set $W(j) = W(j-4) \oplus U.$

TABLE 11.4 AES S-Box Used in the Byte Sub Transformation.

	0	1	2	3	4	5	6	7	8	9	A	B	C	D	E	F
0	63	7C	77	7B	F2	6B	6F	C5	30	01	67	2B	FE	D7	AB	76
1	CA	82	C9	7D	FA	59	47	F0	AD	D4	A2	AF	9C	A4	72	C0
2	B7	FD	93	26	36	3F	F7	CC	34	A5	E5	F1	71	D8	31	15
3	04	C7	23	C3	18	96	05	9A	07	12	80	E2	EB	27	B2	75
4	09	83	2C	1A	1B	6E	5A	A0	52	3B	D6	B3	29	E3	2F	84
5	53	D1	00	ED	20	FC	B1	5B	6A	CB	BE	39	4A	4C	58	CF
6	D0	EF	AA	FB	43	4D	33	85	45	F9	02	7F	50	3C	9F	A8
7	51	A3	40	8F	92	9D	38	F5	BC	B6	DA	21	10	FF	F3	D2
8	CD	0C	13	EC	5F	97	44	17	C4	A7	7E	3D	64	5D	19	73
9	60	81	4F	DC	22	2A	90	88	46	EE	B8	14	DE	5E	0B	DB
A	E0	32	3A	0A	49	06	24	5C	C2	D3	AC	62	91	95	E4	79
B	E7	C8	37	6D	8D	D5	4E	A9	6C	56	F4	EA	65	7A	AE	08
C	BA	78	25	2E	1C	A6	B4	C6	E8	DD	74	1F	4B	BD	8B	8A
D	70	3E	B5	66	48	03	F6	0E	61	35	57	B9	86	C1	1D	9E
E	E1	F8	98	11	69	D9	8E	94	9B	1E	87	E9	CE	55	28	DF
F	8C	A1	89	0D	BF	E6	42	68	41	99	2D	0F	B0	54	BB	16

Note: The shaded row and column show how the byte 9A gets transformed to the byte B8.

The AES Decryption Algorithm

The decryption of AES is accomplished by reversing the order of the rounds' state transformations in the AES algorithm and replacing each transformation by its inverse transformation. A schematic diagram for it can be created from Figure 11.6 by reversing arrows, just as Figure 11.5 was obtained from Figure 11.3 for the scaled-down AES. We briefly summarize the inverses of the four AES component mappings that are needed in the decryption algorithm. Just as before, the **Add Round Key mapping** is its own inverse.

1. The inverse of the **Nibble Sub** mapping is the **Inv Nibble Sub** mapping defined by

$$\begin{bmatrix} c_{0,0} & c_{0,1} \\ c_{1,0} & c_{1,1} \end{bmatrix} \rightarrow \begin{bmatrix} S^{-1}(c_{0,0}) & S^{-1}(c_{0,1}) \\ S^{-1}(c_{1,0}) & S^{-1}(c_{1,1}) \end{bmatrix} \triangleq \begin{bmatrix} d_{0,0} & d_{0,1} \\ d_{1,0} & d_{1,1} \end{bmatrix}$$

where S^{-1} denotes the inverse S-box function. The creation of the table (corresponding to Table 11.4 for the inverse function) will be left to the computer implementation material at the end of this chapter.

2. The inverse of the Shift Row mapping is the Inv Shift Row mapping defined by cyclically shifting row i of a state matrix i units to the right (instead of the left).

3. The inverse of the **Mix Column** mapping, **Inv Mix Column**, is defined by the following $GF(2^8)$ matrix multiplication:

$$A \cdot \begin{bmatrix} c_{0,0} & c_{0,1} & c_{0,2} & c_{0,3} \\ c_{1,0} & c_{1,1} & c_{1,2} & c_{1,3} \\ c_{2,0} & c_{2,1} & c_{2,2} & c_{2,3} \\ c_{3,0} & c_{3,1} & c_{3,2} & c_{3,3} \end{bmatrix} \equiv \begin{bmatrix} d_{0,0} & d_{0,1} & d_{0,2} & d_{0,3} \\ d_{1,0} & d_{1,1} & d_{1,2} & d_{1,3} \\ d_{2,0} & d_{2,1} & d_{2,2} & d_{2,3} \\ d_{3,0} & d_{3,1} & d_{3,2} & d_{3,3} \end{bmatrix}$$

where A is the following four-by-four matrix of bytes:

$$A = \begin{bmatrix} X^3+X^2+X & X^3+X+1 & X^3+X^2+1 & X^3+1 \\ X^3+1 & X^3+X^2+X & X^3+X+1 & X^3+X^2+1 \\ X^3+X^2+1 & X^3+1 & X^3+X^2+X & X^3+X+1 \\ X^3+X+1 & X^3+X^2+1 & X^3+1 & X^3+X^2+X \end{bmatrix}$$

To show that the **Inv Mix Column** asserted above is really the inverse of the **Mix Column** mapping, it suffices to check that its matrix A is the "inverse" of the corresponding matrix for the **Mix Column** transformation.[*]

Security of the AES

The AES cryptosystem was designed to resist attacks from linear and differential cryptanalysis, and up to now, the only known way to attack it is essentially an exhaustive key search. In fact, this would still be the case if only six of the main nine (Step 2) rounds were used, but not if only five were used. Thus, there is already a sufficient margin for error, but if necessary, the algorithm could easily be expanded to have additional rounds. In contrast with Feistel cryptosystems, such as DES where half the bits are preserved in each round, the AES rounds are much more diffusive. In fact, it has been shown that just two of the main rounds are sufficient to ensure that individual bits of the states depend on each of the 128 bits two states back. Such diffusion makes the AES quite resistant against linear and differential cryptanalysis, as well as all other attacks that have been attempted. Unless an efficient new attack on AES gets discovered, with its capacity of accommodating 256-bit keys (of which there are more than 10^{77}), the AES system will be secure for many years to come.[†]

---

[*] It turns out that the matrix A of the **Inv Mix Column** can be obtained by the corresponding matrix for the **Mix Column** transformation by using the classical adjoint formula (Proposition 4.5), but working in byte arithmetic.

[†] As was explained in Chapter 7, Moore's law's estimated waiting time for computers to be able to perform the same on a 256-bit key search as they do today with a 56-bit key search is about 300 years!

Chapter 11 Exercises

1. Perform the following nibble operations:
   (a) $2 + 2$
   (b) $9 + B$
   (c) $2 \cdot 2$
   (d) $9 \cdot B$

2. Perform the following nibble operations:
   (a) $7 + C$
   (b) $B + B$
   (c) $7 \cdot C$
   (d) $B \cdot B$

3. Compute the following scaled-down AES S-box outputs (using Table 11.2):
   (a) $S(C)$
   (b) $S(0)$
   (c) $S(7)$
   (d) $S(3)$

4. Compute the following scaled-down AES S-box outputs (using Table 11.2):
   (a) $S(F)$
   (b) $S(5)$
   (c) $S(B)$
   (d) $S(1)$

5. Perform the following nibble matrix multiplications:

   (a) $\begin{bmatrix} 1 & X^2 \\ X^2 & 1 \end{bmatrix} \cdot \begin{bmatrix} 1 & 2 \\ 3 & 4 \end{bmatrix}$

   (b) $\begin{bmatrix} 1 & X^2 \\ X^2 & 1 \end{bmatrix} \cdot \begin{bmatrix} A & B \\ C & D \end{bmatrix}$

6. Perform the following nibble matrix multiplications:

   (a) $\begin{bmatrix} 1 & X^2 \\ X^2 & 1 \end{bmatrix} \cdot \begin{bmatrix} 4 & 7 \\ 4 & 3 \end{bmatrix}$

   (b) $\begin{bmatrix} 1 & X^2 \\ X^2 & 1 \end{bmatrix} \cdot \begin{bmatrix} F & 8 \\ 3 & B \end{bmatrix}$

7. *Scaled-Down AES Round Key Generation.* Compute the three scaled-down AES round keys κ^0, κ^1, κ^2 corresponding to each of the following 12-bit system keys. Express your answers as hex strings of four nibbles (i.e., in the same format that the system keys are presented).
   (a) $\kappa = 1234$
   (b) $\kappa = A64B$
   (c) $\kappa = 8888$
   (d) $\kappa = FA5A$

8. *Scaled-Down AES Round Key Generation.* Compute the three scaled-down AES round keys κ^0, κ^1, κ^2 corresponding to each of the following 12-bit system keys. Express your answers as hex strings of four nibbles (i.e., in the same format that the system keys are presented).
    (a) $\kappa = 2468$
    (b) $\kappa = ABCD$
    (c) $\kappa = 5D7F$
    (d) $\kappa = 69AB$

9. *Scaled-Down AES Encryption.* For each part, a 16-bit plaintext message is given (in hex form). Use the scaled-down AES algorithm (Algorithm 11.1) to compute the corresponding ciphertexts, using the system keys given in the corresponding parts of Exercise 7.
    (a) $P = 7777$
    (b) $P = DEED$
    (c) $P = FAD2$
    (d) $P = 49A3$

10. *Scaled-Down AES Encryption.* For each part, a 16-bit plaintext message is given (in hex form). Use the scaled-down AES algorithm (see Figure 11.5) to compute the corresponding ciphertexts, using the system keys given in the corresponding parts of Exercise 8.
    (a) $P = 2332$
    (b) $P = FEED$
    (c) $P = 4CAB$
    (d) $P = 82B4$

11. *Scaled-Down AES Decryption.* For each part, a 16-bit ciphertext message is given (in hex form). Use the scaled-down AES decryption algorithm (Algorithm 11.1) to compute the corresponding ciphertexts, using the system keys given in the corresponding parts of Exercise 7.
    (a) $C = 1963$
    (b) $C = A662$
    (c) $C = 5492$
    (d) $C = BB8D$

12. *Scaled-Down AES Decryption.* For each part, a 16-bit ciphertext message is given (in hex form). Use the scaled-down AES decryption algorithm (Algorithm 11.1) to compute the corresponding ciphertexts, using the system keys given in the corresponding parts of Exercise 8.
    (a) $C = 1975$
    (b) $C = 90A2$
    (c) $C = 6110$
    (d) $C = CCCD$

13. Perform the following byte operations:
     (a) $25 + 42$
     (b) $99 + AB$
     (c) $12 \cdot 12$
     (d) $A9 \cdot BB$

14. Perform the following byte operations:
     (a) $70 + 60$
     (b) $BC + BF$
     (c) $88 \cdot 8C$
     (d) $AC \cdot CD$

15. *Mathematical Description of the AES S-Box.* The AES S-box can be described by the following mathematical algorithm that defines a function F from the set of all bytes to itself. We let B be a byte.

     Step 1. First take $B_1 = [b_7, b_6, b_5, b_4, b_3, b_2, b_1, b_0]$ to be the inverse of B in $GF(256)$, if the latter is invertible; otherwise, take it to be the zero byte (i.e., in the latter case $B_1 = B$).

     Step 2. Define $F(B_1) = [b_7', b_6', b_5', b_4', b_3', b_2', b_1', b_0']$ where the eight bits are specified by the following mod 2 matrix computation:

$$\begin{bmatrix} b_0' \\ b_1' \\ b_2' \\ b_3' \\ b_4' \\ b_5' \\ b_6' \\ b_7' \end{bmatrix} = \begin{bmatrix} 1 & 0 & 0 & 0 & 1 & 1 & 1 & 1 \\ 1 & 1 & 0 & 0 & 0 & 1 & 1 & 1 \\ 1 & 1 & 1 & 0 & 0 & 0 & 1 & 1 \\ 1 & 1 & 1 & 1 & 0 & 0 & 0 & 1 \\ 1 & 1 & 1 & 1 & 1 & 0 & 0 & 0 \\ 0 & 1 & 1 & 1 & 1 & 1 & 0 & 0 \\ 0 & 0 & 1 & 1 & 1 & 1 & 1 & 0 \\ 0 & 0 & 0 & 1 & 1 & 1 & 1 & 1 \end{bmatrix} \cdot \begin{bmatrix} b_0 \\ b_1 \\ b_2 \\ b_3 \\ b_4 \\ b_5 \\ b_6 \\ b_7 \end{bmatrix} \oplus \begin{bmatrix} 1 \\ 1 \\ 0 \\ 0 \\ 0 \\ 1 \\ 1 \\ 0 \end{bmatrix}$$

     Note that the bits are stacked in the order of increasing powers of X, also the matrix operation multiplication is done mod 2, not in $GF(256)$. It can be shown that the above defined function is the same as the S-box function for the AES; see Computer Exercise 18.
     (a) Evaluate $F(00)$ (hex notation) and check with Table 11.2 that this is the same as $S(00)$.
     (b) Repeat part (a) with $F(3F)$ and $F(A6)$.

16. *Alternative Mathematical Description of the AES S-Box.* The description of the full AES S-box function given in Exercise 15 can be reformulated to be given completely in terms of modular polynomial ring operations by means of the following algorithm that defines a function G from the set of all bytes to itself. We let B be a byte.

Step 1. First take $B_1 = [b_7, b_6, b_5, b_4, b_3, b_2, b_1, b_0]$ to be the inverse of B in $GF(256)$, if the latter is invertible; otherwise, take it to be the zero byte (i.e., in the latter case $B_1 = B$). (Same first step as in Exercise 19).

Step 2. We view $B_1 = [b_7, b_6, b_5, b_4, b_3, b_2, b_1, b_0]$ as a polynomial
$$B_1 = b_7 X^7 + b_6 X^6 + b_5 X^5 + b_4 X^4 + b_3 X^3 + b_2 X^2 + b_1 X + b_0$$
in the modular ring $\mathbb{Z}_2[X] \pmod{X^8 + 1}$, and then define $F(B_1) = [b_7', b_6', b_5', b_4', b_3', b_2', b_1', b_0']$ where the eight bits are specified by the polynomial $a \cdot B_1 + b$, where $a = X^4 + X^3 + X^2 + X + 1$, $b = X^6 + X^5 + X + 1$, and the operations are computed in $\mathbb{Z}_2[X] \pmod{X^8 + 1}$. It can be shown that the above-defined function is the same as the S-box function for the scaled-down AES; see Computer Exercise 19.

(a) Evaluate $G(00)$ (hex notation) and check with Table 11.2 that this is the same as $S(00)$.

(b) Repeat part (a) with $G(3F)$ and $G(A6)$.

(c) Show that the polynomial $a = X^4 + X^3 + X^2 + X + 1$ is invertible in the ring $\mathbb{Z}_2[X] \pmod{X^8 + 1}$, but that the latter ring is not a field.

Note: In the language of Chapter 3 (where the ring was $\mathbb{Z}_n$) the mapping $N_1 \to a \cdot N_1 + b$ of Step 2 of the algorithm is an invertible *affine mapping* on the finite ring $\mathbb{Z}_2[X] \pmod{X^8 + 1}$.

Note: The last set of exercises will demonstrate the importance of the Nibble/Byte Sub transformations in the security of the scaled-down/full AES cryptosystems. The following basic definition will be used.

Definition 11.3

Suppose that $F:\{\text{length-}n \text{ bit strings}\} \to \{\text{length-}n \text{ bit strings}\}$ is a function, where n is a positive integer. We say that F has the **equal difference property** if $F(B) \oplus F(C) = F(B') \oplus F(C')$, whenever B, B', C, C' are length-n bit strings that satisfy $B \oplus C = B' \oplus C'$.

---

17. *Affine Mappings on $\mathbb{Z}_2[X] \pmod{m}$ Have the Equal Difference Property.*

(a) Suppose that m is a nonconstant polynomial in $\mathbb{Z}_2[X]$ and that $a, b \in \mathbb{Z}_2[X] \pmod{m}$. Show that the affine mapping $G:\mathbb{Z}_2[X] \pmod{m} \to \mathbb{Z}_2[X] \pmod{m}$ defined by $G(f) = a \cdot f + b$ has the equal difference property.

(b) Show that the composition $F \circ G$ of two mappings $F, G:\{\text{length-}n \text{ bit strings}\} \to \{\text{length-}n \text{ bit strings}\}$ that each has the equal difference property will also have the equal difference property.

18. (a) Show that the Add Round Key, Shift Row, and Mix Column transformations of the scaled-down AES cryptosystem each has the equal difference property.

(b) Show that the Add Round Key, Shift Row, and Mix Column transformations of the AES cryptosystem each has the equal difference property.

19. (a) Provide a specific example to show that the Nibble Sub Transformation of the scaled-down AES cryptosystem does not have the equal difference property.
     (b) Provide a specific example to show that the Byte Sub transformation of the AES cryptosystem does not have the equal difference property.

20. Show that if all Nibble/Byte Sub transformations were simply removed from either the scaled-down or the full AES cryptosystem, then the resulting cryptosystem would be easily vulnerable to a known plaintext attack.

Suggestion: Let the resulting encryption mapping be denoted by E. Use the results of Exercises 21 through 22 to show that E has the equal difference property. Suppose that Eve knows a plaintext/ciphertext pair P_0 and $E(P_0)$, and she wishes to decrypt another ciphertext $E(P)$. Show that $E(P_0) \oplus E(P)$ is the same as the result of encrypting the string $P_0 \oplus P$ with $\tilde{E}$, where $\tilde{E}$ is the result of composing all of the Shift Row and Mix Column transformations of E (i.e., the Add Round Keys are removed), and so is independent of the key. Since the key is not needed, Eve can apply $\tilde{E}^{-1}$ to $E(P_0) \oplus E(P)$ (by taking the inverse of the Shift Rows and Mix Column transformations in the reverse order) to obtain $P_0 \oplus P$, from which she will need only XOR with P_0 to obtain P.

Chapter 11 Computer Implementations and Exercises

Note: We shall avail ourselves of the programs for polynomial addition, multiplication, and the division algorithm that were developed in the computer implementation material of the previous chapter. In each of the computer exercises below involving nibbles (for the scaled-down AES) and bytes (for the AES), we will ask that the input and output syntax use hex notation. Of course, the internal programming will need to convert these to formats that are more suitable for computing. As we mentioned in previous chapters, readers whose computing platforms are not so well suited for hexadecimal strings may opt to instead write their programs so that their inputs/outputs are bit strings or binary vectors.

1. *Program for Nibble Addition.*
    (a) Write a program with syntax `outHex = NibbleXOR (n1Hex, n2Hex)` that inputs two nibbles, `n1Hex, n2Hex`, as single-character hex strings, and outputs their sum, `outHex`, which, as explained in the text, is the hex character corresponding to the XOR of the 4-bit strings representing the nibbles.

(b) Run your program on the sums of Example 11.1 and Chapter Exercise 1(a) and (b).

Note: This important computer exercise was also given in Chapter 10.

2. *Program for Nibble Multiplication.*
   (a) Write a program with syntax `outHex = NibbleMult (n1Hex, n2Hex)` that inputs two nibbles, *n1Hex, n2Hex*, as single-character hex strings, and outputs their product, `outHex`, which, as explained in the text, is the hex character corresponding to the *GF*(16) product of polynomials represented by 4-bit strings corresponding to the nibbles.
   (b) Run your program on the products of Example 11.1 and Chapter Exercise 1(c) and (d).

Note: This important computer exercise was also given in Chapter 10.

Writing Programs for AES

The next four exercises will build up a full program for the scaled-down AES encryption. Taken one at a time, the programming tasks are not difficult. In later computer exercises, a similar outline will be employed to write encryption programs for the full AES cryptosystem. Once the former task has been completed, the latter will be very easy. Most of the programs for the former will be easy to modify for the larger DES system. The large data sets of the S-box data may be downloaded from the book's Web page.

3. *Program for Scaled-Down DES Round Key Generation.*
   (a) Write a program with syntax `RoundKeysHexStr = ScaledDownAESRoundKeys(KeyHexStr)` that will generate the three round keys of the scaled-down AES cryptosystem of Algorithm 11.1. The input is `KeyHexStr`, a length-4 hex string representing the system key. The output, `RoundKeysHexStr`, will be a length-12 hex string whose first, second, and third groups of four hex characters represent the zeroth, first, and second round keys.
   (b) Use your program to obtain the three round keys that were obtained in the solution of Example 11.2.
   (c) Use your program to obtain the three round keys for each Chapter Exercise 7.

4. *Program for Scaled-Down AES S-Box.*
   (a) Write a program with syntax `outHex = ScaledDown AESSBox(inHex)` that will perform S-box evaluation for scaled-down DES cryptosystem (according to S-boxes specified by Table 11.2). The input is `inHex`, a single hex character, and the output, `outHex`, will be the hex character $S(\texttt{inHex})$.
   (b) Use your program to perform the S-box evaluations of Chapter Exercise 3.

In the remaining computer exercises that will lead to the scaled-down AES encryption program, with the exception of the final encryption program of Computer Exercise 9 (whose inputs and output will be length-4 hex strings), we will assume that the programs have inputs/outputs that are two-by-two state matrices of nibbles. This will coincide with the explanations given in the text. Some readers may find it more convenient to work with vectors of strings or just binary vectors.

5. *Program for Scaled-Down DES Add Key.*
   (a) Write a program with syntax `outStateHex = Scaled DownAESStateXOR(SHex,KeyHex)` whose inputs and output are two-by-two matrices of single hex digits (i.e., nibbles) and that will perform the Add Round Key operation; that is, the entries of the output matrix are simply the XOR of the corresponding input matrix entries.
   (b) Check your program on the results of each of the Add Round Key calculations that were done in the solution of Example 11.2.

6. *Program for Scaled-Down AES Nibble Sub.*
   (a) Write a program with syntax `OutStateHex = Scaled DownAESNibbleSub(InStateHex)` whose inputs and output are two-by-two matrices of single hex digits (i.e., nibbles) and that will perform the Sub Nibble operation, i.e., the entries of the output matrix result from running the corresponding input matrix entries through the scaled-down AES S-box.
   (b) Check your program on the results of each of the Sub Nibble calculations that were done in the solution of Example 11.2.

7. *Program for Scaled-Down AES Nibble Sub.*
   (a) Write a program with syntax `OutStateHex = Scaled DownAESShiftRow(InStateHex)` whose inputs and output are two-by-two matrices of single hex digits (i.e., nibbles) and that will perform the Shift Row operation of the scaled-down AES to transform the input state into the output state.
   (b) Check your program on the results of each of the Shift Row calculations that were done in the solution of Example 11.2.

8. *Program for Scaled-Down AES Mix Column.*
   (a) Write a program with syntax `OutStateHex = Scaled DownAESMixColumn(InStateHex)` whose inputs and output are two-by-two matrices of single hex digits (i.e., nibbles) and that will perform the Mix Column operation of the scaled-down AES to transform the input state into the output state.
   (b) Check your program on the results of each of the Mix Column calculation that was done in the solution of Example 11.2.

9. *Program for Scaled-Down AES Encryption.*
    (a) Write a program with syntax `CtextHex = ScaledDown`
        `AES(PtextHex, KeyHex)` that will perform the scaled-
        down AES encryption of Algorithm 11.1. The inputs are
        `PtextHex`, a length-4 hex string (four nibbles) represent-
        ing a 16-bit plaintext, and `KeyHex`, a length-4 hex string
        (four nibbles) representing the system key. The output,
        `CtextHex`, is the length-4 hex string representing the
        ciphertext.
    (b) Check your program with the results of Example 11.2.
    (c) Use your program to obtain the ciphertext for each part of
        Chapter Exercise 9.

10. *Program for Scaled-Down AES Inverse S-Box.*
    (a) Write a program `ScaledDownAESInvSBox`, whose syn-
        tax is identical to the program `ScaledDownAESSBox`
        of Computer Exercise 4, except that it will work with the
        inverse S-box that is specified in Table 11.3.
    (b) Check your program on the answers to each part of
        Chapter Exercise 3. [Thus, for example, if the exer-
        cise answer shows $S(C) = B$, you should check that
        `ScaledDownAESInvSBox(B) = C`.]
    (c) Use a `FOR` loop to verify that the S-box specified by
        Table 11.3 really is the inverse function of that specified
        by Table 11.2.

11. *Program for Scaled-Down AES Inverse Mix Column.*
    (a) Write a program with syntax `OutStateHex = Scaled`
        `DownAESInvMixColumn(InStateHex)` whose inputs
        and output are two-by-two matrices of single hex digits
        (i.e., nibbles) and that will perform the Inv Mix Column
        operation of the scaled-down AES to transform the input
        state into the output state.
    (b) Check your program on the results of each of the Mix Col-
        umn calculation that was done in the solution of Exercise
        for the Reader 11.4.

12. *Program for Scaled-Down AES Decryption.*
    (a) Write a program with syntax `PtextHex = ScaledDown`
        `AESDecrypt(CtextHex,KeyHex)` that will perform
        the scaled-down AES decryption (as in Figure 11.3). The
        inputs are `CtextHex`, a length-4 hex string (four nibbles)
        representing a 16-bit ciphertext, and `KeyHex`, a length-4
        hex string (four nibbles) representing the system key. The
        output, `PtextHex`, is the length-4 hex string representing
        the plaintext.
    (b) Check your program with the results of Exercise for the
        Reader 11.4.
    (c) Use your program to obtain the ciphertext for each part of
        the Chapter Exercise 11.

13. *Program for Byte Addition.*
    (a) Write a program with syntax `OutHex = ByteXOR(n1Hex, n2Hex)` that inputs two bytes, n1Hex, n2Hex, as two-character hex strings, and outputs their sum, `outHex`, which, as explained in the text, is the hex string corresponding to the XOR of the 8-bit strings representing the bytes.
    (b) Run your program on the sums of Example 11.3 and Chapter Exercise 13(a) and (b).

    Note: This important computer exercise was also given in Chapter 10.

14. *Program for Byte Multiplication.*
    (a) Write a program with syntax `OutHex = ByteMult (n1Hex, n2Hex)` that inputs two bytes, n1Hex, n2Hex, as two-character hex strings, and outputs their product, *outHex*, which, as explained in the text, is the hex string corresponding to the *GF*(256) product of polynomials represented by 8-bit strings corresponding to the bytes.
    (b) Run your program on the products of Example 11.4 and Chapter Exercise 13(c) and (d).

    Note: This important computer exercise was also given in Chapter 10.

15. *Program Suite for AES.* Write a suite of programs for AES corresponding to those of Computer Exercises 3 through 12. (Take the same names of the programs, but with "ScaledDown" omitted.) Do a Web search for AES applets (there are some that show all intermediate steps), and compare to check the correctness of your programs.

16. *Alternative Form for AES S-Box.*
    (a) Write a program with syntax `OutHex = AESFunctionF (inHex)` that will evaluate the function *F* from bytes to bytes that was defined in Chapter Exercise 19. The input is `inHex`, a double hex character, and the output, `outHex`, will be the double hex character $F(\text{inHex})$.
    (b) Use this program together with that of the appropriate program of Computer Exercise 15 to verify that the function of part (a) coincides with the scaled-down AES S-box function.

17. *Alternative Form for Scaled-Down AES S-Box.*
    (a) Write a program with syntax `OutHex = ScaledDown AESFunctionG(inHex)` that will evaluate the function *G* from nibbles to nibbles that was defined in Chapter Exercise 19. The input is `inHex`, a single hex character, and the output, `outHex`, will be the hex character $G(\text{inHex})$.
    (b) Use this program together with that of the appropriate program of Computer Exercise 15 to verify that the function of part (a) coincides with the scaled-down AES S-box function.

Elliptic Curve Cryptography

An elliptic curve is the set of solutions ("points") of an algebraic equation of the form $y^2 = x^3 + ax + b$, where a and b are numbers belonging to some field (for example, the real numbers, the rational numbers, or the integers mod p, or any finite field), and the variables x, y belong to the same field. Motivated by their graphs in the plane when the parameters are considered to be real numbers, a geometric procedure can be devised to define an "addition" operation on points on an elliptic curve, and this operation naturally extends to elliptic curves over any field. Endowed with this addition operation, elliptic curves provide a rich variety of new number systems. In the mid-1980s, number theorists Neal Koblitz (a professor at the University of Washington) and Victor Miller (then an IBM researcher) noticed that the discrete logarithm problems could be defined on these number systems and could be used as the basis of powerful public key cryptosystems that, for a given key size, tended to be much more robust and approximately 10 times more secure than all other known public key cryptosystems. This has important ramifications for efficient hardware implementations. Part of the reason for this security is the fact that, unlike for modular integers, there is no notion for "size" of points in modular elliptic curves. For example, the (elliptic curve) sum of two points with very small coordinates may have extremely large coordinates. At about the same time as Koblitz and Miller announced their results, Dutch mathematician Hendrik Lenstra discovered that prime factorization algorithms (for integers) could be created that are based on elliptic curve arithmetic, and these appeared to be more powerful than most integer-based factoring algorithms. These facts have transformed elliptic curves into one of the most extensively studied branches of cryptography. We begin by introducing elliptic curves over the real numbers and rigorously defining their addition operation by means of their graphs. We then discuss modular elliptic curves over $\mathbb{Z}_p$ and introduce the discrete logarithm problem for elliptic curves. This leads us to the development of natural extensions of the Diffie–Hellman key exchange and the ElGamal cryptosystems to the setting of modular elliptic curves. The chapter ends with an example of an elliptic curve-based factorization algorithm.

Elliptic Curves over the Real Numbers

Before studying elliptic curves over modular integers, it will be helpful to introduce them over the real numbers, as the latter possess many rich geometric properties that will provide some useful intuition and motivation.

Definition 12.1

Given a pair of real numbers $a, b \in \mathbb{R}$, the associated **elliptic curve** E **over the real numbers** $\mathbb{R}$ is the set of all **points** represented by ordered pairs (x, y) of real numbers that solve the equation

$$y^2 = x^3 + ax + b \qquad (12.1)$$

together with the **point at infinity**, which for brevity is denoted as ∞.[*] The **discriminant** of the elliptic curve is defined to be the following number:

$$\Delta \triangleq 4a^3 + 27b^2 \qquad (12.2)$$

The elliptic curve is called **nonsingular** if its discriminant is nonzero: $\Delta \neq 0$; otherwise, it is called **singular**.

In order to better understand the soon-to-be-developed arithmetic of elliptic curves, we will take a moment to display the graphs of a few elliptic curves that illustrate the three general sorts of graphs that can arise.

Example 12.1

We consider elliptic curves over the real numbers where the parameter a of Equation 12.1 is taken to be -4: $y^2 = x^3 - 4x + b$.

(a) Determine the value of b for which the discriminant of the above elliptic curve is 0 (and thus the elliptic curve will be singular).

(b) Sketch planar graphs of the real-valued solutions of the above elliptic curve for the following parameter values: $b = 0, 2, 4, 6$.

(c) Sketch a planar graph of the real-valued solutions of the singular elliptic curve $y^2 = x^3 - 4x + b$, with the parameter b as determined in part (a).

Solution: Part (a): Setting the discriminant Δ of Equation 12.2 equal to zero, with $a = -4$, the equation is $4(-4)^3 + 27b^2 = 0 \Rightarrow b = \sqrt{4^4/27} \approx 3.0792\ldots$

---

[*] Geometrically, the point at infinity should be thought of as lying infinitely far out in the plane, as we move away from zero in *any* direction.

Parts (b) and (c): We first make a few general observations about the planar graphs of the equation $y^2 = x^3 - 4x + b$. First, the x intercepts of the graph are determined by the root cubic equation $x^3 - 4x + b = 0$, and by the general facts of roots of polynomial equations, the roots will fall into one of three possibilities:[*]

(i) one real root (and two complex roots)
(ii) two distinct real roots (one of them a double root), or
(iii) three distinct real roots

Also, since $x^3 - 4x + b \to -\infty$ as $x \to -\infty$ and since we need to have $x^3 - 4x + b \geq 0$ for the equation $y^2 = x^3 - 4x + b$ to have a real-valued solution, it follows that there will be no solutions to $y^2 = x^3 - 4x + b$ when x is less than the smallest real root of $x^3 - 4x + b = 0$. Similarly, for each x larger than the largest real root of $x^3 - 4x + b = 0$, there will be exactly two values of y that satisfy $y^2 = x^3 - 4x + b$; these will be opposites and will get large (in absolute value) as x gets large. With this initial analysis, a computer graphing utility can then be used to produce the graphs of the four nonsingular elliptic curves for part (b) that are shown in Figure 12.1, and the singular elliptic curve of part (c) shown in Figure 12.2.

Comparing the graphs of the aforementioned two figures, it appears that when the parameter b is less than the critical value (≈ 3.08) that makes the discriminant zero, the elliptic curve consists of two pieces: a single loop (on the left) and a "boomerang"-shaped curve (on the right) that goes off to infinity.[†] As b approaches the critical value, these two pieces get closer to one another until they finally merge (Figure 12.2) when b reaches the critical value. Notice that at the meeting point of these two pieces (the intersection point of Figure 12.2), the curve does not have a well-defined tangent line.[‡] When b is greater than the critical value, the intersection point vanishes and the elliptic curve consists of a single piece (see the labeled curves with $b = 4$ and $b = 6$ of Figure 12.1).

---

[*] From the *fundamental theorem of algebra,* any degree-n polynomial equation of the form $x^n + a_{n-1}x^{n-1} + \cdots + a_1 x + a_0 = 0$, where the coefficients a_i are real numbers, will have exactly n roots, which may be real or complex numbers, some of which may be repeated. If the distinct roots are $r_1, r_2, \cdots, r_k$, and these are repeated with corresponding multiplicities $d_1, d_2, \cdots, d_k$, then we may write $x^n + a_{n-1}x^{n-1} + \cdots + a_1 x + a_0 = (x - r_1)^{d_1}(x - r_2)^{d_2} \cdots (x - r_k)^{d_k}$. Also, any complex roots necessarily occur in conjugate pairs (with the same multiplicities) $r = \alpha \pm \beta i$, so the number of complex roots is even.

[†] It is clear from the graphs on page 454 that elliptic curves are not ellipses. The name comes from their appearance in so-called *elliptic integrals,* which are used to compute arc length of ellipses.

[‡] Not having a tangent line at a point is the geometric definition for a planar curve to be *singular.* It can be shown that this property is equivalent to the discriminant being zero and to the polynomial equation $x^3 + ax + b = 0$ having at least one repeated root.

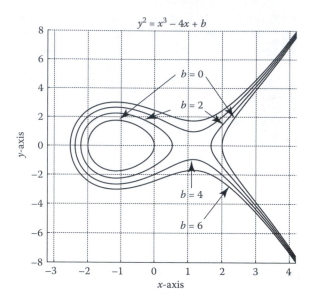

Figure 12.1 Graph of the four nonsingular elliptic curves $y^2 = x^3 - 4x + b$ with $b = 0, 2, 4, 6$. Note that for the values $b = 0, 2$ (that are less than the singular value of $b \approx 3.08$), the elliptic curve has two pieces, the loop on the left and the branch on the right, and three x intercepts, while for $b = 4, 6$ (that are greater than the singular value of $b \approx 3.08$), the elliptic curve has a single piece and a single x intercept.

The Addition Operation for Elliptic Curves

We assume that E is a nonsingular elliptic curve over the real numbers specified by the equation $y^2 = x^3 + ax + b$. We first give a geometric procedure showing how to add points on E, and then we will follow it with an algebraic formula.

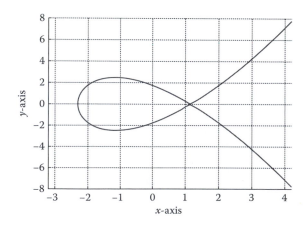

Figure 12.2 Graph of the singular elliptic curve $y^2 = x^3 - 4x + b$ with $b \approx 3.08$.

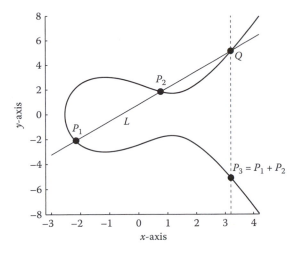

Figure 12.3 Illustration of the addition operation $P_3 = P_1 + P_2$ for two points on an elliptic curve E, in the case that $P_1 \neq P_2, \infty;\ P_2 \neq \infty$.

Algorithm 12.1: Addition of Points on Elliptic Curves over $\mathbb{R}$.
Part I: Geometric Description of Algorithm via the Graph of the Elliptic Curve

Input: Two points, P_1, P_2, on a nonsingular elliptic curve E over the real numbers.

Output: A third point, $P_3 = P_1 + P_2$, on E (the **sum** of the first two points).

Case 1. $P_1 \neq P_2, \infty;\ P_2 \neq \infty.$

> *Step 1.*   Draw the line L through P_1, P_2. The line will intersect E in a (unique) third point Q; see Figure 12.3.

> *Step 2.*   Set the point $P_3 = P_1 + P_2$ to be the reflection of Q about the x axis; i.e., if $Q = (x, y)$, we set $P_3 = (x, -y)$. We make the convention that any vertical line passes through the point at infinity.

Case 2. $P_1 = P_2 \neq \infty.$
Use the procedure of Case 1, but with L taken to be the tangent line of the elliptic curve E at the point P_1. In this case, the line will intersect E in a (unique) second point Q.

Case 3. At least one of the two points $P_i = \infty.$

We use the convention that $P_1 + \infty = P_1,\ \infty + P_2 = P_2.$[*]

The corresponding algebraic formulation of the above algorithm can be derived using the relevant concepts about lines. Before presenting this

---

[*] We caution the reader (especially one who has studied calculus) that the point at infinity for elliptic curves behaves quite differently from the number infinity in the setting of real numbers, where we have $x + \infty = \infty$, for any real number x.

formulation, we give one specific example of an addition that will help to motivate it.[*] Of course, once we give the algebraic algorithm, there will be no need for these sorts of geometric derivations.

Example 12.2

Use the geometric Algorithm 12.1 to find the sum of the points $P_1 = (2, 4)$ and $P_2 = (4, 8)$ on the elliptic curve E defined by $y^2 = x^3 - 4x + 16$.

Solution: The reader should verify that both of the points P_1, P_2 satisfy the equation $y^2 = x^3 - 4x + 16$ and thus belong to E. To compute $P_3 = P_1 + P_2$, we follow the steps of Case 1 in the algorithm. We develop the algebra steps by referring to the entries of the points as follows: $P_1 = (2, 4) \triangleq (x_1, y_1)$ and $P_2 = (4, 8) \triangleq (x_2, y_2)$.

Step 1. The line L through P_1, P_2 has slope $m = (y_2 - y_1)/(x_2 - x_1) = (8 - 4)/(4 - 2) = 2$. The equation of this line, $y = mx + b$, can be determined by substituting one of the points (we will use P_1) and the slope m into this equation to obtain $4 = 2 \cdot 2 + b \Rightarrow b = 0$. Thus, L has equation $y = 2x$. We then substitute this equation into that for E to find the unique third intersection point Q:

$$(2x)^2 = x^3 - 4x + 16 \Rightarrow x^3 - 4x^2 - 4x + 16 = 0$$

Since we already know two roots, $r_1 = 2$, $r_2 = 4$, we can easily determine the third root (that will be the x coordinate of Q) by using the general fact that the sum of the n roots of a degree-n polynomial equation having leading coefficient equal to 1 must always equal the negative of the x^{n-1} coefficient of the polynomial:[†] $r_1 + r_2 + r_3 = -(-4) \Rightarrow r_3 = 4 - r_1 - r_2 = 4 - 2 - 4 = -2$. This is the x-coordinate of Q; to obtain the corresponding y-coordinate, we substitute into the equation for L: $y = 2x = 2 \cdot (-2) = -4$. Thus $Q = (-2, -4)$.

Step 2. The point $P_3 = P_1 + P_2$ will be the reflection of Q about the x-axis: $P_3 = (-2, -(-4)) = (-2, 4)$.

We now present the general algebraic formulation of this addition algorithm.

---

[*] Our example illustrates Case 1 of the algorithm and requires only algebra. Case 2 would require finding the slope of the tangent line, and this can be accomplished by using some calculus (implicit differentiation of the elliptic curve equation).

[†] By the fundamental theorem of algebra (that was described in an earlier footnote), $x^n + a_{n-1}x^{n-1} + \cdots + a_1 x + a_0 = (x - r_1)(x - r_2) \cdots (x - r_n)$, where $r_1, r_2, \cdots, r_n$ are the roots (with multiple roots repeated). When we multiply out the right side, the only way to get monomials of degree $n - 1$ is to choose the term x in all but one of the factors, and to choose $-r_i$ from the remaining factor. It follows that when it is expanded and simplified, the term of degree $n - 1$ on the right will be $-(r_1 + r_2 + \cdots + r_n)x^{n-1}$, and if we equate the coefficient with the corresponding one of the original polynomial, we get that $r_1 + r_2 + \cdots + r_n = -a_{n-1}$.

Algorithm 12.1: Addition of Points on Elliptic Curves over $\mathbb{R}$.
Part II: Algebraic Formulation

Inputs: Two points, $P_1 = (x_1, y_1)$ and $P_2 = (x_2, y_2)$, on a nonsingular elliptic curve E over the real numbers,[*] which is defined by the equation $y^2 = x^3 + ax + b$.

Output: A third point, $P_3 = (x_3, y_3)$, on E that is the sum $P_1 + P_2$.
   If either $P_1, P_2 = \infty$, output $P_3 = $ the other point, and exit the algorithm.

Otherwise, we set (the slope of the line L): $m = \begin{cases} \dfrac{y_2 - y_1}{x_2 - x_1}, & \text{if } P_2 \neq P_2 \\ \dfrac{3x_1^2 + a}{2y_1}, & \text{if } P_2 = P_2 \end{cases}$

   If m is undefined (i.e., if L is a vertical line), output $P_3 = \infty$, and exit the algorithm.

In all remaining cases, we set $\begin{cases} x_3 = m^2 - x_1 - x_2 \\ y_3 = m(x_1 - x_3) - y_1 \end{cases}$, and output $P_3 = (x_3, y_3)$.

Exercise for the Reader 12.1

Use the above algebraic algorithm to redo the computation of Example 12.2.

In earlier chapters, we experienced several sorts of "addition" and "multiplication" operations that were quite different from the usual ones with the same name. The next example shows that elliptic curve addition is different from vector addition even if all of the vectors belong to a given elliptic curve.

Example 12.3

Show that although (2, 4) and (4, 8) belong to the elliptic curve $y^2 = x^3 - 4x + 16$, we have $(2, 4) + (2, 4) \neq (4, 8)$.
   Solution: Using the algebraic form of Algorithm 12.1 with $P_1, P_2 = (2, 4)$, we obtain $m = \dfrac{3 \cdot 2^2 - 4}{2 \cdot 4} = 1$, so we may compute $x_3 = m^2 - x_1 - x_2 = 1^2 - 2 - 2 = -3$, and then $y_3 = m(x_1 - x_3) - y_1 = 1(2 - (-3)) - 4 = 5 - 4 = 1$. Thus $P_3 = (-3, 1)$; that is, $(2, 4) + (2, 4) = (-3, 1) \neq (4, 8)$.

Although this addition operation may appear a bit contrived, it turns out to have some of the properties enjoyed by ordinary addition. For example,

---

[*] In this development, the point at infinity ∞ should be viewed sitting infinitely far up the y-axis yet above the whole x-axis so that every vertical line intersects it (a bit like the North Pole).

since we have (by the algorithm) $P + \infty = P$ for any point P on the curve, the point at infinity behaves like the additive identity zero for ordinary addition of real numbers. Also notice that from (either version of) Algorithm 12.1, it follows that the reflection of any point $P = (x, y)$ of E over the x-axis; that is, the point $(x, -y)$ satisfies $(x, y) + (x, -y) = \infty$—the additive identity for elliptic curve addition. (Notice that this still works when $y = 0$, in which case both points are the same.) The following proposition collects these two facts, along with two other useful properties of elliptic curve addition.

Proposition 12.1: Properties of Elliptic Curve Addition

Suppose that E is a nonsingular elliptic curve defined by $y^2 = x^3 + ax + b$. The addition operation of points of E defined by Algorithm 12.1 has the following properties, where P, Q, and R denote points of E:

(1) *Commutativity.* $P + Q = Q + P$
(2) *Associativity.* $(P + Q) + R = P + (Q + R)$
(3) *Additive Identity.* $P + \infty = P$
(4) *Additive Inverses.* There exists a point $-P$ in E, such that $P + (-P) = \infty$. Moreover, $-(x, y) = (x, -y)$, and $-\infty = \infty$.

We have already proven parts (3) and (4). The commutativity (1) follows easily from either the geometric or the algebraic form of Algorithm 12.1. For example, using the geometric approach, since the line L through P and Q is the same as the line through Q and P, it follows that the third point R (where L intersects the elliptic curve) is the same for both additions, thus, so is its reflection over the x axis, and hence, $P + Q = Q + P$. The proof of the associativity property (2) is much more involved than those that we just gave for the other three properties. Although it can be proved using the algebraic formulation of Algorithm 12.1, this method would be quite messy and require the separation into different cases—we do not recommend that any readers try this (even those who are very adept with algebraic manipulations). There are more elegant geometric approaches to proving the elliptic curve associative law, but such approaches are quite involved; the interested reader may find one such approach in the 11-page Section 2.4 in [Was-03].

Groups

Notice that the four properties of Proposition 12.1 are exactly the first four axioms for ring addition in Definition 10.2, with only a stylistic difference in that we are denoting the additive identity here by ∞ rather than 0. Any nonempty set G with a single binary operation on which these axioms are satisfied composes a very important type of number system in mathematics known as an *abelian group*.

Definition 12.2

An **abelian** (or commutative) **group** is a set G that is endowed with a single binary operation, which in this definition we denote by the symbol $\triangle$, and for which the following axioms hold, where a, b, c denote generic elements of G:

1. *Commutativity.* $a \triangle b = b \triangle a$.
2. *Associativity of Addition.* $(a \triangle b) \triangle c = a \triangle (b \triangle c)$.
3. *Identity.* There exists in G an **identity** element, denoted as e, that satisfies: $a \triangle e = a$.
4. *Inverses.* For each ring element a, there exists a corresponding **inverse**, denoted as a^{-1}, that satisfies $a \triangle a^{-1} = e$.

The adjective *abelian* refers to the commutativity Axiom (1). In more general treatments, nonabelian groups [that satisfy only Axioms (2) through (4) in the above definition] are considered. Since all groups under our consideration will be abelian groups, we often refer to these simply as "groups" (omitting the *abelian* adjective). Groups whose binary operations are denoted as additions (+) are usually called **additive groups**, while those whose binary operations are denoted as multiplications (·) are called **multiplicative groups**. In an additive group, the identity is usually denoted as 0 (zero), and in a multiplicative group it is usually denoted as 1 (one). Also, in an additive group, the inverse of a group element a is usually denoted as $-a$. Note that for elliptic curves, although they are additive groups, the identity ∞ is not denoted as 0.

From Definition 10.2, it follows that any ring with just its addition operation forms an additive group. From Exercise for the Reader 10.1 (and some of the ring axioms of Definition 10.2), it follows that the set of invertible elements $R^{\times}$ in any ring forms a multiplicative group. From Proposition 2.1, it follows that any nonsingular elliptic curve over the real numbers forms an additive group.

Group theory is a vast and very well understood branch of mathematics, and the Diffie–Hellman key exchange as well as the ElGamal cryptosystem can be easily generalized to work with any finite group. Although elliptic curves of the real numbers will always be infinite groups, the soon-to-be-considered elliptic curves over modular integers (of prime modulus) provide us with a rich variety of finite groups on which we will build new cryptosystems.

Notation: For a positive integer n, the additive group analogue for powers $b^n = \underbrace{b \cdot b \cdots b}_{n \text{ times}}$, in multiplicative groups, is a repeated summation of the same element for which we use the following notation:

$$nP \text{ (or } n \cdot P) \triangleq \underbrace{P + P + \cdots + P}_{n \text{ times}} \qquad (12.3)$$

This notation derives from a familiar arithmetic property; for example, $5x = x + x + x + x + x$. We will adhere to this additive notation in all of our work with elliptic curves.[*]

Exercise for the Reader 12.2

Using the notation (Equation 12.3), the result of Example 12.3 can be expressed as $2 \cdot (2, 4) = (-3, 1)$. Compute $3 \cdot (2, 4)$ and $4 \cdot (2, 4)$.

Elliptic Curves over $\mathbb{Z}_p$

Suppose that $p > 3$ is a prime. The definition of an elliptic curve E over $\mathbb{Z}_p$ is the same as the definition of elliptic curves over the real numbers, except that the points on the curve are pairs of mod p integers that satisfy the elliptic curve equation mod p.

Definition 12.3

Given a prime number $p > 3$ and a pair of real numbers $a, b \in \mathbb{R}$, the associated **elliptic curve E over $\mathbb{Z}_p$** (or simply the **elliptic curve mod p**) is the set of all ordered pairs (x, y) of mod p integers that solve the congruence

$$y^2 \equiv x^3 + ax + b \pmod{p} \qquad (12.4)$$

together with the **point at infinity** ∞. The **discriminant** of the elliptic curve is defined to be the following number:

$$\Delta \equiv 4a^3 + 27b^2 \pmod{p} \qquad (12.5)$$

The elliptic curve is called **nonsingular** if its discriminant nonzero $\pmod{p}$: $\Delta \not\equiv 0$; otherwise, it is called **singular**.

Note in contrast to elliptic curves over the real numbers, any *modular elliptic curve E* over $\mathbb{Z}_p$ is a finite set with at most $p^2 + 1$ elements, but we soon will provide a more accurate result.

Example 12.4

Verify that the following modular elliptic curve is nonsingular, and then find all of its points: the elliptic curve E is defined by $y^2 \equiv x^3 - 4x + 16 \equiv x^3 + x + 1 \pmod{5}$.
  Solution: Since $\Delta \equiv 4a^3 + 27b^2 \equiv 4 \cdot 1^3 + 27 \cdot 1^2 \equiv 31 \equiv 1 \not\equiv 0 \pmod 5$, the elliptic curve is nonsingular. A brute-force approach to finding the points of an elliptic curve mod a prime p would be

---

[*] In this regard, from a cryptographic perspective, it would make more sense to use multiplicative notation rather than addition notation in the elliptic curve operation. But addition of points on elliptic curves has been in existence much longer than these cryptographic applications, so the addition notation is here to stay.

TABLE 12.1 Finite Points (x, y) on the Elliptic Curve $y^2 \equiv x^3 + x + 1(\text{mod } 5)$

x	$x^3 + x + 1(\text{mod } 5)$	$y \equiv \sqrt{x^3 + x + 1} \ (\text{mod } 5)$
0	1	1, 4
1	3	None
2	1	1, 4
3	1	1, 4
4	4	2, 3

to simply let x run through the integers mod p: $0, 1, = \cdots, p - 1$, evaluate $x^3 - 4x + 16 \,(\text{mod } p)$ for each such x, and look for any square roots y of this number (mod p). The modular elliptic curve will be the set of all ordered pairs (x, y) that arise in this way together with the point at infinity. These computations are summarized in Table 12.1.

Thus, we can write $E = \{(0, 1), (0, 4), (2, 1), (2, 4), (3, 1), (3, 4), (4, 2), (4, 3), \infty\}$.

Exercise for the Reader 12.3

Verify that the following modular elliptic curve is nonsingular, and then find all of its points: the elliptic curve E is defined by $y^2 \equiv x^3 + x + 1 \,(\text{mod } 7)$.

In our search for square roots in the solution of the above example, we witness that there are either no square roots or exactly two. This is true in general whenever we are dealing with prime moduli, except for the exceptional case that 0 has only one square root (itself). This is summarized in the following proposition. After this, we recall a formula from Proposition 2.14 (proven in the exercises of Chapter 2) that gives a more efficient means of extracting square roots with a prime modulus p, in the case that $p \equiv 3(\text{mod } 4)$.

Proposition 12.2: *Uniqueness of Square Roots Modulo a Prime*

If p is an odd prime number, and $a \not\equiv 0(\text{mod } p)$ is an integer mod p, then the equation $x^2 \equiv a(\text{mod } p)$, that is, $x = \sqrt{a}\,(\text{mod } p)$, has either no solutions or exactly two solutions (mod p). Also, 0 has only one square root mod p: $\sqrt{0} \equiv 0(\text{mod } p)$.

Proof: We first deal with the case when $a = 0$, which is easy but requires a separate treatment. If $x^2 \equiv 0(\text{mod } p)$, then $p \,|\, x^2$, so it follows from Euclid's lemma (Proposition 2.7) that $p \,|\, x$, so that $x \equiv 0(\text{mod } p)$.

For the remainder of the proof, we assume that $a \not\equiv 0(\text{mod } p)$. Let g be a primitive root mod p; such primitive roots exist by Theorem 8.7. Since (by Proposition 8.6) the modular powers $g, g^2, \cdots, g^{p-1}$ make up all nonzero integers mod p, if x is any nonzero integer mod p, we can write $x \equiv g^j$ and $a \equiv g^\ell$ (mod p) for unique integers $j, \ell \in \{1, 2, \cdots, p - 1\}$. Using these representations, the congruence $x^2 \equiv a(\text{mod } p)$ becomes

$g^{2j} \equiv g^{\ell}(\mathrm{mod}\ p)$. By Proposition 8.5(c), this latter congruence is equivalent to $2j \equiv \ell \pmod{p-1}$. Since p is odd, $d = \gcd(2, p-1) = 2$, and by Algorithm 2.3 (and its proof), this latter congruence will have either no solutions (in the case $d \nmid \ell$), in which case a has no square root, or exactly two solutions (in the case $d \mid \ell$), in which case a has exactly two square roots. □

We next put forth the following formula from Proposition 2.15, that provides the two square roots of nonzero modular integer a, whenever they exist, in case of a prime modulus $p \equiv 3(\mathrm{mod}\ 4)$:

$$\sqrt{a} \equiv \pm a^{(p+1)/4} \pmod{p}, \text{ in case } a \not\equiv 0(\mathrm{mod}\ p) \text{ has square roots} \qquad (12.6)$$

Equation 12.6 considerably speeds up the brute-force search of points on elliptic curves mod a prime p, provided that $p \equiv 3(\mathrm{mod}\ 4)$. For each value of x, we need compute only one modular power (with a moderately sized exponent) and check whether it is a square root, rather than squaring all of the $p - 1$ nonzero integers mod p. We demonstrate this faster method in the Example 12.5.

Example 12.5

Verify that the following modular elliptic curve is nonsingular, and then find all of its points: the elliptic curve E is defined by $y^2 \equiv x^3 - 4x + 16 \equiv x^3 + 7x + 5 (\mathrm{mod}\ 11)$.

Solution: Since $\Delta \equiv 4a^3 + 27b^2 \equiv 4 \cdot 7^3 + 27 \cdot 5^2 \equiv 2047$ $1 \not\equiv 0(\mathrm{mod}\,11)$,the elliptic curve is nonsingular. We let x run through the integers mod 11: $0, 1, \cdots, 10$, evaluate $r \equiv x^3 + 7x + 5$ (mod 11) for each such x, and whenever the value of r is nonzero, we use Equation 12.6 to find any square roots of a. The modular elliptic curve will be the set of all ordered pairs (x, y) that arise in this way together with the point at infinity. The computations are summarized in Table 12.2.

From these computations, we may write $E = \{(0, 4), (0, 7), (2, 4), (2, 7), (3, 3), (3, 8), (4, 3), (4, 8), (5, 0), (7, 1), (7, 10), (8, 1), (8, 10), (9, 4), (9, 7), \infty\}$.

Exercise for the Reader 12.4

Verify that the following modular elliptic curve is nonsingular, and then use the method of the above example to find all of its points: the elliptic curve E is defined by $y^2 \equiv x^3 + x + 1 (\mathrm{mod}\ 11)$.

The Variety of Sizes of Modular Elliptic Curves

After working through the previous two examples, the following natural question arises: How many points are in a given elliptic curve modulo a prime $p > 3$? It turns out that there will be roughly p points; the following theorem contains a more precise result dating back to the 1930s.

TABLE 12.2 Determination of All Finite Points (x, y) on the Elliptic Curve $y^2 \equiv x^3 + 7x + 5 \pmod{11}$, Using Equation 12.6 to Compute Square Roots

x	$a \equiv x^3 + 7x + 5$ (mod 11)	a^3 (mod 11)	$(a^3)^2$ (mod 11)	$y \equiv \pm a^3 \pmod{11}$?
0	5	4	5	$4, -4 \equiv 7$
1	2	8	9	No
2	5	4	5	$4, -4 \equiv 7$
3	9	3	9	$3, -3 \equiv 8$
4	9	3	9	$3, -3 \equiv 8$
5	0	–	–	0
6	10	10	1	No
7	1	1	1	$1, -1 \equiv 10$
8	1	1	1	$1, -1 \equiv 10$
9	5	4	5	$4, -4 \equiv 7$
10	8	6	3	No

Theorem 12.3: Hasse's Theorem

If E is an elliptic curve mod p, where $p > 3$ is a prime, then $|E|$ = the number of points in E satisfies the following inequalities:

$$p + 1 - 2\sqrt{p} \leq |E| \leq p + 1 + 2\sqrt{p} \qquad (12.7)$$

A proof of this theorem can be found in Section 2.4 in [Was-03] or Section V.1 in [Sil-86]. It turns out that the range of possible values in Equation 12.7 are essentially all attainable for a given modulus p, for appropriate choices of the coefficients of the elliptic curve. This latter result is more recent; see [Wat-69].[*]

Theorem 12.4: Waterhouse's Theorem

If $p > 3$ is a prime and N is a positive integer that satisfies $p + 1 - 2\sqrt{p} \leq N \leq p + 1 + 2\sqrt{p}$, then there exists a nonsingular elliptic curve E (mod p) with exactly N elements; that is, $|E| = N$.

The Addition Operation for Elliptic Curves over $\mathbb{Z}_p$

The addition operation for modular elliptic curves is defined using exactly the same formulas that were used in Algorithm 12.1, the only difference being that all divisions are performed mod p.[†]

---

[*] Waterhouse originally proved his theorem for N in the range $p + 1 - 2\sqrt{p} < N < p + 1 + 2\sqrt{p}$, but the result was extended in 1987 to hold also at the endpoints by Hans-Georg Rück; see [Rüc-87].

[†] Actually, the same definition works to define elliptic curves over any finite field $GF(p^n)$, but restricting attention to modular integer arithmetic will sufficiently illustrate the richness of elliptic curve cryptography, and the resulting cryptosystems are as secure as those that would result from using other finite fields (of comparable sizes).

Algorithm 12.2: Addition of Points on Elliptic Curves over $\mathbb{Z}_p$

Input: Two points, $P_1 = (x_1, y_1)$ and $P_2 = (x_2, y_2)$, on a nonsingular elliptic curve E over $\mathbb{Z}_p$, which is defined by the congruence $y^2 \equiv x^3 + ax + b$ (mod p) where $p > 3$ is a prime number.

Output: A third point, $P_3 = (x_3, y_3)$, on E that is the **sum** $P_1 + P_2$.

If either $P_1, P_2 = \infty$, output $P_3 =$ the other point, and exit the algorithm. Otherwise, we set (the slope of the line L):

$$m \equiv \begin{cases} (y_2 - y_1) \cdot (x_2 - x_1)^{-1}, & \text{if } P_2 \neq P_2 \\ (3x_1^2 + a) \cdot (2y_1)^{-1}, & \text{if } P_2 = P_2 \end{cases} \pmod{p}.$$

If m is undefined (that is, inverse in the above formula does not exist), output $P_3 = \infty$, and exit the algorithm.

In all remaining cases, we set $\begin{cases} x_3 \equiv m^2 - x_1 - x_2 \\ y_3 \equiv m(x_1 - x_3) - y_1 \end{cases}$ (mod p) and output $P_3 = (x_3, y_3)$.

Example 12.6

(a) Use Algorithm 12.2 to find the sum of the points $P_1 = (2, 4)$ and $P_2 = (5, 0)$ on the modular elliptic curve E defined by $y^2 \equiv x^3 + 7x + 5 \pmod{11}$.

(b) Use Algorithm 12.2 to compute $2P$, where $P = (44, 29)$ is on the modular elliptic curve E defined by $y^2 \equiv x^3 + 9$ (mod 907).

Solution: Part (a): We use the notation of the algorithm $P_1 = (2, 4) \triangleq (x_1, y_1)$, $P_2 = (5, 0) \triangleq (x_2, y_2)$, and $P_3 \triangleq (x_3, y_3)$ (that will be determined). Since working mod 11 we have $(x_2 - x_1)^{-1} \equiv (5-2)^{-1} \equiv 3^{-1} \equiv 4$, we may compute $m = (y_2 - y_1) \cdot (x_2 - x_1)^{-1} \equiv (0-4) \cdot 4 \equiv -16 \equiv 6$. Next, we may compute the coordinates of the sum:

$$\begin{cases} x_3 = m^2 - x_1 - x_2 \equiv 6^2 - 2 - 5 \equiv 29 \equiv 7 \\ y_3 = m(x_1 - x_3) - y_1 \equiv 6(2-7) - 4 \equiv -30 - 4 \equiv -34 \equiv 10 \end{cases}$$

(mod 11) so that $P_3 \equiv (7, 10)$.

Part (b): We set $P = (44, 29) \triangleq (x_1, y_1) \triangleq (x_2, y_2)$ and $P_3 \triangleq (x_3, y_3)$ (that will be determined). We use the extended Euclidean algorithm (Algorithm 2.2) to compute the inverse of $2y_1 \equiv 2 \cdot 29 \equiv 58$ to be 735(mod 907). Thus $m \equiv (3x_1^2 + a) \cdot (2y_1)^{-1} \equiv (3 \cdot 44^2 + 0) \cdot 735 \equiv 538 \pmod{p}$, and hence

$$\begin{cases} x_3 = m^2 - x_1 - x_2 \equiv 6^2 - 2 - 5 \equiv 29 \equiv 7 \\ y_3 = m(x_1 - x_3) - y_1 \equiv 6(2-7) - 4 \equiv -30 - 4 \equiv -34 \equiv 10 \end{cases}$$

(mod 11) so that $P_3 \equiv (7, 10)$.

Exercise for the Reader 12.5

Perform the following modular elliptic curve operations (using Algorithm 12.2), where the ambient elliptic curve is defined by $y^2 \equiv x^3 + x + 1 \pmod{11}$:

(a) $(4, 5) + (0, 10)$
(b) $2(4, 5)$

Since there is no geometric analogue for addition of points on a modular curve, it is no longer obvious that the sum of two points (as defined by this algorithm) will really belong to the same elliptic curve. But this indeed turns out to be the case; moreover, all of the nice properties of Proposition 12.1 remain valid in this setting (as do their corresponding algebraic proofs).

Theorem 12.5

Any nonsingular modular elliptic curve E is a finite (abelian) group under its addition operation.

We next discuss a very useful general group theoretic concept that we experienced in Chapter 8 in the setting of the multiplicative groups $\mathbb{Z}_n^{\times}$.

Definition 12.4

If G is a finite abelian group and $a \in G$, then the **order** of a in G, denoted as $\operatorname{ord}_G(a)$, is the smallest positive integer k such that:

- In additive group notation: $ka = 0$.
- In multiplicative group notation: $a^k = 1$.

Although it is not so obvious that such positive integers exist, the following general theorem provides assurance of this fact, as well as some precisions on the possible values of orders.

Theorem 12.6

If G is a finite group and $a \in G$, then the order of a in G must divide the number of elements in G; that is, $\operatorname{ord}_G(a) \mid |G|$.

A proof of this theorem can be found in any book on abstract algebra; see, for example, [Her-96] or [Hun-96]. Note that in the case $G = \mathbb{Z}_n^{\times}$, this theorem becomes $\operatorname{ord}_{\mathbb{Z}_n^{\times}}(a) \mid \phi(n)$, which we proved in Chapter 8 [except we denoted $\operatorname{ord}_{\mathbb{Z}_n^{\times}}(a)$ simply as $\operatorname{ord}_n(a)$].

Example 12.7

Compute $\operatorname{ord}_E(0, 4)$ where E is the elliptic curve defined by $y^2 \equiv x^3 + 7x + 5 \pmod{11}$ of Example 12.5.

Solution: In Example 12.5, it was found that $|E| = 16$. Thus, by Theorem 12.6, the only possibilities for $\operatorname{ord}_E(0, 4)$ are 2, 4,

8, or 16. Letting $P = (0, 4)$, we compute (using Algorithm 12.2): $2P = (3, 3)$, $4P = 2P + 2P = (9, 7)$, and $8P = 4P + 4P = (5, 0)$, and so it follows that $\text{ord}_E(0, 4) = 16$. (The reader may wish to check that $16P = \infty$.)

Exercise for the Reader 12.6

Compute the orders of the points $(0, 4)$ and $(2, 4)$ of the modular elliptic curve E of Example 12.4.

The point $P = (0, 4)$ on the elliptic curve of Example 12.6 had the maximum possible order; such a point is thus the analogue of primitive root in $\mathbb{Z}_p^\times$. The same proof that we gave in Chapter 8 (Proposition 8.6) shows that if P is any point on a nonsingular modular elliptic curve, and $k = \text{ord}_E(P)$, then the points $P, 2P, 3P, \cdots, kP = \infty$ are distinct points of E, and all higher multiples of P will cycle back through these multiples. Thus, if P has maximum possible order $|E|$, then $\{P, 2P, 3P, \cdots, |E|P = \infty\} = E$. While these analogues of primitive roots do not always exist in elliptic curves, it is always feasible to find points of high order, and such points will be the basis for generalizing the ElGamal types of cryptosystems into the setting of elliptic curves.

The Discrete Logarithm Problem on Modular Elliptic Curves

We first recall the **discrete logarithm problem for** $\mathbb{Z}_p^\times$: Given a primitive root $g \pmod{p}$, and given an integer $a \in \mathbb{Z}_p^\times$, the problem asks for the exponent j (that will be unique mod $p - 1$) for which $g^j \equiv a \pmod{p}$. Since modular elliptic curves do not always have analogues of primitive roots, we use the following more general formulation:

Definition 12.5

If $p > 3$ is a prime, E is a nonsingular elliptic curve mod p, and $P, Q \in E$, the (**elliptic curve**) **discrete logarithm problem** is to find a positive integer m for which $Q = m \cdot P$, if such an integer exists. In case a solution exists, any positive integer that works is called a **discrete logarithm** of Q in the **base** P.

Just as with the modular integer version, the discrete logarithm problem for elliptic curves is an intractable problem. In fact, the elliptic curve problem is more difficult than the mod p version due to the bizarre behavior of elliptic curve addition. One important difference is that when working with integers, there is a notion of size—for example, the number of modular digits. When two small modular integers are multiplied (say, two 12-digit integers mod a 1000-digit number), the result will be small. But with elliptic curve addition, there is no way to distinguish between small and large points. One can have, for example, two points with small coordinates on an elliptic curve whose sum has very large coordinates. In short, modular integer multiplication is much easier to understand than elliptic curve point addition. All that is needed is to have an elliptic curve E and a point of high order; such parameters can always be constructed, as we discuss later.

Example 12.8

Letting E be the elliptic curve defined by $y^2 \equiv x^3 + 7x + 5$ (mod 11), solve the discrete logarithm problem $(7, 1) = m \cdot (0, 4)$.

Solution: We use the brute-force approach: we compute $2 \cdot (0, 4)$, $3 \cdot (0, 4)$, and so forth, until we obtain the element $(7, 1)$. From the previous example, we know that $\text{ord}_E((0, 4)) = 16$, so there will be a solution. In the worst case, we will need to compute up to $15 \cdot (0, 4)$ [since $16 \cdot (0, 4) = \infty$]. Using Algorithm 12.2, we obtain: $2 \cdot (0, 4) = (3, 3)$, $3 \cdot (0, 4) = 2 \cdot (0, 4) + (0, 4) = (2, 4)$, $4 \cdot (0, 4) = 3 \cdot (0, 4) + (0, 4) = (9, 7)$, $5 \cdot (0, 4) = 5 \cdot (0, 4) + (0, 4) = (7, 1)$, so we can stop; $m = 5$ is the desired discrete logarithm.

An Elliptic Curve Version of the Diffie–Hellman Key Exchange

We first review the original Diffie–Hellman key exchange algorithm (Algorithm 9.1) and then illustrate the corresponding modifications needed for the elliptic curve version using a table. To create a common secret key, Alice and Bob choose a large prime number p and a primitive root g (mod p); these can be made public. Then Alice and Bob each choose (preferably randomly) a private integer, a or b, respectively, in the range $1 \le a, b < p - 1$. Alice and Bob each compute their modular power $A \equiv g^a$ or $B \equiv g^b$ (mod p) and send it to each other. Alice and Bob then compute the common key K now as $K \equiv B^a$ and $K \equiv A^b$, respectively. Table 12.3 illustrates the changes that are needed to turn this into an elliptic curve algorithm.

Before giving an example, we make a few comments about the elliptic curve version. As with the traditional version, the large primes can be randomly generated, as explained in Chapter 8. The elliptic curve E, $y^2 \equiv x^3 + ax + b$ (mod p), can then be randomly generated by generating mod p integers for its coefficients a and b.* The nonsingular condition $4a^3 + 27b^2 \not\equiv 0$ (mod p) needs to be checked. If it fails, just generate a new set of parameters for E—but with a large prime p; chances are very small this condition will fail. Since elliptic curves need not have points of maximum possible order $|E|$, we aim only to generate a point G of high order. One quick way to generate a point G on E is to first randomly generate a mod p integer as a potential x coordinate x_1. Then we compute $x_1^3 + ax_1 + b \pmod{p}$ and check to see if it has a square root y_1. If we

---

* We have an inevitable collision of notation: the a and b in the elliptic curve version are coefficients of the elliptic curve, while in the mod p version, they represent Alice's and Bob's secret parameters.

TABLE 12.3 Corresponding Elements for Turning the Standard ($\mathbb{Z}_p$) Diffie–Hellman Key Exchange (Algorithm 9.1) into an Elliptic Curve Version

	$\mathbb{Z}_p$ **Version**	**Elliptic Curve Version**
Base parameters (public)	$p =$ a large prime number	$p =$ a large prime number
	$g =$ a primitive root (mod p)	$E =$ an elliptic curve (mod p)
		$G =$ a point of E of high order
Secret parameters:	(preferably random)	(preferably random)
Alice:	$1 \leq a < p-1$	$\sqrt{p} < n_A < p - \sqrt{p}$
Bob:	$1 \leq b < p-1$	$\sqrt{p} < n_B < p - \sqrt{p}$
Nonsecret parameters:		
Alice:	$A \equiv g^a$ (mod p)	$A \equiv n_A G$
Bob:	$B \equiv g^a$ (mod p)	$B \equiv n_B G$
Obtain common key:		(key is x coordinate of K)
Alice:	$K \equiv B^a$ (mod p)	$K \equiv n_A B$
Bob:	$K \equiv A^b$ (mod p)	$K \equiv n_B A$

restrict our prime p to satisfy $p \equiv 3 \pmod 4$, this square root check can be done quickly using Equation 12.6. Since any integer can have at most two square roots (mod p), it follows from Hasse's theorem that roughly half the time, there will be a square root y_1. We take $G = (x_1, y_1)$. We would like G to have high order so that the corresponding elliptic curve discrete logarithm problem will be hard to solve (this will make the system secure). Specific implementation details on how to find points of high order on an elliptic curve are a bit beyond the scope of this chapter, but we provide an outline of one such scheme. Without knowing the number of points on an elliptic curve, it is generally difficult to compute orders of specific points. If we know the number of points on the elliptic curve E, then the problem can be made simpler by Theorem 12.6. In particular, in case there is a prime number of points on E, then it follows that every point different from the point at infinity will have the maximum possible order equal to $|E|$. There are efficient algorithms for counting points on elliptic curves. The first such polynomial time algorithm is known as *Schoof's algorithm* and was discovered in 1985. It was later embellished by other mathematicians. So one possible scheme for generating elliptic curves with points of high order is to first fix a large prime p (of desired size) and continue the process of randomly generating elliptic curves mod p and checking their orders using a Schoof-type algorithm. (The number of points on a randomly generated elliptic curve mod p tend to be uniformly distributed in the range specified by Waterhouse's theorem.) These orders can be tested for primality using one of the tests of Chapter 8. Once an elliptic curve of prime order is found, any point different from the point at infinity will do. We refer the reader to [HaMeVa-04] and to [Was-03] for details. In the selection of secret parameters, the bounds on n_A, n_B were used to ensure they would be less than $|E|$ by Hasse's theorem, and randomly selected from a large range of numbers. The reason that Alice's and Bob's common keys are equal is because $n_A B \equiv n_A n_B G \equiv n_B n_A G \equiv n_B A \pmod p$. The resulting key K will be a point of E, and thus a pair of modular integers. We used the x coordinate (modular) integer of this vector as the actual key, but any other deterministic choice could work—for example, the y coordinate or the sum (or product) of these two coordinates.

We next give a "small" parameter example to help better illustrate the concepts. The exercises in the computer implementation material at the end of the chapter will allow the reader to work with much larger examples.

Example 12.9

Let p be the prime number 11027.

(a) Randomly generate the coefficients a and b to determine an elliptic curve E defined by $y^2 \equiv x^3 + ax + b \pmod{p}$, and check that it is nonsingular. (Repeat if necessary.)

(b) Randomly generate a point G on E. (For this example, do not be concerned about the order G).

(c) Suppose that Alice chooses her secret parameter to be $n_A = 32$, and Bob takes his to be $n_B = 21$. (These are intentionally made smaller than what was recommended in Table 12.3 for ease of illustration.) Compute their resulting private Diffie–Hellman key created by the elliptic curve protocol of Table 12.3 in both ways.

Solution: Part (a): We randomly generated $a = 4601$ and $b = 548$. We compute $4a^3 + 27b^2 \equiv 9142 \not\equiv 0 \pmod{p}$ so that the elliptic curve $y^2 \equiv x^3 + 4601x + 548 \pmod{p}$ is nonsingular.

Part (b): We randomly generate an x coordinate as a mod p integer: $x_1 = 9954$. We then substitute this into the right side of the elliptic equation to obtain $r \triangleq x_1^3 + 4601x_1 + 548 \equiv 4618$ (mod p). Since $p \equiv 3 \pmod 4$, we use Equation 12.6 to compute a possible square root of r: $y_1 \equiv r^{(p+1)/4} \equiv 8879 \pmod{p}$ (we used fast modular exponentiation). We then check if this is actually a square root of r by squaring it (mod p). Since this turns out to be the case, we have obtained a point $G = (x_1, y_1) = (9954, 8879) \in E$. By the way, $\mathrm{ord}_E(G) = 1099$.

Part (c): We first compute $A \equiv n_A G \equiv 32 \cdot G$. The fastest approach is to keep doubling:

$$2G \equiv G + G \equiv (4023, 9690)$$

$$4G \equiv 2G + 2G \equiv (9395, 4193)$$

$$8G \equiv 4G + 4G \equiv (10151, 6531)$$

$$16G \equiv 8G + 8G \equiv (8344, 5801)$$

$$A \equiv 32G \equiv 16G + 16G \equiv (2652, 8449)$$

These computations can help us now to compute (although Alice should not share them with Bob—Why?):

$$B \equiv n_B G \equiv 21 \cdot G \equiv 16G + 4G + G \equiv (530, 2745) + G \equiv (202, 3553)$$

To compute the shared private key, Alice would compute $K \equiv n_A B \equiv 32 \cdot B$. We use the same approach as before:

$$2B \equiv B + B \equiv (943, 3104)$$

$$4B \equiv 2B + 2B \equiv (933, 2778)$$

$$8B \equiv 4B + 4B \equiv (2960, 45)$$

$$16B \equiv 8B + 8B \equiv (1009, 7768)$$

$$K \equiv 32B \equiv 16B + 16B \equiv (8814, 8359)$$

On Bob's end, he would compute $K \equiv n_B A \equiv 21 \cdot A$, and he could proceed as follows:

$$2A \equiv A + A \equiv (7962, 7796)$$

$$4A \equiv 2A + 2A \equiv (6913, 5918)$$

$$8A \equiv 4A + 4A \equiv (4308, 3378)$$

$$16A \equiv 8A + 8A \equiv (3892, 10524)$$

$$K \equiv 21A \equiv 16A + 4A + A \equiv (8979, 9865) + A \equiv (8814, 8359)$$

Both Alice and Bob have obtained the same vector K, from which they can read off their secret key as the first component: key = 8814.

Fast Integer Multiplication of Points on Modular Elliptic Curves

Notice that in performing the integer multiplications of points in the preceding example, we used an efficient strategy that was analogous to the fast modular exponentiation algorithm of Chapter 6. Indeed, this fast modular exponentiation algorithm (Algorithm 2.2) for computing large modular powers is easily adapted into the following algorithm (by simply converting the multiplicative notation to additive notation). Just as fast modular exponentiation was important in the efficient implementation of the mod integer versions of the Diffie–Hellman and ElGamal cryptosystems, this algorithm will serve as the computational basis for this and other elliptic curve cryptosystems.

Algorithm 12.3: Fast Integer Multiples of Points on Elliptic Curves

Input: A nonsingular modular elliptic curve E, a point $P \in E$, and a positive integer multiplier x.

Output: The point $Q = x \cdot P \in E$.

Step 1.　Use Algorithm 6.1 to create the binary expansion of the exponent x: $x \sim [d_K \ d_{K-1} \ \cdots d_1 \ d_0]$ (base 2)

Step 2.　<Repeatedly double the point P as we run through the binary digits d_k of x, including the result in the cumulative sum only when $d_k = 1$>

Set $Q = \infty$ <Initialize cumulative sum Q>
Set $D = P \pmod{m}$ <Initialize doubling>
FOR $k = 0$ TO K

 IF $d_k = 1$
  Update $Q \rightarrow Q + D$
 END <IF>
  Update $D \rightarrow 2D$ <Doubling need not be done when $k = K$>
 END <k FOR>

Step 3. Output: Q.

Exercise for the Reader 12.7

Let p be the prime number 251.

(a) Let E be the elliptic curve defined by $y^2 \equiv x^3 + 196x + 98 \pmod{p}$. Verify that E is nonsingular. Then, noting that $p \equiv 3 \pmod 4$, use Equation 12.6 to see whether the x coordinate $x_1 = 28$ (which we randomly generated) gives rise to a point on E. If it does, let y_1 denote the resulting y coordinate with the positive sign in Equation 12.6, and let $G = (x_1, y_1) \in E$.

(b) Suppose that Alice chooses her secret parameter to be $n_A = 9$, and Bob takes his to be $n_B = 16$. (These are intentionally made smaller than what was recommended in Table 12.3 for ease of illustration.) Compute their resulting private Diffie–Hellman key created by the elliptic curve protocol of Table 12.3 in both ways.

Representing Plaintexts on Modular Elliptic Curves

The ideas that were used to extend the Diffie–Hellman key exchange to the elliptic curve setting allow us to do the same for the ElGamal cryptosystem of Chapter 9. But there is one issue that first needs to be ironed out: How do we represent plaintexts on an elliptic curve? This issue was completely straightforward in the traditional ElGamal setting since the plaintext space was $\mathbb{Z}_p$, and any plaintext can be represented using blocks of mod p integers. The points of a given elliptic curve mod p, on the other hand, have coordinates that are much less predictable and need not include any given specific mod p integer that appears in our plaintext. Moreover, it would be inefficient to compute all of the points on a given elliptic curve to help us determine a reasonable plaintext space. We will give a very efficient probabilistic algorithm that will be able to effectively represent plaintext through points on an elliptic curve. The algorithm succeeds with probability less than any specified value; for example, we could stipulate that we wanted the failure rate to be less than 10^{-20} percent, which is adequate for all practical purposes. The following probabilistic algorithm is due to Neal Koblitz; it is based on the fact that, given a nonsingular elliptic curve E defined by $y^2 \equiv x^3 + ax + b$ modulo a prime $p > 3$, roughly half of the integers mod p will have square roots, so if we want to represent a given plaintext message* m as an x coordinate of a point on E, there is only about a 50 percent chance that this will be possible, that is, that $m^3 + am + b$ will have a square root (mod p). For a given positive integer K that satisfies $(m+1)K < p$, we can check through the list of integers $Km, Km+1, Km+2, \cdots, Km+(K-1)$ to see if any one can appear as an x coordinate of a point in E. If any such integer so appears, we can unambiguously use it to represent our plaintext m, because $m = \text{floor}([Km+i]/K)$, whenever $0 \le i < K$. Since the probability that any one of these integers will fail is 1/2, the probability that all such attempts will fail may be estimated as $(1/2)^K$. This can be made as small as we like by choosing K to be sufficiently large. Of course, larger values of K necessitate larger values for the prime p, but this fact will not be important in real applications since the primes used will typically have hundreds of digits. The algorithm assumes that $p \equiv 3 \pmod 4$, so that Equation 12.6 may be used to efficiently compute square roots.

Algorithm 12.4: Koblitz's Algorithm for Plaintext Representations on an Elliptic Curve mod p

Input: A prime $p > 3$ satisfying $p \equiv 3 \pmod 4$, an elliptic curve E (mod p) with equation $y^2 \equiv x^3 + ax + b$, a positive integer K (error tolerance parameter), and a nonnegative integer m (the plaintext) that satisfies $(m+1)K < p$.

Output: The point $P \in E$, whose x coordinate is of the form $Km+i$, with $0 \le i < K$, or a failure message.

Note: The estimated failure rate for this algorithm is $(1/2)^K$.

Step 1.   Initialize $i = 0$.

Step 2.   Set $x = Km+i$, $r \equiv x^3 + ax + b \pmod p$, and use fast modular exponentiation (Algorithm 6.5) to compute $y \equiv r^{(p+1)/4}$ (mod p).

Step 3.   Check whether $y^2 \equiv r \pmod p$. If it is, output $P = (x, y)$, and exit the algorithm. If it is not, but $i < K - 1$, update $i \to i+1$ and return to Step 2. If it is not and $i = K - 1$, output failure message and exit algorithm.

Example 12.10

Let p be the prime number 307, and let E be the elliptic curve $y^2 \equiv x^3 + 22x + 153 \pmod p$.

---

* In previous chapters we usually represented plaintext messages with an uppercase P. But since in the context of elliptic curves, P typically is used to represent a generic point on an elliptic curve, we will use a lowercase m as the generic symbol for a plaintext message for the remainder of this chapter.

(a) Suppose that we wish to use Koblitz's Algorithm 12.4 to represent a positive integer plaintext m and that we would like the failure rate to be less than $1/1000$. Find the smallest parameter K in the algorithm that will achieve this. For this value of K, what is the range of plaintext values for m for which the algorithm can be applied?

(b) Apply Algorithm 12.4 using the value of K found in part (a), and with $m = 22$.

Solution: Part (a): Since $2^{-9} \approx 0.0020$, and $2^{-10} \approx 0.00098$, we would use $K = 10$. The equation $(m+1)K < p$ thus implies $(m+1)10 < 307 \Rightarrow 10m < 297 \Rightarrow m < 29.7$, so the range of admissible values for m is $0 \le m \le 29$.

Part (b):

Step 1. Initialize $i = 0$.

Step 2. Set $x = Km + i = 10 \cdot 22 + 0 = 220$, $r \equiv 220^3 + 22 \cdot 220 + 153 \equiv 93 \pmod{p}$, and use fast modular exponentiation (Algorithm 6.5) to compute $y \equiv r^{(p+1)/4} \equiv 93^{77} \equiv 287 \pmod{p}$.

Step 3. Since $y^2 \equiv 93 \equiv r \pmod{p}$, we have found a plaintext representative on E in just one iteration; we output $P = (x, y) = (220, 287)$, and exit the algorithm.

Exercise for the Reader 12.8

Repeat the instructions tasks of Example 12.10, but with the desired failure rate to be less than $1/500$, and with the prime changed to $p = 523$, but with all other parameters (including the elliptic curve equation) the same as in the example.

An Elliptic Curve Version of the ElGamal Cryptosystem

We first review the original ElGamal algorithm (Algorithm 9.4) and then illustrate the corresponding modifications needed for the elliptic curve version using a table. Alice needs to send Bob a message m. They publicly agree on the system parameters consisting of a large prime p and a primitive root g mod p. Alice and Bob each choose (preferably randomly) a private key integer, a or b, respectively, in the range $1 \le a, b < p - 1$. Alice and Bob each compute the modular power, $A \equiv g^a$ or $B \equiv g^b \pmod{p}$, which will be his or her public key. The ciphertext that Alice sends to Bob will be the ordered pair (A, C), where $C \equiv B^a m \pmod{p}$, which Alice computes using Bob's public key B. Bob may decrypt this message by computing $A^{p-1-b}C \pmod{p}$, which will be the original plaintext message m. Table 12.4 illustrates the changes that are needed to turn this into an elliptic curve algorithm.

TABLE 12.4 Corresponding Elements for Turning the Standard ($\mathbb{Z}_p$) ElGamal Cryptosystem (Algorithm 9.4) into an Elliptic Curve Version

	$\mathbb{Z}_p$ **Version**	**Elliptic Curve Version**
Base parameters (public)	$p =$ a large prime number	$p =$ a large prime number
	$g =$ a primitive root (mod p)	$E =$ an elliptic curve (mod p)
		$G =$ a point of E of high order
Secret parameters:	(preferably random)	(preferably random)
Alice:	$1 \leq a < p-1$	$\sqrt{p} < n_A < p - \sqrt{p}$
Bob:	$1 \leq b < p-1$	$\sqrt{p} < n_B < p - \sqrt{p}$
Nonsecret parameters:		
Alice:	$A \equiv g^a \pmod{p}$	$A \equiv n_A G$
Bob:	$B \equiv g^a \pmod{p}$	$B \equiv n_B G$
Encryption of plaintext message m from Alice to Bob:	Ciphertext: (A, C), where $C \equiv B^a m \pmod{p}$	First use Algorithm 12.4 to find a point $P \in E$ representing m. Ciphertext: (A, C), where $C = P + n_A B$
Decryption at Bob's end:	$m \equiv A^{p-1-b} C \pmod{p}$	$P = C - n_B A$ As explained in Algorithm 12.4, m can be easily obtained from P

Similar comments regarding the parameters that were made about the elliptic curve version of the Diffie–Hellman key exchange apply here. The following computation shows that the decryption scheme actually works:

$$C - n_B A = P + n_A B - n_B A = P + K - K = P$$

where we have used the fact that $n_A B$ and $n_B A$ both equal the Diffie–Hellman key K. The following is an example with small parameters to illustrate this concept.

Example 12.11

Let $p = 307$, let E be the (nonsingular) elliptic curve $y^2 \equiv x^3 + 22x + 153 \pmod{p}$, and let $P = (220, 287)$ be the plaintext representative point that was found in Example 12.10.

(a) Randomly generate a point G on E. (For this example, do not be concerned about the order G.)

(b) Suppose that Alice chooses her secret parameter to be $n_A = 32$, and Bob takes his to be $n_B = 54$. Go through the ElGamal encryption process that Alice would need to do to send Bob her message that is represented by the point P. What is the ciphertext?

(c) Go through the ElGamal decryption procedure that would need to be done at Bob's end to decrypt Alice's message.

Solution: Part (a): We randomly generate an x coordinate as a mod p integer: $x_1 = 167$. We then substitute this into the right side of the elliptic curve equation to obtain $r \triangleq x_1^3 + 22x_1 + 153 \equiv 109 \, (\text{mod} \, p)$. Since $p \equiv 3 (\text{mod} \, 4)$, we use Equation 12.6 to compute a possible square root of r: $y_1 \equiv r^{(p+1)/4} \equiv 118 (\text{mod} \, p)$ (we used fast modular exponentiation). We then check if this actually is a square root of r by squaring it (mod p). Since this turns out to be the case, we have obtained a point $G = (x_1, y_1) = (167, 118) \in E$.

Part (b): To encrypt, Alice needs Bob's public key, which he has been able to compute using Algorithm 12.3 to be $B = n_B G = 54G = (188, 55)$. Following the encryption procedure of Table 12.4, Alice multiplies Bob's public key number B by her secret parameter n_A (using Algorithm 12.3) to obtain $n_A B = (204, 265)$. Then (using Algorithm 12.2) she adds her plaintext point to this point to obtain $C = P + n_A B = (24, 136)$. Alice also needs to compute her public key, which could be computed using Algorithm 12.3 to be $A = n_A G = 32G = (30, 45)$. The resulting elliptic curve ElGamal ciphertext that Alice sends to Bob is the pair of points $(A, C) = ((30, 45), (24, 136))$.

Part (c): To decrypt, Bob would first compute (using Algorithm 12.3) $n_B A = (204, 265)$ (this is the Diffie–Hellman key). He would then subtract this from C (the second point of Alice's ciphertext), by using Algorithm 12.2 to add $-n_B A = (204, -265) \equiv (204, 42)$ to C: $C - n_B A = (204, -265) \equiv (220, 287) = P$, as was anticipated.

Exercise for the Reader 12.9

Repeat the instructions tasks of Example 12.11, but with the prime changed to $p = 523$, the plaintext representative point P to be what was found in the solution of Exercise for the Reader 12.8, but with all other parameters (including the elliptic curve equation) the same as in the example. For part (a), try this sequence of values of x_1 (in this order): 395, 402, 195 in the "random" generation of the point G. (These three mod p integers were randomly generated; at least one will work.)

This program of extending integer-based cryptosystems whose security is based on the discrete logarithm problem can be continued in the elliptic curve setting. For example, the ElGamal digital signature scheme (Algorithm 9.5) can also be adapted for elliptic curves. The interested reader may find further such examples in [Sti-06] and [TrWa-06].

A Factoring Algorithm Based on Elliptic Curves

Elliptic curves can be used in a very elegant and effective way in a factoring algorithm. This algorithm nicely complements and is more robust and effective than Pollard's $p - 1$ algorithm (Algorithm 8.5). We first explain how and why this ingenious algorithm of Hendrik Lenstra works, then we state it formally and give a specific example.

Suppose that we wish to factor a positive integer n that we suspect is composite. Although the algorithm works under general circumstances, to simplify this explanation we will assume that $n = pq$ is a product of distinct (large) primes, a particularly notorious sort of composite integer that is used in RSA. Although modular elliptic curves were defined only for prime moduli, we will temporarily "pretend" that n is prime, and we randomly generate a nonsingular elliptic curve E mod n. An easy way to do this would be to first randomly generate three mod n integers x, y, and a, and then determine the mod n integer b by the equation $y^2 \equiv x^3 + ax + b \pmod{n}$. When we are working with large values of p, q, and n (as is typical in difficult factorization problems), it will almost always be the case that such randomly generated curves are nonsingular, so we will not concern ourselves with such verifications. Thus, we have randomly generated an elliptic curve E mod n and a point $P = (x, y)$ on E. Actually, using the Chinese remainder theorem, E can be defined as a pair of elliptic curves mod p and mod q. We temporarily denote these latter two (true) elliptic curves as E_p and E_q. The algorithm involves "attempting" to compute the scalar multiple $B!P$ (using Algorithm 12.3), where B is a suitable positive integer. The reason why this might lead us to a factor of n is because the point P may have different orders in E_p and in E_q. If these orders are indeed different, and one of them does not have any large prime factors, then in the computation of multiples kP (in the chain of computations leading to $B!P$), it will probably happen that we will get $kP = \infty$ in one of E_p or E_q, and a finite point in the other. Now, although we are only doing our calculations mod n, this means that in the corresponding addition process (with Algorithm 12.2) we will have a nontrivial gcd in trying to compute the (slope) parameter m. This corresponds to m being infinite in one of E_p or E_q, and finite in the other, and in this case the denominator used to compute m will be a nontrivial factor (either p or q) of n. Of course, if the orders of P in both E_p and E_q each have large prime factors, then this method will not produce a factorization—but it can be repeated any number of times using newly generated elliptic curves. Herein lies the main (theoretical and practical) advantage over Pollard's $p - 1$ algorithm. In order to work, the latter algorithm needed $n - 1$ to be free of large prime factors. By Waterhouse's theorem (Theorem 12.4), the number of points on randomly generated elliptic curves can lie anywhere in the range $p + 1 \pm 2\sqrt{p}$, and it can furthermore be shown that these numbers tend to be uniformly distributed in this range when the curves are randomly generated. From these facts and from Theorem 12.6, it follows that this elliptic curve factorization algorithm requires only that n be near some integers that do not have any large prime factors. This is a much less stringent requirement than was needed in Pollard's $p - 1$ algorithm. We formally state this algorithm, and then give a specific example.

Algorithm 12.5: Lenstra's Algorithm for Factorization Using Elliptic Curves

Input: An odd composite integer $n > 3$, a positive integer B, and a positive integer K (number of trials).

Output: Either a nontrivial factor of n or no output in case the algorithm does not find one.

Step 0.   *Initialize Trial Counter.* Initialize $i = 1$.

Step 1.   *Generate Elliptic Curve and Point.* Randomly generate three mod n integers x, y, and a, and then choose the mod n integer b by the equation for an elliptic curve E: $y^2 \equiv x^3 + ax + b \pmod{n}$. Set $P = (x, y)$.

Step 2.   Use Algorithm 12.3 in conjunction with Algorithm 12.2 to compute $B!P \pmod{n}$ through the following sequence of steps: $1!P = P$, $2!P = 2(1!P), 3!P = 3(2!P), \cdots B!P = B([B-1]!P)$. At each intermediate application of Algorithm 12.2 in this process, keep track of $d = \gcd(\mathrm{Den}(m), n)$, where $\mathrm{Den}(m)$ is the denominator of m in the algorithm [so either $(x_2 - x_1)$ or $2y_1$]. If this ever turns out to give a nontrivial factor of n, i.e., if $1 < d < n$, then output d as a nontrivial factor and exit the algorithm.

Step 3.   Update $i \rightarrow i + 1$. If $i < K$, return to Step 2; otherwise, exit the algorithm.

We give a "small" example to illustrate this algorithm.

Example 12.12

Apply Lenstra's elliptic curve factoring algorithm with $B = 10$, and $K = 3$ to the integer $n = 345, 283$.

Solution: With $i = 1$, we randomly generate the following integers (mod n): $x = 325604$, $y = 236075$, $a = 275656$. Solving $y^2 \equiv x^3 + ax + b$ for b (mod n) produces $b = 290844$. We now proceed through the sequential computation of $10!P$, where $P = (x, y)$: We use Algorithm 12.3, but we stop if one of the denominators $\mathrm{Den}(m)$ is ever strictly between 1 and n. The computations $2!P, 3!P, 4!P$ all go through, but a nontrivial factor will be detected in the process of computing $5!P$ form $4!P = (201323, 330647)$. This factor 487 is a factor of n, and allows us to factor $n = 487 \cdot 709$.

Computing Note: Computer implementations of Lenstra's algorithm would be needed to effectively demonstrate its superiority over that of Pollard, or even a brute-force algorithm. Due to limitations of floating point arithmetic accuracy, it would be necessary to have symbolic capabilities on the computing platform in order to apply it to integers with more than, say, seven digits. Before applying any factoring algorithm to a given integer n, it is best to first apply one of the efficient primality tests of Chapter 8 (such as Miller–Rabin) to test whether n is prime (and hence already factored).

Chapter 12 Exercises

Note: In a few of the computational exercises involving elliptic curves over real numbers, some answers may involve noninteger real numbers.

In such cases, display all answers to four decimals, but use the default storage for subsequent machine computations.

1. Let E be the elliptic curve over the real numbers defined by $y^2 = x^3 + 8$, and let $P_1 = (1, 3)$ and $P_2 = (2, 4)$ (two points on E).
   (a) Compute $P_1 + P_2$ by using the algebraic formulation of Algorithm 12.1 (as in Example 12.3).
   (b) Compute $P_1 + P_2$ by using the geometric formulation of Algorithm 12.1 (as in Example 12.2).

2. Let E be the elliptic curve over the real numbers defined by $y^2 = x^3 + 2x + 3$, and let $P_1 = (-1, 0)$ and $P_2 = (3, 6)$ (two points on E).
   (a) Compute $P_1 + P_2$ by using the algebraic formulation of Algorithm 12.1 (as in Example 12.3).
   (b) Compute $P_1 + P_2$ by using the geometric formulation of Algorithm 12.1 (as in Example 12.2).

3. Let E be the elliptic curve over the real numbers defined by $y^2 = x^3 + 1$, and let $P = (0, 1)$ (a point on E).
   (a) Compute $2P$.
   (b) Compute $3P, 4P, 5P, 6P, 7P$.
   (c) Based on your calculations in parts (a) and (b), give a formula for nP, whenever n is a positive integer.

4. Let E be the elliptic curve over the real numbers defined by $y^2 = x^3 + 1$, and let $P = (2, 3)$ (a point on E).
   (a) Compute $2P$.
   (b) Compute $3P, 4P$.

5. For each prime modulus given below, let E be the elliptic curve defined by $y^2 = x^3 + 8 \pmod{p}$. Find all points on E:
   (a) $p = 5$
   (b) $p = 7$
   (c) $p = 11$

6. For each prime modulus given below, let E be the elliptic curve defined by $y^2 = x^3 + 3x + 3 \pmod{p}$. Find all points on E:
   (a) $p = 5$
   (b) $p = 7$
   (c) $p = 11$

7. Let E be the modular elliptic curve defined by $y^2 = x^3 + 6x \pmod{13}$. Show that E is nonsingular and then perform the following arithmetic operations on E:
   (a) $(4, 7) + (0, 0)$
   (b) $(4, 7) + (5, 5)$
   (c) $-(4, 7)$
   (d) $(4, 7) + (4, 7)$

8. Let E be the modular elliptic curve defined by $y^2 = x^3 + 3x \pmod{17}$. Show that E is nonsingular and then perform the following arithmetic operations on E:
    (a) $(4, 5) + (0, 0)$
    (b) $(6, 9) + (11, 15)$
    (c) $-(4, 12)$
    (d) $(8, 14) + (8, 14)$

9. Let E be the modular elliptic curve defined by $y^2 = x^3 + 3x + 3 \pmod{5}$.
    (a) Show that E is nonsingular.
    (b) Find all points of E (including the point at infinity).
    (c) Create an addition table for the points of E.

10. Let E be the modular elliptic curve defined by $y^2 = x^3 + 3 \pmod{5}$.
    (a) Show that E is nonsingular.
    (b) Find all points of E (including the point at infinity).
    (c) Create an addition table for the points of E.

11. Let E be the modular elliptic curve defined by $y^2 = x^3 + 2x + 1 \pmod{11}$. Show that E is nonsingular and then perform the following arithmetic operations on E.
    (a) $2(0, 1)$
    (b) $3(0, 1)$
    (c) $6(0, 1)$
    (d) $9(0, 1)$

12. Let E be the modular elliptic curve defined by $y^2 = x^3 + 2x + 1 \pmod{11}$. Show that E is nonsingular and then perform the following arithmetic operations on E.
    (a) $2(6, 8)$
    (b) $4(6, 8)$
    (c) $8(6, 8)$
    (d) $12(6, 8)$

13. Let E be the modular elliptic curve defined by $y^2 = x^3 + 2x + 1 \pmod{11}$. Compute the following orders:
    (a) $\mathrm{ord}_E((0, 10))$
    (b) $\mathrm{ord}_E((3, 1))$
    (c) $\mathrm{ord}_E(\infty)$
    (d) $\mathrm{ord}_E((8, 10))$

14. Let E be the modular elliptic curve defined by $y^2 = x^3 + 2x + 1 \pmod{11}$. Compute the following orders:
    (a) $\mathrm{ord}_E((3, 1))$
    (b) $\mathrm{ord}_E((1, 2))$
    (c) $\mathrm{ord}_E(\infty)$
    (d) $\mathrm{ord}_E((0, 10))$

15. Let E be the modular elliptic curve defined by $y^2 = x^3 + 6x + 3 \pmod 7$. Let $P = (2, 3)$, and then solve the following discrete logarithm problems on E:
    (a) $(2, 4) = m \cdot P$
    (b) $(4, 0) = m \cdot P$
    (c) $(5, 2) = m \cdot P$
    (d) $(5, 5) = m \cdot P$

16. Let E be the modular elliptic curve defined by $y^2 = x^3 + x + 6 \pmod 7$. Let $P = (1, 1)$, and then solve the following discrete logarithm problems on E:
    (a) $(2, 4) = m \cdot P$
    (b) $(4, 5) = m \cdot P$
    (c) $(6, 2) = m \cdot P$
    (d) $(1, 6) = m \cdot P$

17. Let E be the nonsingular modular elliptic curve defined by $y^2 = x^3 + 84x \pmod{269}$. Compute the following scalar multiples of the point $P = (18, 9) \in E$:
    (a) $10 \cdot P$
    (b) $56 \cdot P$
    (c) $135 \cdot P$
    (d) $402 \cdot P$

18. Let E be the nonsingular modular elliptic curve defined by $y^2 = x^3 + 84 \pmod{223}$. Compute the following scalar multiples of the point $P = (9, 12) \in E$:
    (a) $7 \cdot P$
    (b) $46 \cdot P$
    (c) $101 \cdot P$
    (d) $368 \cdot P$

19. Let p be the prime number 163, and let E be the elliptic curve defined by $y^2 \equiv x^3 + 22x + 153 \pmod p$.
    (a) Show that E is nonsingular. Then, noting that $p \equiv 3 \pmod 4$, use Equation 12.6 to see whether the x coordinate $x_1 = 28$ (which we randomly generated) gives rise to a point on E. If it does, y_1 denotes the resulting y coordinate with the positive sign in Equation 12.6, and let $G = (x_1, y_1) \in E$. If it does not, repeat with the values $x_1 = 94, 10$ until this construction produces such a point $G = (x_1, y_1) \in E$. (One of these will work.)
    (b) Suppose that Alice chooses her secret parameter to be $n_A = 19$, and Bob takes his to be $n_B = 41$. (These are intentionally made smaller than what was recommended in Table 12.3 for ease of illustration.) Compute their resulting private Diffie–Hellman key created by the elliptic curve protocol of Table 12.3 in both ways.
    (c) Repeat part (b) with $n_A = 192$ and $n_B = 94$.

20. Let p be the prime number 239, and let E be the elliptic curve defined by $y^2 \equiv x^3 + 39x + 58 \pmod{p}$.
    (a) Show that E is nonsingular. Then, noting that $p \equiv 3 \pmod 4$, use Equation 12.6 to see whether the x coordinate $x_1 = 134$ (which we randomly generated) gives rise to a point on E. If it does, y_1 denotes the resulting y coordinate with the positive sign in Equation 12.6, and let $G = (x_1, y_1) \in E$. If it does not, repeat with the values $x_1 = 3, 96$, until this construction produces such a point $G = (x_1, y_1) \in E$. (One of these will work.)
    (b) Suppose that Alice chooses her secret parameter to be $n_A = 18$, and Bob takes his to be $n_B = 52$. (These are intentionally made smaller than what was recommended in Table 12.3 for ease of illustration.) Compute their resulting private Diffie–Hellman key created by the elliptic curve protocol of Table 12.3 in both ways.
    (c) Repeat part (b) with $n_A = 78$, and $n_B = 152$.

21. (a) Let E be the elliptic curve $y^2 \equiv x^3 + 39x + 58 \pmod{p}$, where $p = 431$. Use Koblitz's algorithm (Algorithm 12.4) to represent the plaintext message $m = 13$, using $K = 10$.
    (b) Repeat part (a) changing p to be 5431, m to 89, and using $K = 50$.
    (c) Repeat part (a) changing p to be 72379, m to 244, and using $K = 100$.

22. (a) Let E be the elliptic curve $y^2 \equiv x^3 + 62x + 9 \pmod{p}$, where $p = 311$. Use Koblitz's algorithm (Algorithm 12.4) to represent the plaintext message $m = 30$, using $K = 10$.
    (b) Repeat part (a) changing p to be 5431, m to 46, and using $K = 50$.
    (c) Repeat part (a) changing p to be 72379, m to 356, and using $K = 100$.

23. Let $p = 439$, let E be the (nonsingular) elliptic curve $y^2 \equiv x^3 + 6x + 167 \pmod{p}$, and let $P = (312, 65)$ be the plaintext representative point.
    (a) Noting that $p \equiv 3 \pmod 4$, generate a point G on E by running through the x coordinates $x_1 = 38, 276, 61$, making use of Equation 12.6 (use the positive square root sign) until one is first found.
    (b) Suppose that Alice chooses her secret parameter to be $n_A = 24$, and Bob takes his to be $n_B = 71$. Go through the ElGamal encryption process that Alice would need to do to send Bob her message that is represented by the point P. What is the ciphertext?
    (c) Go through the ElGamal decryption procedure that would need to be done at Bob's end to decrypt Alice's message.
    (d) Repeat parts (a) and (b) using the following changes in secret parameters: $n_A = 89$, $n_B = 193$.

24. Let $p = 547$, let E be the (nonsingular) elliptic curve $y^2 \equiv x^3 + 6x + 167 \pmod{p}$, and let $P = (316, 521)$ be the plaintext representative point.

    (a) Noting that $p \equiv 3 \pmod 4$, generate a point G on E by running through the x coordinates $x_1 = 284, 341, 61$, making use of Equation 12.6 (use the positive square root sign) until one is first found.

    (b) Suppose that Alice chooses her secret parameter to be $n_A = 19$, and Bob takes his to be $n_B = 57$. Go through the ElGamal encryption process that Alice would need to do to send Bob her message that is represented by the point P. What is the ciphertext?

    (c) Go through the ElGamal decryption procedure that would need to be done at Bob's end to decrypt Alice's message.

    (d) Repeat parts (a) and (b) using the following changes in secret parameters: $n_A = 107$, $n_B = 150$.

25. In each part, a composite number n is specified. Apply Lenstra's elliptic curve factoring algorithm (Algorithm 12.5) using the parameters $B = 10$ and $K = 3$ with the aim of factoring n.

    (a) $n = 295{,}891$

    (b) $n = 1{,}544{,}927$

    (c) $n = 8{,}574{,}421$

    Note: It would be instructive for the reader to ignore the given fact that these integers are composite, and to apply one of the primality tests of Chapter 8 to ascertain these facts.

26. In each part, a composite number n is specified. Apply Lenstra's elliptic curve factoring algorithm (Algorithm 12.5) using the parameters $B = 10$ and $K = 3$ with the aim of factoring n.

    (a) $n = 288{,}619$

    (b) $n = 1{,}728{,}931$

    (c) $n = 11{,}064{,}199$

    Note: It would be instructive for the reader to ignore the given fact that these integers are composite, and to apply one of the primality tests of Chapter 8 to ascertain these facts.

27. Suppose that p is an odd prime number, $p > 3$ that satisfies $p \equiv 2 \pmod 3$.

    (a) Show that the function $f : \mathbb{Z}_p \to \mathbb{Z}_p$ defined by $f(x) \equiv x^3 \pmod p$ is a bijection.

    (b) Show that the number of points $|E|$ on the modular elliptic curve defined by $y^2 = x^3 + 1 \pmod p$ is exactly $p + 1$.

    Suggestion: For part (a), see Exercise 60 of Chapter 2. Use part (a) to prove part (b).

28. Without using Hasse's theorem, show that the number of points on $|E|$ on any modular elliptic curve E defined by $y^2 = x^3 + ax + b \pmod p$ is at most $2p + 1$.

29. Suppose that E is a nonsingular elliptic curve modulo a prime $p > 3$, and that the number of elements $|E|$ of E, is also a prime

number q. Show that for any $P \in E$, with $P \neq \infty$, we have $\operatorname{ord}_E(P) = q$.

30. Suppose that E is a nonsingular elliptic curve over the real numbers.
    (a) Give a geometric condition that is equivalent to $\operatorname{ord}_E(P) = 2$.
    (b) Give a geometric condition that is equivalent to $\operatorname{ord}_E(P) = 3$.
    (c) Give a geometric condition that is equivalent to $\operatorname{ord}_E(P) = 4$.

31. *Reflection Is Needed for Associativity.* Recall that in the geometric version of Algorithm 12.1, for adding two points $P_3 = P_1 + P_2$ on a nonsingular elliptic curve E over the real numbers, we first found the third point Q on the line L passing through P_1, P_2 and then reflected this point over the x axis to obtain the point P_3 (see Figure 12.3). Suppose that we define a new binary operation $\boxed{+}$ on E by using the same procedure, but skipping the reflection; i.e., we define $P_1 \boxed{+} P_2 \triangleq Q$.

    (a) Show this new binary operation $\boxed{+}$ is commutative.
    (b) Give an example to show that $\boxed{+}$ need not be associative. That is, find a specific nonsingular modular elliptic curve E and three points $P, Q, R \in E$ such that $(P \boxed{+} Q) \boxed{+} R \neq P \boxed{+} (Q \boxed{+} R)$.

Chapter 12 Computer Implementations and Exercises

Note: We reiterate a relevant principle that was stated in the computer implementation sections of Chapters 8 and 9: If your computing platform is a floating point arithmetic system, it may allow you only up to 15 or so significant digits of accuracy. Symbolic systems allow for much greater precision, being able to handle hundreds of significant digits. Some platforms allow users to choose if they wish to work in floating point or symbolic arithmetic. If you are working on such a dual-use platform, you may wish to create two separate programs for those who might work with large integers: an ordinary version and a symbolic version (perhaps attaching a Sym suffix to the names of those of the latter type). In case you do not have access to a symbolic system, some particular questions below may need to be skipped or modified so the numbers are of a manageable size.

1. *Adding Points on Nonsingular Elliptic Curves over the Real Numbers.*
   (a) Write a program with syntax SumPoints = Elliptic CurvePointAddition(P1, P2, a, b) whose inputs a, b are integers that determine a nonsingular elliptic curve $y^2 = x^3 + ax + b$ over the real numbers (so that $4a^3 + 27b^2 \neq 0$), and the first two inputs P1, P2 are length-2 vectors representing two points on this elliptic curve. The output, SumPoints, will be the length-2

vector representing the sum P1 + P2 on the elliptic curve computed using the algebraic formulation of Algorithm 12.1.

(b) Check the correctness of the program by testing it on Examples 12.2 and 12.3.

(c) Run your program on the calculations of Chapter Exercises 1 and 3.

(d) On the elliptic curve E determined by $y^2 = x^3 + 8$, find the smallest integer $x > 2$ such that there is another integer y with $P = (x, y) \in E$. Next, use your program to compute $2P$, $3P$, $5P$, and $100P$.

2. *Computing and Listing All Points on an Elliptic Curve mod p.*

(a) Write a program with syntax `Points = EllipticCurve PointsModp(a, b, p)` whose inputs a, b, and p are non-negative integers with $p > 3$ a prime and a, b $< p$. The output, `Points`, is a two-column matrix whose rows list all of the points (x,y) on the elliptic curve $y^2 \equiv x^3 + ax + b$ (mod p). The program will find the points by a brute-force search, running through all integers x mod p and trying to solve for corresponding y's. The final point in the list should be the point at infinity, listed, for example, as Inf, Inf (or replacing Inf by whatever the reader's computing platform uses to represent infinity).

(b) Check the correctness of the program by testing it on Examples 12.4 and 12.5.

(c) Run your program on the calculations of Chapter Exercises 1 and 3.

(d) Use your program to determine the number of points on the modular elliptic curve E defined by $y^2 = x^3 + 8$ (mod p), when
   (i) $p = 211$
   (ii) $p = 2003$
   (iii) $p = 20011$

3. *Computing and Listing all Points on an Elliptic Curve mod p, when $p \equiv 3 \pmod 4$.*

(a) Write a program with syntax `Points = EllipticCurve PointsModp3Mod4(a, b, p)` whose inputs/outputs and functionality are the same as that for the program of the previous computer exercise, except that now it is assumed that $p \equiv 3 \pmod 4$, and instead of using the brute-force approach, Equation 12.6 will be used (as in the solution of Example 12.5).

(b) Check the correctness of the program by testing it on Example 12.5.

(c) Use your program to determine the number of points on the modular elliptic curve E defined by $y^2 = x^3 + 8$ (mod p), when
   (i) $p = 2003$
   (ii) $p = 20011$
   (iii) $p = 200003$

4. *Adding Points on Nonsingular Modular Elliptic Curves.*

   (a) Write a program with syntax `SumPoints = Elliptic CurvePointAdditionModp(P1, P2, a, b, p)` whose inputs a, b, and p are nonnegative integers with $p > 3$ prime and b, $c < p$, and `P1`, `P2` are length-2 vectors giving the coordinates of two points on the elliptic curve $y^2 \equiv x^3 + ax + b$ (mod p). The output, `SumPoints`, is a length-2 vector of mod p integers giving the coordinates of the sum `P1` + `P2` (computed using Algorithm 12.2). In case either of the inputted points does not belong to the specified elliptic curve, the program should produce an error message.

   (b) Check the correctness of the program by testing it on Example 12.6.

   (c) Run your program on the calculations of Chapter Exercises 7 and 11.

   (d) On the modular elliptic curve E determined by $y^2 \equiv x^3 + 8 \pmod{907}$, let $P = (84, 21) \in E$, and use your program to compute $2P$, $3P$, $5P$, and $100P$.

5. *Program for Computing Orders on Modular Elliptic Curves.*

   (a) Write a program with syntax `Order = EllipticCurve PointsModpOrder(P, a, b, p)` whose inputs a, b, and p are nonnegative integers with $p > 3$ prime corresponding to a nonsingular modular elliptic curve E defined by $y^2 \equiv x^3 + ax + b$ (mod p). The input P is a length-2 vector of mod p integers corresponding to a point on E. The output, `Order`, is a positive integer representing the order of P in the abelian group E (mod p). The order is computed by brute force: simply continuing to compute the sum $P + P + P + \ldots$ (iteratively using the program of the previous computer exercise), until we obtain the identity (inf, inf).

   (b) Check the correctness of the program by testing it on Example 12.6.

   (c) Run your program on the calculations of Chapter Exercise 13.

   (d) On the modular elliptic curve E determined by $y^2 = x^3 + 42x + 35 \pmod{1201}$ find an element of maximum order. How many such elements does E have?

6. *Program for Computing Fast Integer Multiples of Points on Elliptic Curves.*

   (a) Write a program with syntax `Q = EllipticCurveFast ScalMult(P, x, a, b, p)` whose inputs a, b, and p are nonnegative integers with $p > 3$ prime corresponding to a nonsingular modular elliptic curve E defined by $y^2 \equiv x^3 + ax + b$ (mod p). The first input P is a length-2 vector of mod p integers corresponding to a point on E, and the second input x is a positive integer representing the scalar multiplier. The output Q is a length-2 vector of mod p integers representing the point $x \cdot P$ on the elliptic curve. The program uses the fast integer multiples algorithm (Algorithm 12.3)

   (b) Check the correctness of the program by testing it on Example 12.9.

(c) Run your program on the calculations of Chapter Exercise 17.

(d) Consider the point $P = (14615, 94962)$ that lies on the modular elliptic curve E determined by $y^2 = x^3 + 36621x + 61200 \pmod{151523}$. Compute $506985 \cdot P$.

7. *Program for the Elliptic Curve Diffie–Hellman Key Exchange.*

(a) Write a program that implements the elliptic curve-based Diffie–Hellman key exchange algorithm (as presented in Table 12.3).

(b) Run your program on Example 12.9 and on Chapter Exercise 19.

(c) Will your program effectively run on elliptic curves randomly generated with a 50-digit prime? Explain, and provide some experimental evidence to support your statement.

8. *Program for Plaintext Representations on an Elliptic Curve mod p Using Koblitz's Algorithm 12.4.*

(a) Write a program with syntax `PtextRep = EllipticCurvePlaintextRepModp(m, a, b, p, K)` whose inputs a, b, and p are nonnegative integers with $p > 3$ is prime satisfying $p \equiv 3 \pmod 4$, and $a, b < p$, specify a nonsingular elliptic curve $y^2 = x^3 + bx + c \pmod p$. The first variable, m, is a nonnegative integer specifying a plaintext message, and the last input variable, K, is a positive integer that specifies the number of trials to search for a representation point. It is assumed that $(m+1)K < p$. The output, `PtextRep`, is a length-2 vector giving the coordinates of a point on the elliptic curve that represents the plaintext, in the sense that $m = \text{floor}(x/K)$, where x denotes the first coordinate of `PtextRep`. The point is found using the Koblitz algorithm (Algorithm 12.4). In case the size condition $(m+1)K < p$ is not met, an error message should be produced indicating to use a larger prime p.

(b) Check the correctness of the program by testing it on some known examples, and examples that you have worked out or created by hand.

9. *Program for the Elliptic Curve ElGamal Cryptosystem.*

(a) Write programs that implement the elliptic curve-based ElGamal encryption and decryption (as presented in Table 12.4).

(b) Run your program on Example 12.11 and on Chapter Exercise 23.

(c) Will your program effectively run on elliptic curves randomly generated with a 50-digit prime? Explain, and provide some experimental evidence to support your statement.

10. *Program for the Lenstra's Elliptic Curve Factoring Algorithm.*

(a) Write a program that will perform Lenstra's elliptic curve factoring algorithm (Algorithm 12.5).

(b) Run your program on Example 12.12 and on Chapter Exercise 25.

(c) Do some experiments by applying your algorithm to products of two probable primes with d digits each (produce the primes using random generation with the Miller–Rabin test, as explained in Chapter 8, with error tolerance less than 1 in 1 billion), starting with $d = 5$ and working your way up through larger values in increments of 5: $d =$ 10, 15, 20, ... to test the effectiveness of this algorithm. Experiment with different values of B. Summarize and analyze the results of your experiments.

Note: This computer exercise is more substantial than average.

11. *Empirical Comparison of Pollard's $p - 1$ Factorization Algorithm with Lenstra's Elliptic Curve Factorization Algorithm.*

(a) Randomly generate 10 pairs of 10-digit probable primes (using the Miller–Rabin test, as explained in Chapter 8 with error tolerance less than 1 in 1 billion). Apply both Pollard's $p - 1$ algorithm (from Chapter 8) and Lenstra's elliptic curve algorithm (from this chapter) to each of the 10 composite numbers resulting from multiplying the two primes, for each pair generated. Use comparable values of the parameter B. Compare the success rate and run times. Summarize and analyze the results of your experiments.

(b) Repeat part (a), except that the probable primes generated should have 20 digits, rather than 10.

Note: This computer exercise is more substantial than average.

Appendices

Appendix A: Sets and Basic Counting Principles

A set is simply a collection of objects. The purpose of this appendix is to establish some standard notation of some basic set theoretic concepts and a few principles that are useful in counting the number of objects in a finite set. The treatment here is deliberately brief. More details and proofs concerning such topics can be found in any book on discrete structures, such as [Sta-11].

Concepts and Notations for Sets

A **set** S is a collection of objects; the objects in a set are called **elements** (or **members**) of the set.* The objects making up a set can be of any sort. For example, the collection of all students in this cryptography class (or in any particular class at any school) is a set. Sets are the building blocks of all discrete structures, including all cryptosystems.

We write $x \in S$ to denote that object x belongs to the set S, and $x \notin S$ to mean x does not belong to the set S. For example, if S denotes the set of all states of the United States, then Hawaii $\in S$, but Guam $\notin S$. Braces ({ }) are a common notation used to describe a set between which we give either a listing of all of the elements or a description of them. For example, we can write the set of the colors of the flag of the United States as {Red, White, Blue}. Listing the elements in a large incongruous set such as {all words in *Webster's Dictionary*} would not be feasible. An ellipsis (…) can be used (repeatedly) to replace obvious patterns; for example, the set of all *positive integers* can be written as {1, 2, 3, …}. Sets such as the positive integers that have a never-ending list of elements are called **infinite sets**, whereas sets like {Red, White, Blue} that contain a finite (ending) list of elements are called **finite sets**.

We stress that the order in which the elements of a set are listed is immaterial. To decide whether two sets are the same, one need only compare elements. Thus, {1, 2, 3, 4} = {2, 1, 4, 3} = {1, 2, 3, 1, 4}, but the duplicate listing as in the last version should be avoided. We next introduce two operations on pairs of sets that produce new sets.

Definition A.1

Let A and B be two sets.

(i)   The **union** of A and B, denoted $A \cup B$, is the set of all elements that belong to A or B (or both). In symbols, we can write $A \cup B = \{x \mid x \in A \text{ or } x \in B\}$.

---

* The reason that we do not enunciate this definition is that it is rather informal. Although it will serve all of our purposes, it can lead to subtle *paradoxes* (contradictions). For the formal rigorous definition of a set (and the related axiomatics), the interested reader would do well to consult the book by Enderton [End-77].

(ii)   The **intersection** of A and B, denoted $A \cap B$, is the set of all elements that belong to A and to B. In symbols, we can write $A \cap B = \{x \mid x \in A$ and $x \in B\}$.

Example A.1

The union and intersection of the two sets $A = \{2, 4, 6, 8, 10\}$ and $B = \{3, 6, 9, 12\}$ are $A \cup B = \{2, 3, 4, 6, 8, 9, 10, 12\}$ and $A \cap B = \{6\}$, respectively. We point out that a set with just one element in it, such as $A \cap B = \{6\}$, is called a *singleton set*.

The union and intersection are the set-theoretical equivalents of the logical disjunction ("OR") and conjunction ("AND") operators. In contrast with logical ideas, many concepts about sets can be visualized using so-called **Venn diagrams**. Venn diagrams represent sets by simple geometric shapes (such as circles) and use shading to represent any particular set in question. Such diagrams are created in Figure A.1 for the union and intersection.

Example A.2

Let A be the set of all female (undergraduates) at your (or some other) college, and let B denote the set of all mathematics majors at this same college. Describe in plain English the two sets $A \cup B$ and $A \cap B$. Next, suppose that Joey and Amy are mathematics majors at this college, Linda is an economics major also at this college, and Jane is a mathematics major at a different college (say Dartmouth College, assuming you are not at Dartmouth). Indicate the truth value of each set inclusion statement below and illustrate with a Venn diagram.

(a) Joey $\in A \cup B$
(b) Amy $\in A \cap B$
(c) Linda $\in A \cap B$
(d) Jane $\in A \cup B$

Solution: $A \cup B = \{$students at your college that are either females or math majors (or both)$\}$. $A \cap B = \{$female mathematics

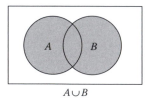

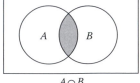

$A \cup B$ $A \cap B$

Figure A.1 Venn diagrams for the union ($A \cup B$) and intersection ($A \cap B$) of two sets A and B.

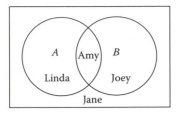

Figure A.2 A Venn diagram for Example A.2.

majors at your college}. The Venn diagram in Figure A.2 (compared with Figure A.1) justifies the following answers:

(a) True
(b) True
(c) False
(d) False

The next definition introduces a simple yet very important possible relationship between two sets.

Definition A.2

Let A and B be two sets. We say that A is a **subset** of B, written $A \subseteq B$, if every element of A is also an element of B; that is, $x \in A \Rightarrow x \in B$. The symbols $A \nsubseteq B$ denote that A is not a subset of B. Note that the subset relation allows the two sets to be equal; that is, $A \subseteq A$. If $A \subseteq B$, and moreover $A \neq B$ (B really has more elements than A), then A is called a **proper subset** of B, and we denote this by $A \subset B$.

Figure A.3 gives a Venn diagram that illustrates the concept of a subset.

Definition A.3

The **empty set**, denoted $\varnothing$, is the set that contains no elements; that is, $\varnothing = \{ \ \}$. Two sets A and B for which $A \cap B = \varnothing$ are said to be **disjoint** (or **mutually exclusive**).

Note that for any set A, we (always) have $\varnothing \subseteq A$. The reason for this is simple: the corresponding implication $x \in \varnothing \Rightarrow x \in A$ is **vacuously true**, meaning that the hypothesis of the implication is never true, so the logical implication is true "by default."

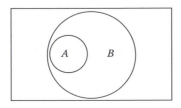

Figure A.3 A Venn diagram illustrating the subset relation $A \subseteq B$.

Definition A.4

In a given context, the **universal set** U is the set of all objects under consideration. In such a context, if A is any set, we define the **complement of** A, denoted $\sim A$, to be the set of all elements of the universal set that are not in A; that is, $\sim A = \{x \in U \mid x \notin A\}$. Outside the context of universal sets, the notion of *set differences* (*relative complements*) of two sets can be formulated as follows: If A and B are two sets, the **set difference** $A \sim B$ is defined to be the set of all elements of A that do not belong to B; that is, $A \sim B = \{x \in A \mid x \notin B\}$.

In any Venn diagram, the universal set is represented by the bounding rectangle; see Figures A.1–A.3. Venn diagrams for the complement as well as the set difference are given in Figure A.4.

As a simple example, to compute $\{2, 4, 6, 8, 10\} \sim \{3, 6, 9, 12\}$, we simply remove any elements from the first set that appear in the second set (that is, only the number 6), to get $\{2, 4, 8, 10\}$.

Venn diagrams are a convenient means of establishing an assortment of *set theoretic identities*, which equate different combinations of generic sets as always being equal. The next example illustrates this concept.

Example A.3

Use Venn diagrams to show that for any three sets A, B, and C, the following *distributive law* is valid: $A \cap (B \cup C) = (A \cap B) \cup (A \cap C)$. *Solution:* It is easily verified that the shaded region in Figure A.5 represents both sets of the asserted equation. (The figure can be obtained in two steps for each side by graphing the parenthesized sets first.)

The following theorem presents several useful identities involving sets, and each can be established using Venn diagrams as in Example A.3.

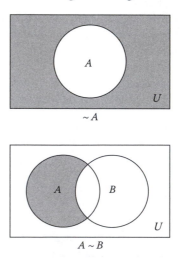

Figure A.4 Venn diagrams illustrating the complement $\sim A$ of a set A (left) and the set difference (relative complement) $A \sim B$.

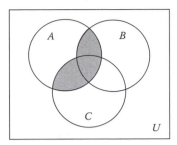

Figure A.5 A Venn diagram for the sets $A \cap (B \cup C)$ and $(A \cap B) \cup (A \cap C)$ of Example A.3.

Theorem A.1: Some Set Theoretic Identities

Let A, B, and C be sets contained in some universal set U. The following identities are then valid:

(a) *Commutativity:* $A \cup B = B \cup A$, $A \cap B = B \cap A$

(b) *Associativity:* $(A \cup B) \cup C = A \cup (B \cup C)$, $(A \cap B) \cap C = A \cap (B \cap C)$

(c) *Distributivity:* $A \cap (B \cup C) = (A \cap B) \cup (A \cap C)$, $A \cup (B \cap C) = (A \cup B) \cap (A \cup C)$

(d) *De Morgan's Laws:* $\sim(A \cup B) = \sim A \cap \sim B$, $\sim(A \cap B) = \sim A \cup \sim B$

(e) *Double Complementation:* $\sim(\sim A) = A$

(f) *Absorption:* $A \cup (A \cap B) = A$, $A \cap (A \cup B) = A$

(g) *Identity Laws:* $A \cap U = A$, $A \cup \varnothing = A$

(h) *Dominance Laws:* $A \cup U = U$, $A \cap \varnothing = \varnothing$

(i) *Complement Laws:* $A \cup \sim A = U$, $A \cap \sim A = \varnothing$

Two Basic Counting Principles

The subject of sophisticated counting methods has evolved into an important branch of mathematics called *combinatorics*. It is often important in cryptography to know exactly or to obtain good estimates for the number of objects in a certain finite set, or to be able to count the number of mathematical operations required to execute a specified algorithm. We present two simple but very useful counting principles that will sufficiently meet our needs in this book.

Notation: If S is any finite set, the symbol $|S|$, which can be read as the **cardinality** of S, denotes the number of elements in the set S.

Example A.4: Motivating Example for the Multiplication Principle

Arlo packs three shirts, two ties, and three pairs of pants for a business trip. How many different outfits can Arlo put together during this trip? Assume that an outfit consists of one choice each of a shirt, tie, and a pair of pants, and that any differences in the choices lead to different outfits.

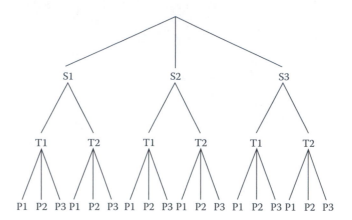

Figure A.6 A tree diagram for the counting problem of Example A.4. To put together an outfit, we start at the top (root) of the tree and first choose one of three shirts {S1, S2, S3}, next we choose a tie from {T1, T2}, and finally we choose a pair of pants from {P1, P2, P3}. Each path from top to bottom represents a different permissible outfit, and no other outfits can be put together.

Solution: One approach is to represent the sequence of choices by a so-called *tree diagram*; such a diagram is shown in Figure A.6. Notice that each outfit corresponds to a unique sequence of choices of a shirt (from S1, S2, and S3), tie (from T1 and T2), and pants (from P1, P2, and P3), and this in turn corresponds to a unique path down the tree, which is completely determined by where it lands on the bottom. Notice that at each stage, the new total number of choices results in the previous total number, multiplied by the new number of choices that one needs to make (since each previous result branches off with the latter number of branches). Thus we have shown that Arlo is able to put together a total of $3 \cdot 2 \cdot 3 = 18$ outfits.

Using a set theoretic approach, the outfits can be viewed to correspond to elements in the **Cartesian product set** $S \times T \times P$ of *vectors* (or *ordered lists*) of the form (s, t, p), where $s \in S, t \in T, p \in P$, and where S denotes the set of all shirts, $S = \{S1, S2, S3\}$, and similarly for T and P. For example, the ordered triple (S2, T1, P1) corresponds to the outfit with the second shirt, the first tie, and the first pair of pants. Since from the following simple set-theoretic proposition $|S \times T \times P| = |S| \cdot |T| \cdot |P|$, we have another way to count the different outfits.

Proposition A.2

If $S_1, S_2, \cdots, S_k$ are finite sets, then the cardinality of their Cartesian product $S_1 \times S_2 \times \cdots \times S_k$ consisting of all vectors of the form $(s_1, s_2, \cdots, s_k)$, where $s_1 \in S_1, \cdots, s_k \in S_k$, is given by

$$|S_1 \times S_2 \times \cdots \times S_k| = |S_1| \cdot |S_2| \cdot \cdots \cdot |S_k|$$

From Proposition A.2 we may obtain the general Multiplication Principle.

> **Multiplication Principle**
>
> Suppose that a sequence of choices is to be made and that there are m_1 options for the first choice, m_2 options for the second choice, and so on, up to the kth choice. If these choices can be combined freely, then the total number of possible outcomes for the whole set of choices equals $m_1 \cdot m_2 \cdots m_k$.

The multiplication principle is extremely useful. To apply it to a counting problem, one must be able to recast the problem at hand into a sequence of unrestricted choices. Examples A.5 and A.6 demonstrate this technique.

Examples

Example A.5

A standard Hawaii license plate consists of a group of three letters followed by a group of three digits; see Figure A.7.

(a) How many (standard) Hawaii license plates can the state produce?
(b) If on the island of Maui, the first letter of the plate must be "M," how many (standard) Maui plates can be produced?

Solution: Part (a): We view creating a Hawaii plate as making a sequence of six unrestricted choices; for each letter slot, we have 26 choices, while for the digit slots we have 10 choices. Hence, by the multiplication principle, the total number of Hawaii plates is $26 \cdot 26 \cdot 26 \cdot 10 \cdot 10 \cdot 10 = 17{,}576{,}000$.

Part (b): Since the first letter is already specified, making a Maui plate can be viewed as a sequence of five choices, with the total number being $26^2 \cdot 10^3 = 676{,}000$.

Example A.6

A three-member committee is to be formed from the U.S. Senate, which has 100 members (2 from each state). The committee will have a chairperson, a vice-chair, and a spokesperson.

(a) How many different such committees can be formed?
(b) How many if Senator A must be on it?
(c) How many if Senators B and C will serve together or not at all?

Figure A.7 A standard Hawaii license plate.

Solution: Part (a): We break up the formation of the committee into the following sequence of three choices: first choose a chair (100 senators to choose from); next, after a chair has been chosen, choose a vice-chair (from the 99 senators remaining); finally, from the 98 senators remaining, we choose the spokesperson.* The multiplication principle tells us that there can be a total of $100 \cdot 99 \cdot 98 = 970{,}200$ such committees.

Part (b): We give two different methods.

Method 1: Separate into disjoint cases first. Many difficult problems can often be reduced to simpler ones using cases. For the problem at hand, there are three natural cases: either A serves as chair, vice-chair, or spokesperson. These three cases are disjoint (no matter how the rest of the committees are formed). (Why?) Using the multiplication principle to fill the remaining slots, by disjointness, we may add up the results to get the answer to part (b): $1 \cdot 99 \cdot 98 + 99 \cdot 1 \cdot 98 + 99 \cdot 98 \cdot 1 = 3 \cdot 99 \cdot 98 = 29{,}106$ (the factor 1 in each of the three terms represents the fact that there is only one choice for the corresponding slot, since Senator A will occupy that slot in each case).

Method 2: Use the multiplication principle directly.

$$\underset{\substack{\text{places to put} \\ \text{Senator A}}}{3} \quad \cdot \quad \underset{\substack{\text{choices to fill} \\ \text{first remaining} \\ \text{slot}}}{99} \quad \cdot \quad \underset{\substack{\text{choices to fill} \\ \text{second remaining} \\ \text{slot}}}{99}$$

Part (c): Separating into the two natural cases: (i) neither B nor C serves and (ii) both B and C serve (which give rise to disjoint sets of committees) seems like the only way to go here. Each of the two cases is amenable to the multiplication principle:

$$\underbrace{\underset{\substack{\text{choices for} \\ \text{the chair}}}{98} \cdot \underset{\substack{\text{choices for} \\ \text{the vice-chair}}}{97} \cdot \underset{\substack{\text{choices for the} \\ \text{spokesperson}}}{96}}_{\text{neither B nor C serve}} + \underbrace{\underset{\substack{\text{positions} \\ \text{for B}}}{3} \cdot \underset{\substack{\text{positions} \\ \text{for C}}}{2} \cdot \underset{\substack{\text{choices for the} \\ \text{remaining position}}}{98}}_{\text{both B and C serve}} = 913{,}164$$

The multiplication principle has both practical and theoretical utility. We use it next to give a proof of an important fact concerning subsets of a finite set.

Proposition A.3

A set S with a finite number n of elements has 2^n subsets.

Proof: We list the elements of the S as $\{a_1, a_2, \cdots, a_n\}$. We can view the formation of a subset $B \subseteq S$ as a sequence of n choices, the ith choice being whether to include the element a_i in the subset B. Since each of these n steps

---

* Of course, this method of choosing a committee has no bearing on the process of how the Senate might actually put together such a committee (usually by nominations and voting); we cast the task as a sequence of choices solely as a mathematical device to solve the counting problem.

has two choices (that is, either $a_i \in B$ or $a_i \notin B$), it follows from the multiplication principle that there are a total of $\underbrace{2 \cdot 2 \cdots \cdots 2}_{n \text{ factors}} = 2^n$ subsets of S. $\square$

Another simple yet often useful rule is a consequence of the basic fact that for any subset $S \subseteq U$ (the universal set), U is the disjoint union of S and its complement $\sim S$. If U is a finite set, this implies that $|U| = |S| + |\sim S| \Rightarrow |S| = |U| - |\sim S|$. We reiterate this in words:

> ### Complement Principle
>
> The number of elements in a set equals the number of elements in the (finite) universal set, less the number of elements that are not in the set.

Example A.7

For security reasons, a university's finance office requires students to create a six-character password to log in to their accounts that contain at least one digit and at least one letter.

(a) How many passwords are possible if the protocol is not case-sensitive?
(b) What if the protocol is case-sensitive?

Solution: Let D denote the set of all six-character strings that contain at least one digit, and L the set of all six-character strings that contain at least one letter. We wish to count the number of passwords in the set $D \cap L$. The sets D and L are difficult to count directly, but their complements are easy. For example, $\sim D$ is the set of all six-character passwords that contain no digits, and therefore consists only of letters. By the multiplication principle, the number of such passwords is 26^6 for part (a) and 52^6 for part (b). In the same fashion, $|\sim L| = 10^6$ [for both parts (a) and (b)]. Also, letting S denote the (universal) set of all six-character passwords, the multiplication principle gives that $|S| = 36^6$ for part (a) and $|S| = 62^6$ for part (b).

The complement principle and then De Morgan's law allows us to write

$$|D \cap L| = |S| - |\sim(D \cap L)| = |S| - |\sim D \cup \sim L|$$

Now (fortunately) the sets $\sim D$ and $\sim L$ are disjoint, so $|\sim D \cup \sim L| = |\sim D| + |\sim L|$. We now have all the information we need to answer the questions.

Part (a): $|S| - (|\sim D| + |\sim L|) = 36^6 - 26^6 - 10^6 = $ 1,866,866,560.
Part (b): $|S| - (|\sim D| + |\sim L|) = 62^6 - 52^6 - 10^6 = $ 37,028,625,920.

Certainly either protocol should be sufficient to accommodate any university.

Appendix B: Randomness and Probability

Randomness is a natural phenomenon that refers to the unpredictability of certain events. Examples of natural events whose outcomes are *random* include flipping a coin (and noting if heads or tails shows), throwing a die (and counting the number of pips that show), and the number of clicks that a Geiger counter makes per unit of time in the presence of a certain radioactive substance. Although by their very nature it is impossible to predict the outcome of a random event, the subject of *probability* aims to shed light on long-term trends or chances of certain outcomes. Randomness is important in cryptography because in many cryptosystems and related algorithms, it is most secure and/or effective to randomly generate certain parameters. This appendix provides some basic results of probability that are used in a few places in the book. The computer implementation material of Chapter 1 provides some details on the generation of random numbers using the computer. Although such computer-generated random numbers are obtained by deterministic schemes and so technically are not random, they possess all of the required statistical properties. More details and proofs concerning the results presented in this appendix can be found in [Ros-02] and also in [Sta-11].

Probability Terminology and Axioms

Probability is the branch of mathematics having to do with rigorously assigning percentages of likelihoods for certain events that are subject to chance. The following questions are typical of those for which probability can be used to answer: What are the chances that a poker hand is a full house? What are the chances that 1000 random guesses of an unknown six-character password will determine the password? We first need some general terminology.

Definition B.1

An **experiment** is any process that has an **outcome**. Each repetition of an experiment is called a **trial**. The **sample space**, S, of an experiment is the set of all possible outcomes.

  We stipulate that exactly one outcome in the sample space occurs each time an experiment is performed. The following example describes some basic experiments.

Example B.1

We describe four experiments along with the associated sample spaces.

(a) Flip a penny and record if it lands on heads or tails: $S = \{H, T\}$.

(b) Throw a die and record the number of pips showing: $S = \{1, 2, 3, 4, 5, 6\}$.

(c) Make 1000 six-character random guesses at an unknown six-character password, where the characters can be any of the uppercase or lowercase English letters, or any of the 10 digits (0, 1, ..., 9). Find out if we correctly hit the password. $S = \{\text{Yes, No}\}$.

(d) Randomly select 40 students and ask them their birthdays. Count the maximum number of people who all share the same birthday. $S = \{1, 2, \cdots, 40\}$.

Definition B.2

An **event** E is a subset of the sample space S; that is, $E \subseteq S$. If the outcome of an experiment belongs to E, we say that the **event E has occurred**.

For any experiment, one goal of probability theory is to create a *probability function* P that assigns to each event E the likelihood $P(E)$ that E will occur. These numbers $P(E)$—the *probabilities*—should have certain properties that we will discuss momentarily. In case of a finite sample space for which each outcome is equally likely [such as each of the experiments in parts (a) and (b) of the preceding example], the following definition of probability should seem quite reasonable.

Definition B.3   Probabilities in Case of a Finite Sample Space with Each Outcome Being Equally Likely

If E is an event in an experiment with finitely many, equally likely outcomes, then

$$P(E) = \frac{|E|}{|S|} \tag{B.1}$$

where (recall) $|E|$ denotes the number of elements in the set E.

With this definition, the computation of probabilities is reduced to counting the number of elements in events. We usually describe a coin (or a die) as *fair* if it is equally likely to turn out any outcome when flipped (or tossed). Thus, for example, when we flip a fair coin, $P(H) = 1/2 = P(T)$. Coins or die that are not fair are usually called *biased* or *loaded*. Unless indicated otherwise, we will assume that all coins, dice, or other gambling devices are fair.

Example B.2

Compute the probability $P(E)$ for each event E described below:

(a) In the experiment of flipping a penny and a nickel, E is the event "at least one head."

(b) In the experiment of throwing a pair of dice, E is the event "the pips add up to 7 or 11" (i.e., the event of winning a basic game of *craps*).

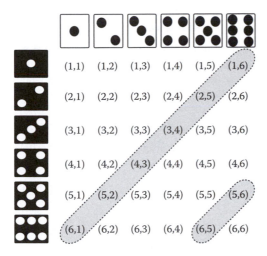

Figure B.1 The sample space of the 36 outcomes in the experiment of rolling two dice. The shaded outcomes are those that will win a game of craps (7 or 11).

Solution: Part (a): Using natural notation, the event E can be expressed as {HT, TH, HH}, and thus $P(E) = |E|/|S| = 3/4$.

Part (b): The 36 elements of the sample space S are shown in Figure B.1, and those that correspond to the sum of the pips being 7 or 11 (that is, those in the event E) are shaded. Note that the elements of the sample space have been displayed in a fashion that make it quite simple to locate the outcomes in E. Thus, since E consists of eight outcomes, we may conclude that $P(E) = |E|/|S| = 8/36 = 2/9 \approx 22.2\%$.

The underlying theory of probability depends heavily on set theory, since, after all, events in probability are subsets of the sample space. During the early 20th century, while modern set theory was developing in a form that would resolve many of the paradoxes that had previously arisen, there became a need to lay down a new foundation for probability. This was accomplished in 1933 by Russian mathematician Andrey Kolmogorov. Kolmogorov based his theory on three very intuitive axioms from which all of probability theory could be developed. His elegant development has been likened to Euclid's axiomatic development of geometry in the latter's time-enduring textbook, *The Elements*. In order to state the Kolmogorov axioms, we will need the following definitions.

Definitions

Definition B.4

Two events, E and F, are called **mutually exclusive** if they are disjoint as sets; that is, if $E \cap F = \emptyset$. A sequence of events, $E_1, E_2, \cdots$ is **pairwise mutually exclusive** if any pair of events taken from this sequence is mutually exclusive; that is, if $i \neq j$ are indexes of two sets from this sequence, then E_i and E_j are mutually exclusive.

Definition B.5: Kolmogorov's Axioms of Probability

Given any set S (a sample space), a **probability** on S is a set function P that assigns to each event E (subset of S) its corresponding probability $P(E)$ in such a way that the following axioms are satisfied:

(1) For any event E, $0 \leq P(E) \leq 1$. (*Any probability lies between 0% and 100%*)

(2) $P(S) = 1$. (*It is a sure thing that any outcome will be in the sample space.*)

(3) If a sequence of events E_1, E_2, $\cdots$ is pairwise mutually exclusive, then

$$P(E_1 \cup E_2 \cup E_3 \cup \cdots) = P(E_1) + P(E_2) + P(E_3) + \cdots \qquad (B.2)$$

(*Probabilities of disjoint events can be added.*)

We point out that this definition does not even require the sample space to be a finite set. From these simple (and hopefully intuitive) axioms, many interesting consequences can be derived. Note that in the third axiom (Equation B.2), if we just take a pair of mutually exclusive events $E_1 = A$ and $E_2 = B$, the axiom states that $P(A \cup B) = P(A) + P(B)$.

The following theorem collects some useful rules of probability that can all be derived from Kolmogorov's axioms. Unless stated otherwise, we assume that we are in the context of a particular sample space S on which a probability function P is defined that satisfies Kolmogorov's axioms.

Theorem B.1: Probability Rules

Given events E and F, the following formulas are valid:

(a) *Complementary Probability Rule.* $P(E) = 1 - P(\sim E)$.
(b) *Monotonicity.* If $E \subseteq F$, then $P(E) \leq P(F)$.
(c) *Inclusion Exclusion Principle.* $P(E \cup F) = P(E) + P(F) - P(E \cap F)$.

Proof: We prove only part (a); the proofs of the other two can be accomplished in a similar fashion. Note that the events E and its complement $\sim E$ are certainly mutually exclusive, and their union equals the sample space S. Using the axioms, we may thus obtain:

$$1 \underset{\text{Axiom 2}}{=} P(S) = P(E \cup \sim E) \underset{\text{Axiom 3}}{=} P(E) + P(\sim E) \implies 1 - P(\sim E) = P(E)$$

as desired.

Example B.3

Suppose that the weather forecast estimates the probability of rain tomorrow as 70%, the probability of lightning tomorrow as 40%, and the probability of rain or lightning as 95%. Compute each of the following probabilities concerning tomorrow's weather:

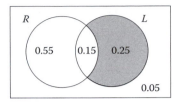

Figure B.2 Venn diagram with probabilities for the solution of Example B.3.

   (a) No lightning.
   (b) Rain and lightning.
   (c) Lightning but no rain.
   (d) Neither rain nor lightning.

   Solution: Using obvious notations, we can write P(R) = 0.7, P(L) = 0.4, and P($R \cup L$) = 0.95.
   Part (a): Using the complementary probability rule, we obtain P($\sim L$) = 1 − P($\sim (\sim L)$) = 1 − P(L) = 1 − 0.4 = 0.6.
   Part (b): Invoking part (c) of Theorem B.1 (and solving it for the probability of the intersection), we can write

$$P(R \cap L) = P(R) + P(L) - P(R \cup L) = 0.7 + 0.4 - 0.95 = 0.15$$

   Part (c): This could also be done with the probability rules, but let us instead show an alternative (more intuitive) approach that uses a Venn diagram. Using the probabilities that were given along with the one just found in part (b), Kolmogorov's Axiom 3 lets us fill in the three probabilities within the circles shown in Figure B.2. The final probability is that of the complement of the union. It can thus be obtained either by the complement rule (since we were given the probability of $R \cup L$), or we can just read it off from the Venn diagram: P($L \sim R$) = 0.25.
   Part (d): The event "neither rain nor lightning" is just $\sim R \cap \sim L = \sim (L \cup R)$ (we have used De Morgan's law), so we can read the corresponding probability off the Venn diagram to be 0.05.

Our next example is a famous problem in probability; most people tend to greatly overestimate its answer when they first hear about it, and many are stupefied by the results.

Example B.4: The Birthday Problem

We let $B(n)$ denote the probability that at least two people will share a common birthday inside a room containing n people with randomly distributed birthdays.* It is clear that $B(1) = 0$, $B(366) = 1$, and for all values of n in between, $B(n)$ must increase as n increases (if more people are in the room, there is

---

* We ignore leap year birthdays (on February 29).

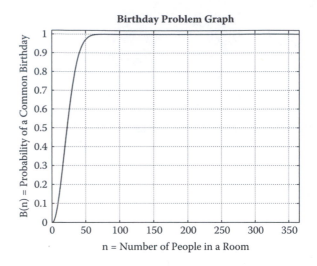

Figure B.3 Plot of the probabilities for the birthday problem of Example B.4.

a better chance of a common birthday). The problem is to find the first value of n for which $B(n) > 0.5$. In other words, how many people would be needed in a room for there to be a better than 50% chance of a common birthday?

Solution: For each value of n, the sample space of this experiment can be viewed as the set of all vectors (ordered lists) $(b_1, b_2, \cdots, b_n)$ of possible birthdays of the n people in the room, and thus has 365^n elements. Let E denote the event that there is a common birthday in the list (that is, $b_i = b_j$, for some $i \neq j$). Describing (or counting) the elements of E is complicated, but the complementary event $\sim E$ is simply the event that there are no common birthdays in the list, and this is easy to count (using the multiplication principle). Thus, using the complementary probability rule, we obtain

$$B(n) \equiv P(E) = 1 - P(\sim E) = 1 - \frac{|\sim E|}{|S|}$$

$$= 1 - \frac{365 \cdot 364 \cdots (365 - n + 1)}{365 \cdot 365 \cdots 365}$$

The numerator was obtained using the multiplication principle: for the first person's birthday there are 365 possibilities; for the second, since there are no common birthdays in any $\sim E$ outcome, there are 364 possibilities; for the third, there are 363, and so on.

It is now easily checked with a simple computer loop that $B(n)$ first exceeds 0.5 when $n = 23$. The plot of all of the values of $B(n)$ is shown in Figure B.3.

Note: We point out that if we attempted to directly count the number of outcomes in E (rather than using the complementary probability rule), there would be an inordinate number of cases to consider.

Probability functions can easily be constructed on any sample space S as follows: Let $p : S \rightarrow [0,1]$ be any function; that is, for each $s_i \in S$, $p(s_i)$ is a real number between 0 and 1 (inclusive). If this function satisfies $\sum_i p(s_i) = 1$, then for any event E, the set function defined by $P(E) = \sum_{s_i \in E} p(s_i)$ is a probability function that satisfies Kolmogorov's axioms. For example, if we apply this construction to the set $\{1, 2, 3, 4, 5\}$ using the function p specified by $p(1) = p(3) = p(5) = 1/9$, and $p(2) = p(4) = 1/3$, the construction would lead to a Kolmogorov probability function that assigns triple the probability to an even number as it does to an odd number; for example, $P(\{1,2,3\}) = p(1) + p(2) + p(3) = 1/9 + 1/3 + 1/9 = 5/9$. Of course, for such a construction to be useful, it must model some actual experiment. We next move on to the important concepts of conditional probability and independence of events. These will lead us to some interesting probability models.

Conditional Probability

Conditional probabilities are probabilities based on additional information being known. We will motivate this important definition with the following simple example: Suppose that two balls are drawn (without replacement) from an urn containing three red and two black balls. Let E be the event "the second ball drawn is black." Then, *a priori* (not knowing anything about the first ball being drawn), $P(E) = 2/5$. However, suppose that we knew that the event F, "the first ball is black" has occurred. This decreases the chance of the second ball being black since at the second draw, the urn has only one black ball along with three red balls. Effectively, knowing that the event F has occurred has reduced the sample space for the computation of the event E. The *reduced sample space* is now F, and the number of outcomes in which E can occur (in the reduced sample space) is the number of outcomes in $E \cap F$. This allows us to conclude that the probability of E given that F has occurred is 1/4. This probability is called the conditional probability of E given F and is written as $P(E \mid F)$. The reader should check that $P(E \mid F)$ will coincide with the ratio of the probabilities $P(E \cap F) / P(F)$. We take this as our general definition of conditional probability.

Definition B.6

Suppose that E and F are events with $P(F) > 0$. The **conditional probability of E given that F has occurred** is defined to be

$$P(E \mid F) = \frac{P(E \cap F)}{P(F)} \qquad \text{(B.3)}$$

We point out that in the setting of a finite sample space with equally likely outcomes, Equation B.1 allows us to rewrite Equation B.3 in the form

$$P(E \mid F) = \frac{|E \cap F|}{|F|}$$

Example B.5

Compute the conditional probabilities in each of the following questions:

(a) Three dice are rolled. What is the conditional probability that (exactly) one of the dice shows a one, given that no two of the dice show the same number?

(b) The conditions at the (fictitious) *Galley Prison* are abysmal: 15% of the inmates have AIDS, 20% have TB, and 10% have both AIDS and TB. A prisoner is randomly chosen and is shown to have TB. What is the probability that he also has AIDS?

Solution: Part (a): Let D represent the event that no two of the dice show the same number, and E be the event that exactly one die shows a one. Viewing the sample space as the set of all ordered triples (a, b, c) of possible outcomes of the three dice, from the multiplication principle we see that $|D| = 6 \cdot 5 \cdot 4$, and $|E \cap D| = 3 \cdot 5 \cdot 4$ (the factor 3 represents the choices of where the one can go in the ordered triple). Since all outcomes are equally likely, we conclude that

$$P(E \mid D) = \frac{|E \cap D|}{|D|} = \frac{3 \cdot \cancel{5} \cdot \cancel{4}}{6 \cdot \cancel{5} \cdot \cancel{4}} = \frac{1}{2}$$

Part (b): With the obvious notations for the events, we can easily compute the desired conditional probability:

$$P(A \mid T) = \frac{P(A \cap T)}{P(T)} = \frac{0.10}{0.20} = \frac{1}{2}$$

Thus, the additional information that the prisoner has TB greatly increases the chances that he has AIDS (from 15% to 50%).

Note that in part (a) of the above example, the corresponding (unconditional) probability could be computed as $P(E) = \dfrac{|E|}{|S|} = \dfrac{3 \cdot 5 \cdot 5}{6 \cdot 6 \cdot 6} = \dfrac{25}{72} \approx$ 0.3742. Thus, in both of the examples above, as well as in the motivating example, the conditional probability $P(E \mid F)$ turned out to be different from the corresponding (unconditional) probability $P(E)$. In other words, the additional knowledge that the event F has occurred changed the probability that event E will occur. In cases where both events have positive (unconditional) probabilities, such pairs of events are called *dependent*, whereas the events E and F will be *independent* if $P(E \mid F) = P(E)$; that is, the knowledge that event F has occurred does not change the probability that F has occurred. Here is a simple example of two independent events. Suppose that we flip a fair coin 10 times, and let F be the event that the first nine flips turned up heads, and E be the event that the tenth flip turns up heads. Since knowing that we landed on heads the first nine times does not influence the 50% chance of getting heads on the tenth flip, we have $P(E \mid F) = P(E)$, and the events are independent.

Conditional probabilities are at the core of all of the (in)famous card-counting strategies that were developed to beat dealers in blackjack games in casinos. The rules of blackjack are set up so that the dealer always has a higher probability of winning (the house advantage). But these probabilities are based on unconditional probabilities, assuming that each new card that comes up from the deck is equally likely. In a card-counting strategy, players mentally keep a tally of the cards that have shown up in a game (until the decks get reshuffled or replaced); for example, a system might assign a "+1" to each high card with value greater than 8, and a "–1" to each card with value less than 5. If at a given point in the game the running total is, say, –9, this means that nine more low cards have been used than high cards, and thus high cards will have a higher conditional probability of showing up next than a low card. Such additional information has been successfully used to give players advantages over dealers, which, in turn, has resulted in casinos adopting certain countermeasures (reshuffling or replacing decks often, video surveillance to spot and promptly "escort out" card counters, and so forth), but it is still possible even at the time of the writing of this book to beat casinos at certain blackjack betting tables.

The definition of conditional probability (Equation B.3) can be easily solved for $P(E \cap F)$ by simply multiplying both sides by $P(F)$:

$$P(E \mid F) = \frac{P(E \cap F)}{P(F)} \quad \Rightarrow \quad P(E \cap F) = P(E \mid F) \cdot P(F) \qquad \text{(B.4)}$$

The resulting Equation (B.4) is known as the **multiplication rule**, and it turns out to be particularly useful both in theory and in problem solving.

Example B.6

Compute the probability of drawing two hearts from a well-shuffled deck of playing cards using the multiplication rule.

Solution: We let H_1 be the event of drawing a heart with the first card, and H_2 be the event of drawing a heart with the second card. The multiplication rule (Equation B.3) tells us that $P(H_1 \cap H_2) = P(H_2 \mid H_1)P(H_1)$. Now since there are 13 hearts, $P(H_1) = 13/52$. After a heart has been removed, there will be only 12 hearts and 51 cards total, so $P(H_2 \mid H_1) = 12/51$. Hence, $P(H_1 \cap H_2) = (13/52) \cdot (12/51) = 1/17$.

Conditioning and Bayes' Formula

We next move on to discuss two important and related consequences of the product rule: *conditioning* and *Bayes' formula*. Both of these methods allow us to use known conditional probabilities to compute unknown probabilities, both absolute and conditional.

Suppose that the sample space S can be **partitioned** into a pairwise mutually exclusive union of events $A_1, A_2, \cdots, A_n$; that is, S is the disjoint

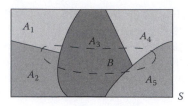

Figure B.4 Mosaic diagram for a partition of a sample space S into disjoint pieces A_i along with an arbitrary event B.

union of these sets (see Figure B.4). Typically, in the setting of probability questions, one would have some good information relating to these events. Now, if B is an arbitrary event, then the events $A_1, A_2, \cdots, A_n$ can also be used to decompose B into pairwise mutually exclusive events:

$$B = (B \cap A_1) \cup (B \cap A_2) \cup \cdots \cup (B \cap A_n)$$

If we use the fact that disjoint probabilities can be added (Kolmogorov's Axiom 3), and then use the multiplication rule, we are led to

$$P(B) = \sum_{i=1}^{n} P(B \cap A_i) = \sum_{i=1}^{n} P(B \mid A_i)P(A_i) \qquad (B.5)$$

Use of Equation B.5 is known as **conditioning the event B on the events** $A_1, A_2, \cdots, A_n$. An important special case is when $n = 2$, so that $A_1 = F$, $A_2 = \sim F$, and Equation B.5 becomes

$$P(B) = P(B \mid F)P(F) + P(B \mid \sim F)P(\sim F)$$

This special case is called "**conditioning on whether or not F has occurred**." Conditioning is used to separate a difficult probability computation into cases that are more manageable. We give an example shortly, but first we show how with one more small yet significant step we can obtain Bayes' formula that allows us to compute (difficult) conditional probabilities in terms of (simpler) reverse conditional probabilities. We continue to suppose that the sample space is partitioned by $A_1, A_2, \cdots, A_n$. By first using the definition of conditional probability and then using conditioning and the product rule, we arrive at **Bayes' formula**:

$$P(A_j \mid B) = \frac{P(A_j \cap B)}{P(B)} = \frac{P(B \mid A_j)P(A_j)}{\sum_{i=1}^{n} P(B \mid A_i)P(A_i)} \qquad (B.6)$$

Example B.7: Quality Control

Three factories A, B, and C all produce the same model of a computer. Their outputs consist of 20%, 35%, and 45%, respectively, of the total computers produced. Factory A has a 6% defect rate for the computers it produces, while the defect rates for factories B and C are 3% and 2%, respectively.

(a) If a computer is randomly selected, compute the probability that it will turn out to be defective.
(b) Suppose that the computer selected did turn out to be defective. Compute the probability that it came from factory A.

Solution: We use the following natural notations for the relevant events: D = the selected computer is defective, A, B, C = the selected computer came from factory A, B, C, respectively.

Part (a): Since A, B, C clearly partition the sample space S, we can condition on whether the computer came from factory A, B, or C:

$$P(D) = P(D \mid A)P(A) + P(D \mid B)P(B) + P(D \mid C)P(C)$$
$$= (.06)(.2) + (.03)(.35) + (.02)(.45)$$
$$= 0.0315$$

Part (b): We are asked to find $P(A \mid D)$. This at first might seem difficult, but since the problem gives us all of the relevant reverse conditional probabilities, we are set up perfectly to use Bayes' formula. Since we have already found $P(D)$ in part (a), we do not need the full formula (Equation B.6):

$$P(A \mid D) = \frac{P(A \cap D)}{P(D)} = \frac{P(D \mid A)P(A)}{P(D)} = \frac{(.06)(.2)}{.0315} = 0.3810$$

Thus, knowing the computer is defective has nearly doubled the chances that it came from factory A.

Random Variables

When performing an experiment, we are often interested in some numerical quantity (or function) of the outcome. Since such a numerical quantity depends on a random outcome (for each trial of an experiment), we call such a function a *random variable*.

Definition B.7

A **(discrete) random variable**, usually denoted as X, Y, Z, and so forth, is a real-valued function with domain being the sample space—that is, $X : S \to \mathbb{R}$—that has either a finite or countably infinite set of outcomes.

For a discrete random variable X, we can enumerate its range: $x_1, x_2, x_3, \cdots$, and since these values cover all possible outcomes of the experiment, their associated probabilities must add up to 1: $\sum_i p(x_i) = 1$, where $p(x_i) = P([X = x_i])$.

Example B.8: Binomial Random Variables

We consider an experiment with only two outcomes, which we refer to as a *success* or a *failure*. Thus, we may write: P(*success*) = p, and P(*failure*) = $q = 1 - p$ (since $p + q = 1$). For example, if the experiment is flipping a coin, and we label "heads" as a *success*, then $p = q = 1/2$. If the experiment is rolling a pair of dice and a *success* is considered winning a game of craps (the dice adding up to either 7 or 11), then $p = 2/9$ and $q = 7/9$. We repeat our experiment for a total of n trials, and we let $X =$ total number of *successes* in the n independent trials.

X is called a **binomial random variable** with parameters n (number of trials) and p (probability of success); it is denoted as $X \sim \mathscr{B}(n, p)$. The range of X is {0, 1, 2, ..., n}.

Definition B.8

Suppose that X is a discrete random variable and let the x_i's be the values of X for which $p(x_i) > 0$. Thus, $\sum p(x_i) = 1$. The **expectation** (or the **expected value**) of the random variable X is given by

$$E[X] = \sum x_i p(x_i) = \sum x_i P([X = x_i]) \qquad (B.7)$$

The expectation is thus a weighted average of the different values that a random variable can assume, with the weights equaling the corresponding probabilities that the values will occur. Although the outcomes of a random variable are "random," the expectation gives an estimate for the long-term average outcome of the random variable, if the experiment were repeated a large number of times. For example, in the setting of X being the payoff in a gambling bet, the expected value gives the very important long-term expected payoff of the bet. The following result is often useful in computing expectations.

Theorem B.2: Linearity of Expectation

If $X_1, X_2, \cdots, X_n$ are (discrete) random variables and $a_1, a_2, \cdots, a_n$ are real numbers, then

$$E[a_1 X_1 + a_2 X_2 + \cdots + a_n X_n] = a_1 E[X_1] + a_2 E[X_2] + \cdots + a_n E[X_n] \qquad (B.8)$$

Example B.9

Compute the expected value of the binomial random variable $X \sim \mathscr{B}(n, p)$.

Solution: Proceeding directly with the definition would involve a laborious technical computation. A much more natural way to compute $E[X]$ is to use linearity. We let $X_i (1 \leq i \leq n)$ denote the outcome on the ith trial; that is, $X_i = 1$ if the ith trial is a success, and otherwise $X_i = 0$. Then clearly $X = \sum_{i=1}^{n} X_i$. Furthermore, the X_i's all have the same expected values: $E[X_i] = 0 \cdot P[X_i = 0] + 1 \cdot P[X_i = 1]$

$= 0 \cdot q + 1 \cdot p = p$. Thus, by Theorem B.2, we may conclude that $E[X] = E\left[\sum_{i=1}^{n} X_i\right] = \sum_{i=1}^{n} E[X_i] = \sum_{i=1}^{n} p = np$.

The result of this example has numerous practical realizations. Here is a simple one: If we flip a fair coin 1000 times, and let X = the total number of flips that turn up heads, then $X \sim \mathscr{B}(1000, 0.5)$, so by the result of the previous example, we have $E[X] = np = 1000(0.5) = 500$. In other words, if we flip a fair coin 1000 times, we would expect that about 500 of these flips (that is, half of them) would land on heads.

Appendix C: Solutions to All Exercises for the Reader

Chapter 1: An Overview of the Subject

EFR 1.1

(a) The function C **is not onto** since any string whose first two bits are different from 10 will not be in the range. The function C **is one-to-one**, since $C(\sigma) = C(\sigma')$ implies $1010 \cdot \sigma = 1010 \cdot \sigma'$ and by ignoring the first four bits, we get that $\sigma = \sigma'$.

(b) Suppose that $D(b_1 b_2 b_3) = D(b_1' b_2' b_3')$. Let $d_1 d_2 d_3 = D(b_1 b_2 b_3)$, $d_1' d_2' d_3' = D(b_1' b_2' b_3')$. Equating first bits: $d_1 = d_1'$, the definition of D tells us that $b_1 = b_1'$. Next, since $d_2 = d_2'$, the definition of D tells us that $b_1 + b_2$, $b_1' + b_2'$ are either both even, or both odd. But since we already know that $b_1 = b_1'$, this means that b_2, b_2' are either both even, or both odd. Since these bits can only be 0 or 1, this forces them to be equal, i.e., $b_2 = b_2'$. Finally, since $d_3 = d_3'$, a similar argument shows $b_3 = b_3'$. We have thus shown that $D(b_1 b_2 b_3) = D(b_1' b_2' b_3')$ implies $b_1 b_2 b_3 = b_1' b_2' b_3'$, i.e., **$D$ is one-to-one**.

EFR 1.2

(a) YQQFFTQUOQYMZMFZAAZ

(b) Jenkins is a turncoat

EFR 1.3

(a) FWMSOOCNOYHZC

(b) Break out at midnight

EFR 1.4

(a) First we state the procedure using the Vigenère tableau (Table 1.2) and then we explain why it works.

Procedure: For each keyword letter, look in the corresponding *row* of the Vigenère tableau for the ciphertext letter A; the column letter where A is found will be the corresponding letter for the decryption keyword, if Vigenère encryption is used. For example, the first letter of the Vigenère keyword *money* is m, and we find that in the m-row of the Vigenère tableau, the letter A appears in the o-column. So the first letter of the Vigenère decryption keyword is o.

Why This Works: In the Vigenère encryption process, each letter of the keyword corresponds to a substitution shift cipher where a gets shifted to the keyword letter. For example, if the first keyword letter is m, then the corresponding shift would shift the plaintext letter a to the ciphertext letter M, and all letters are shifted 12 letters down (looking at Table 1.3 will be helpful). In order to reverse this shift, we could either shift the ciphertext letters 12 units up the alphabet, or shift them the complementary number $26 - 12 = 14$ units down, corresponding to the shift where a goes to O. (Because with the latter option, applying both the original shift and the latter shift would result in a shift of $12 + (26 - 12) = 26$ letters down the alphabet, which simply brings the plaintext letters back to themselves.) In summary, Vigenère decryption can be achieved by using the Vigenère encryption process on the modified keyword by taking each letter of the original keyword, and using instead the letter that is obtained by shifting a in the opposite direction by the same amount, or the complementary number of letters down. The Vigenère tableau is organized in such a way that these reverse shifts are readily obtained by the indicated lookup procedure.

(b)   Using the procedure of part (a), the corresponding Vigenère decryption keyword would be *omnwc*.

Note: Here is an explanation in terms of shift ciphers: The Vigenère encryption keyword *money* corresponds to shifts of 12, 14, 13, 4, 24 down the alphabet (looking at Figure 1.3 of the text might be helpful), the corresponding inverse shifts would be $26 - 12, 26 - 14, 26 - 13, 26 - 4, 26 - 24 = 14, 12, 13, 22, 2$, which correspond to the keyword *omnwc*.

EFR 1.5

(a)   Removing the duplicated letter b, the modified keyword *barcelon* results in the Playfair array:

b	a	r	c	e
l	o	n	d	f
g	h	ij	k	m
p	q	s	t	u
v	w	x	y	z

Inserting x's between double letters of the plaintext, and pairing off the letters gives us:

me et ag en ty ul lo va tx th ea ub er ge re st au ra nt

Encrypting each pair according to the applicable case 1, 2, or 3 produces:

uf cu bh rf yc pf on wb sy qk br pe bc mb cb tu eq cr ds

and thus the following ciphertext:

(b)   Breaking off the ciphertext into pairs (and putting it in lowercase) gives:

ma xh nv gl be rc cx si hb xs gb bc ac mr de rq rz

Using the array of part (a), and reversing each of appropriate cases 1, 2, or 3 of the Playfair encryption cipher produces the following:

> he wi lx lb ec ar ry in ga si lv er br ie fc as ex

Putting the words together and removing redundant x's gives the original message: "He will be carrying a silver briefcase."

Chapter 2: Divisibility and Modular Arithmetic

EFR 2.1

   (a) $16,000 = 2^7 \cdot 5^3$, $42,757 = 11 \cdot 13^2 \cdot 23$

   (b) $\gcd(100, 76) = 4$, $\gcd(16000, 960) = 320$

EFR 2.2

   (a) $\mathrm{lcm}(12, 28) = 84$, $\mathrm{lcm}(100, 76) = 1900$

   (b)   Using the indicated notation, we let $g = p_1^{\sigma_1} \cdot p_2^{\sigma_2} \cdots p_n^{\sigma_n}$. Since for each index i, we have $\alpha_i \geq \sigma_i$, it follows that $c \equiv p_1^{\alpha_1 - \sigma_1} p_2^{\alpha_2 - \sigma_2} \cdots p_n^{\alpha_n - \sigma_n}$ is a nonnegative integer that satisfies

$$g \cdot c = p_1^{\sigma_1} p_2^{\sigma_2} \cdots p_n^{\sigma_n} \cdot p_1^{\alpha_1 - \sigma_1} p_2^{\alpha_2 - \sigma_2} \cdots p_n^{\alpha_n - \sigma_n} = p_1^{\alpha_1} p_2^{\alpha_2} \cdots p_n^{\alpha_n} = a$$

and thus $g \mid a$. A similar argument shows that $g \mid b$, and thus $g \mid \gcd(a, b)$. Conversely, if k is any common divisor of a and b, then the prime factorization of k can only contain primes from the list $p_1, p_2, \cdots, p_k$. For any index i, the exponent of p_i in the prime factorization of k cannot exceed α_i (since $k \mid a$) and it cannot exceed β_i (since $k \mid b$), so it follows that this exponent cannot exceed $\min(\alpha_i, \beta_i) = \sigma_i$. From these facts it follows (just as above) that $k \mid g$, and so $g = \gcd(a, b)$.

     Next we let $\ell = p_1^{\mu_1} \cdot p_2^{\mu_2} \cdots p_n^{\mu_n}$. Since for each index i, we have $\alpha_i \leq \mu_i$, it follows that $c \equiv p_1^{\mu_1 - \alpha_1} p_2^{\mu_2 - \alpha_2} \cdots p_n^{\mu_n - \alpha_n}$ is a nonnegative integer that satisfies

$$a \cdot c = p_1^{\alpha_1} p_2^{\alpha_2} \cdots p_n^{\alpha_n} \cdot p_1^{\mu_1 - \alpha_1} p_2^{\mu_2 - \alpha_2} \cdots p_n^{\mu_n - \alpha_n} = p_1^{\mu_1} p_2^{\mu_2} \cdots p_n^{\mu_n} = \ell$$

and thus $a \mid \ell$. A similar argument shows that $b \mid \ell$, and thus $\mathrm{lcm}(a, b) \mid \ell$. Conversely, if k is any common multiple of a and b, then the prime factorization of k must contain all primes from the list $p_1, p_2, \cdots, p_k$. For any index i, the exponent of p_i in the prime factorization of k must be at least α_i (since $a \mid k$) and it must be at least β_i (since $b \mid k$), so it follows that this exponent must be at least $\max(\alpha_i, \beta_i) = \mu_i$. From these facts it follows (just as above) that $\ell \mid k$, and so $\ell = \mathrm{lcm}(a, b)$. □

   (c)   The result follows from multiplying the expressions g and ℓ for $\gcd(a, b)$ and $\mathrm{lcm}(a, b)$ of part (b), and using the fact that $\max(\alpha_i, \beta_i) + \min(\alpha_i, \beta_i) = \alpha_i + \beta_i$.

EFR 2.3

(a) Letting $q = \text{floor}(a/d)$ and $r = a - qd$, we certainly have that q and r are integers that satisfy $dq + r = a$. Since $q = \text{floor}(a/d) \leq a/d$, it follows that $qd \leq a$, and so $r = a - qd \geq 0$. Also, since $q = \text{floor}(a/d) > a/d - 1$, it follows that $qd > (a/d - 1)d = a - d$, so that $r = a - qd < d$.

(b) (i) quotient: $q = \text{floor}(123/5) = 24$, remainder: $r = a - qd$ $= 123 - 24 \cdot 5 = 3$.

(ii) quotient: $q = \text{floor}(-874/15) = -59$, remainder: $r = a - qd$ $= -874 - (-59) \cdot 15 = 11$.

EFR 2.4

$$\begin{cases} 91 = 1 \cdot 65 + 26 \\ 65 = 2 \cdot 26 + 13 \\ 26 = 2 \cdot 13 + 0 \end{cases} \text{ so gcd}(65,91) = 13$$

$$\begin{cases} 1665 = 1 \cdot 910 + 755 \\ 910 \ = 1 \cdot 755 + 155 \\ 755 \ = 4 \cdot 155 + 135 \\ 155 \ = 1 \cdot 135 + 20 \\ 135 \ = 6 \cdot 20 \ + 15 \\ 20 \ = 1 \cdot 15 \ + 5 \\ 15 \ = 3 \cdot 5 \ \ +0 \end{cases} \text{ so gcd}(1665, 755) = 5$$

EFR 2.5

(a) Solving the second-to-last equation of the first set in the preceding solution for the remainder 13 gives $13 = 65 - 2 \cdot 26$. Next, we substitute into this the expression for 26 obtained by solving the first equation, $26 = 91 - 1 \cdot 65$, to obtain $13 = 65 - 2 \cdot 26 = 65 - 2 \cdot (91 - 1 \cdot 65) = 65 - 2 \cdot 91 + 2 \cdot 65 = 3 \cdot 65 - 2 \cdot 91$. In summary: $13 = \text{gcd}(65, 91) = 3 \cdot 65 - 2 \cdot 91$.

Similarly, we start with the second-to-last equation of the second set of the solution to EFR 2.4, solving for the remainder $5 = 20 - 1 \cdot 5$. We then successively work our way up the list of equations, substituting each remainder in turn, and expressing in terms of the two most recent dividends:

$$5 = 20 - 1 \cdot 15 = 20 - 1 \cdot (135 - 6 \cdot 20) = 7 \cdot 20 - 1 \cdot 135$$

$$= 7 \cdot (155 - 1 \cdot 135) - 1 \cdot 135 = 7 \cdot 155 - 8 \cdot 135$$

$$= 7 \cdot 155 - 8 \cdot (755 - 4 \cdot 155) = -8 \cdot 755 + 39 \cdot 155$$

$$= -8 \cdot 755 + 39 \cdot (910 - 1 \cdot 755) = 39 \cdot 910 - 47 \cdot 755$$

$$= 39 \cdot 910 - 47 \cdot (1665 - 1 \cdot 910) = -47 \cdot 1665 + 86 \cdot 910$$

In summary: $5 = \text{gcd}(1665, 910) = -47 \cdot 1665 + 86 \cdot 910$.

EFR 2.6

Apply the division algorithm to write $a = 2q_a + r_a$ and $b = 2q_b + r_b$, where $0 \le r_a, r_b < 2$. Thus, a is even(odd) if $r_a = 0(1)$, and similarly for b. Since $a - b = 2(q_a - q_b) + r_a - r_b$, we see that $a \equiv b \pmod 2 \Leftrightarrow 2 \,|\, (r_a - r_b)$, but since r_a, r_b can only take on the values 0 or 1, it follows that $2 \,|\, (r_a - r_b)$ if, and only if, r_a, r_b are both 0 or both 1; i.e., $a \equiv b \pmod 2$ if, and only if, a and b have the same parity. □

There are two congruence classes (mod 2): the even integers and the odd integers.

EFR 2.7

(a) $88 + 1234 + 82645 \equiv 8 + 4 + 5 \equiv 17 \equiv 7 \pmod{10}$.

(b)   Since 10 is a factor of 11!, we have $11! \equiv 0$, and so $(11!) \equiv 0^2 \equiv 0 \pmod{10}$.

Any integer is congruent to its one's digit mod 10 (since 10 divides the difference of the integer and its one's digit).

EFR 2.8

For a fixed modulus m, the mod function $a \to \mathrm{mod}(a, m)$ is never one-to-one since, for example, $\mathrm{mod}(0, m) = \mathrm{mod}(m, m)$. The function maps the integers onto the set of possible remainders $\{0, 1, 2, \cdots, m\}$, mod m, since the image of each of these remainders is itself.

EFR 2.9

Working mod 12, we have $11 + 8 \equiv 19 \equiv 7$, $5 \cdot 8 = 40 \equiv 4$ and $11^2 \equiv (-1)^2 \equiv 1$, and these computations give the following equalities in $\mathbb{Z}_{12}$: $11 + 8 = 7$, $5 \cdot 8 = 4$, and $11^2 = 1$. We will check to see whether an element $b \in \mathbb{Z}_{12}$ exists such that $5b = 1$ by multiplying 5 by all elements until (if ever) we find one that works. Working in $\mathbb{Z}_{12}$, we have $5 \cdot 1 = 5$, $5 \cdot 2 = 10$, $5 \cdot 3 (\equiv 15) = 3$, $5 \cdot 4 (\equiv 20) = 8$, $5 \cdot 5 (\equiv 25) = 1$. So $b = 5$ has the indicated property.

EFR 2.10

(a)   Initial Vectors: $U = [1155, 1, 0]$, $V = [862, 0, 1]$.

Since $V(1) = 862$ is positive, we update the vectors:
$W = U - \mathrm{floor}(U(1)/V(1))V = [293, 1, -1]$, $U = V = [862, 0, 1]$,
    $V = W = [293, 1, -1]$.

Since $V(1) = 293$ is positive, we update the vectors:
$W = U - \mathrm{floor}(U(1)/V(1))V = [276, -2, 3]$, $U = V = [293, 1, -1]$,
    $V = W = [276, -2, 3]$.

Since $V(1) = 276$ is positive, we update the vectors:
$W = U - \mathrm{floor}(U(1)/V(1))V = [17, 3, -4]$, $U = V = [276, -2, 3]$,
    $V = W = [17, 3, -4]$.

Since $V(1) = 17$ is positive, we update the vectors:
$W = U - \mathrm{floor}(U(1)/V(1))V = [4, -50, 67]$, $U = V = [17, 3, -4]$,
    $V = W = [4, -50, 67]$.

Since $V(1) = 4$ is positive, we update the vectors:
$W = U - \mathrm{floor}(U(1)/V(1))V = [1, 203, -272]$, $U = V = [4, -50, 67]$,
    $V = W = [1, 203, -272]$.

Since $V(1) = 1$ is positive, we update the vectors:
$W = U - \text{floor}(U(1)/V(1))V = [0, -862, 1155]$, $U = V = [1, 203, -272]$, $V = W = [0, -862, 1155]$.
Since $V(1) = 0$, the algorithm terminates.

We can read off the answer of the last updated vector U: $d = 1$, $x = 203$, and $y = -272$; it can be readily checked that these numbers satisfy $d = 1155x + 862y$.

(b) In light of Proposition 2.11, since gcd(862, 1155) = 1 and since $y = -272 \equiv 883 \pmod{1155}$, we have $862^{-1} = 883$ in $\mathbb{Z}_{1155}$.

EFR 2.11

(a) Since $d = \gcd(123, 456) = 3 \mid 12$, we can follow Algorithm 2.3 to obtain the $d = 3$ solutions of the congruence.

Step 1. We solve the modified congruence $(123/3)y \equiv (12/3) \pmod{456/3}$, i.e., $41y \equiv 4 \pmod{152}$. Using the extended Euclidean algorithm (Algorithm 2.2) as in the solution to EFR 2.10, we compute $41^{-1} = 89$. We multiply both sides of the modified congruence by this inverse to solve it:

$$y \equiv 41^{-1} \cdot 41y \equiv 89 \cdot 4 \equiv 52 \pmod{152}$$

Step 2. We may now list the $d = 3$ solutions of the original congruence: 52, 52 + 456/3, 52 + 2 · 456 / 3 = {52, 204, 356}. (The reader may wish to check each of these.)

(b) Since $15x + 4 \equiv 20 \pmod{25}$ is equivalent to $15x \equiv 16 \pmod{25}$, and since $d = \gcd(15, 25) = 5 \nmid 16$, we know that the congruence has no solution.

EFR 2.12

(a) Since $6x + 2 \equiv 5 \pmod 9$ is equivalent to $6x \equiv 3 \pmod 9$, and since $d = \gcd(6, 9) = 3 \mid 3$, we know that the congruence has $d = 3$ solutions. We may obtain them using Algorithm 2.3.

Step 1. We solve the modified congruence: $(6/3)y \equiv (3/3) \pmod{9/3}$, i.e., $2y \equiv 1 \pmod 3$. Since $2 \cdot 2 = 4 \equiv 1 \pmod 4$, we know $2^{-1} = 2 \pmod 3$, and we may solve this modified congruence by multiplying both sides by this inverse to obtain $y \equiv 2^{-1} \cdot 1 \equiv 2 \cdot 1 \equiv 2 \pmod 9$.

Step 2. The three solutions of the original congruence are {2, 2 + 9/3, 2 + 2 · 9 / 3} = {2, 5, 8}.

(b) Since $6x + 2 \equiv 3 \pmod 9$ is equivalent to $6x \equiv 1 \pmod 9$, and since $d = \gcd(6, 9) = 3 \nmid 1$, we know that the congruence has no solution.

(c) Since gcd(5,9) = 1, we know the congruence has a unique solution. Since $5 \cdot 2 \equiv 10 \equiv 1 \pmod 9$, we know $5^{-1} = 2$ in $\mathbb{Z}_9$, and this inverse can be used to solve the congruence

$5x + 2 \equiv 3 \pmod 9 \Rightarrow 5x \equiv 1 \pmod 9 \Rightarrow 5^{-1} \cdot 5x \equiv 2 \cdot 1 \pmod 9 \Rightarrow x \equiv 2 \pmod 9$

EFR 2.13

In order to apply the Chinese remainder theorem, we first need to put the third congruence of the system, $3x \equiv 4(\text{mod } 7)$, into proper form. Since $3 \cdot 5 \equiv 15 \equiv 1(\text{mod } 7)$ and $3^{-1} = 5(\text{mod } 7)$, we can multiply both sides of this third congruence by this inverse to convert it into $x \equiv 6(\text{mod } 7)$. We may now apply the solution scheme of the Chinese remainder theorem to the equivalent system of congruences:

$$\begin{cases} x \equiv 0(\text{mod } 2) \\ x \equiv 2(\text{mod } 5) \\ x \equiv 6(\text{mod } 7) \end{cases}$$

Since the moduli are pairwise relatively prime, Equation 2.8 (in the proof of the Chinese remainder theorem) provides us with a scheme for obtaining a simultaneous solution. We first set $N = 2 \cdot 5 \cdot 7 = 70$. With $b_1, b_2, b_3 = 0, 2, 6$ and $n_1, n_2, n_3 = 2, 5, 7$, in order to use Equation 2.8, we must first determine e_1, e_2, e_3 by their defining equations: $e_i(N / n_i) \equiv 1 \ (\text{mod } n_i)$.

For e_1: $e_1 \cdot 35 \equiv 1(\text{mod } 2) \Leftrightarrow e_1 \cdot 1 \equiv 1(\text{mod } 2) \Leftrightarrow e_1 \equiv 1(\text{mod } 2)$.
For e_2: $e_2 \cdot 14 \equiv 1(\text{mod } 5) \Leftrightarrow e_2 \cdot 4 \equiv 1(\text{mod } 5) \Leftrightarrow e_2 \equiv 4(\text{mod } 5)$
[since $2^{-1} = 3(\text{mod } 5)$].
For e_3: $e_3 \cdot 10 \equiv 1(\text{mod } 7) \Leftrightarrow e_3 \cdot 3 \equiv 1(\text{mod } 7) \Leftrightarrow e_3 \equiv 5(\text{mod } 7)$.

Now we have all that we need to apply Equation 2.8 to obtain the desired solution:

$$x = \sum_{i=1}^{3} b_i e_i (M / n_i) = 0 \cdot 1 \cdot (35) + 2 \cdot 4 \cdot (14) + 6 \cdot 5 \cdot (10)$$

$$= 412 \equiv 62(\text{mod } 70)$$

The general solution to the original system of congruences is the set of all integers that are congruent to 62 (mod 70).

EFR 2.14

(a) Since for each index j, $m_j | \text{lcm}(m_1, m_2, \cdots, m_k)$, the implication $\text{lcm}(m_1, m_2, \cdots, m_k) | b \Rightarrow m_j | b$ follows from Proposition 2.12. We write $\text{lcm}(m_1, m_2, \cdots, m_k) = p_1^{\alpha_1} p_2^{\alpha_2} \cdots p_n^{\alpha_n}$, where $p_1 < p_2 < \cdots < p_n$ are the distinct primes appearing in the unique factorization. For each index i, we must have $p_i^{\alpha_i}$ appearing as a factor of at least one m_j (since otherwise a lower exponent could be used in the lcm). Thus, from the assumptions $m_2 | b, \cdots, m_k | b$, we may infer that $p_i^{\alpha_i} | b$, for each index i, $1 \le i \le n$. Thus each prime p_i appears in the prime factorization of b, with exponent at least α_i. This means that $\text{lcm}(m_1, m_2, \cdots, m_k) = p_1^{\alpha_1} p_2^{\alpha_2} \cdots p_n^{\alpha_n}$ must divide b. □

(b) This is a direct consequence of the result of part (a) with $b = ax - c$ (using the definition of a congruence). □

EFR 2.15

We had reduced the Hindu puzzle to finding the smallest positive solution of Equation 2.6:

$$\begin{cases} x \equiv 1(\text{mod } 4) \\ x \equiv 1(\text{mod } 5) \\ x \equiv 1(\text{mod } 6) \\ x \equiv 0(\text{mod } 7) \end{cases}$$

We may apply Proposition 2.14 to the first three of these congruences to convert them into the single equivalent congruence: $x \equiv 1 \pmod{\text{lcm}(4,5,6)}$. Since $\text{lcm}(4,5,6) = \text{lcm}(2^2, 5, 2 \cdot 3) = 2^2 \cdot 5 \cdot 3 = 60$, the original system is equivalent to: $\begin{cases} x \equiv 1(\text{mod } 60) \\ x \equiv 0(\text{mod } 7) \end{cases}$ which is now amenable to the algorithm of the Chinese remainder theorem (since the moduli are relatively prime).

We first set $N = 60 \cdot 7 = 420$. With $b_1, b_2 = 1, 0$ and $n_1, n_2 = 60, 7$, in order to use Equation 2.8, we must first determine e_1, e_2 by their defining equations: $e_i(N / n_i) \equiv 1(\text{mod } n_i)$.
For $e_1: e_1 \cdot 7 \equiv 1(\text{mod } 60) \Leftrightarrow e_1 \equiv 43(\text{mod } 60)$ (using the extended Euclidean algorithm).
For $e_2: e_2 \cdot 60 \equiv 1(\text{mod } 7) \Leftrightarrow e_2 \cdot 4 \equiv 1(\text{mod } 7) \Leftrightarrow e_2 \equiv 2(\text{mod } 7)$ [since $4^{-1} = 2(\text{mod } 7)$].

Now we have all that we need to apply Equation 2.8 to get a desired solution:

$$x = \sum_{i=1}^{2} b_i e_i (M / n_i) = 1 \cdot 43 \cdot (7) + 0 \cdot 2 \cdot (60) = 301(\text{mod } 420)$$

Putting this solution in the context of the original problem (of Example 2.9), we conclude that the smallest number of eggs that the woman could have had is 301.

Chapter 3: The Evolution of Codemaking until the Computer Era

EFR 3.1

(a)   Let $\psi(y) \equiv \alpha^{-1}(y - \beta)(\text{mod } m)$ be the asserted inverse function $\psi : \mathbb{Z}_m \to \mathbb{Z}_m$. For any x in $\mathbb{Z}_m$, we have $\psi(\phi_{\alpha,\beta}(x)) \equiv \psi(\alpha x + \beta) \equiv \alpha^{-1}(\alpha x + \beta - \beta) \equiv \alpha^{-1}(\alpha x) \equiv x \pmod{m}$.

Also, $\phi_{\alpha,\beta}(\psi(x)) \equiv \phi_{\alpha,\beta}(\alpha^{-1}(x - \beta)) \equiv \alpha\alpha^{-1}(x - \beta) + \beta \equiv (x - \beta) + \beta \equiv x$ This shows $\psi = \phi_{\alpha,\beta}^{-1}$. □

(b) Since $5 \cdot 21 = 105 \equiv 1 \pmod{26}$, $5^{-1} = 21 \pmod{26}$, by the result of part (a), $\phi_{5,21}^{-1}(y) \equiv 21(y - 11) \equiv 21y + 3 \pmod{26}$.

EFR 3.2

(a) As in the solution to Example 3.2, the plaintext message corresponds to the following vector of integers [2, 14, 3, 4, 1, 11, 20, 4, 0, 11, 4, 17, 19]. To each integer in this vector, we must apply the affine mapping $\phi_{5,8}(x) \equiv 5x + 8 \pmod{26}$ to obtain the corresponding integers for the ciphertext. Since $\phi_{5,8}(2) \equiv 5 \cdot 2 + 8 \equiv 18 \pmod{26}$, $\phi_{5,8}(14) \equiv 5 \cdot 14 + 8 \equiv 78 \equiv 0 \pmod{26}$, the first two integers in the ciphertext vector are 18 and 0. Continuing in the fashion, we obtain the entire vector of ciphertext integers: [18, 0, 23, 2, 13, 11, 4, 2, 8, 11, 2, 15, 25], and by using Table 3.1, this corresponds to the ciphertext SAXCNLECILCPZ.

(b) Since $\phi_{5,8}^{-1}(x) \equiv 5^{-1}(x - 8) \pmod{26}$, and since (as in the preceding Exercise for the Reader) $5^{-1} = 21 \pmod{26}$, we have $\phi_{5,8}^{-1}(x) \equiv 21(x - 8) \equiv 21x - 168 \equiv 21x + 14 \equiv \phi_{21,14}(x) \pmod{26}$. The ciphertext CJISEIZCZRCFPCUWXCVZ corresponds via Table 3.1 to following vector [2, 9, 8, 18, 4, 8, 25, 2, 25, 17, 2, 5, 15, 2, 20, 22, 23, 2, 21, 25]. To recover the corresponding integers for the plaintext, we apply the inverse mapping just found to each integer in the above vector. So the first two plaintext integer values would be $\phi_{21,14}(2) \equiv 21 \cdot 2 + 14 \equiv 56 \equiv 4 \pmod{26}$ and $\phi_{21,14}(9) \equiv 21 \cdot 9 + 14 \equiv 203 \equiv 21 \pmod{26}$. Continuing in this fashion we obtain the entire vector of ciphertext integers: [4, 21, 0, 2, 20, 0, 19, 4, 19, 7, 4, 15, 17, 4, 18, 8, 3, 4, 13, 19]. Next we use Table 3.1 to recover the plaintext: "evacuatethepresident."

EFR 3.3

Since the scytale of Figure 3.4 has septagon cross sections when the cloth is wrapped four times around, the plaintext will be filled in as indicated below and unraveling the cloth would correspond to transposing the plaintext letters so they are read from top to bottom (and then left to right). This produces the ciphertext

R	E	S	C
U	E	T	H
E	P	R	I
N	C	E	F
R	O	M	T
H	E	P	A
L	A	C	E

RUENRHLEEPCOEASTREMPCCHIFTAE.

EFR 3.4

(a) The homophones are naturally ordered by the interval of integers that represent them. For reference, we first form a length-26 vector Vec whose entries give the number of homophones of each of the 26 letters in order. So (from Table 1.1) Vec = (82, 15, 28, ..., 20, 1). The strongest security of such a system (if we are as unpredictable as possible) would be achieved by randomly choosing a homophone to represent each plaintext character and then applying the affine mapping to obtain the corresponding ciphertext character. One way to see this is to imagine that we have 26 spinners corresponding to the 26 numbers in Vec.

Each spinner corresponds to a plaintext letter and has the circle divided into equal sectors, the number of which is the number of homophones for the plaintext letter. The sectors are labeled with the plaintext integer homophones. For example, the spinner for the plaintext letter a would have 82 sectors labeled 0–81. Each time we need to encrypt a plaintext letter a, we would spin the corresponding spinner, not the integer homophone on which the spinner lands, and apply the affine mapping to it to obtain the corresponding ciphertext character.

(b)   The deterministic scheme given here is less secure than the randomized scheme that we gave in part (a). Here is one reason: In this deterministic scheme, the homophone cipher integers will for a given plaintext letter always recycle in the same order. Such imbedded recycled sequences could be detected in long ciphertexts, greatly aiding potential hackers. For example, the length of the cycles for any plaintext character is the number of its homophones.

EFR 3.5

(a)   One way to build such a device would be to create double wheels. Each double wheel consists of two wheels labeled (in the same order) with the letters A to Z (just like a single Jefferson wheel). But, each double wheel can be pulled apart and locked together so that A on the left wheel lines up with any other letter on the second wheel. Notice that all other letters on the right wheel would be shifted by an equal amount from the letters on the left wheel. In this way, such a double wheel can conveniently represent any shift cipher. Now to realize the Vigenère cipher, with a keyword—say, *columbia*—we simply take eight such double disks (the same as there are keyword letters), line up the A's in the left disk with the keyword letters (in order). So in the first disk, A on the left disk lines up with C on the right disk, and A on the second left disk lines up with O on the second right disk, and so on. We load these eight double disks (in order) on the spindle. To encode a plaintext—say, *meetinchurch*—with this Vigenère cipher device, we would simply line up the left disks with the first eight plaintext characters: *meetinch*, and the corresponding ciphertext characters would appear right next to them on the right disks. Then we would line up the final four plaintext characters *urch* on the left disks of the first four disks (to recycle the keyword), and the right disks would give us the final four ciphertext characters. Decryption is equally straightforward (just line up the ciphertext letters in order on the right disks to read off the plaintext letters from the left disks). The left and right disks could even be painted different colors for further ease of use.

EFR 3.6

The table for τ tells us that $\tau(1) = 3$. Using this output with the table for σ, we infer that $(\sigma \circ \tau)(1) = \sigma(\tau(1)) = \sigma(3) = 2$. In the same fashion, we use the tables to find that

$(\sigma \circ \tau)(2) = \sigma(\tau(2)) = \sigma(4) = 1$. We continue this to find $(\sigma \circ \tau)(3) = 5$, $(\sigma \circ \tau)(4) = 3$, $(\sigma \circ \tau)(5) = 5$, and $(\sigma \circ \tau)(6) = 6$. We can now write out the tabular form for $\sigma \circ \tau$:

$\begin{pmatrix} 1 & 2 & 3 & 4 & 5 & 6 \\ 2 & 1 & 5 & 3 & 4 & 6 \end{pmatrix}$. In practice, one would simply directly write out the tabular form for a composition permutation by looking at the tabular forms of the permutations that are being composed and tracing the elements one at a time. To get the tabular form for the inverse $(\sigma \circ \tau)^{-1}$, we simply turn the one just found "upside down" and reorder the columns so the first row is sorted from 1 to 6:

$\begin{pmatrix} 1 & 2 & 3 & 4 & 5 & 6 \\ 2 & 1 & 4 & 5 & 3 & 6 \end{pmatrix}$. If we turn upside down the tabular forms for τ and σ, and then compose them, we obtain:

$$\tau^{-1} \circ \sigma^{-1} = \begin{pmatrix} 1 & 2 & 3 & 4 & 5 & 6 \\ 5 & 6 & 1 & 2 & 3 & 4 \end{pmatrix} \circ \begin{pmatrix} 1 & 2 & 3 & 4 & 5 & 6 \\ 4 & 3 & 6 & 1 & 5 & 2 \end{pmatrix}$$

$$= \begin{pmatrix} 1 & 2 & 3 & 4 & 5 & 6 \\ 2 & 1 & 4 & 5 & 3 & 6 \end{pmatrix}$$

Notice that this is the same as the tabular form for $(\sigma \circ \tau)^{-1}$.

EFR 3.7

Let $\psi = f^{-1} \circ g^{-1}$, which is a function from C to A. We need to show that ψ is the inverse of the composition function $g \circ f : A \to C$. First, for any $a \in A$, we have

$$\psi((g \circ f)(a)) = (f^{-1} \circ g^{-1})((g(f(a))) = f^{-1}(g^{-1}(g(f(a))) = f^{-1}(f(a)) = a$$

In the same fashion, for any $c \in C$, we have

$$(g \circ f)(\psi(c)) = (g \circ f)((f^{-1} \circ g^{-1})(c)) = g(f(f^{-1}(g^{-1}(c)))) = g(g^{-1}(c)) = c$$

This proves that $\psi = (g \circ f)^{-1}$. $\square$

EFR 3.8

Applying the constructive procedure in the proof of part (d) of Proposition 3.1, we detect the following nontrivial cycles (in order, starting with smallest elements): (3,7), (2,4), (6,10), (8,9). Forming their product yields: $\sigma = (3,7)(2,4)(6,10)(8,9)$.

EFR 3.9

Since $\sigma(i) \equiv i + 1 (\mathrm{mod}\ 6)$, it follows that the k-fold composition σ^k would thus amount to adding 1 k times, or just adding k (mod 6), and this is the k-unit shift. Since $k + (6 - k) \equiv 0 (\mathrm{mod}\ 6)$, it follows that $(\sigma^k)^{-1} = \sigma^{6-k}$, so $(\sigma)^{-1}[= (\sigma^1)^{-1}] = \sigma^5$, $(\sigma^2)^{-1} = \sigma^4$, and $(\sigma^3)^{-1} = \sigma^3$. By composing permutations, we first obtain $\sigma^2 = \sigma \circ \sigma = (0, 1, 2, 3, 4, 5) \circ (0, 1, 2, 3, 4, 5) = (0, 2, 4)(1, 3, 5)$.

From this result, we can compose again to obtain $\sigma^3 = \sigma^2 \circ \sigma = (0,2,4)(1,3,5) \circ (0,1,2,3,4,5) = (0,3)(1,4)(2,5)$. The cycle decompositions for $\sigma^4 = (\sigma^2)^{-1}$ and $\sigma^5 = (\sigma)^{-1}$ follow from inverting the ones we have [via Proposition 3.1(a) and (b)].

EFR 3.10

With the given initial settings, the position of the rotors for the first plaintext letter encryption will be: Rotor 1: 5, Rotor 2: 2, Rotor 3: 5 (Rotor 1 advanced one notch from the initial settings). For the second plaintext letter the rotor settings will be: Rotor 1: 0, Rotor 2: 3, Rotor 3: 5 (since Rotor 1 advanced to 0, Rotor 2 also advanced on notch from 2 to 3). For the third plaintext encryption, the rotor settings will be: Rotor 1: 1, Rotor 2: 3, Rotor 3: 5, and for the fourth they will be: Rotor 1: 2, Rotor 2: 3, Rotor 3: 5. We now provide the sequence of transformations that produces the ciphertext, letter by letter (as the solution of Example 3.8):

$$\begin{array}{ccccccccccccc} \beta & F & C_5[\rho_1] & C_2[\rho_2] & C_5[\rho_3] & F^{-1} & \tau & F & C_5[\rho_3^{-1}] & C_2[\rho_2^{-1}] & C_5[\rho_1^{-1}] & F^{-1} & \beta \\ b\mapsto d\mapsto 3 & \mapsto & 1 & \mapsto & 0 & \mapsto & 2\mapsto c\mapsto e\mapsto 4 & \mapsto & 2 & \mapsto & 3 & \mapsto & 1\mapsto b\mapsto d \end{array}$$

$$\begin{array}{ccccccccccccc} \beta & F & C_0[\rho_1] & C_3[\rho_2] & C_5[\rho_3] & F^{-1} & \tau & F & C_5[\rho_3^{-1}] & C_3[\rho_2^{-1}] & C_0[\rho_1^{-1}] & F^{-1} & \beta \\ e\mapsto e\mapsto 4 & \mapsto & 5 & \mapsto & 0 & \mapsto & 2\mapsto c\mapsto e\mapsto 4 & \mapsto & 2 & \mapsto & 1 & \mapsto & 1\mapsto b\mapsto d \end{array}$$

$$\begin{array}{ccccccccccccc} \beta & F & C_1[\rho_1] & C_3[\rho_2] & C_5[\rho_3] & F^{-1} & \tau & F & C_5[\rho_3^{-1}] & C_3[\rho_2^{-1}] & C_1[\rho_1^{-1}] & F^{-1} & \beta \\ e\mapsto e\mapsto 4 & \mapsto & 2 & \mapsto & 1 & \mapsto & 3\mapsto d\mapsto b\mapsto 1 & \mapsto & 5 & \mapsto & 0 & \mapsto & 0\mapsto a\mapsto f \end{array}$$

$$\begin{array}{ccccccccccccc} \beta & F & C_2[\rho_1] & C_3[\rho_2] & C_5[\rho_3] & F^{-1} & \tau & F & C_5[\rho_3^{-1}] & C_3[\rho_2^{-1}] & C_2[\rho_1^{-1}] & F^{-1} & \beta \\ f\mapsto a\mapsto 0 & \mapsto & 4 & \mapsto & 3 & \mapsto & 5\mapsto f\mapsto a\mapsto 0 & \mapsto & 4 & \mapsto & 3 & \mapsto & 2\mapsto c\mapsto c \end{array}$$

We have thus demonstrated the ciphertext to be *DDFC*.

Chapter 4: Matrices and the Hill Cryptosystem

EFR 4.1

$$(AB)C = \left(\begin{bmatrix} 2 & -4 \\ 1 & 6 \end{bmatrix}\begin{bmatrix} 8 & 0 \\ -4 & 1 \end{bmatrix}\right)\begin{bmatrix} 3 & 7 \\ 5 & 5 \end{bmatrix} = \left(\begin{bmatrix} 32 & -4 \\ -16 & 6 \end{bmatrix}\right)\begin{bmatrix} 3 & 7 \\ 5 & 5 \end{bmatrix}$$

$$= \begin{bmatrix} 76 & 204 \\ -18 & -82 \end{bmatrix}$$

$$A(BC) = \begin{bmatrix} 2 & -4 \\ 1 & 6 \end{bmatrix}\left(\begin{bmatrix} 8 & 0 \\ -4 & 1 \end{bmatrix}\begin{bmatrix} 3 & 7 \\ 5 & 5 \end{bmatrix}\right) = \begin{bmatrix} 2 & -4 \\ 1 & 6 \end{bmatrix}\left(\begin{bmatrix} 24 & 56 \\ -7 & -23 \end{bmatrix}\right)$$

$$= \begin{bmatrix} 76 & 204 \\ -18 & -82 \end{bmatrix} = (AB)C$$

EFR 4.2

In the course of proving both identities, we assume that the sizes of the matrices involved are compatible so that both sides of the identities are defined.

(i) The (i,j) entry of $A(B+C)$ is the dot product of the ith row of A, $[a_{i1} \ a_{i2} \cdots a_{im}]$, and the jth column of $B+C$, $\begin{bmatrix} b_{1j}+c_{1j} \\ b_{2j}+c_{2j} \\ \vdots \\ b_{mj}+c_{mj} \end{bmatrix}$, and so equals

$$\sum_{k=1}^{m} a_{ik}(b_{kj}+c_{kj}) = \sum_{k=1}^{m} a_{ik}b_{kj} + \sum_{k=1}^{m} a_{ik}c_{kj}.$$ Since the last two sums give the (i,j) entries of AB and AC, respectively, the identity $A(B+C) = AB + AC$ is thus established. □

(ii) The (i,j) entry of $\alpha(A+B)$ is $\alpha(a_{ij}+b_{ij}) = \alpha a_{ij} + \alpha b_{ij}$. Since the first term on the right is the (i,j) entry of αA, and the second term is the (i,j) entry of αB, the identity $\alpha(A+B) = \alpha A + \alpha B$ is proved. □

EFR 4.3

We assume that $A = [a_{ij}]$ is an $n \times m$ matrix.

(i) The (i,j) entries of AI_m is the dot product of the ith row of A, $[a_{i1} \ a_{i2} \cdots a_{im}]$, with the jth column of I_m, $\begin{bmatrix} 0 \\ \vdots \\ 0 \\ 1 \\ 0 \\ \vdots \\ 0 \end{bmatrix} \leftarrow \text{row } j,$

and so equals a_{ij}, which is the (i,j) entries of A. This proves that $AI_m = A$.

(ii) The (i,j) entries of $I_n A$ is the dot product of the ith row of I_n, $[0 \ \cdots \ 0 \ \underset{\underset{\text{column } i}{\uparrow}}{1} \ 0 \ \cdots \ 0]$, with the jth column of A, $\begin{bmatrix} a_{1j} \\ a_{2j} \\ \vdots \\ a_{nj} \end{bmatrix}$, and so equals a_{ij}, which is the (i,j) entries of A. This proves that $I_n A = A$. □

EFR 4.4

(a) If the ith row of A is the zero vector (all entries are zero), and B is any matrix with the same number of rows as A has rows, then, since any entry of the ith row of the matrix product AB will be obtained as a dot product using the ith row of A (which is the zero vector), the result will be zero. This proves that the ith row of AB is also the zero vector. In particular, if A were invertible and we took $B = A^{-1}$, this would imply that the ith row of AA^{-1} is the zero vector. But this would contradict the property that $AA^{-1} = I$ (since the identity matrix does not have any rows of zeros). This contradiction shows that no matrix with a row of zeros can be invertible. □

(b) If the jth column of the matrix A is all zeros, and B is any matrix with the same number of columns as A has rows, then, since any entry of the jth column of the matrix product BA will be obtained as a dot product using the jth column of A (which is the zero vector), the result will be zero. This proves that the jth column of BA is also the zero vector. In particular, if A were invertible and we took $B = A^{-1}$, this would imply that the jth column of $A^{-1}A$ is the zero vector. But this would contradict the property that $A^{-1}A = I$ (since the identity matrix does not have any rows of zeros). This contradiction shows that no matrix with a column of zeros can be invertible. □

EFR 4.5

$$\det\begin{bmatrix} 5 & -6 & 9 \\ -12 & 2 & 7 \\ 2 & 3 & -7 \end{bmatrix} = 5\det\begin{bmatrix} 2 & 7 \\ 3 & -7 \end{bmatrix} - (-6)\det\begin{bmatrix} -12 & 7 \\ 2 & -7 \end{bmatrix}$$

$$+ 9\det\begin{bmatrix} -12 & 2 \\ 2 & 3 \end{bmatrix}$$

$$= 5(2\cdot(-7) - 7\cdot3) + 6((-12)\cdot(-7) - 7\cdot2)$$

$$+ 9((-12)\cdot3 - 2\cdot2)$$

$$= -175 + 420 - 360$$

$$= -115$$

EFR 4.6

Since $\det(M) = \det\left(\begin{bmatrix} 2 & 6 \\ 3 & -9 \end{bmatrix}\right) = 2\cdot(-9) - 6\cdot3 = -36 \neq 0,$

the matrix M is invertible and its inverse is (by Theorem 4.3):

$$M^{-1} = \frac{1}{\det(M)}\begin{bmatrix} -9 & -6 \\ -3 & 2 \end{bmatrix} = -\frac{1}{36}\begin{bmatrix} -9 & -6 \\ -3 & 2 \end{bmatrix} = \frac{1}{36}\begin{bmatrix} 9 & 6 \\ 3 & -2 \end{bmatrix}$$

or $\begin{bmatrix} 1/4 & 1/6 \\ 1/12 & -1/18 \end{bmatrix}$

Since $\det(N) = \det\left(\begin{bmatrix} 2 & 6 \\ 3 & 9 \end{bmatrix}\right) = 2\cdot9 - 6\cdot3 = 0$, the matrix N is not invertible.

EFR 4.7

(a) Working mod 3, we have $A \equiv \begin{bmatrix} 2 & 7 \\ 4 & 1 \end{bmatrix} \equiv \begin{bmatrix} 2 & 1 \\ 1 & 1 \end{bmatrix}$, B

$\equiv \begin{bmatrix} 1 & 2 \\ 9 & 8 \end{bmatrix} \equiv \begin{bmatrix} 1 & 2 \\ 0 & 2 \end{bmatrix}$, and we compute

$$A + B \equiv \begin{bmatrix} 2 & 1 \\ 1 & 1 \end{bmatrix} + \begin{bmatrix} 1 & 2 \\ 0 & 2 \end{bmatrix} \equiv \begin{bmatrix} 3 & 3 \\ 1 & 3 \end{bmatrix} \equiv \begin{bmatrix} 0 & 0 \\ 1 & 0 \end{bmatrix},$$

$$AB \equiv \begin{bmatrix} 2 & 1 \\ 1 & 1 \end{bmatrix} \cdot \begin{bmatrix} 1 & 2 \\ 0 & 2 \end{bmatrix} \equiv \begin{bmatrix} 2 & 6 \\ 1 & 4 \end{bmatrix} \equiv \begin{bmatrix} 2 & 0 \\ 1 & 1 \end{bmatrix} \pmod 3$$

Since $\det(B) \equiv 1 \cdot 2 - 2 \cdot 0 \equiv 2$ is invertible mod 3 (with $2^{-1} = 2$), it follows from Theorem 4.4 that

$$B^{-1} \equiv \det(B)^{-1} \cdot \begin{bmatrix} 2 & -2 \\ -0 & 1 \end{bmatrix} \equiv 2 \cdot \begin{bmatrix} 2 & 1 \\ 0 & 1 \end{bmatrix} \equiv \begin{bmatrix} 4 & 2 \\ 0 & 2 \end{bmatrix} \equiv \begin{bmatrix} 1 & 2 \\ 0 & 2 \end{bmatrix} \pmod 3$$

(b)   Working mod 10, we compute

$$A + B \equiv \begin{bmatrix} 2 & 7 \\ 4 & 1 \end{bmatrix} + \begin{bmatrix} 1 & 2 \\ 9 & 8 \end{bmatrix} \equiv \begin{bmatrix} 3 & 9 \\ 13 & 1 \end{bmatrix} \equiv \begin{bmatrix} 3 & 9 \\ 3 & 1 \end{bmatrix},$$

$$AB \equiv \begin{bmatrix} 2 & 7 \\ 4 & 1 \end{bmatrix} \cdot \begin{bmatrix} 1 & 2 \\ 9 & 8 \end{bmatrix} \equiv \begin{bmatrix} 65 & 60 \\ 13 & 16 \end{bmatrix} \equiv \begin{bmatrix} 5 & 0 \\ 3 & 6 \end{bmatrix} \pmod{10}$$

Since $\det(B) \equiv 1 \cdot 8 - 2 \cdot 9 \equiv 0 \pmod{10}$, it follows from Theorem 4.4 that B^{-1} does not exist (mod 10).

EFR 4.8

Using cofactor expansion (Algorithm 4.1), we compute $\det(A) = -39$.

(a)   Since $-39 \equiv 6 \pmod{15}$ and $\gcd(6,15) = 3$, it follows from Theorem 4.4(1) that A is not invertible (mod 15).

(b)   Since $-39 \equiv 9 \pmod{16}$ and $\gcd(9,16) = 1$, it follows from Theorem 4.4(1) that A is invertible (mod 16). The classical adjoint matrix is

$$\mathrm{adj}(A) = \begin{bmatrix} \det\begin{pmatrix} 4 & 5 \\ 0 & 1 \end{pmatrix} & -\det\begin{pmatrix} 1 & 3 \\ 0 & 1 \end{pmatrix} & \det\begin{pmatrix} 1 & 3 \\ 4 & 5 \end{pmatrix} \\ -\det\begin{pmatrix} 8 & 5 \\ 5 & 1 \end{pmatrix} & \det\begin{pmatrix} 1 & 3 \\ 5 & 1 \end{pmatrix} & -\det\begin{pmatrix} 1 & 3 \\ 8 & 5 \end{pmatrix} \\ \det\begin{pmatrix} 8 & 4 \\ 5 & 0 \end{pmatrix} & -\det\begin{pmatrix} 1 & 1 \\ 5 & 0 \end{pmatrix} & \det\begin{pmatrix} 1 & 1 \\ 8 & 4 \end{pmatrix} \end{bmatrix}$$

$$= \begin{bmatrix} 4 & -1 & -7 \\ 17 & -14 & 19 \\ -20 & 5 & -4 \end{bmatrix} \equiv \begin{bmatrix} 4 & 15 & 9 \\ 1 & 2 & 3 \\ 12 & 5 & 12 \end{bmatrix}$$

Since $\det(A)^{-1} \equiv 9^{-1} \equiv 9 \pmod{16}$, it follows that

$$A^{-1} = \det(A)^{-1} \cdot \mathrm{adj}(A) = 9 \cdot \begin{bmatrix} 4 & 15 & 9 \\ 1 & 2 & 3 \\ 12 & 5 & 12 \end{bmatrix} \equiv \begin{bmatrix} 4 & 7 & 1 \\ 9 & 2 & 11 \\ 12 & 13 & 12 \end{bmatrix} \pmod{16}$$

EFR 4.9

For a 2×2 matrix $A = \begin{bmatrix} a & b \\ c & d \end{bmatrix}$, each of the submatrices A_{ij}

is the 1×1 submatrix of A obtained by deleting row i and column j, and since the determinant of a 1×1 matrix is simply the number (inside the brackets), we obtain

$$\text{adj}(A) = \begin{bmatrix} \det(A_{11}) & -\det(A_{21}) \\ -\det(A_{12}) & \det(A_{22}) \end{bmatrix} = \begin{bmatrix} d & -b \\ -c & a \end{bmatrix}$$

The special cases now readily follow from Proposition 4.5.

EFR 4.10

The $\mathbb{Z}_{26}$ vector corresponding to the plaintext is (see Table 3.1):

[19 0 17 8 3 22 23 6 23 22 13 20 0 13 5 7 7 20]

Regrouping it into a three-row ciphertext matrix gives

$C = \begin{bmatrix} 19 & 8 & 23 & 22 & 0 & 7 \\ 0 & 3 & 6 & 13 & 13 & 7 \\ 7 & 22 & 23 & 20 & 5 & 20 \end{bmatrix}$. The corresponding

plaintext (uncoded) matrix U is given by $U = A^{-1}C$. We use

Proposition 4.5 to compute $A^{-1} \equiv \begin{bmatrix} 1 & 1 & 25 \\ 0 & 25 & 1 \\ 25 & 0 & 1 \end{bmatrix} \pmod{26}$,

so that

$$U = A^{-1}C \equiv \begin{bmatrix} 1 & 1 & 25 \\ 0 & 25 & 1 \\ 25 & 0 & 1 \end{bmatrix} \cdot \begin{bmatrix} 19 & 8 & 23 & 22 & 0 & 7 \\ 0 & 3 & 6 & 13 & 13 & 7 \\ 7 & 22 & 23 & 20 & 5 & 20 \end{bmatrix}$$

$$\equiv \begin{bmatrix} 2 & 15 & 6 & 15 & 8 & 20 \\ 17 & 19 & 17 & 7 & 18 & 13 \\ 24 & 14 & 0 & 24 & 5 & 13 \end{bmatrix} \pmod{26}$$

Using Table 3.1 to process the entries of U (in the prescribed order) produces the plaintext message: "Cryptography is fun." (Spaces were inserted and the additional n was deleted.)

Chapter 5: The Evolution of Codebreaking until the Computer Era

EFR 5.1

As in the solution of Example 5.1, we first tabulate the frequencies of most often occurring ciphertext letters, along with the counts of the number of ciphertext letters that *do not* appear directly adjacent to it in any *cipherword*.

CIPHERTEXT LETTER	OBSERVED FREQUENCY	NON-ADJACENT COUNT	CIPHERTEXT LETTER	OBSERVED FREQUENCY	NON-ADJACENT COUNT
W	.129	4	G	.069	12
D	.090	7	U	.066	16
A	.089	15	E	.059	8
Y	.080	9	S	.051	12
R	.073	13	O	.036	19
N	.069	11	B	.034	21

Comparing with the predicted frequencies of Table 5.1, it seems quite likely that (the ciphertext letter) $W \leftrightarrow e$. To ascertain the next highest frequency of vowels a, o, and i (according to Table 5.1), we note that D, Y, and E have high frequencies and are very social letters (due to their low nonadjacent count numbers in the table). The higher frequencies of D and Y tell us that these should represent a and o (in some order). The single letter cipherword Y (which appears several times) shows that Y represents either a or i, so we are led to $Y \leftrightarrow a$, $D \leftrightarrow o$, and $E \leftrightarrow i$. The cipherword, AUW, which appears many times, thus becomes AUe, and probably represents *the* or *she*, but because of the high frequency of A, we postulate $A \leftrightarrow t$, and thus $U \leftrightarrow h$. Remaining higher frequency consonants are: n, s, r, d, l. There are three cipherwords that end in SS, and S has high frequency. Since only one of these consonants often gets doubled as a word ending, this leads us to $S \leftrightarrow l$. Making these three vowel substitutions into the ciphertext produces the following (first stage):

thNoQHhoQt ReGioN KeaN Rhe aGMioQRlK NelateO all heN eMJeNiXeGtR aGO JaNtial RQTTeRReR to a TaNeeN OailK oG the liLNaNK RteJR oN iG the hall oP the XaiG LQilOiGH the ToeOR talVeO oP Bhat Rhall Be Oo BheG Be PiGiRh TolleHe eFeG the HiNlR Bho VGeB that theK BeNe HoiGH to Le XaNNieO JNeteGOeO to Le ToGRiOeNiGH iXJoNtaGt LQRiGeRR JoRitioGR eFeG theK Bho VGeB that theK BoQlO haFe to BoNV hiGteO aLoQt PaLQloQR RQitoNR aR PoN TaNol Rhe BaR aG oNJhaG heN oGlK GeaN NelatiFe BaR a FaGilla PlaFoNeO RiRteN XaNNieO to aG oJtiTiaG iG Rt JaQl Rhe haO QReO XoRt oP the XoGeK PNoX heN PatheNR eRtate Rhe BaR Got iG loFe that iR Got oPteG GoN eFeN loGH at a tiXe Rhe BoQlO eaNG heN liFiGH LQt hoB Rhe BaR to eaNG it hoB Rhe BaR to ToGZQeN the BoNlO alXoRt eGtiNelK PoN the BoNlOR oBG HooO Rhe OiO Got Ree

The above partially translated ciphertext is beginning to reveal itself. The first word thNoQHhoQt most probably represents "throughout," leading us to the correspondences: $N \leftrightarrow r$, $Q \leftrightarrow u$, and $H \leftrightarrow g$ (and the frequency correspondence is encouraging). The three cipherwords thus becomes *ReGior Kear Rhe*, and because of their high frequencies, it is quite probable that R and G correspond to two of *n*, *s*, and *d*. The only second word that would make sense is *senior*, leading us to $R \leftrightarrow s$ and $G \leftrightarrow n$, and the third word leads us to conjecture that $K \leftrightarrow y$. Only two higher frequency cipherletters, B and O, appear in the table. Since the cipherwords *Be* and *BaR = Bas* appear, we are led to the higher frequency consonant *w*: $B \leftrightarrow w$. With what we have so far, the partially translated cipherwords *worlOs* and *woulO* lead us to $O \leftrightarrow d$. Making these eight additional substitutions into the Stage 1 partial ciphertext produces the following (second stage):

```
throughout senior year she anMiously relateO
all her eMJeriXents anO Jartial suTTesses
to a Tareer Oaily on the liLrary steJs or
in the hall oP the Xain LuilOing the ToeOs
talVeO oP what shall we Oo when we Pinish
Tollege eFen the girls who Vnew that they
were going to Le XarrieO JretenOeO to Le
TonsiOering iXJortant Lusiness Jositions
eFen they who Vnew that they woulO haFe to
worV hinteO aLout PaLulous suitors as Por
Tarol she was an orJhan her only near relat-
iFe was a Fanilla PlaForeO sister XarrieO
to an oJtiTian in st Jaul she haO useO Xost
oP the Xoney ProX her Pathers estate she
was not in loFe that is not oPten nor eFer
long at a tiXe she woulO earn her liFing
Lut how she was to earn it how she was to
TonZuer the worlO alXost entirely Por the
worlOs own gooO she OiO not see
```

It is quite easy now to complete the rest of the decryption. The original formatted plaintext is the following passage from the novel *Main Street*, written by Sinclair Lewis in 1920:

> Throughout Senior year she anxiously related all her experiments and partial successes to a career. Daily, on the library steps or in the hall of the Main Building, the co-eds talked of "What shall we do when we finish college?" Even the girls who knew that they were going to be married pretended to be considering important business positions; even they who knew that they would have to work hinted about fabulous suitors. As for Carol, she was an orphan; her only near relative was a vanilla-flavored sister married to an optician in St. Paul. She had used most of the money from her father's estate. She was not in love—that is, not often, nor ever long at a time. She would earn her living.
>
> But how she was to earn it, how she was to conquer the world— almost entirely for the world's own good—she did not see.

EFR 5.2

The ciphertext contains the following five matched 5-grams (i.e., each has one duplication): *IPGBI, AIHBK, IHBKV, HBKVV, BKVVK, CUBJX, IBJXC*, with separation distances: 426, 204, 204, 204, 204, 60, 390. The gcd's of pairs of these separation distances are 6, 12, and 30 (with 6 repeated most often). In particular, the length of the codeword should divide the overall gcd, which is 6. Thus, the possible codeword lengths are 2, 3, or 6. The frequency distribution of all ciphertext characters whose positions are congruent to 1 (mod 6) is shown in the figure below:

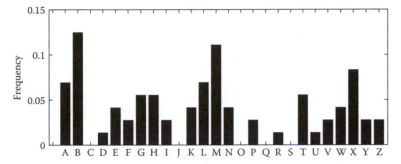

The highest peaks are candidates for the ciphertext characters representing *e*. If *B* represented *e*, then *X* (which is four letters back from *B*) would represent *a* (which is four letters back from *e*), and similarly *U* and *W* would have to represent *x* and *z*, respectively. The latter two representations do not seem likely (it would mean, according to the frequency plot that *z* occurred with frequency about 5% in the plaintext, compared to its 0.1% usual frequency). It would also entail that *Q* represents *t*, so that *t* does not appear in the plaintext (also unlikely since Table 5.1 tells us it should occur about 9.1% of the time). So we rule out this possibility and move on to consider the next: what if *M* represented *e*? This possibility would also lead to unreasonable consequences: For example, we would need to have *z* occurring twice as often as *a* in the plaintext. What about the third possibility of *X* representing *e*? Then *T*, *S*, and *Q* would need to represent *a*, *z*, and *x*, respectively. The frequencies corroborate well. Also, the most frequent cipherletter *B* would have to represent *i*, which, although not typical, is possible if the passage is a personal narrative (often using the personal pronoun *I*). Thus we have good evidence that the first keyword letter is *t*. Similar analysis on the remaining positions of the keyword letter lead us to the codeword being *trepid*. The decryption can now be achieved as taught in Chapter 1. The plaintext, in its original format, is the following passage from *The Time Machine*, written by H. G. Wells in 1895:

Already I saw other vast shapes—huge buildings with intricate parapets and tall columns, with a wooded hill-side dimly creeping in upon me through the lessening storm. I was seized with a panic fear. I turned frantically to the Time Machine, and strove hard to readjust it. As I did so the shafts of the sun smote through the thunderstorm. The grey downpour was swept aside and vanished like the trailing garments of a ghost. Above me, in the intense blue of the summer sky, some faint brown shreds of cloud whirled into nothingness.

EFR 5.3

Step 1. *Determine the length of the keyword.* We first count (using the appropriate applet from this chapter) the number of character matches when one copy of the ciphertext (that is placed directly below another copy) is horizontally shifted different numbers of units to the right of the top copy:

Horizontal shift parameter i:	1	2	3	4	5	6	7	8	9	10
Number of exact matches:	19	19	15	23	21	20	31	43	27	25

The keyword length is the shift parameter giving rise to the largest number of exact matches, so it is 8.

Step 2. *Determine the keyword.* Now we determine the ith keyword letter (for $i = 1$ to $i = 8$): Start with $i = 1$.

We form the length frequency vector F_1 for the ciphertext letters in all positions that are congruent to 1 (mod 8).

Then we compute the 26 dot products of F_1 with the shifted English distribution vectors V_j (for $j = 0$ to 25). The index j for which the dot product is maximum corresponds to the first keyword letter (by the basic integer/letter correspondence of Table 3.1).

When $i = 1$, these 26 dot products are:

j	0	1	2	3	4	5	6	7	8	9	10	11	12
$F_1 \cdot V_j$	0.033	0.036	0.046	0.038	0.038	0.043	0.051	0.036	0.028	0.038	0.041	0.036	0.032

j	13	14	15	16	17	18	19	20	21	22	23	24	25
$F_1 \cdot V_j$	0.044	0.028	0.026	0.044	0.069	0.041	0.034	0.031	0.042	0.039	0.039	0.038	0.033

Since $j = 17$ gives the clear maximum dot product, the first keyword letter should be *R* (corresponding to 17 in Table 3.1).

After forming the vector F_2, the frequencies of ciphertext letters in all positions that are congruent to 2(mod 8), the 26 dot products are computed to be:

j	0	1	2	3	4	5	6	7	8	9	10	11	12	13
$F_2 \cdot V_j$	0.047	0.036	0.030	0.037	0.068	0.040	0.033	0.036	0.045	0.035	0.033	0.035	0.031	0.033

j	14	15	16	17	18	19	20	21	22	23	24	25
$F_2 \cdot V_j$	0.034	0.048	0.039	0.039	0.036	0.050	0.042	0.038	0.031	0.040	0.034	0.032

Since $j = 4$ gives the clear maximum dot product, the second keyword letter should be *E* (corresponding to 4 in Table 3.1).

In a similar fashion, we obtain the remaining six keyword letters to reveal the keyword is *REPUBLIC*. The decryption may now be achieved as taught in Chapter 1. The plaintext, in its original format, is as follows:

> I would not have it imagined, however, that he was one of those cruel potentates of the school who joy in the smart of their subjects; on the contrary, he administered justice with discrimination rather than severity, taking the burthen off the backs of the weak, and laying it on those of the strong. Your mere puny stripling that winced at the least flourish of the rod was passed by with indulgence; but the claims of justice were satisfied by inflicting a double portion on some little, tough, wrong-headed, broad-skirted Dutch urchin, who sulked and swelled and grew dogged and sullen beneath the birch. All this he called "doing his duty by their parents"; and he never inflicted a chastisement without following it by the assurance, so consolatory to the smarting urchin, that "he would remember it, and thank him for it the longest day he had to live."

EFR 5.4

(a)   We separate into cases depending on the number of elements in the intersection of the cycles $|\sigma \cap \tau|$. Note that since τ is a 2-cycle, it is its own inverse [Proposition 3.1(b)] so $\tau\tau = \tau^2 = id$.

Case 1. $|\sigma \cap \tau| = 0$. In this case, the cycles are disjoint, so they commute (by Proposition 3.1), so we have $\tau\sigma\tau = \tau\tau\sigma = id\sigma = \sigma$. So $\tau\sigma\tau$ is a 3-cycle because σ is.

Case 2. $|\sigma \cap \tau| = 1$. Since cycles may be cyclically ordered, we may assume that $a = i$. Thus, by tracing each element through the composition $\tau\sigma\tau$, we obtain $\tau\sigma\tau = (a,j)(a,b,c)(a,j) = (a)(b,c,j) = (b,c,j)$, once again a 3-cycle.

Case 3. $|\sigma \cap \tau| = 2$. Since cycles may be cyclically ordered, we may assume that $a = i$, and ($b = j$, or $c = j$). If $b = j$, $\tau\sigma\tau = (a,b)(a,b,c)(a,b) = (a,c,b)$ while if $c = j$, $\tau\sigma\tau = (a,c)(a,b,c)(a,c) = (a,c,b)$.

(b)   The cases where the number of elements in the intersection of the cycles $|\sigma \cap \tau|$ is either 0 or 1 are handled identically to what was done in part (a), so we have left to consider the case where $|\sigma \cap \tau| = 2$. Since cycles may be cyclically ordered, we may assume that $a_1 = i$, and we let $\ell > 1$ be the index for which $a_\ell = j$. For greater clarity, we separate into two cases:

Case 1. $\ell < k$: By tracing each element through the composition $\tau\sigma\tau$, we obtain $\tau\sigma\tau = (a_1, a_\ell)(a_1, a_2, \cdots, a_{\ell-1}, a_\ell, a_{\ell+1}, \cdots, a_k)(a_1, a_\ell) = (a_1, a_{\ell+1}, \cdots, a_k, a_\ell, a_2, \cdots, a_{\ell-1})$.

Case 2. $\ell = k$: $\tau\sigma\tau = (a_1, a_k)(a_1, a_2, \cdots, a_{k-1}, a_k)(a_1, a_k) = (a_1, a_k, a_2, a \cdots, a_{k-1})$.

(c)   *Proof of Theorem 5.4:* Using the fact that τ is a 2-cycle, it is its own inverse [Proposition 3.1(b)] so $\tau\tau = \tau^2 = id$, we may write

$$\tau\sigma\tau = \tau\sigma_1\sigma_2 \cdots \sigma_k \tau = \tau\sigma_1\tau^2\sigma_2\tau^2 \cdots \tau^2\sigma_k\tau$$

$$= (\tau\sigma_1\tau)(\tau\sigma_2\tau)\cdots(\tau\sigma_k\tau) \equiv \sigma_1'\sigma_2' \cdots \sigma_k'$$

By part (b), each of the above defined permutations σ_i' is a cycle with the same length as its corresponding σ_i. The proof will be complete if we can show that the cycles σ_i' are disjoint. But this follows by observing from the above proof of part (b) that the elements permuted by σ_i' and σ_i are identical as sets. □

Chapter 6: Representation and Arithmetic of Integers in Different Bases

EFR 6.1

(a) [1101001111] (base 2) ~ $2^9 + 2^8 + 2^6 + 2^3 + 2^2 + 2^1 + 2^0 = 847$. [777] (base 8) ~ $7 \cdot 8^2 + 7 \cdot 8^1 + 7 \cdot 8^0 = 511$. [123ABC] (hex) ~ $1 \cdot 16^5 + 2 \cdot 16^4 + 3 \cdot 16^3 + 10 \cdot 16^2 + 11 \cdot 16^1 + 12 \cdot 16^0 = 1,194,684$.

(b) (i) *Step 1:* $R = 122$. *Step 2:* Largest power of 2 not exceeding $R = 122$ is $2^6 = 64$, so ($K = 6$) and $c_6 = 1$. Update $R \to 122 - 64 = 58$. Step 2 is repeated: Now, $k = 5$ is the largest exponent such that $2^k \leq 58(= R)$, so we set $c_5 = 1$, and update $R \to 58 - 32 = 26$.

Step 2 is repeated: $k = 4$, $c_4 = 1$, $R \to 26 - 16 = 10$.
Step 2 is repeated: $k = 3$, $c_3 = 1$, $R \to 10 - 8 = 2$.
Step 2 is repeated: $k = 1$, $c_1 = 1$, $R \to 2 - 2 = 0$.
So 122 ~ [1111010] (base 2).

(ii) *Step 1:* $R = 9675$. *Step 2:* Largest power of 32 not exceeding $R = 9675$ is $32^2 = 1024$, so ($K = 2$) and $c_2 = \text{floor}(9675/1024) = 9$. Update $R \to 9675 - 9.32^2 = 459$. Step 2 is repeated: Now, $k = 1$ is the largest exponent such that $32^k \leq 459(= R)$, so we set $c_1 = \text{floor}(459/32) = 14$ and update $R \to 459 - 14 \cdot 32 = 11$. Step 2 is repeated: $k = 0$, $c_0 = 11$, $R \to 11 - 11 = 0$.
So 9675 ~ [9 14 11] (base 32).

(iii) *Step 1:* $R = 52,396$. *Step 2:* Largest power of 16 not exceeding $R = 52,396$ is $16^3 = 4096$, so ($K = 3$) and $c_3 = \text{floor}(52396/4096 = 12$.) Update $R \to 52,396 - 12 \cdot 16^3 = 3244$. Step 2 is repeated: Now, $k = 2$ is the largest exponent such that $16^k \leq 3244(= R)$, so we set $c_2 = \text{floor}(3244/16^2) = 12$ and update $R \to 3244 - 12 \cdot 16 = 172$.
Step 2 is repeated: $k = 1$, $c_1 = 10$, $R \to 172 - 10 \cdot 16 = 12$.
Step 2 is repeated: $k = 0$, $c_0 = 12$, $R \to 12 - 12 = 0$.
So 52,396 ~ [12 12 10 12] = [CCAC] (hex).

EFR 6.2

(a) The (binary) addition algorithm is summarized in the diagram. The resulting addition [101111] + [001111] = [111110] (base 2) can be translated term by term into an ordinary integer addition [with the same method used in part (a) of the solution to EFR 6.1]: 47 + 15 = 62.

	1	*1*	*1*	*1*		
1	0	1	1	1	1
0	0	1	1	1	1
1	1	1	1	1	0

(b) The (hexadecimal) addition algorithm is summarized in the diagram. In each step, the letters are converted to their integer equivalents to perform the required addition. Here is a summary: E + 2 = 14 + 2 = 16, so the rightmost entry in the table is 0 and the carry 1 is put over 4 (in italics) in the next column to the left. Next, we add 1(the carry) + 4 + A = 1 + 4 + 10 = 15 = F. F gets entered into the second-from-right entry, and there is no carry. The third-from-right slot addition is now D + A = 13 + 10 = 23, so we have a carry of 1, and enter 23 − 16 = 7. The final addition is now 1(the carry) + 7 + 1 = 9.

   The resulting addition [7D4E] + [1AA2] = [97F0] (base 16) can be translated term by term into an ordinary integer addition [with the same method used in part (a) of the solution to EFR 6.1]: 32,078 + 6818 = 38,896.

	1		*1*	
7	D	4	E
1	A	A	2
9	7	F	0

EFR 6.3

(a) The (binary) subtraction algorithm is summarized in the diagram. Note that the first borrow was needed on the second-from-right digit subtraction 0 − 1. We borrowed 1 from the next preceding digit which gave us 2 + 0 − 1 = 1. The resulting subtraction [101101] + [001111] = [011110] (base 2) translates into the following integer subtraction: 45 − 15 = 30.

−1	*−1*	*−1*	*−1*		
1	0	1	1	0	1
0	0	1	1	1	1
0	1	1	1	1	0

(b) The (hexadecimal) subtraction algorithm is summarized in the diagram. In each step, the letters are converted to their integer equivalents to perform the required subtraction. Here is a summary: E − 2 = 14 − 2 = 12 = C, so the rightmost entry in the table is C, and there is no borrow to put over 4 in next column to the left. Next, we need to subtract 4 − A = 4 − 10, so we need to borrow 1 from the previous digit, and this adds 16 so the subtraction is now feasible: 16 + 4 − 10 = 10 = A. So A gets entered into the second-from-right entry, and there is no borrow; −1 is put over the preceding column. The third-from-right slot subtraction is now −1(the borrow) + D − A = −1 + 13 − 10 = 2, so we enter 2, and there is no borrow. The final subtraction is now 7 − 1 = 6. The resulting subtraction

[7D4E] − [1AA2] = [62AC] (hex) translates into the following integer subtraction: $32{,}078 − 6818 = 25{,}260$.

$$
\begin{array}{ccccc}
 & ^{-1} & & & \\
7 & D & 4 & E \\
1 & A & A & 2 \\
\hline
6 & 2 & A & C
\end{array}
$$

EFR 6.4

(a) The (binary) multiplication algorithm is summarized in the diagram. Note that as with all base 2 multiplications there are no carries. In the additions of the digits, there were carries, which we omitted. The resulting multiplication $[1111] \times [1111] = [11100001]$ (base 2) translates into the following integer multiplication: $15 \times 15 = 225$.

$$
\begin{array}{ccccccccc}
 & & & & 1 & 1 & 1 & 1 \\
 & & & \times & 1 & 1 & 1 & 1 \\
\hline
 & & & & 1 & 1 & 1 & 1 \\
 & & & 1 & 1 & 1 & 1 \\
 & & 1 & 1 & 1 & 1 \\
 & 1 & 1 & 1 & 1 \\
\hline
1 & 1 & 1 & 0 & 0 & 0 & 0 & 1
\end{array}
$$

(b) The base 7 multiplication algorithm is summarized in the diagram below. Here is how the computations went: We start by multiplying the last digit ($i = 0$) of the second number, 0, with the first number, and we get all zeros, so this is the first (unshifted) row in the multiplication table results. Next, we multiply the second-to-middle digit ($i = 1$) of the second number, 2, with the first number (262). Starting with the last digit ($j = 0$), we have $2 \cdot 2 = 4$(base 7) and there is no carry, so we simply put 4 in at the right of the second row in the multiplication table results, which is padded with one 0 (one unit shift). Next ($j = 1$), we multiply $2 \cdot 6 = 12 = 1 \cdot 7 + 5$ (base 7) so we put 5 to the left of 4 in the second row of the multiplication table results, and put the carry above the 2 in the next column (in the figure we have 4/1, where the 4 pertains to a carry that will occur in the next digit multiplication). We now add our carry to the last multiplication for this digit ($j = 2$), to get $1 + 2 \cdot 2 = 5$(base 7), so we place a 5 to the left of the last 5 that was entered. We now move on to the left digit ($i = 2$) of the second number, 5. First we multiply ($j = 0$) $5 \cdot 2 = 10 = 1 \cdot 7 + 3$(base 7), so we enter 3 at the right of the third row in the multiplication table results, which is padded with two 0's at the right (two-unit shift), and put the carry (1) on top of the column for the next digit (carries for this multiplication are written before the "/"). We add the carry (1) to the next multiplication: ($j = 1$) $1 + 5 \cdot 6 = 31 = 4 \cdot 7 + 3$(base 7), so we put 3 to the left of the just-entered 3 in the third row of the multiplication table results, and put the carry above 4 in the next column. In the final ($j = 2$) multiplication, we add the carry (4) $4 + 5 \cdot 2 = 14 = 2 \cdot 7 + 0$, so we enter 0 to the left of the 3 that was just entered, and since this was the last digit, we enter the final carry 2 to the left of this 3. Finally, we add up (in base 7) these three computed (and shifted rows), which can be done in two steps using Algorithm 6.2. The resulting multiplication $[262] \times [520] = [212140]$ (base 7) translates into the following integer multiplication: $142 \times 259 = 36,778$.

				4/1	1/	
				2	6	2
			$\times$	5	2	0
				0	0	0
			5	5	4	
	2	0	3	3		
	2	1	2	1	4	0

EFR 6.5

Using Algorithm 6.1, we obtain the binary expansion for the exponent: $225 \sim [11100001]$ (base 2). We initialize cumulative product, $a = 1$, and the cumulative square, $s = 289$. We now proceed through the iterations of Step 2:

With $k = 0$, the corresponding binary digit of the exponent is 1, so we multiply the cumulative product with s to get 289(mod 311): $a = 289$. We update cumulative square (mod 311) for the next iteration: $s \to s^2 = 289^2 \equiv 173$.

With $k = 1$, the corresponding binary digit of the exponent is 0, so we do not multiply the cumulative product with s. We update cumulative square (mod 311) for the next iteration: $s \to s^2 = 173^2 \equiv 73$.

With $k = 2$, the corresponding binary digit of the exponent is 0, so we do not multiply the cumulative product with s. We update cumulative square (mod 311) for the next iteration: $s \to s^2 = 73^2 \equiv 42$

With $k = 3$, the corresponding binary digit of the exponent is 0, so we do not multiply the cumulative product with s. We update cumulative square (mod 311) for the next iteration: $s \to s^2 = 42^2 \equiv 209$.

With $k = 4$, the corresponding binary digit of the exponent is 0, so we do not multiply the cumulative product with s. We update cumulative square (mod 311) for the next iteration: $s \to s^2 = 209^2 \equiv 141$.

With $k = 5$, the corresponding binary digit of the exponent is 1, so we multiply the cumulative product with s to get $a = 40749 \equiv 8 \pmod{311}$. We update cumulative square (mod 311) for the next iteration: $s \to s^2 = 141^2 \equiv 288$.

With $k = 6$, the corresponding binary digit of the exponent is 1, so we multiply the cumulative product with s to get $a = 2304 \equiv 127 \pmod{311}$. We update cumulative square $\pmod{311}$ for the next iteration: $s \to s^2 = 288^2 \equiv 218$. With $k = 7$ (last iteration) the corresponding binary digit of the exponent is 1, so we multiply the cumulative product with s to get $a = 27{,}686 \equiv 7 \pmod{311}$.

So we may conclude final answer to be $289^{225} \equiv 7 \pmod{311}$.

Chapter 7: Block Cryptosystems and the Data Encryption Standard (DES)

EFR 7.1

(a) According to the indicated key scheduling rule, the three round keys are given by (see Table 6.1): $\kappa^1 = 0 \oplus \text{E} = 0000 \oplus 1110 = 1110$, $\kappa^2 = \text{E} \oplus 6 = 1110 \oplus 0110 = 1000$, and $\kappa^3 = 6 \oplus 0 = 0110 \oplus 0000 = 0110$. The left and right halves of P are just D and E: $L_0 = \text{D} = 1101$, $R_0 = \text{E} = 1110$.

$$L_1 = R_0 = 1110$$

Round 1: $R_1 = L_0 \oplus f_{\kappa^1}(R_0) = L_0 \oplus [\kappa^1 \oplus R_0]$
$$= 1101 \oplus [1110 \oplus 1110] = 1101$$
$$L_2 = R_1 = 1101$$

Round 2: $R_2 = L_1 \oplus f_{\kappa^2}(R_1) = L_1 \oplus [\kappa^2 \oplus R_1]$
$$= 1110 \oplus [1000 \oplus 1101] = 1011$$
$$L_3 = R_2 = 1011$$

Round 3: $R_3 = L_2 \oplus f_{\kappa^3}(R_2) = L_2 \oplus [\kappa^3 \oplus R_2]$
$$= 1101 \oplus [0110 \oplus 1011] = 0000$$

The ciphertext is now obtained by switching the left and right halves of the final (third) round: $C = (R_3, L_3) = 00001011 = 0\text{B}$ (in hex format).

(b) To decrypt, we proceed just as above, but we reverse the key sequence: For greater clarity we will denote all left/right intermediate strings using primes. The left and right halves of C are just 0 and B: $L_0' = 0 = 0000$, $R_0' = \text{B} = 1011$.

$$L_1' = R_0' = 1011$$

Round 1: $R_1' = L_0' \oplus f_{\kappa^3}(R_0') = L_0' \oplus [\kappa^3 \oplus R_0']$
$$= 0000 \oplus [0110 \oplus 1011] = 1101$$
$$L_2' = R_1' = 1101$$

Round 2: $R_2' = L_1' \oplus f_{\kappa^2}(R_1') = L_1' \oplus [\kappa^2 \oplus R_1']$
$$= 1011 \oplus [1000 \oplus 1101] = 1110$$

$$L_3' = R_2' = 1110$$

Round 3: $R_3' = L_2' \oplus f_{\kappa^1}(R_2') = L_2' \oplus [\kappa^1 \oplus R_2']$
$$= 1101 \oplus [1110 \oplus 1110] = 1101$$

The plaintext is now obtained by switching the left and right halves of the final (third) round: $C = (R_3', L_3') = 11011110 = DE$ (in hex format), which is, as expected, the original plaintext.

EFR 7.2

(a) *Method 1:* For a one-round Feistel cryptosystem, if the plaintext is $L_0 R_0$, then the ciphertext will be $R_1 L_1$, where $L_1 = R_0$, $R_1 = L_0 \oplus f_{\kappa^1}(R_0)$. The decryption of the ciphertext $L_0' R_0' \triangleq R_1 L_1$ produces $R_1' L_1'$, where $L_1' = R_0'$, $R_1' = L_0' \oplus f_{\kappa^1}(R_0')$. We need to show this is the plaintext $L_0 R_0$, i.e., that $R_1' = L_0$ and that $L_1' = R_0$. Indeed, making use of the previous equations, we obtain $L_1' = R_0' = L_1 = R_0$. Similarly, $R_1' = L_0' \oplus f_{\kappa^1}(R_0') = R_1 \oplus f_{\kappa^1}(L_1) = L_0 \oplus f_{\kappa^1}(R_0) \oplus f_{\kappa^1}(R_0)$ $= L_0 \oplus 0 = L_0$.

Method 2: This approach uses compositions and will be more sophisticated than that of Method 1, but more elegant and easier to generalize. We consider functions on pairs of bit strings (L,R). We define the switch function s by $s(L,R) = (R,L)$. Also given a round function f (so here we will have $f = f_{\kappa^1}$) we define $G(L,R) = (R, L + f(R))$. Thus, $(s \circ G)(L,R) = (L + f(R), R)$, and so $(s \circ G) \circ (s \circ G)(L + f(R), R) = s(R, L \oplus f(R) \oplus f(R)) = s(R, L \oplus 0) = s(R, L) = (L,R)$. This shows that $(s \circ G) \circ (s \circ G) = id$, so $(s \circ G)$ is its own inverse. But $(s \circ G)$ is the one-round Feistel encryption mapping! This proves what we want. For future reference, we note that since s is clearly its own inverse, the equation $(s \circ G) \circ (s \circ G) = id$ also implies that $s = G \circ s \circ G$.

(b) We now move on to the two-round system. Using the prime notation for the decryption strings as per the suggestion, it suffices to show that (i) $L_2' = R_0$ and (ii) $R_2' = L_0$. We start with the given relations $L_0' = R_2$, $R_0' = L_2$, and then use the decryption algorithm to express all primed strings in terms of L_2, R_2:

From the first round of decryption:

$$L_1' = R_0' = L_2,$$

$$R_1' = L_0' \oplus f_{\kappa^2}(R_0') = R_2 \oplus f_{\kappa^2}(L_2).$$

From the second round of decryption:

$$L_2' = R_1' = R_2 \oplus f_{\kappa^2}(L_2),$$

$$R_2' = L_1' \oplus f_{\kappa^1}(R_1') = L_2 \oplus f_{\kappa^1}(R_2 \oplus f_{\kappa^2}(L_2)).$$

Next, we will prove (i) and (ii) by starting with the last two equations and repeatedly substituting the encryption equations until all occurrences of the second round encryption strings L_2, R_2 are replaced by zeroth round encryption strings L_0, R_0.

Proof of (i):

$$L_2' = R_2 \oplus f_{\kappa^2}(L_2) = L_1 \oplus f_{\kappa^2}(R_1) \oplus f_{\kappa^2}(R_1) \quad \text{(since } R_2 = L_1 \oplus f_{\kappa^2}(R_1),$$
$$L_2 = R_1)$$

$$= L_1 \oplus 0 \qquad\qquad\qquad \text{(since } A \oplus A = 0)$$

$$= R_0 \qquad\qquad\qquad \text{(since } L_1 = R_0, \; A \oplus 0 = A)$$

Proof of (ii):

$$R_2' = L_2 \oplus f_{\kappa^1}(R_2 \oplus f_{\kappa^2}(L_2)) = L_2 \oplus f_{\kappa^1}(L_1 \oplus f_{\kappa^2}(R_1) \oplus f_{\kappa^2}(R_1))$$

$$\text{(since } R_2 = L_1 \oplus f_{\kappa^2}(R_1), L_2 = R_1)$$

$$= R_1 \oplus f_{\kappa^1}(L_1 \oplus 0) \qquad\qquad \text{(since } L_2 = R_1, \; A \oplus A = 0)$$

$$= L_0 \oplus f_{\kappa^1}(R_0) \oplus f_{\kappa^1}(R_0 \oplus 0) \qquad \text{(since } R_1 = L_0 \oplus f_{\kappa^1}(R_0), \; L_1 = R_0)$$

$$= L_0 \qquad\qquad\qquad \text{(since } A \oplus 0 = A)$$

Method 3: Using the notation of Method 2 of the one-round Feistel cryptosystem, it follows that the two-round encryption mapping can be written as $E = s \circ G_2 \circ G_1$, and the decryption mapping as $D = s \circ G_1 \circ G_2$ (since the keys are used in the reverse order). Thus, using what we learned in part (a), we have

$$D \circ E = s \circ G_1 \circ (G_2 \circ s \circ G_2) \circ G_1 = s \circ G_1 \circ (s) \circ G_1$$

$$= s \circ (G_1 \circ s \circ G_1) = s \circ (s) = id$$

It should be quite clear now how Method 2 can be used to show the self-encrypting property of the Feistel cryptosystem with any number of rounds.

EFR 7.3

(i) To compute $S_1(1110)$, since the two outer bits are 10 (the binary expansion of 2), we use row 2 of the S-box 1 table, and since the two inner bits are 11 (the binary expansion of 3), we use column 3. Since the row 2, column 3 entry of the table is 0, we have $S_1(1110) = 0$.

(ii) As explained in (i), the answer will be the row 2, column 3 entry of the S-box 2 table, which is 1, i.e., $S_2(1110) = 0$.

EFR 7.4

We start with the ciphertext that was obtained in Example 7.2: $C = 10001011$.

Step 1. We apply initial permutation (from Equation 7.1 $p \mapsto b_2 b_5 b_1 b_3 b_8 b_4 b_7 b_6$): $P = 10001011 \mapsto 01101010$.

Step 2. We apply the two-round Feistel system:

The two keys κ^1, κ^2 have been prepared in the example $\kappa^1 = 00001111$, $\kappa^2 = 01010110$ (but in this decryption algorithm, they get used in the reverse order).

We enter the Feistel system with the left and right halves extracted from the result of Step 1: $01101010 \mapsto L_0 = 0110$, $R_0 = 1010$.

Feistel Round 1:

We first evaluate $f_{\kappa^2}(R_0)$, by following the outline of Figure 7.5:

- Expansion: $E(R_0) = 11000011$ (see Table 7.4)
- XOR with the round key: $\kappa^2 \oplus E(R_0) = 10010101 \triangleq B_1 B_2$
- Apply S-boxes: $C_1 = S_1(B_1) = S_1(1001) = 01$, $C_2 = S_2(B_2) = S_2(0101) = 10$
- Apply final permutation (Table 7.6): $C_1 C_2 = 0110 \to 1001$

So we have $f_{\kappa^2}(R_0) = 1001$.

The results of Round 1 are thus $L_1 = R_0 = 1010$, $R_1 = L_0 \oplus f_{\kappa^2}(R_0) = 0110 \oplus 1001 = 1111$.

Feistel Round 2:

We first evaluate $f_{\kappa^1}(R_1)$, by following the outline of Figure 7.5:

- Expansion: $E(R_1) = 11111111$ (see Table 7.4)

- XOR with the round key: $\kappa^1 \oplus E(R_1) = 11110000 \triangleq B_1 B_2$
- Apply S-boxes: $C_1 = S_1(B_1) = S_1(1111) = 00$, $C_2 = S_2(B_2) = S_2(0000) = 01$
- Apply final permutation (Table 7.6): $C_1 C_2 = 0001 \to 0100$

So we have $f_{\kappa^1}(R_1) = 0100$.

The results of Round 2 are thus $L_2 = R_1 = 1111$, $R_2 = L_1 \oplus f_{\kappa^1}(R_1) = 1010 \oplus 0100 = 1110$.

The result of the Feistel system is $R_2 L_2 = 11101111$.

Step 3. Apply inverse initial permutation (from Equation 7.2 $c_1 c_2 c_3 c_4 c_5 c_6 c_7 c_8 \mapsto c_3 c_1 c_4 c_6 c_2 c_8 c_7 c_5$) to the result of Step 2 to obtain the original plaintext of Example 7.2: $11101111 \mapsto 11011111$.

EFR 7.5

Usually not. Since individual bits are permuted, and hex characters represent four bits, the hex characters in the original plaintext do not get permuted unless the bit permutation is done on adjacent groups of four bits. For example, the hex representation of the IP is 365B26EB85E5DF25, whereas the original plaintext hex representation BB5DEE19586739A9 contained different hex symbols (for example, A appeared in the original plaintext, but not in the IP of it).

EFR 7.6

There are 5 plaintext blocks: $1010100011 \rightarrow P_1 = 10, P_2 = 10,$
$P_3 = 1.0, P_4 = 00, P_5 = 11$

(a) In the electronic codebook mode, the corresponding cipherblocks are just the encryptions of the plaintext blocks using
Table 7.20: $C_1 = E(P_1) = 11, C_2 = E(P_2) = 11, C_3 = E(P_3) = 11,$
$C_4 = E(P_4) = 10, C_5 = E(P_5) = 01 \rightarrow 1111111001.$

(b) We use Equation 7.7 and Table 7.20 to recursively determine the cipherblocks:

$$C_1 = E(P_1 \oplus C_0) = E(10 \oplus 10) = E(00) = 10,$$
$$C_2 = E(P_2 \oplus C_1) = E(10 \oplus 10) = E(00) = 10$$
$$C_3 = E(P_3 \oplus C_2) = E(10 \oplus 10) = E(00) = 10,$$
$$C_4 = E(P_4 \oplus C_3) = E(00 \oplus 10) = E(10) = 11$$
$$C_5 = E(P_5 \oplus C_4) = E(11 \oplus 11) = E(00) = 10 \rightarrow 1010101110$$

EFR 7.7

Since $k = 1$, the plaintext gets processed one bit at a time:
$011110 \rightarrow c_1 = 0, c_2 = 1, c_3 = 1, c_4 = 1, c_5 = 1, c_6 = 0.$ Thus, we
will need to run through Steps 2 and 3 of Algorithm 7.4 six
times:

$i = 1$: $E(S_1) = E(10) = 11.$ So $T_1 = 1, R_1 = 0,$ $p_1 = c_1 \oplus T_1 = 0 \oplus 1 = 1,$
and $S_2 = R_1 c_1 = 00.$

$i = 2$: $E(S_2) = E(00) = 10.$ So $T_2 = 1, R_2 = 0,$ $p_2 = c_2 \oplus T_2 = 1 \oplus 1 = 0,$
and $S_3 = R_2 c_2 = 01.$

$i = 3$: $E(S_3) = E(01) = 00.$ So $T_3 = 0, R_1 = 1,$ $p_3 = c_3 \oplus T_3 = 1 \oplus 0 = 1,$
and $S_4 = R_3 c_3 = 11.$

$i = 4$: $E(S_4) = E(11) = 01.$ So $T_4 = 0, R_4 = 1,$ $p_4 = c_4 \oplus T_4 = 1 \oplus 0 = 1,$
and $S_5 = R_4 c_4 = 11.$

$i = 5$: $E(S_5) = E(11) = 01.$ So $T_5 = 0, R_5 = 1,$ $p_5 = c_5 \oplus T_5 = 1 \oplus 0 = 1,$
and $S_6 = R_5 c_5 = 11.$

$i = 6$: $E(S_6) = E(11) = 01.$ So $T_6 = 0,$ $p_6 = c_6 \oplus T_6 = 0 \oplus 0 = 0.$
Thus, we have recovered the original plaintext:
$p_1 p_2 p_3 p_4 p_5 p_6 = 011110.$

EFR 7.8

We temporarily denote the asserted plaintext subblocks of Step 3
in Algorithm 7.5 as p_i'. The other common strings of Algorithms
7.4 and 7.5 are clearly identical. We must show that $p_i' = p_i$.
Using the definitions of these two quantities in (Step 3 of) the
algorithms, we obtain $p_i' = c_i \oplus T_i = (p_i \oplus T_i) \oplus T_i = p_i.$

EFR 7.9

As in the solution of Example 7.5, the encryption is done bit
by bit with $p_1 = 1, p_2 = 0, p_3 = 1, p_4 = 1, p_5 = 1, p_6 = 0.$

$i = 1$: $E(S_1) = E(10) = 11.$ So $T_1 = 1, R_1 = 0,$ $c_1 = p_1 \oplus T_1 = 1 \oplus 1 = 0,$
and $S_2 = R_1 T_1 = 01.$

$i = 2$: $E(S_2) = E(01) = 00.$ So $T_2 = 0, R_2 = 1,$ $c_2 = p_2 \oplus T_2 = 0 \oplus 0 = 0,$
and $S_3 = R_2 T_2 = 10.$

$i = 3$: $E(S_3) = E(10) = 11$. So $T_3 = 1$, $R_3 = 0$, $c_3 = p_3 \oplus T_3 = 1 \oplus 1 = 0$, and $S_4 = R_3 T_3 = 01$.

$i = 4$: $E(S_4) = E(01) = 00$. So $T_4 = 0$, $R_4 = 1$, $c_4 = p_4 \oplus T_4 = 1 \oplus 0 = 1$, and $S_5 = R_4 T_4 = 10$.

$i = 5$: $E(S_5) = E(10) = 11$. So $T_5 = 1$, $R_5 = 0$, $c_5 = p_5 \oplus T_5 = 1 \oplus 1 = 0$, and $S_6 = R_5 T_5 = 01$.

$i = 6$: $E(S_6) = E(01) = 00$. So $T_6 = 0$, and $c_6 = p_6 \oplus T_6 = 0 \oplus 0 = 0$.

Thus, the ciphertext that gets transmitted is $c_1 c_2 c_3 c_4 c_5 c_6 =$ `000100`.

EFR 7.10

To decrypt in the OFB mode, we need only XOR the ciphertext subblocks with the leftmost subblocks of the encrypted shift registers: $p_i = c_i \oplus T_i$.

Chapter 8: Some Number Theory and Algorithms

EFR 8.1

(a) Since the first bit must be 1, there are $2^{299} \approx 1.019 \times 10^{90}$ 300-bit numbers. The largest integer that can be represented by 300 bits (corresponding to the bit string of 300 1's) is $2^{300-1} \approx 2.037 \times 10^{90}$. Similarly, the largest integer that can be represented by 299 bits is $2^{299-1} \approx 5.093 \times 10^{89}$. Thus there are $\pi(2^{299}) - \pi(2^{298})$ 300-bit prime numbers. The prime number theorem estimates this number to be $2^{299}/\ln(2^{299}) - 2^{298}/\ln(2^{298}) \approx 2.449 \times 10^{87}$.

(b) The number of 300-bit odd numbers is (since the first and last bits must be 1) $2^{298} \approx 5.093 \times 10^{89}$. By part (a), the proportion of 300-bit odd numbers that are prime is approximately $2.449 \times 10^{87} / 5.093 \times 10^{89} \approx 1/2080$, and this is the probability that a randomly selected 300-bit odd number turns out to be prime.

EFR 8.2

Method 1: Fast Modular Exponentiation. We begin with $18^2 \equiv 5 \, (\mathrm{mod} \, 29)$ and continue to square both sides until the exponents exceed at least half of the desired exponent:

$$18^4 \equiv 5^2 \equiv 25$$

$$18^8 \equiv 25^2 \equiv 625 \equiv 16$$

$$18^{16} \equiv 16^2 \equiv 256 \equiv 24$$

$$18^{32} \equiv 25$$

$$18^{64} \equiv 16$$

$$18^{128} \equiv 24$$

$$18^{256} \equiv 25$$

$$18^{512} \equiv 16$$

We are now able to use the aforementioned powers to compute the desired power of 18 (mod 29). This is because of the binary expansion $802 \sim [1100100010]$ (base 2), which is equivalent to $8052 = 512 + 256 + 32 + 2$, as the reader can easily check. It follows that we may compute

$$18^{802} = 18^{512} \cdot 18^{256} \cdot 18^{32} \cdot 18^{2} \equiv 16 \cdot 25 \cdot 25 \cdot 5 \equiv 4 (\text{mod } 29)$$

Method 2: By Fermat's little theorem, $18^{28} \equiv 1 (\text{mod } 29)$. If we apply the division algorithm to the integer division of 802 by 28, we obtain $802 = 28 \cdot 28 + 18$. It follows that $18^{802} \equiv (18^{28})^{28} \cdot 18^{18} \equiv 1^{80} \cdot 4 \equiv 4 \ (\text{mod } 29)$.

EFR 8.3

By factoring each input into primes and then using Equation 8.1 of Proposition 8.3, we obtain

$$\phi(15) = \phi(3 \cdot 5) = (3-1) \cdot 3^{0} \cdot (5-1) \cdot 5^{0} = 8$$

$$\phi(20) = \phi(2^{2} \cdot 5) = (2-1) \cdot 2^{1} \cdot (5-1) \cdot 5^{0} = 8$$

$$\phi(208) = \phi(2^{4} \cdot 13) = (2-1) \cdot 2^{3} \cdot (13-1) \cdot 5^{0} = 96$$

$$\phi(2208) = \phi(2^{5} \cdot 3 \cdot 23) = (2-1) \cdot 2^{4} \cdot (3-1) \cdot 3^{0} \cdot (23-1) \cdot 23^{0} = 704$$

$$\phi(6624) = \phi(2^{5} \cdot 3^{2} \cdot 23) = (2-1) \cdot 2^{4} \cdot (3-1) \cdot 3^{1} \cdot (23-1) \cdot 23^{0} = 2112$$

EFR 8.4

Proof of Euler's Theorem: As in the proof of Fermat's little theorem, we will denote elements of $\mathbb{Z}_m$ (the integers mod m) using square brackets: $[k]$ represents the set of all integers that are congruent to k (mod m). We will construct a function with domain and codomain both being the set A of mod m integers that are relatively prime to m: $f : A \to A$, defined by $f([x]) = [ax]$. By Proposition 2.10, this definition will give the same output no matter which representative we use of $[x]$, so it is a well-defined function on elements of A. But we still need to check that the images, which *a priori* could be any mod m integers, are actually in the set A (i.e., so the codomain of the function can be taken to be A.) Indeed, $[ax]$ did not belong to A; this would mean that ax is not relatively prime to m. But since both a and x are relatively prime to m, their product ax also must be, so indeed $[ax] \in A$.

Next we will show that f is one-to-one. Suppose that $f([x]) = f([y])$. This means that $[ax] = [ay]$, or $ax \equiv ay \ (\text{mod } m)$. By definition, this means that $m \mid (ax - ay)$ or $m \mid a(x - y)$. But since $\gcd(m, a) = 1$, it follows that each of the prime powers in the factorization of m must divide into the latter factor $x - y$, and thus, so must their product m, so $m \mid x - y$, which means that $[x] = [y]$, so f is one-to-one.

Since f is a one-to-one function of the set $A = \{[a_1], [a_2], \cdots,$ $[a_K]\}$ (note that $K = \phi(m)$) to itself, it follows that the images of f: $f([a_1])$, $f([a_2])$, $\cdots$, $f([a_K])$, are simply a relisting of the elements of A, in perhaps a different order. It follows that if we multiply representatives from these to listings of the set A, we will get the same result (mod n):

$$a_1 \cdot a_2 \cdot \ \ldots \ \cdot a_K \equiv (a \cdot a_1) \cdot (a \cdot a_2) \cdot \ \ldots \ \cdot (a \cdot a_K) \equiv a^{K-1}(a_1 \cdot a_2 \cdot \ \ldots \ \cdot a_K)(\bmod\ m)$$

This equation implies that $p \mid [a^{K-1}(a_1 \cdot a_2 \cdot \ \ldots \ \cdot a_K) - a_1 \cdot a_2 \cdot \ \ldots \ \cdot a_K]$ or $m \mid [(a^{\phi(m)} - 1)(a_1 \cdot a_2 \cdot \ \cdots \ \cdot a_K)]$, and the same argument used above shows that since m has no common factors with $a_1 \cdot a_2 \cdot \ \ldots \ \cdot a_K$, we must have $m \mid (a^{\phi(m)} - 1)$, so that $a^{\phi(m)} \equiv 1$ (mod m), as we wished to prove. $\square$

EFR 8.5

(a) Since $\gcd(7,58) = 1$, and $\phi(58) = \phi(2 \cdot 29) = (2-1) \cdot 2^0 \cdot (29-1) \cdot 29^0 = 28$, Euler's theorem tells us that $7^{28} \equiv 1 (\bmod\ 58)$. Using the division algorithm for the integer division of 8486 by 28 gives $8486 = 303 \cdot 28 + 2$, and consequently $7^{8486} \equiv (7^{28})^{303} \cdot 7^2 \equiv 1 \cdot 49 \equiv 49 (\bmod\ 58)$.

(b) Finding the last three digits of any number is the same as the answer we would get by converting it to an integer modulo 1000. Thus, we wish to find 13^{2017} (mod 1000). Since $\gcd(13,1000) = 1$, and $\phi(1000) = \phi(2^3 \cdot 5^3) = (2-1) \cdot 2^2 \cdot (5-1) \cdot 5^2 = 400$, Euler's theorem tells us that $13^{400} \equiv 1$ (mod 1000). Applying the division algorithm to the given exponent divided by 400 gives $2017 = 5 \cdot 4 + 17$; hence, $13^{2017} \equiv (13^{400})^5 \cdot 13^{17} \equiv 1 \cdot 13 \equiv 933 (\bmod\ 1000)$. (In the last computation we used Algorithm 6.5.)

EFR 8.6

The tables below illustrate all of the modular powers [up to the $\phi(n)$th] for $n = 4$ [with $\phi(4) = 2$)] and $n = 9$ [with $\phi(9) = 6$]. The column entry giving the order of each element a is highlighted. From the tables we see that there is one primitive root mod 4: 3, and there are two primitive roots mod 9: 2 and 5.

(a)

	a^k (mod 4)	
	$k = 1$	$k = 2$
$a = 1$	1	1
$a = 3$	3	1

(b)

	a^k (mod 9)					
	$k = 1$	$k = 2$	$k = 3$	$k = 4$	$k = 5$	$k = 6$
$a = 1$	1	1	1	1	1	1
$a = 2$	2	4	8	7	5	1
$a = 4$	4	7	1	4	1	1
$a = 5$	5	7	8	4	2	1
$a = 7$	7	4	1	7	4	1
$a = 8$	8	1	8	1	8	1

EFR 8.7

(a) The prime factorization of 334 is $2 \cdot 167$, so by Theorem 8.7, 334 has $\phi(334) = 166$ primitive roots.

(b) By Proposition 8.5(b), the orders of any integer relatively prime to 334(mod 334) must divide $\phi(334) = 166 = 2 \cdot 83$, so the only possible orders are 1, 2, 83, and 166 (in which case we have a primitive root). To find the smallest primitive root, we go through positive integers a relatively prime to 334, starting with 3, computing (using Algorithm 6.5) the modular powers a^2, a^{83} (mod 334) until we find an integer whose order is 166: $3^2 \equiv 9, 3^{83} \equiv 1 \pmod{334}$ [so $\text{ord}_{334}(3) = 83$], $5^2 \equiv 25, 5^{83} \equiv 333 \pmod{334}$ so $\text{ord}_{334}(5) = 166$, and 5 is the smallest primitive root (mod 334).

(c) First of all, the order is only defined for the $\phi(334) = 166$ that are relatively prime to 334. These 166 modular integers are accounted for as powers (from 1 to 166) of our primitive root 5 by Proposition 8.6. But Proposition 8.9 tells us that $\text{ord}_n(5^j) = \dfrac{\text{ord}_n(5)}{\gcd(j, \text{ord}_n(5))} = \dfrac{166}{\gcd(j, 166)}$. The only way this can equal 2 is if the denominator is 83, i.e., $j = 83$. So there is only one element of order 2. The same argument shows that there are 83 integers (mod 334) that have order 83, and the remaining 82 are primitive roots [they will correspond to powers 5^j with $\gcd(j, 166) = 1$].

EFR 8.8

(a) $n = 2581$

Step 1. Initialize the trial counter $i = 0$.

Step 2. Randomly generate a base: $a = 2102$. We use Algorithm 6.5 to compute the modular power $a^{n-1} \equiv 2102^{2580} \equiv 2137$ (mod 2581).

Step 3. Since $a^{n-1} \not\equiv 1 \pmod{n}$, Fermat's test has proved that $n = 2581$ is composite with witness $a = 2102$. The reader may check this by factoring $2581 = 29 \cdot 89$.

(b) $n = 1889$

Step 1. Initialize the trial counter $i = 0$.

Step 2. Randomly generate a base: $a = 1710$. We use Algorithm 6.5 to compute the modular power $a^{n-1} \equiv 1710^{1888} \equiv 1 \pmod{1889}$. So 1889 has passed Fermat's primality test with this value of a; since i is now 1, we repeat Step 2.

Step 2. (Second repetition) Randomly generate a base: $a = 241$. We use Algorithm 6.5 to compute the modular power $a^{n-1} \equiv 241^{1888} \equiv 1 \pmod{1889}$. So 1889 has passed Fermat's primality test with this value of a; since i is now 2, we repeat Step 2.

Step 2. (Third repetition) Randomly generate a base: $a = 1724$. We use Algorithm 6.5 to compute the modular power

$a^{n-1} \equiv 1724^{1888} \equiv 1 \pmod{1889}$. So 1889 has passed Fermat's primality test with this value of a; since i is now 3, we repeat Step 2.

Step 2. (Fourth repetition) Randomly generate a base: $a = 1194$. We use Algorithm 6.5 to compute the modular power $a^{n-1} \equiv 1194^{1888} \equiv 1 \pmod{1889}$. So 1889 has passed Fermat's primality test with this value of a; since i is now 4, this was the final iteration of Step 2.

Step 3. Declare 1889 as probably prime. (The reader may check that 1889 is indeed prime, so the test worked.)

EFR 8.9

(a) $n = 2581$

Step 1. First we express $n - 1 = 2580$ as $2^f \cdot m$, where $f = 2$, and $m = 645$. Initialize counter $i = 0$.

Step 2. Randomly generate a base: $a = 253$. We compute $A_0 = a^m \equiv 253^{645} \equiv 2370 \pmod{2581}$. Since $A_0 \not\equiv \pm 1 \pmod{n}$, we enter into the FOR loop, but the FOR loop runs FOR $j = 1$ TO $f - 2 = 0$, which means it is an empty loop. Since the (empty) FOR loop has completed (without finding n composite or probably prime), we do one last squaring: $A_1 \equiv A_0^2 \equiv 2370^2 \equiv 644 \pmod{721}$. Since $A_1 \not\equiv -1 \pmod{2581}$, this proves that n is composite, with witness $a = 645$.

(b) $n = 1889$

• *Step 1.* First we express $n - 1 = 1888$ as $2^f \cdot m$, where $f = 5$, and $m = 59$. Initialize counter $i = 0$.

Step 2. Randomly generate a base: $a = 527$. We compute $A_0 = a^m \equiv 527^{59} \equiv 1648 \pmod{1889}$. Since $A_0 \not\equiv \pm 1 \pmod{n}$, we enter into the FOR loop of Step 2, in which we compute $A_1 \equiv A_0^2 \equiv 1411 \pmod{1889}$, $A_2 \equiv A_1^2 \equiv 1804 \pmod{1889}$, $A_3 \equiv A_2^2 \equiv 1558 \pmod{1889}$. Since the FOR loop has completed (without finding n composite or probably prime), we do one last squaring: $A_4 \equiv A_3^2 \equiv 1888 \pmod{1889}$. Since $A_4 \equiv -1$, this trial finds n to be probably prime. Since i is now 1, we repeat Step 2.

Step 2. (Second repetition) Randomly generate a base: $a = 1033$. Since $A_0 \not\equiv \pm 1 \pmod{n}$, we enter into the FOR loop of Step 2, in which we compute $A_1 \equiv A_0^2 \equiv 1411 \pmod{1889}$ (*Note*: At this point we could realize that the rest of Step 2 must be the same as that for the previous iteration—Why?), $A_2 \equiv A_1^2 \equiv 1804 \pmod{1889}$, $A_3 \equiv A_2^2 \equiv 1558 \pmod{1889}$. Since the FOR loop has completed (without finding n composite or probably prime), we do one last squaring: $A_4 \equiv A_3^2 \equiv 1888 \pmod{1889}$. Since $A_4 \equiv -1$, this trial finds n to be probably prime. Since i is now 2, this was the final iteration of Step 2.

Step 3. Declare 1889 as probably prime. (The reader may check that 409 is indeed prime so the test worked.)

EFR 8.10

We apply Pollard's $p - 1$ test to $n = 12637211$

Step 1. We compute $a^{B!} \pmod{n}$ by using fast modular exponentiation $B = 15$ times:

$$2^{1!} \equiv 2, \, 2^{2!} \equiv (2^{1!})^2 \equiv 4, \, 2^{3!} \equiv (2^{2!})^3 \equiv 64, \, 2^{4!} \equiv (2^{3!})^4 \equiv 4{,}140{,}005,$$

$$2^{5!} \equiv (2^{4!})^5 \equiv 1{,}682{,}197, \qquad\qquad 2^{6!} \equiv (2^{5!})^6 \equiv 2{,}045{,}693,$$

$$2^{7!} \equiv (2^{6!})^7 \equiv 9{,}957{,}770, \qquad\qquad 2^{8!} \equiv (2^{7!})^8 \equiv 7{,}623{,}286,$$

$$2^{9!} \equiv (2^{8!})^9 \equiv 4{,}873{,}635, \qquad\qquad 2^{10!} \equiv (2^{9!})^{10} \equiv 1{,}816{,}279,$$

$$2^{11!} \equiv (2^{10!})^{11} \equiv 4{,}226{,}082, \qquad\qquad 2^{12!} \equiv (2^{11!})^{12} \equiv 454{,}276,$$

$$2^{13!} \equiv (2^{12!})^{13} \equiv 2{,}248{,}423, \qquad 2^{14!} \equiv (2^{13!})^{14} \equiv 2{,}796{,}135,$$

$$2^{15!} \equiv (2^{14!})^{15} \equiv 2{,}796{,}135$$

Step 2. We use the Euclidean algorithm (Algorithm 2.1) to compute $d = \gcd(a^{B!} - 1, n) = \gcd(2{,}796{,}135, 12{,}637{,}211) = 3001$. This is the outputted nontrivial factor of n.

Chapter 9: Public Key Cryptography

EFR 9.1

(a) By Proposition 8.5(b), for any integer $a \in \mathbb{Z}_{79}^{\times}$, $\mathrm{ord}_{79}(a) \mid \phi(79) = 78 = 2 \cdot 3 \cdot 13$, so the only possibilities for $\mathrm{ord}_{79}(a)$ are $(1,) 2, 3, 6, 13, 26, 39$, and 78, and we need only look at these modular powers of a to check if a is a primitive root. Starting with $a = 2$, we have $2^2 \equiv 4$, $2^3 \equiv 8$, $2^6 \equiv 64$, $2^{13} \equiv 23$, $2^{39} \equiv 1 \pmod{79}$, so $\mathrm{ord}_{79}(2) = 39$. Next with $a = 3$, we have $3^2 \equiv 9$, $3^3 \equiv 27$, $3^6 \equiv 18$, $3^{13} \equiv 24$, $3^{39} \equiv 78 \pmod{79}$, so $\mathrm{ord}_{79}(3) = 78$, and $g = 3$ is the smallest primitive root.

(b) Bob (computes and) sends Alice the number $B = g^b \equiv 3^{33} \equiv 57 \pmod{79}$, and Alice sends Bob the number $A = g^a \equiv 3^{51} \equiv 71 \pmod{79}$. On her end, Alice computes the common (secret) key as $K = B^a \equiv 57^{51} \equiv 61 \pmod{79}$, and Bob computes it as $K = A^b \equiv 71^{33} \equiv 61 \pmod{79}$.

EFR 9.2

(a) Since $\phi(n) = (p-1)(q-1) = 66 \cdot 36 = 2376$, e will be the smallest integer greater than 1000 that is relatively prime to $\phi(n) = 2376$. Since e must be odd, we begin with 1001 and compute $\gcd(1001, \phi(n)) = 11$, and then move on to check $\gcd(1003, \phi(n)) = 1$, so $e = 1003$ is a legitimate encryption exponent. To encrypt the plaintext $P = 2012$, we raise P to the power of $e \pmod{n}$ (using Algorithm 6.5) to obtain $C \equiv P^e \equiv 2012^{1003} \equiv 2095 \pmod{n}$.

(b)   We use the extended Euclidean Algorithm 2.2 to compute $d \equiv e^{-1} \equiv 1843 \,(\mathrm{mod}\,\phi(n))$, which is the decryption exponent. To decrypt the ciphertext, we need to raise it to the power of decryption exponent $d = 1843 \,(\mathrm{mod}\,n = 2479)$: $2095^{1843} \equiv 2012$ (mod 2479)—the original plaintext message.

EFR 9.3

(a)   From Algorithm 9.2, Alice's (private key) decryption mapping is $D(m) \equiv m^d \,(\mathrm{mod}\,n)$, where $n = pq = 6887$. Thus, by Algorithm 9.3, her digital signature will be $s = D(P) \equiv P^d \equiv 5^{139} \equiv 1903 \,(\mathrm{mod}\,6887)$.

(b)   To verify Alice's signature, according to Algorithm 9.3, Bob must apply Alice's (public key) encryption mapping $E(m) \equiv m^e \,(\mathrm{mod}\,n)$, to the purported signature $s = 1903$: $s^e \equiv 1903^{2569} \equiv 5 \,(\mathrm{mod}\,6887)$. Since this is the original document P, Alice's signature has been verified.

(c)   Alice would use Bob's public key to encrypt her digitally signed document, and thus send him $E_B(s) \equiv s^{e_B} \equiv 1903^{1007} \equiv 409 \,(\mathrm{mod}\,6887)$.

(d)   Bob's private key can be computed as $d_B \equiv e_B^{-1} \equiv 3023$ $(\mathrm{mod}\,\phi(n))$. He would first use this to decrypt Alice's digital signature: $s \equiv 409^{3023} \equiv 1903 \,(\mathrm{mod}\,6887)$, and he would then proceed as in the solution to part (b) to verify this signature.

EFR 9.4

(a)   Since $n = pq$ and $\phi(n) = (p-1)(q-1) = n - p - q + 1$ are known, we can solve the first equation for $q = n/p$ and substitute this into the latter equation to obtain:

$$\phi(n) = (p-1)(q-1) = n - p - \frac{n}{p} + 1 \;\Rightarrow\; p + \frac{n}{p}$$

$$= n + 1 - \phi(n) \;\Rightarrow\; p^2 - [n+1-\phi(n)]p + n = 0$$

The last equation is a quadratic equation in p (with all other quantities being known), so we can solve it for p using the quadratic formula

$$p = \frac{[n+1-\phi(n)] \pm \sqrt{[n+1-\phi(n)]^2 - 4n}}{2}$$

One of the solutions is p and the other is q (since the roles of p and q are interchangeable).

(b)   Using the formula of part (a) with $n = 628883$ and $\phi(n) = 627288$, since the bracketed expression is $n + 1 - \phi(n) = 1596$, we obtain

$$p,q = \frac{[1596] \pm \sqrt{[1596]^2 - 4 \cdot 628883}}{2}$$

$$= \frac{1596 \pm \sqrt{31,684}}{2} = \frac{1596 \pm 178}{2} = 709,\; 887$$

EFR 9.5

(a) By Proposition 8.5(b), for any integer $k \in \mathbb{Z}_{1231}^{\times}$, $\mathrm{ord}_{1231}(k) \mid$ $\phi(1231) = 1230 = 2 \cdot 3 \cdot 5 \cdot 41$, so the only possibilities for $\mathrm{ord}_{1231}(k)$ are (1,) 2, 3, 5, 6, 10, 15, 41, 82, 123, 205, 246, 410, and 1230, and we need only look at these modular powers of k to check if k is a primitive root. Starting with $k = 701$, a series of simple exponentiations shows that all but the last of these modular powers is different from 1, so $\mathrm{ord}_{1231}(k) = 1230$, and so $g = 701$ is the desired primitive root.

(b) Bob's public key is $B \equiv g^b \equiv 701^{954} \equiv 143 (\mathrm{mod}\, 1231)$, and Alice's is $A \equiv g^a \equiv 701^{212} \equiv 990 (\mathrm{mod}\, 1231)$. Alice computes $C \equiv B^a P \equiv 144 \cdot 44 \equiv 181 (\mathrm{mod}\, 1231)$. Thus, the entire ciphertext would be $(A, C) = (990, 181)$.

(c) The decryption exponent is $p - 1 - b = 276$ and $A^{276} \equiv 143^{276}$ $\equiv 966 (\mathrm{mod}\, 1231)$, and so $d_\kappa((A,C)) \equiv A^{p-1-b} C \equiv 966 \cdot 181 \equiv$ ⌐ $44 (\mathrm{mod}\, 1231)$, as expected.

EFR 9.6

(a) We first use the extended Euclidean algorithm (Algorithm 2.2) to compute $d^{-1} \equiv 73 (\mathrm{mod}\, 1230)$. Now we may compute both $r \equiv g^d \equiv 701^{337} \equiv 250 (\mathrm{mod}\, 1231)$ and $s \equiv d^{-1}(P - ar) \equiv 73 \cdot$ $(44 - 212 \cdot 250) \equiv 102 (\mathrm{mod}\, 1230)$. Thus, the digitally signed document that Alice sends to Bob is $(P, r, s) = (44, 250, 102)$.

(b) To authenticate that the document 44 (the first component of the vector received from Alice) was really signed by her, Bob uses the second two components of the vector received, along with Alice's public ElGamal key $A = 990$, computes $A^r r^s \equiv 290 \cdot 69 \equiv 314 (\mathrm{mod}\, 1231)$, and compares this with $g^P \equiv 701^{44} \equiv 314 (\mathrm{mod}\, 1231)$. Since the results agree, Alice's signature has been authenticated.

EFR 9.7

Suppose that Mallory has obtained two ElGamal digital signatures, (P, r, s) and (P', r, s'), that were created by Alice for different plaintexts ($P \neq P'$), but using the same signature exponent d [and hence sharing the same value of $r \equiv g^d (\mathrm{mod}\, p)$]. If we subtract the equations $s \equiv d^{-1}$ $(P - ar)$ and $s' \equiv d^{-1}(P' - ar)(\mathrm{mod}\, p - 1)$, we obtain that $s - s' \equiv d^{-1}(P - P')(\mathrm{mod}\, p - 1)$. Now, if it turns out that $P - P'$ (which Mallory knows) is relatively prime to $p - 1$, then she could use the extended Euclidean algorithm to compute $t \equiv (P - P')^{-1}(\mathrm{mod}\, p - 1)$, and multiply the previous congruence by this to obtain $s - s' \equiv d^{-1}(P - P') \Rightarrow$ $(s - s')t \equiv d^{-1}(P - P')(P - P')^{-1}(\mathrm{mod}\, p - 1)$, or $d^{-1} \equiv (s - s')t$ $(\mathrm{mod}\, p - 1)$. From this, she could use the extended Euclidean algorithm once more to recover Alice's secret decryption exponent d. Next, if r is also invertible (mod $p - 1$), then Mallory could compute $r^{-1}(\mathrm{mod}\, p - 1)$, and use this to solve for Alice's private key a as follows:

$$s \equiv d^{-1}(P - ar) \Rightarrow ds \equiv P - ar \Rightarrow ar$$

$$\equiv P - ds \Rightarrow a \equiv (P - ds)r^{-1} \pmod{p-1}$$

She would now be able to read all of Alice's confidential messages (that the latter receives on her ElGamal cryptosystem) and be able to send out bogus documents with Alice's authentic ElGamal digital signature!

EFR 9.8

(a) $a_2 = 5 > 3 = a_1$, $a_3 = 9 > 8 = a_1 + a_2$, $a_4 = 18 > 17 = a_1 + a_2 + a_3$, $a_5 = 36 > 35 = a_1 + a_2 + a_3 + a_4$, $a_6 = 100 > 71 = a_1 + a_2 + a_3 + a_4 + a_5$. This shows that $[a_1,\ a_2,\ a_3,\ a_5,\ a_5,\ a_6]$ is superincreasing.

(b) We use Algorithm 9.6:

Step 1. $S = 27$, Index $= 6$.

Step 2. Since $S < a_6 = 100$, we set $x_6 = 0$, and update Index $\to 6 - 1 = 5$.

Step 2. (Second iteration) Since $S < a_5 = 36$, we set $x_5 = 0$, and update Index $\to 5 - 1 = 4$.

Step 2. (Third iteration) Since $S \geq a_4 = 18$, we set $x_4 = 1$, update $S \to 27 - 18 = 9$ and update Index $\to 4 - 1 = 3$.

Step 2. (Fourth iteration) Since $S \geq a_3 = 9$, we set $x_3 = 1$, update $S \to 9 - 9 = 0$ and update Index $\to 3 - 1 = 2$.

Since S is now 0, the remaining two steps clearly set $x_2 = x_1 = 0$, and thus the unique solution to the given superincreasing knapsack problem is $[x_1,\ x_2,\ x_3,\ x_5,\ x_5,\ x_6] = [0,\ 0,\ 1,\ 1,\ 0,\ 0]$.

EFR 9.9

(a) Since the binary expansion of $\sum_{i=1}^{k} 2^i$ is a vector of k 1's, it follows (from the binary expansion development of Chapter 6) that this integer is one less than the next binary integer, which is a single 1 followed by k 0's. Since this latter binary vector represents the integer 2^{k+1}, it follows that the given sequence is superincreasing (but just barely, since each term is exactly one more than the sum of the previous terms).

(b) This can be accomplished using (strong) mathematical induction. Basis Step: $i = 1$: Since a_1 is a positive integer, it must be at least $1 = 2^{i-1}$. Inductive Step: We assume $a_i \geq 2^{i-1}$ for all indices i up to some fixed index k. It suffices to show that the relation holds for the index $k + 1$: $a_{k+1} \geq 2^{(k+1)-1}$. But the superincreasing assumption tells us that $a_{k+1} > a_k + \cdots + a_2 + a_1$. By our inductive hypothesis, $a_k + \cdots + a_2 + a_1 \geq \sum_{i=1}^{k} 2^{i-1}$, and by the solution of part (a), this latter sum equals $2^k - 1$. It therefore follows that $a_{k+1} \geq 2^k = 2^{(k+1)-1}$, as we needed to show. $\square$

EFR 9.10

(a) Working mod $m = 201$, we have $w \cdot [a_1 \ a_2 \ a_3 \ a_4 \ a_5 \ a_6] \equiv 77 \cdot [3, 5, 9, 18, 36, 100] \equiv [30, 184, 90, 180, 159, 62]$. Thus, this latter vector is Bob's public key $[b_1 \ b_2 \ b_3 \ \cdots \ b_n]$.

(b) The ciphertext is $f_b([1, 1, 1, 0, 0, 0]) = x_1 b_1 + x_2 b_2 + x_3 b_3 + x_4 b_4 + x_5 b_5 + x_6 b_6 = 1 \cdot 30 + 1 \cdot 184 + 1 \cdot 90 + 0 \cdot 180 + 0 \cdot 159 + 0 \cdot 62 = 304 = s$.

(c) Using the extended Euclidean algorithm (Algorithm 2.2), we compute $w^{-1} \equiv 47$. (This need be computed only once and could be supplied with the rest of Bob's private key.) Since $w^{-1} \cdot s \equiv 47 \cdot 304 \equiv 17 (\mathrm{mod}\, m)$, the plaintext will be the solution of the superincreasing knapsack problem with weight vector $[a_1 \ a_2 \ a_3 \ a_4 \ a_5 \ a_6]$ and knapsack weight $s' = 17$. Algorithm 9.6 quickly produces the original plaintext.

EFR 9.11

The values of b and D turn out exactly as in the solution of Example 9.10, except that the factor 557 is found one iteration sooner (when $t = 3$, instead of $t = 2$).

Chapter 10: Finite Fields in General, and $GF(2^8)$ in Particular

EFR 10.1

Proof:

(1) If a and b have multiplicative inverses a^{-1} and b^{-1}, respectively, then $(ab) \cdot (b^{-1} a^{-1}) = a(bb^{-1})a^{-1} = a \cdot 1 \cdot a^{-1} = a \cdot a^{-1} = 1$, thus $(ab)^{-1} = b^{-1} a^{-1}$, so $ab \in R^{\times}$.

(2) Since $(a^{-1})^{-1} = a$, it follows that the inverse of an invertible element is invertible.

(3) Since $a \cdot 1 = a$ for all $a \in R$, this will still, of course, hold when a is an element of the smaller set $R^{\times}$. $\square$

EFR 10.2

(a) The addition and multiplication tables are shown below:

+	0	1	a		·	0	1	a
0	0	1	a		**0**	0	0	0
1	1	a	0		**1**	0	1	a
a	a	0	1		**a**	0	a	1

(b) There are only two nonzero elements in F: 1 and a. The multiplicative identity is always its own inverse (in any ring), and the rule $a \cdot a = 1$ shows that a is also its own inverse. Thus, all nonzero elements have inverses, so F is a field.

EFR 10.3

Using the XOR addition and the multiplication table, we may compute:

$$10 \cdot (01 \oplus 10) = 10 \cdot 11 = 01 = 10 \oplus 11 = 10 \cdot 01 \oplus 10 \cdot 10,$$
$$10 \cdot (01 \oplus 01) = 10 \cdot 00 = 00 = 10 \cdot 01 \oplus 10 \cdot 01,$$
$$10 \cdot (01 \oplus 11) = 10 \cdot 10 = 11 = 10 \oplus 01 = 10 \cdot 01 \oplus 10 \cdot 11,$$
$$10 \cdot (10 \oplus 10) = 10 \cdot 00 = 00 = 10 \cdot 10 \oplus 10 \cdot 10,$$
$$10 \cdot (10 \oplus 11) = 10 \cdot 01 = 10 = 11 \oplus 01 = 10 \cdot 10 \oplus 10 \cdot 11,$$
$$10 \cdot (11 \oplus 11) = 10 \cdot 00 = 00 = 10 \cdot 11 \oplus 10 \cdot 11,$$
$$11 \cdot (01 \oplus 10) = 11 \cdot 11 = 10 = 11 \oplus 01 = 11 \cdot 01 \oplus 11 \cdot 10,$$
$$11 \cdot (01 \oplus 01) = 11 \cdot 00 = 00 = 11 \cdot 01 \oplus 11 \cdot 01,$$
$$11 \cdot (01 \oplus 11) = 11 \cdot 10 = 01 = 11 \oplus 10 = 11 \cdot 01 \oplus 11 \cdot 11,$$
$$11 \cdot (10 \oplus 10) = 11 \cdot 00 = 00 = 11 \cdot 10 \oplus 11 \cdot 10,$$
$$11 \cdot (10 \oplus 11) = 11 \cdot 01 = 11 = 01 \oplus 10 = 11 \cdot 10 \oplus 11 \cdot 11,$$
$$11 \cdot (11 \oplus 11) = 11 \cdot 00 = 00 = 11 \cdot 11 \oplus 11 \cdot 11$$

EFR 10.4

(a) Working mod 5 on the coefficients, we obtain

$$(X^5 + 4X^3 + 2) + (3X^3 + 2X) \equiv X^5 + 7X^3 + 2X + 2 \equiv X^5 + 2X^3 + 2X + 2$$

$$(X^5 + 4X^3 + 2) \cdot (3X^3 + 2X) \equiv (X^5 + 4X^3 + 2) \cdot 3X^3 + (X^5 + 4X^3 + 2) \cdot 2X$$

$$\equiv 3X^8 + 2X^6 + X^3 + 2X^6 + 3X^4 + 4X$$

$$\equiv 3X^8 + 4X^6 + 3X^4 + X^3 + 4X$$

(b) If we distribute the multiplication among the terms of the second factor, it becomes evident that most of the terms cancel:

$$(X^n + X^{n-1} + X^{n-2} + \cdots + X + 1) \cdot (X - 1)$$

$$\equiv (X^n + X^{n-1} + X^{n-2} + \cdots + X + 1) \cdot X - (X^n + X^{n-1} + X^{n-2} + \cdots + X + 1) \cdot 1$$

$$\equiv [X^{n+1} + X^n + X^{n-1} + \cdots + X^2 + X] - [X^n + X^{n-1} + X^{n-2} + \cdots + X + 1]$$

$$\equiv X^{n+1} - 1 \equiv X^{n+1} + p - 1$$

EFR 10.5

(a) Since (as was pointed out in the solution of Example 10.8) $X^2 - 1$ factors as $(X + 1)(X - 1)$ in any $\mathbb{Z}_p[X]$, it follows that $X^2 - 1$ is not irreducible in any $\mathbb{Z}_p[X]$.

(b) The solution of Example 10.8 showed that $X^2 + 1 = (X + 1)^2$, so $X^2 + 1$ is not irreducible in $\mathbb{Z}_2[X]$. We show that $X^2 + 1$ is irreducible in $\mathbb{Z}_p[X]$ for any prime $p > 3$. To do this, we need to show that $X^2 + 1$ can have no linear factor $X + a$ (with leading coefficient 1), where $a \in \mathbb{Z}_p$. Indeed, if there were such a linear factor, then we would necessarily have $X^2 + 1 = (X + a)(bX + c)$, for some modular integers $b, c \in \mathbb{Z}_p$. Multiplying out the right side produces

$$X^2 + 1 = bX^2 + (ab + c)X + ac \implies b \equiv 1,\ ab + c \equiv 0,\ ac \equiv 1$$

Substituting the first equation into the second gives $a + c \equiv 0 \Rightarrow c \equiv -a$, so the third equation now becomes $-a^2 \equiv 1 \Rightarrow a^2 \equiv -1 \pmod{p}$. All of this has shown that $X^2 + 1$ is irreducible in $\mathbb{Z}_p[X] \Leftrightarrow -1$ does not have a mod p square root.

Since $-1 \equiv 1 \pmod 2$ is its own square root (mod 2), this confirms the first result that $X^2 + 1$ is not irreducible in $\mathbb{Z}_2[X]$. It can easily be checked whether -1 has mod p square root, for the remaining values $p = 3, 5, 7$. For $p = 3$, since $1^2 \equiv 1, 2^2 \equiv 1$, we see that -1 does not have a mod 3 square root, so that $X^2 + 1$ is irreducible in $\mathbb{Z}_3[X]$. Similarly, one finds that -1 does not have a mod 7, so that $X^2 + 1$ is irreducible in $\mathbb{Z}_7[X]$. On the other hand, since $2^2 \equiv 4 \equiv -1 \pmod 5$, -1 *does* have a mod 5 square root, so that $X^2 + 1$ is not irreducible in $\mathbb{Z}_5[X]$.

Note: In general, it can be shown (using primitive roots) that -1 has a mod p square root if, and only if, $p = 2$, or $p \equiv 1 \pmod 4$; see Exercise 30.

EFR 10.6

The two divisions are performed according to the schematic diagrams below.

Coefficient Work Is mod 2:	Coefficient Work Is mod 7:

Coefficient Work Is mod 2:

$$
\begin{array}{r}
X^2 \qquad\qquad\qquad \\
X^3 + 1 \overline{)\ X^5 + 0X^4 + 0X^3 + 0X^2 + 0X + 1} \\
\underline{-X^5 \qquad\qquad -X^2 \qquad\qquad} \\
X^2 \qquad + 1
\end{array}
$$

Coefficient Work Is mod 7:

$$
\begin{array}{r}
X^3 \qquad + 2X + 4 \\
X^2 + 5 \overline{)\ X^5 + 0X^4 + 0X^3 + 4X^2 + 0X + 0} \\
\underline{-X^5 \qquad -5X^3 \qquad\qquad} \\
2X^3 + 4X^2 \qquad \\
\underline{-2X^3 \qquad -10X} \\
4X^2 + 4X \\
\underline{-4X^2 \qquad -20} \\
4X + 1
\end{array}
$$

(a) From the first schematic diagram, we read off the quotient $= X^2$, and the remainder $= X^2 + 1$, so that $X^5 + 1 = X^2 \cdot (X^3 + 1) + X^2 + 1$ in $\mathbb{Z}_2[X]$.

(b) From the second schematic diagram, we read off the quotient $= X^3 + 2X + 4$, and the remainder $= 4X + 1$, so that $X^5 + 4X^2 = (X^3 + 2X + 4) \cdot (X^2 + 5) + 4X + 1$ in $\mathbb{Z}_7[X]$.

EFR 10.7

If the remainder in the division algorithm of f by g is zero, this means that $f = q \cdot g + 0 = q \cdot g$ in $\mathbb{Z}_p[X]$, which means that $g \mid f$ [note that since $\deg(g) > 0$, g cannot be 0]. Conversely, from $g \mid f$ we get that $f = q \cdot g = q \cdot g + 0$, for some $q \in \mathbb{Z}_p[X]$, and by uniqueness of the quotient and remainder in the division algorithm (Proposition 10.6), it follows that the remainder in the division algorithm of f by g is zero.

EFR 10.8

(a) Since floor(3/2) = 1, Algorithm 10.2 tells us that to determine whether $X^3 + X + 1$ is irreducible in $\mathbb{Z}_2[X]$, we need only check for degree 1 (linear factors): X or $X + 1$. Certainly X cannot be a factor (because when X is multiplied by any non-zero polynomial, the lowest degree term will have degree at least 1). To check whether X is a factor, we apply the division algorithm to the division of $X^3 + X + 1$ by $X + 1$. This produces a remainder 1 $(X^3 + X + 1 = (X^2 + X) \cdot (X + 1) + 1)$, so the algorithm tells us that $X^3 + X + 1$ is irreducible in $\mathbb{Z}_2[X]$.

(b) *Step 1.* We initialize deg = 1.

Step 2. (First iteration) We need to check if any of the degree-1 polynomials in $\mathbb{Z}_2[X]$ with leading coefficient 1 are factors of $X^4 + X^3 + X + 1$. These polynomials are X and $X + 1$. As in part (a), X clearly cannot be a factor. If we apply the division algorithm to $X^4 + X^3 + X + 1$ divided by $X + 1$, we obtain $X^4 + X^3 + X + 1 = (X^3 + 1)(X + 1)$, showing that $X^4 + X^3 + X + 1$ is not irreducible in $\mathbb{Z}_2[X]$.

Note: In case the first iteration of Step 1 failed to produce a factorization, since floor(4/2) = 2, we would need to go through one more iteration to check whether any of the following degree-2 polynomials is a factor of a degree-4 polynomial: X^2, $X^2 + 1$, $X^2 + X$, $X^2 + X + 1$. Of these four, since the first two obviously factor into smaller degree polynomials, they would not need to be checked [or we could see that they could not be factors as in part (a)].

EFR 10.9

Since $X^4 + X + 1 \equiv 0 \Rightarrow X^4 \equiv -X - 1 \equiv X + 1 \pmod{X^4 + X + 1}$, we use the reduction method by substituting $X + 1$ for X^4 in any powers of X higher than 3 (repeatedly, if necessary), until the polynomial has degree 3:

$$X^7 + X^4 + X^2 + X \equiv X^3 \cdot X^4 + X^4 + X^2 + X$$

$$\equiv X^3 \cdot (X + 1) + (X + 1) + X^2 + X$$

$$\equiv X^4 + X^3 + (X + 1) + X^2 + X$$

$$\equiv (X + 1) + X^3 + X^2 + 1 \qquad \text{(since } 2X \equiv 0 \text{ in } \mathbb{Z}_2[X])$$

$$\equiv X^3 + X^2 + X \pmod{X^4 + X + 1}$$

We could have also used the division algorithm on the division of $X^7 + X^4 + X^2 + X$ by $X^4 + X + 1$ to find a quotient of X^3, and a remainder of $X^3 + X^2 + X$, the latter being the same answer we obtained above.

EFR 10.10

The $2^2 = 4$ elements of $\mathbb{Z}_2[X](\mathrm{mod}\, X^2)$ are [by (4)]: $\mathbb{Z}_2[X](\mathrm{mod}\, X^2) = \{0, 1, X, X+1\}$. Adding polynomials in $\mathbb{Z}_2[X](\mathrm{mod}\, m)$ (for any m) is always the same as ordinary polynomial addition $\mathbb{Z}_2[X]$, which simply amounts to XORing the coefficients. We multiply as usual in $\mathbb{Z}_2[X]$, but then replace any powers that are (at least 2) using the identity $X^2 \equiv 0(\mathrm{mod}\, X^2)$. Thus, for example, $X(X+1) \equiv X^2 + X \equiv X(\mathrm{mod}\, X^2)$ and $XX \equiv X^2 \equiv 0(\mathrm{mod}\, X^2)$.

+	0	1	X	X+1
0	0	1	X	X+1
1	1	0	X+1	X
X	X	X+1	0	1
X+1	X+1	X	1	0

.	0	1	X	X+1
0	0	0	0	0
1	0	1	X	X+1
X	0	X	0	X
X+1	0	X+1	X	1

From the multiplication table, we see that the (nonzero) element X does not have an inverse, thus $\mathbb{Z}_2[X](\mathrm{mod}\, X^2)$ is not a field. (But we already knew this from Theorem 10.7 since $X^2 = X \cdot X$ is not irreducible.)

EFR 10.11

(a) Additions in $GF(16)$ are just ordinary polynomial additions in $\mathbb{Z}_2[X]$, but since we do not have an addition table, we need to convert the hex terms into polynomials (using Table 10.6), do the addition, and then convert back to hex:

$$6 + D \rightarrow (X^2 + X) + (X^3 + X^2 + 1) \equiv X^3 + X + 1 \rightarrow B$$

(b) Since, from Table 10.7, $A^2 = 8$, we have $4 + A^2 = 4 + 8 \rightarrow (X^2) + (X^3) \equiv X^3 + X^2 \rightarrow C$. We can use Table 10.7 to multiply this by D to get the final answer: 3.

(c) We use Table 10.7 three times: $(A \cdot B) \cdot C \cdot D = (2) \cdot C \cdot D = (2 \cdot C) \cdot D = (B) \cdot D = 6$.

EFR 10.12

(a) $64 + CB \rightarrow (X^6 + X^5 + X^2) + (X^7 + X^6 + X^3 + X + 1) = X^7 + X^5 + X^3 + X^2 + X + 1 \rightarrow AF$.

(b) $AA \cdot AA \rightarrow (X^7 + X^5 + X^3 + X) \cdot (X^7 + X^5 + X^3 + X) = X^{14} + X^{10} + X^6 + X^2$. If we divide this latter polynomial by $g = X^8 + X^4 + X^3 + X + 1$, the remainder is $X^7 + X^5 + X^4 + X \rightarrow B2$.

(c) $B9 + 12 \rightarrow (X^7 + X^5 + X^4 + X^3 + 1) + (X^4 + X) = X^7 + X^5 + X^3 + X + 1 \rightarrow AB$. $A4(B9 + 12) = A4 \cdot AB \rightarrow (X^7 + X^5 + X^2) \cdot (X^7 + X^5 + X^3 + X + 1) = X^{14} + X^9 + X^3 + X^2$. If we divide this latter polynomial by $g = X^8 + X^4 + X^3 + X + 1$, the remainder is $X^7 + X^6 + X^5 \rightarrow E0$.

EFR 10.13

(a) We apply the polynomial Euclidean algorithm to compute the gcd of defining polynomial $g = X^4 + X + 1$ and the polynomial $A \rightarrow X^3 + X$:

$$X^4 + X + 1 = (X)(X^3 + X) + (X^2 + X + 1)$$

$$X^3 + X = (X + 1)(X^2 + X + 1) + (X + 1)$$

$$X^2 + X + 1 = (X)(X + 1) + 1$$

We already knew the gcd would be 1 since all nonzero elements in a field are invertible. We now work backwards through the above equations to express 1 as a combination of the original polynomials. Keep in mind that all coefficient arithmetic is mod 2:

$$X^2 + X + 1 = (X)(X + 1) + 1 \Rightarrow 1 = X^2 + X + 1 + (X)(X + 1)$$

$$X^3 + X = (X + 1)(X^2 + X + 1) + (X + 1)$$

$$\Rightarrow X + 1 = (X^3 + X) + (X + 1)(X^2 + X + 1)$$

Substituting this into the first equation produces:

$$1 = (X^2 + X + 1) + (X)((X^3 + X) + (X + 1)(X^2 + X + 1))$$

$$= (X^2 + X + 1)(X^2 + X + 1) + (X)(X^3 + X)$$

In this latter expression, we substitute $X^2 + X + 1 = (X^4 + X + 1) + (X)(X^3 + X)$ (from the very first equation of the Euclidean algorithm) to obtain:

$$1 = (X^2 + X + 1)(X^2 + X + 1) + (X)(X^3 + X)$$

$$= (X^2 + X + 1)((X^4 + X + 1) + (X)(X^3 + X)) + (X)(X^3 + X)$$

$$= (X^2 + X + 1)(X^4 + X + 1) + (X^3 + X^2)(X^3 + X)$$

The last equation tells us that $(X^3 + X)^{-1} = X^3 + X^2$ (mod $X^4 + X + 1$); i.e., $\mathbf{A}^{-1} = \mathbf{C}$.

(b) We apply the polynomial Euclidean algorithm to compute the gcd of the defining polynomial $g = X^8 + X^4 + X^3 + X + 1$ and the polynomial $\mathbf{1A} \rightarrow X^4 + X^3 + X$:

$$X^8 + X^4 + X^3 + X + 1 = (X^4 + X^3 + X^2)(X^4 + X^3 + X) + (X + 1)$$

$$X^4 + X^3 + X = (X^3 + 1)(X + 1) + 1$$

We already knew the gcd would be 1 since all nonzero elements in a field are invertible. We now work backwards through the above equations to express 1 as a combination of the original polynomials. Keep in mind that all coefficient arithmetic is mod 2:

$$X^4 + X^3 + X = (X^3 + 1)(X + 1) + 1 \Rightarrow 1 = (X^4 + X^3 + X) + (X^3 + 1)(X + 1)$$

$$X^8 + X^4 + X^3 + X + 1 = (X^4 + X^3 + X^2)(X^4 + X^3 + X) + (X + 1)$$

$$\Rightarrow X + 1 = (X^8 + X^4 + X^3 + X + 1)$$

$$+ (X^4 + X^3 + X^2)(X^4 + X^3 + X)$$

We substitute the latter obtained equation into the former to produce:

$$1 = (X^4 + X^3 + X) + (X^3 + 1)(X + 1)$$

$$= (X^4 + X^3 + X) + (X^3 + 1)((X^8 + X^4 + X^3 + X + 1)$$

$$+ (X^4 + X^3 + X^2)(X^4 + X^3 + X))$$

$$= (X^3 + 1)(X^8 + X^4 + X^3 + X + 1)$$

$$+ [(X^3 + 1)(X^4 + X^3 + X^2) + 1](X^4 + X^3 + X)$$

$$= (X^3 + 1)(X^8 + X^4 + X^3 + X + 1)$$

$$+ [X^7 + X^6 + X^5 + X^4 + X^3 + X^2 + 1](X^4 + X^3 + X)$$

The last equation tells us that $(X^4 + X^3 + X)^{-1} = X^7 + X^6 + X^5 + X^4 + X^3 + X^2 + 1 \pmod{X^8 + X^4 + X^3 + X + 1}$; i.e., $1A^{-1} = FD$.

Chapter 11: The Advanced Encryption Standard (AES) Protocol

EFR 11.1

(a) Answer: 0. The XOR of a binary string with itself is always the string of zeros, and 0000(binary) = 0(hex).

(b) $9 \cdot 9 \to 1001 \cdot 1001 \to (X^3 + 1) \cdot (X^3 + 1) \equiv (X^3 + 1) \cdot X^3 + (X^3 + 1) \cdot 1 \equiv X^6 + X^3 + X^3 + 1 \underset{X^4 \equiv X+1}{\equiv} (X + 1)X^2 + X \equiv X^3 + X^2 + 1$

(mod $X^4 + X + 1$). (Recall the coefficient arithmetic is mod 2) $\to 1101 \to$ D.

EFR 11.2

Here is a summary of each of the steps:

Creation of Round Keys: $W(0) = \begin{bmatrix} 4 \\ 6 \end{bmatrix}$, $W(1) = \begin{bmatrix} 8 \\ C \end{bmatrix}$.

To create $W(2)$, we XOR $W(0)$ with RCON(1) to get $\begin{bmatrix} C \\ 6 \end{bmatrix}$, we

take the SubNib or the RotNib of $W(1)$ to get $\begin{bmatrix} B \\ A \end{bmatrix}$, and $W(2)$

will be the XOR of the last two vectors: $W(2) = \begin{bmatrix} 7 \\ C \end{bmatrix}$. Now $W(3) = W(1) \oplus W(2) = \begin{bmatrix} F \\ 0 \end{bmatrix}$.

To create $W(4)$, we XOR $W(2)$ with RCON(2) to get $\begin{bmatrix} 4 \\ C \end{bmatrix}$, we take the SubNib or the RotNib of W(3) to get $\begin{bmatrix} 9 \\ 7 \end{bmatrix}$, and $W(4)$ will be the XOR of the last two vectors: $W(4) = \begin{bmatrix} D \\ B \end{bmatrix}$. Now $W(5) = W(3) \oplus W(4) = \begin{bmatrix} 2 \\ B \end{bmatrix}$.

The round keys thus are $\kappa^0 = 468C$, $\kappa^1 = 7CF0$, and $\kappa^2 = DB2B$.

Convert plaintext to nibble array: $P \to \begin{bmatrix} 4 & B \\ 7 & 0 \end{bmatrix}$

++

Round 0: Add Round Key: $\begin{bmatrix} 4 & B \\ 7 & 0 \end{bmatrix} \oplus \kappa^0 \to \begin{bmatrix} 0 & 3 \\ 1 & C \end{bmatrix}$

++

Round 1: Nibble Sub: $\begin{bmatrix} 0 & 3 \\ 1 & C \end{bmatrix} \to \begin{bmatrix} 9 & C \\ D & B \end{bmatrix}$

Shift Row: $\begin{bmatrix} 9 & C \\ D & B \end{bmatrix} \to \begin{bmatrix} 9 & C \\ B & D \end{bmatrix}$

Mix Column: $\begin{bmatrix} 9 & C \\ B & D \end{bmatrix} \to \begin{bmatrix} 3 & D \\ 9 & 8 \end{bmatrix}$

Add Round Key: $\begin{bmatrix} 3 & D \\ 9 & 8 \end{bmatrix} \oplus \kappa^1 \to \begin{bmatrix} 4 & 2 \\ 5 & 8 \end{bmatrix}$

++

Round 2: Nibble Sub: $\begin{bmatrix} 4 & 2 \\ 5 & 8 \end{bmatrix} \to \begin{bmatrix} 4 & 6 \\ 1 & A \end{bmatrix}$

Shift Row: $\begin{bmatrix} 4 & 6 \\ 1 & A \end{bmatrix} \to \begin{bmatrix} 4 & 6 \\ A & 1 \end{bmatrix}$

Add Round Key: $\begin{bmatrix} 4 & 6 \\ A & 1 \end{bmatrix} \oplus \kappa^2 \to \begin{bmatrix} 9 & 4 \\ 1 & A \end{bmatrix}$

Ciphertext: $\to 914A$

EFR 11.3

We first use the S-box Table 11.2 to compute all of 16 images:

$S(0) = 9(\text{hex}) = 1001(\text{binary})$
$S(1) = D(\text{hex}) = 1101(\text{binary})$
$S(2) = 6(\text{hex}) = 0110(\text{binary})$
$S(3) = C(\text{hex}) = 1100(\text{binary})$
$S(4) = 4(\text{hex}) = 0100(\text{binary})$

$S(5) = 1(\text{hex}) = 0001(\text{binary})$
$S(6) = 2(\text{hex}) = 0010(\text{binary})$
$S(7) = E(\text{hex}) = 1110(\text{binary})$
$S(8) = A(\text{hex}) = 1010(\text{binary})$
$S(9) = 8(\text{hex}) = 1000(\text{binary})$
$S(A) = 0(\text{hex}) = 0000(\text{binary})$
$S(B) = F(\text{hex}) = 1111(\text{binary})$
$S(C) = B(\text{hex}) = 1011(\text{binary})$
$S(D) = 5(\text{hex}) = 0101(\text{binary})$
$S(E) = 3(\text{hex}) = 0011(\text{binary})$
$S(F) = 7(\text{hex}) = 0111(\text{binary})$

If we apply the inverse S-box using Table 11.3 to each of these 16 images, we find that in each case we will get back to the corresponding input that was used in the S-box. For example, $S(9) = 1000$, using the second row and the zeroth column of Table 11.3, we see that $S^{-1}(8) = 9$, i.e., $S^{-1}(S(9)) = 9$. Thus the inverse S-box really is the inverse function of the S-box function.

EFR 11.4

Here is a summary of each of the steps:
The round keys are $\kappa^0 = 2A74$, $\kappa^1 = E490$, and $\kappa^2 = 4CDC$ (just as in Example 11.2).

Convert ciphertext to nibble array: $C \rightarrow \begin{bmatrix} 2 & B \\ 9 & 9 \end{bmatrix}$

+++

Round 2: Add Round Key: $\begin{bmatrix} 2 & B \\ 9 & 9 \end{bmatrix} \oplus \kappa^2 \rightarrow \begin{bmatrix} 6 & 6 \\ 5 & 5 \end{bmatrix}$

Shift Row: $\begin{bmatrix} 6 & 6 \\ 5 & 5 \end{bmatrix} \rightarrow \begin{bmatrix} 6 & 6 \\ 5 & 5 \end{bmatrix}$

Inv Nibble Sub: $\begin{bmatrix} 6 & 6 \\ 5 & 5 \end{bmatrix} \rightarrow \begin{bmatrix} 2 & 2 \\ D & D \end{bmatrix}$

+++

Round 1: Add Round Key: $\begin{bmatrix} 2 & 2 \\ D & D \end{bmatrix} \oplus \kappa^1 \rightarrow \begin{bmatrix} C & B \\ 9 & D \end{bmatrix}$

Inv Mix Column: $\begin{bmatrix} C & B \\ 9 & D \end{bmatrix} \rightarrow \begin{bmatrix} 7 & 5 \\ 6 & A \end{bmatrix}$

Shift Row: $\begin{bmatrix} 7 & 5 \\ 6 & A \end{bmatrix} \rightarrow \begin{bmatrix} 7 & 5 \\ A & 6 \end{bmatrix}$

Inv Nibble Sub: $\begin{bmatrix} 7 & 5 \\ A & 6 \end{bmatrix} \rightarrow \begin{bmatrix} F & D \\ 8 & 2 \end{bmatrix}$

+++

Round 0: Add Round Key: $\begin{bmatrix} F & D \\ 8 & 2 \end{bmatrix} \oplus \kappa^0 \rightarrow \begin{bmatrix} D & A \\ 2 & 6 \end{bmatrix}$

Plaintext: $\rightarrow$ D2A6

EFR 11.5

$77 + 99 \rightarrow [0111][0111] \oplus [1001][1001] = [1110][1110] \rightarrow$ EE

D7·2F $\rightarrow [1101][0111] \cdot [0010][1111]$

$$\rightarrow (X^7 + X^6 + X^4 + X^2 + X + 1)(X^5 + X^3 + X^2 + X + 1)$$

Multiplying this product and simplifying (working mod 2 on the coefficients), we arrive at $X^{12} + X^{11} + X^{10} + X^9 + X^6 + X^5 + X^4 + X^3 + X^2 + 1$. We need to reduce this polynomial so that there are no powers higher than X^7; we use the division algorithm to divide this polynomial by $g = X^8 + X^4 + X^3 + X + 1$. When this is done, we obtain a quotient $q = X^4 + X^3 + X^2 + X + 1$ and a remainder $r = X^6 + X^4 + X^2$, such that $X^{12} + X^{11} + X^{10} + X^9 + X^6 + X^5 + X^4 + X^3 + X^2 + 1 = qg + r$. The remainder will be the product we want: D7·2F $\rightarrow r \rightarrow$ $[0101][0100] \rightarrow$ 54.

Chapter 12: Elliptic Curve Cryptography

EFR 12.1

The slope is given by $m = \dfrac{y_2 - y_1}{x_2 - x_1} = \dfrac{8 - 4}{4 - 2} = 2$. Substituting this into the coordinate formulas, we obtain

$$\begin{cases} x_3 = m^2 - x_1 - x_2 = 2^2 - 2 - 4 = -2 \\ y_3 = m(x_1 - x_3) - y_1 = 2(2 - (-2)) - 4 = 4 \end{cases}$$

so that $P_3 = (x_3, y_3) = (-2, 4)$.

EFR 12.2

Since $3 \cdot (2, 4) = (2, 4) + 2 \cdot (2, 4) = (2, 4) + (-3, 1)$, we apply Algorithm 12.1 with $P_1 = (x_1, y_1) = (2, 4)$ and $P_2 = (x_2, y_2)$ $= (-3, 1)$: The slope is given by $m = \dfrac{y_2 - y_1}{x_2 - x_1} = \dfrac{1 - 4}{-3 - 2} = \dfrac{3}{5}$.

Substituting this into the coordinate formulas, we obtain

$$\begin{cases} x_3 = m^2 - x_1 - x_2 = (3/5)^2 - 2 - (-3) = 1.36 \\ y_3 = m(x_1 - x_3) - y_1 = (3/5)(2 - 1.36) - 4 = -3.616 \end{cases}$$

so that $3 \cdot (2, 4) = (1.36, -3.616)$.

There are now two ways to use Algorithm 12.1 to compute $4 \cdot (2, 4)$: either as $2 \cdot (2, 4) + 2 \cdot (2, 4)$ or as $(2, 4) + 3 \cdot (2, 4)$. To use the first scheme, we write $P_1 = (x_1, y_1) = (-3, 1)$ and $P_2 = (x_2, y_2) = (-3, 1)$. Since the points being added are equal, Algorithm 12.1 tells us to set $m = \dfrac{3x_1^2 + a}{2y_1} = \dfrac{3 \cdot (-3)^2 + (-4)}{2 \cdot 1} =$

$\frac{23}{2}$. Substituting this into the coordinate formulas, we obtain

$$\begin{cases} x_3 = m^2 - x_1 - x_2 = (11.5)^2 - (-3) - (-3) = 138.25 \\ y_3 = m(x_1 - x_3) - y_1 = (11.5)(-3 - 138.25) - 1 = -1625.375 \end{cases}$$

so that $4 \cdot (2, \ 4) = (138.25, \ -1625.375)$.

EFR 12.3

Since $\Delta \equiv 4a^3 + 27b^2 \equiv 4 \cdot 1^3 + 27 \cdot 1^2 \equiv 31 \equiv 3 \not\equiv 0 \pmod{7}$, the elliptic curve is nonsingular. The computations used to compute the points of the elliptic curve are summarized in the table below:

x	$x^3 + x + 1$ (mod 7)	$y \equiv \sqrt{x^3 + x + 1}$ (mod 7
0	0	1,6
1	3	None
2	4	2,5
3	3	None
4	6	None
5	5	None
6	6	None

Thus, we can write $E = \{(0, \ 1), (0, \ 6), (2, \ 2), (2, \ 5), \infty\}$.

EFR 12.4

Since $\Delta \equiv 4a^3 + 27b^2 \equiv 4 \cdot 1^3 + 27 \cdot 1^2 \equiv 31 \equiv 9 \not\equiv 0 \pmod{11}$, the elliptic curve is nonsingular. We let x run through the integers mod 11: $0, 1, \cdots, 10$, evaluate $r \equiv x^3 + 7x + 5$ (mod 11) for each such x, and whenever the value of r is nonzero, we use Equation 12.6 to find any square roots of a. The modular elliptic curve will be the set of all ordered pairs (x, y) that arise in this way together with the point at infinity. The computations are summarized in the following table:

x	$a \equiv x^3 + x + 1$ (mod 11)	a^3 (mod 11)	$(a^3)^2$ (mod 11)	$y \equiv \pm a^3$ (mod 11)
0	1	1	1	$1, -10 \equiv 9$
1	3	5	3	$5, -5 \equiv 6$
2	0	–	–	0
3	9	3	9	$3, -3 \equiv 8$
4	3	5	3	$5, -5 \equiv 6$
5	10	10	1	No
6	3	5	3	$5, -5 \equiv 6$
7	10	10	1	No
8	4	9	4	$9, -9 \equiv 2$
9	2	8	9	No
10	10	10	1	No

From these computations, we may write: $E = \{(0, 1), (0, 9), (1, 5), (1, 6),$
$(2, 0), (3, 3), (3, 8), (4, 5), (4, 6), (6, 5), (6, 6), (8, 2), (8, 9), \infty\}$.

EFR 12.5

(a) We use the notation of Algorithm 12.3: $P_1 = (4, 5) \triangleq (x_1, y_1)$,
$P_2 = (0, 10) \triangleq (x_2, y_2)$, and $P_3 \triangleq (x_3, y_3)$ (that will be
determined). Since working mod 11 we have $(x_2 - x_1)^{-1} \equiv$
$(0 - 4)^{-1} \equiv 7^{-1} \equiv 8$, we may compute $m = (y_2 - y_1) \cdot (x_2 - x_1)^{-1}$
$\equiv (10 - 5) \cdot 8 \equiv 40 \equiv 7$. Next, we may compute the coordinates
of the sum $\begin{cases} x_3 = m^2 - x_1 - x_2 \equiv 7^2 - 4 - 0 \equiv 45 \equiv 1 \\ y_3 = m(x_1 - x_3) - y_1 \equiv 7(4 - 1) - 5 \equiv 16 \equiv 5 \end{cases}$
(mod 11), so that $P_3 \equiv (1, 5)$.

(b) Here we set $P_1 = (4, 5) \triangleq (x_1, y_1)$, $P_2 = (4, 5) \triangleq (x_2, y_2)$,
and $P_3 \triangleq (x_3, y_3)$ (that will be determined). Since the
inverse of $2y_1 \equiv 2 \cdot 5 \equiv 10$ is 10 (mod 11), we compute
$m \equiv (3x_1^2 + a) \cdot (2y_1)^{-1} \equiv (3 \cdot 4^2 + 1) \cdot 10 \equiv 490 \equiv 6 \,(\text{mod } 11)$,
and hence $\begin{cases} x_3 = m^2 - x_1 - x_2 \equiv 6^2 - 4 - 4 \equiv 28 \equiv 6 \\ y_3 = m(x_1 - x_3) - y_1 \equiv 6(4 - 6) - 5 \equiv -17 \equiv 5 \end{cases}$
(mod 11), so that $P_3 \equiv (6, 5)$.

EFR 12.6

Since we found in Example 12.4 that the elliptic curve E has
nine points, it follows from Theorem 12.6 that the only possible orders for points are 1 (only for the point at infinity),
3, and 9. So to check the order of a point P, we need only
compute $3P = 2P + P$. If this is the point at infinity, then the
order is 3; otherwise, it must be 9.

For $P = (0, 4)$, we first use Algorithm 12.2 to compute
$2P = P + P = (4, 3)$, then we use this result and the same
algorithm to compute $3P = 2P + P = (4, 3) + (0, 4) = (2, 4)$.
It follows that $\text{ord}_E(0, 4) = 9$.

From this computation, it follows that $3 \cdot (2, 4) = 3 \cdot (3P) =$
$9P = \infty$, so $\text{ord}_E(2, 4) = 3$.

We set $P = (44, 29) \triangleq (x_1, y_1) \triangleq (x_2, y_2)$, and $P_3 \triangleq (x_3, y_3)$
(that will be determined). We use the extended Euclidean
algorithm (Algorithm 2.2) to compute the inverse of
$2y_1 \equiv 2 \cdot 29 \equiv 58$ to be 735 (mod 907). Thus $m \equiv (3x_1^2$
$+ a) \cdot (2y_1)^{-1} \equiv (3 \cdot 44^2 + 0) \cdot 735 \equiv 538 (\text{mod } p)$, and hence
$\begin{cases} x_3 = m^2 - x_1 - x_2 \equiv 6^2 - 2 - 5 \equiv 29 \equiv 7 \\ y_3 = m(x_1 - x_3) - y_1 \equiv 6(2 - 7) - 4 \equiv -30 - 4 \equiv -34 \equiv 10 \end{cases}$
(mod 11), so that $P_3 \equiv (7, 10)$.

EFR 12.7

(a) Since $\Delta \equiv 4a^3 + 27b^2 \equiv 4 \cdot 196^3 + 27 \cdot 98^2 \equiv 171 \not\equiv 0 (\text{mod } p)$,
the elliptic curve E is nonsingular. We first compute
$r \triangleq x_1^3 + 196x_1 + 98 \equiv 28^3 + 196 \cdot 28 + 98 \equiv 179$ and then use
fast modular exponentiation to compute $r^{(p+1)/4} \equiv 179^{(251+1)/4}$

$\equiv 179^{63} \equiv 207(\text{mod } p)$. Since the square of the latter integer is $207^2 \equiv 179 \equiv r(\text{mod } p)$, it follows (from Equation 12.6) that this latter number, $y_1 = 207$, is a square root of r and thus gives rise to a point $G = (x_1, y_1) = (28, 207) \in E$.

(b)   We use Algorithm 12.3 to compute $A \equiv n_A G \equiv 9 \cdot G$.

Step 1. The binary expansion of 9 is $[d_3 d_2 d_1 d_0] = [1001]$.

Step 2. We initialize $Q = \infty$ (cumulative sum) and $D = G$ (doubling point).

$k = 0$: Since $d_0 = 1$, we update $Q \to Q + D = \infty + G = G = (28, 207)$, and we double D for the next round: $D \to D + D = 2 \cdot (28, 207) = (61, 121)$. (Algorithm 12.2 was used twice.)

$k = 1$: Since $d_1 = 0$, we leave Q alone, and we double D for the next round: $D \to D + D = 2 \cdot (61, 121) = (99, 205)$.

$k = 2$: Since $d_2 = 0$, we leave Q alone, and we double D for the next round: $D \to D + D = 2 \cdot (99, 205) = (159, 110)$.

$k = 3$: Since $d_3 = 1$, we update $Q \to Q + D = (28, 207) + (159, 110) = (95, 176) = 9G = A$.

In a slightly more abbreviated fashion, the computation for $B \equiv n_B G \equiv 16 \cdot G$ proceeds as follows: The binary expansion of 16 is $[d_4 d_3 d_2 d_1 d_0] = [10000]$. Initialize $Q = \infty$ and $D = G$.

$k = 0$: $D \to D + D = (61, 121)$. (From what was just done for Alice.) $k = 1$: $D \to D + D = (99, 205)$. $k = 2$: $D \to D + D = (159, 110)$. $k = 3$: $D \to D + D = (16, 157)$. $k = 4$: $Q \to Q + D = \infty + (16, 157) = (16, 157) = 16G = B$.

To compute the shared private key, Alice would compute $K \equiv n_A B \equiv 9B$. The abbreviated computation runs as follows: Initialize $Q = \infty$ and $D = B$. $k = 0$: $Q \to Q + D = (16, 157)$. $D \to D + D = (217, 126)$. $k = 1$: $D \to D + D = (229, 82)$. $k = 2$: $D \to D + D = (162, 126)$. $k = 3$: $Q \to Q + D = (247, 76) = 9B = K$.

Bob would compute $K \equiv n_B A \equiv 16A$ using Algorithm 12.3 as follows: Initialize $Q = \infty$ and $D = A$.
$k = 0$: $D \to D + D = (14, 243)$. $k = 1$: $D \to D + D = (114, 139)$. $k = 2$: $D \to D + D = (74, 107)$. $k = 3$: $D \to D + D = (247, 76)$. $k = 4$: $Q \to Q + D = (247, 76) = 16A = K$.

EFR 12.8

(a)   Since $2^{-9} = 0.00195...$, we would use $K = 9$. The equation $(m+1)K < p$ thus implies $(m+1)9 < 523 \Rightarrow 9m < 514 \Rightarrow m < 57.1$, so the range of admissible values for m is $0 \le m \le 57$.

(b)

Step 1. Initialize $i = 0$.

Step 2. Set $x = Km + i = 9 \cdot 22 + 0 = 198$, $r \equiv 198^3 + 22 \cdot 198 + 153 \equiv 351(\text{mod } p)$, and use fast modular exponentiation (Algorithm 6.5) to compute $y \equiv r^{(p+1)/4} \equiv 351^{77} \equiv 457(\text{mod } p)$.

Step 3. Since $y^2 \equiv 172 \not\equiv r(\text{mod } p)$, we update $i \to i+1 = 1$, and return to Step 2.

Step 2. (Second iteration) Set $x = Km + i = 9 \cdot 22 + 1 = 199$, $r \equiv 199^3 + 22 \cdot 199 + 153 \equiv 382(\text{mod } p)$, and use fast modular exponentiation (Algorithm 6.5) to compute $y \equiv r^{(p+1)/4} \equiv 382^{77} \equiv 229(\text{mod } p)$.

Step 3. (Second iteration) Since $y^2 \equiv 141 \not\equiv r(\text{mod } p)$, we update $i \to i+1 = 2$, and return to Step 2.

Step 2. (Third iteration) Set $x = Km + i = 9 \cdot 22 + 2 = 200$, $r \equiv 200^3 + 22 \cdot 200 + 153 \equiv 38(\text{mod } p)$, and use fast modular exponentiation (Algorithm 6.5) to compute $y \equiv r^{(p+1)/4} \equiv 38^{77} \equiv 123(\text{mod } p)$.

Step 3. (Third iteration) Since $y^2 \equiv 485 \not\equiv r(\text{mod } p)$, we update $i \to i+1 = 3$, and return to Step 2.

Step 2. (Fourth iteration) Set $x = Km + i = 9 \cdot 22 + 2 = 201$, $r \equiv 201^3 + 22 \cdot 201 + 153 \equiv 371(\text{mod } p)$, and use fast modular exponentiation (Algorithm 6.5) to compute $y \equiv r^{(p+1)/4} \equiv 371^{77} \equiv 246(\text{mod } p)$.

Step 3. (Fourth iteration) Since $y^2 \equiv 371 \equiv r(\text{mod } p)$, we have found a plaintext representative on E; we output $P = (x, y) = (201, 246)$, and exit the algorithm.

EFR 12.9

(a) The first two values for x_1 do not lead to a point on the elliptic curve because in each case, formula (12.6) $y_1 \equiv r^{(p+1)/4} (\text{mod } p)$ for a square root of $r \triangleq x_1^3 + 22x_1 + 153$ results in a number that is not a square root of r. However, the third number, $x_1 = 195$, does lead to a point on E: $r \triangleq x_1^3 + 22x_1 + 153 \equiv 40$, $y_1 \equiv r^{(p+1)/4} \equiv 119(\text{mod } p)$, and $y_1^2 \equiv r \ (\text{mod } p)$. We have thus obtained the point $G = (x_1, y_1) = (195, 119) \in E$.

(b) To encrypt, Alice will need Bob's public key, which he will have been able to compute using Algorithm 12.3 to be $B = n_B G = 54G \equiv (105, 86)$. Following the encryption procedure of Table 12.4, Alice multiplies Bob's public key number B by her secret parameter n_A (using Algorithm 12.3) to obtain $n_A B = 32B = (296, 308)$. Then (using Algorithm 12.2) she adds her plaintext point $P = (201, 246)$ (from the solution of the previous exercise for the reader) to this point to obtain $C = P + n_A B = (203, 72)$. Alice will also need to compute her public key, which could be computed using Algorithm 12.3 to be $A \equiv n_A G \equiv 32G \equiv (8, 494)$. The resulting elliptic curve ElGamal ciphertext that Alice sends to Bob is the pair of points $(A, C) = ((8, 494), (203, 72))$.

(c): To decrypt, Bob would first compute (using Algorithm 12.3) $n_B A = 54A = (296, 308)$ (this is the Diffie–Hellman key). He would then subtract this from C (the second point of Alice's ciphertext), by using Algorithm 12.2 to add $-n_B A = (296, -308) \equiv (296, 215)$ to C: $C - n_B A = (201, 2246) = P$, as was anticipated.

Appendix D: Answers and Brief Solutions to Selected Odd-Numbered Exercises

Chapter 1

1. All three are functions.
   (a) Domains of F and G are $\{a, b, c, d\}$, domain of H is $\{a, b, c\}$. Codomains of F and H are $\{1, 2, 3, 4\}$, codomain of G is $\{1, 2, 3\}$. Range of F is $\{1, 2, 3, 4\}$, range of G is $\{1, 2, 3\}$, range of H is $\{1, 2, 4\}$,
   (b) F and H are one-to-one, G is not.
   (c) F and G are onto, H is not.
   (d) Only F is bijective (both one-to-one and onto).

3. (a) Yes
   (b) Yes
   (c) 00010111
   (d) cabby

5. (a) *Such a function need not be onto.* For example, the set $A = \{1, 2\}$ has more elements than the set $B = \{1\}$, but the function f from A to B defined by $f(1) = 1 = f(2)$ is not onto. *Such a function can never be one-to-one.* Reason: Since one-to-one functions can never have duplicated outputs, the range must be the same size as the domain. But the range is a subset of the codomain, so for a one-to-one function, the size of the codomain must be at least as large as the domain.
   (b) *Such a function need not be one-to-one.* For example, the set $A = \{1, 2\}$ has fewer elements than the set $B = \{1, 2, 3\}$, but the function f from A to B defined by $f(1) = 1 = f(2)$ is not one-to-one. *Such a function can never be onto.* Reason: The range is at most as large as the domain A, which is assumed to be smaller than the codomain B.

7. In each of the examples below, we will specify a function $f : \{1, 2, 3, \cdots\} \to \{1, 2, 3, \cdots\}$.
   (a) Any constant function, such as $f(i) = 1$, for each $i \in \{1, 2, 3, \cdots\}$ is neither one-to-one nor onto.
   (b) The right shift function $f(i) = i + 1$ for each $i \in \{1, 2, 3, \cdots\}$ is one-to-one but not onto since 1 is not in the range.
   (c) The function defined by $f(1) = 1$ and $f(i) = i - 1$, for each $i \in \{1, 2, 3, \cdots\}$, is onto but not one-to-one since $f(1) = 1 = f(2)$.
   (d) The function defined by taking each even integer to the odd integer right before it, and each odd integer to the even integer right after it, is bijective and satisfies the indicated condition

(an output never equals its input). Here is a formula for this function: $f(i) = i - 1$ if i is even, and $f(i) = i + 1$ if i is odd.

9. (a) Yes

(b) No. Reason: No string ending in a zero is in the range.

(c) No

11. (a) Yes. Reason: If $f(b_1 b_2 b_3 b_4 b_5 b_6 b_7 b_8) = f(b_1' b_2' b_3' b_4' b_5' b_6' b_7' b_8')$, this means that $b_2 b_4 b_6 b_8 b_1 b_3 b_5 b^* = b_2' b_4' b_6' b_8' b_1' b_3' b_5' b'^*$ so equating bits gives us that $b_i = b_i'$ for all indices i except $i = 7$. But since $b^* = b'^*$, we also must have (according to the definition of f) that $b_6 + b_7 + b_8$, $b_6' + b_7' + b_8'$ are both even or both odd, and since we already know that the first and third of these three terms are the same, it follows that b_7, b_7' are both even or both odd. Since they can only be 0 or 1, they must be the same.

(b) Yes. Reason: Given any length-8 string $d_1 d_2 d_3 d_4 d_5 d_6 d_7 d_8$, we need to find an input string $b_1 b_2 b_3 b_4 b_5 b_6 b_7 b_8$ that will give us this output under f, i.e., satisfying $b_2 b_4 b_6 b_8 b_1 b_3 b_5 b^* = d_1 d_2 d_3 d_4 d_5 d_6 d_7 d_8$. From this latter equation, it follows that we must set $b_2 = d_1$, $b_4 = d_2$, $b_6 = d_3$, $b_8 = d_4$, $b_1 = d_5$, $b_3 = d_6$, $b_5 = d_7$, so the only bit left to specify in the input is b_7. Since b_6, b_8 are already specified as d_3, d_4, in order to have $b^* = d_8$, we will need to choose b_7 so that if $d_8 = 1$, then $b_7 + d_3 + d_4$ ($= b_6 + b_7 + b_8$) is even, whereas if $d_8 = 0$, then $b_7 + d_3 + d_4$ is odd. In either case, notice that we must have $b_7 + d_3 + d_4 + d_8$ be odd. This can clearly be done (in only one way) as follows: if $d_3 + d_4 + d_8$ is odd, we must have $b_7 = 0$, while if $d_3 + d_4 + d_8$ is even, we must have $b_7 = 1$.

(c) The inverse function's formula was determined in part (b) in the process of showing f is onto. Here is the summary formula of the inverse function: $f^{-1}(d_1 d_2 d_3 d_4 d_5 d_6 d_7 d_8) = d_5 d_1 d_6 d_2 d_7 d_3 d^{**} d_4$ ($= b_1 b_2 b_3 b_4 b_5 b_6 b_7 b_8$), where $d^{**} = 0$, if $d_3 + d_4 + d_8$ is odd, and otherwise $d^{**} = 1$.

13. (a) (i)   WKHVKLSPHQWZLOODUULYHDWQRRQ

(ii)   ODBORZXQWLOIULGDB

(iii)   DOZDBVXVHWKHEDFNGRRU

(iv)   WKHSKRQHLVEXJJHG

(b) (i)   Bring the item to Jenkins

(ii)   Send Agent Polk a signal

(iii)   Intercept their case worker

(iv)   Check in to the hotel

15. (a) (i)   PDAODELIAJPSEHHWNNERAWPJKKJ

(ii)   HWUHKSQJPEHBNEZWU

(iii)   WHSWUOQOAPDAXWYGZKKN

(iv)   PDALDKJAEOXQCCAZ

(b) (i)   Waiting for instructions

(ii)   Subject has boarded plane

    (iii)   Make initial contact as a businessman
    (iv)   Operation has been compromised

17. (a) (i)   KVGCLBGAGXXPZZNKVKZJGKXGFCP
    (ii)   COAVSPLBVSPYIWFKC
    (iii)   RZYKCLLGGDLXSOEUHHFF
    (iv)   KVGZLHESKCFNXUGN

    (b) (i)   Harrell will be waiting for you
    (ii)   The meeting with Watson is as set up
    (iii)   Come alone but bring your piece
    (iv)   Rent a room in the Hotel Marignon in the Fifth Arrondisement

19. (a) (i)   QMYIGSTGCPQZKGRLXKPKIGBRTDDT
    (ii)   RLAHSZDUPOMGPKKCAW
    (iii)   LRXYHYWDBPGYCBADIDKT
    (iv)   QMGVMSPCSKCZLVPGKU

    (b) (i)   Take cover now
    (ii)   The money is buried underneath
    (iii)   Pay off the watchman
    (iv)   Pretend you are a professor; once inside, copy the files

21. If we have any string of plaintext and the corresponding string of ciphertext, for each matched letters in the strings, the corresponding keyword shift letter is specified by the number of letters that the plaintext letter gets shifted down the alphabet to get the corresponding ciphertext letter. If we shift 0 letters down, the keyword letter is a; if we shift 1 letter down, the keyword letter is b, etc. (see Table 1.3). Thus, for example, if we knew that a Vigenère cipher was used to convert the plaintext "theyhavegrenades" into the ciphertext POMQRETANZWXEBAZ, since the first letter t goes to P, which is $26 - (19 - 15) = 22$ letters down the alphabet, we get the first keyword letter must be w. (To see this, refer to Table 1.3, and use the fact that since P is to the left of t, the shift must have cycled back after passing z.) Similarly, since the next plaintext letter h goes to O, which is $14 - 7 = 7$ letters down the alphabet, the second keyword letter must be h. Continuing in this fashion, the given plaintext/ciphertext correspondence produces the following keyword sequence "whiskeywhiskeywh," so it appears that the keyword is *whiskey*. For such an attack to completely determine the keyword, the known strings must be at least as long as the (unknown) keyword.

23. (a) (i)   DVGXDDVVDAGXDDAGDVVVDXDDVXDGDVX
        DDDDADGAVGDXAXXGXVAVFDG
    (ii)   ADGDAVVVDDVVAFDFGXDVDDAGDXGXXG
        AGVG
    (iii)   VDDXXDDDAXDDDFGVDGVDDFFGDVGDXV
        AGGFVDFGDV
    (iv)   DAGVFGGXDXVDXDDDDFVADDAGGVXXXDGD

   (b) (i)   retreat

     (ii)   more munitions needed in Normandy

    (iii)   strike tomorrow at 4 am

    (iv)   Metz is a lost battle redeploy in Lyon

25. Each plaintext letter gives rise to two ciphertext letters. In Step 1, the plaintext letters are assigned to unique pairs of ciphertext, but into Step 2 these pairs are broken as the letters are put row by row into an array, and after mixing up the columns, the letters are processed column by column. So it is very unlikely that identical adjacent pairs of letters will give rise to the same four-letter ciphertext passages. Here is a specific example. Under the keyword PARIS, the ADFGVX ciphertexts for "abc" and "cba" are VFDGFF and GFFVFD, respectively. In the first, "ab" corresponds to VFDG, while in the second it corresponds to FVFD.

27. No. Shifting to the left by k letters is the same as shifting to the right by $26 - k$ letters.

29. Generally (i) is more secure. This makes sense since the key needed to describe (i) has length nm, while the key needed to describe (ii) has length $n + m$ (which is usually smaller than nm). For an extreme case, consider what happens when m is a factor of n (or the other way around). Then the cipher (ii) is equivalent to a Vigenère cipher with keylength n, so the additional Vigenère cipher of keylength m adds no additional security. Here is a specific example: if in (ii) $n = 4$ and $m = 2$, and the corresponding keywords are *gold* and *be*, then the cipher (ii) is just the Vigenère cipher whose keyword is *hsmh* (this is simply the ciphertext when the Vigenère cipher with keyword *be* is applied to the plaintext *gold*), which is much less secure than a Vigenère cipher with a key of length $nm = 8$. In general, the effective keylength of the cipher (ii) will be the least common multiple of n and m, so (ii) will compare better with (i), in terms of security, in cases where n and m do not share many common factors. But even when n and m have no common factors (other than 1), so that the effective single Vigenère cipher keylength in (ii) is nm, the system (i) has many more actual Vigenère cipher keys than the effective length mn keys resulting from (ii). In particular, the individual characters in (ii) can be chosen randomly, but those in (ii) cannot be made to have a random pattern.

31. Such a system has perfect secrecy. With the knowledge of just one letter of ciphertext, any of the 26 possible plaintext letters is equally possible.

Chapter 2

1. (a) False

   (b) True

   (c) False

(d) True, since $a = (-1)(-a)$

(e) False, only when $a = 0$ since by Definition 2.1, 0 cannot divide any integer, but if $a \neq 0$ the relation is true

(f) False

3. Since $a \mid b$, there is an integer k such that $b = ak$, and since $b \mid a$, there is an integer ℓ such that $a = b\ell$. Substituting the first equation into the second gives $a = (ak)\ell$, but since $a \neq 0$ (since it divides b; see Definition 2.1), we can divide both sides by a to obtain $1 = k\ell$, and this forces $\ell = \pm 1$. Thus $a = \ell b = \pm b$. □

5. Both (a) and (e) are primes.

7. (a) $24 = 2^3 \cdot 3$

(b) $88 = 2^3 \cdot 11$

(c) $675 = 3^3 \cdot 5^2$

(d) $6400 = 2^8 \cdot 5^2$

(e) $74,529 = 3^2 \cdot 7^2 \cdot 13^2$

(f) $183,495,637 = 13^3 \cdot 17^4$

9. (a) quotient: $q = 22$, remainder: $r = 1$

(b) quotient: $q = 21$, remainder: $r = 3$

(c) quotient: $q = -39$, remainder: $r = 1$

(d) quotient: $q = 67$, remainder: $r = 11$

(e) quotient: $q = 49$, remainder: $r = 3$

(f) quotient: $q = -44$, remainder: $r = 2$

11. (a) 12

(b) 100

(c) 4

(d) 280

(e) 600

(f) 11025

13. (a) $36 = 3 \cdot 12 + 0$ (Euclidean algorithm terminates after one step.) So $\gcd(36, 12) = 12$.

(b) $\begin{cases} 25 = 1 \cdot 20 + 5 \\ 20 = 4 \cdot 5 + 0 \end{cases} \Rightarrow \gcd(20, \ 25) = 5$
$$\Rightarrow \operatorname{lcm}(20, \ 25) = 20 \cdot 25 / 5 = 100$$

(c) $\begin{cases} 100 = 1 \cdot 56 + 44 \\ 56 \ = 1 \cdot 44 + 12 \\ 44 \ = 3 \cdot 12 + 8 \\ 12 \ = 1 \cdot 8 + 4 \\ 8 \ \ = 2 \cdot 4 + 0 \end{cases} \Rightarrow \gcd(100, 56) = 4$

(d) $\begin{cases} 1400 = 2 \cdot 560 + 280 \\ 560 = 2 \cdot 280 + 0 \end{cases} \Rightarrow \gcd(560, 1400) = 280$

(e) $\begin{cases} 120 = 2 \cdot 50 + 20 \\ 50 = 2 \cdot 20 + 10 & \Rightarrow \gcd(120,\ 50) = 10 \\ & \Rightarrow \mathrm{lcm}(120,\ 50) = 120 \cdot 50 / 10 = 600 \\ 20 = 2 \cdot 10 + 0 \end{cases}$

(f) $\begin{cases} 5788125 = 47 \cdot 121275 + 88200 \\ 121275 = 1 \cdot 88200 + 33075 \\ 88200 = 2 \cdot 33075 + 22050 & \Rightarrow \gcd(5788125,\ 121275) = \\ 33075 = 1 \cdot 22050 + 11025 & \qquad 11025. \\ 22050 = 2 \cdot 11025 + 0 \end{cases}$

15. (a) Since one of the numbers is a factor of the other, we can do this by inspection: $12 = \gcd(36,\ 12) = 36 \cdot 0 + 12 \cdot 1$.

(b) Starting with the equation $12 = 1 \cdot 8 + 4 \Rightarrow 4 = 12 - 1 \cdot 8$, from the Euclidean algorithm computation of Exercise 13(c), and working our way up the list to perform substitutions of the new remainders, we obtain $4 = 12 - 1 \cdot 8 = 12 - 1 \cdot 4 \cdot 12 - 1 \cdot 44$ $\Rightarrow 4 = 4 \cdot 12 - 1 \cdot 44 = 4 \cdot (56 - 1 \cdot 44) - 1 \cdot 44 = 4 \cdot 56 - 5 \cdot 44 \Rightarrow$ $4 = 4 \cdot 56 - 5 \cdot 44 = 4 \cdot 56 - 5 \cdot (100 - 1 \cdot 56) = -5 \cdot 100 + \cdot 9 \cdot 56$.

(c) From the Euclidean algorithm computation of Exercise 13(d), since it terminated after two iterations, the first equation leads to the desired expression: $1400 = 2 \cdot 560 + 280 \Rightarrow$ $280 = -2 \cdot 560 + 1400$.

(d) Starting with the equation $33075 = 1 \cdot 22050 + 11025 \Rightarrow$ $11025 = 33075 - 1 \cdot 22050$, from the Euclidean algorithm computation of Exercise 13(e), and working our way up the list to perform substitutions of the new remainders, we obtain $11025 = 33075 - 1 \cdot 22050 = 33075 - 1 \cdot (88200 - 2 \cdot 33075)$ $= 3 \cdot 33075 - 1 \cdot 88200 \Rightarrow 11025 = 3 \cdot (121275 - 1 \cdot 88200) -$ $1 \cdot 88200 = 3 \cdot 121275 - 4 \cdot 88200 \Rightarrow 11025 = 3 \cdot 121275 - 4 \cdot$ $(5788125 - 47 \cdot 121275) = 191 \cdot 121275 - 4 \cdot 5788125$.

17. (a) True
(b) False
(c) False
(d) True
(e) False
(f) False

19. (a) 14
(b) 8
(c) 16
(d) 7
(e) First notice that (working in $\mathbb{Z}_{24}$) $21^2 = 9$, so $21^4 = (21^2)^2 = 9^2 = 9$, and it follows that any power of 9, and hence any even power of 21, is 9. Since $223 = 2 \cdot 111 + 1$, we have $21^{223} = 21^{2 \cdot 111 + 1} = (21^2)^{111} \cdot 21^1 = (9)^{111} \cdot 21 = 9 \cdot 21 = 21$.

21.

$\mathbb{Z}_2$			$\mathbb{Z}_2$			$\mathbb{Z}_4$					$\mathbb{Z}_4$			
+	0	1	×	0	1	+	0	1	2	3	+	0	1	2	3
0	0	1	0	0	0	0	0	1	2	3	0	0	0	0	0
1	1	0	1	0	1	1	1	2	3	0	1	0	1	2	3
						2	2	3	0	1	2	0	2	0	2
						3	3	0	1	2	3	0	3	3	1

In $\mathbb{Z}_2$ only 1 is invertible, while in $\mathbb{Z}_4$ the invertible elements are 1 and 3.

23. The invertible elements in $\mathbb{Z}_8$ are $\{1, 3, 5, 7\}$ and each of these elements is its own inverse.

25. All nonzero elements of $\mathbb{Z}_7$ are invertible; 2 and 4 are inverses; 3 and 5 are inverses; 1 and 6 are self-inverses.

27. Using the fact (see the solution to Exercise 23) that every odd number in $\mathbb{Z}_8$ is its own inverse, each of the given congruences can be solved:

(a) $3x \equiv 5 \Rightarrow x \equiv 1 \cdot x \equiv 3^{-1} \cdot 3x \equiv 3^{-1} \cdot 5 \equiv 3 \cdot 5 \equiv 15 \equiv 7 \pmod 8$

(b) $7x + 2 \equiv 3 \Rightarrow 7x \equiv 3 - 2 \equiv 7 \Rightarrow x \equiv (7^{-1} \cdot 7 \equiv 7 \cdot 7 \equiv 49 \equiv)\ 1$ $\pmod 8$ (cf. footnote to Example 2.8)

(c) $5x - 2 \equiv 2 \Rightarrow 5x \equiv 2 + 2 = 4 \Rightarrow x \equiv 1 \cdot x \equiv 3^{-1} \cdot 3x \equiv 5^{-1} \cdot 4 \equiv$ $5 \cdot 4 \equiv 20 \equiv 4 \pmod 8$

29. (a) Initial Vectors: $U = [388, 1, 0]$, $V = [3, 0, 1]$. Since $V(1) = 3$ is positive, we update the vectors: $W = U - \text{floor}(U(1)/V(1))V = [1, 1, -129]$, $U = V = [3, 0, 1]$, $V = W = [1, 1, -129]$. Since $V(1) = 1$ is positive, we update the vectors: $W = U - \text{floor}(U(1)/V(1))V = [0, -3, 388]$, $U = V = [1, 1, -129]$, $V = W = [0, -3, 388]$. Since $V(1) = 0$, the algorithm terminates. From the last updated vector U we obtain: $d = 1$, $x = 1$, and $y = -129$; it can be readily checked that these numbers satisfy: $d = 388x + 3y$. In light of Proposition 2.11, since $\gcd(388, 3) = 1$ and since $y = -129 \equiv 259 \pmod{388}$, we have $3^{-1} = 259$ in $\mathbb{Z}_{388}$.

(b) Initial Vectors: $U = [388, 1, 0]$, $V = [55, 0, 1]$. Since $V(1) = 55$ is positive, we update the vectors: $W = U - \text{floor}(U(1)/V(1))V = [3, 1, -7]$, $U = V = [55, 0, 1]$, $V = W = [3, 1, -7]$. Since $V(1) = 3$ is positive, we update the vectors: $W = U - \text{floor}(U(1)/V(1))$ $V = [1, -18, 127]$, $U = V = [3, 1, -7]$, $V = W = [1, -18, 127]$. Since $V(1) = 1$ is positive, we update the vectors: $W = U - \text{floor}(U(1)/V(1))V = [0, 55, -388]$, $U = V = [1, -18, 127]$, $V = W = [0, 55, -388]$. Since $V(1) = 0$, the algorithm terminates.

From the last updated vector U we obtain $d = 1$, $x = -18$, and $y = 127$; it can be readily checked that these numbers satisfy $d = 388x + 55y$. In light of Proposition 2.11, since $\gcd(388, 55) = 1$, we have $55^{-1} = y = 127$ in $\mathbb{Z}_{388}$.

(c) Computing as above, we obtain $149^{-1} = 125$ in $\mathbb{Z}_{388}$.

(d) Since (as the extended Euclidean algorithm will show) $\gcd(97, 388) = 97 > 1$, (by Proposition 2.11) 97 is not invertible in $\mathbb{Z}_{388}$.

31. (a) Initial Vectors: $U = [1353, 1, 0]$, $V = [2, 0, 1]$. Since $V(1) = 2$ is positive, we update the vectors: $W = U - \text{floor}(U(1)/V(1))V = [1, 1, -676]$, $U = V = [2, 0, 1]$, $V = W = [1, 1, -676]$. Since $V(1) = 1$ is positive, we update the vectors: $W = U - \text{floor}(U(1)/V(1))V = [0, -2, 1353]$, $U = V = [1, 1, -676]$, $V = W = [0, -2, 1353]$. Since $V(1) = 0$, the algorithm terminates. From the last updated vector U, we obtain $d = 1$, $x = 1$, and $y = -676$; it can be readily checked that these numbers satisfy $d = 1353x + 2y$. In light of Proposition 2.11, since $\gcd(1353, 2) = 1$ and since $y = -676 \equiv 677 \,(\text{mod } 1353)$, we have $2^{-1} = 677$ in $\mathbb{Z}_{1353}$.

(b) Since (as the extended Euclidean algorithm will show) $\gcd(44, 1353) = 11 > 1$, (by Proposition 2.11) 44 is not invertible in $\mathbb{Z}_{1353}$.

(c) The extended Euclidean algorithm terminates after six iterations to show that $d = 1353x + 886y$, where $d = 1$, $x = -203$, and $y = 310$; so it follows from Proposition 2.11 that $886^{-1} = 310$ in $\mathbb{Z}_{1353}$.

(d) The extended Euclidean algorithm terminates after nine iterations to show that $d = 1353x + 350y$, where $d = 1$, $x = 76$, and $y = -259$ ($\equiv 1094$); so it follows from Proposition 2.11 that $350^{-1} = 1094$ in $\mathbb{Z}_{1353}$.

33. Making use of the inverse information obtained in Exercises 29 and 31, we can easily solve each of the indicated congruences:

(a) $3x \equiv 59 \Rightarrow x \equiv 1 \cdot x \equiv 3^{-1} \cdot 3x \equiv 3^{-1} \cdot 59 \equiv 259 \cdot 59 \equiv 149 \,(\text{mod } 388)$

(b) $149x \equiv 225 \Rightarrow x \equiv 1 \cdot x \equiv 149^{-1} \cdot 149x \equiv 149^{-1} \cdot 225 \equiv 125 \cdot 225 \equiv 189 \,(\text{mod } 388)$

(c) $2x \equiv 1225 \Rightarrow x \equiv 1 \cdot x \equiv 2^{-1} \cdot 2x \equiv 2^{-1} \cdot 1225 \equiv 677 \cdot 225 \equiv 789 \,(\text{mod } 1353)$

(d) $886x \equiv 35 \Rightarrow x \equiv 1 \cdot x \equiv 886^{-1} \cdot 886x \equiv 886^{-1} \cdot 35 \equiv 310 \cdot 35 \equiv 26 \,(\text{mod } 1353)$

35. (a) Since $d = \gcd(3, 18) = 3 \,|\, 6$, we can follow Algorithm 2.3 to obtain the $d = 3$ solutions of the congruence.

Step 1. We solve the modified congruence: $(3/3)y \equiv (6/3) \,(\text{mod } 18/3)$; i.e., $y \equiv 2 \,(\text{mod } 6)$. Plainly, the unique solution is $2 \,(\text{mod } 6)$.

Step 2. We may now list the $d = 3$ solutions of the original congruence: $2, 2 + 18/3, 2 + 2 \cdot 18/3 = \{2, 8, 14\}$. (The reader may wish to check each of these.)

(b) Since $d = \gcd(6, 27) = 3 \not| \, 16$, we know that the congruence has no solution.

(c) Since $d = \gcd(8,\ 28) = 4\,|\,12$, we can follow Algorithm 2.3 to obtain the $d = 4$ solutions of the congruence.

Step 1. We solve the modified congruence: $(8/4)y \equiv (12/4)$ $(\bmod\ 28/4)$; i.e., $2y \equiv 3(\bmod\ 7)$. Since $2^{-1} = 4\,(\bmod\ 7)$, we obtain $y \equiv 2^{-1} \cdot 2y \equiv 2^{-1} \cdot 3 \equiv 4 \cdot 3 \equiv 5(\bmod\ 7)$.

Step 2. We may now list the $d = 4$ solutions of the original congruence: $5, 5 + 28/4, 5 + 2 \cdot 28/4, 5 + 3 \cdot 28/4 = \{5, 12, 19, 26\}$.

(d) Since $d = \gcd(8,\ 28) = 4\,I\,6$, we know that the congruence has no solution.

37. (a) Since $d = \gcd(6,\ 776) = 2\,|\,6$, we can follow Algorithm 2.3 to obtain the $d = 2$ solutions of the congruence.

Step 1. We solve the modified congruence: $(6/2)y \equiv (28/2)$ $(\bmod\ 776/2)$, i.e., $3y \equiv 14(\bmod\ 388)$. In Exercise 29(a) we computed $3^{-1} = 259(\bmod\ 388)$, and we can use this to solve for y: $y \equiv 3^{-1} \cdot 14 \equiv 259 \cdot 14 \equiv 134(\bmod\ 388)$.

Step 2. We may now list the $d = 2$ solutions of the original congruence: $134, 134 + 776/2 = \{134, 522\}$.

(b) Since $d = \gcd(15,\ 1940) = 5\,I\,21$, we know that the congruence has no solution.

(c) Solution: $\{189, 577, 965, 1353\}$ [The answer to Exercise 29(c) will make this easy.]

(d) Solution: $\{747, 2100, 3453, 4806\}$ [The answer to Exercise 31(c) will make this easy.]

39. (a) By the Chinese remainder theorem, the solution obtained will be unique modulo $N = 35$. The inverse of $N/5 = 7(\bmod\ 5)$ is $e_1 = 3$, and the inverse of $N/7 = 5(\bmod\ 7)$ is $e_2 = 3$; thus, by Equation 2.8 the solution is $x = 3 \cdot 3 \cdot 7 + 4 \cdot 3 \cdot 5 \equiv 18(\bmod\ 35)$.

(b) By the Chinese remainder theorem, the solution obtained will be unique modulo $N = 165$. The inverse of $N/3 = 55 \equiv 1$ $(\bmod\ 3)$ is $e_1 = 1$; the inverse of $N/5 = 33 \equiv 3(\bmod\ 5)$ is $e_2 = 2$; the inverse of $N/11 = 15 \equiv 4(\bmod\ 11)$ is $e_3 = 3$; thus, by Equation 2.8, the solution is $x = 2 \cdot 1 \cdot 55 + 1 \cdot 2 \cdot 33 + 3 \cdot 3 \cdot 15 = 311 \equiv 146(\bmod\ 165)$.

(c) Solution: $x \equiv 3446(\bmod\ 4290)$.

41. Letting x denote the number of coins, we have the following system:

$$
\begin{cases} x & \equiv 8(\bmod\ 15) \\ x - 8 & \equiv 11(\bmod\ 14) \\ x - 8 - 11 \equiv 5(\bmod\ 13) \end{cases} \Rightarrow \begin{cases} x \equiv 8(\bmod\ 15) \\ x \equiv 5(\bmod\ 14) \\ x \equiv 11(\bmod\ 13) \end{cases}
$$

The Chinese remainder theorem is applicable and yields the following solution: $x \equiv 2273(\bmod\ 2730)$, and thus 2273 is the smallest possible number of coins.

43. (a) 6, 1, 2

(b) Suppose that $x_1' x_2' \cdots x_{13}'$ differs from $x_1 x_2 \cdots x_{13}$ in exactly one digit. If the differing digits x_k', x_k shared an odd index, then since all odd index digits have coefficient 1 in Equation 2.9, it would follow from Equation 2.9 that the two odd digits would be equal (this could be seen by subtracting one version of Equation 2.9 from the other)—a contradiction. So assume the differing digits have an even index. Subtracting Equation 2.9 for $x_1 x_2 \cdots x_{13}$ from Equation 2.9 for $x_1' x_2' \cdots x_{13}'$ would result (after removing common terms and simplifying) in the equation $3x_k' \equiv 3x_k \pmod{10}$. But since 3 is invertible (mod 10), we could multiply both sides by the inverse to obtain $x_k' \equiv x_k \pmod{10}$, so in any case, we are led to a contradiction, thus proving that two valid ISBN-13 numbers cannot differ in only a single digit. $\square$

(c) We assume that $x_1 x_2 \cdots x_{13}$ was incorrectly typed as $y_1 y_2 \cdots y_{13}$, where each $y_i = x_i$, with exactly two adjacent permuted exceptions, $y_j = x_{j+1}$, $y_{j+1} = x_j$, with the two digits being different. Assume that both ISBNs checked with Equation 2.9. Then, by subtracting the corresponding two equations (and simplifying), we would be left with $2(x_{i+1} - x_i) \equiv 0 \pmod{10}$. Since 2 is not invertible, we cannot conclude that $x_{i+1} \equiv x_i \pmod{10}$. More precisely, since $2(x_{i+1} - x_i) \equiv 0 \pmod{10}$ is equivalent to $10 \mid 2(x_{i+1} - x_i)$, which in turn (divide the divisibility by 2) is equivalent to $5 \mid (x_{i+1} - x_i)$, the error will not be detected if $x_{i+1} \equiv x_i \pmod 5$. (But otherwise it would be detected.)

(d) If $x_1 x_2 \cdots x_{13}$ violated Equation 2.9, we could change just the check digit x_{13} so that Equation 2.9 would hold. We could similarly easily change any of the other odd-indexed digits (that each has coefficient ± 1 in Equation 2.9) to ensure that the modified sequence would satisfy Equation 2.9.

47. (a) True. *Proof:* Since $a \mid b$ and $a \mid (b+1)$, Theorem 2.1(b) tells us that $a \mid [(b+1) - b]$, or $a \mid 1$, which forces $a = \pm 1$. $\square$

(b) True. *Proof:* Since n is even, we can write $n = 2k$, for some integer k. Squaring gives $n^2 = 4k^2$, showing that $4 \mid n^2$. $\square$

(c) True. *Proof:* Since a and b have the same parity, we have $a \equiv b \pmod 2$, or that $2 \mid a - b$. It follows that $2 \mid (a-b)(a+b) = a^2 - b^2$, showing that $a^2 - b^2$ is even. $\square$

(*Note:* With a little more work, it can be shown that $4 \mid a^2 - b^2$.)

49. (a) False: $4 \nmid 1 \cdot 2 \cdot 3$.

(b) True. *Proof:* First we factor $a^4 - a^2 = a^2(a^2 - 1) = a^2(a+1)(a-1)$. We separate into two cases: *Case 1: a* is even; i.e., $2 \mid a$. Then $4 \mid a^2$, so (by the factorization above) $4 \mid (a^4 - a^2)$. *Case 2: a* is odd; then $a+1$ and $a-1$ must both be even; i.e., $2 \mid a \pm 1$. Thus $4 = 2 \cdot 2 \mid (a+1)(a-1)$, and so (by the factorization above) $4 \mid (a^4 - a^2)$. $\square$

(c) False. Counterexample: $4^9 - 3 = 11 \cdot 23831$.

51. (a) Since $d \mid b$ and $d \mid c$ if, and only if, $ad \mid ab$ and $ad \mid ac$, it follows that the set $\{ad : d$ is a common divisor of a and $b\}$ is exactly the set of common divisors of ab and ac.

Therefore, $\gcd(ab, ac)$ is the greatest element of the set $\{ad : d \text{ is a common divisor of } a \text{ and } b\}$, which is a $\gcd(a, c)$.

(b) From part (a), it follows that $d \gcd(b/d, c/d) = \gcd(d \cdot (b/d), d \cdot (c/d)) = \gcd(b, c) = d$, and this forces $\gcd(b/d, c/d) = 1$. This latter fact implies that the numerator and denominator of $\dfrac{b/d}{c/d}$ have no common factors, so is the lowest terms representation of $\dfrac{b}{c}$. (This fact corresponds to the arithmetical fact that to put a fraction in lowest terms, we need to cancel out the greatest common factor of the numerator and denominator.)

(c) Use the Euclidean algorithm to obtain $d = \gcd(b, c)$, and then apply part (b). For example, (the Euclidean algorithm gives) $\gcd(1474, 39463) = 67$, so the lowest terms form of $1474/39463$ is $(1474/67)/(39463/67) = 22/589$.

53. If n had a nontrivial factorization $n = ab$, then the indicated identity would produce a nontrivial factorization of $2^{ab} - 1$. This proves that if $2^n - 1$ is composite, then so is n, which is the contrapositive, and so logically equivalent to the indicated statement.

55. (a) We will prove that $n = \displaystyle\sum_{k=0}^{D} d_k \cdot 10^k$ is always congruent to $(d_0 + d_2 + d_4 + \cdots - d_1 - d_3 - d_5 - \cdots) \pmod{11}$. [Note that this is more than we needed to show, since the indicated statement simply states that if one of these quantities is congruent to $0 \pmod{11}$, then so is the other.] Collecting like terms in the difference, we obtain:

$$n - (d_0 + d_2 + d_4 + \cdots - d_1 - d_3 - d_5 - \cdots)$$

$$= [d_0 \cdot 10^0 + d_1 \cdot 10^1 + d_2 \cdot 10^2 + d_3 \cdot 10^3 + \cdots + d_D \cdot 10^D]$$

$$- (d_0 + d_2 + d_4 + \cdots - d_1 - d_3 - d_5 - \cdots)$$

$$= [d_0 \cdot (10^0 - 1) + d_1 \cdot (10^1 + 1) + d_2 \cdot (10^2 - 1)$$

$$+ d_3 \cdot (10^3 + 1) + \cdots + d_D \cdot (10^D - (-1)^D)]$$

But since $10 \equiv -1 \pmod{11}$, it follows that $(10^{2k} - 1) \equiv ((-1)^{2k} - 1) \equiv 0$ and $(10^{2k+1} + 1) \equiv ((-1)^{2k+1} + 1) \equiv 0 \pmod{11}$, so that all of the terms in the last bracketed expression above must be congruent to $0 \pmod{11}$. □

(b) Collecting like terms in the difference, we obtain:

$$n - (d_0 + 10d_1 + 100d_2 - d_3 - 10d_4 - 100d_5 + d_6 + 10d_7 + 100d_8 - \cdots)$$

$$= [d_0 \cdot 10^0 + d_1 \cdot 10^1 + d_2 \cdot 10^2 + d_3 \cdot 10^3 + \cdots + d_D \cdot 10^D]$$

$$- (d_0 + 10d_1 + 100d_2 - d_3 - 10d_4 - 100d_5 + d_6 + 10d_7 + 100d_8 - \cdots)$$

$$= [d_0 \cdot (10^0 - 1) + d_1 \cdot (10^1 - 10) + d_2 \cdot (10^2 - 100) + d_3 \cdot (10^3 + 1)$$

$$+ d_4 \cdot (10^4 + 10) + d_5 \cdot (10^5 + 100)d_6 \cdot (10^6 - 1) + d_7 \cdot (10^7 - 10)$$

$$+ d_8 \cdot (10^8 - 100) - \cdots]$$

Now each expression in parentheses in the last expression is (when the greatest common factor of 10 is factored out) one of these two forms: $10^j \cdot (10^{6n} - 1)$ or $10^j \cdot (10^{6n+3} + 1)$, where $j = 0$, 1, or 2, and n is a nonnegative integer. But since $10^3 \equiv 3^3 \equiv -1 \pmod 7$, it follows that $10^{6n} \equiv 1 \pmod 7$, and thus each of these two expressions is congruent to $0 \pmod 7$. The assertion now follows, just as it did in part (a). □

57. If $m > 1$ is even, then the sum $\displaystyle\sum_{k=1}^{m-1} k$ contains an odd number of terms. Apart from the middle term $m/2$, the remaining terms can be paired off as $m/2 + j$ and $m/2 - j$, as j runs from 1 to $m/2 - 1$. Each such pair adds up to $(m/2 + j) + (m/2 - j) = m \equiv 0 \pmod m$, so the entire sum must equal this remaining middle term: $m/2$. □

59. (a) This follows directly from Proposition 2.14(b).
   (b) The statement of part (a) fails if the assumption that a and b are relatively prime is omitted. Counterexample: $6 \equiv 2 \pmod 4$ and $6 \equiv 2 \pmod 2$, but $6 \not\equiv 2 \pmod{2 \cdot 4}$.

61. (a) This follows directly from Proposition 2.14(a), with $b = x^a - a$ (using the definition of a congruence). □
   (b) From the result just proved, it follows that an integer a will have a square root x mod pq if, and only if, x is both a square root of a mod p and mod q. From Exercise 60, this will happen if, and only if, a belongs to both of the sets $\{0^2, 1^2, 2^2, \cdots, [(p-1)/2]^2\}$ $\pmod p$ and $\{0^2, 1^2, 2^2, \cdots, [(q-1)/2]^2\}$ $\pmod q$. Furthermore, since each such *nonzero* element (of either set) has two distinct square roots under the corresponding modulus, and zero has only one square root for each of the one, two, or four possibilities, the Chinese remainder theorem can be used to compute all of the distinct square roots.
   (c) (i) With $p = 5$ and $q = 7$, since $a = 9$ belongs to both sets $\{0^2, 1^2, 2^2, \cdots, [(p-1)/2]^2\} = \{0, 1, 4\} \pmod p$ and $\{0^2, 1^2, 2^2, \cdots, [(q-1)/2]^2\} = \{0, 1, 4, 2\} \pmod q$, it follows that a does indeed have a square root—in fact, four of them. Since $\sqrt{a} \equiv \sqrt{4} \equiv \pm 2 \equiv 2$, $3 \pmod 5$ and $\sqrt{a} \equiv \sqrt{2} \equiv \pm 3 \equiv 3$, $4 \pmod 7$ with any of the four combinations of these square roots, we can use the Chinese remainder theorem to obtain a square root of a mod pq. For example, using the smaller roots, the Chinese remainder theorem applied to the system $\begin{cases} x \equiv 2 \pmod 5 \\ x \equiv 3 \pmod 7 \end{cases}$ produces the unique solution $x = 17 \pmod{35}$. The other three combinations produce the following three solutions: $x = 3$, 18, $32 \pmod{35}$, and so $\sqrt{a} \equiv \sqrt{9} \equiv \{3, 17, 18, 32\} \pmod{35}$. [The reader may wish to verify that the squares of each of these numbers are congruent to $9 \pmod{35}$.]
       (ii) $\sqrt{51} \equiv \{102, 391\} \pmod{493}$
   (d) Part (a) would remain true (for the same reason as before). The result of part (b) would be simplified somewhat since all

integers have a unique square root mod 2. To see whether a has a square root mod $2q$, we need only check whether a belongs to $\{0^2, 1^2, 2^2, \cdots, [(q-1)/2]^2\} \pmod{q}$. If a appears on the list as zero [i.e., $\sqrt{a} \equiv 0 \pmod{q}$], then a will have a unique square root mod $2q$, obtained by using the Chinese remainder theorem to solve the system $x \equiv \sqrt{a} \pmod 2$, $x \equiv 0 \pmod q$, whereas if a appears on the list as a nonzero square, then a will have two square roots mod $2q$, each obtainable using the Chinese remainder theorem using one of the two square roots of a mod q.

(e) (i)   11 has no square roots mod 26 (since it has none mod 13).

    (ii) $\sqrt{68} \equiv \{5, 38\} \pmod{86}$.

63. (a) By Exercise 61, $x^2 \equiv 46 \pmod{413}$ is equivalent to
$$\begin{cases} x^2 \equiv 46 \equiv 4 \pmod 7 \\ x^2 \equiv 46 \pmod{59} \end{cases}.$$

The first congruence plainly has two solutions $x \equiv \pm 2 \pmod 7$. Proposition 2.14 tells us that the latter will have a solution if, and only if, $46^{(59+1)/4} \equiv 46^{15}$ is a square root of $46 \pmod{59}$. We work our way up to this power: $46^2 \equiv 51$, $46^4 \equiv 51^2 \equiv 5$, $46^8 \equiv 5^2 \equiv 25$, so $46^{15} \equiv 46^8 \cdot 46^4 \cdot 46^2 \cdot 46 \equiv (25 \cdot 5) \cdot (51 \cdot 46) \equiv 7 \cdot 45 \equiv 20$. Since $20^2 \equiv 46 \pmod{59}$, we know there are two square roots of 46: $x \equiv \pm 20 \pmod{59}$. Applying the Chinese remainder theorem to the four possible square root combinations mod 7 and mod 59 produces the four square roots mod 413: $\sqrt{46} \equiv \{79, 138, 275, 334\} \pmod{413}$.

(b) There are no such square roots.

(c) $\sqrt{34} = \{6568, 8711, 14862, 17005\} \pmod{23573}$.

65. Working mod 2, if $a \equiv 0$, or if $a \equiv 1$, the expression $a^{4n+1} - a$ is clearly congruent to zero, so $2 \mid (a^{4n+1} - a)$. In order to work mod 5, we factor $a^{4n+1} - a = a(a^{4n} - 1) = a(a^{2n} + 1)(a^n + 1)(a^n - 1)$. If $a \equiv 0$, the first factor is zero, if $a \equiv \pm 1$, one of the last two factors is 0, and finally, if $a \equiv \pm 2$, then $a^2 \equiv 4 \equiv -1$, so $a^{4n} - 1 \equiv 0$. This proves that $5 \mid (a^{4n+1} - a)$, and when combined with $2 \mid (a^{4n+1} - a)$, produces $10 \mid (a^{4n+1} - a)$. This means that $a^{4n+1} \equiv a \pmod{10}$. □

Chapter 3

1. (a) IYPKDLQMCTNDIBFDSCTQI

   (b) HYBQJYBQPBFTQQTCBPVDFWDXCTKK

   (c) TSSPCIYPNDKKBNNLQBIX

   (d) IYPNTQPXBFLQMPCIYPSKTTCBQV

3. (a) youmustexitbeforeholtarrives

   (b) weararedshirtandcap

   (c) lookforatallchinesewoman

   (d) togainaccessusethewordcleveland

5. (a) TALHTPEIRAOOSNCSWEIIENLD

(b) More troops needed.

7. (a) The three known plaintext $\rightarrow$ ciphertext correspondences are r(17) $\rightarrow$ M(12), c(2) $\rightarrow$ P(15), and t(19) $\rightarrow$ W(22). In terms of the affine mapping, this translates into the following three congruences: $17\alpha + \beta \equiv 12$, $2\alpha + \beta \equiv 15$, and $19\alpha + \beta \equiv 22 \pmod{26}$. Subtracting the second from the first produces $15\alpha \equiv -3 \pmod{26}$, which implies $\alpha \equiv 15^{-1}(-3) \equiv 7 \cdot (-3) \equiv -21 \equiv 5 \pmod{26}$. Substituting this into the second equation, we obtain $2\alpha + 5 \equiv 15 \Rightarrow 2\alpha \equiv 10 \Rightarrow \alpha \equiv 5 \pmod{26}$. This gives the key = (5, 5), from which we can decrypt the plaintext: americanfootball

(b) The two known plaintext $\rightarrow$ ciphertext correspondences lead us to the system $13\alpha + \beta \equiv 1$, $18\alpha + \beta \equiv 20 \pmod{26}$. Subtracting the first from the second gives $5\alpha \equiv 19 \Rightarrow \alpha \equiv 5^{-1} \cdot 19 \equiv 21 \cdot 19 \equiv 9 \pmod{26}$ Then substitute this into either of the original equations to obtain the key = (9, 14), from which we can decrypt the plaintext: gunsandbutter

(c) The two known plaintext $\rightarrow$ ciphertext correspondences lead us to the system: $\beta \equiv 9$, $13\alpha + \beta \equiv 22 \pmod{26}$. This system is easily solved to produce the key = (7, 9), from which we may decrypt the plaintext: grandcentralstation

(d) In a similar fashion, it is possible to solve for the key = (21, 22), and use it to decrypt: ferrarisandlamborghinis

9. (a) nulls: 23 0 6

(b) no nulls: [5, 20, 13, 28]; with some nulls: [23, 5, 11, 20, 0, 13, 28]

(c) Take the train to Grenoble.

11. (a) $\begin{pmatrix} 1 & 2 & 3 & 4 & 5 & 6 & 7 \\ 2 & 3 & 5 & 7 & 1 & 4 & 6 \end{pmatrix}$

(b) $\begin{pmatrix} 1 & 2 & 3 & 4 & 5 & 6 & 7 \\ 1 & 2 & 5 & 4 & 6 & 7 & 3 \end{pmatrix}$

(c) $\begin{pmatrix} 1 & 2 & 3 & 4 & 5 & 6 & 7 \\ 7 & 3 & 5 & 1 & 2 & 6 & 4 \end{pmatrix}$

(d) $\begin{pmatrix} 1 & 2 & 3 & 4 & 5 & 6 & 7 \\ 1 & 2 & 5 & 4 & 6 & 7 & 3 \end{pmatrix}$

13. (a) (2, 3, 4, 5) (6, 9) (7, 8)

(b) (1, 10) (2, 9) (3, 8) (4, 7) (5, 6)

(c) (1, 2, 7, 4, 5)

(d) (1, 2, 3, 4)

(e) $(1, 2, 5) \circ (7, 4, 2)^{-1} = (1, 2, 5) \circ (2, 4, 7) = (1, 2, 4, 7, 5)$

(f) (1, 3, 5, 9, 1)

15. Neither. It is definitely not a transposition cipher since the location of the ciphertext characters within a ciphertext is identical to the location of the corresponding plaintext character within the plaintext. The permutation changes each time we encrypt a plaintext letter, so it is not a simple substitution cipher, but a similar system to the Vigenère cryptosystem, (typically) with a much larger period.

17. (a) BDBE
(b) CDFCEA
(c) DFECBACFDF

19. It was shown in the solution of EFR 3.1 that $\phi_{\alpha,\beta}^{-1}(x) \equiv \alpha^{-1}$ $(x - \beta)$ (mod m). But $\alpha^{-1}(x - \beta) \equiv \alpha^{-1}x - \alpha^{-1} \cdot \beta \equiv \phi_{\alpha^{-1},-\alpha^{-1}\beta}(x)$.

21. None. Since a shift cipher $x \mapsto x + \kappa \pmod{26}$ is the affine cipher $\phi_{1,\kappa}$, the indicated cryptosystem is just a composition of three affine ciphers, and by Exercise 20, compositions of affine mappings are again affine mappings. So such a triple composition of affine ciphers would be the same as a single affine cipher.

23. (a) [13, 39, 36, 22, 28, 39, 2, 30, 22, 20, 13, 1, 19, 12]
(b) themeetingissetup

25. Each plaintext/ciphertext character correspondence $a_i \leftrightarrow A_i$ gives rise to a congruence $a_i\alpha + \beta \equiv A_i \pmod{p}$ in the three unknowns (the key parameters) α, β, p. This is equivalent to $p \mid a_i\alpha + \beta - A_i$. There may be several solutions for however many plaintext/ciphertext correspondences are known (since there are infinitely many primes), so it is not feasible to solve for the actual key. However, since the cipher is really a substitution cipher, it would be more effective to use the known information together with frequency analysis of the ciphertext. If p is known, any two plaintext/ciphertext correspondences will suffice for a known plaintext attack. We could simply subtract the resulting two congruences (so β will be eliminated) to obtain one of the form $a'\alpha \equiv A' \pmod{p}$, where $a' \not\equiv 0 \pmod{p}$. Since p is prime, we can multiply both sides by $(a')^{-1}$ to solve for α, and then substitute this into either of the original equations to recover β, thus determining the key.

27. (a) BORKSNUCOYFNUO
(b) evacuatetheprisoners
(c) It is not a simple substitution cipher since individual plaintext characters are not encrypted. This is an example of a block cipher (with block length 2), since encryption takes place on blocks of two plaintext letters at a time.
(d) Since the system processes characters in blocks of two, we could only gain useful information by feeding pairs of ciphertext characters in the decryption machine. There are 676 ordered pairs of ciphertext letters, and each corresponds to an integer (mod 676) (as explained in Exercise 26), and similarly for the 676 pairs of plaintext letters. Thus each corresponding pair of ciphertext/plaintext letters gives rise to a corresponding pair of mod 676 integers A_i/a_i, which, in turn, gives rise

to a congruence $a_i\alpha + \beta \equiv A_i \pmod{676}$. We need at least two such congruences to solve for the unknown key (α, β). Once we have two of them $a_i\alpha + \beta \equiv A_i$, $a_i'\alpha + \beta \equiv A_i' \pmod{676}$, such that the difference $a_i - a_i'$ is relatively prime to 676, we can subtract them to eliminate β, then solve for α, and then substitute this back into one of the original congruences to determine β and, hence, the key. So how can we obtain two such equations? One way is to continue to decrypt ciphertext pairs corresponding to the mod 676 integers 0, 1, 2, ..., until such a pair is found.

(e) This is addressed in our solution to (d).

29. (a) When the cycle $\sigma = (a_1, a_2, \cdots, a_\ell)$ is applied to any of its elements, the element gets shifted one unit down the list (and the last element a_ℓ gets cycled back to a_1); it follows that σ^2 moves each element two units down the list (with the last two elements $a_{\ell-1}, a_\ell$ getting cycled back to a_1, a_2), σ^3 moves each element three units down the list, and so on. Since the list has ℓ elements, if follows that $\sigma^\ell = id$ and $\sigma^k \neq id$, for any positive integer $k < \ell$. This proves that the order of $\sigma = (a_1, a_2, \cdots, a_\ell)$ is ℓ. □

(b) It will be helpful to first note the following observation:
Claim: If k is any positive integer and $\sigma = (a_1, a_2, \cdots, a_\ell)$ is an ℓ cycle, then $\sigma^k = \sigma^r$, where r is the remainder when k is divided by ℓ.

Proof of Claim: We use the division algorithm to write $k = q\ell + r$. Since k is nonnegative, so must be q, and since [by part (a)] $\sigma^\ell = id$, we obtain $\sigma^k = \sigma^{q\ell+r} = \sigma^{q\ell}\sigma^r = (\sigma^\ell)^q \sigma^r = (id)^q \sigma^r = \sigma^r$, as asserted. □

If $\sigma_1\sigma_2 \cdots \sigma_t$ is the disjoint cycle decomposition of σ, then, since disjoint cycles commute, it follows that for any positive integer k, we have $\sigma^k = \sigma_1^k \sigma_2^k \cdots \sigma_t^k$. Since each of the power cycles on the right side of this equation acts on disjoint elements, it follows that the whole product will be the identity permutation exactly when each separate power cycle is the identity—i.e., $\sigma_j^k = id$; but by the claim, $\sigma_j^k = \sigma_j^{r_j}$, r_j is the remainder when k is divided by ℓ. Thus, it follows that $\sigma^k = id \Leftrightarrow r_j = 0$, for all indices $j \Leftrightarrow \ell_j \mid k$, for all indices j. It follows from the definition of order that the order of σ is $\text{lcm}(\ell_1, \ell_2, \cdots, \ell_t)$. □

(c) Each of the cycles σ_j appearing in the disjoint cycle decomposition of σ has order ℓ_j, which is a positive integer that is at most n (since the permutation acts on n objects). Since the lcm of a set of integers divides their product, the result now follows from part (b). □

31. (a) We use notation similar to what was used on the smaller scale in Example 3.8. Here, σ denotes the unit shift on $\mathbb{Z}_{26}$, and $F : \{a, b, c, \cdots, z\} \to \mathbb{Z}_{26}$ is the "full" bijection indicated in Table 1.3. Because of the given rotor settings, it follows that the encryption of the four plaintext letters will correspond to the following rotor settings:

First plaintext letter rotor settings: Rotor 1: 25, Rotor 2: 0, Rotor 3: 13
Second plaintext letter rotor settings: Rotor 1: 0, Rotor 2: 1, Rotor 3: 13
Third plaintext letter rotor settings: Rotor 1: 1, Rotor 2: 1, Rotor 3: 13
Fourth plaintext letter rotor settings: Rotor 1: 2, Rotor 2: 1, Rotor 3: 13

We now describe the encryption, one letter at a time. One difference with Example 3.8 is that the rotor permutations are given here in terms of letters rather than mod 26 integers. As we perform the compositions below, we mentally identify each letter with its integer representative.

$$a \overset{\beta}{\mapsto} f \overset{F}{\mapsto} 5 \overset{C_{25}[\rho_1]}{\mapsto} 12 \overset{C_0[\rho_2]}{\mapsto} 22 \overset{C_{13}[\rho_3]}{\mapsto} 6 \overset{F^{-1}}{\mapsto} g \overset{\tau}{\mapsto} \ell \overset{F}{\mapsto} 11 \overset{C_{13}[\rho_3^{-1}]}{\mapsto} 1 \overset{C_0[\rho_2^{-1}]}{\mapsto} 9 \overset{C_{25}[\rho_1^{-1}]}{\mapsto}$$

$$22 \overset{F^{-1}}{\mapsto} w \overset{\beta}{\mapsto} w$$

$$u \overset{\beta}{\mapsto} g \overset{F}{\mapsto} 6 \overset{C_0[\rho_1]}{\mapsto} 3 \overset{C_1[\rho_2]}{\mapsto} 17 \overset{C_{13}[\rho_3]}{\mapsto} 22 \overset{F^{-1}}{\mapsto} w \overset{\tau}{\mapsto} v \overset{F}{\mapsto} 21 \overset{C_{13}[\rho_3^{-1}]}{\mapsto} 3 \overset{C_1[\rho_2^{-1}]}{\mapsto} 24 \overset{C_0[\rho_1^{-1}]}{\mapsto}$$

$$14 \overset{F^{-1}}{\mapsto} o \overset{\beta}{\mapsto} o$$

$$c \overset{\beta}{\mapsto} z \overset{F}{\mapsto} 25 \overset{C_1[\rho_1]}{\mapsto} 3 \overset{C_1[\rho_2]}{\mapsto} 17 \overset{C_{13}[\rho_3]}{\mapsto} 22 \overset{F^{-1}}{\mapsto} w \overset{\tau}{\mapsto} v \overset{F}{\mapsto} 21 \overset{C_{13}[\rho_3^{-1}]}{\mapsto} 3 \overset{C_1[\rho_2^{-1}]}{\mapsto} 24$$

$$\overset{C_1[\rho_1^{-1}]}{\mapsto} 8 \overset{F^{-1}}{\mapsto} i \overset{\beta}{\mapsto} h$$

$$h \overset{\beta}{\mapsto} i \overset{F}{\mapsto} 8 \overset{C_2[\rho_1]}{\mapsto} 11 \overset{C_1[\rho_2]}{\mapsto} 21 \overset{C_{13}[\rho_3]}{\mapsto} 4 \overset{F^{-1}}{\mapsto} e \overset{\tau}{\mapsto} q \overset{F}{\mapsto} 16 \overset{C_{13}[\rho_3^{-1}]}{\mapsto} 14 \overset{C_1[\rho_2^{-1}]}{\mapsto} 19$$

$$\overset{C_2[\rho_1^{-1}]}{\mapsto} 6 \overset{F^{-1}}{\mapsto} g \overset{\beta}{\mapsto} u$$

The complete ciphertext (in the usual uppercase) is thus *WOHU*.

(b) To decrypt, we now trace each of the ciphertext letters (in order) with the same composition of permutations:

$$w \overset{\beta}{\mapsto} w \overset{F}{\mapsto} 22 \overset{C_{25}[\rho_1]}{\mapsto} 9 \overset{C_0[\rho_2]}{\mapsto} 1 \overset{C_{13}[\rho_3]}{\mapsto} 11 \overset{F^{-1}}{\mapsto} \ell \overset{\tau}{\mapsto} g \overset{F}{\mapsto} 6 \overset{C_{13}[\rho_3^{-1}]}{\mapsto} 22 \overset{C_0[\rho_2^{-1}]}{\mapsto} 12 \overset{C_{25}[\rho_1^{-1}]}{\mapsto}$$

$$5 \overset{F^{-1}}{\mapsto} f \overset{\beta}{\mapsto} a$$

$$o \overset{\beta}{\mapsto} o \overset{F}{\mapsto} 14 \overset{C_0[\rho_1]}{\mapsto} 24 \overset{C_1[\rho_2]}{\mapsto} 3 \overset{C_{13}[\rho_3]}{\mapsto} 21 \overset{F^{-1}}{\mapsto} v \overset{\tau}{\mapsto} w \overset{F}{\mapsto} 22 \overset{C_{13}[\rho_3^{-1}]}{\mapsto} 17 \overset{C_1[\rho_2^{-1}]}{\mapsto} 3 \overset{C_0[\rho_1^{-1}]}{\mapsto}$$

$$6 \overset{F^{-1}}{\mapsto} g \overset{\beta}{\mapsto} u$$

$$h \overset{\beta}{\mapsto} i \overset{F}{\mapsto} 8 \overset{C_1[\rho_1]}{\mapsto} 24 \overset{C_1[\rho_2]}{\mapsto} 3 \overset{C_{13}[\rho_3]}{\mapsto} 21 \overset{F^{-1}}{\mapsto} v \overset{\tau}{\mapsto} w \overset{F}{\mapsto} 22 \overset{C_{13}[\rho_3^{-1}]}{\mapsto} 17 \overset{C_1[\rho_2^{-1}]}{\mapsto} 3 \overset{C_1[\rho_1^{-1}]}{\mapsto}$$

$$25 \overset{F^{-1}}{\mapsto} z \overset{\beta}{\mapsto} c$$

$$u \overset{\beta}{\mapsto} g \overset{F}{\mapsto} 6 \overset{C_2[\rho_1]}{\mapsto} 19 \overset{C_1[\rho_2]}{\mapsto} 14 \overset{C_{13}[\rho_3]}{\mapsto} 16 \overset{F^{-1}}{\mapsto} q \overset{\tau}{\mapsto} e \overset{F}{\mapsto} 4 \overset{C_{13}[\rho_3^{-1}]}{\mapsto} 21 \overset{C_1[\rho_2^{-1}]}{\mapsto} 11 \overset{C_2[\rho_1^{-1}]}{\mapsto}$$

$$8 \overset{F^{-1}}{\mapsto} i \overset{\beta}{\mapsto} h$$

33. Using the notation established in the proof of Proposition 3.3(a), the general Enigma encryption mapping may be expressed as follows: $E = \beta \circ F^{-1} \circ \rho^{-1} \circ F \circ \tau \circ F^{-1} \circ \rho \circ F \circ \beta$. Assume that a plaintext letter, call it a^*, is encrypted to itself. This means that $E(a^*) = a^*$, or in terms of the composition, $\beta \circ F^{-1} \circ \rho^{-1} \circ F \circ \tau \circ F^{-1} \circ \rho \circ F \circ \beta(a^*) = a^*$. Using the fact that the plugboard permutation β is its own inverse (since it is a product of disjoint two-cycles), if we apply the composition $F^{-1} \circ \rho \circ F \circ \beta$ to both sides of this equation, it simplifies to $\tau \circ F^{-1} \circ \rho \circ F \circ \beta(a^*) = F^{-1} \circ \rho \circ F \circ \beta(a^*)$ or $\tau(F^{-1} \circ \rho \circ F \circ \beta(a^*)) = F^{-1} \circ \rho \circ F \circ \beta(a^*)$. This says that the reflector permutation τ fixes the letter $F^{-1} \circ \rho \circ F \circ \beta(a^*)$, but this is impossible since the reflector in the Enigma is chosen so that no letter is fixed (i.e., the permutation is a product of a full 13 disjoint two-cycles). With this contraction, Proposition 3.3(b) is proved. □

35. (a) With this modification, β will still be its own inverse, and since this is the only property of the reflector that was used in the proof of Proposition 3.3(a), the result will still hold with this modification. But if there are fewer than a full set of 13 two-cycles, then some letters will be left fixed by the permutation, and the proof we gave for Exercise 33 shows that these lead to letters that will be fixed by the Enigma machine (for any rotor settings, some letters will always get fixed by the entire Enigma process in this situation, although the letters that get fixed change with the rotor movements). So the property of Proposition 3.3(b) no longer will hold.
   (b) If the reflector permutation has any cycles of at least length 3 in its disjoint cycle decomposition, then it will no longer be its own inverse, and the proof of Proposition 3.3(a) shows that this will destroy the self-decrypting property of the Enigma. From the proof of Exercise 33, we see that the nonfixing property of the Enigma machine might still hold, and this would be equivalent to the reflector permutation not fixing any letters.

37. (a) 15, 45, 15
   (b) 325; 44,850; 3,453,450
   (c) Following the suggestion, we first assume that the k plugboard cables are all different colors: red, green, blue, …. Now, the number of ways that we can hook up the first (red) cable is the number of ways of choosing two of the n empty sockets, so $C(n, 2)$. After we make a choice and hook the first cable up, there will be $n - 2$ open sockets, and we need to select two of them to which we will hook up the second (green) cable. This can be done in $C(n - 2, 2)$ ways. By the same token, there will then be $C(n - 4, 2)$ ways to hook up the third (blue) cable. Continuing in this fashion, the multiplication principle tells us that the number of different ways that we can hook up the k different colored cables equals

$C(n,\ 2)\cdot C(n-2,\ 2)\ \cdot C(n-4,\ 2)\cdot C(n-6,\ 2)\cdots\cdots C(n-2k,\ 2)$

$$=\frac{n!}{2!(n-2)!}\cdot\frac{(n-2)!}{2!(n-4)!}\cdot\frac{(n-4)!}{2!(n-6)!}\cdot\frac{(n-6)!}{2!(n-8)!}\cdots\cdots\frac{(n-2k+2)!}{2!(n-2k)!}$$

$$=\frac{n!}{2!\,\cancel{(n-2)!}}\cdot\frac{\cancel{(n-2)!}}{2!\,\cancel{(n-4)!}}\cdot\frac{\cancel{(n-4)!}}{2!\,\cancel{(n-6)!}}\cdot\frac{\cancel{(n-6)!}}{2!\,\cancel{(n-8)!}}\cdots\cdots\frac{\cancel{(n-2k+2)!}}{2!(n-2k)!}$$

$$=\frac{n!}{2^k(n-2k)!}$$

However, since the colors of the cables are unimportant, the above count needs to be corrected by dividing the number of ways of permuting the k cables, i.e., $k!$. This gives the answer $\dfrac{n!}{(n-2k)!\,k!\,2^k}$ that was asserted in part (c). The reader should verify that this formula gives the answers for parts (a) and (b).

Chapter 4

1. (a) 3×2

   (b) 3

   (c) $\begin{bmatrix} 10 & 0 \\ 9 & 3 \end{bmatrix}$

   (d) $\begin{bmatrix} -8 & 7 & -1 \\ 2 & 4 & -5 \\ 13 & 13 & 3 \end{bmatrix}$

   (e) Undefined

   (f) $\begin{bmatrix} 19 & -5 & 8 \\ -4 & 4 & 16 \\ -2 & -5 & 21 \end{bmatrix}$

3. (a) $\begin{bmatrix} 4 & 16 \\ 63 & 14 \end{bmatrix}$

   (b) $\begin{bmatrix} 40 & -12 \\ -6 & -22 \end{bmatrix}$

   (c) $A^2 = A\cdot A = \begin{bmatrix} -20 & -28 \\ 42 & 1 \end{bmatrix}$

   (d) $\begin{bmatrix} 5 & 56 \\ 31 & 30 \\ -22 & -18 \end{bmatrix}$

   (e) Undefined

(f) $\begin{bmatrix} -7 & -16 & 36 \\ -18 & -12 & 40 \\ 13 & -86 & 127 \end{bmatrix}$

5. $A' = \begin{bmatrix} 2 & 6 \\ -4 & 5 \end{bmatrix}$, $C' = \begin{bmatrix} 9 & -2 & -5 \\ -4 & 0 & -6 \\ 3 & 7 & 6 \end{bmatrix}$,

$D' = \begin{bmatrix} 2 & 7 & 5 \\ 6 & 4 & 6 \end{bmatrix}$

7. $A^{-1} = \dfrac{1}{34}\begin{bmatrix} 5 & -6 \\ 4 & 2 \end{bmatrix}$, $B^{-1} = \dfrac{1}{28}\begin{bmatrix} 2 & 4 \\ 3 & -8 \end{bmatrix}$

9. As pointed out in the text, multiplying an $n \times m$ matrix with an $m \times r$ matrix requires a grand total of $nr(2m-1)$ mathematical operations (additions and multiplications). Since A has size 100×2 and B has size 2×100, computing the product AB will thus require $100 \cdot 100 \cdot (2 \cdot 2 - 1) = 30,000$ mathematical operations to produce a 100×100 matrix D. Then, since C has size 100×2, computing the product DC will require an additional $100 \cdot 2 \, (2 \cdot 100 - 1) = 39,800$ mathematical operations for a grand total of 69,800 mathematical operations in computing the triple product $(AB)C$. On the other hand, if we were to compute $A(BC)$, we would first compute BC with $2 \cdot 2 \cdot (2 \cdot 100 - 1) = 796$ mathematical operations to produce a 2×2 matrix E, and then compute the product AE with a total of $100 \cdot 2 \cdot (2 \cdot 2 - 1) = 600$ mathematical operations, yielding a grand total of only 1396 mathematical operations. Thus, the first method would have required 5000 percent times as many mathematical operations!

11. (a) $\begin{bmatrix} 1 & 5 \\ 3 & 0 \end{bmatrix}$

(b) $\begin{bmatrix} 3 & 5 \\ 3 & 2 \end{bmatrix}$

(c) $\begin{bmatrix} 0 & 1 \\ 3 & 3 \end{bmatrix}$

(d) $\begin{bmatrix} 7 & 7 \\ 1 & 4 \end{bmatrix}$

13. (a) $AB \equiv \begin{bmatrix} 1 & 5 \\ 3 & 0 \end{bmatrix}$, $BA \equiv \begin{bmatrix} 0 & 3 \\ 4 & 3 \end{bmatrix}$, $A^2 \equiv \begin{bmatrix} 0 & 2 \\ 2 & 1 \end{bmatrix}$,

$CD \equiv \begin{bmatrix} 0 & 1 \\ 1 & 0 \\ 3 & 2 \end{bmatrix}$, DE is undefined, $EC \equiv \begin{bmatrix} 3 & 4 & 1 \\ 2 & 3 & 0 \\ 3 & 4 & 2 \end{bmatrix}$

(b) $AB \equiv \begin{bmatrix} 4 & 6 \\ 3 & 4 \end{bmatrix}$, $BA \equiv \begin{bmatrix} 0 & 8 \\ 4 & 8 \end{bmatrix}$, $A^2 \equiv \begin{bmatrix} 0 & 2 \\ 2 & 1 \end{bmatrix}$,

$CD \equiv \begin{bmatrix} 5 & 6 \\ 1 & 0 \\ 8 & 2 \end{bmatrix}$, DE is undefined, $EC \equiv \begin{bmatrix} 3 & 4 & 6 \\ 2 & 8 & 0 \\ 3 & 4 & 7 \end{bmatrix}$

Note: Each of these answers should agree with the corresponding answers of Exercise 3, after the latter are converted into mod 5 matrices.

15. $\det(A) \equiv 7 \,(\text{mod } 9)$, $A^{-1} \equiv 7^{-1} \cdot \begin{bmatrix} 5 & -1 \\ -3 & 2 \end{bmatrix} \equiv 4 \cdot \begin{bmatrix} 5 & 8 \\ 6 & 2 \end{bmatrix} \equiv$

$\begin{bmatrix} 2 & 5 \\ 6 & 8 \end{bmatrix} (\text{mod } 9)$; $\det(B) \equiv 3 \,(\text{mod } 9)$, since 3 is not relatively

prime to 9, B is not invertible (mod 9).

17. (a) 46
   (b) 4
   (c) 0
   (d) $A^{-1} = \dfrac{1}{46} \begin{bmatrix} -1 & 23 & -19 \\ -15 & 23 & -9 \\ 14 & -46 & 36 \end{bmatrix}$

   (e) Since $4^{-1} \equiv 16 \,(\text{mod } 21)$, $A^{-1} \equiv 4^{-1} \cdot \begin{bmatrix} -1 & 23 & -19 \\ -15 & 23 & -9 \\ 14 & -46 & 36 \end{bmatrix}$

   $\equiv 16 \cdot \begin{bmatrix} 20 & 2 & 2 \\ 6 & 2 & 12 \\ 14 & 17 & 15 \end{bmatrix} \equiv \begin{bmatrix} 5 & 11 & 11 \\ 12 & 11 & 3 \\ 14 & 20 & 9 \end{bmatrix} (\text{mod } 21)$

   (f) Since $\det(A) \equiv 0$ is not invertible (mod 23), the matrix A is not invertible (mod 23).

19. (a) (i)   FKEEZNVRUFJHKNDRPMELUSROUQNA
        (ii)   DRMHKEWZHPNQXTPHQR
        (iii) WDGIAQGYGJRTFLEURZAF
        (iv)   FKYPLRVHYWTHQSAH

   (b) (i)   Check into the Mayflower.
        (ii)   Take out the ceiling panel by window.
        (iii) Assemble the unit.
        (iv)   Bring and leave hotel immediately.

21. (a) (i)   AXEZAHBTFGJCTTERRIDMHTNGCBP
        (ii)   LJJZHVHNATNYZUCYNL
        (iii) LWHYSQMYQAXEBDDNYBFBS
        (iv)   AXEWDKRVZTMNMKQQQD

(b) Note that $A^{-1} \equiv \begin{bmatrix} 1 & 1 & 25 \\ 0 & 25 & 1 \\ 25 & 0 & 1 \end{bmatrix}$ (mod 26).

   (i)   Math is fun.
  (ii)   Hill patented his cipher.
 (iii)   The FBI keeps files on all cryptographers.
 (iv)   Celebrity gets delayed in cryptography.

23. (a) The given information can be translated (using Table 3.1) into the following matrix equation: $A \begin{bmatrix} 19 & 4 & 13 \\ 22 & 14 & 1 \end{bmatrix} = \begin{bmatrix} 25 & 6 & 16 \\ 2 & 2 & 25 \end{bmatrix}$. If we restrict this equation to using only the first and third columns of the partial plaintext and ciphertext matrices, the equation becomes $A \begin{bmatrix} 19 & 13 \\ 22 & 1 \end{bmatrix} = \begin{bmatrix} 25 & 16 \\ 2 & 25 \end{bmatrix}$, and since $\det\left(\begin{bmatrix} 19 & 13 \\ 22 & 1 \end{bmatrix} \right) \equiv 19 \pmod{26}$, the latter matrix equation can be solved by right multiplying

by $\begin{bmatrix} 19 & 13 \\ 22 & 1 \end{bmatrix}^{-1} \equiv 19^{-1} \cdot \begin{bmatrix} 1 & -13 \\ -22 & 19 \end{bmatrix} \equiv 19^{-1} \cdot \begin{bmatrix} 1 & -13 \\ -22 & 19 \end{bmatrix} \equiv$

$11 \cdot \begin{bmatrix} 1 & 13 \\ 4 & 19 \end{bmatrix} \equiv \begin{bmatrix} 11 & 13 \\ 18 & 1 \end{bmatrix}$ (mod 26). We point out that the other three options for selecting two of the three columns in the matrix equation result in noninvertible matrices. This yields $A \equiv \begin{bmatrix} 25 & 16 \\ 2 & 25 \end{bmatrix} \begin{bmatrix} 19 & 13 \\ 22 & 1 \end{bmatrix}^{-1} \equiv$

$\begin{bmatrix} 25 & 16 \\ 2 & 25 \end{bmatrix} \begin{bmatrix} 11 & 13 \\ 18 & 1 \end{bmatrix} \equiv \begin{bmatrix} 17 & 3 \\ 4 & 25 \end{bmatrix}$.

We may now decrypt the ciphertext in the usual fashion, revealing the plaintext to be: "The revolt will be on Bastille Day."

(b) Since the character correspondence represented by the plaintext portion "reou" occupies the fourth through the seventh characters, only the "eo" subportion fills an entire column of the plaintext integer matrix. Since all known plaintext columns begin with e, whose integer equivalent is 4, it follows that the determinants of any of the 2×2 matrices formed by known plaintext columns will always be even, and hence not relatively prime to 26. Thus, none of these matrices can be inverted (mod 26), and so none of the corresponding partial encryption equations can be solved for the encryption matrix.

(c) The given information can be translated (using Table 3.1) into the following matrix equation: $A \begin{bmatrix} 17 & 22 & 17 \\ 24 & 8 & 14 \\ 11 & 13 & 21 \end{bmatrix} =$

$$\begin{bmatrix} 1 & 24 & 5 \\ 20 & 21 & 10 \\ 2 & 24 & 24 \end{bmatrix}. \quad \text{Since} \quad \det\left(\begin{bmatrix} 17 & 22 & 17 \\ 24 & 8 & 14 \\ 11 & 13 & 21 \end{bmatrix}\right) \equiv$$

$4 \pmod{26}$ is not relatively prime to 26, it follows that this matrix cannot be inverted, so the matrix equation cannot be solved for A. The same problem occurs if we try to permute the columns of this matrix. Thus the given information is not sufficient to decrypt the plaintext.

25. Any choice that would render the integer matrix (obtained using Table 3.1) to be invertible mod 26 would work. One such example, and a particularly convenient one at that, would be to use the strings $abb \cdots bb$, $bab \cdots bb$, $bba \cdots bb$, $\cdots$, $bbb \cdots ab$, $bbb \cdots ba$. This would result in the matrix equation $AI = C$, which directly gives the encoding matrix.

27. *Proof:* We assume that the sizes of the matrices involved are compatible so that both sides of the identities are defined. The (i,j) entry of $(A+B)C$ is the dot product of the ith row of $A+B$: $[a_{i1}+b_{i1} \quad a_{i2}+b_{i2} \quad \cdots \quad a_{im}+b_{im}]$, and the jth column of

$$C, \begin{bmatrix} c_{1j} \\ c_{2j} \\ \vdots \\ c_{mj} \end{bmatrix}, \text{ and so equals } \sum_{k=1}^{m}(a_{ik}+b_{kj})c_{kj} = \sum_{k=1}^{m}a_{ik}c_{kj} + \sum_{k=1}^{m}b_{ik}c_{kj}.$$

Since the last two sums give the (i,j) entries of AC and BC, respectively, the identity $(A+B)C = AC + BC$ is thus established. □

29. *Proof:* The (i,j) entry of $(A+B)'$ is the (j,i) of $A+B$, which is $a_{ji}+b_{ji}$. Since the first term is the (i,j) entry of A', and the second term is the (i,j) entry of B', it follow that their sum is the (i,j) entry of $A'+B'$, proving that $(A+B)' = A'+B'$. □

31. (a) *Proof:* Let $C = B^{-1}A^{-1}$. In order to show that $C = (AB)^{-1}$, we must show that $C(AB) = I$ and that $(AB)C = I$. Using associativity of matrix multiplication, we compute $C(AB) = (B^{-1}A^{-1})(AB) = B^{-1}(A^{-1}A)B = B^{-1}(I)B = B^{-1}B = I$. (We have used the definition of inverse for the matrices A and B, and the multiplicative identity property of the identity matrix that $IM = MI = M$.) Similarly, $(AB)C = (AB)(B^{-1}A^{-1}) = A(B^{-1}B)A^{-1} = A(I)A^{-1} = AA^{-1} = I$. □

(b) *Proof:* Let $C = (A^{-1})'$. In order to show that $C = (A')^{-1}$, we must show that $CA' = I$ and that $A'C = I$. This can be accomplished by repeatedly collapsing product of A and its inverse t times, in a similar fashion to what was done in the proof of part (a). For example:

$$CA^t = \underbrace{A^{-1}A^{-1}\cdots(A^{-1}}_{t \text{ factors}} \underbrace{A)A\cdots A}_{t \text{ factors}} = \underbrace{A^{-1}A^{-1}\cdots A^{-1}}_{t-1 \text{ factors}} I \underbrace{AA\cdots A}_{t-1 \text{ factors}}$$

$$= \underbrace{A^{-1}A^{-1}\cdots A^{-1}}_{t-1 \text{ factors}} \underbrace{AA\cdots A}_{t-1 \text{ factors}} = \cdots A^{-1}A = I \,\square$$

Chapter 5

5. Key:

a	b	c	d	e	f	g	h	i	j	k	l	m	n	o	p	q	r	s	t	u	v	w	x	y	z
Q	X	J	T	N	S	D	F	A	B	W	E	R	P	K	M	I	L	Z	V	Y	O	H	G	C	U

7. Key:

a	b	c	d	e	f	g	h	i	j	k	l	m	n	o	p	q	r	s	t	u	v	w	x	y	z
P	Q	W	C	E	D	A	M	I	K	Y	N	O	T	V	U	F	Z	H	S	L	J	G	R	B	X

9. Key:

a	b	c	d	e	f	g	h	i	j	k	l	m	n	o	p	q	r	s	t	u	v	w	x	y	z
V	Y	R	P	N	M	C	F	E	T	L	X	S	W	Q	H	K	B	I	U	D	Z	J	A	O	G

11. Key:

a	b	c	d	e	f	g	h	i	j	k	l	m	n	o	p	q	r	s	t	u	v	w	x	y	z
P	A	Y	U	Q	X	H	D	F	B	V	S	W	N	J	Z	R	L	I	K	O	M	C	G	E	T

13. Vigenère keyword: JARGON

15. Vigenère keyword: CAROUSEL

17. Vigenère keyword: HYBRIDS

19. Vigenère keyword: TYPHOONS (*Note:* The plaintext passages from Exercises 13–18 came from novels by Stephen Crane that were written in the late 19th century.)

21. (a) 0
(b) $3/12 = 0.25$
(c) 0.25

23. (a) By Theorem 5.3, $E[\mathscr{A}(\text{STR}1, \text{STR}2)] = [0.2, 0.75, 0.05] \cdot [0.2, 0.75, 0.05] = 0.605$.
(b) Again by Theorem 5.3, we have $E[\mathscr{A}(\text{STR}1, \text{STR}2)] = [0.2, 0.75, 0.05] \cdot [0.05, 0.2, 0.75] = 0.1975$ for the single unit shift, and $E[\mathscr{A}(\text{STR}1, \text{STR}2)] = [0.2, 0.75, 0.05] \cdot [0.75, 0.05, 0.2] = 0.1975$ (also) for the two-unit shift.

25. (a) The distribution vector is Vec = [Vec0, .50], where Vec0 is the length-10 vector [0.05, 0.05, …, 0.05]. By Theorem 5.3,

E $[\mathscr{I}(\text{STR}1, \text{STR}2)]$ = Vec • Vec = $(0.05)^2 + \square + (0.05)^2 + (0.5)^2 = 10$ • $(0.05)^2 + (0.5)^2 = 0.275$.

(b) For any nontrivial shift of the 11 letters, the corresponding distribution will be the corresponding nontrivial shift VecShift of Vec. Although there are 10 different such shifted vectors, their dot products with Vec are all the same since the single 0.5 entries of each will not be matched. By Theorem 5.3, $E[\mathscr{I}(\text{STR}1, \text{STR}2)]$ = Vec • ShiftVec = 9 $\cdot (0.05)^2 + 2 \cdot (0.05)(0.5) = 0.0725$.

(c) Given a ciphertext, if we form the length-11 vector VecCiph of frequencies of the letters and take 11 dot products with all possible shifted vectors of Vec [in part (a)] of the ordinary language distribution, then, from parts (a) and (b), one of these dot products should theoretically be nearly 400 percent larger than all of the others. The corresponding shift will be the key.

27. (a) Since the characters of the strings are randomly selected from the 12 ciphertext letters, the corresponding distribution vector is the length-12 vector Vec = (1/12, 1/12, ..., 1/12). By Theorem 5.3, $E[\mathscr{I}(\text{STR}1, \text{STR}2)]$ = Vec • Vec = $(1/12)^2 + \cdots + (1/12)^2 = 2 \cdot (1/12)^2 = 1/12$.

(b) 0.14

(c) 3, Kalakaua

29. Vigenère keyword: JARGON

31. Vigenère keyword: CAROUSEL

33. Vigenère keyword: HYBRIDS

35. Vigenère keyword: TYPHOONS

37. (a) FALSE: Here is the simplest counterexample: Take STR1 = aa, STR2 = ab, STR3 = bb. We have $\mathscr{I}(\text{STR}1, \text{STR}2) = \mathscr{I}(\text{STR}2, \text{STR}3) = 1/2$, but $\mathscr{I}(\text{STR}1, \text{STR}3) = 0$. (b) This is always false because Corollary 5.2(a) tells us both expectations equal 1/26.

39. (a) YES: The administrator will work with the frequency distribution for the letters in all English dictionary words (possibly different from Table 5.1 for passages of written English, but obtainable from the literature). A simple frequency analysis would be effective and proceed quickly (at 1000 passwords/second, it would only take 10 seconds to process the 10,000 encrypted passwords). The total ciphertext length is expected to be $5 \cdot 10,000 = 50,000$, which is certainly of ample size for a frequency analysis. (b) YES: To proceed, the administrator could first view each English letter as having two homophones (its uppercase letter and its lowercase letter). The size of the plaintext/ciphertext alphabets are thus doubled from what they were in part (a), but the data is sufficiently large to

ascertain all of the homophones. In the next step, the administrator would need only check a few passwords using trial and error to test the different case choices. For example, in a word of length 5 (the average length), at most $2^5 = 32$ trials would be needed, and once the upper/lowercase cipherletters for a given plaintext letter are determined, they need not be checked again. Thus, a table of the complete key could quickly be constructed. (c) YES: The plaintext characters are the 26 letters and the 10 digits. The key step is to first determine which 10 ciphercharacters correspond to the 10 digits. Once this is done, the word portion of each cipherpassword can easily be split from the digit portions and a frequency analyses can be performed to match the cipherletters with plaintext letters. The plaintext digits all have the same frequencies so would not be amenable to frequency analysis, but there are only 10 of them, so the resulting $10! \approx 3.6$ million permutations could be checked by trail and error. There are a number of ways to accomplish the key step; here is one such method: Assuming that the digits are uniformly distributed with frequency 1/10 each, if d is any cipherdigit, the expected number of cipherpasswords that begin with d is given by:
$n_d = 10,000 \cdot P(W_{\text{digit}}) \cdot P(W_d \mid W_{\text{digit}}) > 10,000 \cdot (0.5) \cdot (1/10) =$
500, where W_{digit} is the event that a randomly selected password begins with a digit and W_d is the event that a randomly selected password begins with the cipherdigit d. (We conservatively estimated $P(W_{\text{digit}}) > .5$, it would be more realistic to estimate $P(W_{\text{digit}}) = .80$, and this would further simplify the analysis that follows.) Similarly, if e is any cipherletter, the expected number of cipherpasswords that begin with e is given by $n_d < 10,000 \cdot (0.5) \cdot P(e) = 5000 \, P(e)$, where $P(e)$ is the frequency of the cipherletter e among all cipherletters. Since all but about 10 cipherletters will have frequencies of less than 1/20 (cf. Table 5.1) and thus will result in $n_d < 250$, we may weed out all but about 20 ciphercharacters as candidates for cipherdigits. But, if the first character of a cipherpassword were a cipherletter (rather than a cipherdigit) then all of the last four characters of such a cipherpassword would have to be cipherdigits, and, in particular, could not be the same as the first cipherletter. This fact could be used to screen out all remaining ciphercharacters that are not cipherdigits: simply collect all (> 250) cipherpasswords that begin with a given ciphercharacter to be tested. For each, check to see of any of the four last ciphercharacters is the same as the first ciphercharacter. If there is a match, the ciphercharacter is eliminated from being a cipherdigit. Since there will be $> 4 \cdot 250 = 1000$ such four ciphercharacter strings to check, the probability is very high that all cipherletters will be eliminated in this fashion. (d) Believe it or not, the answer is still YES! If the administrator randomly generates a password of 8 lowercase characters, the probability of hitting an actual

password is $p = 10{,}000/26^8 = 4.788... \times 10^{-8}$. Since she/he can test 1000 passwords/second, in one week, the number of passwords that could be tested is $N = 1000 \cdot 60$ (sec/min) $\cdot 60$ (min/hr) $\cdot 24 \cdot 7 = 6.048 \times 10^8$. This means that the expected number of actual passwords that is actually be discovered would be $p \cdot N = 28.96$. Each of these could promptly be compared with their corresponding cipherpassword to reveal the partial correspondence of the codebook (for all 8 plaintext letters). The process could be halted when about 80 percent of the key is ascertained. The remaining portion of the key could then be obtained by choosing cipherwords and using trial and error for any unknown corresponding plaintext letters. This example nicely illustrates why it is recommended that passwords meet certain seemingly bizarre and inconvenient requirements: e.g., must have at least one uppercase letter, at least one digit, and at least one special character (e.g., &, %, $).

41. The only change in the proof of Proposition 5.1 given in the text needed to convert it into a proof of Proposition 5.3 is to change the three sequences of A, B, C, $\cdots$, Z that appear to sequences x_1, x_2, $\cdots$, x_ℓ.

43. (a) The entries in each of the vectors V_i, V_j, V_0, $V_{|i-j|}$ are the same, but cyclically shifted different amounts. To compute the dot product, $v_i \square V_j$, we multiply entries of V_0, shifted i units to the right, by corresponding entries of V_0, shifted j units to the right. But this means that the cyclic distance (i.e., if we equally place the 26 English letters on a circle in alphabetical order) between the entries that are being multiplied to form if dot product is $|i-j|$. The result is thus the same as the shifted dot product $V_{|i-j|} \square V_0$.

(b) Although part (a) tells us that the number of different dot products of the 26 computed in the Freidman attack is only 14, only one of them will (theoretically) be the maximum, and it could occur as any of the 26, so no essential reduction in work could be achieved that is solely based on the result of part (a).

Chapter 6

Note: For simplicity of notation we will write all base b expansions as strings whenever $b \le 10$ or $b = 16$ (hex format).

1. (a) (As in Table 6.2): 0, 1, 10, 11, 100, 101, 110, 111, 1000, 1001, 1010, 1011, 1100, 1101, 1110, 1111, 10000, 10001, 10010, 10011, 10100, 10101, 10110, 10111, 11000, 11001

(b) 0, 1, 2, 3, 4, 5, 6, 7, 10, 11, 12, 13, 14, 15, 16, 17, 20, 21, 22, 23, 24, 25, 26, 27, 30, 31

(c) See Table 6.2.

3. (a) 1100100, 1100101, 1100110, 1100111, 1101000, 1101001, 1101010, 1101011, 1101100, 1101101, 1101110, 1101111, 1110000, 1110001, 1110010, 1110011, 1110100, 1110101, 1110110, 1110111, 1111000, 1111001, 1111010, 1111011, 1111100, 1111101

   (b) 144, 145, 146, 147, 150, 151, 152, 153, 154, 155, 156, 157, 160, 161, 162, 163, 164, 165, 166, 167, 171, 172, 173, 174, 175

   (c) 64, 65, 66, 67, 68, 69, 6A, 6B, 6C, 6D, 6E, 6F, 70, 71, 72, 73, 74, 75, 76, 77, 78, 79, 7A, 7B, 7C, 7D

5. (a) 42

   (b) 14,043

   (c) 11,259,375

   (d) 703

   (e) 4,886,735,530

   (f) 17,343,427

7. (a) 66 = 1000010 (base 2) = 102 (base 8) = 42 (hex)

   (b) 237 = 11101101 (base 2) = 355 (base 8) = ED (hex)

   (c) 1925 = 11110000101 (base 2) = 3605 (base 8) = 785 (hex)

   (d) 12587 = 11000100101011 (base 2) = 30453 (base 8) = 312B (hex)

   (e) 28,000 = 110110101100000 (base 2) = 66540 (base 8) = 6D60 (hex)

   (f) 150,269 = 100100101011111101 (base 2) = 445375 (base 8) = 24AFD (hex)

9. (a) 66 = 2110 (base 3) = 73 (base 9) = [2 12] (base 27)

   (b) 237 = 22210 (base 3) = 283 (base 9) = [8 21] (base 27)

   (c) 1925 = 2122022 (base 3) = 2568 (base 9) = [2 17 8] (base 27)

   (d) 12,587 = 12202101 (base 3) = 18235 (base 9) = [17 7 5] (base 27)

   (e) 28,000 = 11021020 (base 3) = 42361 (base 9) = [1 11 11 1] (base 27)

   (f) 150,269 = 21122010 (base 3) = 248115 (base 9) = [7 17 3 14] (base 27)

11. (a)   (i) agent = 00000 00110 00100 01101 10011

      (ii) met = 01100 00100 10011

      (iii) liaison = 01011 01000 00000 01000 10010 01110 01101

   (b)   (i) help

      (ii) keller

      (iii) now

13. (a) 11011; 12 + 15 = 27

   (b) 11104; 2921 + 1755 = 4676

   (c) 14464; 43,724 + 39,320 = 83,044

   (d) [24 18]; 558 + 60 = 618

15. (a) 1001100111; 428 + 187 = 615

   (b) 56100010; 11,979,065 + 112,338 = 12,091,400

(c) A4444CB; 11,189,196 + 161,057,023 = 172,246,219
(d) [1 0 4 6 21]; 355,233 + 38,063 = 393,296

17. (a) 0001 or 1; 12 − 11 = 1
(b) 6062; 3721 − 599 = 3122
(c) 1099CD; 11150028 − 10062079 = 1087949
(d) 355,233 − 38,063 = 317,170

19. (a) 101010; $6 \times 7 = 42$
(b) 23177; $365 \times 27 = 9855$
(c) 679FEC; $2764 \times 2457 = 6,791,148$
(d) [2 3 14 5]; $558 \times 60 = 33,480$

21. (a) 1100000111; $25 \times 31 = 775$
(b) 23413214; $2916 \times 1755 = 5,117,580$
(c) 66BE0D34; $43,724 \times 39,423 = 1,723,731,252$
(d) [2 15 23 17 11 17]; $14,209 \times 1813 = 25,760,917$

23. (a) 4
(b) 2
(c) 24
(d) 63

25. Each hexadecimal digit is equivalent to a string of four binary digits via Table 6.1. Since 16 is a power of 2, the digit placements correspond so that the adjacent groups of four binary digits can be converted to their single hex equivalents, and the expansions will represent the same number. The binary expansion should initially be padded on the left with redundant zeros, so that the number of binary digits is a multiple of 4. Here is a simple example: To convert the binary expansion: [10011] to hex, we first pad with three zeros on the left: [0001 0011]. Then we use Table 6.1 to look up the hex equivalents of the groups of four binary digits: [13] (hex). The reader should check that both expansions represent 19.

27. Each corresponding pair of digits must be added with the current carry (in the formation of NewCar in Step 2); this amounts to two additions for each of the n digits. Then, in computing s_i in the same step, we need to subtract a certain quantity from the number just computed, giving a total of n subtractions. Thus we have to do a total of $3n$ additions/subtractions. In case certain carries are zero, they need not be added, cutting down the number of additions by n. In total, there are thus between $2n$ and $3n$ additions/subtractions. (Actually, we can possibly cut this down a bit further if we avoid subtracting NewCar terms or single digits that are zero.)

29. This follows from Theorem 6.1 with $b = 2$: Any positive integer W can be written as $W = c_K 2^K + c_{K-1} 2^{K-1} + \cdots + c_1 2^1 + c_0$, where K is a nonnegative integer, and each coefficient c_i is either 0 or 1. If $W \le 2^{n-1}$, then we may assume in this expansion that $K < n$ (because 2^K would by itself be greater than W). This means that the weight W is the sum of all the weights 2^i whose coefficients c_i are 1.

31. (a) (i) Since a lies in the range $-32 = -2^{6-1} \leq a < 2^{6-1} = 32$, we can use length 6 (or larger) two's complement representation. Since 17 is nonnegative, the first digit is 0. The remaining digits are given by the binary expansion of 17, which is (by Algorithm 6.1) 10001; so we have the two's complement representation $17 \sim [010001]$. [Note: it would also be correct to use any larger number of digits by tacking on additional zeros after the first (sign) digit, in this and any two's complement representation.]

   (ii) Since -22 lies in the range $-32 = -2^{6-1} \leq a < 2^{6-1} = 32$, we can use length 6 (or larger) two's complement representation. Since -22 is negative, the first digit is 1, and the remaining 5 digits will be the binary expansion of $32 - |-22| = 10$, which (by Algorithm 6.1) is 01010. Thus we have the two's complement representation $-22 \sim [101010]$.

   (iii) Since -32 lies in the same range, we can use the same procedure to obtain the following two's complement representation: $-32 \sim [100000]$.

   (b) (i) Since the first digit is 1, the sign of the number a being represented is zero. The remaining five digits 10011, which is the binary representation of $16 + 2 + 1 = 19$, represents $32 - |a|$. Solving for $|a|$ gives $|a| = 13$, so the number being represented is -13.

   (ii) 12

   (iii) 31

33. (a) When a is nonnegative, the remainder of this division is just a, while if a is negative, then since $a = -|a|$, it follows that $a = (-1) \cdot 2^{n-1} + (2^{n-1} - |a|)$ so the remainder of the division of a by 2^{n-1}. In either case, it follows that the two's complement representation of a uses (after the first digit that determines the sign) precisely the binary expansion of this remainder as the remaining digits.

   (b) By the result of part (a), here is a simple algorithm for converting an integer in the range $-2^{n-1} \leq a < 2^{n-1}$, to its two's complement representation $a \sim [b_{n-1} \, b_{n-2} \, \ldots b_1 \, b_0]$

   Step 1. If $a < 0$, set $b_{n-1} = 1$, otherwise set $b_{n-1} = 0$.

   Step 2. Compute the remainder r of the division of a by 2^{n-1}.

   Step 3. Take $[b_{n-2} \, b_{n-3} \cdots b_1 \, b_0]$ to be the binary expansion of r. (Padding with zeros on the left, as needed so there will be $n-1$ bits.)

   (c) An algorithm for the reverse process can also be based on the result of part (a). We assume that we are given an n bit two's complement representation $[b_{n-1} \, b_{n-2} \, \ldots b_1 \, b_0]$ The following steps will compute the integer a that is represented:

   Step 1. If $b_{n-1} = 1$, set the sign of a to be negative, otherwise take it to be positive.

Step 2. Convert the binary expansion of $[b_{n-2} \; b_{n-3} \cdots b_1 \; b_0]$ into an integer r. Then set $|a| = 2^{n-1} - r$.

Step 3. Attach the sign found in Step 1 to the absolute value found in Step 2 to recover a.

Chapter 7

1. (a) 100110
   (b) 101101
   (c) B
   (d) C

3. (a) 101010
   (b) 001100
   (c) Since XORing a bit with 1 toggles the bit to its opposite value, it follows that the bit values of $S \oplus \mathbf{1}$ will be the opposite of the corresponding bit values of S.

5. (a) 2A (hex)
   (b) 0D (hex)

7. (a) 72 (hex)
   (b) EA (hex)

9. (a) 11
   (b) 11
   (c) 01
   (d) 01

11. (a) 0010
    (b) 1011
    (c) 0001
    (d) 1101

13. (a) $01110110 = 76$ (hex)
    (b) $01000001 = 41$ (hex)

15. (a) 1111
    (b) 0001
    (c) 0101
    (d) 0100

25. *Proof:* We let $\hat{C} = \mathrm{DES}_{\bar{\kappa}}(\bar{P})$. We must show that $\hat{C} = \bar{C}$, where $C = \mathrm{DES}_{\kappa}(P)$. As in Exercise 4, we let $\mathbf{1}$ denote the string of 1's (whose size changes when needed). We will show that after each of the 16 DES rounds, the left and right strings produced in the encryption of $\mathrm{DES}_{\bar{\kappa}}(\bar{P})$, which we denote as $\hat{L}_i, \hat{R}_i$, are just the complements $\bar{L}_i, \bar{R}_i$, of the left and right strings L_i, R_i, produced after the same round of the encryption of $\mathrm{DES}_{\kappa}(P)$. Since this is certainly true when $i = 0$ (before we enter into the Feistel rounds), we assume that $\hat{L}_{i-1} = \bar{L}_{i-1}, \hat{R}_{i-1} = \bar{R}_{i-1}$, and our goal is to show that $\hat{L}_i = \bar{L}_i, \hat{R}_i = \bar{R}_i$. We first observe that for any (32-bit) string R, we have $E(\bar{R}) = E(R) \oplus \mathbf{1}$, where E denotes the DES expansion function. From this, and the fact that $\bar{\kappa}^i = \kappa^i \oplus \mathbf{1}$, it

follows that $E(\hat{R}_i) \oplus \overline{\kappa^i} = E(\hat{R}_i) \oplus 1 \oplus \kappa^i \oplus 1 = E(R) \oplus \kappa^i$, so that the inputs into the S-boxes (in the round being considered) in the encryption of $\text{DES}_\kappa(P)$ are identical to those in the encryption of $\text{DES}_{\bar{\kappa}}(\bar{P})$. In completing the round in the encryption of $\text{DES}_{\bar{\kappa}}(\bar{P})$, the new left and right vectors are formed by the Feistel scheme (of Definition 7.2): $\hat{L}_i = \hat{R}_{i-1}, \hat{R}_i = \hat{L}_i \oplus f_{\kappa^i}(\hat{R}_{i-1})$. If we make the substitutions $\hat{L}_{i-1} = \bar{L}_{i-1}, \hat{R}_{i-1} = \bar{R}_{i-1}$, these become $\hat{L}_i = \bar{R}_{i-1}, \hat{R}_i = \bar{L}_{i-1} \oplus f_{\kappa^i}(\bar{R}_{i-1})$. Since $L_i = R_{i-1} \Rightarrow \bar{L}_i = \bar{R}_{i-1}$, the first of the previous two equations shows that $\hat{L}_i = \bar{L}_i$. Also, since $f_{\kappa^i}(\bar{R}_{i-1}) = f_{\kappa^i}(R_{i-1})$, the second of the previous two equations shows $\hat{R}_i = \bar{L}_{i-1} \oplus f_{\kappa^i}(R_{i-1}) = 1 \oplus L_{i-1} \oplus f_{\kappa^i}(R_{i-1}) = 1 \oplus R_i = \bar{R}_i$, as was needed to show. $\square$

31. Although the suggestion will work (and it seems, *a priori*, like the most straightforward way to proceed), the alternative approach using compositions that was outlined in the solution of Exercise for the Reader 7.2 is going to be much more transparent. Continuing with what was developed in that example, we can write the Feistel encryption function as the composition $E = s \circ G_{NR} \circ G_{NR-1} \circ \cdots \circ G_2 \circ G_1$ where s denotes the left/right switch function $s(L,R) = (R,L)$, and $G_i(L,R) = (R, L \oplus f_{\kappa^i}(R))$ is the ith round function. We proved that $G_i \circ s \circ G_i = s$. We wish to show that the inverse of E is the similar mapping with the round keys used in the reverse order, i.e., the mapping $D = s \circ G_1 \circ G_2 \circ \cdots \circ G_{NR-1} \circ G_{NR}$. We can use this identity to telescope the composition $E \circ E$ to show we get the identity mapping as follows:

$$D \circ E = (s \circ G_1 \circ G_2 \circ \cdots \circ G_{NR-1} \circ G_{NR}) \circ (s \circ G_{NR} \circ G_{NR-1} \circ \cdots \circ G_2 \circ G_1)$$

$$= (s \circ G_1 \circ G_2 \circ \cdots \circ G_{NR-1}) \circ (G_{NR} \circ s \circ G_{NR}) \circ (G_{NR-1} \circ \cdots \circ G_2 \circ G_1)$$

$$= (s \circ G_1 \circ G_2 \circ \cdots \circ G_{NR-1}) \circ (s) \circ (G_{NR-1} \circ \cdots \circ G_2 \circ G_1)$$

$$= (s \circ G_1 \circ G_2 \circ \cdots \circ (G_{NR-1} \circ s \circ G_{NR-1}) \circ \cdots \circ G_2 \circ G_1)$$

$$= (s \circ G_1 \circ G_2 \circ \cdots \circ (s) \circ \cdots \circ G_2 \circ G_1)$$

$$\vdots$$

$$= s \circ (G_1 \circ s \circ G_1)$$

$$= s \circ (s)$$

$$= id \quad \square$$

We now compare this with the following much less intuitive (albeit more direct) approach:

Second Proof: Following the suggestion, we will use mathematical induction to prove that $L_i' = R_{NR-i}, R_i' = L_{NR-i}$, for each $i = 0, 1, \ldots, NR$.

Basis Step: $i = 0$: By assumption $L_0' = R_{NR}, R_0' = L_{NR}$, because $[R_{NR}, L_{NR}]$, the ciphertext that is outputted from the Feistel system is the same as $[L_0', R_0']$, which is the input of the decryption program.

Inductive Step: We assume the identity $L_i' = R_{NR-i}$, $R_i' = L_{NR-i}$ is true for an index $i < NR$. From this, we need to show the identity is true for the index $i + 1$.

Indeed, by definition of the $(i + 1)$th Feistel round (on the decryption), we have $L_{i+1}' = R_i'$ By the inductive hypothesis, $R_i' = L_{NR-i}$. But by definition of the $(NR - i)$th Feistel round (on the encryption), we have $L_{NR-i} = R_{NR-i-1} = R_{NR-(i+1)}$. Combining these three equations yields $L_{i+1}' = R_{NR-(i+1)}$.

We deal with R_{i+1}' in a similar fashion, but it will be a bit more complicated since the round keys will enter into the picture. For ease of notation we will denote the ith round key function f_{k^i} simply as f_i. We point out that since decryption uses the keys in the opposite order, the ith round key function in decryption is f_{NR+1-i}. The definition of the $(i + 1)$th Feistel round (on the decryption) implies that $R_{i+1} = L_i' \oplus f_{NR+1-(i+1)}(R_i') = L_i' \oplus f_{NR-i}(R_i')$. By the induction hypothesis, this can be rewritten as $R_{i+1}' = R_{NR-i} \oplus f_{NR-i}(L_{NR-i})$ If in this latter expression we make the substitutions $R_{NR-i} = L_{NR-i-1} \oplus f_{NR-i}(R_{NR-i-1})$, $L_{NR-i} = R_{NR-i-1}$ (which come from definition of the $(NR-i)$th Feistel round on the encryption), we are led to $R_{i+1}' = L_{NR-i-1} \oplus f_{NR-i}(R_{NR-i-1}) \oplus f_{NR-i}(R_{NR-i-1}) = L_{NR-i-1} \oplus 0 = L_{NR-i-1}$, and the proof is complete. □

Chapter 8

1. (a) $\pi(10^9) \rightarrow 10^9/\ln(10^9) \approx 48,254,942$

   (b) $\pi(10^{10}) - \pi(10^9) \rightarrow 10^{10}/\ln(10^{10}) - 10^9/\ln(10^9) \approx 386,039,539$

   (c) $\pi(10^{12}) - \pi(10^9) \rightarrow 10^{12}/\ln(10^{12}) - 10^9/\ln(10^9) \approx$
   36,142,951,883

3. (a) $\pi(2^{100}) - \pi(2^{99}) \rightarrow 2^{100}/\ln(2^{100}) - 2^{99}/\ln(2^{99}) \approx 9.052 \times 10^{27}$

   (b) Probability $\approx 9.052 \times 10^{27}/2^{98} \approx 1/35$

   (c) $\pi(2^{1000}) - \pi(2^{999}) \rightarrow 2^{1000}/\ln(2^{1000}) - 2^{999}/\ln(2^{999}) \approx$
   7216×10^{297}

   (d) Probability $\approx 7.7216 \times 10^{297}/2^{998} \approx 1/346$

5. (a) 3
   (b) 1
   (c) 4
   (d) 37

7. (a) 16
   (b) 120
   (c) 384
   (d) 2400

9. (a) Since n is even, we can write $n = 2^f m$, where m is odd and f is a positive integer. From Equation 8.1 of Proposition 8.3, it follows that $\phi(n) = 2^{f-1}\phi(m)$ and $\phi(2n) = 2^f \phi(m)$, and from these two equations it follows that $\phi(2n) = 2\phi(n)$.

(b) If n is odd, then it follows from Proposition 8.3 that $\phi(2n) = (2-1) \cdot 2^0 \cdot \phi(n) = \phi(n)$.

11. This can be done using Equation 8.1 of Proposition 8.3.

(a) $\{2\}$

(b) $\{5, 8, 10, 12\}$

(c) $\varnothing$ (no solutions)

(d) $\{13, 21, 26, 28, 36, 42\}$

13. (a) 1

(b) 7

(c) 24

(d) 8

15. (a)

$a^k \pmod 6$		
	$k=1$	$k=2$
$a=1$	1	1
$a=5$	5	1

(b)

$a^k \pmod{12}$				
	$k=1$	$k=2$	$k=3$	$k=4$
$a=1$	1	1	1	1
$a=5$	5	1	5	1
$a=7$	7	1	7	1
$a=11$	11	1	11	1

From the table of modular powers, we see that $\mathrm{ord}_6(1) = \mathrm{ord}_{12}(1) = 1$.

(*Note*: It is trivial that $\mathrm{ord}_n(1) = 1$ for any modulus n.) The orders of all other elements are 2, and 5 is a primitive root (mod 6), but there are no primitive roots mod 12.

17. (a) $\mathrm{ord}_{10}(3) = 4$

(b) $\mathrm{ord}_{21}(6)$ is undefined since 6 is not relatively prime to 21.

(c) $\mathrm{ord}_{304}(21) = 36$

19. (a) (i) $12 = 2^2 \cdot 3$ is not of any of the forms listed in Theorem 8.7, so there are no primitive roots.

(b) (i) 13 is prime, so by Theorem 8.7 there are $\phi(\phi(13)) = \phi(12) = 4$ primitive roots (mod 13).

(ii) Since $\mathrm{ord}_{12}(2) = 12 = \phi(13)$, 2 is a primitive root mod 13.

(c) (i) Since $14 = 2 \cdot p$, with $p = 7$, Theorem 8.7 tells us that there are $\phi(\phi(14)) = \phi(6) = 2$ primitive roots mod 14.

(ii) Since $\mathrm{ord}_{14}(3) = 6 = \phi(14)$, 3 is a primitive root mod 14.

21. (a) (i) $25 = p^2$ with $p = 5$; so by Theorem 8.7 we know there are $\phi(\phi(25)) = \phi(20) = 8$ primitive roots mod 25.

(ii) The possible mod 25 orders (for integers relatively prime to 25) must divide [by Proposition 8.5(b)] $\phi(25) = 20 = 2^2 \cdot 5$, and so can only be one of the following values: 1 (only for $a = 1$), 2, 4, 5, 10, or 20 (in which case we have a primitive root). To find the smallest primitive root mod 25, go through all integers relatively prime to 25 starting with $a = 2$, and compute a^2, a^4, a^5, a^{10} (mod 25) until you find the smallest for which these powers are all different from 1; this will be the smallest primitive root. $2^2 \equiv 4$, $2^4 \equiv 16$, $2^5 \equiv 7$, $2^{10} \equiv 24$ (mod 25). So we can stop: $g = 2$ is the smallest primitive root. *Alternatively:* A more sophisticated approach would be to use Theorem 8.8(b). Since it is easy to check that $g = 2$ is a primitive root mod 5, the theorem tells us that $2 + 5 = 7$ will be a primitive root mod 25.

(iii) By Corollary 8.10, if j is relatively prime to $\mathrm{ord}_{25}(g) = 20$, then g^j will also be a primitive root. Using $j = 3$, we get that $2^3 \equiv 8$ is another primitive root mod 25.

(b) (i) Since $39 = 3 \cdot 13$ is not of any of the forms listed in Theorem 8.7, there are no primitive roots mod 39.
(ii) and (iii) are not applicable.

(c) (i) Since 31 is prime, Theorem 8.7 tells us that there are $\phi(\phi(31)) = \phi(30) = 8$ primitive roots mod 31.

(ii) The possible mod 31 orders (for integers relatively prime to 31) must divide [by Proposition 8.5(b)] $\phi(31) = 30 = 2 \cdot 3 \cdot 5$ and so can only be one of the following values: 1 (only for $a = 1$), 2, 3, 5, 6, 10, 15, or 30 (in which case we have a primitive root). To find the smallest primitive root mod 31, go through all integers relatively prime to 31 starting with $a = 2$, and compute a^2, a^3, a^5, a^6, a^{10}, a^{15} (mod 25) until you find the smallest for which these powers are all different from 1; this will be the smallest primitive root. $2^2 \equiv 4, 2^3 \equiv 8, 2^5 \equiv 1$ (mod 31), so 2 does not work, and we move on to $a = 3$: $3^2 \equiv 9, 3^3 \equiv 27, 3^5 \equiv 26, 3^6 \equiv 16, 3^{10} \equiv 25, 3^{15} \equiv 30$ (mod 31), and we can stop: $g = 3$ is the smallest primitive root.

(iii) By Corollary 8.10, if j is relatively prime to $\mathrm{ord}_{25}(g) = 30$, then g^j will also be a primitive root. Using $j = 7$, we get that $3^7 \equiv 17$ is another primitive root mod 31.

(d) (i) Since $50 = 2 \cdot p^2$, with $p = 5$, Theorem 8.7 tells us that there are $\phi(\phi(50)) = \phi(20) = 8$ primitive roots mod 50.

(ii) Since we know from the solution of part (a) that $g = 2$ is a primitive root of $p^2 = 25$, Theorem 8.8(b) tells us that $G = g + p^2 = 2 + 25 = 27$ (since it is odd) will be a primitive root of 50.

(iii) By Corollary 8.10, if j is relatively prime to $\phi(50) = 20$, then G^j will also be a primitive root. Using $j = 3$, we get that $27^3 \equiv 33$ is another primitive root mod 50.

(e) (i) Since $52 = 2^2 \cdot 13$ is not of any of the forms listed in Theorem 8.7, there are no primitive roots mod 52.
(ii) and (iii) are not applicable.

(f) (i) Since $961 = p^2$, with $p = 31$, Theorem 8.7 tells us that there are $\phi(\phi(31^2)) = \phi(930) = 240$ primitive roots mod 961.

(ii) Since we know from the solution of part (c) that $g = 3$ is a primitive root of $p = 31$, Theorem 8.8(b) tells us that g (since it is odd) will be a primitive root of 961.

(iii) By Corollary 8.10, if j is relatively prime to $\phi(961) = 930$, then G^j will also be a primitive root. Using $j = 7$, we get that $3^7 \equiv 265$ is another primitive root mod 961.

23. All of the integers mod 223 for which order is defined (i.e., which are relatively prime to 223) are powers of the primitive root $g = 3$ of part (a) (by Theorem 8.7). By Proposition 8.5(b), the order of any such element must divide $\phi(223) = 222 = 2 \cdot 3 \cdot 37$. Since 10 is not a factor, there are no integers with order 10, so the answer to part (d) is zero. But both 6 and 74 are factors. By Equation 8.2 of Proposition 8.9 with a taken to be the primitive root 3: $\operatorname{ord}_{223}(3^j) = \dfrac{\operatorname{ord}_{223}(3)}{\gcd(j, \operatorname{ord}_{223}(3))} = \dfrac{222}{\gcd(j, 222)}$. Thus, the mod 223 integers with order = 6 correspond to the modular powers $3^j (1 \le j \le 222)$ that will make the latter fraction = 6, i.e., $\dfrac{222}{\gcd(j, 222)} = 6 \Rightarrow \gcd(j, 222) = 222/6 = 37$. The corresponding integers 2 are clearly $j = 37, 5 \cdot 37$ $(2 \cdot 37, 3 \cdot 37,$ and $4 \cdot 37$ are out since they have other common factors with 222, $7 \cdot 37$ is already too large). Thus there are two elements of order 6 (mod 223). Part (c) is done in the same fashion. Elements of order 74 must be of the form $3^j (1 \le j \le 222)$ with $\dfrac{222}{\gcd(j, 222)} = 74 \Rightarrow \gcd(j, 222) = 222/74 = 3$. A computation shows that there are 36 such elements.

Chapter 9

1. (a) Since the prime factorization of $p - 1$ is $772 = 2^2 \cdot 193$, the only possibilities for $\operatorname{ord}_p(2)$ are [by Proposition 8.5(b)]: 2, 4, 193, 386, and 772 (for a primitive root). Since $2^{193} \equiv 317$, and $2^{386} \equiv 772$, it follows that $\operatorname{ord}_p(2) = 772$, and thus $g = 2$ is a primitive root mod p.

(b) $B = g^b \equiv 2^{603} \equiv 122 (\bmod\ 773)$, $A = g^a \equiv 2^{333} \equiv 277 (\bmod\ 773)$

(c) $K = B^a \equiv 122^{333} \equiv 75 (\bmod\ 773)$, $K = A^b \equiv 277^{333} \equiv 75 (\bmod\ 773)$

3. (a) $g = 307$ (*Note:* $\operatorname{ord}_p(301) = 97, \operatorname{ord}_p(303) = 388, \operatorname{ord}_p(305) = 97$)

(b) $A = 1143, B = 300$

(c) $K = 245$

5. (a) $C \equiv P^e \equiv 1234^{143} \equiv 4347 (\bmod\ n)$

(b) $C^d \equiv 4347^{47} \equiv 1234 \equiv P (\bmod\ n)$

(c) $s \equiv P^{d_A} \equiv 1234^{367} \equiv 273 (\bmod\ 6887)$

(d) $s^{e_A} \equiv 273^{4303} \equiv 1234 \equiv P (\bmod\ 6887)$

7. (a) $C = 58684$

   (b) $C^d \equiv 58684^{29401} \equiv 12345 \equiv P(\bmod n)$

   (c) $s = 11824$

   (d) $s^{e_A} \equiv 11824^{50443} \equiv 12345 \equiv P(\bmod 69353)$

9. (a) $\phi(n) = (p-1)(q-1) = 2376$, $\gcd(169, \phi(n)) = 1$, so $e = 169$ is a legitimate encryption exponent. We use the extended Euclidean Algorithm 2.2 to compute $d \equiv e^{-1} \equiv 2137(\bmod \phi(n))$, which is the decryption exponent.

   (b) $C = 1744$

   (c) $C^d \equiv 1744^{2137} \equiv 1234 \equiv P \ (\bmod n)$

   (d) $s = 69$

   (e) We use the extended Euclidean Algorithm 2.2 to compute Alice's private key: $e_A \equiv d_A^{-1} \equiv 751(\bmod \phi(n))$. Authentication: $s^{e_A} \equiv 69^{751} \equiv 1234 \equiv P(\bmod 2376)$

11. (a) Bob's public key is $B \equiv g^b \equiv 2^{603} \equiv 122(\bmod 773)$, and Alice's is $A \equiv g^a \equiv 2^{333} \equiv 277(\bmod 773)$. Alice computes $C \equiv B^a P \equiv 75 \cdot 321 \equiv 112(\bmod 773)$. Thus, the entire ciphertext would be $(A, C) = (277, 112)$.

    (b) The decryption exponent is $p - 1 - b = 169$ and $A^{169} \equiv 277^{169} \equiv 134(\bmod 1231)$, and so $d_\kappa((A,C)) \equiv A^{p-1-b}C \equiv 134 \cdot 112 \equiv 321 \equiv P(\bmod 1231)$.

    (c) We first use the extended Euclidean algorithm (Algorithm 2.2) to compute $d^{-1} \equiv 79(\bmod p - 1)$. Now we may compute both $r \equiv g^d \equiv 2^{215} \equiv 356(\bmod 1231)$ and $s \equiv d^{-1}(P - ar) \equiv 79 \cdot (321 - 333 \cdot 356) \equiv 495(\bmod 1230)$. Thus, the digitally signed document that Alice sends to Bob is $(P, r, s) = (321, 356, 495)$.

    (d) To authenticate that the document 321 (the first component of the vector received from Alice) was really signed by her, Bob uses the second two components of the vector received along with Alice's public ElGamal key $A = 277$ and computes $A^r r^s \equiv 629 \cdot 731 \equiv 637(\bmod 1231)$, and compares this with $g^P \equiv 2^{321} \equiv 637(\bmod 1231)$. Since the results agree, Alice's signature has been authenticated.

13. (a) $g = 5055$

    (b) $(A, C) = (3533, 2030)$

    (c) $d_\kappa((A, C)) \equiv A^{p-1-b}C \equiv 3533^{2341} \cdot 2030 \equiv 1758 \cdot 2030 \equiv 4321 \equiv P \ (\bmod p)$

    (d) $(P, r, s) = (4321, 1870, 5215)$

    (e) Since $A^r r^s \equiv 860 \equiv g^P(\bmod p)$, the signature is authenticated.

15. (i) (a) The sequence is easily checked to be superincreasing, since $m = 175 > 171 =$ the sum of the weights, and $\gcd(w, m) = 1$, the key is legitimate.

        (b) The public key is the mod m vector: $w \cdot [a_1 \ a_2 \cdots a_6] \equiv 88 \cdot [3 \ 5 \ 8 \ 18 \ 36 \ 100] \equiv [89 \ 90 \ 929 \ 18 \ 50] \ (\bmod m)$.

        (c) The ciphertext is $f_b([1, 0, 1, 0, 1, 0]) = x_1 b_1 + x_2 b_2 + x_3 b_3 + x_4 b_4 + x_5 b_5 + x_6 b_6 = 1 \cdot 89 + 0 \cdot 90 + 1 \cdot 92 + 0 \cdot 9 + 1 \cdot 18 + 0 \cdot 50 = 199 = s$.

(d) Using the extended Euclidean algorithm (Algorithm 2.2), we compute $w^{-1} \equiv 2$. Since $w^{-1} \cdot s \equiv 2 \cdot 199 \equiv 48 (\mathrm{mod}\, m)$, the plaintext will be the solution of the superincreasing knapsack problem with weight vector $[a_1\ a_2\ a_3\ a_4\ a_5\ a_6]$ and knapsack weight $s' = 17$. Algorithm 9.6 quickly produces the original plaintext.

(ii) (b) The public key is the mod m vector $w \cdot [a_1\ a_2\ \cdots\ a_6] \equiv 371 \cdot [18\ 36\ 100\ 184\ 360\ 750] \equiv [878\ 306\ 850\ 114\ 160\ 1300]\, (\mathrm{mod}\, m)$.

(c) $s = 1888$ (ciphertext)

(d) Using the extended Euclidean algorithm (Algorithm 2.2) we compute $w^{-1} \equiv 981$. Since $w^{-1} \cdot s \equiv 981 \cdot 1888 \equiv 478$ $(\mathrm{mod}\, m)$, the plaintext will be the solution of the superincreasing knapsack problem with weight vector $[a_1\ a_2\ a_3\ a_4\ a_5\ a_6]$ and knapsack weight $s' = 17$. Algorithm 9.6 quickly produces the original plaintext.

(iii) (b) The public key is the mod m vector: $w \cdot [a_1\ a_2\ \cdots\ a_6]$ $\equiv 205 \cdot [5\ 9\ 18\ 34\ 72\ 144] \equiv [167\ 129\ 258\ 106\ 174\ 62]\, (\mathrm{mod}\, m)$.

(c) $s = 599$ (ciphertext)

(d) Using the extended Euclidean algorithm (Algorithm 2.2) we compute $w^{-1} \equiv 173$. Since $w^{-1} \cdot s \equiv 173 \cdot 599 \equiv 95 (\mathrm{mod}\, m)$, the plaintext will be the solution of the superincreasing knapsack problem with weight vector $[a_1\ a_2\ a_3\ a_4\ a_5\ a_6]$ and knapsack weight $s' = 17$. Algorithm 9.6 quickly produces the original plaintext.

17. (a) The public key is the mod m vector: $w \cdot [a_1\ a_2\ \cdots a_9] \equiv 365 \cdot$ $[5\ 9\ 18\ 34\ 72\ 144] \equiv [1095\ 326\ 287\ 574\ 1148\ 524\ 1204\ 987\ 932](\mathrm{mod}\, m)$.

(b) (i) 4831
(ii) 4666
(iii) 4025

(c) The main parameters needed for the decryptions are $w^{-1} \equiv 538$ (for each decryption), and

(i) $w^{-1} \cdot s \equiv 1311 (\mathrm{mod}\, m)$

(ii) $w^{-1} \cdot s \equiv 982 (\mathrm{mod}\, m)$

(iii) $w^{-1} \cdot s \equiv 894 (\mathrm{mod}\, m)$

19. (a) If either C or $\tilde{C}$ were not relatively prime with n, then the corresponding gcd (which could be found with the Euclidean algorithm) would be one of the two (secret) prime factors of $n = pq$; thus, the factorization of n could be found, and from this Eve could obtain the (secret) decryption exponent. She would thus be able to decrypt not only these two but all future messages sent to either Bob or Ben (by anyone, not just Alice). (We point out that this case is typically quite rare.) Henceforth we assume that both C and $\tilde{C}$ are relatively prime to n. Since $\gcd(e, \tilde{e}) = 1$, Eve could use the extended Euclidean algorithm (Algorithm 2.2) to determine integers x and y such that $ex + \tilde{e}y = 1$. From this and the given ciphertext equations, we deduce that $C^x \tilde{C}^y \equiv (P^e)^x (P^{\tilde{e}})^y \equiv P^{ex + \tilde{e}y} \equiv P \pmod{n}$. From the equation $ex + \tilde{e}y = 1$, it follows that one of x or y is a positive integer, and the other is negative. Without loss of generality, we assume that $x < 0$. Now since $C^x \equiv (C^{-1})^{|x|}$, Eve needs only first compute $C^{-1} \pmod{n}$, using the extended Euclidean algorithm, then raise the result and $\tilde{C}$ to the appropriate (positive) modular powers, and multiply them (mod n) to determine the plaintext P.

(b) The extended Euclidean algorithm (Algorithm 2.2) tells us that with $x = -587$ and $y = 38$, we have $ex + \tilde{e}y = 1$. Another application of the extended Euclidean algorithm tells us that $C^{-1} = 5179 \pmod{n}$. Using fast modular exponentiation (Algorithm 6.5), we compute $C^x \equiv (C^{-1})^{587} \equiv 5179^{587} \equiv 1538 \pmod{6887}$ and also $\tilde{C}^y \equiv 1902^{38} \equiv 2594$. The plaintext P will be the product of these two numbers (mod 6887): $P = 1538 \cdot 2594 \equiv 1999$. (As a check, the reader should verify that this number encrypts to C and $\tilde{C}$, when the RSA encryption exponents e and $\tilde{e}$ are used, respectively.

(c) As in part (b), we now obtain $ex + \tilde{e}y = 1$, with $x = -578$ and $y = 2917$. Also, $C^{-1} = 405390 \pmod{n}$. Hence $P \equiv C^x \tilde{C}^y \equiv (C^{-1})^{578} (\tilde{C})^{2917} \equiv 856876 \cdot 766399 \equiv 888888 \pmod{n}$.

21. (a) $D_e : \mathbb{Z}_p \to \mathbb{Z}_p :: D_e(P) \equiv P^d \pmod{p}$, where d is the inverse of $e \pmod{p-1}$. In the given setting, we compute $d \equiv 275^{-1} \equiv 11 \pmod{p-1}$. Thus, $D_e(777) \equiv 777^{11} \equiv 600 \equiv P \pmod{p}$. (The reader may wish to verify that $E_e(P) \equiv 777$.)

(b) This system is totally insecure. Eve could simply use the extended Euclidean algorithm to compute d (the private key) but inverting e (the public key) (mod $p-1$).

23. Following the suggestion, Eve could use the extended Euclidean algorithm to compute $x^{-1} \pmod{n}$. When Bob decrypts $\tilde{C} \equiv Cx^e \equiv P^e x^e \equiv (Px)^e \pmod{n}$, the result (that he gives to Eve) will be $\tilde{C}^d \equiv (Px)^{ed} \equiv Px \pmod{n}$, and Eve need only multiply the latter number by $x^{-1} \pmod{n}$ to obtain the plaintext.

25. (a) Since $g^{L_g(a) + L_g(b)} \equiv g^{L_g(a)} g^{L_g(b)} \equiv ab \equiv g^{L_g(ab)} \pmod{p}$, it follows that [see Proposition 8.5(c)] $L_g(ab) \equiv L_g(a) + L_g(b) \pmod{p-1}$.

(b) Since $g^{-L_g(a)} \equiv (g^{L_g(a)})^{(-1)} \equiv a^{-1} \pmod{p} \equiv ab \equiv g^{L_g(ab)}$, it follows that $L_g(a^{-1}) = -L_g(a) \pmod{p-1}$.

(c) Since $g^{kL_g(a)} \equiv (g^{L_g(a)})^k \equiv a^k \pmod{p}$, it follows that $L_g(a^k) = kL_g(a) \pmod{p-1}$.

(d) Since $h^{L_h(g) \cdot L_g(a)} \equiv (h^{L_h(g)})^{L_g(a)} \equiv (g)^{L_g(a)} \equiv a \pmod{p}$, it follows that $L_h(a) \equiv L_h(g) \cdot L_g(a) \pmod{p-1}$.

29. For each part, $s = 10$ and $f = 74277$

(a) Prime factor 787 found at the iteration when $t = 2$.

(b) Prime factor 787 found at the iteration when $t = 0$.

(c) Algorithm did not find a prime factor with this seed.

Chapter 10

1. (a) S is a ring with 0 serving as both the additive and multiplicative identity. S is sometimes called the trivial ring.

(b) Since S is using the binary operations of the ring $\mathbb{Z}_6$, the ring axioms (1), (2), (5), (6), and (8) are all inherited. We do need to check that S is *closed* under these operations, i.e., when we add/multiply any two elements of S, the result lands in S. But this clearly holds because $3 + 3 = 0$, and $3 \cdot 3 = 3$ (in S). S contains the additive identity 0, and 3 is its own additive inverse. Although S does not contain the multiplicative identity 1 of $\mathbb{Z}_6$, since $3 \cdot 3 = 3$ (in S), it follows that 3 serves as a multiplicative identity for S. This all shows that S is a ring.

(c) Since the sum/product of even numbers is even, it follows that S is closed under the addition/multiplication operations of $\mathbb{Z}_{10}$. Also S has the additive identity 0 and is closed under additive inverses. It suffices to verify that 6 serves as a multiplicative identity for S: $6 \cdot 2 = 2$, $6 \cdot 4 = 4$, $6 \cdot 6 = 6$, $6 \cdot 8 = 8$.

3. (a) The trivial ring $S = \{0\}$ is also trivially a field since there are no nonzero elements. So the requirement that every nonzero element have multiplicative inverse is automatically true.

(b) S is a field since there is only one nonzero element (3), which is that multiplicative identity, and the multiplicative identity is always its own inverse in any ring.

(c) Since $4 \cdot 4 = 6$ in S and 6 is the multiplicative identity, we have $4^{-1} = 4$. Since $2 \cdot 8 = 6$ in S, 2 and 8 are inverses of each other. Thus all nonzero elements have multiplicative inverses, so S is a field.

5. In each part, since the operations are done separately in each component using arithmetic in the ring $\mathbb{Z}_n$ it follows that all of the ring axioms hold for these systems of vectors. The additive identity is the zero vector, the multiplicative identity is the vector of 1's.

7. (a) $X^5 + X$

(b) $X^5 + 3X^2 + 5X + 3$

(c) $3X^2 + 6$

9. (a) $X^6 + X^5 + X^4 + X^3 + X + 1$

(b) $10X^6 + 2X^5 + 9X^4 + 10X^3 + X$

(c) $5X^6 + 2X^5 + 6X^4 + 3X^3 + X^2 + 5X + 5$

11. (a) Quotient $= X^3 + 1$, remainder $= 0$.

(b) Quotient $= X^3 + X + 1$, remainder $= 0$.

(c) Quotient $= X^3 + X^2 + X + 1$, remainder $= X + 1$.

(d) Quotient $= 2X^3 + 2X + 1$, remainder $= X$.

13. (a) irreducible

(b) irreducible

15. (a) $X^5 + X + 1$

(b) $2X^3 + 6X^2 + 5$

17. (a) Since $X^3 + X + 1 = (X+1)(X^2 + X + 2)$ in $\mathbb{Z}_3[X]$, the ring $\mathbb{Z}_3[X](\bmod X^3 + X + 1)$ is not a field.
   (b) $2X^2, X^2 + X + 1$

19. (a) C5
   (b) 1F
   (c) B3

21. For each part, the polynomial version of the extended Euclidean algorithm should be used as in Example 10.14 and Exercise for the Reader 10.13. [But for part (c), Table 10.7 could be used.]
   (a) Inverse $= X + 1$
   (b) Inverse $= X^2 + 2X + 2$
   (c) Inverse $=$ A
   (d) Inverse $=$ A2

29. Impossible. *Proof:* If a is any nonzero element in the ring (such an element exists since the ring has at least two elements), since the ring axioms tell us that $0 \cdot a = 0$ and $1 \cdot a = a$, if 1 were the same as 0, the two left sides of these equations would be the same and it would follow that the two right sides would also be the same—i.e., $0 = a$—which is a contradiction. □

Chapter 11

1. (a) 0
   (b) 2
   (c) 4
   (d) C

3. (a) B
   (b) 9
   (c) E
   (d) C

5. (a) $\begin{bmatrix} 1 & X^2 \\ X^2 & 1 \end{bmatrix} \cdot \begin{bmatrix} 1 & 2 \\ 3 & 4 \end{bmatrix} = \begin{bmatrix} 1 & 4 \\ 4 & 1 \end{bmatrix}.$

$\begin{bmatrix} 1 & 2 \\ 3 & 4 \end{bmatrix} = \begin{bmatrix} 1 \cdot 1 + 4 \cdot 3 & 1 \cdot 2 + 4 \cdot 4 \\ 4 \cdot 1 + 1 \cdot 3 & 4 \cdot 2 + 1 \cdot 4 \end{bmatrix} = \begin{bmatrix} D & 1 \\ 3 & 7 \end{bmatrix}$

   (b) $\begin{bmatrix} 1 & X^2 \\ X^2 & 1 \end{bmatrix} \cdot \begin{bmatrix} A & B \\ C & D \end{bmatrix} = \begin{bmatrix} 1 & 4 \\ 4 & 1 \end{bmatrix} \cdot \begin{bmatrix} A & B \\ C & D \end{bmatrix} =$

$\begin{bmatrix} 1 \cdot A + 4 \cdot C & 1 \cdot B + 4 \cdot D \\ 4 \cdot A + 1 \cdot C & 4 \cdot B + 1 \cdot D \end{bmatrix} = \begin{bmatrix} F & A \\ 2 & 7 \end{bmatrix}$

7. (a) $\kappa^0 = 1234, \kappa^1 = \text{DEEA}, \kappa^2 = \text{ED07}$

   (b) $\kappa^0 = \text{A64B}, \kappa^1 = \text{D299}, \kappa^2 = \text{6AF3}$

(c) $\kappa^0 = 8888$, $\kappa^1 = A22A$, $\kappa^2 = 94BE$

(d) $\kappa^0 = FA5A$, $\kappa^1 = 7B21$, $\kappa^2 = 9DBC$

9. (a) E737
  (b) 9616
  (c) DE19
  (d) AB2C

11. (a) 5134
  (b) A2C6
  (c) 31E1
  (d) D4F7

13. (a) 67
  (b) 32
  (c) 1F
  (d) 80

15. (a) *Step 1.* $B_1 = [0, 0, 0, 0, 0, 0, 0, 0]$
  Step 2. Since matrix multiplication with the zero vector is zero, the XOR on the right is just the right side vector, which give the bits of the output in the reverse order, so $F(B_1) = [0, 1, 1, 0, 0, 0, 1, 1] \sim 63$. By Table 11.2, this agrees with $S(00)$.

  (b) *Step 1.* $B_1 = [0, 0, 0, 1, 1, 0, 0, 1]$ (the polynomial extended Euclidean algorithm could be used to get this).
  Step 2.

$$
\begin{bmatrix} b_0' \\ b_1' \\ b_2' \\ b_3' \\ b_4' \\ b_5' \\ b_6' \\ b_7' \end{bmatrix}
=
\begin{bmatrix}
1 & 0 & 0 & 0 & 1 & 1 & 1 & 1 \\
1 & 1 & 0 & 0 & 0 & 1 & 1 & 1 \\
1 & 1 & 1 & 0 & 0 & 0 & 1 & 1 \\
1 & 1 & 1 & 1 & 0 & 0 & 0 & 1 \\
1 & 1 & 1 & 1 & 1 & 0 & 0 & 0 \\
0 & 1 & 1 & 1 & 1 & 1 & 0 & 0 \\
0 & 0 & 1 & 1 & 1 & 1 & 1 & 0 \\
0 & 0 & 0 & 1 & 1 & 1 & 1 & 1
\end{bmatrix}
\cdot
\begin{bmatrix} b_0 \\ b_1 \\ b_2 \\ b_3 \\ b_4 \\ b_5 \\ b_6 \\ b_7 \end{bmatrix}
\oplus
$$

$$
\begin{bmatrix} 1 \\ 1 \\ 0 \\ 0 \\ 0 \\ 1 \\ 1 \\ 0 \end{bmatrix}
=
\begin{bmatrix} 0 \\ 1 \\ 1 \\ 0 \\ 1 \\ 0 \\ 0 \\ 0 \end{bmatrix}
\oplus
\begin{bmatrix} 1 \\ 1 \\ 0 \\ 0 \\ 0 \\ 1 \\ 1 \\ 0 \end{bmatrix}
=
\begin{bmatrix} 1 \\ 0 \\ 1 \\ 0 \\ 1 \\ 1 \\ 1 \\ 0 \end{bmatrix}
$$

$$F(N_1) = [b'_7, b'_6, b'_5, b'_4, b'_3, b'_2, b'_1, b'_0] = [0, 1, 1, 1, 0, 1, 0, 1]$$
~ 75. By Table 11.2, this agrees with $S(3F)$.

Chapter 12

1. (a)
   (b) $(-2, 0)$

3. (a) $2P = (0, -1)$
   (b) $3P = \infty$, $4P = (0, 1)$, $5P = (0, -1) = 2P$,
       so $6P = 5P + P = 2P + P = 3P = \infty$,
       $7P = 6P + P = 3P + P = 4P = (0, 1)$.
   (c) If we continue the progression in part (b), we are led to
       $nP = (n-3)P = (n-6)P = \cdots$, and it follows that $nP = rP$,
       where r is the remainder in the division of n by 3. This yields

       the formula $nP = \begin{cases} \infty, & \text{if } n \equiv 0 (\text{mod } 3) \\ (0,\ 1), & \text{if } n \equiv 1 (\text{mod } 3) \\ (0,\ -1), & \text{if } n \equiv 2 (\text{mod } 3) \end{cases}$

5. (a) $E = \{(1, 2), (1, 3), (2, 1), (2, 4), (3, 0), \infty\}$
   (b) $E = \{(0, 1), (0, 6), (1, 3), (1, 4), (2, 3), (2, 4), (3, 0),$
       $(4, 3), (4, 4), (5, 0), (6, 0), \infty\}$
   (c) $E = \{(1, 3), (1, 8), (2, 4), (2, 7), (5, 1), (5, 10), (6, 2),$
       $(6, 9), (8, 5), (8, 6), (9, 0), \infty\}$

7. Since the discriminant is nonzero, $4a^3 + 27b^2 \equiv 4 \cdot 6^3 + 27 \cdot 0^2 \equiv$
   $6 \not\equiv 0 (\text{mod } 13)$, the elliptic curve is nonsingular.
   (a) $(8, 12)$
   (b) $(8, 1)$
   (c) $-(4, 7) = (4, -7) \equiv (4, 6) \ (\text{mod } 13)$
   (d) $(9, 9)$

9. (a) Since the discriminant is nonzero, $4a^3 + 27b^2 \equiv 4 \cdot 3^3 +$
       $27 \cdot 3^2 \equiv 1 \not\equiv 0 (\text{mod } 5)$, the elliptic curve is nonsingular.
   (b) $E = \{(3, 2), (3, 3), (4, 2), (4, 3), \infty\}$
   (c) We point out that the group axioms (see Proposition 12.1)
       allow many shortcuts in forming the addition table shown
       here. For example, since addition is commutative, additions
       need only be computed in one order. Also, the inverse $(a, -b)$
       (mod 5) of a point (a, b) always results in the point at infinity
       when added to the point.

+	∞	(3,2)	(3,3)	(4,2)	(4,3)
∞	∞	(3,2)	(3,3)	(4,2)	(4,3)
(3,2)	(3,2)	(4,3)	∞	(3,3)	(4,2)
(3,3)	(3,3)	∞	(4,2)	(4,3)	(3,2)
(4,2)	(4,2)	(3,3)	(4,3)	(3,2)	∞
(4,3)	(4,3)	(4,2)	(3,2)	∞	(3,3)

11. Since the discriminant is nonzero, $4a^3 + 27b^2 \equiv 4 \cdot 2^3 + 27 \cdot 1^2 \equiv 4 \not\equiv 0 \pmod{11}$, the elliptic curve is nonsingular.
   (a) $2(0, 1) = (0, 1) + (0, 1) = (1, 9)$
   (b) $3(0, 1) = 2(0, 1) + (0, 1) = (1, 9) + (0, 1) = (8, 1)$
   (c) $6(0, 1) = 3(0, 1) + 3(0, 1) = (8, 1) + (8, 1) = (10, 3)$
   (d) $9(0, 1) = 6(0, 1) + 3(0, 1) = (10, 3) + (8, 1) = (5, 2)$

13. (a) 16
   (b) 4
   (c) $\operatorname{ord}_E(\infty) = 1$ (always on any nonsingular elliptic curve)
   (d) 16

15. (a) 5
   (b) 3
   (c) 2
   (d) 4

17. (a) (148, 20)
   (b) (256, 132)
   (c) (15, 84)
   (d) (218, 229)

19. (a) Since the discriminant is nonzero: $4a^3 + 27b^2 \equiv 4 \cdot 22^3 + 27 \cdot 153^2 \equiv 141 \not\equiv 0 \pmod{163}$, the elliptic curve is nonsingular. To check for a square root of $r \triangleq x_1^3 + 22x_1 + 153 \equiv 28^3 + 22 \cdot 28 + 153 \equiv 64$, since $p \equiv 3 \pmod 4$, we may use Equation 12.6 $y_1 \equiv r^{(p+1)/4} \equiv 64^{41} \equiv 155 \pmod p$. (We used fast modular exponentiation.) Since $y_1^2 \equiv 64 \equiv r \pmod p$, this leads to a point $G = (x_1, y_1) = (28, 155) \in E$.
   (b) Using Algorithm 12.3, Alice and Bob would respectively compute:
$$A \equiv n_A G \equiv 19 \cdot (28, 155) = (28, 155)$$
$$B \equiv n_B G \equiv 41 \cdot (28, 155) = (28, 8)$$

   Alice would compute the shared Diffie–Hellman key as $K = n_A B = 19 \cdot (28, 8) = (28, 8)$.
   Bob would compute it as $K = n_B A = 41(28, 155) = (28, 155)$.

   Note: It was a bit of an accident that Bob's public key B coincides with the private key K. The technical reason that this happened is that $\operatorname{ord}_E(G) = 246$, and $19 \cdot 41 = 779 \equiv 41 \equiv n_B \pmod{\operatorname{ord}_E(G)}$. Just as was proved in Chapter 8 in the setting of integers, it is not hard to show that if i and j are positive integers with $i \equiv j \pmod{\operatorname{ord}_E(G)}$, then $iG = jG$.
   (c) $A = \infty$, $B = (28, 155)$, $K = \infty$

21. (a) (130, 48)
   (b) (4450, 127)
   (c) (24401, 44434)

23. (a) The first number given results in the point $G = (38, 145)$.

(b) Alice will need to get from Bob his public key $B = n_B G = (7, 119)$. From this, she computes $n_A B = (329, 179)$ and adds her plaintext to this to produce $C = P + n_A B = (45, 331)$. The ciphertext she sends Bob will be the pair of points $(A, C) = ((356, 273), (45, 331))$, where she will have needed to compute her public key $B = n_A G = (356, 273)$.

(c) To decrypt, Bob first multiplies the first point A of the ciphertext (Alice's public key) by his private parameter $K = n_B A = (329, 179)$, he then needs to subtract this point from the second point C of the ciphertext to recover the plaintext: $C - (329, 179) = C + (329, -179) \equiv (45, 331) + (329, 260) = (312, 65)$.

(d) Ciphertext: $(A, C) = ((148, 148), (260, 301))$.

25. Results will vary, depending on the random choices being made in the execution of Lenstra's algorithm. Here are the prime factorizations:

(a) $449 \cdot 659$

(b) $929 \cdot 1663$

(c) $2237 \cdot 3833$

29. This follows directly from Theorem 12.6.

31. We first observe that $P_1 \boxplus P_2 = -(P + Q)$, where "+" is the addition operation on the elliptic curve (under which it is a group).

(a) "$\boxplus$" is commutative because "+" is commutative.

(b) Since $\left(P \boxplus Q\right) \boxplus R = -(-(P + Q) + R) = P + Q - R$, and $P \boxplus \left(Q \boxplus R\right) = -(P - (Q + R)) = -P + Q + R$, it follows that $2(P - R) = \left[\left(P \boxplus Q\right) \boxplus R\right] - \left[P \boxplus \left(Q \boxplus R\right)\right]$.

Thus, if "$\boxplus$" is associative, then $2(P - R) = \infty$ for all points P, R. Just about any randomly chosen example will show this statement is false. For example, if $P = (2, 4)$ on the curve $y^2 = x^3 - 4x + 16$ of Example 12.2, and $R = -P$, then the preceding equation becomes:

$$\infty = 2(P - R) = 2(2P) = 2(-3, 1) = (553/4, -13003/8),$$

a contradiction!

Appendix E: Suggestions for Further Reading

Synopsis

The purpose of this appendix is to recommend a few selected books that will provide additional information on various topics relating to cryptography. These books, in turn, contain their own useful collections of references. In addition to offering guidance for readers wishing to delve deeper into the subject, another goal is to provide sources for special cryptography projects and papers that professors might wish to offer to some of their particularly motivated and/or strong students.

History of Cryptography

The most comprehensive book that was ever written on the history of cryptography is that of Kahn [Kah-96]. In this 1200 page tome, Kahn has done an admirable job researching numerous elements of cryptography and how they influenced history, particularly during the 20th century. The first edition of Kahn's book came out in 1967, and the subsequent 1996 edition contains an additional chapter on public key cryptography. The book by Singh [Sin-00] provides a lean and lively treatment of the subject, and very nicely complements Kahn's book in the more recent history of cryptography. Although both books were written for the general public, Singh's book contains some well done technical explanations of many of the cryptographic concepts. We also point out two good references that describe the celebrated breaking of the notorious German Enigma machines: The book by Kozaczuk [Koz-84] (translated from Polish) details how the Polish codebreakers were able to achieve their initial (and significant) contributions. The book *Codebreakers: The Inside Story of Bletchley Park* [HiSt-01] consists of 27 narratives written by cryptographers at Bletchley Park after it became legal for them to talk about their work.

Mathematical Foundations

Many of the mathematical concepts of cryptography fall under the general headings of discrete structures and number theory. Rosen has written an excellent text on discrete structures, which is now in its sixth edition: [Ros-07]. The author (of this cryptography book) is currently finishing a book [Sta-11] on discrete structures that will be written in the same format as this book (with computer implementations). Rosen has also written an excellent introductory text on number theory [Ros-05]. Number theory is one of the oldest subjects of mathematics, and it has received a renewed boost in interest due in no small part to its uses in cryptography. An exciting historical account of the more recent history and groundbreaking achievements of number theory can be found in the book by du Sautoy [Sau-03]. Much of the number theory that gets used in cryptography is algorithmic. A gem of an introduction to the subject of algorithmic number theory is the book by Bressoud and Wagon [BrWa-99]. Although it is written with snippets of code from a particular software package (Mathematica), it should be accessible to any reader who is experienced with any related computing platform. Two other fine books on this subject that are more advanced are the book by Bach and Shallit [BaSh-96], and the book by Cohen [Coh-93]. For readers wishing to learn more about finite fields, the book by Lidl and Niederreiter [LiNi-83] provides an excellent and comprehensive treatment. The study of abstract number systems such as fields, rings, and groups falls under an area of mathematics known as abstract algebra. For readers wishing to learn more about some of the general theory in this area, either of the books [Her-96] by Herstein or [Hun-96] by Hungerford would be fine places to start.

Computer Implementations

Schneier's book [Sch-96] covers a great deal of cryptographic algorithms and provides corresponding code in the C programming language. Bishop's book [Bis-02] shows how to create Java applets for an assortment of cryptographically related algorithms. Although it was written under the title of a book on elliptic curve cryptography, the book [HaMeVa-03] by Hankerson, Menezes, and Vanstone contains some very useful information on building efficient algorithms for finite field arithmetic (useful in AES and related cryptosystems as well as for elliptic curve cryptosystems), including useful tables of industrial irreducible polynomials of high degree that are well suited for cryptographic systems.

Elliptic Curves

Elliptic curves have some beautiful connections in several diverse branches of mathematics and have been the subject of intensive research for hundreds of years. For more details and mathematical proofs of some of the number theoretic results relating to elliptic curves, see the books by Silverman [Sil-86], and Washington [Was-03]. The book [HaMeVa-03] by Hankerson, et. al, that was cited in the preceeding paragraph contains a thorough algorithmic development of elliptic curve cryptography over finite fields, along with extensive implementation ideas.

Additional Topics in Cryptography

The subject of cryptography has long ago become too vast for it to be possible to write a single book (that would fit into one's backpack and) that decently covers all of its topics. The sizable handbook by Menezes, van Oorschot, and Vansctone [MeOoVa-96] has become a standard reference on computer-era cryptography; it contains a wealth of information, along with over 50 pages of references to numerous seminal scholarly papers. Our book has not devoted much attention to the area of network security, in part, because this area is still under a continuous flux of extensive adaptation and development. A good (and very accessible) introduction to the underlying concepts of network security can be found in the book by Forouzan [For-08]. Daemen and Rijmen, the famed creators of AES (Chapter 11), have coauthored a book [DaRi-02] that thoroughly describes the details, insights, and resistance to cryptanalysis concepts that led to the design of their cryptosystem. Four good textbooks on general cryptography that are written at a higher level than our book and which contain coverage of some additional topics are the following: Stinson's book [Sti-06] has become a standard text for many upper-level cryptography classes, and is now in its third edition. The book by Trappe and Washington [TrWa-03] discusses many applications and implementation protocols. Mollin's book [Mol-06] covers a great deal of number theory and shows how it has been implemented in cryptography. Finally, the relatively short book by Buchmann [Buc-04] is a good read for people with more advanced mathematical backgrounds (e.g., those who have already completed a decent course in abstract algebra).

References

[AgKaSa-04]   Agrawal, Manindra, Neeraj Kayal, and Nitin Saxena, PRIMES is in P, *Annals of Mathematics* 38, 781–893 (2004).

[AlGrPo-94]   Alford, William R., Andrew Granville, and Carl Pomerance, There are infinitely many Carmichael numbers, *Annals of Mathematics* 140(3), 703–722 (1994).

[BaSh-96]   Bach, Eric, and Jeffrey Shallit, *Algorithmic Number Theory*, Vol. 1: *Efficient Algorithms*, MIT Press, Cambridge, MA (1996).

[Bis-02] Bishop, David, *Introduction to Cryptography with Java Applets,* Jones and Bartlett, Sudbuny, MA (2002).

[BiSh-91]   Biham, Eli, and Adi Shamir, Differential cryptanalysis of DES-like cryptosystems, *Journal of Cryptography* (1), 3–72 (1991).

[BrWa-99]   Bressoud, David, and Stan Wagon, *A Course in Computational Number Theory*, John Wiley & Sons, Hoboken, NJ (1999).

[Buc-04]   Buchmann, Johannes A., *Introduction to Cryptography, Second Edition*, Springer-Verlag, NY (2004).

[CaWi-93]   Campbell, Keith W., and Michael J. Wiener, DES is not a group, *Proceedings of CRYPTO '93, Advances in Cryptography* (Lecture Notes in Computer Science), Springer-Verlag, New York 30–43 (1976).

[Coh-93]   Cohen, Henri, *A Course in Computational Number Theory*, Springer-Verlag, New York (1993).

[Cop-93]   Coppersmith, Don, The Data Encryption Standard (DES) and its strength against attacks, *IBM Journal of Research and Development* 38(3), 243–250 (1993).

[CoWi-90]   Coppersmith, Don, and Shmuel Winograd, Matrix multiplication via arithmetic progressions, *Journal of Symbolic Computation* 9(3), 251–280 (1990).

[DaLaPo-93]   Damgård, Ivan, Peter Landrock, and Carl Pomerance, Average case error estimates for the strong probable prime test, *Mathematics of Computation* 61(203), 177–194 (1993).

[DaRi-02]   Daemen, Joan, and Vincent Rijmen, *A Design of Rijndael: AES—The Advanced Encryption Standard*, Springer-Verlag, New York (2002).

[DiHe-76]   Diffie, Whitfield, and Martin Hellman, New directions in cryptography, *IEEE Transactions on Information Theory* 22, 644–654 (1976).

[End-77]   Enderton, Herbert B., *Elements of Set Theory*, Academic Press, New York (1977).

[For-08]   Forouzan, Behrouz A., Cryptography and Network Security, McGraw-Hill, New York (2008).

[Gai-89]   Helen F. Gaines, *Cryptanalysis*, Dover Publications, New York (1989).

[GaJo-79]   Garey, Michael R., and David S. Johnson, *Computers and Intractability: A Guide to the Theory of NP-Completeness*, W.H. Freeman and Company, New York (1979).

[Gor-05]   Gordon, Raymond G., *Ethnologue: Languages of the World*, 15th ed., SIL International, Dallas, TX (2005).

[HaMeVa-04]   Hankerson, Darrel, Alfred Menezes, and Scott Vanstone, *Guide to Elliptic Curve Cryptography*, Springer-Verlag, New York (2004).

[HaWr-80]   Hardy, Godfrey H., and Edward M. Wright, *An Introduction to the Theory of Numbers*, Oxford University Press, Oxford (1980).

[Her-96]   Herstein, Israel N., *Abstract Algebra*, 3rd ed., John Wiley & Sons, Hoboken, NJ (1996).

[HiSt-01]   Hinsley, F.H., and Alan Stripp (eds.), *Codebreakers: The Inside Story of Bletchley Park*, Oxford University Press, Oxford (2001).

[Hun-96]   Hungerford, Thomas W., *Abstract Algebra: An Introduction*, 2nd ed., Brooks Cole, Independence, KY (1996).

[Kah-96]   Kahn, David, *The Codebreakers: The Comprehensive History of Secret Communication from Ancient Times to the Internet*, Scribner, New York (1996).

[KaLaMo-07]   Kaye, Phillip, Raymond Laflamme, and Michele Mosca, *An Introduction to Quantum Computing*, Oxford University Press, New York (2007).

[KePfPi-04]   Kellerer, Hans, Ulrich Pferschy, and David Pisinger, *Knapsack Problems*, Scribner Press, Princeton, NJ (2004).

[Knu-98]   Knuth, Donald E., *The Art of Computer Programming*, Vol. 2, 3rd ed., Addison Wesley, Upper Saddle River, NJ (1998).

[Kob-94]   Koblitz, Neal, *A Course in Number Theory and Cryptography*, 2nd ed., Springer-Verlag, New York (1994).

[KoHi-99]   Kolman, Bernard, and David R. Hill, *Introductory Linear Algebra: An Applied First Course*, 7th ed., Prentice-Hall, Upper Saddle River, NJ (1999).

[Koz-84]   Kozaczuk, Władysław, *ENIGMA: How the German Machine Cipher Was Broken, and How It Was Read by the Allies in World War Two*, University Publications of America, Bethesda, MD (1984).

[LePa-06]   Leemis, Lawrence M., and Stephen K. Park, *Discrete-Event Simulation, A First Course*, Prentice-Hall, Upper Saddle River, NJ (2006).

[LiNi-83]   Lidl, Rudolf, and Harald Niederreiter, *Encyclopedia of Mathematics and Its Applications*, Vol. 20: *Finite Fields*, Addison-Wesley, Reading, MA (1983).

[LiXi-04]   Ling, San, and Chaoping Xing, *Coding Theory: A First Course*, Cambridge University Press, Cambridge (2004).

[Mat-94]   Matsui, Mitsuru, Linear cryptanalysis method for the DES cipher, *EUROCRYPT '93* (Lecture Notes in Computer Science #765), Springer-Verlag, New York 386–397 (1994).

[MeOoVa-96]   Menezes, Alfred, Paul van Oorschot, and Scott Vanstone, *Handbook of Applied Cryptography*, Chapman & Hall/CRC Press, Boca Raton, FL (1996).

[MeHe-78]   Merkle, Ralph C., and Martin E. Hellman, Hiding information and signatures in knapsack trapdoors, *IEEE Transactions on Information Theory*, IT-24, 525–530 (1978).

[MeHe-81]   Merkle, Ralph C., and Martin E. Hellman, On the security of multiple encryption, *Communications of the ACM* 24, 465–467 (1981).

[Mil-76]   Miller, Gary L., Riemann's hypothesis and tests for primality, *Journal of Computer System Sciences* 13(3), 300–317 (1976).

[Mol-03]   Mollin, Richard A., *RSA and Public Key Cryptography*, Chapman & Hall/CRC, Boca Raton, FL (2003).

[Mol-06]   Mollin, Richard A., *An Introduction to Cryptography, Second Edition*, Chapman & Hall, Boca Raton, FL (2006).

[Moo-05]   Moon, Todd, K., *Error Correction Coding: Mathematical Methods and Algorithms*, John Wiley & Sons, Hoboken, NJ (2005).

[MuScWe-03]   Musa, Mohammad, Edward Schaefer, and Stephen Wedig, A simplified Rijndael algorithm and its linear and differential cryptanalyses, *Cryptologia* 27, 148–177 (2003).

[Nag-01]   Nagell, Trygve, *Introduction to Number Theory*, 2nd ed., American Mathematical Society, Providence, RI (2001).

[Pol-74]   Pollard, John M., Theorems on factorization and primality testing, *Proceedings of the Cambridge Philosophical Society* 76, 521–528 (1974).

[Poo-05]   Poole, David, *Linear Algebra: A Modern Introduction*, Cengage Learning, Florence, KY (2005).

[Rab-76]   Rabin, Michael O., Probabilistic algorithm for testing primality, *Journal of Number Theory* 12(1), 128–138 (1976).

[Rob-05]   Robinson, Sara, Toward an optimal algorithm for matrix multiplication, *SIAM News* 38(9), Society of Industrial and Applied Math (SIAM), Philadelphia, PA (2005).

[Ros-05]   Rosen, Kenneth H., *Elementary Number Theory and Its Applications*, 5th ed., Addison-Wesley, New York (2005).

[Ros-07]   Rosen, Kenneth H., *Discrete Mathematics and Its Applications*, 6th Edition, McGraw-Hill, NY (2007).

[Ros-02]   Ross, Sheldon M., *A First Course in Probability*, 6th ed., Prentice-Hall, Upper Saddle River, NJ (2002).

[Rüc-87]   Rück, Hans-Georg, A note on elliptic curves over finite fields, *Mathematics of Computation* 49(179), 301–304 (1987).

[Sau-03]   du Sautoy, Marcus, *The Music of Primes*, Harper Collins, NY (2003).

[Sch-96]   Schneier, Bruce, *Applied Cryptography: Protocols, Algorithms, and Source Code in C*, 2nd ed., John Wiley & Sons, Hoboken, NJ (1996).

[Sil-86]   Silverman, Joseph H., *The Arithmetic of Elliptic Curves*, Springer-Verlag, New York (1986).

[Sin-99]   Singh, Simon, *The Code Book: Evolution of Secrecy from Many Queen of Scots to Quantum Cryptography*, Doubleday, NY (1999).

[Sta-11]   Stanoyevitch, Alexander, *Discrete Structures with Contemporary Applications*, Chapman & Hall/CRC Press, Boca Raton, FL (Forthcoming).

[Sti-06]   Stinson, R. Douglas, *Cryptography, Theory and Practice*, 3rd ed., Chapman & Hall/CRC, Boca Raton, FL (2006).

[Str-69]   Strassen, Volker, Gaussian elimination is not optimal, *Numerishe Mathematik* 13, 354–356 (1969).

[TrWa-06]   Trappe, Wade, and Lawrence C. Washington, *An Introduction to Cryptography with Coding Theory*, 2nd ed., Prentice-Hall, Upper Saddle River, NJ (2006).

[Vau-01]   Vaudenay, Serge, Cryptanalysis of the Chor-Rivest cryptosystem, *Journal of Cryptology* 14(2), 87–100 (2001).

[Was-03]   Washington, Lawrence C., *Elliptic Curves, Number Theory and Cryptography*, Chapman & Hall/CRC Press, Boca Raton, FL (2003).

[Wat-69]   Waterhouse, William C., Abelian varieties over finite fields, *Annals scientifiques de l'École Normale Supérieure* 4(4), 521–560 (1969).

Index of Corollaries, Lemmas, Propositions, and Theories

Corollary 5.2, 194
Corollary 8.10, 305
Corollary 9.2, 342

Lemma 8.11: Square Root of One mod p, 312

Proposition 2.4: The Division Algorithm, 47
Proposition 2.5, 49
Proposition 2.7: Euclid's Lemma, 51
Proposition 2.8: Basic Properties of Congruences, 54
Proposition 2.9: Congruences and Remainders, 55
Proposition 2.10: Validity of Congruent Substitutions in Modular Arithmetic, 56
Proposition 2.11: Inverses in $\mathbb{Z}_m$, 60
Proposition 2.12, 68
Proposition 2.14, 70
Proposition 3.1, 116
Proposition 3.2: General One-Unit Shift Permutations and Their Powers, 119
Proposition 3.3: General Properties of Enigma Machines, 126
Proposition 4.1: Some Properties of Matrix Arithmetic, 149
Proposition 4.5: Classic Adjoint Formula for the Inverse of an $n \times n$ Matrix, 159
Proposition 4.6: General Cofactor Expansion for Determinants of Square
             Matrices, 171
Proposition 5.1: Expected Values of Indices of Coincidence, 193
Proposition 7.1: Properties of the XOR Operator, 255
Proposition 8.3, 299
Proposition 8.5, 302
Proposition 8.6, 303
Proposition 8.9: Order of Powers Formula, 305
Proposition 8.12, 312
Contraposititve of Proposition 8.12, 313
Proposition 9.1, 341
Proposition 9.3, 351
Proposition 9.4: One-Way Functions for Knapsack Cryptosystems, 353
Proposition 10.1, 380
Proposition 10.3, 387
Proposition 10.6, 393
Proposition 12.1: Properties of Elliptic Curve Addition, 458
Proposition 12.2: Uniqueness of Square Roots Modulo a Prime, 461
Proposition A.2, 496
Proposition A.3, 498

Theorem 2.1, 44
Theorem 2.2: Fundamental Theorem of Arithmetic, 44
Theorem 2.3: Euclid, 46
Theorem 2.6, 50
Theorem 2.13: The Chinese Remainder Theorem, 69

Theorem 4.2: On the Invertibility of Square Matrices, 155
Theorem 4.3: On the Invertibility of 2×2 Matrices, 155
Theorem 4.4: On the Invertibility of Square Modular Integer Matrices, 157
Theorem 5.3: Expected Values of Indices of Coincidence, 196
Theorem 5.4, 205
Theorem 6.1: Base b Representations of Integers, 222
Theorem 8.1: Prime Number Theorem, 294
Theorem 8.2: Fermat's Little Theorem, 297
Theorem 8.4: Euler's Theorem, 300
Theorem 8.7: Existence and Number of Primitive Roots, 304
Theorem 8.8: Determination of Primitive Roots, 305
Theorem 8.13: Performance Guarantee for the Miller–Rabin Primality Test, 314
Theorem 9.5: Security Guarantee for the RSA Cryptosystem, 357
Theorem 10.2: Inventory of Finite Fields, 382
Theorem 10.4, 389
Theorem 10.5, 389
Theorem 10.7, 397
Theorem 12.3: Hasse's Theorem, 463
Theorem 12.4: Waterhouse's Theorem, 463
Theorem 12.5, 465
Theorem 12.6, 465
Theorem A.1: Some Set Theoretic Identities, 495
Theorem B.1: Probability Rules, 504
Theorem B.2: Linearity of Expectation, 512

Index of Algorithms

Algorithm 1.1: Generating Random Integers Using rand, 40

Algorithm 2.1: The Euclidean Algorithm, 49

Algorithm 2.2: The Extended Euclidean Algorithm, 62

Algorithm 2.3: (Procedures for Solving $ax \equiv c \pmod{m}$ in the Case $d = \gcd(a,m) > 1$ and $d \mid c$), 65

Algorithm 4.1: Cofactor Expansion Algorithm for Computing Determinants, 153

Algorithm 5.1: The Friedman Ciphertext-Only Attack on the Vigenère Cipher, 200

Algorithm 5.2: Generating Random Permutations of a Vector $[1, 2, 3, \ldots, k]$, 219

Algorithm 6.1: Construction of the Base b Expansion of a Positive Integer n, 222

Algorithm 6.2: Addition of Two Base b Integers, 230

Algorithm 6.3: Subtraction of Two Base b Integers, 232

Algorithm 6.4: Multiplication of Two Base b Integers, 235

Algorithm 6.5: Fast Modular Exponentiation, 239

Algorithm 7.1: Scaled-Down DES Encryption
    Part I: Outline, 259
    Part II: Details, 259

Algorithm 7.2: Scaled-Down DES Decryption, 264

Algorithm 7.3: Data Encryption Standard (DES)
    Part I: Outline, 265
    Part II: Details, 266

Algorithm 7.4: Cipher Feedback (CFB) Mode Encryption, 276

Algorithm 7.5: Cipher Feedback (CFB) Mode Decryption, 278

Algorithm 8.1: Gauss's Algorithm for Finding a Primitive Root Modula a Prime p, 307

Algorithm 8.2: Randomized Fermat Primality Test, 310

Algorithm 8.3: The Miller–Rabin Primality Test, 313

Algorithm 8.4: The Miller–Rabin Primality Test with a Factoring Enhancement, 315

Algorithm 8.5: Pollard's $p - 1$ Factorization Algorithm, 317

Algorithm 9.1: The Diffie–Hellman Key Exchange Protocol, 336

Algorithm 9.2: The RSA Public Key Cryptosystem, 340

Algorithm 9.3: A General Digital Signature for Public Key Cryptosystems, 343

Algorithm 9.4: The ElGamal Public Key Cryptosystem, 346

Algorithm 9.5: The ElGamal Digital Signature Protocol, 347

Algorithm 9.6: Solving a Knapsack Problem with Superincreasing Weights, 351

Algorithm 9.7: Merkle–Hellman Knapsack Cryptosystem, 354

Algorithm 9.8: Probabilistic Factoring Algorithm for an RSA Modulus n Given the Decryption Exponent d, 358

Algorithm 10.1: Division Algorithm in $\mathbb{Z}_p[X]$, 392

Algorithm 10.2: Brute-Force Irreducibility Test for Polynomials in $\mathbb{Z}_p[X]$, 394

Algorithm 10.3: Ben Or's Irreducibility Test, 410

Algorithm 11.1: Scaled-Down AES Encryption
    Part I: Outline, 421
    Part II: Details, 423

Algorithm 11.2: Scaled-Down AES Encryption
    Part I: Outline, 435
    Advanced Encryption Standard (AES) with 128 Bit Keys
    Part II: Details, 437

Algorithm 12.1: Addition of Points on Elliptical Curves over $\mathbb{R}$
Part I: Geometric Description of Algorithm via the Graph of the Elliptical Curve, 455
Part II: Algebraic Formulation, 457
Algorithm 12.2: Addition of Points on Elliptic Curves over $\mathbb{Z}_p$, 464
Algorithm 12.3: Fast Integer Multiple of Points on Elliptic Curves, 470
Algorithm 12.4: Koblitz's Algorithm for Plaintext Representations on an Elliptic Curve mod p, 472
Algorithm 12.5: Lenstra's Algorithm for Factorization Using Elliptic Curves, 476

Subject Index

A

Abelian groups, 458–460, 465
Absorption, set theoretic identities, 495
Active attacks; *See also specific attacks*
  basic concepts, 12
  on OFB mode, 285
Addition
  AES
    AES algorithm operations, 433
    computer program for, 449
    encryption, 423
    Galois fields, 413, 423
    nibble, 419, 445–446
  algorithm complexity analysis, assessing work required to execute, 246, 247
  elliptic curve, 454–458
    algebraic algorithm, 455–458
    computer implementations and exercises, 483–484, 485
    curves over $\mathbb{Z}_p$, 463–466
    geometric algorithm, 455, 456
  finite fields, 383, 384
    Galois fields, 398, 413, 423
    rings, 378–379, 381, 410
  matrix, 146–147, 149, 150, 175
  modular integer systems, 59
  polynomial; *See* Polynomials, addition
  vector, 457
Addition algorithm with base b expansions, 229–231
Additive groups, 459
Additive identity, 378, 383, 410
  elliptic curve addition, 458
  modular arithmetic, 59
Additive inverses, 379
Add Round Key operation, AES
  computer program for, 447
  decryption, 431
  encryption, 422, 424, 429, 437
  exercises, 444–445
  full (128 bit) AES, 438
ADFGVX cipher, 32–35, 102
Adjacent digits, 78, 80
Adjacent letters, 182, 183, 215
Adjacent pairs, 34, 571

Adleman, Leonard, 22, 331, 338, 339
Advanced encryption standard (AES) protocol, 21, 417–449
  byte representation and arithmetic, 432–434
  computer implementations and exercises, 445–449
  decryption algorithm, 439–440
  development of, 254, 417–419
  encryption algorithm, 437–438
  exercises, 441–445
  exercise solutions, 560–563, 609–611
  full (128 bit key) AES, 432, 437–438
  Galois fields, 399, 400
  nibbles, 419–421
  scaled-down version, 421–429
    computer programs for, 446–449
    decryption, 429–432
    encryption, 421–429
    encryption algorithm, 435–437, 438
  security of, 440
Affine ciphers
  computer programs for, 136, 137–140
    with homophones, 138–139
    with homophones and nulls, 139–140
    with nulls, 137–138
  evolution of codemaking, 96–100
  passive attacks on, 98–100
Affine function/mapping, 96
  AES, 444
  composition, 109–110
  digraph block cryptosystems, 132–133
  homophones, 105–106
  nulls, 103–104
Agrawal, Manindra, 309
AKS test, 309
Algebraic algorithm, elliptic curve addition, 455–458
Algorithm complexity analysis, assessing work required to execute, 246–247
Algorithms, defined, 3–4
Alice (literature convention), 2, 22, 23, 339, 340

Alphabets
  basic concepts, 3
  cryptosystem components, 94
  English, 13–14, 95
  monoalphabetic and polyalphabetic
      ciphers, 12–15
  number of characters, 95
  plaintext conversion to numerical
      equivalents, 225–228
American Standard Code for
      Information Interchange
      (ASCII), 227, 254, 340
Ancient codes, 91–94
Arab culture, cryptography in, 102
Arithmetic
  algorithm complexity analysis,
      assessing work required to
      execute, 246–247
  elliptic curve, 451
  integers in different bases
      addition algorithm with base b
          expansions, 229–231
      computer implementations and
          exercises, 248–250
      exercises, 241–247
      exercise solutions, 536–540
      large integers, 237–239
      multiplication algorithm
          with base b expansions,
          234–237
      subtraction algorithm with base
          b expansions, 231–234
  matrix, 175
      addition, subtraction, and scalar
          multiplication, 146–147
      multiplication, 147–151;
          See also Matrix
          multiplication
      properties of, 149–150
  modular integer systems, 59; See
      also Divisibility and modular
      arithmetic
  nibble
      addition and multiplication,
          419–420
      computer implementations and
          exercises, 445–446
ASCII, 227, 254, 340
Assissi, Benicio de, 102
Associativity
  abelian group, 459
  addition, 378, 383, 459
  elliptic curve, 458, 483
  matrix arithmetic, 149, 150
  multiplication, 379, 383
  rings, 379
  set theoretic identities, 495
Asymmetric key cryptography,
      21–22; See also Public key
      cryptography

Attacks on cryptosystems, 2; See also
      specific attacks
  affine ciphers, 98–100
  evolution of codebreaking; See
      Evolution of codebreaking
      till computer era
  overview, 12–15
Authentication
  basic concepts, 1
  features of public key
      cryptosystems, 25
  public key cryptography, 343–345
      digital signatures, 343–345
      ElGamal cryptosystem,
          347–349, 373
      RSA digital signatures, 371
Avalanche condition, strong, 419
Avalanche effect, 272, 290–291

B

Babbage, Charles, 187, 207
Babbage/Kasiski attack, 108
      computer programs to aid in,
          216–218
      Vignière cipher demise, 188–192
Bases; See also Integers in different
      bases
  elliptic curve discrete logarithm
      problem, 466
Bayes' formula, 510–511
Belaso, Giovanni Battista, 15
Ben-Or's irreducibility determination
      algorithm, 410–411, 414–415
Biased, probabilities, 502
Big-O notation, 247
Bijections
  finite fields, 382
  overview, 5–7
Binary alphabet, basic concepts, 3
Binary expansions
  AES algorithm operations, 434
  integers in different bases, 221,
      224–227
      addition algorithm with base b
          expansions, 231–234
      multiplication algorithm in base
          b expansions, 234–237
Binary operations
  abelian group, 458–460
  algorithm complexity analysis,
      assessing work required to
      execute, 246–247
  elliptic curve, 483
  finite fields, 377–378
  rings, 378–379, 381, 406–407
Binary strings
  basic concepts, 3
  plaintext conversion to numerical
      equivalents, 225–228

Binary vectors
    knapsack problem reformulation, 350
    nibble addition and multiplication, 419
    rings, 406–407
Binomial random variables, 511–513
Birthday problem, 505–507
Bit operations, work required to execute
        algorithm complexity
        analysis, 246–247
Bits, word size, 238
Bit strings, 238
    conversion programs, 286
    nibble addition, 420
    plaintext conversion to numerical
        equivalents, 225–228
Bitwise representation, AES algorithm,
        432, 433
Bletchley Park, 22, 202, 206–208, 252
Block ciphers, 20, 26
    evolution of codebreaking, 190
    Hill cryptosystem, 162–166; *See
        also* Hill cryptosystem
    Playfair cipher as, 18
Block cryptosystems, 132–133, 251–292
    computer implementations and
        exercises, 286–292
    DES, 265–272
        adoption of, 252–254
        fall of, 272–273
        scaled-down version, 258–265
        triple, 273–274
    evolution of computers into
        cryptosystems, 251–252
    exercises, 279–286
        public key cryptography,
            367–368
        solutions, 540–545, 599–601
    Feistel cryptosystems, 255–258
    modes of operation for, 274–279
        block mode, 274–276
        cipherblock chaining (CBC)
            mode, 275–276
        cipher feedback (CFB) mode,
            276–278
        electronic codebook (ECB)
            mode, 274–275
        output feedback (OFB) mode,
            278–279
    XOR operation, 254–255
Block matrix multiplication, 172–174
Block mode operations, block
        cryptosystem, 274–276
Block size
    AES
        versus Rijndael, 419
        scaled-down versus 128-bit key,
            421
    DES, scaled-down, 258
    Feistel cryptosystems, 255

Bob (literature convention), 2, 22, 339,
        340
Broadcast attack, RSA cryptosystem,
        366
Brute-force approach
    DES attacks, 273
    elliptic curve discrete logarithm
        problem, 467
    irreducibility test for polynomials in
        $\mathbb{Z}_p[X]$, 394, 395
    knapsack problem, 374
    modular inverses, 87
    passive attacks on substitution
        cipher, 13
    points on modular elliptic curve,
        452–456, 462
Byte, definition of, 276
Byte arithmetic/operations, AES,
        432–434
    AES algorithm operations, 432, 433
    computer program for, 449
    exercises, 443
    nibble-byte subtransformations,
        444–445
    sub transformation, encryption
        algorithm, 424, 436, 437,
        439

C

Caesar cipher, 9–11, 94
    evolution of codemaking, 101
    shift ciphers, 95
Cardinality, 495, 496
Carmichael, Robert, 311
Carmichael numbers, 311–312
Carries
    addition algorithm with base b
        expansions, 229, 230
    multiplication algorithm with base
        b expansion, 234, 235, 236,
        238
Cartesian product set, 496
Cauchy, Augustus, 382
CBC (cipherblock chaining) mode,
        275–276
Ceiling function, definition of, 48
Certification, primes, 309
CFB (cipher feedback) mode, 276–278
Chain matrix multiplication, 167–168
Champollion, Jean-Francois, 92
Change of base formula, 224
Chiffre indéchiffrable, le, 15, 108
Chinese remainder theorem, 67–71,
        359
    computer implementations and
        exercises, 89
    elliptic curve-based factoring
        algorithm, 476
    RSA cryptosystem, 341

Chor-Rivest cryptosystem, 356
Chosen ciphertext attacks, 12, 32
    affine ciphers, 99–100
    exercises, 133
    Hill cryptosystem, 164, 170
    RSA cryptosystem, 366
Chosen plaintext attacks, 12, 13, 99
    differential cryptanalysis, 272
    Hill cryptosystem, 164
    linear cryptanalysis, 273
Church, Alonzo, 207
Cipherblock chaining (CBC) mode,
        block cryptosystems,
        275–276
Cipher feedback (CFB) mode, block
        cryptosystems, 276–278
Ciphergram, computer program
        for extracting data from
        ciphertext string, 216–218
Ciphers
    ADFGVX, 32–35
    versus code, 91
    Playfair, 18–25
    programming with integer
        arithmetic, 38–39
    standards, 2
    substitution, 8–11
    terminology, 94
    Vignière, 15–18
Ciphertext
    basic concepts, 2
    partial substitutions, program for,
        215
    substitution ciphers, 8–11
Ciphertext attacks
    affine ciphers, 99–100
    chosen; See Chosen ciphertext
        attacks
    types of, 12
Ciphertext-only attacks, 12, 13
    affine ciphers, 98, 136
    frequency analysis-based, 186
    homophonic cryptosystems and,
        106–107
    on shift cipher, 38
    Vignière cipher, 200–201
Classical adjoint formula for matrix
        inversions, 159–162,
        171–172, 176
Clay foundation, 24
Closure, ring, 380
Cocks, Clifford, 22, 23
Code, versus cipher, 91
Code-book attacks, DES, 274–275
Codebreaking; See Decryption;
        Evolution of codebreaking
        till computer era
Codemaking; See Encryption;
        Evolution of codemaking till
        computer era

Coding theory
    congruency applications, 77–79
    Shannon's contributions to, 25
Codomain, basic concepts, 4, 5
Coefficient formula, polynomials, 387
Coefficients, 385, 390
Cofactor expansion algorithm, 153–154,
        157, 160, 171–172, 529
    classical adjoint formula with, 159
    computer platform caveat, 161
    computer programs, 176
Cogitata Physica-Mathematica
        (Mersenne), 81
Cohen, Henri, 314
Coincidence, index of, 193–201
Column index, AES encryption
        algorithm, 437–438
Column matrix, 146
Combinatorics, 495
Common modulus attack, RSA
        cryptosystem, 365
Commutation, composition of functions
        and, 110
Commutative rings, 58
Commutativity
    addition, 378, 458
    elliptic curve
        abelian group, 458–460
        addition, 458
        exercises, 483
    matrix arithmetic, 148–149, 150
    ring multiplication, 379
    set theoretic identities, 495
Complementarity probability rule, 504
Complementary keys, DES, 284
Complementary plaintext, DES, 284
Complement bit strings, exercise,
        279–280
Complements, set, 494, 495, 499
Complexity analysis of algorithms,
        assessing work required to
        execute, 246–247
Complexity of polynomials, RSA
        security guarantees, 357
Complex roots, elliptic curves over real
        numbers, 453
Composite integers, defined, 44
Compositeness
    Carmichael numbers, 311–312
    Miller–Rabin test, 314
    witness to, 309–310
Composite numbers, Lenstra's
        algorithm application, 482
Composition of functions, 332–333
    dissection of Enigma machine into
        permutations, 119–120
    evolution of codemaking, 109–110
    inverse of, 429
    permutations, computer program
        for, 141

repeated, 117
scaled-down Enigma machine, 120–121
triple, 122
Computational number theory, 309
Computation issues
algorithm complexity analysis, assessing work required to execute, 246–247
floating point platform limitations, 85, 87, 161, 237, 240, 296, 314, 317–318, 325, 369, 483
elliptic curve operations, 483
Lenstra's algorithm, 477
RSA cryptosystem, 341
Moore's law, 440
public key cryptography, 334
vector representation of polynomials, 387–388
Computation of orders, 303
Computer-generated random numbers, 40, 41
Computer implementations and exercises
AES, 445–449
block cryptosystems, 286–292
codebreaking evolution, 214–220
Babbage/Kasiski attack, programs to aid in, 216–218
frequency analysis, programs to aid in, 214–215
Friedman attack, programs related to, 218–220
index of coincidence, 218
codemaking evolution, 136–143
cofactor expansion method, 89, 161, 176
DES, 287–292
division algorithm, 86
elliptic curve cryptography, 483–487
modular elliptic curves, 484, 485
nonsingular elliptic curve, 483–484, 485
fast modular exponentiation, 240
Feistel cryptosystems, 287
finite fields, 411–415
Hill cryptosystem, 177–178
integers in different bases, 224, 248–250
matrices and Hill cryptosystem, 174–179
fast matrix multiplication, 179
modular matrices, 175–177, 178–179
scalar multiplication, 175
square (invertible) matrix, 175–176
Strassen's algorithm, 179

modular arithmetic, 85–89
Chinese remainder theorem, 89
congruences, 88
Euclidean algorithm, 86–88
prime factorialization, 85–86
number theory and algorithms, 325–329
overview, 35–41
computer-generated random numbers, 39–41
integer/text conversions, 36–37
programming basic ciphers with integer arithmetic, 38–39
vector/string conversions, 35–36
public key cryptography, 369–375
random substitution ciphers, 220
readings in, 616
three-round Feistel systems, 287
XOR program, 287
Concatenation, 7
Conditional probability, 507–509
Conditioning, 195, 509–511
Confederate cipher disk, 11
Confusion
one-time pad, 25–26
Shannon's properties of, 272, 419
Congruence classes, 54–55
Congruences
addition of elliptic curves over $\mathbb{Z}_p$, 464
basic properties, 54
Chinese remainder theorem, 67–71
computer implementations and exercises, 88
congruent mod m, 53
divisibility and modular arithmetic, 52–58
exercises
credit card error detecting codes, 79–80
divisibility criteria, 82–83
ISBN error detecting codes, 77–79
round robin tournaments, 80
modular elliptic curves, 461–462
solving, 61, 64–66
validity of congruent substitutions in modular arithmetic, 56–57
in $\mathbb{Z}_p[X]$ modulo, 395–396
Conjugates, of permutation, 123
Constant polynomial, 385
Continuous infinite sets, 4
Contrapositive, Fermat's little theorem, 309
Convergence, Gauss's primitive root finding algorithm, 325
Conversions
integer/text, 36–37
vector/string and string/vector, 35–36, 286

Coppersmith, Don, 150, 272
Correspondence, English alphabet, 95
Counter mode of operation, block
    cryptosystems, 285–286
Counting principles, 495–499
Credit card error-detecting codes,
    79–80, 89
Cryptanalysis
    basic concepts, 3
    linear and differential, 272–273
Cryptography, 1–2
Cryptosystems
    basic concepts, 1–2
    block; See Block cryptosystems
    formal definition, 94–96
Cycle decomposition form invariance,
    205–206
Cyclic permutations/cycles, evolution of
    codemaking, 114–119

D

Daemen, Joan, 418
Data encryption standard (DES),
    20–21, 265–272
    adoption of, 252–254
    AES development, 417–419
    computer programs, 287–292
    exercises, 282–283
    fall of, 272–273
    public key cryptography, 333
    scaled-down version, 258–265
    self-decryption proof, 285
    triple, 273–274
Decimal expansion, integers in
    different bases, 222
Decomposition, disjoint cycle, 115–116,
    117, 124
Decryption; See also specific systems
    basic concepts, 2–3
    codebreaking; See Evolution of
        codebreaking till computer
        era
    Playfair and Vignière ciphers, 39
Decryption algorithm
    AES, 439–440
    self-decryption proof, three-round
        Feistel systems, 285
Decryption exponent, 551, 552, 605,
    606
    ElGamal cryptosystem, 345–346,
        347
    RSA cryptosystem
        computer programs for, 370,
            371, 372
        probabilistic factoring
            algorithms for RSA modulus,
            358
        public key, 340–341, 342
        security guarantee, 357

Decryption functions
    cryptosystem components, 94
    substitution ciphers, English
        alphabet, 96
Definitions of basic concepts, 1–4
De Morgan's Laws, 495
Density, primes, 308
Dependent events, 508
DES; See Data encryption standard
DES algorithm, 262, 264, 265, 267
    computer programs for, 290
    scaled-down DES, 258–259
Descartes, René, 295
Determinant, square (invertible)
    matrix, 153–155
Differential cryptanalysis, 272, 273
Diffie, Whit, 21, 22, 331, 333
Diffie–Hellman key exchange, 21, 22,
    331, 346
    computer program for, 369–370
    discrete logarithms, 334
    elliptic curve version, 467–468, 474
        computer implementations and
            exercises, 486
        exercises, 481
    exercises, 360–361, 366–367
    with groups, 459
    public key cryptography, 336–337
Diffusion
    one-time pad, 26
    Shannon's properties of, 272, 419
Digital signatures and authentication,
    25
    ElGamal cryptosystem, 347–349,
        373
    public key cryptography, 343–345
    RSA cryptosystem, 340, 370–371
Digital Signature Standard (DSS), 345
Digraphs, 107, 132–133
Dimensions, matrix, 145
Direct method, modular exponentiation,
    247
Discrete infinite sets, 4
Discrete logarithm problem, 303, 306
    exercises, 367
    on modular elliptic curves, 466–467
    modular elliptic curves, 480
    public key cryptosystems, 338
    review of, 334–335
Discrete random variable, defined, 511
Discriminant, elliptic curve, 452
Disjoint cases, multiplication principle,
    498
Disjoint ciphertext character sets, 189
Disjoint cycle decomposition, 115–116,
    124, 205–206
Disjoint probabilities, addition to
    Kolmogorov's axiom, 510
Disjoint sets, 492
Disjoint union, sets, 509–510

Distinct primes, square root modulo
        m, 84
Distributive laws
    finite fields, 384
        division algorithm, 392
        polynomial multiplication, 387,
            388
        rings, 379, 380, 384
    matrix arithmetic, 149, 171
    multiplication algorithm in base b
            expansions, 236
    Venn diagrams, 494
Distributivity, set theoretic identities,
        495
Dividend
    definition of, 47
    division algorithm for $\mathbb{Z}_p[X]$, 391,
        392
Divisibility and modular arithmetic,
        43–89
    Chinese remainder theorem, 67–71
    divisibility definition and examples,
        43–44
    division algorithm, 47–48
    Euclidean algorithm, 48–52
    exercises, 71–85
    exercise solutions, 517–522,
        572–581
    extended Euclidean algorithm,
        61–64
    greatest common divisors and
            relatively prime integers,
            46–47
    modular arithmetic and
            congruences, 52–58
    modular integer systems, 58–60
    modular inverses, 60–61
    primes, 44–46
    solving linear congruences, 64–66
Divisibility criteria, application of
        congruences, 82–83
Division, polynomial; *See* Polynomials,
        division
Division algorithm, 519, 556–557, 563,
        584
    AES, 421, 434, 445
    computer implementations and
            exercises, 86
    congruences, 55–56
    conversions among bases and
            integer equivalents, 223
        addition algorithm with base b
            expansions, 229
        subtraction algorithm with base
            b expansions, 232
    Euclidean algorithm and, 48–50
    extended, 158
    Fermat's little theorem, 297, 298,
        546
    matrix arithmetic, 158

modular arithmetic, 47–48, 87, 547
    computer programs for, 80, 86
    congruences and remainders, 55,
        56, 80
    Euclidean algorithm, 49, 50–51,
        64
    exercises, 72
    nibbles, 421, 445
    polynomial, 391–395, 421, 434
        computer programs for, 412
        Euclidean algorithm, 404, 405
        exercises, 407, 408, 411
Divisor
    definition of, 47
    division algorithm for $\mathbb{Z}_p[X]$, 391,
        392
Domain, basic concepts, 4, 5
Dominance laws, set theoretic
        identities, 495
Dot product
    horizontal shifted, computer
            program for, 218
    matrix operations, 146, 148
    vectors, 199–200
Double complementation, set theoretic
        identities, 495
Double DES, 273

E

ECB (electronic codebook) mode,
        274–275
Eckert, J. Presper, 252
Egyptian hieroglyphics, 92, 93, 95
Electronic codebook (ECB) mode,
        block cryptosystems,
        274–275
Electronic Numerical Integrator and
        Calculator (ENIAC), 252
Elements
    matrix, 145
    sets, 491
Elements, The (Euclid), 45–46, 503
Elgamal, Taher, 345
ElGamal cryptosystem, 345–347
    computer programs, 372–373
    digital signatures with, 347–349
    discrete logarithms, 334
    elliptic curve addition, 466
    elliptic curve version, 481
        computer implementations and
            exercises, 486
        plaintext representation,
            471–473
        procedure, 473–475
    exercises, 363–364, 366–367
    with groups, 459
    mathematical problems providing
            security, 338
    modular exponentiation, 301

Elliptic curve cryptography, 25, 451–487
  addition of elliptic curves over $\mathbb{Z}_p$, 463–466
  addition operation for, 454–458
  computer implementations and exercises, 483–487
  Diffie–Hellman key exchange version, 467–468
  ElGamal cryptosystem version, 473–475
  elliptic curves over finite fields, 463
  elliptic curves over real numbers, 452–454
  elliptic curves over $\mathbb{Z}_p$, 460–462
  exercises, 477–483
  exercise solutions, 563–567, 611–613
  factoring algorithm based on, 475–477
  groups, 458–462
  modular
    discrete logarithm problem on, 466–467
    fast integer multiplication of points on, 470–471
    plaintext representation on, 471–473
    sizes of, 462–463
  readings in, 616
  selections for further reading, 616
Ellis, James, 22–23
Empty sets, 493
Empty strings, 3, 7
Encryption; *See also specific systems*
  basic concepts, 2–3
  codemaking evolution; *See Evolution of codemaking till computer era*
  cryptosystem components, 94
Encryption algorithm, AES, 435–439
  128 bit keys, 437–439
  scaled-down, 435–437
Encryption exercises, block cryptosystems, 282–283
Encryption key, basic concepts, 2
Encryption mapping, two-round, 541–543
Encryption programs
  AES, scaled-down, 421–425
  DES, scaled-down, 288
  public key cryptography
    ElGamal cryptosystem, 372–373
    Merkle–Hellman knapsack cryptosystem, 374–375
  RSA cryptosystem, 370
  three-round Feistel systems, 287
English alphabet, 13–14, 95

ENIAC (Electronic Numerical Integrator and Calculator), 252
Enigma machines
  attack methods, 201–205
    German usage protocols, 202–203
    Polish codebreakers, 203, 204
    Rejewski's attack, 203–205
  evolution of codemaking, 111–114
    computer programs, 141–143
    dissection into permutations, 119–126
    scaled-down, 120–121
    special properties of, 126–127
Entropy, 21
Entry, matrix, 145
Equal difference property, 444–445
Equality, polynomials in $\mathbb{Z}_p[X]$, 385
Equivalence relations, 54
Error-detecting codes
  credit card, 79–80, 89
  ISBN, 77–79, 88–89
Error propagation, block cryptosystems, 285
Euclid, 45, 503
Euclidean algorithm
  computer implementations and exercises, 86–88
  divisibility and modular arithmetic, 48–52
  extended, 61–64, 347, 552
    addition of elliptic curves over $\mathbb{Z}_p$, 464
    computer implementations and exercises, 414
    polynomials, 404
    RSA cryptosystem, 342
  polynomials, 404–405, 408–409, 414
  RSA security guarantees, 360
Euclid's lemma, 51, 312, 461
Euler, Leonhard, 298, 299
Euler's little theorem, 297–298
Euler's phi function, 298–299, 303, 320, 326
Euler's theorem, 300–301, 302, 359
  exercises, 320
  proof of, 546–547
Eve (literature convention), 2, 23
Event, sample space subset, 502
Evolution of codebreaking till computer era, 181–200
  computer implementations and exercises, 214–220
    Babbage/Kasiski attack, programs to aid in, 216–218
    frequency analysis, programs to aid in, 214–215
    Friedman attack, programs related to, 218–220

Enigmas, attack methods, 201–205
    German usage protocols,
        202–203
        Polish codebreakers, 203, 204
        Rejewski's attack, 203–205
    exercises, 208–214
    exercise solutions, 530–536,
        592–595
    frequency analysis attacks, 181–186
    index of coincidence, 193–201
    invariance of cycle decomposition
        form, 205–208
    Turing and Bletchley Park,
        206–208
    Vignière cipher demise, 187–192
        Babbage/Kasiski attack,
            188–192
        Friedman attack, 192
Evolution of codemaking till computer
        era, 91–143
    affine ciphers, 96–100
    ancient codes, 91–94
    composition of functions, 109–110
    computer implementations and
        exercises, 136–143
    cyclic permutations/cycles, 114–119
    enigma machines, 111–114
        dissection into permutations,
            119–126
        special properties of, 126–127
    exercises, 127–136
    exercise solutions, 522–526,
        581–587
    formal definition of cryptosystem,
        94–96
    homophones, 105–109
    nulls, 102–105
    permutations
        computer representations of,
            140–143
        cyclic, 114–119
        enigma machine dissection into,
            119–126
        tabular form notation for,
            110–111
    steganography, 100–102
    tabular form notation for
        permutations, 110–111
Exercise solutions, 451–487, 515–567
Expansion function, DES, 266, 267
Expansions
    DES, 261, 269
    integers in different bases, 221, 222,
        223, 224–227
        addition algorithm with base b
            expansions, 229–231
        multiplication algorithm with
            base b expansions, 234–237
        subtraction algorithm with base
            b expansions, 231–234

Expected value, binomial random
        variable, 512–513
Experiment, defined, 501
Exponentiation
    algorithm complexity analysis,
        assessing work required to
        execute, 247
    discrete logarithms, 334, 335
    fast modular, 239–240, 545–546
    squaring algorithm for, 250
Exponents
    decryption; See Decryption
        exponent
    magic, Fermat's little theorem, 297,
        298, 300
    modular exponentiation; See Fast
        modular exponentiation;
        Modular exponentiation
    RSA cryptosystem, 340, 341, 342
    signature, ElGamal cryptosystem,
        347
Extended Euclidean algorithm, 88,
        347, 552
    addition of elliptic curves over $\mathbb{Z}_p$,
        464
    divisibility and modular arithmetic,
        61–64
    polynomials, 404

F

Factorization, prime; See Prime
        factorization
Factorials, 13
Factoring
    elliptic curve arithmetic-based, 482
    elliptic curve cryptography-based
        algorithm, 451, 475–477
    Miller–Rabin test with factoring
        enhancement, 315–316,
        328–329
    Pollard p-1 factoring algorithm,
        316–319
    public key cryptosystems
        computer implementations and
            exercises, 371–372
        elementary factoring method,
            368
        one-way functions, 333
        RSA security guarantees, 358
        spread 331–368, 338
Factoring problem, 309
Factorization
    fundamental theorem of arithmetic,
        44
    primes, 44, 45, 85–86, 357, 358
    RSA cryptosystem, 342
    RSA security guarantees, 342, 357,
        358
Factors, divisibility, 43, 389

Fair, probability concepts, 502
Fast integer multiplication of points,
        elliptic curve, 470–471,
        485–486
Fast matrix multiplication, 150, 179
Fast modular exponentiation, 239–240,
        296–297, 545–546
    Diffie–Hellman key exchange,
        elliptic curve protocol, 469
    discrete logarithms, 335
    Koblitz's algorithm, 472, 473
Feedback modes, block cryptosystems
    cipher feedback (CFB) mode,
        276–278
    output feedback (OFB) mode,
        278–279
Feistel, Horst, 253
Feistel cryptosystems, 253, 255–258,
        259, 260, 263, 264, 440,
        542–543
    computer implementations and
        exercises, 287
    DES, 265
    exercises, 280–281
    self-decryption proof, 285
Fermat, Pierre de, 295, 296
Fermat's little theorem, 295–298, 546
    exercises, 319, 320
    Pollard p-1 factoring algorithm
        basis, 317
Fermat's primality test, 309–311
    computer programs for, 328
    exercises, 323
Feynman, Richard, 357
Field isomorphism, 382
Finite fields, 377–415
    AES; See Advanced encryption
        standard protocol
    binary operations, 377–378
    building from $\mathbb{Z}_p[X]$, 396–399
    computer implementations and
        exercises, 411–415
    definition of, 381
    elliptic curves over, 463
    exercises, 406–411
    exercise solutions, 554–560,
        608–609
    fields, 381–384
        addition and multiplication
            tables, 384
        definition of, 381
        inventory of, 382
    Galois fields, 382, 399–403
    polynomials
        Euclidean algorithm for,
            404–406
        vector representation of,
            387–388

polynomials in $\mathbb{Z}_p[X]$
    addition and multiplication of,
        386–387
    congruences in modulo as fixed
        polynomial, 395–396
    divisibility in, 389–390
    division algorithm for, 391–395
    as ring, 388–389
    polynomials with coefficients in
        $\mathbb{Z}_p$, 385
    rings, 378–380
Finite sets, 4, 452, 491
Finite strings, 7
First on, first off, 333
Fixed elements, cyclic permutation,
        115
Floating point platform limitations,
        240, 325; See also
        Computation issues
Floor function, 40, 47–48
Flowers, Tommy, 252
FORTRAN, 252
Frequency analysis
    computer program for modular
        frequency counts, 216
    computer programs to aid in,
        214–215
Frequency analysis attacks
    evolution of codebreaking, 181–186
    homophonic cryptosystems and,
        106–107
    Vignière cipher, 189–190
Frequency vector, Friedman attack,
        199
Friedman, William F., 188
Friedman attack, 197–201
    computer programs related to,
        218–220
    index of coincidence, 194
    Vignière cipher demise, 192
Functions; See also Mapping
    basic concepts, 3
    composition of, evolution of
        codemaking, 109–110
    cryptosystem components, 94
    overview, 4–8
        inverse, 7–8
        one-to-one and onto, bijections,
            5–7
    substitution ciphers, 8–11
Fundamental theorem of algebra, 453
Fundamental theorem of arithmetic, 44,
        46, 51–52

G

Gadsby (Wright), 14
Galois, Evariste, 382, 383

Galois fields, 254, 382, 399–403, 404
    AES; *See also* Advanced encryption
        standard protocol
        AES algorithm operations, 432,
            433
        encryption, 423–424, 432
        Mix Column mapping, 430
        nibble addition and
            multiplication, 419, 420
    building finite fields from $\mathbb{Z}_p[X]$,
        396–399
    computer programs for
        addition/multiplication, 413
    computation of inverses, 414
Gauss, Carl Friedrich, 52–53, 382
Gaussian elimination, 159
Gauss's algorithm
    computer program for, 326
    exercises, 322, 325
    primitive roots, 307–308
General substitution cipher, known
        plaintext attack, 13
Geometric algorithm, elliptic curve
        addition, 455, 456
German usage protocols for Enigmas,
        202–203
Government Communications
        Headquarters (GCHQ), 22,
        23
Governments, 3, 356–357
Gram, 190
Graphs, elliptic curve, 453, 454, 455
Greatest common divisors and
        relatively prime integers,
        46–47
Great Internet Mersenne Prime Search
        (GIMPS), 82
Groups
    DES, 273
    elliptic curve cryptography,
        458–462
Group theory, 459–460

H

Hackers, 2
Hadamard, Jacques, 294
Hardy, Godfrey, 294
Hasse's Theorem, 463, 468
Hawaiian alphabet, 210
Hellman, Martin, 21, 22, 273, 331, 333,
        352, 353
Hexadecimal form
    AES algorithm operations, 432,
        433, 434
    DES, 282
        computer programs, 290
        decryption program, 291

Galois field computations, 400, 401,
        402, 403, 408
integers in different bases, 221,
        224–227
        addition algorithm with base b
            expansions, 231–234
        multiplication algorithm in base
            b expansions, 234–237
    nibble operations, 420
Hieroglyphics, 92, 93, 95
Hill, Lester, 162
Hill cryptosystem, 162–166, 169
    computer programs, 177–178
    exercises, 169–171
Hindu puzzle, 67–71
History of cryptography
    ADFGVX cipher, 33–34
    Caesar cipher, 9–11
    codebreaking; *See* Evolution of
        codebreaking till computer
        era
    codemaking; *See* Evolution of
        codemaking till computer era
    communications technology,
        108–109
    Mersenne primes, 81–82
    one-time pad, 25–28
    public key cryptography, 21–25
    readings in, 615
    selections for further reading, 615
Homophones, 523–524, 593–594
    affine ciphers with, 138–140
    evolution of codemaking, 105–109
    randomized encryption system,
        106–107
Horizontal shifted dot products, 218
Horizontal shifted match counts, 218

I

IBM, 252, 418, 419
Identity
    abelian group, 459
    additive, 379, 383, 410
        elliptic curve addition, 458
    multiplicative, 379, 380, 383, 410
    polynomial, 390
Identity function, 110, 123
Identity matrix, 151, 152
Identity permutation, 96
Image, basic concepts, 4
Inclusion Exclusion principle,
        probability rules, 504
Independent events, 508
Indeterminate X, 385
Index of coincidence, 193–201, 218
Indian culture, cryptography in, 102
Industrial-grade primes, 314, 372

Infinite sets, 4, 491
Infinity
    elliptic curves over modular
        integers, 460, 462
    elliptic curves over real numbers, 452
Initial permutation, DES, 265, 266,
    270, 271
    computer program for, 289
    inverse, 264, 265, 289
    scaled-down, 259, 260, 263, 264
Input set, basic concepts, 5
Institute of Electrical and Electronics
    Engineers (IEEE), 238
Integer arithmetic, overview, 38–39
Integers
    alphabets, 95–96
    divisibility and modular arithmetic;
        See Divisibility and modular
        arithmetic
    floor function, 40
    modular orders of invertible
        modular integers, 301–302
    number theory, 43
Integers in different bases, 221–250
    arithmetic with large integers,
        237–239
    computer implementations and
        exercises, 248–250
    exercises, 241–247
    exercise solutions, 536–540,
        595–599
    fast modular exponentiation,
        239–240
    hexadecimal and binary expansions,
        224–227
        addition algorithm with base b
            expansions, 231–234
        multiplication algorithm in base
            b expansions, 234–237
    representation of, 221–224
Integer size
    RSA cryptosystem, 341
    symbolic versus floating point
        systems, 240, 314
Integers modulo m, 58
Integer systems
    modular, 58–60
    relatively prime integers, 46–47
Integer/text conversions, 36–37
Integral domains, 409–410
Integrity, basic concepts, 1
Intersection, sets, 492–495
Invariance of cycle decomposition
    form, 205–208
Inverse functions
    overview, 7–8
    S-box, 430
    shift permutation, 10
    substitution ciphers, English
        alphabet, 96

Inverse permutation
    computer program for, 141
    cycle, 116
    DES, 289
    substitution ciphers, English
        alphabet, 96
Inverse problem, 24
Inverses/inversion/invertibility
    abelian group, 459
    AES
        computer programs for, 448
        S-box, 444, 448
    composition of functions, 332–333,
        429
    elliptic curve addition, 458
    finite fields
        Galois fields, 414
        polynomial Euclidean algorithm
            for determination of,
            408–409
        rings, 379–380, 407
    matrices, 176–177, 430
        classical adjoint for, 159–162
        computer implementations and
            exercises, 174–176, 178–179
        definition of, 151–153
        definition of invertible matrix,
            151–152
        determinant of, 153–155
        Hill cryptosystem, 162–166
        square (invertible), 155–156
        square modular integer,
            157–158
    modular, 60–61
        brute-force approach, 87
        extended Euclidean algorithm,
            88
        modular orders of invertible
            modular integers, 301–302
    notation for, 332
Invertible affine mapping, AES S-box
    description, 444
Inv Mix Column, 440
Inv Nibble Sub mapping, 430, 431,
    439–440
Inv Shift Row, 440
Irreducible polynomials; See
    Polynomials, irreducible/
    irreducibility
ISBN error detecting codes, 77–79,
    88–89
Isomorphism, field, 382

J

Jacobi, Carl Gustav, 382
Japan, Enigma machine, 112
Jefferson, Thomas, 107–108
j-fold composition, 117
j-unit shift, 123

K

Kasiski, Friederich W., 187
Kayal, Neeraj, 309
k-cycle, 116
Keyboard, Enigma machine elements,
        112, 113, 121
Key exchange
    Diffie–Hellman, 336–337
    secure, quest for, 332–333
Key exchange protocols, 331
Key extraction permutation, DES, 261
Key generation matrix, AES
        encryption, 425, 426
Key κ, AES encryption, 424
Keylength
    AES, 417, 419
    DES cryptosystem, 253
    one-time pad, 27, 40
    Vignière cipher, 189, 190, 191, 198,
        572
        Babbage/Kasiski attack, 216
        Friedman attack, 201
Key permutation, English alphabet
        substitution cyphers, 96
Keys
    basic concepts, 2
    cryptosystem components, 94
    one-time pad, program for creating,
        40
    private key cryptosystems, 21
    public key cryptography, 23
    substitution ciphers, 9
Key schedule, Feistel cryptosystems, 255
Key search, Moore's law, 440
Key size
    AES, 417, 421, 432
    DES, 254
        scaled-down, 258
        triple, 273–274
Keyspace
    DES, 265
    Diffie–Hellman key exchange, 336,
        337
    RSA cryptosystem, 340
Knapsack problems/cryptosystems,
        349–352
    computer programs for, 374–375
    mathematical problems providing
        security, 338
    Merkle–Hellman, 352–356
    public key cryptosystems, 338
Known plaintext attacks, 12, 13, 32,
        132, 583
    AES Nibble/Byte Sub
        Transformations and, 445
    affine ciphers, 98–99
    ElGamal cryptosystem, 367
    Hill cipher, 177–178
Koblitz, Neal, 158, 451

Koblitz's algorithm, 472, 473, 481, 486
Kolmogorov, Andrey, 503
Kolmogorov axioms, 503–504, 505, 510
Kolmogorov probability functions, 507
Kronecker delta, 192

L

Lampboard, Enigma machine elements,
        112, 113
Large integers, arithmetic with, 24,
        237–239
Leading term, polynomials in $\mathbb{Z}_p[X]$,
        387
Lenstra, Hendrik, 451
Lenstra's algorithm, 476–477, 482, 487
Letter frequency, English alphabet,
        13–14, 107
Linear congruences
    Chinese remainder theorem, 67–71
    solving, 64–66
Linear cryptanalysis, 272, 273
Linguistic properties of language, and
        frequency-based attacks, 182
Logarithms, discrete, 334–335
Lorenz cipher, 252
Lorenz encryption machines, 252
Lucifer system, 95

M

Magic exponent, Fermat's little
        theorem, 297, 298, 300
Mallory (literature convention), 2, 23
Mapping, 4; *See also* Functions
    AES
        decryption, 429–432
        encryption, 422–423, 424,
            436–437, 439–440
        affine function; *See* Affine function/
            mapping
        two-round, 541–543
MARS, 418
Match counts, horizontal shifted, 218
Mathematical description, AES S-box,
        443–444
Mathematical foundations of
        cryptography, 2, 3
    readings in, 615–616
    selections for further reading,
        615–616
Matrices, 145–179
    AES, 424–425, 437, 444
    anatomy of matrix, 145–146
    arithmetic operations, 149–151
        addition, subtraction, and scalar
            multiplication, 146–147, 149,
            150, 175
        multiplication; *See also* Matrix
            multiplication

classical adjoint for matrix
   inversions, 159–162
computer implementations and
   exercises, 174–179
definition of, 147–148
exercises, 166–174
exercise solutions, 526–530,
   587–592
Hill cryptosystem, 162–166
modular integer systems, 156–158,
   161
multiplication, 147–149
nibble, 424–425, 427
   exercises, 441
   scaled-down AES encryption,
      422
noncommutative ring, 379
noncommutativity of, 148–149
square (invertible) matrix
   definition of, 151–153
   determinant of, 153–155
   inverses of 2x2 matrices,
      155–156
transpose of matrix, 156
Matrix distributive law, 171
Matrix multiplication, 147–149
   AES decryption, 440
   AES encryption, 423
   associativity property, 149
   block, 172–173
   chain, 167–168
   computer implementations and
      exercises, 179
   definition of, 147–148
   Mix Column mapping, 430
   nibble, 441
   noncommutativity of, 148–149
   ring axioms and, 379
   scalar; See Scalar multiplication
   Strassen's algorithm, 173–174
Matsui, Mitsuru, 273
Mauchly, John W., 252
Members, set, 491
Menezes, Alfred, 273, 616, 617
Merkle, Ralph, 22, 273, 331, 352, 353
Merkle–Hellman knapsack
      cryptosystem, 352–356
   computer program for, 374–375
   exercises, 364–365
Mersenne, Marin, 81
Mersenne primes, 342
Microdots, 100
Miller, Gary, 312
Miller, Victor, 451
Miller–Rabin test, 312–314
   computer program for, 327–329
   exercises, 323
   with factoring enhancement,
      315–316, 323
Minoan script, 93, 94

Mix Column Transformation, AES, 440
   computer programs for, 447, 448
   decryption, 431, 440, 448
   encryption, 422–423, 424, 428,
      436–437
   exercises, 444–445
   inverse, 431, 440
Modes of operation, block
      cryptosystems, 274–279, 285
mod function, computer, 57, 86–87, 161
Mod n primitive roots, exercises,
      321–322
Modular arithmetic; *See also*
      Divisibility and modular
      arithmetic
   AES algorithm operations, 433, 434
   Chinese remainder theorem, 67–71
   computer implementations and
      exercises, 175–179
   and congruences, 52–58
   elliptic curve-based factoring
      algorithm, 476
   exercises, 321
   integer systems, 58–60
   inverses, 60–61
   matrix, 175
   Mix Column mapping, 430
   public key cryptography; *See* Public
      key cryptography
   solving linear congruences, 64–66
   square root modulo m, 83–84
Modular elliptic curves
   addition of elliptic curves over $\mathbb{Z}_p$,
      463–466
   computer implementations and
      exercises, 484, 485
   Diffie–Hellman key exchange,
      467–470
   discrete logarithm problem on,
      466–467
   exercises, 478, 479, 480, 481,
      482–483
   fast integer multiplication of points
      on, 470–471
   plaintext representation on, 471–473
   properties of, 460–462
   sizes of, 462–463
Modular exponentiation
   algorithm complexity analysis,
      assessing work required to
      execute, 247
   discrete logarithms, 334, 335
   Euler's theorem, 300–301
   exercises, 319, 320
   fast, 239–240, 296–297, 545–546;
      See also Fast modular
      exponentiation
   squaring algorithm for, 250
Modular frequency counts, computer
      program for, 216

Modular integer matrices, 156–158
    computer implementations and
            exercises, 175–177, 178–179
        addition and scalar
            multiplication, 175
        determinant of, computing using
            cofactor expansion, 176
    invertibility, 157–158, 161, 175–176,
            178–179
Modular integers
    alphabets, 95–96
    elliptic curve-based factoring
            algorithm, 476
    elliptic curves over, 459, 460–462
    invertible, modular orders of,
            301–302
    rings, 379
Modular inverses
    brute-force approach, 87
    Hill cryptosystem decryption, 164
Modular orders of invertible modular
            integers, 301–302
Modular polynomials, 402–403, 406,
            411, 443
Modular powers, 321
Modulus attacks, RSA cryptosystem,
            342, 365
Monoalphabetic ciphers, passive attacks
            on substitution cipher, 12–15
Monotonicity, probability rules, 504
Moore, Gordon, 356
Moore's law, 356–357, 440
Multiples, divisibility, 43, 389
Multiple solutions, knapsack problems,
            349–350
Multiplication
    AES algorithm operations, 433,
            434
    algorithm complexity analysis,
            assessing work required to
            execute, 246, 247
    algorithm with base b expansions,
            234–237
    counting principles, 495–499
    fast integer multiplication of points
            on modular elliptic curves,
            470–471
    fields, 383
    finite fields, 384
    Galois fields, 399, 400, 401, 402
        AES encryption, scaled-down
            version, 423
        AES security, 417
        computer program for, 413
    matrix; See Matrix multiplication
    modular integer systems, 59
    mutativity of, 380
    nibble, 419, 446
    polynomials; See Polynomials,
            multiplication

rings, 378, 380, 381, 406–407, 410
    scalar; See Scalar multiplication
    vector, polynomials in $\mathbb{Z}_p[X]$,
            388
Multiplication principle, 13, 495–499
Multiplication rule, 509
Multiplicative functions, exercises,
            324–325
Multiplicative groups, 459
Multiplicative identity, 379, 383, 410
Multiplicative inverse, rings, 379–380
Mutativity of multiplication, and
            distributive law, 380
Mutually exclusive events, 503,
            509–510
Mutually exclusive (disjoint) sets, 493

N

Nagell, Tryqve, 294
National Bureau of Standards (NB),
            253
National Institute of Standards and
            Technology (NIST), 251,
            253, 254, 345, 417, 418
National Security Agency (NSA), 3,
            252–253, 357
Native American languages, 93–94
Navajo speakers in WW II, 93
n-gram, 190
Nibbles, AES, 419–421
    computer implementations and
            exercises, 445–446
    encryption, 424–425, 427
    exercises, 441, 444–445
Nibble Sub mapping, inverse of, 430,
            439–440
Nibble Sub Transformation, AES
    computer programs for, 447
    decryption, 431
    encryption, 422, 424, 428
    exercises, 445
Nicolas, Jean Gustave, Baron de la
            Vallée Poussin, 294
Noncommutative ring, 379
Nonrepudiation, 25, 340
Nonsingular elliptic curve
    as abelian group under addition
            operation, 465
    computer implementations and
            exercises, 483–484, 485
    definition of, 452
    exercises, 478, 479, 480, 481, 483
    fast integer multiplication of points
            on, 470–471
    graphs, 453, 454
    over modular integers, 460–461
    over real numbers, 452
    Waterhouse's Theorem, 463
NP complete problems, 24, 350

Nulls
    affine ciphers with, computer
        programs, 137–138
    evolution of codemaking, 102–105
    homophones combined with, 107
Number of rounds
    DES, scaled-down, 258
    Feistel cryptosystems, 255
Numbers, matrix terminology, 146
Number systems, abelian group,
        458–460
Number theory and algorithms, 43,
        293–329
    Carmichael numbers, 311–312
    computer implementations and
        exercises, 325–329
    divisibility and modular arithmetic;
        See Divisibility and modular
        arithmetic
    Euler phi function, 298–299
    Euler's theorem, 300–301
    exercises, 319–325
    exercise solutions, 545–550,
        601–604
    Fermat's little theorem, 295–298
    Fermat's primality test, 309–311
    Miller–Rabin test, 312–316
        with factoring enhancement,
            315–316
    modular orders of invertible
        modular integers, 301–302
    order of powers formula, 305–308
    Pollard p-1 factoring algorithm,
        316–319
    prime number generation,
        308–309
    prime number theorem, 293–295
    primitive roots, 302–305
        determination of, 304–305
        existence of, 304

O

Object weights, knapsack problems,
        349, 350–352
    computer programs for, 374
    Merkle–Hellman knapsack
        cryptosystem, 352–356
Octal expansions, 225
OFB (output feedback) mode, 278–279
One, multiplicative identity in R, 379
One-time pad, 25–28, 40
One-to-one functions
    overview, 5–7
    substitution ciphers, 8–11
One-unit shift permutations, 119
One-way functions
    Merkle–Hellman knapsack
        cryptosystem, 353
    public key cryptography, 333–334

Onto functions
    overview, 5–7
    substitution ciphers, 8–11
Ordered lists, Cartesian product set,
        496
Ordered pairs, 20, 378, 564, 583
    binary operations, 377, 378
    elliptic curves
        modular, 460, 461, 462
        over real numbers, 452
Order of powers formula, 305–308
Orders, 293
    computer program for, 326
    computing, 303
    elliptic curve
        addition of elliptic curves over
            $\mathbb{Z}_p$, 465, 466
        computer implementations and
            exercises, 485
    exercises, 321
    modular, of invertible modular
        integers, 301–302
Outcome, experiment definition, 501
Output feedback (OFB) mode
    active attack on, 285
    block cryptosystems, 278–279
Output target set, basic concepts, 5
Overview, 1–41
    attacks on cryptosystems, 12–15
    computer implementations and
        exercises, 35–41
        computer-generated random
            numbers, 39–41
        integer/text conversions, 36–37
        programming basic ciphers with
            integer arithmetic, 38–39
        vector/string conversions, 35–36
    definitions of basic concepts, 1–4
    exercises, 28–35
        ADFGVX cipher, 32–35
        solutions, 515–517, 569–572
    functions, 4–8
        inverse, 7–8
        one-to-one and onto, bijections,
            5–7
    one-time pad, perfect secrecy,
        25–28
    Playfair cipher, 18–25
    substitution ciphers, 8–11
    Vignière cipher, 15–18

P

P = NP question, 24
Painvin, Georges, 33–34
Pairwise mutually exclusive events,
        503, 509–510
Paradoxes, set definition, 491
Partial substitutions, computer program
        for, 215

Pascal, Blaise, 295
Passive attacks
    on affine ciphers, 98–100
    basic concepts, 12
    on substitution cipher, 12–15
Perfect secrecy, 26
Performance guarantee, Miller–Rabin
        test, 314
Periodicity, powers of mod integers,
        293
Periodic substitution ciphers, Friedman
        attack, 192
Permutation ciphers, 101
Permutations
    conjugates of, 123
    evolution of codemaking
        computer representations of,
            140–143
        cyclic, 114–119
        enigma machine dissection into,
            119–126
        tabular form notation for,
            110–111
    random, computer program for
        generating, 219–220
    substitution ciphers, 9
Phaistos disk, 92–93, 94
Phi function, Euler's, 298–299, 303,
        320
Plaintext
    basic concepts, 2, 3
    conversion to numerical equivalents,
        225–228
    cryptosystem components, 94
    Enigma machine properties, 126
    monoalphabetic and polyalphabetic
        ciphers, 12–13
    representation on modular elliptic
        curves, 471–473
        computer implementations and
            exercises, 486
        exercises, 481, 482
    scytale cipher, 101–102
    substitution ciphers, 8–11
Plaintext attacks, 12
    affine ciphers, 98–99
    chosen; See Chosen plaintext
        attacks
    known; See Known plaintext attacks
Playfair, Lyon, 18
Playfair cipher
    overview, 18–25
    programming with integer
        arithmetic, 39
Plugboard, Enigma machine elements,
        112, 113, 121
Points, elliptic curve, 451
    addition, 455
    computer implementations and
        exercises, 484, 485–486

Diffie–Hellman key exchange, 468
elliptic curves over real numbers,
        452
    modular, determination of number
        of, 462, 463
Polish codebreakers, Enigma attack
        methods, 203, 204
Pollard, John, 317
Pollard p-1 factoring algorithm,
        316–319
    comparison with Lenstra's
        algorithm, 487
    computer program for, 329
    exercises, 323
Polyalphabetic ciphers, passive attacks
        on substitution cipher, 12–15
Polynomial complexity, RSA security
        guarantees, 357
Polynomials
    addition, 388, 398
        computer program for, 411
        exercises, 407
        nibble, 419, 420
        polynomials in $\mathbb{Z}_p[X]$, 386–387
    AES algorithm operations, 432,
        433–434
    Ben-Or's irreducibility determination
        algorithm, 410–411
    building finite fields from, 396–399
    with coefficients in $\mathbb{Z}_p$, 385
    computer programs
        for checking irreducibility, 412
        for extended and regular
            Euclidean algorithm for, 414
        for multiplication, 413
    congruences in $\mathbb{Z}_p[X]$ modulo as
        fixed polynomial, 395–396
    constant, 385
    divisibility in, 389–390
    division, 407, 408
        computer program for, 412
        division algorithm for, 391–395
        nibble operations, 421
    elliptic curves over, 460–462
    Euclidean algorithm, 404–406,
        408–409
    fundamental theorem of algebra, 453
    Galois fields, 382, 399–403
    irreducible/irreducibility, 405
        Ben-Or's irreducibility
            determination algorithm,
            410–411, 414–415
        computer program for checking,
            412
        computer programs for
            checking, 412
        defined, 390
        exercises, 408
        test of, 394, 395
    modular, 402–403, 406, 411, 443

multiplication, 386–387, 388, 398, 407
    AES algorithm operations, 433, 434
    computer programs for, 412, 413
    nibble, 419, 420
    in $\mathbb{Z}_p[X]$, 386–387
    in $\mathbb{Z}_p[X](\bmod\ m)$, 413
    nibble addition and multiplication, 419, 420
    as ring, 388–389
    vector representation of, 387–388
Polynomial time algorithms, 309, 355–356
    Schoof's, 468
Polynomial time prime factorization algorithm, 357
Positive integers, number theory, 43
Positive integer solutions, 70, 295, 320
Powers
    exercises, 321
    modular orders of invertible modular integers, 301–302
    order of powers formula, 305–308
    periodicity in, 293
P problems, 24
Prime certification tests, 309
Prime factorialization, 24, 309, 357
    computer implementations and exercises, 85–86
    elliptic curve arithmetic-based algorithms, 451
Prime factors, 45, 46
    elliptic curves, 476
    modular inverses, 60–61
    Pollard p-1 factoring algorithm, 317, 318
    prime factorization program, 85–86
    prime number theorem, 294
    public key cryptography, 605
    RSA cryptosystem, 347, 368
Prime modulus
    elliptic curve points, 478
    elliptic curves over modular integers, 459, 460–462
Prime numbers
    Diffie–Hellman key exchange, 336, 337
    ElGamal cryptosystem, 347
    Fermat's primality test, 309–311
    finite fields, 377
    generation of, 308–309
    industrial-grade, 314
    modular arithmetic, 44–46
        computer implementations and exercises, 85
        factorizations, 44, 45
        fundamental theorem of arithmetic, 44
        Mersenne primes, 81–82

    relatively prime integers, 46–47
    square root modulo, 83–84
    Wilson's theorem, 84–85
    modular powers, 321
    Pollard p-1 factoring algorithm, 316–319
    primitive roots, 303
    RSA cryptosystem, 340, 342
    Sophie Germain primes, 337
    tests of primality
        Carmichael numbers, 311–312
        computer programs for, 327–329
        exercises, 323–324
        Fermat's little theorem, 309–311
        Fermat's primality test, 309–311, 327
        Miller–Rabin test, 312–316, 327–329
        Pollard p-1 factoring algorithm, 316–319
Prime number theorem, 293–295
    exercises, 319, 545
    prime number generation, 308
Primitive roots
    elliptic curve analogues, 466
    modular elliptic curves, 461
    number theory, 293, 302–305, 547–548
    public key cryptography
        computer programs for, 326
        determination of, 304–305
        Diffie–Hellman key exchange, 336, 337
        exercises, 321–322
        existence of, 304
        Gauss's algorithm, 307–308
        number theory concepts, 302–305
Private key
    Diffie–Hellman key exchange, 469, 470
    public key cryptography, 23, 24, 338
        ElGamal cryptosystem, 346
        Merkle–Hellman knapsack cryptosystem, 353
Private key cryptosystems, 21
Probabilistic factoring algorithm, RSA security guarantees, 358
Probabilistic primality test, 308
Probability, 295; See also Randomness and probability
Probability function, 502, 504, 507
Probability rules, 504
Product
    matrix multiplication, 148
    nibble, 419
    polynomials in $\mathbb{Z}_p[X]$, 386
    rings, 410
Proper subsets, 493
Pseudoprime generating program, 329

Pseudorandom numbers, 27
Public key
  ElGamal cryptosystem, 346
  Merkle–Hellman knapsack
      cryptosystem, 353
  public key cryptography, 23, 338
  RSA security guarantees, 357
Public key cryptography, 21–22,
      331–375
  computer implementations and
      exercises, 369–375
  definition of, 94
  Diffie–Hellman key exchange,
      336–337
  digital signatures and
      authentication, 343–345
  discrete logarithm problem, review
      of, 334–335
  ElGamal cryptosystem, 345–349
    digital signatures with, 347–349
  exercises, 360–369
  exercise solutions, 550–554,
      604–607
  features of cryptosystems, 24–25
  government controls on
      cryptography, 356–357
  informal analogy for cryptosystem,
      331–332
  knapsack problems, 349–356
    Merkle–Hellman knapsack
      cryptosystem, 352–356
  number theory concepts
    orders, 301–302
    primitive roots, 302–305
  one-way functions, 333–334
  quest for complete public key
      cryptosystem, 337–338
  quest for secure electronic key
      exchange, 332–333
  RSA cryptosystem, 338–343
  RSA security guarantees, 357–360
Puzzles, Chinese remainder theorem,
      67–71

Q

Quality control, 510–511
Quantum computers, 357
Quotient
  definition of, 47
  division algorithm for $\mathbb{Z}_p[X]$, 391,
      392

R

Rabin, Michael, 312
rand, random integer generation, 40
Randomized encryption
  homophones, 106–107
  nulls, 104–105

Randomly generated matrix,
      computation of invertibility
      probability, 178–179
Randomness and probability, 501–513
  binomial random variables,
      511–513
  birthday problem, 505–507
  conditional probability, 507–509
  conditioning and Bayes' formula,
      509–511
  pseudorandom number generation
      algorithm, 27
  random variables, 511–513
  terminology and axioms, 501–507
Random numbers, computer-generated,
      28, 39–41
Random permutations, computer
      program for generating,
      219–220
Random substitution ciphers, 220
Random variables
  binomial, 511–513
  discrete, 511
Range, functions, 4, 5
$RC6$, 418
Real numbers
  elliptic curves over, 452–454, 478,
      483–484
  floor function, 40
Rearrangement, substitution ciphers, 9
Reflection, and associativity, 483
Reflector, Enigma machine elements,
      112, 113
Reflexivity, congruency properties, 54
Rejewski, Marian, 203, 204
Rejewski's attack, 203–205
Relative complements, set, 493
Relatively prime integers, 50, 60, 61, 96
  exercises, 83, 324
  modular arithmetic, 46–47
  pairwise, 68, 69, 70, 71
  passive attacks on affine cipher,
      98, 99
  programs, 89
Remainder(s)
  congruences and, 55–56
  definition of, 47
  division algorithm for $\mathbb{Z}_p[X]$, 391,
      392
Rijmen, Vincent, 418
Rijndael, 418–419
Rings
  AES S-box, 444
  building finite fields from $\mathbb{Z}_p[X]$,
      396–399
  commutative, 58
  congruences in $\mathbb{Z}_p[X]$ modulo as
      fixed polynomial, 395–396
  exercises, 406–407, 408
  finite fields, 378–380, 381, 383, 384

integral domains, 409–410
polynomials in $\mathbb{Z}_p[X]$ as, 388–389
Ritter, Richard, 111
Rivest, Ronald, 22, 331, 338, 339
Root cubic equation, elliptic curve graphs, 453
Roots
    elliptic curves over real numbers, 453
    Gauss's algorithm, 325
    matrix, computer implementations and exercises, 174
    modular elliptic curves, 461
    polynomials in $\mathbb{Z}_p[X]$, 409
    primitive; *See* Primitive roots
Rosetta stone, 92
rot13 cipher shift, 10
Rotate Nibble operator, 425
Rotors, Enigma machine elements, 112, 113, 120, 121–122
Rotor window, Enigma machine elements, 113
Round constants, AES encryption, 425, 439
Round key function
    DES, 267, 269
        computer programs for, 288, 289–290
        scaled-down, 263, 281
    Feistel cryptosystems, 255
Round keys
    AES, 422, 424
        computer program for, 446
        exercises, 441, 442
    DES, 259, 265, 271, 282
        computer programs for, 287–288, 289
        generation of, 259
Round-off errors, 161
Round robin tournaments, application of congruences, 80
Rounds
    AES, 421, 422, 440
    DES, 258, 260, 261, 264
    Feistel cryptosystems, 255
Row matrix, 146
Rózycki, Jerzy, 203, 204
RSA (Rivest, Shamir, Adleman) cryptosystem, 24, 273, 339
    computer programs for, 370–371, 372
    development of, 22
    digital signatures, 344–345
    exercises, 361–363, 365–366, 367–368
    mathematical problems providing security, 338
    Public key cryptography, 338–343
    security guarantees, 357–360
RSA *RC6*, 418

RSA Security, 45, 294, 345
RSA-640, 327, 372
Russian alphabet, 95

S

Sample space, experiment, 501–503
    partitioned, 509–510
    reduced, conditional probability, 507
Saxena, Nitin, 309
S-box
    AES
        computer programs for, 446–447, 448, 449
        encryption, 423–424, 428
        encryption algorithm, 437, 439
        exercises, 441, 443–444
        inverse, 430, 448
    DES, 267, 268
        computer programs for, 288, 289
        exercise, 284
        scaled-down, 261–262, 281, 282
S-box table, AES, 423
Scalar multiplication
    computer implementations and exercises, 175
    elliptic curve exercises, 480
    matrix, 146–147
    polynomials in $\mathbb{Z}_p[X]$, 388
Scalars, defined, 146, 147
Scaled-down AES; *See* Advanced encryption standard protocol
Scaled-down DES
    computer programs for, 287–289
    exercises, 281–282
Scaled-down Enigma machines
    composition of functions, 120–121
    computer programs, 141–143
Scherbius, Arthur, 111
Schoof's algorithm, 468
Scytale, 101
Scytale cipher, 101–102, 128, 136–137
Second quotient, division algorithm for $\mathbb{Z}_p[X]$, 391
Self-cancelling properties, XOR, 255, 429
Self-decryption proof, DES and Feistel cryptosystems, 285
Serpent, 419
Set differences, 494
Sets
    basic concepts, 4, 5
    basic counting principles, 495–499
    binary operations, 377
    concepts and notations, 491–495
    finite fields, 377
    modular elliptic curves, 452
Set theory, probability theory and, 503
Shamir, Adi, 22, 338, 339, 355–356

Shannon, Claude, 25, 26
Shannon's properties of diffusion and confusion, 272, 419
Shift cipher, 38, 95
Shift permutation, 189
    Caesar cipher, 10
    one-unit, 119
Shift register, cipher feedback (CFB) mode, 276
Shift Row mapping
    inverse of, 440
    reverse order, 429
Shift Row Transformation, AES
    decryption, 431
    encryption, 422, 424, 428, 429, 436
    exercises, 444–445
Shor, Peter, 357
Signature exponent, ElGamal cryptosystem, 347
Significant digits, computing platforms and, 325; *See also* Computation issues
Simultaneous congruences, Hindu puzzle, 67–71
Single linear congruence, solving, 66
Singleton set, 492
Singular elliptic curve
    definition of, 452
    graphs, 454
    over modular integers, 460
    over real numbers, 452
Sizes
    matrix, 145
    of modular elliptic curves, 462–463
Sophie Germain primes, 337
Spaces
    frequency analysis-based attacks, 183
    RSA cryptosystem, 340
    substitution ciphers, 10
Spinner, randomized encryption, 104–105
Square (invertible) matrix
    computer implementations and exercises, 175–176
    definition of, 146, 151–153
    determinant computation, 159
    determinant of, 153–155
    general cofactor expansions, 171–172
    inverses of 2x2 matrices, 155–156, 174–175
Square roots
    modular elliptic curves, 461, 462
    modulo m, 83–84
Standards, 2
    Digital Signature Standard (DSS), 345
    encryption; *See* Advanced encryption standard protocol; Data encryption standard

State matrix, Mix Column mapping, 430
State transformations (mappings); *See* Mapping
Statistical frequency counts, 13–14
Steganography, 100–102
Storage
    two's complement representation scheme, 245–246
    as vectors or strings, 248
Strassen, Volker, 150
Strassen's algorithm, 150–151, 173–174, 179
Stream modes, block cryptosystems, 276–279
Strings, 254
    basic concepts, 3
    computer programs
        for extracting ciphertext data from ciphertext string, 216–218
        XOR operation, 287
    integers in different bases, 248–250
    vector/string conversions, 35–36
String size, AES, 417
Strong avalanche condition, AES, 419
Subblocks, cipher feedback (CFB) mode, 276
Submatrix, 154
Sub Nibble operator, 425
Subsets, 5, 492
Substitution box, DES, 261–262, 267, 268
Substitution ciphers
    Caesar cipher, 9–11
    evolution of codemaking, 102
        cryptosystem components, 95
        homophonic, 107
        steganography, 100–101
    frequency analysis-based attacks, 183–186
    overview, 8–11
    passive attack example, 12–15
    random, computer implementations and exercises, 220
Substitution permutation network, 419
Substitutions
    congruent, 56–57
    partial, computer program for, 215
Subtraction
    algorithm complexity analysis, assessing work required to execute, 246–247
    matrix, 146–147
    rings, 379
Subtraction algorithm with base b expansions, 231–234
Sum
    addition of elliptic curves over $\mathbb{Z}_p$, 464
    elliptic curve addition, 455

nibble, 419
polynomials in $\mathbb{Z}_p[X]$, 386
Superincreasing weights, knapsack
    problem, 350–352
  computer programs for, 374
  exercises, 364–365
  Merkle–Hellman knapsack
    cryptosystem, 352–356
*Symbolic Analysis of Relay and
    Switching Circuits, A.*
    (Shannon), 25
Symbolic computing platforms, 296,
    314, 325, 334, 369
  elliptic curve operations, 483
  Lenstra's algorithm, 477
  public key cryptography, 334
  RSA cryptosystem, 341
Symmetric key cryptosystems, 21, 23,
    24; *See also* Private key
    cryptosystems
  definition of, 94
  DES development, 95
  substitution ciphers, English
    alphabet, 96
Symmetry
  congruency properties, 54
  matrix, 156

T

Tables
  basic concepts, 3–4
  tabular form notation for
    permutations, 110–111, 220
Tangent line, elliptic curve properties,
    453
T-attack, 272–273
Tempest devices, 357
Ternary expansions, 225
Text
  integer/text conversions, 36–37
  plaintext; *See* Plaintext
Three-round Feistel systems,
    280–281
  computer implementations and
    exercises, 287
  self-decryption proof, 285
Time algorithm, Schoof's, 468
*Traicté des Chiffres ou Secrètes
    Manières d'Escrire*
    (Vignière), 15
Transitivity
  congruency properties, 54
  divisibility, 44, 68, 389
Transpose of matrix, 156, 171
Transposition ciphers, 101–102
Trapdoor (one-way) function, 333–334,
    353
*Treatise on Numerals and Secret Ways
    of Writing* (Vignière), 15

Tree diagram, counting principles, 496
Trial (experiment), defined, 501
Trigram, 190
Trigraphs, 107
Triple composition, 122
Triple DES, 273–274, 291–292, 333
Trithemius, Johannes, 15
Trivial cycle, 115, 116
Turing, Alan, 206–208
Twofish, 419
Two-round encryption mapping,
    541–543
Two-round Feistel systems, 258, 259,
    263, 280
Two's complement representation
    scheme, 245–246

U

Union, sets, 491, 492–495
Unique factorization, in $\mathbb{Z}_p[X]$, 405
Universal set, 494

V

Vacuously true, 493
van Oorschot, Paul, 273, 617
Vanstone, Scott, 273, 616, 617
Vatican ciphers, 102
Vaudenay, Serge, 356
Vector addition, 457
Vector multiplication, polynomials in
    $\mathbb{Z}_p[X]$, 388
Vectors, 254
  Cartesian product set, 496
  conversion programs, 286
  dot product formula, 199–200
  integers in different bases,
    248–250
  knapsack problem reformulation,
    350
  nibble addition and multiplication,
    419
  polynomial representations,
    387–388
  rings, 406–407
  XOR program, 287
Vector/string conversions, 35–36
Venn diagrams, 492–495, 505
Vernam, Gilbert S., 26
Vernam cipher, 26
Verser, Rocke, 273
Vignière, Blaise de, 15
Vignière cipher, 107
  demise of, 187–192
    Babbage/Kasiski attack,
      188–192
    Friedman attack, 192
    Friedman attack, 197–201
    ciphertext-only, 200–201

Hill cryptosystem with, 166
one-time pad as, 28
overview, 15–18
programming with integer
  arithmetic, 38–39
Vignière tableau, 16, 17

W

Waterhouse's theorem, 463, 476
Weak keys, DES, 284
Weights, object; *See* Object weights,
  knapsack problems
Wheatstone, Charles, 18
Wilson's theorem, 84–85
Winograd, Shmuel, 150
Witness, primality test, 309–311, 314
Word length, 10, 240
Word size, 238
Wright, Edward V., 294
Wright, Ernest Vincent, 14

X

XOR operation, 254–255, 383, 407
  AES, 428, 445
    computer implementations and
      exercises, 447, 449
    encryption, 427
  exercise, 285
  nibble addition, 420
  self-cancelling properties, 255, 429

Y

Young, Thomas, 92

Z

Zero, 378
Zero polynomial, 385, 387, 394
Zuse, Konrad, 251, 252
Zygalski, Henryk, 203, 204